Organic Nanorea

From Molecular to Supram

Organic Nanoreactors

From Molecular to Supramolecular Organic Compounds

Samahe Sadjadi
Gas Conversion Department
Iran Polymer and Petrochemical Institute
Faculty of Petrochemicals, Tehran, Iran

ELSEVIER

AMSTERDAM • BOSTON • HEIDELBERG • LONDON
NEW YORK • OXFORD • PARIS • SAN DIEGO
SAN FRANCISCO • SINGAPORE • SYDNEY • TOKYO
Academic Press is an Imprint of Elsevier

Academic Press is an imprint of Elsevier
125 London Wall, London EC2Y 5AS, UK
525 B Street, Suite 1800, San Diego, CA 92101-4495, USA
50 Hampshire Street, 5th Floor, Cambridge, MA 02139, USA
The Boulevard, Langford Lane, Kidlington, Oxford OX5 1GB, UK

British Library Cataloguing-in-Publication Data
A catalogue record for this book is available from the British Library

Library of Congress Cataloging-in-Publication Data
A catalog record for this book is available from the Library of Congress

ISBN: 978-0-12-801713-5

For information on all Academic Press publications
visit our website at https://www.elsevier.com/

Publisher: Cathleen Sether
Acquisition Editor: Katey Birtcher
Editorial Project Manager: Jill Cetel
Production Project Manager: Paul Prasad Chandramohan
Designer: Mathew Limbert

Typeset by Thomson Digital

Contents

12. Supercritical Fluids in Nanoreactor Technology

Sodeh Sadjadi

13. Pyrene: The Guest of Honor

Nicolas P.E. Barry, Bruno Therrien

14. Nanoreactors Based on Porphyrin-Functionalized Carbon Compounds

Pietro Tagliatesta, Barbara Floris, Stefano Bellucci

15. Therapeutic Nanoreactors: Toward a Better Blood Substitute

Agnes Ostafin, Kyu Bum Han

List of Contributors

Numbers in Parentheses indicate the pages on which the author's contributions begin.

Nicolas P.E. Barry (421), Department of Chemistry, University of Warwick, Coventry, UK

Stefano Bellucci (463), INFN - Istituto Nazionale di Fisica Nucleare, Laboratori Nazionali di Frascati LNF, Frascati, Latium, Rome, Italy

Giuseppe Borsato (203), Dipartimento di Scienze Molecolari e Nanosistemi, Università Ca' Foscari di Venezia, Mestre (VE), Italy

Kyu Bum Han (519), Department of Materials Science and Chemical Engineering, University of Utah, Nano Institute, Salt Lake City, UT, United States of America

F.G. Cirujano (305), Institute of Chemical Technology, Polytechnic university of Valencia, Superior Council for Scientific Research, Valencia, Spain

Ionel A. Dinu (341), Department of Chemistry, University of Basel, Basel, Switzerland

Tomaz Einfalt (341), Department of Chemistry, University of Basel, Basel, Switzerland

Barbara Floris (463), Dipartimento di Scienze e Tecnologie Chimiche, Università Tor Vergata, via della ricerca scientifica, Roma, Rome, Italy

George W. Gokel (235), Department of Biology, Center for Nanoscience, University of Missouri, St. Louis, MO, United States of America

Frédéric Hapiot (15), University of Artois, Laboratory of Catalysis and Solid State Chemistry (UCCS), Faculty of Sciences Jean Perrin, Lens, Cédex, France

F.X. Llabrés i Xamena (305), Institute of Chemical Technology, Polytechnic University of Valencia, Superior Council for Scientific Research, Valencia, Spain

Mihai Lomora (341), Department of Chemistry, University of Basel, Basel, Switzerland

Stéphane Menuel (15), University of Artois, Laboratory of Catalysis and Solid State Chemistry (UCCS), Faculty of Sciences Jean Perrin, Lens, Cédex, France

Eric Monflier (15), University of Artois, Laboratory of Catalysis and Solid State Chemistry (UCCS), Faculty of Sciences Jean Perrin, Lens, Cédex, France

Saeedeh Negin (235), Departments of Chemistry & Biochemistry, Center for Nanoscience, University of Missouri, St. Louis, MO, United States of America

Agnes Ostafin (519), Department of Materials Science and Chemical Engineering, University of Utah, Nano Institute, Salt Lake City, UT, United States of America

Cornelia G. Palivan (341), Department of Chemistry, University of Basel, Basel, Switzerland

Viktoriia Postupalenko (341), Department of Chemistry, University of Basel, Basel, Switzerland

Joshua B. Puplampu (85), Organic Chemistry Department, A.M. Butlerov Chemical Institute, Kazan (Volga Region) Federal University, Kazan, Russian Federation

M. Rosa Axet (111), Coordination Chemistry Laboratory, University of Toulouse, Emile Monso, Toulouse Cedex, France

Jolanta Rousseau (15), University of Artois, Laboratory of Catalysis and Solid State Chemistry (UCCS), Faculty of Sciences Jean Perrin, Lens, Cédex, France

Cyril Rousseau (15), University of Artois, Laboratory of Catalysis and Solid State Chemistry (UCCS), Faculty of Sciences Jean Perrin, Lens, Cédex, France

Samahe Sadjadi (1, 159, 257), Gas Conversion Department, Iran Polymer and Petrochemical Institute, Faculty of Petrochemicals, Tehran, Iran

Sodeh Sadjadi (373), Nuclear Science and Technology Research Institute, Tehran, Iran

Alessandro Scarso (203), Dipartimento di Scienze Molecolari e Nanosistemi, Università Ca' Foscari di Venezia, Mestre (VE), Italy

Philippe Serp (111), Coordination Chemistry Laboratory, University of Toulouse, Emile Monso, Toulouse Cedex, France

Ivan I. Stoikov (85), Organic Chemistry Department, A.M. Butlerov Chemical Institute, Kazan (Volga Region) Federal University, Kazan, Russian Federation

Bruno Therrien (421), Institut de Chimie, Université de Neuchâtel, Neuchâtel, Switzerland

Alena A. Vavilova (85), Organic Chemistry Department, A.M. Butlerov Chemical Institute, Kazan (Volga Region) Federal University, Kazan, Russian Federation

Brian D. Wagner (43), Department of Chemistry, University of Prince Edward Island, Charlottetown, Canada

Luidmila S. Yakimova (85), Organic Chemistry Department, A.M. Butlerov Chemical Institute, Kazan (Volga Region) Federal University, Kazan, Russian Federation

Pietro Tagliatesta (463), Dipartimento di Scienze e Tecnologie Chimiche, Università Tor Vergata, via della ricerca scientifica, Roma, Rome, Italy

Chapter 1

Introduction to Nanoreactors

Samahe Sadjadi
Gas Conversion Department, Iran Polymer and Petrochemical Institute, Faculty of Petrochemicals, Tehran, Iran

1 APPROACHES TO ARTIFICIAL ENZYMES

Nature provides us with the most efficient catalysts, enzymes that catalyze reactions under mild and green conditions (ie, atmospheric pressure, temperature, and aqueous solution). Efficiency and chemo-, regio-, and stereoselectivity in enzyme-catalyzed reactions are so remarkable that they inspire scientists to design synthetic systems with comparable activity and selectivity [1,2]. Enzymes made of proteins are more than just highly evolved catalysts. They recognize and respond to molecules other than their specific substrate and product [3]. The evolution of artificial enzymes is in its infancy and its main goal is efficient catalysis [1,3].

Using enzyme principles, efficient and selective synthetic catalysts can be designed that may be more practical than enzymes in the industrial section, due to their more robust entities and greater tolerance to reaction conditions such as temperature and pH.

Two key features of enzymes are their binding ability and product release. Enzymes can stabilize the transition state of a reaction more than the ground state of the substrate is stabilized. This can be achieved through additional binding interactions between enzyme functionalities and the substrate within its cavity. This concept is illustrated in Fig. 1.1 with a unimolecular example where the enzyme-substrate complex is stabilized relative to the free species in solution. The activation barrier to reaction is represented by the difference ΔG_{cat} and ΔG_{uncat} for the enzyme catalyzed and the unanalyzed reactions, respectively [4]. It is clear from this picture that, for catalysis to work, the difference $\Delta G_{ETS\#}$ must be larger than ΔG_{ES}.

The microenvironment within the enzyme cavity has a great influence on the reaction catalyzed by enzymes. Most of the effective parameters on enzyme operation—such as substrate preorganization, restricted substrate motion, protein dynamics, covalent binding of the transition state, and desolvation of the substrate—are induced by this microenvironment [5]. Therefore, designing a binding cavity in the structure of artificial enzymes is of great importance [2,6].

One of the most important challenges of development of artificial enzymes is product inhibition, which prevents a catalytic turnover or decreases the reaction

Organic Nanoreactors. http://dx.doi.org/10.1016/B978-0-12-801713-5.00001-X

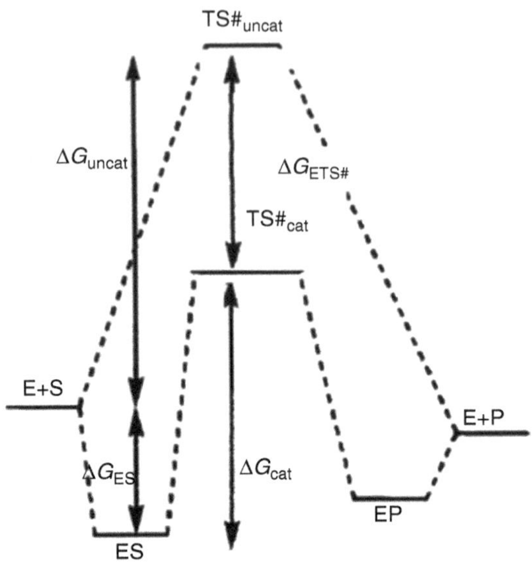

FIGURE 1.1 **Enzymes ability to stabilize the transition state of a reaction relative to that of the ground state.** S, substrate; E, enzyme; P, product; ES, enzyme–substrate complex; TS#, transition state; EP, enzyme–product complex. *(Adapted from Ref. [4] with permission of Elsevier.)*

rate [7]. Product inhibition can occur due to higher stabilization of product by attractive interactions within the binding cavity compared to the reactants [7].

Pauling's enzyme catalysis theory, which addressed transition state stabilization and molecular recognition, was the pioneering work in the field of artificial enzymes [4,8]. Thereafter, remarkable efforts have been performed to realize this theory. Much of the research has focused on the design and synthesis of complex molecules and has been very time-consuming and complicated. Taking advantage of recent advances in molecular biology, biochemistry, and combinatorial, polymer, and supramolecular chemistry, various approaches based on selection have been developed.

These approaches include: (1) *design approach,* which is the designing of host molecules with functional groups that are expected to be involved in catalysis of the chosen reaction, (2) *transition state analog-selection approach,* which consists of generation of a library of hosts in the presence of a transition state analogue and selecting the best host, and (3) *catalytic activity-selection approach,* in which combinatorial chemistry has been used to generate and screen possible catalysts for enzyme-like activity [4].

In this regard, a process that is of growing interest is the synthesis of nanometer-size reaction chambers, that is, nanoreactors (NR), with a confined microenvironment providing a cavity effect that can encapsulate guest molecule(s) through binding interactions [7].

2 NANOREACTORS

2.1 Nanoreactor Definition

Impelled by fascination with natural NRs such as mitochondria, nucleus, Golgi apparatus, lysosomes, and the pores of channel proteins, and against the background of host–guest chemistry, the concept of the NR emerged in the late 1990s [9]. A NR is not a simple nanosize reaction vessel like bench-top or microreactor [10]. The size, shape, and microenvironment within a NR have a remarkable role in the chemical process and can dictate new activity and selectivity [7,11]. The yield, kinetic, and mechanism of reaction taking place in the NR may be different from what is expected in the bulk solvent due to encapsulation effects (vide infra) [10].

Nanoreactors also can be equipped with active sites such as transitional metal species. In this case, the restricted space in the NR can affect reaction rate and mechanism through additional interactions of the reactants and active sites and modification of the concentration of the reactant close to the active site [9]. In this context, immobilization of chiral active centers within porous materials has attracted much attention and there are numerous reports confirming the merit of nanospace confinement in the chemical reactions. Molecular dynamics simulations have been used to clarify the precise way in which enantioselectivities are enhanced [12].

2.2 Encapsulation Effects

Encapsulation of the guest molecule(s) within the hydrophobic microenvironment of the NR results in guest isolation from the bulk. This microenvironment can induce a specific conformation of the guest. Molecular recognition and selective encapsulation of the guest(s) is one of the key features of NRs, and the guest(s) of complementary shape, size, and chemical surface are expected to be encapsulated more easily than other molecules [5,7].

The properties of the NR interior wall and the interactions between encapsulated guest and walls can affect the guest's redox potential and Gibb's free energy, changing the guest's reactivity. Additionally, these interactions influence the molecular alignment, rotational dynamics, and evolution of the reaction transition state. In the case of bimolecular reaction within the confined environment of a NR, the yields of products can alter if two molecules cannot align themselves adequately to achieve the transition state or to relax fully once it is formed [10]. Furthermore, encapsulation can result in stabilization of a specific transition state and favoring a reaction pathway that is not favored in the bulk solution.

Nanoreactor walls with high surface-to-volume ratios generate a confined reaction space, which can result in increase of concentration of the reactants and reaction rate by reducing the mean free path [10]. More frequent collision with molecules can be expected in the restricted environment of a NR due to repulsion of walls' surfaces and molecules [9,10]. Therefore, reaction pathways that are unfavorable in terms of energy in solution might be accessible by performing reaction inside a NR [7]. Reaction rate of reactants inside NRs can also be affected by absorption of reactants on the surface which can slow down the diffusion.

Finally, the confined microenvironment of a NR might induce segregation or phase separation of solvents and reactants inside. For example, the displacement of a given solvent with the absorbed reactants can result in a different reaction mechanism or kinetics [9].

Among the discussed encapsulation effects, the dimension of the confined space, the number of molecules inside, the interaction between the wall and the reactants, and the presence of catalytic functionalities determine the dominant encapsulation effects [9,10].

NMR spectroscopy is a useful and informative method for the characterization of molecular location within NRs and investigation of the effects of the shape and size of molecular capsules [10,13]. Using spectroscopic methods for detection of molecular location inside a NR, one should keep in mind that encapsulation within the NR and the presence of other molecules in NRs can change spectral properties of the guest molecule(s).

X-ray crystallography provides the clearest structural evidence of the often-subtle intermolecular interactions. Combination of solution-state methods (NMR spectroscopy, mass spectrometry) and solid-state methods (X-ray structure analysis) is necessary for an overall understanding of the host–guest behavior. In the case of self-assembled NRs, mechanisms of guest inclusion, guest exchange, and host–guest interactions are not simple, due to the dynamic, flexible nature of self-assembled NRs [13].

2.3 Reaction Kinetics Inside Nanoreactors

Rate-equation for a simple bimolecular reaction in which reactants A and B lead to product C can be calculated using Eq. 1.1.

$$v = \mathrm{d}[\mathrm{C}]/\mathrm{d}t = k_\mathrm{a}\,[\mathrm{A}][\mathrm{B}] \qquad (1.1)$$

EQUATION 1.1 *Reproduced from Ref. [5] by permission of The Royal Society of Chemistry (RSC).*

For the reaction inside the NR, a more complicated rate equation is expected due to encapsulation effects as well as product release in-and-out of the NR [5].

When the reaction between A and B within the NR is the rate-determining step, the rate equation simplifies to Michaelis–Menten kinetics and depends solely on the rate-constant of this step, that is, k_b, and on the NR concentration with the encapsulated substrates (NR⊃A.B) (Eq. 1.2) [5]. In Fig. 1.2a, energy diagrams of these equations have been represented.

$$v = \mathrm{d}[\mathrm{C}]/\mathrm{d}t = k_\mathrm{b}\,[(\mathrm{NR} \supset \mathrm{A.B})] \qquad (1.2)$$

EQUATION 1.2 *Reproduced from Ref. [5] by permission of The Royal Society of Chemistry (RSC).*

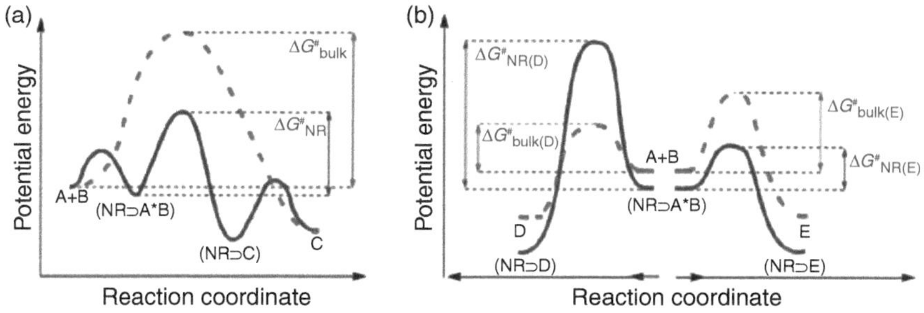

FIGURE 1.2 (a) Simplified reaction profiles of a reaction in the bulk solution (dashed red line) and of a reaction within a NR (blue line). (b) Simplified reaction profiles of a reaction leading to product D that is destabilized by the NR (blue line), compared to the bulk solution (dashed red line); and of a reaction leading to product E that is stabilized by the NR (blue line), compared to the bulk solution (dashed red line), *(Reproduced from Ref. [5] with permission of The Royal Society of Chemistry (RSC).)*

The rate-constant k is a function of the thermodynamic activation parameters, that is, the Gibbs free energy of activation ($\Delta G^{\#}$), and hence to the activation enthalpy ($\Delta H^{\#}$) and the activation entropy ($\Delta S^{\#}$) via the Eyring and Arrhenius equations: $\Delta G^{\#} = \Delta H^{\#} - T.\Delta S^{\#} = -RT(\ln k) + c$ (T, temperature; R, gas constant; c, a constant). [5].

For reactions within NRs, two scenarios can be conceived, though in practice they might be combined:

1. The NR effect is just bringing the substrates together without affecting the activation parameters (ie, $\Delta G^{\#}_{NR} = \Delta G^{\#}_{bulk}$). In this case, the reaction pathway changes from bimolecular to intramolecular. Since the activation parameters in Eqs. 1.1 and 1.2 are the same, the rate constant remains the same, $k_a = k_b$, and the difference in the rate is due to the fact that: $[(NR \supset A.B)] > [A][B]$[7].
2. The NR changes activation parameters (ie, $\Delta G^{\#}_{NR} \neq \Delta G^{\#}_{bulk}$) and reduces the reaction activation energy barrier (ie, $\Delta G^{\#}_{NR} < \Delta G^{\#}_{bulk}$) through stabilization of transition states ($\Delta G^{\#}$) (Fig. 1.2a). Stabilization can stem from noncovalent interactions between the transition state and the functional groups in NR microenvironment, enthalpic stabilization ($\Delta H^{\#}$), and/or entropic factors [7]. As mentioned before, encapsulation of substrate within the confined cavity of NR with definite size and shape can result in substrate preorganization toward the transition state and decrease the potential negative entropy of a reaction.

In a bimolecular reaction, where two or more products are expected (reagents A and B give products D and E), selectivity and reactivity of reaction in NR may be different from bulk solvent.

This observation can be attributed to the changes in the reaction activation energy barrier $\Delta G^{\#}$ of products D and E compared to the bulk. Encapsulation can stabilize the transition state, which leads to product E, thereby favoring its formation. In contrast, formation of product E can become unfavorable through encapsulation and destabilization of the transition state that lead to it (Fig. 1.2b) [7].

2.4 Product Inhibition

Product release out of the NR is an essential parameter for catalytic systems based on NRs. It can influence the catalytic turnover and/or the reaction rate inside the NR and prevent its utility as a true catalyst [5].

Enthalpic effects, such as multiple attractive interactions of product with the NR, as well as entropic ones, can influence product inhibition. A strategy to suppress this problem is engaging product in a follow-up reaction that leads to a compound with lower affinity for the NR.

Other approaches are based on designing NRs in such a way that they have a more binding affinity to reactants than products, or aiming for the reaction products with a size or shape that doesn't match the cavity [2].

Largeness of product in some NRs may prevent their release and result in product inhibition. In this case, designing of less constrained NRs with larger portals, and/or aiming for smaller products, can furnish a solution to the problem [7].

2.5 Nanoreactor Classification

According to a widely accepted classification, there are three main categories of NRs: biological, self-assembled, and natural or synthetic NRs [2,9].

2.5.1 Natural or Synthetic Nanoreactors

These can be divided into three subclasses of (1) inorganic solids such as synthetic zeolites [14,15], clays, ordered mesoporous silicas [16,17], and metal oxide frameworks; (2) organic nanostructures such as covalent organic frameworks or microporous polymers; and (3) hybrid organic–inorganic NRs such as metal organic frameworks [18–20], zeolite with organic groups as lattice (ZOL), aluminophosphate (AlPO), silico-alumino-phosphate (SAPO), and periodic mesoporous organosilicas (PMOs). The interest in inorganic NRs stems from the fact that they are resistant against high temperature and pressure and can be considered as good candidates for industrial applications [9].

2.5.2 Biological Nanoreactors

The use of biological molecules, assemblies, and systems in the synthesis of nanomaterials, which has emerged from investigations at the interfaces among biology, chemistry, and materials science, has provided novel approaches as alternatives to conventional synthetic methods [2,21,22].

Material templating, assembly, and crystallization can be achieved using organized biomolecular systems such as protein cages [23–26], lipid assemblies [27], bacterial S-layers [28], and DNA [29,30]. This research field has attracted growing attention, and several reports and review articles have been published in this regard [31,32].

The current interest in these biosystems stems from the fact that they display a high degree of organization, are often easy to modify, and occur in well-defined self-assembly motifs [2].

2.5.3 *Self-Assembled Nanoreactors*

Self-assembled molecular capsules are three-dimensional structures generated through reversible, noncovalent interactions such as hydrogen bonding between complementary subunits. They circumvent many problems of covalent capsules, such as multistep and complicated synthetic methods and lack of versatility. Therefore, recent researches are focused on self-assembled systems [2]. Micelles, emulsions, vesicles, and noncovalent molecular capsules are examples of self-assembled NRs [2].

As encapsulation within noncovalent self-assembled capsules is reversible. Depending on capsule entity, guest exchange inside and outside of it can occur via diffusion of the guest or (partial) dissociation of the capsule [5,7]. Encapsulation within molecular capsules is more than a physical entrapment. Interactions such as hydrophobic interactions, π–π interactions, or weak coordinative interactions play important roles in the selective encapsulation of particular guest molecules [7].

Self-assembly can also be induced by metal–ligand interactions [33,34]. In this case, supramolecular coordination complexes that are thermodynamically favored products can be obtained from spontaneous metal—ligand bond formation in a mixture of soluble metal and ligand precursors [19]. These self-assembled capsules, which are known as coordination-driven self-assembly, are more robust than hydrogen bond capsules [13,35].

Initially, efforts were centered on encapsulation of small substrate, but recently encapsulation of larger substrate such as adamantine and fullerenes has been reported using NRs [36,37].

3 NANOREACTOR POTENTIAL APPLICATIONS

Nanoreactors hold promise for applications ranging from selective and size-constrained organic synthesis to biomedical advances, and as analytical tools to study reaction mechanisms [38]. The applications of each type of NR are discussed in separate chapters, so here we will briefly discuss some of their most important applications only.

3.1 Catalysis

One of the most important and promising applications of NRs is catalysis. Nanoreactors can be used to address the major problems in the field of catalysis, including catalyst recovery, selectivity, and efficiency. They have the potential to approach catalysts benefiting from high activity, high selectivity, high stability, and easy separation [39]. Various types of NRs—including dendrimers [39], self-assembled coordination cages [40], cyclodextrines [41], cucurbiturils [42], calixarenes, resorcinarenes, and so forth [2],—have been used to catalyze a wide range of chemical reactions.

Selective encapsulation of substrate(s), preorganization of substrates in the confined space of the NR, and an increase of local concentration within the NR can result in unusual regio- and stereoselectivity, reaction rate, and yield, and an alteration in the reaction pathway. It is worth nothing that NRs can also be used in photochemical reactions and development of photocatalysts. As an example, dendrimer-protected TiO_2 nanoparticles were synthesized and their photocatalytic activity was compared to that of TiO_2 nanoparticles in photodegradation of 2,4-dichlorophenoxyacetic acid. The results indicated the superior activity of dendritic photocatalysts, which was attributed to stabilization of nanoparticles and reinforcement of photocatalytic activity [43].

3.2 Protection and Stabilization

Encapsulation of the guest molecule within NR can result in guest protection from the bulk solution. This can be exploited for the stabilization, isolation, and identification of unstable and highly reactive intermediates, such as radicals, carbenes, and nitrenes, and labile products [13,44]. Warmuth reviewed inner-phase stabilization of cyclobutadiene, *o*-benzyne, and cycloheptatetraene and their properties in solution using molecular containers [44]. In this context, Raymond et al. reported encapsulation and stabilization of various iminium cations [45], aromatic diazonium cations, and tropylium cations [46] within self-assembled coordination NRs. Furthermore, this type of NR was exploited for encapsulation and stabilization of reactive tetrahedral P_4 molecules [47] and coordinatively unsaturated transition-metal complex [48]. Molecular encapsulation within NRs has also been used for protection of organic dyes from aggregation effects and photochemical degradation [49]. Encapsulation not only results in physical protection but also influences the absorption wavelength or the fluorescence quantum yield [7]. As an example, study of encapsulation of fluorescent dye within fluorescent dendritic NRs demonstrated that encapsulation can result in quenching/sensitization processes that can be exploited for a variety of purposes, including development of chemosensors, and construction of light harvesting antennas [50].

Nanoreactors can also be used for protecting one reaction site or blocking a reaction through encapsulation of a reactant. As an example, the regioselectivity of nucleophilic substitution of aryl-substituted allylic chlorides can be controlled through encapsulation and shielding of the internal reaction site of the substrate within the cavity of coordination capsule [51].

3.3 Templating and Stabilizing of Nanomaterials

Nanoreactors' microenvironment can be exploited for the synthesis of nanomaterials with definite shape and size such as nanoparticles [25,52–54] and nanotubes [55]. Entrapment of nanoparticles within NR cavity can be used for stabilization of nanoparticle and preventing them from agglomeration. Additionally, encapsulation of nanoparticles can influence their properties through modification of their orientation and configuration [56,57]. Various NRs such as dendrimers and coordination cages have been employed for this purpose. As examples, synthesis of silica nanoparticles with diameter of 2–4 nm [58] and highly monodispersed core-shell nanoparticles with monodisperse silica core nanoparticle and a metal oxide (TiO_2 or ZrO_2) layer with a diameter of 3 nm [59] have been synthesized by using self-assembled hollow spherical complexes as "endo-templates." There are also numerous reports on the templating and stabilizing of various mono- and bimetallic dendrimer encapsulated nanoparticles such as Ag [60], Au [61], Co [62], Pd [63], Ni [64], Cu [65], Ag–Pd [66], Au–Ag [67], Pd–Au [68], etc. The dendrimer-encapsulated nanoparticles have a wide range of applications, especially in catalysis. It is worth mentioning that the application of NRs for templating and stabilizing nanomaterials is not limited to metal oxides or metallic nanoparticles. As an example, dendritic NRs were used for synthesizing and promoting the properties of quantum dots [69].

3.4 Polymer Science

Another application of NRs is in polymer science. Polymerization in the interior cavity of NR can be used for controlling the rate of polymerization and molecular weight of the resultant polymer. Furthermore, the development of self-healing polymers can be achieved by using NRs. Herein, three representative examples have been presented to disclose the utility of NRs in this research field.

Polymerization of aniline within the confined environment of spherical polyelectrolyte brushes was performed [70]. Encapsulation of aniline monomers inside a NR and controlled delivery of the oxidizing reagent to the reaction volume provide optimal conditions for matrix polymerization. The excellent kinetic stability of the resulting core–shell particles, as well as the high macroscopic conductivity of the respective composite, open up perspectives for novel materials [70].

A spherical self-assembled complex bearing 24 methyl methacrylate units in its interior was prepared by Fujita et al. It was demonstrated that endohedral radical polymerization can be promoted due to the increased concentration of monomer units at the core of the complex [71].

Using NRs, White and coworkers developed self-repaired polymers. This approach was based on incorporation of microcapsules loaded with healing agents, dicyclopentadiene monomers, and using a proper catalyst in epoxy matrix. Formation of cracks ruptured the microcapsules and released healing agents through capillary forces that were polymerized by contact with the catalyst, result in bonding of the crack faces [72].

3.5 Development of Nanomedicines

Nanoreactors lead to a remarkable development in the field of nanomedicine [73–75]. There are numerous reports and review articles regarding the applications of various NRs, such as dendritic NRs and polymersomes, as nanocarriers and nanovectors for therapy, diagnosis, and drug and gene delivery [76–80]. A high loading capability, biocompatibility and biodegradability, controlled release, stability, and robustness are important factors in this regard.

3.6 Sensors

Nanoreactors can be used as a platform for developing sensing and stimuli-responsive systems. For example, a NR based on a biocompatible calcium phosphate cage has been synthesized that responds to reactive oxygen species and emits fluorescence at near-infrared region [10]. Stimuli-responsive NRs such as polymer vesicles, which also refers to smart polymer vesicles, responds to stimulus such as changes of pH, temperature, and light, and can be considered as useful carriers for encapsulation and controlled delivery of drug and other biological compounds [81–85].

4 CONCLUSIONS

Nanoreactors are nanoscale containers with the ability of selective encapsulation of guest(s). The confined microenvironment within a NR can remarkably influence the process inside. There are various kinds of NRs, including biological, self-assembled, and natural or synthetic NRs. Up to now, numerous reports have been published on the synthesis and applications of NRs. It is expected that this research field will develop even more rapidly and offer solutions to many challenging topics in material science, catalysis, biomedicine, sensors, etc. It is hoped that the concepts presented in this opening chapter provide information to fundamental concepts of NRs and their potential applications. In the following chapters, various kinds of organic NRs, as well as their potential applications, will be discussed in detail.

REFERENCES

[1] Kirby AJ, Hollfelder F. From enzyme models to model enzymes. Cambridge: RSC Publishing; 2009.
[2] Vriezema DM, et al. Self-assembled nanoreactors. Chem Rev 2005;105:1445–89.
[3] Kirby AJ. Enzyme mechanisms, models, and mimics. Angew Chem Int Ed Engl 1996;35:707–24.
[4] Motherwell WB, Bingham MJ, Six Y. Recent progress in the design and synthesis of artificial enzymes. Tetrahedron 2001;57:4663–86.

[5] Koblenz TS, Wassenaar J, Reek JNH. Reactivity within a confined self-assembled nanospace. Chem Soc Rev 2008;37:247–62.
[6] van Leeuwen PWNM. Supramolecular catalysis. Weinheim, Germany: WILEY-VCH Verlag GmbH & Co; 2008.
[7] Brinker UH. Molecular encapsulation. Weinheim, Germany: WILEY-VCH Verlag GmbH & Co; 2010.
[8] Pauling L. Molecular architecture and biological reactions. Chem Eng News 1946;24(10):1375–7.
[9] Llabres i Xamena FX. Metal organic frameworks as heterogeneous catalysts, 12. Cambridge: RSC Publishers; 2013.
[10] Ostafin A, Landfester K. Nanoreactor engineering for life sciences and medicine. Boston, MA: Artech House; 2009.
[11] Takaoka K, et al. Crystallographic observation of an olefin photodimerization reaction that takes place via thermal molecular tumbling within a self-assembled host. Chem Commun 2006;15:1625–7.
[12] Thomas JM, Raja R. Exploiting nanospace for asymmetric catalysis: confinement of immobilized, single-site chiral catalysts enhances enantioselectivity. Acc Chem Res 2008;41:708–20.
[13] Yoshizawa M, Klosterman JK, Fujita M. Functional molecular flasks: new properties and reactions within discrete, self-assembled hosts. Angew Chem Int Ed 2009;48:3418–38.
[14] Salavati-Niasari M. Synthesis, characterization of cobalt(II) complex nanoparticles encapsulated within nanoreactors of zeolite-Y and their catalytic activities. J Mol Catal A Chem 2009;310:51–8.
[15] Salavati-Niasari M. Template synthesis and characterization of hexaaza macrocycles containing pyridine iron(II) complex nanoparticles dispersed within nanoreactors of zeolite-Y. Chem Commun 2009;12:359–63.
[16] Karimi B, Zareyee D. Design of a highly efficient and water-tolerant sulfonic acid nanoreactor based on tunable ordered porous silica for the von Pechmann reaction. Org Lett 2008;10:3989–92.
[17] Rostamnia S, Xin H. $Pd(OAc)_2$@SBA-15/PrEn nanoreactor: a highly active, reusable and selective phosphine-free catalyst for Suzuki–Miyaura cross-coupling reaction in aqueous media. Appl Organometal Chem 2013;27:348–52.
[18] Raynal M, et al. Supramolecular catalysis. Part 2: artificial enzyme mimics. Chem Soc Rev 2014;43:1734–87.
[19] Cook TR, Zheng YR, Stang PJ. Metal–organic frameworks and self-assembled supramolecular coordination complexes: comparing and contrasting the design, synthesis, and functionality of metal–organic materials. Chem Rev 2013;113:734–77.
[20] Yoon M, Srirambalaji R, Kim K. Homochiral metal–organic frameworks for asymmetric heterogeneous catalysis. Chem Rev 2012;112:1196–231.
[21] Douglas T, Young M. Virus particles as templates for materials synthesis. Adv Mater 1999;11:679–81.
[22] Shenton W, et al. Inorganic-organic nanotube composites from template mineralization of tobacco mosaic virus. Adv Mater 1999;11:253–6.
[23] Douglas T, Young MJ. Host–guest encapsulation of materials by assembled virus protein cages. Nature 1998;393:152–5.
[24] Meldrum FC, et al. Synthesis of inorganic nanophase materials in supramolecular protein cages. Nature 1991;349:684–7.
[25] Shin Y, Dohnalkova A, Lin Y. Preparation of homogeneous gold-silver alloy nanoparticles using the apoferritin cavity as a nanoreactor. J Phys Chem C 2010;114:5985–9.

[26] Ueno T, et al. Size-selective olefin hydrogenation by a Pd nanocluster provided in an apo-ferritin cage. Angew Chem Int Ed 2004;43:2527–30.
[27] Archibald DD, Mann S. Template mineralization of self-assembled anisotropic lipid micro-structures. Nature 1993;364:430–3.
[28] Shenton W, et al. Synthesis of cadmium sulphide superlattices using self-assembled bacterial S-layers. Nature 1997;389:585–7.
[29] Coffer J, et al. Dictation of the shape of mesoscale semiconductor nanoparticle assemblies by plasmid DNA. Appl Phys Lett 1996;69:3851–3.
[30] Alivisatos AP, et al. Organization of "nanocrystal molecules" using DNA. Nature 1996;382:609–11.
[31] Arora PS, Kirshenbaum K. Nano-tailoring: stitching alterations on viral coats. Chem Biol 2004;11:418–20.
[32] Patterson DP, Prevelige PE, Douglas T. Nanoreactors by programmed enzyme encapsulation inside the capsid of the bacteriophage P22. ACS Nano 2012;26:5000–9.
[33] Hastings CJ, et al. Enzyme-like catalysis of the Nazarov cyclization by supramolecular encapsulation. J Am Chem Soc 2010;132:6938–40.
[34] Pluth MD, Bergman RG, Raymond KN. Proton-mediated chemistry and catalysis in a self-assembled supramolecular host. Acc Chem Res 2009;42:1650–9.
[35] Fujita M. Metal-directed self-assembly of two- and three-dimensional synthetic receptors. Chem Soc Rev 1998;27:417–25.
[36] Slagt VF, et al. Assembly of encapsulated transition metal catalysts. Angew Chem Int Ed 2001;40:4271–4.
[37] Yanase M, Haino T, Fukazawa Y. A self-assembling molecular container for fullerenes. Tetrahedron Lett 1999;40:2781–4.
[38] Renggli K, et al. Selective and responsive nanoreactors. Adv Funct Mater 2011;21:1241–59.
[39] Wang D, Astruc D. Dendritic catalysis: basic concepts and recent trends. Coord Chem Rev 2013;257:2317–34.
[40] Horiuchi S, Murase T, Fujita M. Diels–Alder reactions of inert aromatic compounds within a self-assembled coordination cage. Chem Asian J 2011;6:1839–47.
[41] Hapiot F, et al. Recent breakthroughs in aqueous cyclodextrinassisted supramolecular catalysis. Catal Sci Technol 2014;4:1899–908.
[42] Assaf KI, Nau WM. Cucurbiturils: from synthesis to high-affinity binding and catalysis. Chem Soc Rev 2015;44:394–418.
[43] Nakanishi Y, Imae T. Synthesis of dendrimer-protected TiO2 nanoparticles and photodegradation of organic molecules in an aqueous nanoparticle suspension. J Colloid Interface Sci 2005;285:158–62.
[44] Warmuth R. Inner-phase stabilization of reactive intermediates. Eur J Org Chem 2001;423–37.
[45] Dong VM, et al. Molecular recognition and stabilization of iminium ions in water. J Am Chem Soc 2006;128:14464–5.
[46] Brumaghim JL, et al. Encapsulation and stabilization of reactive aromatic diazonium ions and the tropylium ion within a supramolecular host. Eur J Org Chem 2004;2004:5115–8.
[47] Mal P, et al. White phosphorus is air-stable within a self-assembled tetrahedral capsule. Science 2009;324:1697–9.
[48] Kawano M, et al. Direct crystallographic observation of a coordinatively unsaturated transition-metal complex in situ generated within a self-assembled cage. J Am Chem Soc 2006;128:6558–9.
[49] Arunkumar E, Forbes CC, Smith BD. Improving the properties of organic dyes by molecular encapsulation. Eur J Org Chem 2005;2005:4051–9.

[50] Balzani V, et al. Fluorescent guests hosted in fluorescent dendrimers. Tetrahedron 2002;58:629–37.
[51] Kohyama Y, Murase T, Fujita M. A self-assembled cage as a non-covalent protective group: regioselectivity control in the nucleophilic substitution of aryl-substituted allylic chlorides. Chem Commun 2012;48:7811–3.
[52] Sun G, et al. Controlling assembly of paired gold clusters within apoferritin nanoreactor for in vivo kidney targeting and biomedical imaging. J Am Chem Soc 2011;133:8617–24.
[53] Zhu Z, et al. Preparation of nickel nanoparticles in spherical polyelectrolyte brush nanoreactor and their catalytic activity. Ind Eng Chem Res 2011;50:13848–53.
[54] Chung YM, Rhee HK. Pt–Pd bimetallic nanoparticles encapsulated in dendrimer nanoreactor. Catal Lett 2003;85:159–64.
[55] Lu M, et al. Formation of carbon nanotubes in silicon-coated alumina nanoreactor. Carbon 2004;42:1846–9.
[56] Serp P, Philippot K. Nanomaterials in catalysis. Weinheim, Germany: Wiley-VCH Verlag & Co; 2013.
[57] Zhan B-Z, et al. Zeolite-confined nano-RuO_2: a green, selective, and efficient catalyst for aerobic alcohol oxidation. J Am Chem Soc 2003;125:2195–9.
[58] Suzuki K, Sato S, Fujita M. Template synthesis of precisely monodisperse silica nanoparticles within self-assembled organometallic spheres. Nature Chem 2010;2:25–9.
[59] Suzuki K, et al. The precise synthesis and growth of core–shell nanoparticles within a self-assembled spherical template. Angew Chem Int Ed 2011;50:4858–61.
[60] Tao L, et al. Modification of multi-wall carbon nanotube surfaces with poly(amidoamine) dendrons: synthesis and metal templating. Chem Commun 2006;47:4949–51.
[61] D'Aleo A, et al. Oligothia dendrimers for the formation of gold nanoparticles. Adv Funct Mater 2004;14:1167–77.
[62] Aranishi K, Zhu QL, Xu Q. Dendrimer-encapsulated cobalt nanoparticles as high- performance catalysts for the hydrolysis of ammonia borane. Chem Cat Chem 2014;6:1375–9.
[63] Mizugaki T, et al. PAMAM dendron-stabilised palladium nanoparticles: effect of generation and peripheral groups on particle size and hydrogenation activity. Chem Commun 2008;2:241–3.
[64] Knecht MR, Garcia-Martinez JC, Crooks RM. Synthesis, characterization, and magnetic properties of dendrimer-encapsulated nickel nanoparticles containing <150 atoms. Chem Mater 2006;18:5039–44.
[65] Gao H, et al. Dendrimer-encapsulated copper as an oveloligonucleotides label for sensitive electrochemical stripping detection of DNA hybridization. Biosens Bioelectron 2013;48:210–5.
[66] Chung Y-M, Rhee H-K. Dendrimer-templated Ag–Pd bimetallic nanoparticles. J Colloid Interface Sci 2004;271:131–5.
[67] Liu H, et al. Tunable synthesis and acetylation of dendrimer-entrapped or dendrimer-stabilized gold–silver alloy nanoparticles. Colloids Surf B 2012;94:58–67.
[68] Knecht MR, et al. Structural rearrangement of bimetallic alloy PdAu nanoparticles within dendrimer templates to yield core/shell configurations. Chem Mater 2008;20:1019–28.
[69] Zhao Y, et al. Architecture of stable and water-soluble CdSe/ZnS core–shell dendron nanocrystals via ligand exchange. J Colloid Interf Sci 2009;339:336–43.
[70] Korovin AN, et al. Nanoreactor-assisted polymerization toward stable dispersions of conductive composite particles. Macromol Rapid Commun 2011;32:462–7.
[71] Murase T, Sato S, Fujita M. Nanometer-sized shell molecules that confine endohedral polymerizing units. Angew Chem Int Ed 2007;46:1083–5.

[72] White SR, et al. Autonomic healing of polymer composites. Nature 2001;409:794–7.
[73] De Oliveira H, Thevenot J, Lecommandoux S. Smart polymersomes for therapy and diagnosis: fast progress toward multifunctional biomimetic nanomedicines. Wiley Interdiscip Rev Nanomed Nanobiotechnol 2012;4:525–46.
[74] Liu G-Y, Chen C-J, Ji J. Biocompatible and biodegradable polymersomes as delivery vehicles in biomedical applications. Soft Matter 2012;8:8811–21.
[75] Cheng Z, Thorek DLJ, Tsourkas A. Porous polymersomes with encapsulated Gd-labeled dendrimers as highly efficient MRI contrast agents. Adv Funct Mater 2009;19:3753–9.
[76] Kesharwani P, Jain K, Jain NK. Dendrimer as nanocarrier for drug delivery. Prog Polym Sci 2014;39:268–307.
[77] Kesharwani P, Iyer AK. Recent advances in dendrimer-based nanovectors for tumor-targeted drug and gene delivery. Drug Discovery Today 2015;20:536–47.
[78] De Oliveira H, Thevenot J, Lecommandoux S. Smart polymersomes for therapy and diagnosis: fast progress toward multifunctional biomimetic nanomedicines. WIREs Nanomed Nanobiotechnol 2012;4(5):525–46.
[79] Liu GY, Chen CJ, Ji J. Biocompatible and biodegradable polymersomes as delivery vehicles in biomedical applications. Soft Matter 2012;8:8811–21.
[80] Qiao Z, Shi X. Dendrimer-based molecular imaging contrast agents. Prog Polym Sci 2015;44:1–27.
[81] Kim KT, et al. Smart nanocontainers and nanoreactors. Nanoscale 2010;2:844–58.
[82] Cabane E, Malinova V, Meier W. Synthesis of photocleavable amphiphilic block copolymers: toward the design of photosensitive nanocarriers. Macromol Chem Phys 2010;211:1847–56.
[83] Cabane E, et al. Photoresponsive polymersomes as smart, triggerable nanocarriers. Soft Matter 2011;7:9167–76.
[84] Pangu GD, et al. Ultrasonically induced release from nanosized polymer vesicles. Macromol Biosci 2010;10:546–54.
[85] Dua J, O'Reilly RK. Advances and challenges in smart and functional polymer vesicles. Soft Matter 2009;5:3544–61.

Chapter 2

Cyclodextrins as Porous Material for Catalysis

Jolanta Rousseau, Stéphane Menuel, Cyril Rousseau, Frédéric Hapiot, Eric Monflier
University of Artois, Laboratory of Catalysis and Solid State Chemistry (UCCS), Faculty of Sciences Jean Perrin, Lens, Cédex, France

1 CYCLODEXTRINS: A BRIEF OVERVIEW

1.1 Structure and Supramolecular Properties

Cyclodextrins (CDs) have never ceased to fascinate scientists around the world due to their structures and properties [1]. These inexpensive natural cyclic oligosaccharides could be obtained from the enzymatic degradation of a renewable material, namely starch. This family of water-soluble macrocycles are built up from D-(+)-glucopyranosyl units linked by α-1,4-glycosidicbonds (Fig. 2.1). In the enzymatically produced CD family, three derivatives are mainly known: α-, β-, and γ-CD, constituted of six, seven, and eight glucose units, respectively (Fig. 2.1).

Due to the rigid $^4C_1$-chair conformation of glucose units, CDs possess a truncated cone shape with a hydrophilic external surface and a hydrophobic cavity (Fig. 2.2). The internal hydrophobicity mainly results from the inner hydrogens H-3 and H-5 pointing into the cavity. The external hydrophilicity originates from the hydroxyl groups on both sides of the conical structure, pointing away from the cavity and toward the aqueous environment. The narrowest primary side bears the 6-OH groups and the largest secondary side bears the 2- and 3-OH groups (Fig. 2.2). The different reactivity of the CD hydroxyl groups permits us to envisage a large range of selective substitutions on the primary and secondary rims, including different polymerization reactions.

The particular three-dimensional CD structure with its hydrophilic surface and apolar cavity, impacts water solubility and the ability to partially or totally encapsulate hydrophobic molecules of appropriated size and shape in aqueous solution as well as in solid-state through the formation of a reversible "host–guest" complex (Fig. 2.3) [2].

Organic Nanoreactors. http://dx.doi.org/10.1016/B978-0-12-801713-5.00002-1

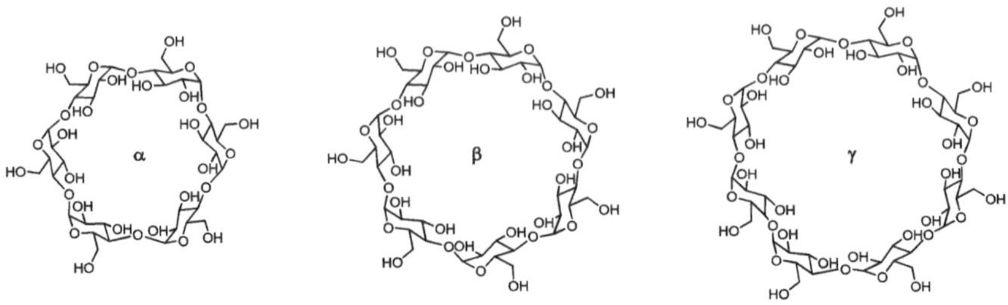

FIGURE 2.1 **Structures of the native α-, β-, and γ-CD.**

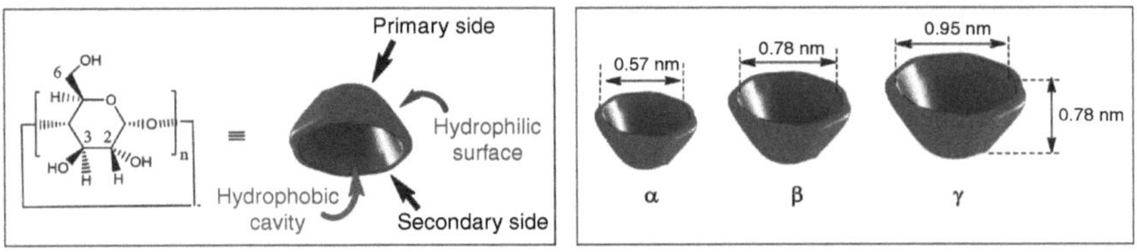

FIGURE 2.2 **3D structures with geometric dimensions of α-, β-, and γ-CD.**

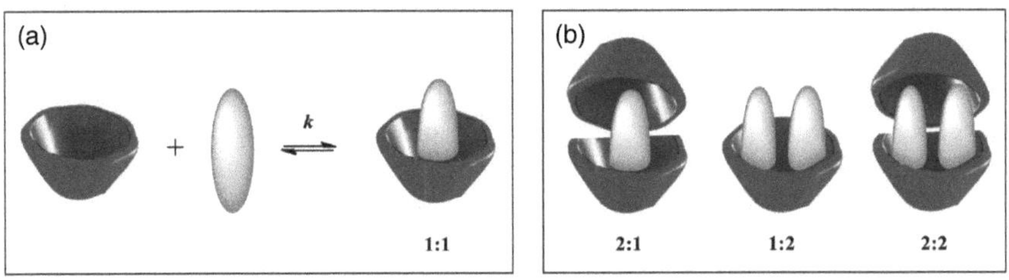

FIGURE 2.3 **Schematic illustration of the supramolecular association existing between CDs as "host" and a "guest" molecule (a) and examples of stoichiometries for "host-guest" complexes (b).**

The driving forces resulting in these inclusion complexes are attributed to various factors such as hydrophobic interactions, van der Waals forces, electronic effects, and steric factors. These interactions make them behave like enzymes to yield high specificity and reaction rates, especially in water.

1.2 CD-Based Polymers

While native CDs are appropriate for the molecule recognition of a wide range of substrates, specific applications require the use of more elaborated CD-structures. Concurrently to mono- and polysubstituted CDs, CD-based polymers appear as a powerful tool. For example, covalent polymer networks containing CDs are of great interest because their cross-linked macromolecular structures reveal a cooperative action between the CD cavities or between the cavities

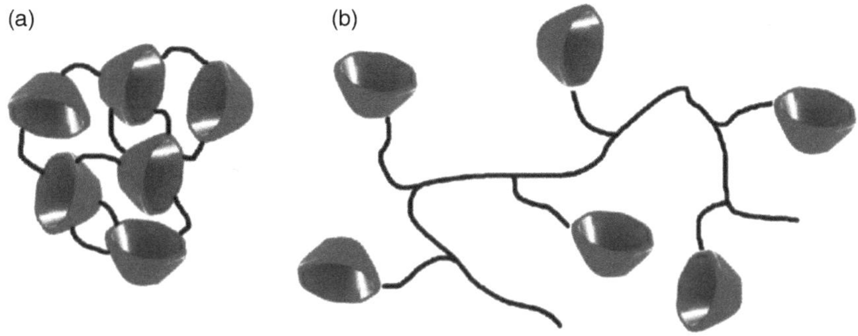

FIGURE 2.4 **Cross-linked (a) and linear (b) CD-based polymers.**

and the polymeric network. Several CD-based polymers have already been described. They can be classified in two categories according to the polymer type, namely reticulated cross-linked CD-based polymers or linear polymers bearing pendant linked CD (Fig. 2.4).

Different parameters can be changed for the modification of the properties of the CD-based polymers: type of the CDs, the nature of the cross-linker or the polymer and the ratio between hydrophobic (cross-linker or polymer) and hydrophilic part (CDs) used for the synthesis. The tuning of these parameters gives polymers with different characteristics (chemical and mechanical properties, surface area, etc.) [3].

1.2.1 Cross-Linked CD-Based Polymers

- Urethane cross-linker: Urethane CD-based polymers are obtained by reaction of the CD hydroxyl groups with diisocyanates via polycondensation reaction. The most widely used diisocyanate cross-linkers are tolylene-2,4-diisocyanate (TDI) and hexamethylene diisocyanate (HDI) [4]. It is also possible to synthesize polymers with specific properties with exotic diisocyanates such as isophorone diisocyanate (IPDI) (Scheme 2.1) [5].
  A major limitation in the use of diisocyanates as cross-linkers is their toxicity. In this context, they cannot be used for food, pharmaceuticals, or cosmetics.
- Carbonate cross-linker: CD-based polycarbonates are obtained by reaction of the hydroxyl groups of the CDs with *N,N'*-carbonyldiimidazole (CDI) or diphenylcarbonate (DPC). During the carbonate cross-linking formation, phenol or imidazole are produced as byproducts and are generally removed by Soxhlet extraction (Scheme 2.2) [6].
- Ether cross-linker: Polyether CD-based polymers are mostly prepared from epoxides opening in basic conditions [7]. The most widely used epoxides are epichlorohydrin (EPI) and its nontoxic equivalent, the ethylene glycol diglycidyl ether (EGDE) (Scheme 2.3).

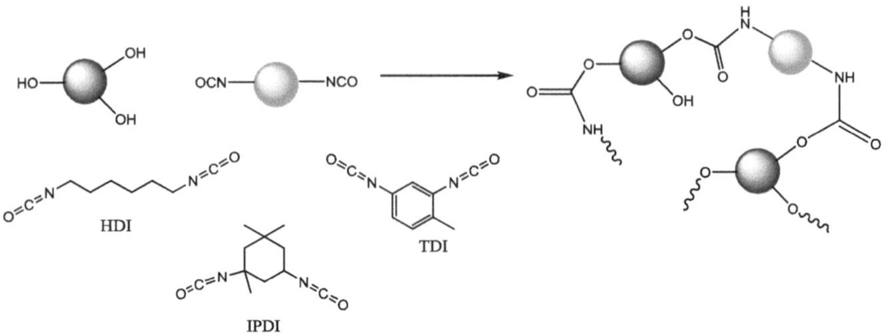

SCHEME 2.1 **CD-based polyurethanes synthesis from diisocyanate and polyols.** *(Adapted from Ref. [4,5].)*

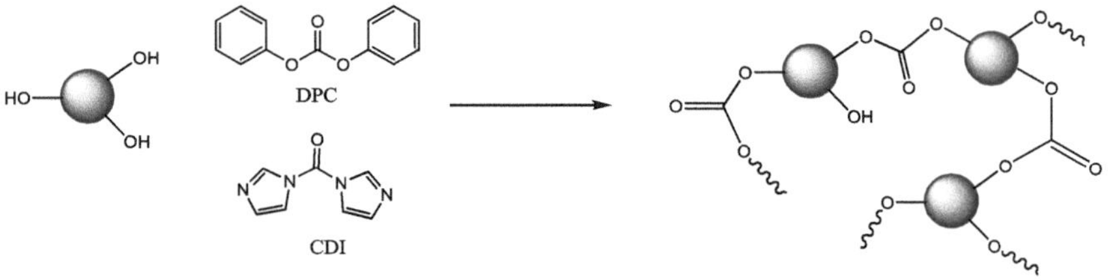

SCHEME 2.2 **CD-based polycarbonates synthesis from carbonates and polyols.** *(Adapted from Ref. [6].)*

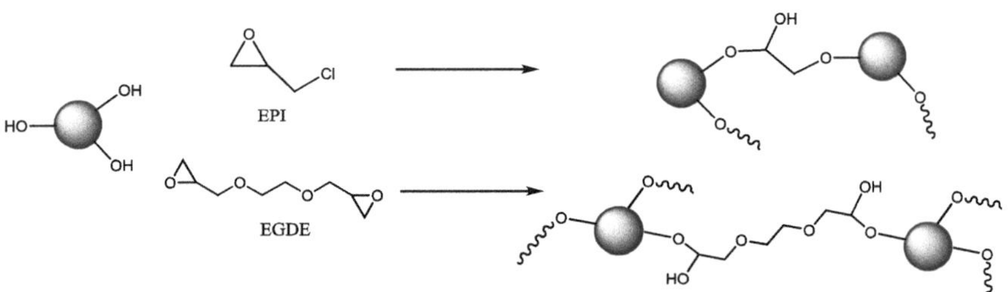

SCHEME 2.3 **Polyethers synthesis from EPI or EGDE and polyols.** *(Adapted from Ref. [7].)*

1.2.2 Linear CD-Based Polymers

Two different approaches can be used to synthesize linear CD-based polymers. The first one consists in initially synthesizing CD-monomers that can be engaged in a polymerization process to obtain CD-based polymers. Classic CD functionalizations are generally made with acrylates or vinyl derivatives (see next section) [8]. The second method consists in reacting native or modified CDs with an already synthesized polymer to functionalize it (see next section).

1.3 Applications of CDs

Because of their unique features, CDs are powerful tools for many industrial products, technologies, and analytical methods. The negligible cytotoxic effects of CDs are one of their essential characteristics for industrial applications, such as drug carriers, foods and flavors, cosmetics, packing, textiles, formulation, separation processes, environment protection, and fermentation, particularly in the area of catalysis. As such, the key target in the development of new technologies through catalysis is to find more highly selective processes that occur in mild conditions, involve fewer auxiliary stages (separation, purification), and produce less in waste and byproducts [9]. Homogeneous catalytic systems offer advantages of high activity and selectivity and mild reaction conditions. Wide use of homogeneous catalysis is restricted first of all by the high cost of separation of catalysts from reaction products, which prevents catalyst recycling [10]. A possible approach to go around these problems consists in the immobilization of active metal complex onto an organic or inorganic support. However, there are certain problems associated with this approach, such as a much restricted mobility of catalysts (which strongly affects their properties), aggregation processes, and washing out of the products from catalyst.

Biphasic catalysis is an elegant alternative to recover the water-soluble catalyst at the end of the reaction by simple decantation (Fig. 2.5). Once the reaction is complete, the catalyst is retained in the aqueous compartment while reaction products are recovered in the organic phase. As a result, the catalyst is readily separated and can be recycled.

As organic substrates are generally poorly soluble in water, a phase transfer agent is needed in addition to the use of a water-soluble catalyst. In this context, CDs have been efficiently used as mass transfer promoters and are still the subject of intensive research [11]. It is known that CDs and modified CDs can be efficiently used as mass transfer agents because an inclusion complex can be formed between the CDs and the organic substrate [12]. This has been done with a broad scope of reactions as alkylation [13], metathesis [14],

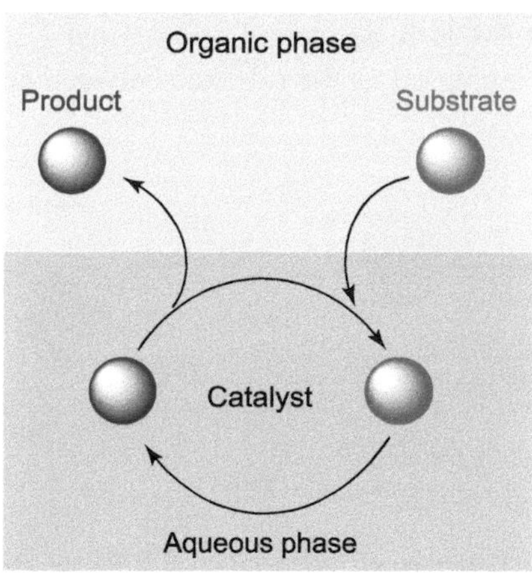

FIGURE 2.5 **Biphasic catalysis process.**

telomerization [15] or polymerization reactions [16], and in rhodium-catalyzed biphasic hydroformylation of higher olefins [17].

2 CD-BASED POLYMERS AS MASS-TRANSFER PROMOTERS

In this part, we will focus on the synthesis of CD-based polymers and their use in mass transfer promoters. Their beneficial impact on the catalytic performance is especially described through relevant examples of the literature.

2.1 Ester Hydrolysis

Pioneered works related to CD-based polymers for catalysis were reported by Nozakura's group. The model reaction was the hydrolysis of various *p*-nitrophenol esters [18]. The CD-based polymers synthesis was realized in a controlled two steps procedure. First, CD-acryloyl monomers were synthesized and purified by gel chromatography through Sephadex G15. The monomers were then polymerized in an H_2O/MeOH mixture (v/v 1/1) using 2,2′-azobis (isobutyronitrile) as an initiator to give polyacryloyl-CD polymers (Scheme 2.4).

These CD-based polymers were engaged in hydrolysis reaction of various *p*-nitrophenol esters derivatives. The reaction was monitored by UV at 400 nm. Among the CD-based polymers, only poly-β-CD-A showed a high hydrolysis activity. The pseudo first-order rate constant k' measured for the ester hydrolysis catalyzed by poly-β-CD-A is compared to the rate constant measured using the native β-CD. The rate enhancements resulting from the use of poly-β-CD-A are presented in Table 2.1.

Rate enhancements ranging from 1.7 to 3.3 were correlated to the binding of the ester with the hydrophobic cavity of the CDs in the polymer structure.

The cooperative action between neighboring β-CD units grafted onto the polymer backbone was also demonstrated by running hydrolysis of *p*-nitrophenyl *p*-nitrobenzoate (*p*NP*p*NB) with different polymers carrying β-CD at changing distances from each other. The polymers tested were poly-β-CD-A, poly-β-CD-NAC, and β-CD-A-acrylamide copolymer (1:6). The acceleration effect of the polymer increased as the distance between neighboring CD units

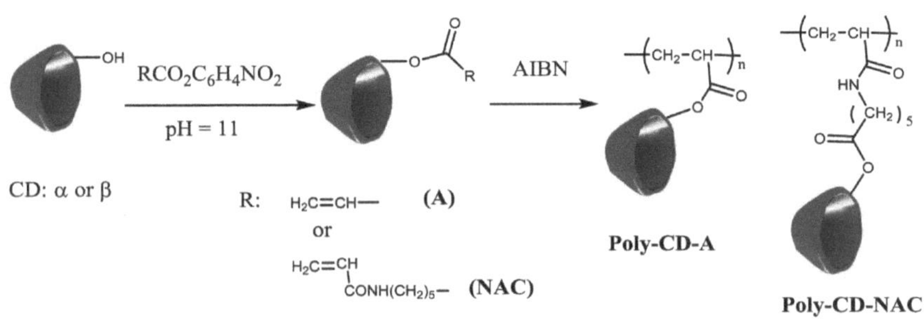

SCHEME 2.4 Two-step synthesis of polyacryloyl CD-based polymers. *(Adapted from Ref. [18a] with permission of Wiley.)*

TABLE 2.1 Hydrolysis Rate Constants of *p*-Nitrophenyl Esters with Native β-CD and Poly-β-CD-A

p-Nitrophenyl esters	k_{uncat} 10^{-4} (s^{-1})	$k'_{\beta\text{-CD}}$ 10^{-4} (s^{-1})	$k'_{polymer}$ 10^{-4} (s^{-1})	$k'_{polymer}/k_{\beta\text{-CD}}$ 10^{-4} (s^{-1})
Acetate	0.90	2.73	4.54	1.7
Propionate	0.77	2.20	3.43	1.6
Valerate	0.53	1.28	2.76	2.1
Benzoate	0.08	0.1	0.31	3.1
p-Nitrobenzoate	1.96	4.64	15.5	3.3
p-Chlorobenzoate	0.18	0.58	1.81	3.1
p-Toluate	0.04	0.11	0.33	3.0

Adapted from Ref. [18] with permission of Wiley and American Chemical Society.

TABLE 2.2 Hydrolysis of *p*NP*p*NB With β-CD Derivatives

Catalyst	k' 10^{-4} (s^{-1})	$k'/k'_{\beta\text{-CD}}$ 10^{-4} (s^{-1})
β-CD	4.65	—
Poly-β-CD-A	15.5	3.3
Poly-β-CD-NAC	8.6	1.8
β-CD-A-acrylamide copolymer (1:6)	3.8	0.8

Adapted from Ref. [18 b] with permission of American Chemical Society.

decreased. Poly-β-CD-A, in which β-CD moieties were located most closely, had an effect 3.3 times larger than β-CD and 4.1 times larger than the β-CD-A-acrylamide copolymer, in which the β-CD units were far from each other. The efficacy of poly-β-CD-NAC was in between (Table 2.2).

Others polymers were also engaged in *p*-nitrophenyl acetate hydrolysis. CD-poly (vinylamine) polymers showed a higher catalytic activity than β-CD alone and a cooperative action between CD residues and NH_2 groups [19]. Seven CD-poly (vinylamine) polymers were synthesized. The substitution degree (SD) of amino functions substituted by β-CD ranged from 0.3 to 5 mol% (Scheme 2.5).

The catalytic activity of all polymers was always higher than that of CD alone and was inversely proportional to the number of CDs on the polymer backbone. The polymer having the lower SD was the best one in terms of catalytic activity. The authors showed that both CD and NH_2 groups are important for the esters hydrolysis. CDs concentrate the ester in the polymer environment, and the NH_2 groups are the active functions involved in the cleavage.

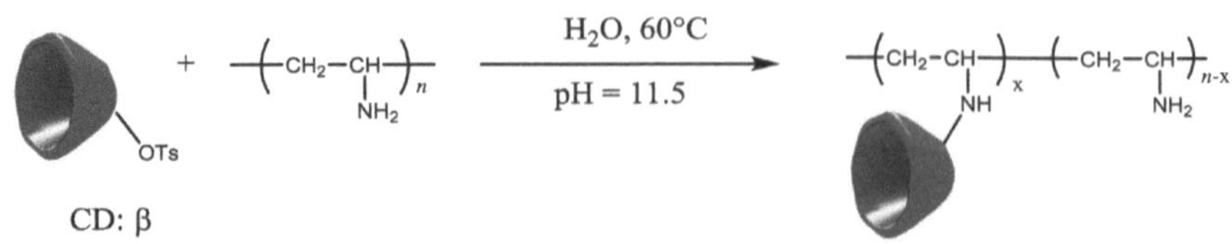

SCHEME 2.5 **Synthesis of CD-poly (vinylamine) polymers [19].**

2.2 Nucleophilic Substitution

Copolymers consisting of β-CD and EPI were used for the nucleophilic substitution of halogeno-alkanes [20]. The ratio between CD and EPI varied from 3 to 10. Methylated copolymers were also synthesized from the previous ones by methylation. The nucleophilic substitution of three bromo-alkanes was considered as a model reaction using sodium iodide as nucleophile in the presence of β-CD alone or with the synthesized CD-EPI copolymers (Scheme 2.6).

In all cases, the reaction was much faster when catalyzed by the CD-EPI copolymers. The rate has increased with a direct relation of the size of the substrate. These results can be explained by the contribution of more than one β-CD that forms a supramolecular complex that acts during the phase-transfer process [20].

2.3 Oxidation

CDs immobilized onto cellulose with EPI as cross-linker were prepared and tested in oxidation of cinnamaldehyde to benzaldehyde (Scheme 2.7) [21].

Different cellulose-supported CD catalysts were used to assess their effectiveness for the oxidation process. CDs immobilized onto the cellulose were α-CD, β-CD, and 2-hydroxypropyl-β-CD (2-HP-β-CD). The amount of CDs immobilized onto the polymers was 0.18 mmol/g. The authors showed the importance of the CD cavity size, especially regarding the reaction yield. The α-CD-cellulose copolymer (cavity size of ~5.2 Å) gave benzaldehyde in 59%

$$\text{R-Br} + \text{NaX} \xrightarrow[\text{or Polymers}]{\beta\text{-CD}} \text{R-X} + \text{NaBr}$$

$$\text{R} = n\text{-}C_8H_{17}\,;\ n\text{-}C_{12}H_{25}\,;\ n\text{-}C_{18}H_{37} \qquad \text{X} = \text{I}\,;\ \text{Cl}\,;\ \text{CN}\,;\ N_3\,;\ \text{SCN}$$

SCHEME 2.6 **Nucleophilic substitution of halogeno-alkane.** *(Adapted from Ref. [20].)*

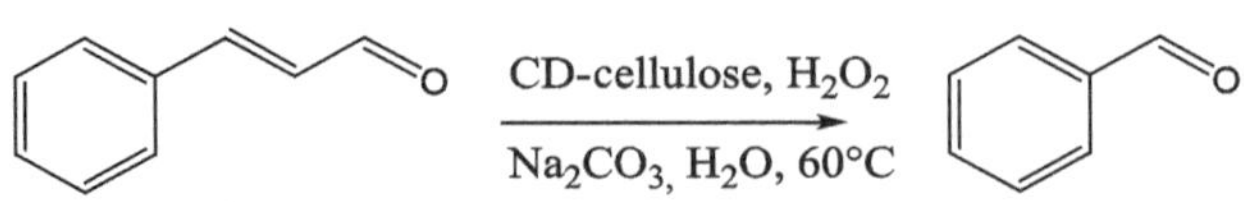

SCHEME 2.7 **Oxidation of cinnamaldehyde to benzaldehyde.** *(Adapted from Ref. [21] with permission of American Chemical Society.)*

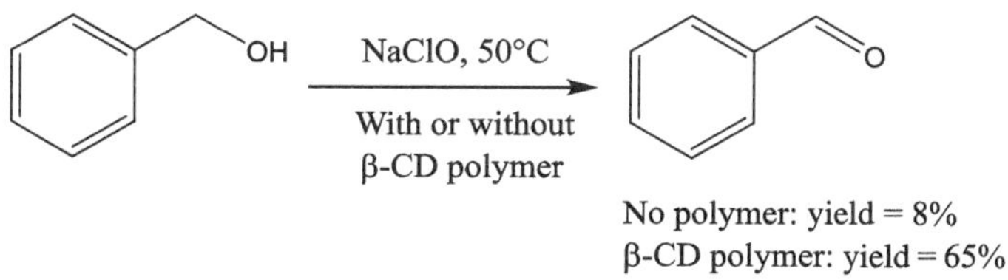

SCHEME 2.8 Benzyl alcohol oxidation in the presence of β-CD-EP copolymer. *(Adapted from Ref. [22] with permission of Elsevier.)*

yield, while 2-HP-CD (cavity size of ~9.6 Å) yielded 81% benzaldehyde under the same conditions.

The authors also proposed a mechanistic explanation, where the hydroxyl group attached to the CD interacted with the carbonyl group of the cinnamaldehyde by weak intermolecular interaction. This significantly changed the chemical environment of the carbon within the inclusion complex and facilitated the oxidation reaction.

Benzyl alcohol oxidation was also realized with a β-CD-EPI copolymer [22]. A EP/CD ratio of 44 was found to be optimal (Scheme 2.8).

Compared with the control experiment realized in the presence of NaClO (oxidation yield of 8%), the yield of the reaction rose up to 65% using the β-CD-EPI copolymer.

2.4 Aldol Condensation

A copolymer bearing proline and permethylated β-CD was used as catalyst in aldol condensation [23]. The linear copolymer bearing both pendant permethylated β-CD and proline groups was designed on the basis that the hydrophobic cavity of the CDs could approach the substrates close to the proline that acted as catalyst through host–guest interactions. The synthesis of the CD monomer was carried out by a copper-catalyzed azide–alkyne cycloaddition. The CD monomer was then polymerized with a protected hydroxyproline methacrylate to give the linear polymer with a monomer ratio proline/CD of 4.The Me-β-CD-Pro polymer was subsequently obtained after acid deprotection of proline (Scheme 2.9).

A different behavior of the Me-β-CD-Pro copolymer was observed upon varying the pH of the solution. It was found that the polymer had an isoelectric point at pH 3.8 (pH at which the two charges neutralize each other as proline is a zwitterion). At this pH, DLS analysis showed that there are no more single chains but only homogeneous nanoaggregates with an average diameter around 100 nm. Conversely, at pH7, the copolymers existed as individual entities with an average diameter of 10–12 nm. The efficacy of the Me-β-CD-Pro polymer was assessed in an aldol reaction using *p*-nitrobenzaldehyde and cyclohexanone as reactants at pH 7 or pH 3 (Scheme 2.10).

The Me-β-CD-Pro copolymer catalyzed the aldol reaction both at pH 7 and 3.8. At pH 7, the reaction was much faster, with 91% yield conversion within 7h, but the reaction was not stereoselective (Table 2.3, entry 1). At pH 3.8, the

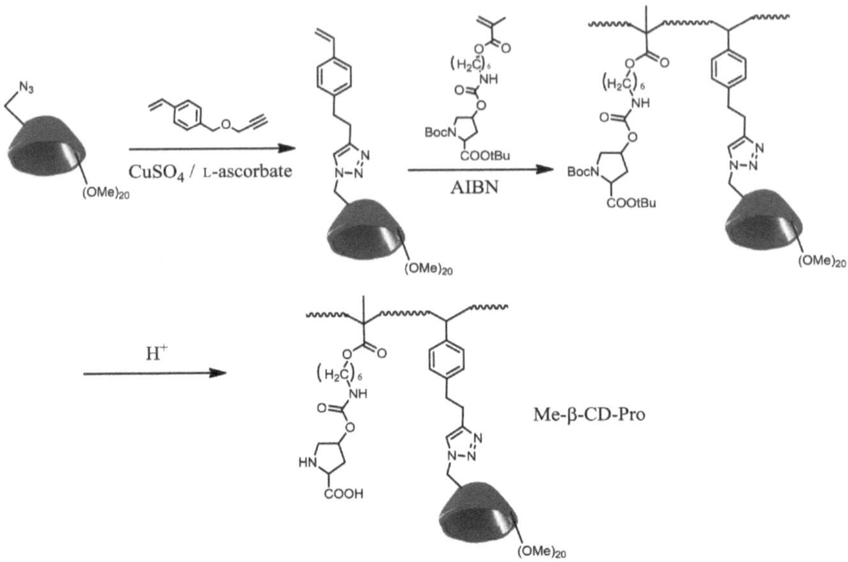

SCHEME 2.9 **Synthesis of the linear copolymer Me-β-CD-Pro.** *(Adapted from Ref. [23] with permission of American Chemical Society.)*

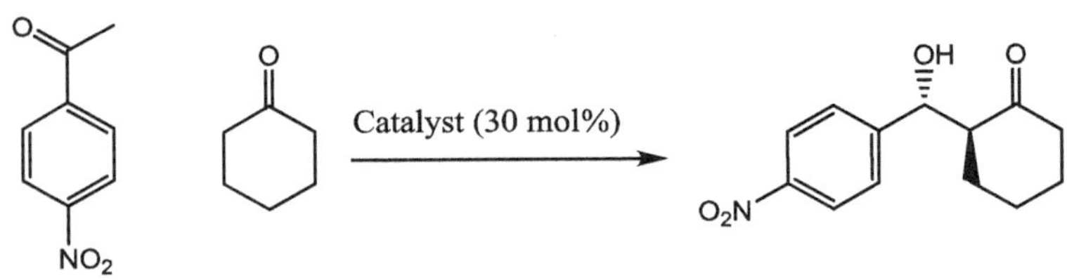

SCHEME 2.10 **Aldol reaction with Me-β-CD-Pro as catalyst.** *(Adapted from Ref. [23] with permission of American Chemical Society.)*

reaction was slower than at pH 7, but the stereoselectivity was much higher (Table 2.3, entry 2).

To explain the catalytic results, an "exclusion of water" mechanism was proposed. Considering that the nonionizable structure of the copolymer backbone was globally hydrophobic, the interpolymer complexation and the formation of larger aggregates were compatible with an expulsion of water from the vicinities of the inner active centers, allowing the reaction to occur in a stereoselective way. The decrease in the reaction rate was in agreement with a poorer accessibility to the active sites.

2.5 Organometallic Catalysis

The catalytic performance of pendant poly (*N*-acryloyloxysuccinimide) (poly-NAS) bearing monoamino randomly methylated β-CD (RAME-β-CD—NH_2) and ethanolamine (Scheme 2.11) was evaluated in aqueous Rh-catalyzed

TABLE 2.3 Aldol Reaction Catalyzed by Me-β-CD-Pro at pH 7 and pH 3.8

Entry	pH	Time (h)	Yield (%)	Anti/syn	ee (%)
1	7.0	4.5	67	1.4/1	6
		7	91	1.4/1	18
2	3.8	24	26	>20/1	>99
		48	41	5/1	>99

Adapted from Ref. [23] with permission of American Chemical Society.

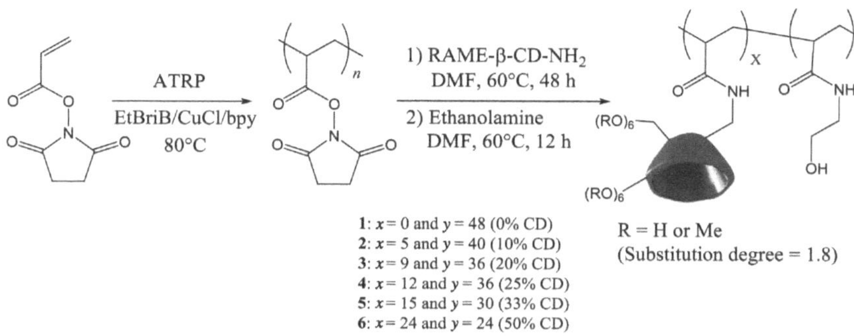

SCHEME 2.11 Synthesis of the RAME-β-CD-based polymers. *(Adapted from Ref. [24] with permission of American Chemical Society.)*

hydroformylation of 1-decene and 1-hexadecene under 50 bar of CO/H_2 at 80°C [24].

The catalytic results showed that high-substituted CD-based polymers were more effective than low-substituted polymers. For 1-decene that could interact with only one CD cavity, polymers **5** and **6** (high CD-substituted polymers) were as efficient as RAME-β-CD alone. In that case, each CD covalently attached to the polymer chain acted more or less as a CD alone. For the longer 1-hexadecane substrate, the low CD-substituted polymer **2** gave approximately the same yield of RAME-β-CD (~40%). In contrast, polymers **5** and **6** gave much higher yields (~80%). This results from the synergic cooperation of two CDs in close vicinity on the polymer backbone.

CD-grafted polymers synthesized from poly(*N*-acryloxysuccinimide) (poly-NAS), mono-amino randomly methylated β-CD (RAME-β-CD—NH_2), and pendant water-soluble phosphane moieties were also used as additives in the aqueous Rh-catalyzed hydroformylation of 1-hexadecene [25]. Within these polymers, both the supramolecular properties of the CD and the coordination ability of the phosphane were combined into the same molecular object. The catalytic results show that increasing the local concentration of the metal–phosphane complex and the CDs at the water/organic interface improves the reaction conversion. The explanation is that—after recognition of the substrate by the CDs, and as they are both graft on the polymer (CD and phosphane)—they can interact and react rapidly due to the proximity of all the partners of the reaction (CD, catalyst, and substrate).

3 IMPRINTED CD-BASED POLYMERS FOR CATALYSIS

Molecular imprinting is a method developed for the design of materials with memory of the shape of the templates. Molecules of similar shape, size, and functional group will fit perfectly into the binding sites created during the synthesis of the materials. Molecularly imprinted polymers (MIPs) constitute a promising material that offers many applications for the separation sciences and purification [26], drug delivery [27], biological antibodies receptors systems [28], and catalysis [29]. This is due to their ability to recognize various molecules like amino acids and proteins [30], nucleotide derivatives [31], pollutants [32], drugs, and food [33].

The general principle of molecular imprinting is presented in Scheme 2.12. A template molecule is mixed with functional monomers that can have specific interaction with the template. This results in a self-preorganized complex between the template molecule and the monomers. A cross-linker is added and this preorganized complex undergoes a polymerization to produce a rigid structure bearing imprinted sites. Eventually, removal of the template liberates cavities that can specifically recognize and bind the target molecule (Scheme 2.12).

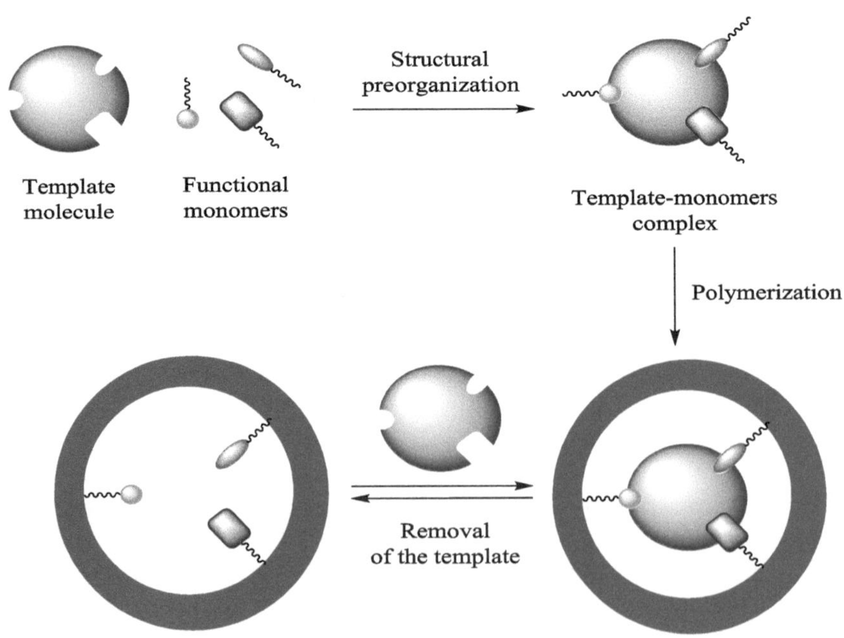

SCHEME 2.12 Molecular imprinting method.

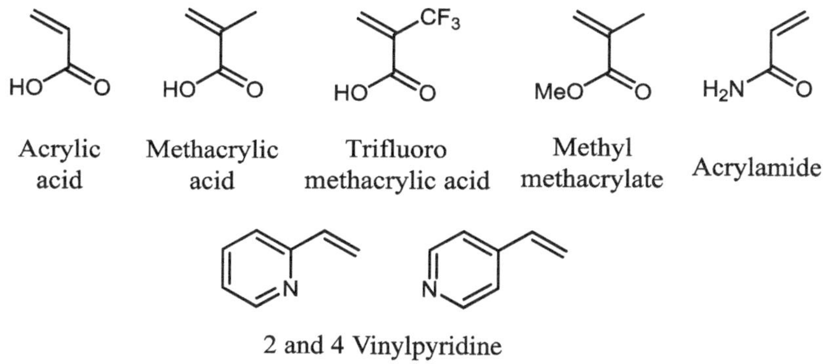

FIGURE 2.6 Structure of common monomers for the synthesis of MIPs. *(Adapted from Ref. [34].)*

Today, the most common method used to obtain CD-based MIPs is the radical polymerization of vinyl monomers. Generally, the polymerization procedure is performed under mild reaction conditions [34]. These processes can be used with a wide range of functional groups and template structures.

The choice of the monomers is very important, as they should associate with the template to create highly specific cavities after removal of the latter. Typical functional monomers are carboxylic acids and their derivatives, sulfonic acids, amides, alcohols, and heteroaromatic bases (Fig. 2.6), as they can interact with the template as hydrogen-bond donors and acceptors and also through ionic interactions.

The role of the cross-linker is also important as it will control the morphology of the polymer matrix and also fix the functional group of the monomers around the template. They also stabilize the binding sites and give mechanical stability to the polymers. The type and amount of the cross-linkers have a deep impact on the selectivity and the binding capability of the MIPs. Common cross-linkers are showed in Fig. 2.7.

Efforts have been made to develop new functional monomers adapted to the synthesis of MIP that are known to possess good binding capacity, such as CDs

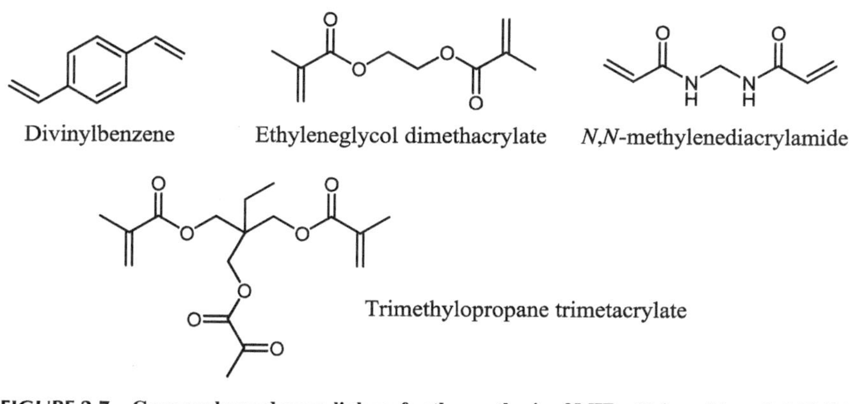

FIGURE 2.7 Commonly used cross-linkers for the synthesis of MIPs. *(Adapted from Ref. [34].)*

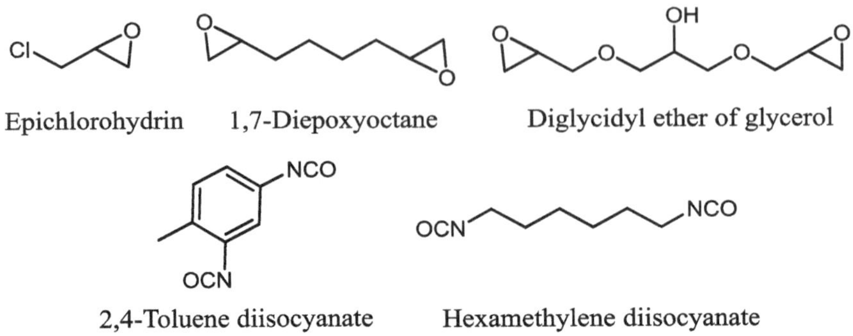

FIGURE 2.8 **Commonly used cross-linker for the synthesis of CD-MIPs.** *(Adapted from Ref. [35b].)*

[35], calixarenes [36], and crown-ethers [37]. MIPs based on these host molecules were used for recognition, waste treatment, and also catalysis.

The synthesis of MIPs from CDs as functional "monomers" is also based on the structural preorganization around a template molecule, but with one difference. In this approach, it is the template that binds inside the CD cavity, thus forming a host–guest complex. Two polymerization methods involving different cross-linkers can be used, depending on the nature of the CD (native or chemically modified).

- The first polymerization method concerns native CDs and diisocyanates cross-linkers such as 2,4-toluene diisocyanate or hexamethylene diisocyanate in DMSO (Fig. 2.8), leading to polyurethanes.
- The second method concerns native CDs or chemically vinylmodified CDs in water. The native CDs are polymerized with epoxides (mainly EPI) as cross-linkers. Vinyl monomers of CDs can also be polymerized by radical polymerization with acrylic derivatives, similar to the elaboration of traditional MIPs (Fig. 2.6).

Although this method requires in some cases the synthesis of vinyl monomer of CDs, it has several advantages over the method in DMSO. Receptors can be prepared for template molecules with various functional groups (carboxylate, amine, and others) to react with the diisocyanate. In water, the formation of inclusion complexes between the template and CDs is much more efficient than in DMSO, making imprinting more effective.

The choice of the template molecule is also very important. It was established that a minimum size is required to have an efficient molecular imprinting. For a good molecular imprinting, the size of the template should be at least 10 Å, which allows the formation of a complex with a CD/template stoichiometry of 2/1. With a template bigger than 10 Å, imprinted dimer or trimer macroligands are obtained, while the functionalized monomer is obtained with a small template or in the absence of a template (Scheme 2.13).

The size of a cross-linker also has a considerable effect on the efficacy of the imprinting. The distance between its reacting groups should correspond to the distance between CD molecules in a complex. A number of binding agents

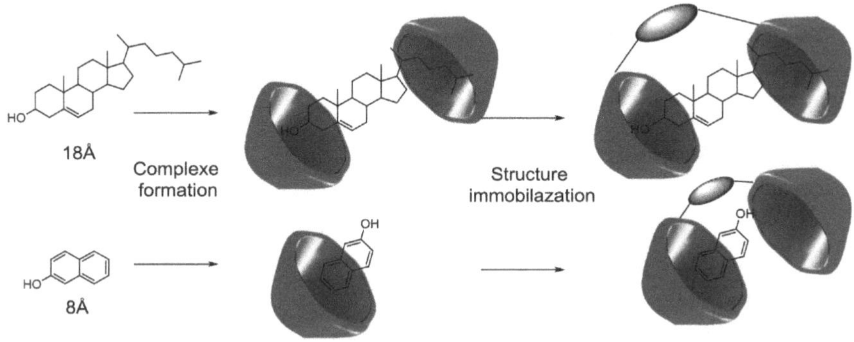

SCHEME 2.13 Binding modes in CD molecular imprinting. *(Adapted from Ref. [35b].)*

was used for the imprinting of 1,1′-bis(2-naphthol), and the efficacy of imprinting was shown to decrease in the following order: EPI, 1,3-diepoxybutane, 1,7-diepoxyoctane, diglycidyl ether of glycerol. The yield in dimeric products was maximum for EPI as a cross-linking agent.

CDs that imprinted macroligand were involved in different catalytic processes such as Wacker oxidation or oxidative coupling. The influence of the template and the linker was studied. A comparison of the reaction performances was established between the imprinted macroligand and the nonimprinted macroligand or CD monomers.

3.1 Wacker Oxidation

Karakhanov group performed a comparative study on the Wacker oxidation of various alkenes with different CD-based imprinted macrocycles [38]. In Table 2.4, the different CD-based MIPs assessed in terms of the oxidation reaction of several terminal alkenes are gathered. The Wacker oxidation was carried out in conditions where the rate of Pd^0 and Cu^0 reoxidation was high compared to the rate of substrate transformation, under oxygen pressure, at 50°C, with a $PdSO_4$:$CuSO_4$:HPA (heteropolyacid) mixture (1:10:10) in a two-phase water/substrate system.

The efficiency of the molecular imprinting and thereby of the ligand ability to recognize a guest molecule could be characterized by the ratio of the turnover frequency (TOF) for the molecularly imprinted macroligand (TOF_{imp}) to the turnover frequency for the ligand synthesized without the template molecule (TOF_{nonimp}):

$$\mathrm{TOF} = \frac{n(\text{product})}{n(\text{cat}) \times t};\ \mathrm{Eff} = \frac{\mathrm{TOF}_{\text{imp}}}{\mathrm{TOF}_{\text{nonimp}}}$$

The authors assessed MIP-CD efficacy using terminal C_7–C_{16} alkenes. The results are presented in Fig. 2.9. For the two MIP-CD families, the imprinted macrocycles are always more efficient than the nonimprinted, except for 1-nonene. The size of the template also seems to be important, as the

TABLE 2.4 Different MIPs-CD Tested in Wacker Oxidation

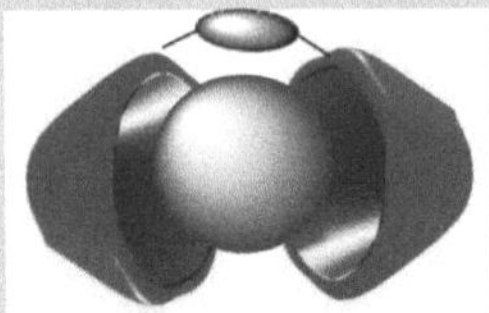			

β-CD	2,4-Toluenediiso-cyanate (TDI)	—	β-CD(TDI)
β-CD	2,4-toluenediiso-cyanate (TDI)	Dodecene	β-CD(TDI)$(C_{12}H_{24})_{imp}$
β-CD	2,4-toluenediiso-cyanate (TDI)	Hexadecene	β-CD(TDI)$(C_{16}H_{32})_{imp}$
β-CD-6-$(CH_2$-CH_2-COOH$)_n$	*N*,*N*-methylenedi-acrylamide (MDA)	—	β-CD-6-$(CH_2$-CH_2-COOH$)_n$(MDA)
β-CD-6-$(CH_2$-CH_2-COOH$)_n$	*N*,*N*-methylenedi-acrylamide (MDA)	Hexadecene	β-CD-6-$(CH_2$-CH_2-COOH$)_n$(MDA) $(C_{16}H_{32})_{imp}$
β-CD	2,4-toluenediiso-cyanate (TDI)	*Tert*-butyl-stryrene	β-CD(TDI)$(tBuSty)_{imp}$

Adapted from Ref. [38].

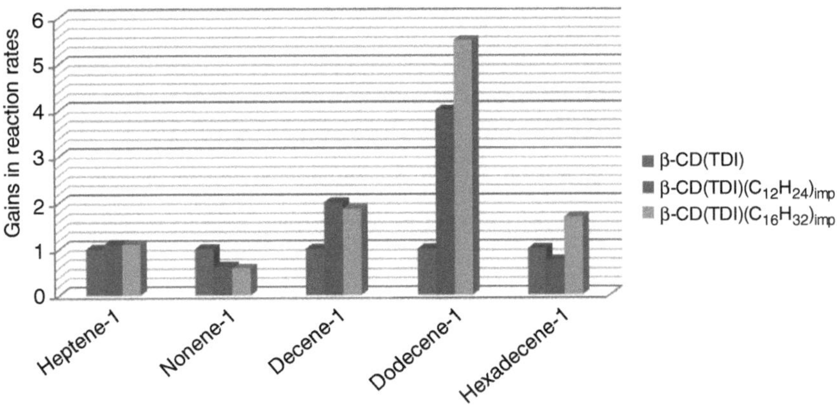

FIGURE 2.9 Efficiency of the Wacker oxidation of *n*-alkenes with various MIP-CD. *(Adapted from Ref. [38].)*

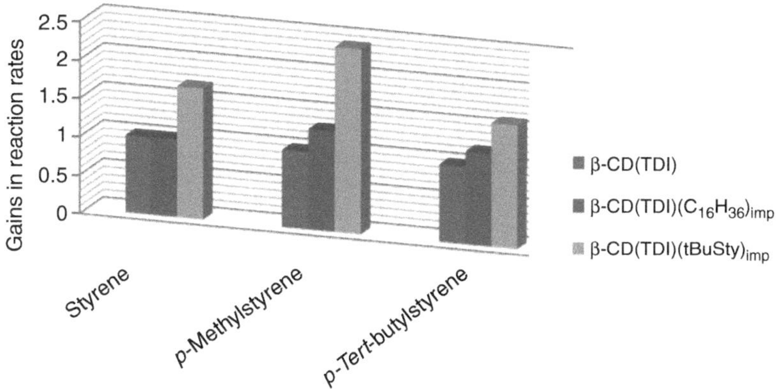

FIGURE 2.10 Efficiency of the Wacker oxidation of styrene derivatives with various MIP-CD. *(Adapted from Ref. [38].)*

macrocycles imprinted with 1-hexadecene give the best results. The authors say that is probably due to the optimal size of 1-hexadecene.

Differences in the rate of *n*-alkenes oxidation were also observed when different cross-linkers were used. Switching from the control macroligand to the one imprinted with 1-hexadecene and *N*,*N*-methylenediacrylamide increased the reaction rate for all the substrates, starting from not only 1-dodecene and 1-hexadecene but also 1-heptene. It should be noted that changing the cross-linking agent on more polar EPI led to a negative effect on the reaction rates. Karakhanov et al. argued that it could probably be due to the hydrophilicity of glycidyl residues that lowered the inclusion complex stability.

The Wacker oxidation was also realized with styrene derivatives instead of linear alkene. This study also confirmed the importance of the template during the imprinting process. Styrene derivatives were submitted to the Wacker oxidation conditions in the presence of CD-based MIPs formed in the presence or in the absence of different templates ($C_{16}H_{32}$ or *p-tert*-butylstyrene) to evaluate the efficacy of the templating approach. The results are presented in Fig. 2.10.

When the MIP-CD was elaborated from a linear alkene (1-hexadecene) as template (β-CD(TDI)$(C_{16}H_{32})_{imp}$), the improvement of the reaction was not significant when compared to the MIP-CD elaborated without template [β-CD(TDI)]. This was probably due to the inappropriate shape of the cavity formed during the imprinting. In contrast, when the macroligand was made with a styrene derivative (*p-tert*-butylstyrene), the oxidation rate was 2.5 times faster than that measured with the reference [β-CD(TDI)].

3.2 Oxidative Coupling

Oxidative coupling was also realized with MIP-CD, especially homocoupling of naphthols and coupling between styrene derivatives and benzene to give stilbene derivatives.

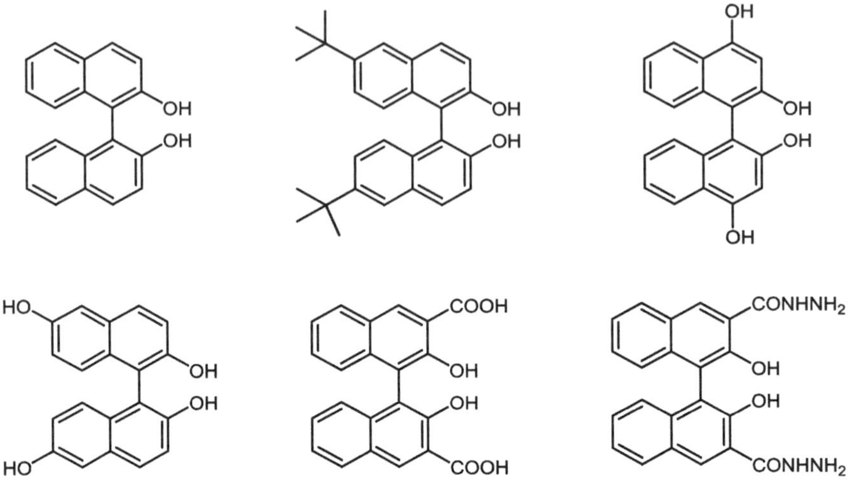

FIGURE 2.11 Bis-naphthols used as template. *(Adapted from Ref. [40] with permission of Wiley.)*

3.2.1 Naphthol Derivatives Homocoupling

The effect of the template on MIP-CD efficiency during the catalytic process has been studied. Diverse bis-naphthols were tested as the template (Fig. 2.11), and the polymerization was carried out using EPI, diepoxyethane, or diglycidyl ether of glycerol [39]. The best MIP-CD was realized using 1,1′-bisnaphtol as template and EPI as cross-linker (β-CD(EPI)(Binol)$_{imp}$).

The β-CD(EPI)(Binol)$_{imp}$ was engaged in the β-naphtol dimerization in the presence of iron III as oxidant in a chloroform/water mixture (1/1) at 60°C. The results are summarized in Table 2.5. The macroligand obtained via imprinting β-CD(EPI)(Binol)$_{imp}$ was considerably more active than systems imprinted with β-CD(EPI) and $FeCl_3$. It should also be pointed out that the catalyst was

TABLE 2.5 β-Naphthol Dimerization Catalyzed by Fe(III) Complexes Using Macroligands

Catalytic system	Time (min)	Yield (%)
$FeCl_3$	30 120	26 43
$FeCl_3$/β-CD(Epy)	30 120	29 66
$FeCl_3$/(β-CD(Epy)(Binol)$_{imp}$	30 120	53 85
$FeCl_3$/(β-CD(Epy)(Binol)$_{imp}$, 2nd cycle	120	79

Adapted from Ref. [40] with permission of Wiley.

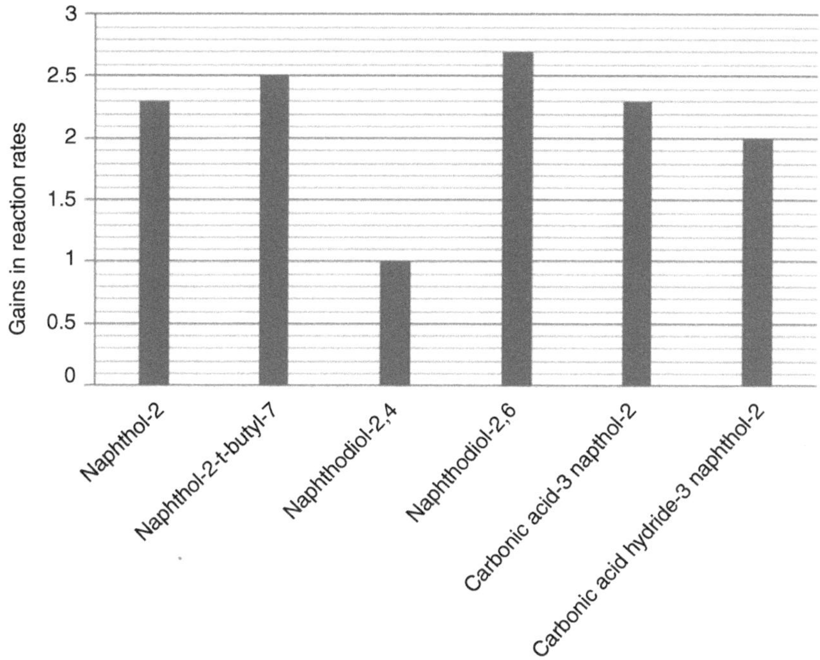

FIGURE 2.12 Fe(III)-catalyzed naphthol dimerization using β-CD(EPI)(Binol). *(Adapted from Ref. [39].)*

easily isolated from the reaction products and could be used repeatedly without significant loss in activity.

Various β-naphthol derivatives were also engaged in the oxidative dimerization reaction [40]. The rate of the oxidative combination for these 2-naphthols was considerably higher using β-CD(Epy)(Binol)$_{imp}$ than for similar catalysts synthesized in the absence of template molecules. The values of efficiency, indicating the reaction acceleration, are shown in Fig. 2.12. In all cases, the improvement with MIP-CD is significant, showing that the macroligand templating facilitates the approach of the reacting naphthalene fragments. However, no specificity of the substrate was observed.

The oxidative coupling of benzene with styrene and some paraderivatives catalyzed by palladium (II) acetate was also investigated in the presence of MIP-CD [41]. Under this condition, the approach of the reacting molecules was also required for a sufficient reaction rate, and the product itself could be a model of the transition state. Trans-stilbene was used as a template for molecular imprinting. Pd-complexes with the synthesized macroligands turned out to be effective catalysts for the oxidative coupling process (Fig. 2.13).

The reaction yield was higher for the macroligand synthesized in the presence of the template compared to the nonimprinted macroligand. It should be noted that, if the cross-linking is performed without the template molecule, the CD hydroxyl groups are substituted randomly, and the structure of the inclusion

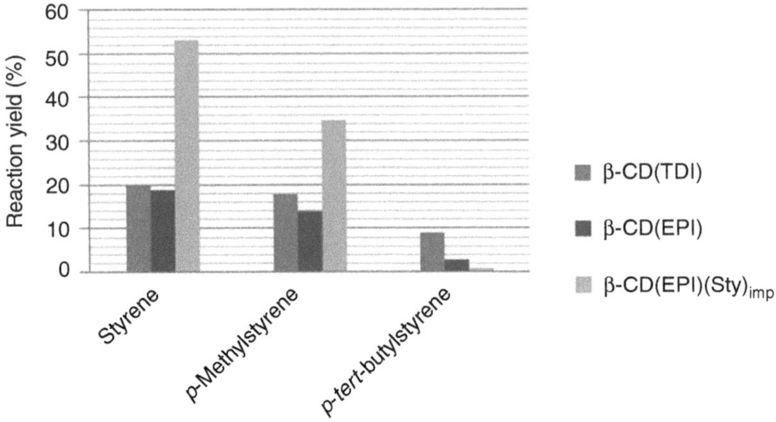

FIGURE 2.13 **Oxidative coupling of styrene derivatives using β-CD(EPI)(Sty)$_{imp}$.** *(Adapted from Ref. [41] with permission of Springer.)*

complexes that form prohibits the approach of the reacting molecules, thus preventing the formation of any C—C bond. Besides, when hydroxyls are substituted, the chaotic arrangement even decreases the reaction yield compared to the process occurring in the presence of nonmodified CD. The effect of the reacting molecule sizes could be clearly understood with the benzene and *p-tert*-butylstyrene combination. The reaction rate was maximal for the catalyst containing nonmodified *β*-CD, the rate being very slow. It appeared that *p-tert*-butylstyrene was too large and could not approach benzene efficiently for coupling using the stilbene-imprinted macroligand.

To conclude this part, molecular imprinting with CDs-based materials could be an effective approach for the synthesis of catalysts capable of converting bulky molecules in two-phase systems, or for overcoming rate limitations for processes requiring an optimal arrangement of several reacting molecules. The possibility of adjusting the catalyst activity and selectivity by a careful selection of a template and cross-linking agents, the suitability of imprinted materials both for two-phase systems and heterogeneous supports for metal complexes and nanoparticles is a striking advantage of the suggested method.

4 CD-BASED NANOSPONGES

CD-based nanosponges are soft materials featuring various physico-chemical proprieties, notably a modulation of the swelling capacity by a judicious choice of the cross-linking agent and CD molar ratio to control the reticulation (Fig. 2.14). The resulting rigidity of the network and the hydrogen-bond dynamics affect the water-holding capacity to produce water confinement in nanosponges [42] and also modify the elastic properties [43]. This class of hyper-cross-linked polymer has a typical nanostructural organization with numerous nanosized cavities and a lot of interconnecting voids to form porous

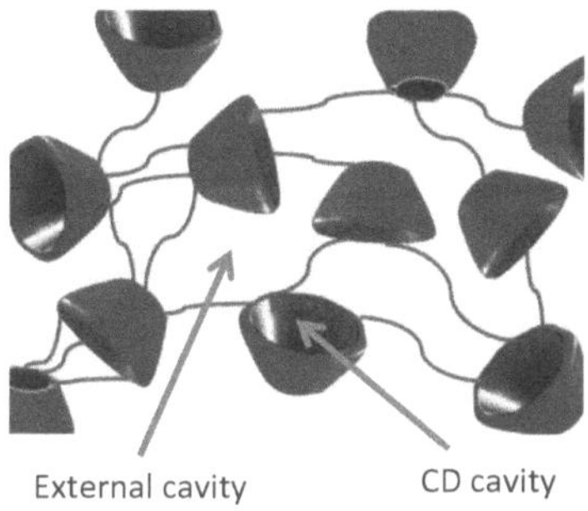

FIGURE 2.14 **Schematic representation of CD-based nanosponges.**

materials. Nanosponges have the ability of physisorption of numerous substances, both hydrophobic and hydrophilic, with a high flexibility [44]. First introduced for the hyper-cross-linked polystyrene [45], the concept of nanosponges was rapidly used for environmental applications, notably for water purification technology [46]. Outstanding properties of CDs allowed improving the removal of organic pollutants from water [47]. Because of their harmlessness, CD-based nanosponges are currently used in pharmaceutical research [48], especially for injectable formulation and oral administration [49]. CD-based nanosponges were also used to improve the solubility of drugs in anticancer therapy [50] and the modulation of drug release [51].

The inherently chiral cavity introduced by CDs was efficiently used for the enantiodifferentiating photoisomerization of cyclooctenes through intermolecular interactions [52]. In this approach, supramolecular effects were expected to control photochirogenesis by nonconvalent interactions between guests and CDs at the ground and excited states. In this strategy, the confinement of the substrate plays a major role for the control of the chirality transfer during the excited step, notably by the increase in activation of free energy between the two transition states to preferentially orient the reaction toward only one stereoisomer (Scheme 2.14) [53].

To implement this concept, the crucial point is the accessibility of the CD cavities, notably after their chemical modification to synthesize nanosponges.

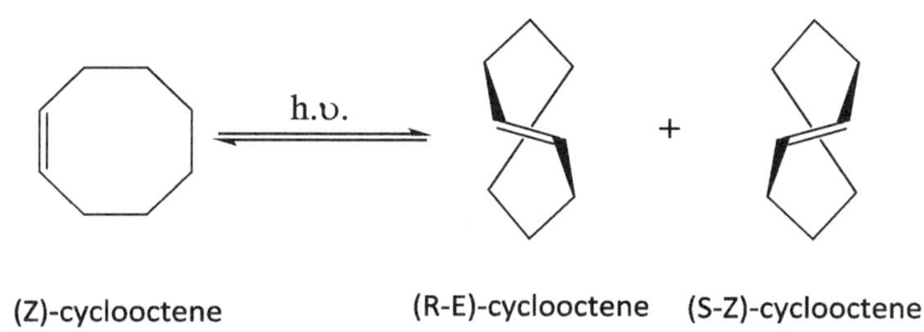

SCHEME 2.14 **Enantiodifferentiating photoisomerization of cyclooctenes.** *(Adapted from Ref. [53] with permission of Royal Society of Chemistry.)*

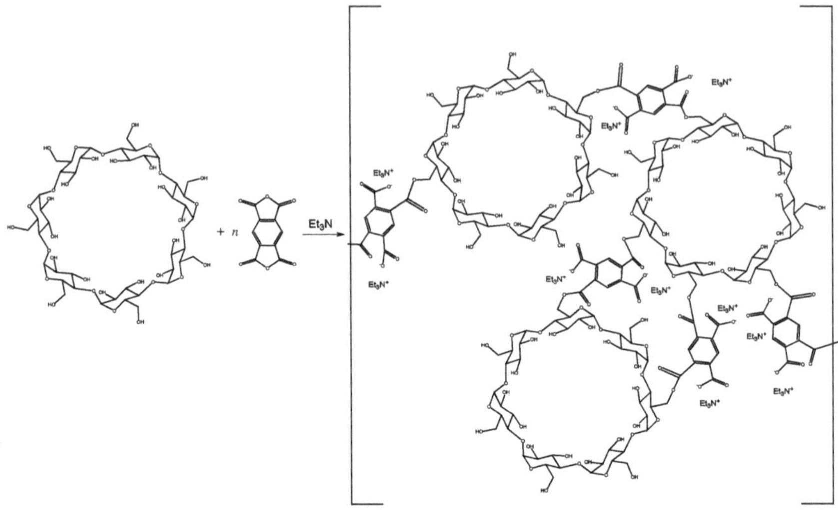

FIGURE 2.15 **Pyromellitate-linked polymeric CD-based nanosponges.** *(Adapted from Ref. [54].)*

This is true if these materials form solid polymer nanoparticles. With pyromellitate-linked polymeric CD-based nanosponges [54], the location of the linker plays a crucial role regarding the cavities' accessibility (Fig. 2.15). Indeed, the aromatic moiety can potentially be included in the CD or at least can cap the cavity. These last points explain the slight modification in the circular dichroism bands upon addition of cyclooctene into the nanosponges by exclusion of the shallowly included linker. However, the pyromellitic parts consist of two carboxylic functions available for an ionic modulation of the complexion properties of these materials. The solution's pH may change the conformation of the cavities' environment and will generate an increase in enantiodifferentiating photoisomerization by the CD nanosponges.

When increasing the aqueous concentration, pyromellitate CD-based nanosponges give highly viscous dispersions and eventually hydrogels network [55,56]. Dynamic light-scattering analysis of aqueous solutions of nanosponges showed an increase in the particle size with a sequential phase evolution. At low concentration, the polymeric solution led to the formation of aggregates evolving toward a precipitation of swelling particles. At higher concentration, a flowing gel (leading to a rigid gel for the most concentrated solution) was observed. This property allowed for a phase-controlled photochirogenesis by supramolecular association. Astonishingly, swelling gel media exhibited enantioselectivity for the cyclooctene photoisomerization [57]. The gelation mechanism was incriminated to rationalize this result. A fraction of pyromellitate residue did not link to CDs and was exposed at the surface of the nanosponge particles. The variation in the circular dichroism bands suggests an inclusion of the surfacing pyromellitate group by surfacing CDs of other particles. The

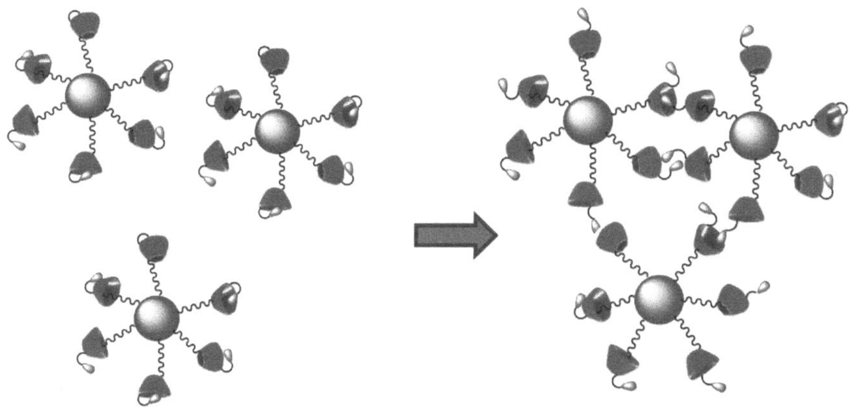

FIGURE 2.16 **Aggregation during swelling gelation process.**

resulting aggregation driven by hydrophobic forces led to the formation of a photochirogenic environment as a result of external cavities formed during the swelling gelation process (Fig. 2.16).

Note that a rigid gel, because of the less dynamic nature of its framework, greatly affects the efficacy of the supramolecular photochirogenesis. The importance of external cavities during the photoisomerization process was also indirectly revealed through investigations on the solvent effect. Indeed, polar solvents such as ethanol are well known to destabilize host/guest complexes [58,59], which results in a decrease in the enantioselectivity. However, such a decrease in the guest affinity upon addition of ethanol was not observed using CD-based nanosponges. This result did not find its justification in the cyclooctene/CD interaction but in the external CD-cavities of the nanosponges.

Another application is related to the ability of cross-linked CD-based polymers to produce metal-loaded solids. Different approaches led to high concentrations of metal salts in polymeric frameworks, among them a reticulation in the presence of metal ions solution [60] and the reflux of a polymer suspension in a metal salt solution [61] or a physical mixture in a planetary ball mill [62].

Taking a green approach, it was also possible to obtain high metal-loaded CD-based polymers by reticulation of CD and diisocyanate under sonication in the presence of Pd(II) and Cu(I) salts, with or without triphenylphosphine (Figure 2.17) [63]. The metal salt-containing cross-linked CD-based polymers led to polymer-nanoparticle composites upon dispersion in water. The resulting heterogeneous system was assessed in a typical Sonogashira reaction with *p*-iodoanisole and phenylacetylene as reactants and TPP as a stabilizing agent. In all cases, the nanocomposite system proved to be much more effective. The most interesting benefit of this polymeric system lies in the very low leaching of both metal and ligand. The efficient metal chelator and ligand encapsulation resulted in an easy separation of the products with no detrimental effect on the catalytic activity. In this context, the system could be recycled. Note that

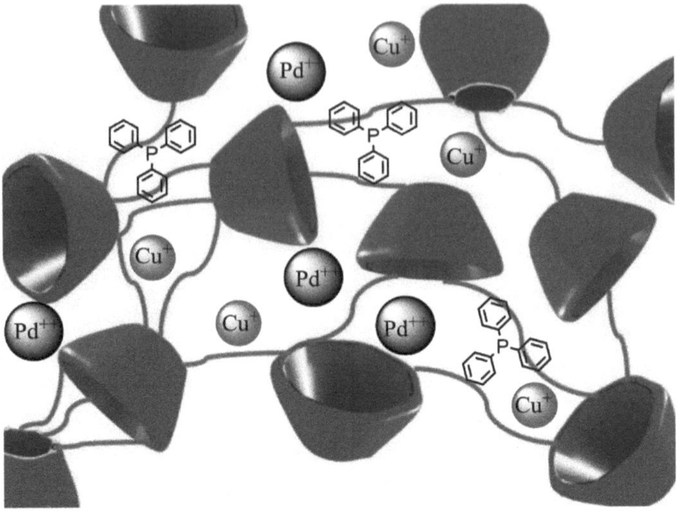

FIGURE 2.17 Metal-loaded CD-based polymers.

the CD-based nanocomposites system was effective in various solvents such as glycerol, DMF, and water (Fig. 2.17).

The polar structure of CDs and cross-linked derivatives is sensitive to dielectric heating. This was exploited in a green approach for cross-coupling reactions and alkyne–azide cycloadditions under microwave irradiation [64]. The solid metal-loaded cross-linked CDs were used without any ligand. In Heck reaction, Pd(II) nanocomposite CD-based polymers led to a complete conversion of styrene with 4-bromoacetophenone within 1 h, while a nearly complete conversion was obtained within 3 h using an oil bath (conventional heating). Similarly, a complete conversion of halogenoaryl with phenylboronic acid as observed in 45 min in a Suzuki reaction, while 3 h were required using an oil bath. These results were extended to the alkyne–azide cycloaddition reaction using Cu (I) nanocomposite cross-linked CD-based polymers as catalysts. High yields were obtained, with benzyl azide and phenylacetylene at 70°C for 10 min under microwave irradiations.

Thus, the use of metal-loaded cross-linked CD-based polymers is part of a green chemistry approach for organic synthesis, as much for the versatility of the potential catalytic applications as for their use under microwave assistance and their recyclability. Furthermore, the high absorption of organic compounds in nanosponges, associated with their chiral environments, place this material for promising applications in sustainable chemistry.

5 CONCLUSIONS

In conclusion, three different types of CD-based polymers (cross-linked polymers, polymers bearing pendant CDs, imprinted CD-based polymers or nanosponges) have been shown to be efficient materials in biphasic or aqueous catalytic processes. An accurate control of the catalytic activity and selectivity can be done by a judicious choice of CDs, cross-linking agent, type of copolymer, and synthetic method used. The complexion properties of CDs are used in a cooperative process, and the amount of hydrophilic CDs connected to the hydrophobic polymer moiety makes it possible to use these materials in aqueous catalysis.

Thus, concurrently with other CD-based materials for catalysis (such as CD-based metallic nanoparticles [65], β-CD-silica hybrid [66], and more recently CD-based metal–organic frameworks (MOF) [67]), CD-based polymers also appear to be very effective materials to improve catalytic performance in aqueous media.

REFERENCES

[1] (a) Villiers A. Compt Rend 1891;112:536–8. (b) Crini G. Chem Rev 2014;114(21):10940–75. and references cited therein.

[2] (a) Szejtli J. Chem Rev 1998;98:1743–53. (b) Martin del Valle EM. Process Biochem 2004;39:1033–46. (c) Cyclodextrins and their complexes. In: Dodziuk H, editor. Chemistry, analytical methods, applications. Germany: Wiley-VCH Verlag GmBH; 2006. p. 507.

[3] Folch-Cano C, Yazdani-Pedram M, Olea-Azar C. Molecules 2014;19:14066–79.

[4] (a) Ma M, Li D-Q. Chem Mater 1999;11:872–4. (b) Mohamed MH, Wilson LD, Headley JV. Carbohyd Res 2011;346:219–29. (c) Mhlanga SD, Marnba BB, Krause RW, Malefetse TJ. J Chem Technol Biotechnol 2007;82:382–8. (d) Xiao P, Dudal Y, Shahgaldian P, Corvini PFX. Polym Chem 2011;2:120–5. (e) Landfester K, Bechthold N, Tiarks F, Antonietti M. Macromolecules 1999;32:5222–8. (f) Landfester K. Angew Chem, Int Ed 2009;48:4488–507.

[5] (a) Tiarks F, Landfester K, Antonietti M. J Polym Sci Polym Chem 2001;39:2520–4. (b) Landfester K, Bechthold N, Tiarks F, Antonietti M. Macromolecules 1999;32:2679–83. (c) Kettel MJ, Hildebrandt H, Schaefer K, Moeller M, Groll J. ACS Nano 2012;6:8087–93.

[6] (a) Antonietti L, Aymonier C, Schlotterbeck U, Garamus VM, Maksimova T, Richtering W, Mecking S. Macromolecules 2005;38:5914–20. (b) Trotta F, Cavalli R. Compos Interfaces 2009;16:39–48. (c) Swaminathan S, Pastero L, Serpe L, Trotta F, Vavia P, Aquilano D, Trotta M, Zara G, Cavalli R. Eur J Pharm Biopharm 2010;74:193–201. (d) Mognetti B, Barberis A, Marino S, Trotta F, Berta G, Francia SD, Cavalli R. J Incl Phenom Macrocycl Chem 2012;74:201–10. (e) Trotta F, Cavalli R, Martina K, Biasizzo M, Vitillo J, Bordiga S, Vavia P, Ansari K. J Incl Phenom Macrocycl Chem 2011;71:189–94.

[7] (a) Fenyvesi E. J Incl Phenom Macrocycl Chem 1988;6:537–43. (b) Szeman J, Fenyvesi E, Szejtli J. J Incl Phenom 1987;5:427–31. (c) Murai S, Imajo S, Maki Y, Takahashi K, Hattori K. J Colloid Interf Sci 1996;183:118–23. (d) Renard E, Deratani A, Volet G, Sebille B. Eur Polym J 1997;33:49–57. (e) Martin R, Scanchez I, Cao R, Rieumont J. Supramol Chem 2006;18:627–31.

[8] (a) Folch-Cano F, Yazdani-Pedram M, Olea-Azar C. Molecules 2014;19:14066–79. (b) Tonelli AE. J Incl Phenom Macrocycl Chem 2007;60:197–202. (c) Munteanu M, Choi S, Ritter H. Macromolecules 2008;41:9619–23.

[9] (a) Anastas PT, Warner JC. Green chemistry: theory and practice. Oxford: Oxford University Press; 1998. (b) Sheldon R. Green chemistry. Weinheim: Wiley–VCH; 2003.

[10] Leadbeater NE, Marco M. Chem Rev 2002;102:3217–24.

[11] (a) Zahalka HA, Januszkiewicz K, Alper H. J Mol Catal 1986;35:249–53. (b) Benyei A, Joo F. J Mol Catal 1990;58:151–63. (c) Anderson JR, Campi EM, Jackson WR. Catal Lett 1991;9:55–8. (d) Monflier E, Blouet E, Barbaux Y, Mortreux A. Angew Chem Int Ed 1994;33:2100–2. (e) Pinel C, Gendreau-Diaz N, Bréhéret A, Lemaire M. J Mol Catal A Chem 1996;112:L157–61. (f) Kirschner D, Green T, Hapiot F, Tilloy S, Leclercq L, Bricout H, Monflier E. Adv Synth Catal 2006;348:379–87. (g) Kirschner D, Jaramillo M, Green T, Hapiot F, Leclercq L, Bricout H, Monflier E. J Mol Catal A Chem 2008;286:11–20.

[12] Leclercq L, Bricout H, Tilloy S, Monflier E. J Colloid Interface Sci 2007;307:481–7.

[13] (a) Hirano K, Yorimitsu H, Oshima K. Adv Synth Catal 2006;348:1543. (b) Hirano K, Yorimitsu H, Oshima K. Chem Commun 2008;28:3234–41.

[14] Brendgen T, Fahlbusch T, Frank M, Schühle DT, Seßler M, Schatz J. Adv Synth Catal 2009;351:303–7.

[15] Behr A, Leschinski J, Awungacha C, Simic S, Knoth T. Chem Sus Chem 2009;2:71–6.

[16] Ding L, Jiao X, Deng J, Zhao W, Yang W. Macromol Rapid Commun 2009;30:120.

[17] Hapiot F, Leclercq L, Azaroual N, Fourmentin S, Tilloy S, Monflier E. Curr Org Synth 2008;5:5162–71.

[18] (a) Furue M, Harada A, Nozakura SI. J. Polym. Science, Polym. Lett. Ed. 1975;13:357–60. (b) Harada A, Furue M, Nozakura SI. Macromol 1976;9:705–10.

[19] Martel B, Morcellet M. Eur Polym J 1995;31:1089–93.

[20] Tanak N, Yamaguchi A, Araki Y, Araki M. Chem Lett 1987;16:715–8.

[21] Zujin Y, Hongbing J. ACS Sustainable Chem Eng. 2013;1:1172–9.

[22] Jiang H, Yang Z, Zhou X, Fang Y, JI H. Chinese J Chem Eng 2012;20:784–92.

[23] Doyagüez EG, Rodriguez-Hernandez J, Corrales G, Fernandez-Mayoralas A, Gallardo A. Macromolecules 2012;45:7676–83.

[24] Potier J, Menuel S, Fournier D, Fourmentin S, Woisel P, Monflier E, Hapiot F. ACS Catal 2012;2:1417–20.

[25] Potier J, Menuel S, Hapiot F, Monflier E. Beilstein J Org Chem 2014;10:2642–8.

[26] (a) Tamayo FG, Turiel E, Martín-Esteban A. J Chromatogr A 2007;1152:32–40. (b) Haginaka J. J Chromatogr B 2008;866:3–13.

[27] Puoci F, Iemma F, Picci N. Curr Drug Deliv 2008;5:85–96.

[28] (a) Longo L, Vasapollo G. Met-Based Drugs 2008;2008:281843 (doi: 10.1155/2008/281843). (b) Ge Y, Turner EPF. Chem Eur J 2009;15:8100–7.

[29] Wulff G, Liu J. Acc Chem Res 2012;45:239–47.

[30] (a) Dong C, Lu R, Huaizhou Z, Chenjia X, Lu Z, Ying Y, Hengzhi W, Yucheng L, Roberts MF, Jeffrey C, Naughton MJ, Zhifeng R, Chiles TC. Nat Nanotechnol 2010;5:597–601. (b) Morelli I, Chiono V, Vozzi G, Ciardelli G, Silvestri D, Giusti P. Sens Actuators B Chem 2010;150:394–401. (c) Scorrano S, Mergola L, Del Sole R, Vasapollo G. Int J Mol Sci 2011;12:1735–43.

[31] Longo L, Vasapollo G. Mini-Rev Org Chem 2008;5:163–70.

[32] (a) Pichon V, Chapuis-Hugon F. Anal Chim Acta 2008;622:48–61.

[33] (a) Puoci F, Cirillo G, Curcio M, Iemma F, Spizzirri UG, Picci N. Anal Chim Acta 2007;593:164–70. (b) Baggiani C, Anfossi L, Giovannoli C. Anal Chim Acta 2007;591: 29–39.

[34] (a) Chen L, Xu S, Li J. Chem Soc Rev 2011;40:2922–42. (b) Vasapollo G, Del Sole R, Mergola L, Rosaria Lazzoi M, Scardino A, Scorrano S, Mele G. Int J Mol Sci 2011;12:5908–45.

[35] (a) Asanuma H, Hishiya T, Komiyama M. Adv Mater 2000;12:1019–30. (b) Karakhanov EA, Maximav AL. Curr Org Chem 2010;14:1284–95.

[36] Lu C-Y, He H-C, He X-W, Zeng X-S. Chinese J Chem 2004;22:275–8.
[37] Luo X, Liu L, Deng F, Luo S. J. Mater Chem A 2013;1:8280–6.
[38] (a) Karakhanov EA, Maksimov AL, Rusova EA, Kardasheva YS, Terenina MV, Buchneva TS, Guchkova AY. Macromol Symp 2003;204:159–73. (b) Karakhanov EA, Karapetyan LM, Kardasheva YS, Maksimov AL, Runova EA, Skorkin VA, Terenina MV. Macromol Symp 2006;235:39–51.
[39] Karakhanov EA, Kardahseva YS, Maksimov AL, Egazar'yants SV, Karapetyan LM, Zatolochnaya O. A. Petrol Chem 2007;47:402–8.
[40] Karakhanov EA, Karapetyan LM, Kardahseva YS, Maksimov AL, Runova EA, Terenina MV, Filippova TY. Macromol Symp 2008;270:106–16.
[41] Karakhanov EA, Maksimov AL, Runova EA, Kardasheva YS, Terenina MV, Kardashev SV, Skorkin VA, Karapetyan LM, Russian TalanovaMY. Chem Bull Int Ed 2008;57:780–92.
[42] Crupi V, Fontana A, Majolino D, Mele A, Melone L, Punta C, Rossi B, Rossi F, Trotta F, Venuti V. J Incl Phenom Macro Chem 2014;80:69–75.
[43] Crupi V, Fontana A, Giarola M, Longeville S, Majolino D, Mariotto G, Mele A, Paciaroni A, Rossi B, Trotta F, Venuti V. J Phys Chem B 2014;118:624–33.
[44] Tejashri G, Amrita B, Darshana J. Acta Pharmaceutica 2013;63:335–58.
[45] Davankov VA, Ilyin MM, Tsyurupa MP, Timofeeva GI, Dubrovina LV. Macromolecules 1996;29:8398–403.
[46] Li DQ, Ma M. Chem Tech 1999;29:31–7.
[47] Arkas M, Allabashi R, Tsiourvas D, Mattausch EM, Perfler R. Environ Sci Technol 2006;40:2771–7.
[48] Chilajwar SV, Pednekar PP, Jadhov KR, Gupta GJC, Kadam VJ. Expert Opin Drug Del 2014;11:111–20.
[49] Selvamuthukumar S, Anandam S, Kannam K, Manavalan R. J Pharm Pharm Sci 2012;15: 103–11.
[50] Ansari KA, Torne SJ, Vavia PR, Trotta F, Cavalli R. Curr Drug Del 2011;8:194–202.
[51] Lembo D, Swaminathan S, Donalisio M, Civra A, Pastero L, Aquilano D, Vavia P, Trotta F, Cavalli R. J Pharmaceutics 2013;443:262–72.
[52] Yang C, Mori T, Wada T, Inoue Y. New J Chem 2007;31:697–702.
[53] Yang C, Inoue Y. Chem Soc Rev 2014;43:4123–43.
[54] Liang WT, Yang C, Nishijima M, Fukuhara G, Mori T, Mele A, Castiglione F, Caldera F, Trotta F, Inoue Y. Beilstein J Org Chem 2012;8:1305–11.
[55] Mele A, Castiglione F, Malpezzi L, Ganazzoli F, Raffaini G, Trotta F, Rossi B, Fontana A, Giunchi G. J Incl Phenom Macro Chem 2011;69:403–9.
[56] Mele A, Castiglione F, Malpezzi L, Ganazzoli F, Raffaini G, Trotta F, Rossi B, Fontana A, Giunchi G. J Incl Phenom Macro Chem 2011;69:403–9.
[57] Liang WT, Yang C, Zhou DY, Haneoka H, Nishijima M, Fukuhara G, Mori T, Castiglione F, Mele A, Caldera F, Trotta F, Inoue Y. Chem Commun 2013;49:3510–2.
[58] Rekharsky MV, Inoue Y. Chem Rev 1998;98:1875–917.
[59] Inoue Y, Dong F, Yamamoto K, Tong LH, Tsuneishi H, Hakushi T, Tai A. J Amer Chem Soc 1995;117:11033–4.
[60] Biswas M, Mukherjee A. Adv Polym Sci 1994;115:89–123.
[61] Warshawsky A, Strikovsky AG, Vilensky MY, Jerabek K. Separ Sci Technol 2002;37: 2607–22.
[62] Huang JX, Moore JA, Acquaye JH, Kaner RB. Macromolecules 2005;38:317–21.
[63] Cintas P, Cravotto G, Gaudino EC, Orio L, Boffa L. Catalysis Sci Technol 2012;2:85–7.

[64] Cravotto G, Gaudino EC, Tagliapietra S, Carnaroglio D, Procopio A. Green Processing Synth 2012;1:269–73.

[65] (a) Strimbu L, Liu J, Kaifer AE. Langmuir 2003;19:483–5. (b) Noël S, Léger B, Herbois R, Ponchel A, Tilloy S, Wenz G, Monflier E. Dalton Trans 2012;41:13359–63.

[66] Kiasat AR, Nazari S. J Incl Phenom Macrocycl Chem 2013;77:429–38.

[67] Han S, Wei Y, Grzybowski BA. Chem Eur J 2013;19:11194–8.

Chapter 3

The Use of Cucurbit[*n*]urils as Organic Nanoreactors

Brian D. Wagner
Department of Chemistry, University of Prince Edward Island, Charlottetown, Canada

1 INTRODUCTION

Cucurbit[*n*]urils (CB[*n*]) are cyclic oligomers of glycoluril, connected by pairs of methylene bridges. The pairs of methylene bridges between the monomers give CB[*n*] molecules very rigid structures (at least up to $n = 8$), in contrast to that of the most popular family of supramolecular hosts, cyclodextrins, or calixarenes. This rigid and well defined internal cavity has resulted in the establishment of CB[*n*]s as versatile, selective, and highly useful supramolecular hosts, and their rapidly growing popularity which is now beginning to rival that of cyclodextrins. This popularity and utility of CB[*n*] has resulted in the publication of numerous general review articles on their physical and host properties and applications over the past 20 years [1–10] There have also been several recent review articles on specific applications of CB[*n*], including as "nanotoys" [11], molecular rotors [12], nanoscale building blocks [13], and drug encapsulation and delivery agents [14], and in chiral applications [15].

The chapter is focused on the use of CB[*n*] molecular hosts as nanoreactors. Nanoreactors are defined as nanoscale reaction vessels, inside of which chemical reactions can occur, and represent an increasingly important focus of supramolecular chemistry [16]. Nanoreactors are useful as templates for controlling reaction pathway, and for stabilizing higher-energy conformations or isomers of guests, allowing for different target products than might be obtained in homogeneous solution reactions. Nanoreactors can be self-assembled structures formed through noncovalent interactions (such as micelles, vesicles, or polymersomes) [17–20], or can be single discrete molecular (covalently bonded) hosts, such as cyclodextrins [21], calixarenes [22], cavitands [23], and of course CB[*n*]. CB[*n*] are especially attractive as nanoreactors, due to their rigid, well-defined cavities (particularly in comparison to those of cyclodextrins and calixarenes), and their strong and selective supramolecular binding of guest molecules. These two aspects of CB[*n*] chemistry, in relationship to their utility as nanoreactors, will be reviewed in detail in Sections 2 and 3. It should be noted that there have been

Organic Nanoreactors. http://dx.doi.org/10.1016/B978-0-12-801713-5.00003-3

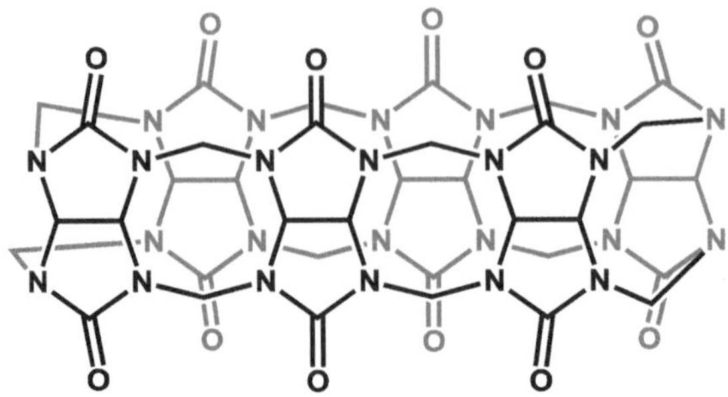

FIGURE 3.1 **Chemical structure of cucurbituril, the original member of the CB[*n*] family (with *n* = 6).**

a few previous reviews of specific aspects of the use CB[*n*] as nanoreactors or nanocapsules, including on applications for time-resolved fluorescence assays [24], photochemistry [25], redox reactions [26,27], and catalysis [28,29]. Ref. [29] in particular is an excellent comprehensive review of the use of CB[*n*] to catalyze reactions. The aim of this chapter is to provide a representative review of the application of CB[*n*] nanoreactors for all types of chemical reactions, with a particular emphasis on resulting chemistry that is unique to the CB[*n*] cavity.

The first CB[*n*] to be synthesized and isolated was the *n* = 6 oligomer cucurbit[6]uril, referred to initially simply as cucurbituril [30,31]. It was first synthesized by Behrend et al. in 1905 [30]. Although the ability of this new compound to bind I_2 molecules was recognized by these researchers in 1905, the exact molecular structure of this unique and interesting hollow cage compound was not determined until 75 years later. In 1981, Freeman, Mock, and Shih were intrigued by this 1905 report, reproduced the synthesis, and elucidated the interesting structure, which is shown in Fig. 3.1. They also named this novel cage compound *cucurbituril*, after its perceived resemblance to the *cucurbitae* family of plants, including pumpkins. They were able to obtain a crystal structure and fully characterize its molecular structure and the physical dimensions of the internal cavity. They also demonstrated the ability of cucurbituril to encapsulate small guest molecules within its internal cavity, forming host–guest inclusion complexes, as will be discussed in detail in Section 3. Following this pioneering work by Mock, and their introduction of this fascinating new molecular host, research into the properties and host–guest inclusion complexes of cucurbituril grew rapidly, with a number of research groups investigating this new host compound.

The first new cucurbituril molecule to be synthesized and characterized was decamethylcucurbit[5]uril, reported by Stoddart et al. in 1992 [32]. However, the small cavity size of the *n* = 5 ring was not particularly useful for the inclusion of guests. The real breakthrough in expanding the CB[*n*] families came in 2000, when Kim et al. [33] reported a useful synthesis method for obtaining the

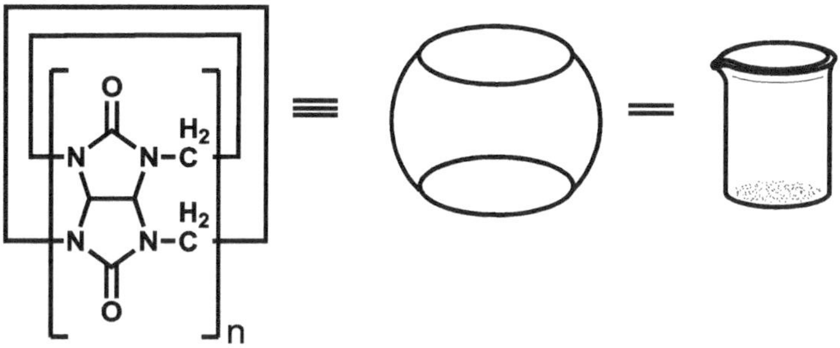

FIGURE 3.2 General chemical structure of the CB[*n*] family of hosts (and the analogy of the CB[*n*] host structure to a molecular beaker or nanoreactor).

n = 5,6,7, and 8 homologs of cucurbituril; these were referred to as CB[*n*]. The general chemical structure of a CB[*n*] host is shown in Fig. 3.2. Other research groups also provided other synthetic strategies, greatly improving the availability of the full suite of CB[*n*] hosts [34,35]. Then in 2005, Isaacs et al. reported the synthesis and characterization of CB[10], a very large molecular host, which can incorporate significant large guests into its cavity [36]. Isaacs has recently published an interesting report on the latest understanding of the mechanism of CB[*n*] formation [37]. These synthetic breakthroughs made available a significant range of cavity sizes, thus greatly expanding the range of guest molecule shapes and sizes which could be incorporated into the cucurbituril cavity, and therefore the range of applications which could be undertaken.

Other cucurbituril synthetic innovations also followed. In 2001, Kim et al. reported the synthesis of CB[5] and CB[6] with cyclohexyl groups around the equators; these showed good water solubility [38]. Day et al. reported a method for synthesizing partially substituted CB[*n*] hosts [39]. Isaacs et al. synthesized and characterized various CB[*n*] analogs, some of which are intrinsically fluorescent hosts [40,41]. Kim et al. provided an excellent review of functionalized cucurbiturils in 2007 [42]. However, the focus in this chapter will be on unsubstituted/underivatized CB[*n*] molecules. Most recently, Ni, Tao et al. reported the largest member of the family yet, CB[14], which shows a twisted, "figure eight" shape, and so in fact exhibits a smaller accessible cavity than smaller CB[*n*] hosts [43].

A number of interesting applications of CB[*n*] molecules have been reported which make use of their rigid shape and cavity (besides their use as nanoreactors, the focus of this chapter); some of these will be briefly listed here for the sake of interest. Mock and Pierdpont reported one of the first practical applications of cucurbituril, in the construction of a pH-activated molecular switch [44]. Blanch et al. have made use to the rigid hollow spherical nature of CB[*n*]s to construct a molecular ball bearing from an inclusion complex of CB[7] [45] and a "molecular gyroscope" from the inclusion of CB[5] within CB[10] [46]. Kim has made extensive studies of the use of cucurbiturils as "beads" in a variety of rotaxanes, polyrotaxanes, and other interlocked structures and molecular machines [47–50]. Isaacs et al. have demonstrated the capacity of the CB[*n*] family for self-sorting applications [51]. Nau et al. have demonstrated the utility of CB[*n*] in drug delivery applications [52], and have reported careful determination of the cavity size and volume [53]. Finally, Keinan et al. have provided a useful recent review of three of the application areas of CB[*n*] hosts, namely their use in sensors, as catalysts, and in molecular motors [54].

In the remaining sections of this chapter, the physical and host properties of CB[*n*] as they relate to their applications as nanoreactors will be discussed in detail, and a detailed representative survey of the specific types of reactions which have been templated or catalyzed by them will be presented. This is not intended to be a comprehensive review, as the number of publications on CB[*n*] both as hosts and as nanoreactors is extensive. Furthermore, the many other various uses of CB[*n*] (briefly described previously) for the formation of rotaxanes and other interlocked species, self-assembled capsules and adducts, and in drug delivery, as inhibitors of reactions, and as components of nanomachines and nanostructures, will not be covered. This chapter will, in a nutshell, explore the uses of CB[*n*] as nanosize reaction flasks.

2 PHYSICAL PROPERTIES OF CUCURBIT[*n*]URILS

The distinguishing features of the CB[*n*] family of hosts, features that have led to their increasingly wide applications as supramolecular hosts and nanoreactors, are their rigid structure defined by the two methylene bridges joining adjacent glycoluril monomer units (as shown in Fig. 3.2), their spherical shape and cavity (at least up to $n = 8$), the equal and opposite carbonyl-lined portals providing access to their internal cavities, and the distinct local environment within their rigid cavities.

Table 3.1 lists the physical properties of the CB[*n*], which are important and relevant to their behavior as supramolecular hosts. As can be seen from Table 3.1, the physical size of the CB[*n*]s and the available cavity volume show a wide and useful range, allowing the matching between the CB[*n*] nanoreactors of choice and the size of the guest reactant(s). However, the height of the CB[*n*] cavity, from portal to portal, is essentially the same for CB][*n*] with $n = 5 - 10$, at 9.1 Å. In terms of access of a guest to the CB[*n*] cavity, the guest size is limited to a degree by the portal size, which is restricted relative to the maximum cavity diameter. The portal diameter ranges from 2.4 to up to 11 Å from $n = 5$ to $n = 10$. It is of interest to note that for the larger CB[*n*] molecules ($n = 10$, 14), the rigidity of the cavity has been lost. While this may have been disadvantageous in terms of the loss of a rigid nanoreactors vessel, in some applications it might present an advantage, as the host flexibility might allow for incorporation of a guest that is larger in fact than the spherical cavity diameter, through flattening of the CB[10] host. In the case of CB[14], the molecule is actually twisted, so does not present a traditional CB[*n*] cavity; investigations

TABLE 3.1 Relevant Physical Properties of the Cucurbit[*n*]uril Family of Hosts

CB[*n*]	Molecular weight (g/mol)	Portal diameter (Å)	Inner cavity diameter (Å)	Cavity volume (Å^3)	# Waters in cavity†	Aqueous solubility (mM)
5	830	2.4	4.4	82	2	20–30
6	996	3.9	5.8	164	4	0.018
7	1163	5.4	7.3	279	8	20–30
8	1329	6.9	8.8	479	12	<0.01
10	1661	9–11	10.7–12.6	—	22	—
14	2324	—	—	—	—	41 (D_2O)

†Calculated based on a 55% packing efficiency [53].
Information compiled from Refs. [6,7,10,43].

of its potential utility as a nanoreactor are as yet unexplored. The binding of guests in these various CB[*n*] hosts, including examples of distortions of the host cavity shape for larger members of the family, will be discussed in detail in Section 3.

One of the most important properties of CB[*n*] as it relates to their utility as hosts and nanoreactors is their solubility, both in water and in organic solvents. The aqueous solubility is, of course, of prime importance and interest, as performing host–guest inclusion in water will yield the maximum driving forces for inclusion, with the polarity differences between the microenvironment within the cavity and that of the bulk solvent being maximized. Thus, binding constants will be maximized by performing the inclusion in aqueous solution. Karcher et al. did extensive studies on the aqueous solubility of the original cucurbituril host [55]. The solubility of CB[6] itself is rather low in water; in fact, it is lower than that of either CB[5] or CB[7], as can be seen from Table 3.1. The aqueous solubility of CB[5] and CB[7], both on the order of 20–30 mM, is sufficient to allow for their easy use in water for most applications. However, in the case of CB[6], CB[8] and CB[10], low aqueous solubility may be a concern for their use as nanoreactors.

In the case of CB[6], it was established early on that the low aqueous solubility could be significantly enhanced in one of two ways: by addition of salts [56,57] or acids [7,35,58]. In both cases, interactions of the cations involved (metal cations in the case of salts; protons in the case of acids) with the partially negative oxygen atoms of the carbonyl groups lining the portals results in complexes with overall positive charge, dramatically increasing the aqueous solubility. Buschmann et al. demonstrated the binding of CB[6] to various metal cations, and the resulting increase in solubility that could be obtained [56]. Kim et al. showed that the binding of sodium cation to CB[6] is reversible and thus provides a highly soluble molecular container for the binding of organic guests [57]. They reported significant solubility of CB[6] in 0.20 M metal salt solutions, such as 37 mM in KCl and 70 mM in $CaCl_2$ [57]. In the case of acidic solutions, an increased solubility of CB[6] of 61 mM in 1:1 formic acid:water is obtained [58]. Even for the moderately aqueous soluble CB[7], enhanced solubility can be obtained to about 700 mM![7].

Another important property of the CB[*n*] family of hosts is their high thermal stability, which is a result of their strong, rigid covalently bonded structure. Buschmann et al. have measured the thermal properties of cucurbituril through various calorimetric studies [59]. Subsequent thermogravimetric studies have confirmed the high stability of the entire family from $n = 5–8$, with decomposition occurring only above 370°C in all four cases [7]. This high stability means that thermal control of reactions in CB[*n*] nanoreactors is possible at relatively high temperatures, which are typically limited by the solvent system and not by the CB[*n*].

Another crucial aspect of CB[*n*] hosts is the microenvironmental nature of the internal cavity [54,60–62]. The polarity within the cavity of hosts is important, both in terms of the driving forces for inclusion of guests (particularly hydrophobic ones), and in terms of the effect that inclusion within the cavity has on the spectroscopic properties of the included guest; these will be discussed in detail in Sections 3 and 4. The polarity within a microenvironment is typically explored using polarity-sensitive or solvatochromic fluorescent or absorption probes. In the case of CB[*n*], it has been established that the effective polarity within the cavity is similar to that of *n*-octanol [60]. Thus, the polarity is significantly lower than that of water, but not as low as that of nonpolar organic solvents such as cyclohexane. This lower-polarity environment will have significant impacts on the emission properties of included fluorescent guests, for example, as will be discussed in Section 4. However, Nau and coworkers have shown that the environmental properties of CB[*n*] cavities, and their impacts on solvatochromic probes, are best characterized and explained by the *incredibly low polarizabilities* within CB[*n*] cavities. In fact, they have shown that the polarizability within a CB[*n*] cavity approaches that of the gas phase [53,60,61]. These results were obtained by studying the effect of inclusion into CB[7] on a particular solvatochromic molecule, 2,3-diazabicyclo-[2.2.2]oct-2-ene [60]. They have determined that the polarizability P of the CB[*n*] cavity microenvironment is equal to 0.12. For comparison, that of perfluorohexane, commonly considered to be one of the least interacting solvents, is 0.159, while that of the gas phase is, by definition, 0 [53]. Nau et al. have been able to ascribe the unique host properties of CB[*n*]s to this very low polarizability within their internal cavities [54]. This makes CB[*n*] essentially nanosize reaction vessels with interiors similar to a bulk gas phase; that is, they provide pockets of gas-phase–like microenvironments for reaction within a bulk-water environment. This unique low-polarizability microenvironment also results in significant shielding of guest NMR signals [62].

Finally, the electrostatic potentials of CB[*n*]s is also an important factor and driving force in their behavior as molecular hosts and nanoreactors [5,7]. CB[*n*] show relatively negative electrostatic potentials, within the cavities and especially at the portals; these potentials are much more negative than is the case for cyclodextrins, for example [5,7]. This means that electrostatic interactions are much more important in the host–guest inclusion processes of CB[*n*]s than for cyclodextrins, and also, correspondingly, that CB[*n*] hosts show high selectivity for cationic guests [5,7]. In the next section, the host properties of CB[*n*], in terms of the range of guests included, and the driving forces for inclusion, will be discussed in detail, to help in the understanding of their utility as nanoreactors.

3 HOST PROPERTIES OF CUCURBIT[*n*]URILS

CB[*n*] act as nanoreactors by forming supramolecular host–guest inclusion complexes with reactant guests. Supramolecular chemistry, or "chemistry beyond the molecule," refers to structures involving two or more individual molecular species held together only by noncovalent interactions, such as dispersion and dipole–dipole forces, and hydrogen bonding [63–65]. Supramolecular host–guest inclusion complexes represent one of the simplest examples of a supramolecular system. These complexes form in solution (usually aqueous) when a relatively small guest molecule becomes included within the interior cavity of a larger, cage-like host molecule, such as a CB[*n*]. This process is illustrated in Fig. 3.3; as indicated in this figure, host–guest inclusion is a dynamic, equilibrium process.

The stronger the host–guest interactions and the driving forces for inclusion (such as the hydrophobic effect), the stronger the resulting host–guest complex will be [64]. This is manifested in the magnitude of the binding constant K, which for the simplest example of a 1:1 host:guest inclusion complex {H:G}, is defined as:

$$K = \frac{[\{H:G\}]}{[H][G]} \tag{3.1}$$

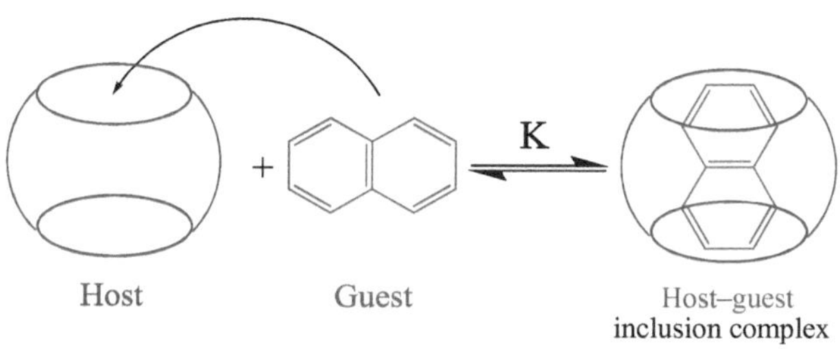

FIGURE 3.3 A cartoon depiction of the formation of a 1:1 CB[*n*]:guest inclusion complex, illustrating the equilibrium nature of the inclusion process and the binding constant K.

Overall, the magnitude of K is determined by the value of the Gibbs Energy of inclusion, $\Delta_{incl}G^{\circ}$, since for any equilibrium:

$$K = e^{-\{\Delta G/RT\}} \tag{3.2}$$

Since $\Delta G_{incl} = \Delta H_{incl} - T\Delta S_{incl}$, there are both enthalpic and entropic possible contributions to the stability of the host–guest complex, which is favored by a negative enthalpy of inclusion and by a positive entropy of inclusion [64]. The various contributions to ΔG are often referred to as the "driving forces" for the inclusion process.

Nau et al. have presented excellent discussions of the mechanism of the inclusion of guests into the CB[6] cavity [66], as well as the important driving forces for the formation of CB[*n*] host–guest inclusion complexes in general [54]. In terms of the mechanism of complexation by CB[6], they showed that the process is dominated by the constrictive binding resulting from the narrow carbonyl-lined portals, and that the high energy barrier to ingression keeps guests that are too large from forming inclusion complexes [66]. They studied the dynamics of this inclusion process using NMR spectroscopy and concluded that it is this barrier to ingression that determines the magnitude of the binding constant. They further showed that the inclusion can be fine-tuned through the choice of pH and/or salt cation, which bind to the portals and change the rate of ingression, and they emphasized that most reported binding constants reflect this presence of H^+ or metal cations associated with the portals, and not the inherent affinity for the guest on the part of the CB[6] cavity. They presented mechanisms for the inclusion of both neutral (direct ingress into the cavity) and organic ammonium guests (initial interaction between the NH_3^+ and the carbonyls, followed by a flip of the organic guest moiety into the cavity) [66]. In terms of driving forces for inclusion into CB[*n*] cavities, they showed that these are dominated by a combination of electrostatic interaction of cationic guests with the carbonyl rims (ion-dipole attractions), and hydrophobic effects [54]. The latter are particularly strong for CB[*n*] hosts, due to the extraordinarily low polarizability within the CB[*n*] cavity, as discussed in Section 2. In terms of the different CB[*n*] hosts, binding of a particular guest is determined by the size match between the guest and the corresponding host cavity volume, as expected. Isaacs et al. have also discussed the mechanism for inclusion of guests into CB[*n*] cavities [7]. They presented a comprehensive mechanistic scheme for the binding of protonated guests by metal cation-complexed CB[6] (Scheme 2 in Ref. [7]), which also involves an initial interaction of the NH_3^+ group with the carbonyls, followed by a flip of the organic component into the cavity.

In addition to the simple 1:1 CB[*n*]:guest inclusion complex depicted in Fig. 3.3, other CB[*n*]:guest stoichiometries are also possible. For example, in the case of bimolecular reactions occurring within a CB[*n*] cavity, a 1:2 host:guest inclusion complex must be able to be formed. In the case of reaction templating by the CB[*n*] nanoreactors, 2:2 complexes are often involved. In the case of large or elongated guests, 2:1 complexes might form in which two CB[*n*] nanoreactors encapsulate the two ends of a single guest. These four most common possible CB[*n*] inclusion complexes are illustrated in Fig. 3.4; all of these have been observed in the application of CB[*n*] as nanoreactors, as will be discussed in Section 5.

CB[*n*] have proven to be exceptionally useful hosts for the encapsulation of a wide range of guest molecules. The first molecule reported to form a complex with a CB[*n*] was I_2, reported in the original 1905 paper [30]. The first guests to be reported as forming *inclusion* complexes with a CB[*n*] were protonated aliphatic amines, reported in the original paper by Mock et al. in 1981 announcing the synthesis and structural determination of cucurbituril [31]. Since then many types, sizes, and shapes of guest molecules have been shown to be included within CB[*n*] host cavities. This section will provide a brief overview of the types of guest molecules that have been encapsulated by CB[*n*] hosts. In general, cationic species are particularly strongly bound, due to polar-electrostatic interactions with the carbonyl portals as mentioned previously; a survey of representative cationic guests included in CB[*n*] will be presented in Section 3.1. However, many neutral molecules are also well bound; these will be reviewed in Section 3.2. In addition, other special types of guest, including metal complexes and other host molecules, can also be included; these will be surveyed in Section 3.3. It is this strong and selective binding of neutral and cationic guests, with very specific complexation geometries and modes, together with their rigid nature, which makes CB[*n*]s incredibly useful and applicable as nanoreactors.

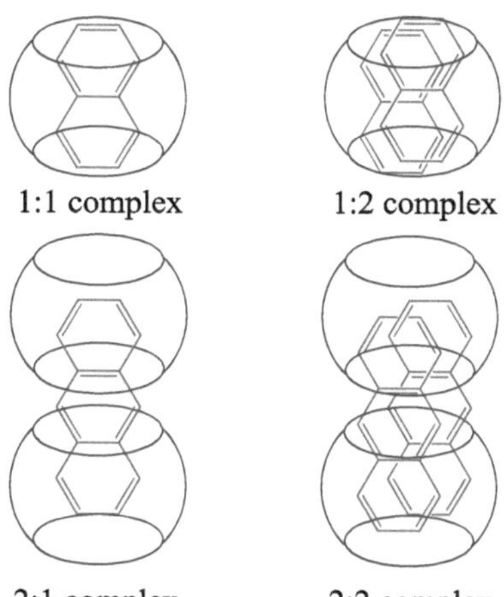

FIGURE 3.4 Depictions of the four host:guest complex stoichiometries encountered in the use of CB[*n*]s as nanoreactors.

3.1 Cationic Guests

As mentioned earlier, protonated aliphatic amines were the first guests shown to be included within the cucurbituril cavity, using NMR, in the original paper by Mock et al. in 1981 [31]. Even in that first paper, the selectivity of cucurbituril as a host was demonstrated: cyclopentanemethyl amine was found to bind to cucurbituril with a binding constant on the order of 10^6 M^{-1} in acidic solution, but cyclohexanemethyl amine was found not to bind at all. Since that initial report in 1981, extensive studies of the binding of cationic guests have been reported, including protonated alkyl amines and diamines [67–71], aryl amines [72,73], amino acids and amino alcohols [74], pyridines [75] and viologens [76,77], and metal cations [56,72,78]. These binding studies have been performed using a variety of techniques, including NMR [67,68,70,76,77], calorimetry [69,71,74], UV–Vis [56,72,73,77,79], and electrochemistry [75,76]. In the case of protonated alkyl and dialkylamines, the strongest binding in CB[6] was found to occur with *n*-hexyldiamine **1** ($K = 2.8 \times 10^6$ M^{-1} in 1:1 formic acid:water solution). The structures of guests **1–4** discussed in this section are shown in Fig. 3.5. This is a result of the presence of two ammonium cation moieties, which can interact with both portals, via inclusion within the cavity of the alkyl chain, which is at the optimum length for proper interaction of both end NH_3^+ groups with the two opposing portals for the 6-membered alkyl chain [67]. By comparison, $K = 2.4 \times 10^4$ M^{-1} for the inclusion of *n*-pentylamine **2** into CB[6] in 1:1 formic acid:water solution, illustrating the increased complex stability when there are two sets of ammonium–carbonyl portal interactions. The extremely strong binding of diammonium guests species has in fact been exploited to generate various highly stable supramolecular 2D and 3D structures, such as cyclic oligorotaxanes [70].

Aryl amines and other cations form strong inclusion complexes with CB[*n*]; in addition to charge–dipole interactions, there can also be interactions between

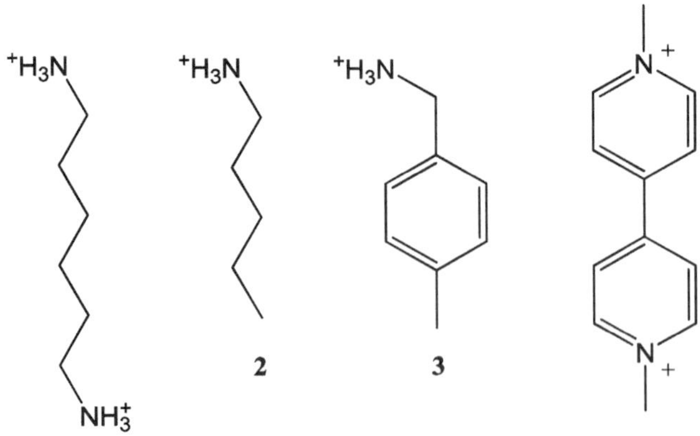

1 2 3 4

FIGURE 3.5 **Representative cationic guests 1–4 included in CB[*n*] host cavities.**

the π-electrons of the aromatic guest with the CB[*n*] cavity walls. These interactions can help to compensate for the larger area of the hydrophobic part of the guest, especially for binding in the original cucurbituril. One of the first aryl ammonium guests to be included in a CB[*n*] was the 4-methylbenzylammonium ion **3** in CB[6], which was studied by Knoche et al. It formed an inclusion complex with a binding constant of $K = 1.2 \times 10^4\ M^{-1}$ [72]. This group also studied various other aryl amines, including azobenzene derivatives, which formed very strong complexes with CB[6] in HCl solutions [73]. Of particular interest and relevance are the studies of Kim et al. [76] and Kaifer et al. [77] on the binding of methylviologen **4**, which is a bipyridyl molecule, capable of forming ion–dipole interactions in its dication form at both portals of a CB[*n*] host. In this case, CB[7] was the best host to match the size of the guest and cavity. Extremely stable host–guest inclusion complexes of the dication with CB[7] were observed, with K on the order of $10^5\ M^{-1}$ [77]. This is in contrast to the case of cyclodextrin hosts, in which the neutral methylviologen is the preferred guest; this again indicated the importance of ion–dipole interactions in CB[7] host chemistry, and the preference for cationic guests. Inclusion also affected the electrochemical properties of the methylviolgen guest; these effects have resulted in the application of CB[7] as a nanoreactor for methylviologen reactions, as will be described in Section 5.

Metal cations show reasonably strong binding to CB[*n*], via a "capping" mechanism in which they sit outside the CB[*n*] cavity and interact with the ring of portal carbonyl groups [56,78,79]. Many different metal cations have been shown to associate with CB[6], for example, including alkali, alkali earth, and lanthanides. As discussed earlier, the association of metal cations with the carbonyl rims results in increased aqueous solubility of the CB[*n*] host (particularly the otherwise nearly insoluble CB[6]), but can also interfere with the inclusion of desired guests. On the other hand, as described by Kim et al. [57], these metal cations can serve as a "lid" helping to encapsulate the guest and keep it in the cavity. Thus, in any use of CB[*n*] hosts as nanoreactors, the involvement (and potential interference) of metal cations present in the reaction solution must be taken into account.

3.2 Neutral Guests

Although the strong charge–dipole interactions seen in the binding of cations by CB[*n*] hosts are not present for neutral guests, the latter can still form strong complexes with CB[*n*] in many cases. The first report of the inclusion of neutral guests in cucurbituril was for a series of organic dyes, including methyl orange **5** [79]. The structures of guests **5–9** discussed in this section are shown in Fig. 3.6. It should be noted that these series of "neutral" organic dyes were all ionic salts, and the reactions were carried out in acidic solution, so in fact the tail end of the guests were anionic sulfate groups. The included part of the dyes was, however, the neutral aromatic moieties. The advantage of using dye molecules as guest

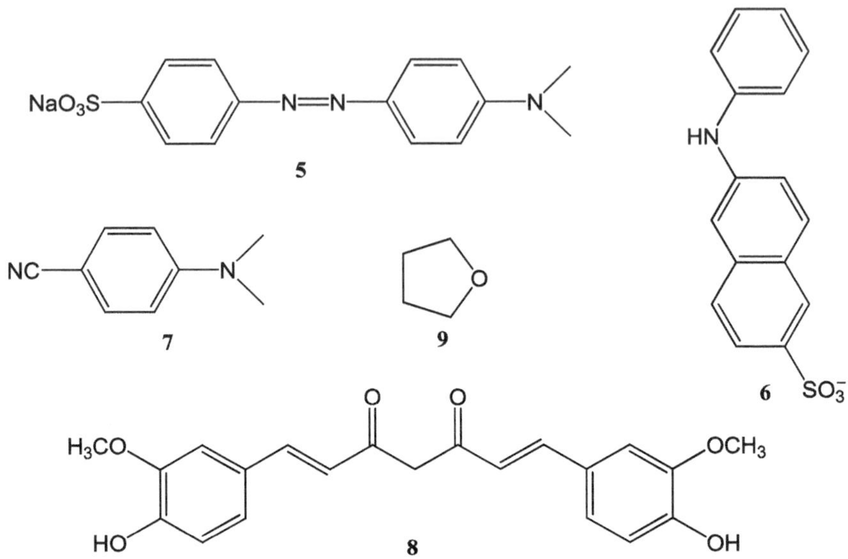

FIGURE 3.6 Representative neutral guests 5–9 included in CB[*n*] host cavities.

is that it allows for the study of the inclusion process by spectroscopic means, either UV-Vis absorption (in this case) or fluorescence. The effect of inclusion on spectroscopic properties will be discussed in Section 4. These authors noted that the binding constants increased with decreasing acid concentration, due to competitive binding of protons by the cucurbituril portals.

The first report that inclusion of a fluorescent guest in a CB[*n*] could result in enhancement of the guest fluorescence was that for the enhancement of the commonly used polarity-sensitive probe 2-anilino-6-naphthalene sulfonic acid **6**, the fluorescence of which was enhanced by a factor of 5.0 upon inclusion into CB[6] [80]. Again, this is an anionic molecule overall, but it was the neutral aniline moiety that was actually included in the cavity, with a relatively low binding constant of 52 M^{-1}. This CB[6]:2,6-ANS complex was described as a *molecular jack o'lantern*, as it is analogous to having a candle (fluorescent molecule) inside the molecular pumpkin cucurbituril. Subsequent studies by this group showed that an even stronger binding and larger enhancement was obtained with this fluorescent guest with CB[7], with a binding constant of 600 M^{-1} and an enhancement of 25 [81]. Other fluorescent dyes and guests were soon studied in CB[*n*] hosts using fluorescence spectroscopy, including naphthalene [82], 4-(dimethylamino)benzonitrile **7** [61], and curcumin **8** [83]. The case of curcumin is of particular interest here. As can be seen from Fig. 3.5, this is a long, symmetrical molecule, which forms 2:1 host:guest complexes with CB[6] in which each of two CB[6] hosts encapsulates each end of the molecule [83]. Koner and Nau have published an excellent summary and review of the encapsulation of fluorescent dyes by CB[*n*] hosts [61].

Besides fluorescent guests, a number of other neutral aliphatic and other molecules have been shown to be encapsulated by CB[*n*] hosts, including aliphatic alcohols and nitriles [84], tetrahydrofuran **9** [57,85,86], and various peptides [87,88]. Interestingly, atomic guests have also been shown to be included within CB[*n*] cavities, specifically the Xenon atom in CB[6] [86,89]. Finally, the very large neutral molecule fullerene has been shown to form inclusions complexes with the large CB[8]; in this case a partial 1:2 host:guest complex forms, in which a single fullerene is bound to two opposing CB[8] cavities, with very minimal penetration into the host cavity and interactions mainly occurring with the host portals [90].

3.3 Other Guests

To conclude this section on the host properties of CB[*n*] nanoreactors, there are a few other types of guests of special interest that do not quite fit the category of simple cations or neutral molecules but which also can be included within CB[*n*] cavities. The structures of guests **10–13** discussed in this section are shown in Fig. 3.7. For example, metal macrocycles have been shown to be included within CB[*n*] cavities, such as Cu(II)cyclam **10** in CB[8] [91]. X-ray crystallography shows that this metal macrocycle is completely included deep within the CB[8] cavity, illustrating the potential usefulness of this relatively large host as a nanoreactor for large reactants. Kim et al. have also shown that ferrocene derivatives, such as hydroxymethylferrocene **11**, can be included within the CB[7] cavity [92] They showed that the neutral and cationic derivatives bind with extremely high binding constants, on the order

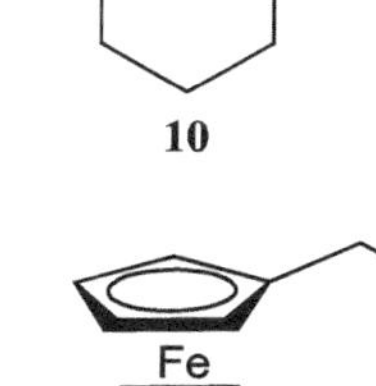

10

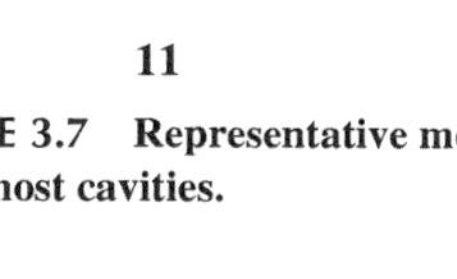

11

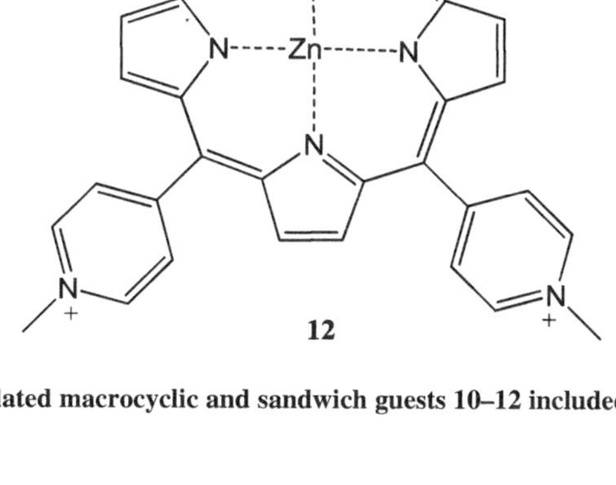

12

FIGURE 3.7 Representative metallated macrocyclic and sandwich guests 10–12 included in CB[*n*] host cavities.

of 10^9–10^{10} and 10^{12}–10^{13} M^{-1} respectively(!), but that anionic derivatives did not bind at all. This is another excellent demonstration of the preference of CB[*n*] cavities for cationic guests, and of their still strong ability to bind neutral guests as well, and therefore their broad utility as nanoreactors. Finally, the very large tetrapyridinyl porphyrin molecules such as **12** were shown to be included within the very large CB[10] cavity [93]. In order to accommodate this large guest, the CB[10] molecule was shown to flatten significantly to present an oval-shaped cavity which can accommodate the porphyrin ring. Once again this illustrates the flexibility of the larger CB[*n*] hosts, and the overall broad range of guests shapes and sizes that can in fact be included in these molecular nanoreactors.

4 EFFECTS OF CUCURBIT[*n*]URIL HOSTS ON GUEST PHYSICAL AND STRUCTURAL PROPERTIES

Inclusion of guest species within the CB[*n*] cavity can have profound and measurable effects on guest physical properties, which can lead to useful applications of CB[*n*] hosts [64,65]. Furthermore, these effects on guest properties provided experimental methods with which to study the specific host–guest inclusion process, and to extract the binding constant, by measuring changes to a physical or spectroscopic property as a function of added host concentration (a host titration). Although the focus of this chapter is on the effects of inclusion into CB[*n*] on the chemical properties and hence reactivities of guest molecules, a brief overview of the effects of CB[*n*] inclusion on guest physical properties is useful here in this section, as these may also have significant impacts on the reactivities observed, or on the methods (eg, spectroscopic) used to study the chemical reactions.

4.1 Effects of CB[*n*] Hosts on Guest Solubility

Since one of the major driving forces for inclusion into CB[*n*] cavities in aqueous solution is the hydrophobic effect, the aqueous solubility of guests that are hydrophobic can be significantly increased by the presence of CB[*n*], particularly the more highly water-soluble members such as CB[5] and CB[7]. The groups of Karcher et al. [55,94] and Buschmann et al. [95,96] studied both the solubility of cucurbituril itself and its ability to solubilize organic compounds and dyes. They proposed the use of cucurbituril for removal of such dyes from wastewater. Huang et al. reported extensive studies on the use of various CB[*n*] hosts to overcome the aqueous solubility problems associated with the kinetin **13** (Fig. 3.8), an important plant hormone [97]. They found that CB[7], as well as derivatized CB[6], substantially enhanced the solubility of kinetin in water. In the case of CB[7], the aqueous solubility of **13** was increased 35-fold at the maximum CB[7] concentration used of 16.75 mM. Nau et al. have studied the increased solubility of neutral organic molecules in the presence of CB[*n*] hosts,

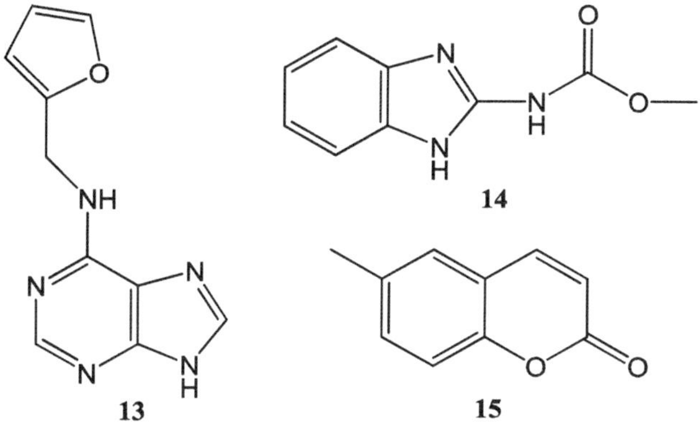

FIGURE 3.8 Representative guests of interest 13–15 that show significant changes in solubility and/or spectroscopic properties upon inclusion within CB[n] host cavities.

for example various drugs of the benzimidazole family using CB[7] [98]. In the case of the drug albendazole **14** (Fig. 3.8), a huge increase of solubility was reported, from 0.003 mmol/L to 0.300 mmol/L upon complexation by CB[7], an increase of a 1000-fold!

The aqueous solubility enhancement of guests with potential medicinal or pharmaceutical applications such as those mentioned previously has led to significant research into the use of CB[n] hosts as drug delivery agents. Macartney has published an excellent recent review of the encapsulation of drug molecules by cucurbiturils, and the effect of inclusion on their physical and chemical properties, including solubility [14].

4.2 Effects of CB[n] Hosts on Guest Spectroscopic Properties

The effects of cucurbituril on guest spectroscopic properties was recognized by Mock et al. at the time of their isolation and characterization of the host molecule in the early 1980s. They used NMR spectroscopy to perform the first binding studies on cucurbituril, for alkyl ammonium guests [31], and concluded that the interior of the cucurbituril cavity provides magnetic shielding of the protons of included guests, resulting in significant upfield shifting of ^{1}H NMR guest signals. This is extremely useful, not only for the NMR-based determination of binding constants through CB[n] titration experiments, but also because the analysis of which specific guest protons are shifted upfield provides important structural information about the nature of the host–guest inclusion complex, and the specific mode of inclusion (eg, which parts, or all, of the guest molecule is residing inside the cavity). In subsequent work on protonated diaminoalkanes, they showed that the multiplets from the methylenes adjacent to the ammonium groups disappear upon inclusion into cucurbituril and reappear significantly upfield [99].

Mock et al. also reported the effect of cucurbituril inclusion on the UV-Vis spectrum of guests [99]. For example, they showed that the UV spectrum of 4-methylbenzylamine **3** undergoes significant changes upon addition of CB[6], including appearance of fine structure and spectral shifting. The measurement of the absorbance at a particular wavelength where there is significant change in the guest spectrum upon addition of the CB[*n*] host provided another experimental method for determination of the binding constant.

The largest spectral changes observed upon CB[*n*] binding is that of the fluorescence spectrum, especially in the case of highly polarity-sensitive guests. As in the case of both the NMR and UV-visible absorption spectrum, these changes are the result of the significantly different microenvironment experience by a guest within the CB[*n*] cavity as compared to free in bulk solution, as described in Section 2. There have been extensive studies reported on the inclusion of fluorescent probes and dyes into CB[6] hosts, with a number of review articles published [13,61,64]. Significant enhancement and spectral shifting of the fluorescence spectra of polarity-sensitive guests have been reported. For example in the case of 2,6-ANS **6**, the fluorescence was increased 25-fold in the presence of CB[7], as mentioned previously. In the case of 6-methylcoumarin **15** (Fig. 3.8), a red-shift from 411 to 443 nm was seen upon encapsulation into CB[8], with an increase in fluorescent quantum yield from 0.008 to 0.040 [100]. A number of factors can lead to these large changes in fluorescence intensity, including the polarity effects on the energy gap between the ground and excited states, the restricted intramolecular rotational motion of the guest included within the cavity, and the protection of the guest from fluorescent quenchers such as O_2, as well as specific solvent–solute interaction [13,64].

4.3 Effects of CB[*n*] Hosts on Guest Structure and Isomerization

There are three aspects of the CB[*n*] cavity which make it potentially useful in controlling the structure, isomerization, speciation, or conformation of a guest molecule. One is the unique microenvironment, as described previously; this is the main factor in the effect of CB[*n*] on guest spectroscopic properties. This unique low-polarity environment has a polarizibility even lower than that of the least interacting solvents; this may allow for the presence of otherwise unstable forms of a guest. The second is the pair of oppositely positioned carbonyl-rimmed portals, located 9.1 Å apart. This means that guest molecules with two interacting groups, in particular two cationic centers, will adopt the conformation or isomer which maximizes the group–carbonyl interactions. The third and perhaps most important aspect in general is the confining nature of the CB[*n*] cavity. Steric interactions, and loss of rotational freedom, may result in the presence inside the cavity of forms of reactants that are thermally unstable in free solution. It is this ability of CB[*n*] hosts to control guest conformation and/or speciation which is at the heart of their incredible and diverse utility as nanoreactors. In this section, some illustrative examples of species uniquely

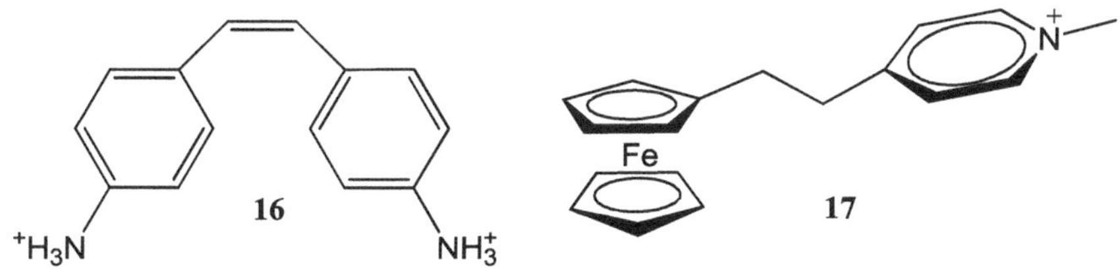

FIGURE 3.9 Representative guests/conformers 16 and 17 that are stabilized within CB[*n*] host cavities.

observed or dominant in CB[*n*] cavities will be described, and Section 4 will present many more examples where such effects of CB[*n*] nanoreactors are used to control guest reactions.

Kim et al. showed that the *cis* isomer of diaminostilbene **16** (Fig. 3.9) (prepared photochemically from the more stable *trans* isomer), is highly stable when prepared within a CB[7] cavity, existing for over 30 days at room temperature [101]. By contrast, it slowly but clearly converts back to the more stable *trans* form in free solution at room temperature. Even upon heating to 60°C, the *cis* isomer included in CB[7] is stable for at least a day, whereas in the absence of CB[7] host isomerization back to the *trans* is complete within 5 h. Using *ab initio* calculations, the authors showed that the C-shaped *cis* isomer (as shown in Fig. 3.9) fits well into the CB[7] cavity, and the two terminal ammonium cations interact well with the two CB[7] portal oxygen atoms. However, the *trans* isomer is too long to allow both ammonium groups to interact with portal oxygen atoms. This greatly stabilizes the *cis* isomer relative to the *trans* in the confined environment of the CB[7] nanoreactor.

Fedin and coworkers [102,103] have isolated a number of distinct species within CB[*n*] nanoreactors. For example, they identified a previously unrecognized hydronium di-cation in a solid-state complex with CB[6] [102]. Instead of the commonly recognized $H_7O_3^+$ monocation (essentially a proton coordinated by three water molecules), they observed a cyclic $H_{14}O_6^+$ species, stabilized and templated by a set of five hydrogen bonds to the CB[6] portal carbonyls, forming an exclusion complex. In another report, they showed that the *trans* form of the metal complex $[Co(en)_2Cl_2]^+$ (en = ethylenediamine) was selectively encapsulated in CB[8] from a mixture of *cis and trans* complexes [103]. Furthermore, the *trans* complex was thermally stabilized in the CB[8] nanoreactor. This shows the potential utility of CB[*n*] nanoreactors to select specific reactant isomers for subsequent selective reactions.

Macartney et al. showed that the (*E*)-1-ferrocenyl-2-(1-methyl-4-pyridinium) ethylene cation **17** (Fig. 3.9) is stabilized by inclusion into CB[7] [104]. They showed that CB[7] inclusion affects the UV-Vis spectrum and electrochemical properties, and most importantly prevents photoisomerization to the (*Z*) form. Isaacs et al. investigated the effect of CB[*n*] hosts on a series of five triazene–arylene oligomers [105]. They demonstrated that the presence of three of the

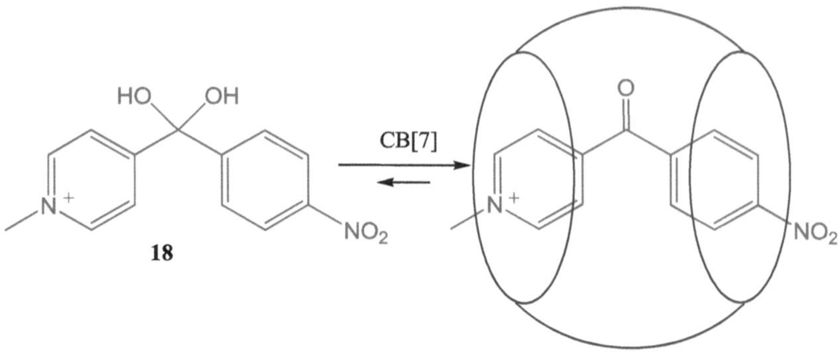

FIGURE 3.10 Stabilization of the ketone form of 18 via inclusion within the CB[7] host cavity. *(Adapted from Ref. [106] with permission of American Chemical Society. Copyright 2008.)*

CB[*n*] hosts, namely CB[7], CB[8] and CB[10], resulted in selective population of specific conformers, and controlled the folding and refolding of these oligomers. Sotiriou-Leventis, Leventis et al. showed that inclusion of *N*-methyl-4-(*p*-nitrobenzoyl)pyridinium cation **18** CB[7] shows a preference for the keto form, whereas in water solution there is an equilibrium between the ketone and *gem*-diol forms [106]. This is illustrated in Fig. 3.10, and showed that CB[7] can be used to force this compound to be in the ketone form; again this would have significant implications for the reactivity of this compound.

As a final example of the control by CB[*n*] hosts of guest structure, Wagner et al. [107] were able to use CB[8] hosts to isolate two rather elusive isomers of Cu(II)cyclam (cyclam **10** is shown in Fig. 3.7), namely the *trans*-I and *trans*-II form, as opposed to the lower energy *trans*-III form. This was the first report of an unsubstituted cyclam existing in either of these two isomeric forms in the solid state; these were previously seen only for substituted cyclams. In this case, the binding mode within the cavity was shown to preferentially be the *trans*-I and *trans*-II forms, through specific hydrogen bonding with the carbonyl portals, which was not optimized for the *trans*-III form.

The isolation of otherwise higher energy forms (or at least exclusive selection of one equilibrium form) within a CB[*n*] cavity, as a result of the confinement within the CB[*n*] nanoreactor and specific host–guest interactions which can result in different stabilities of isomers or conformers within the cavity as opposed to in free solution, well illustrates the potential utility of CB[*n*] nanoreactors for controlling guest reactivity.

4.4 Effects of CB[*n*] Hosts on Guest Aggregation

The final effect of CB[*n*] inclusion on guest properties that is relevant to their use as nanoreactors is the effect of inclusion on guest aggregation. This is somewhat related to solubility, as described in Section 4.1, because increased solubility results in decreased aggregation, but the effect on guest aggregation is worth

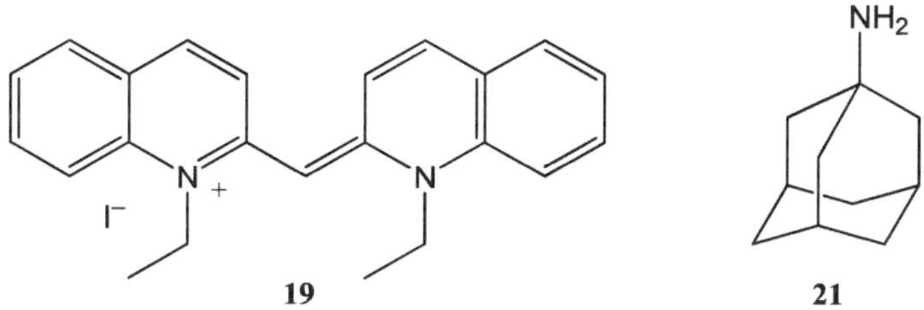

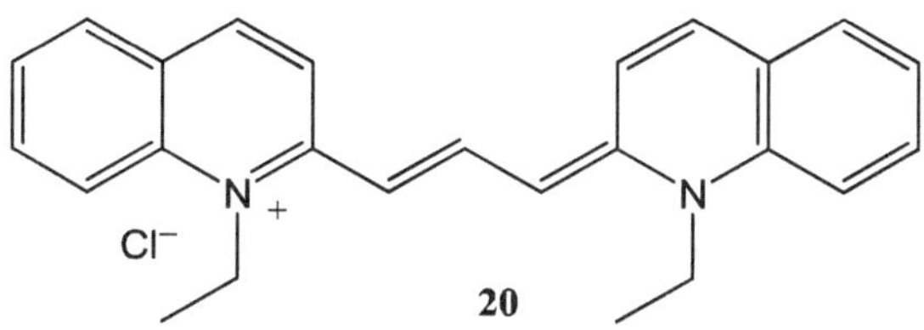

FIGURE 3.11 **Dye guests 19 and 20 deaggregated by CB[7], and competitive guest 21.**

considering on its own merits, as it can have a significant impact on reaction pathways. Kaifer et al. showed that inclusion of the cyanine dyes pseudoisocyanine **19** and pinacyanol **20** (Fig. 3.11) into CB[7] prevents the formation of J- and H-aggregates, respectively, formation that is prevalent in aqueous solution at similar dye concentrations [108]. The presence of CB[6] had no effect on aggregation for either dye, indicating that the CB[6] cavity is too small to thread the dye molecules, illustrating the importance of the size match between guest and host. Furthermore, they showed that a competing guest, 1-aminoadamantane **21** (Fig. 3.11), could be added, with its concentration being adjusted to finetune the relative amount of aggregation occurring.

Isaacs et al. [109] used a novel double-cavity cucurbituril, bis-ns-CB[10], to prepare a supramolecular polymer with poly(diallyldimethylammonium chloride. Interestingly, they found that the double cavity bis-ns-CB[10] played two roles. First, it acted as a de-aggregation agent, keeping individual polymer strands separated by binding the sides of the chains; this allowed the individual polymers to elongate instead of branching. Second, it acted as a "cupping agent". In the latter role, the double cavities served to link together adjacent polymers, thus giving higher overall structure to the resulting material.

In the next section, an extensive survey of the use of CB[*n*] nanoreactors to control guest reactivity, product selectivity, and/or to catalyze guest reactions, via their various effects on guest properties and structure, will be presented.

5 EFFECTS OF CUCURBIT[*n*]URILS ON GUEST REACTIVITY AND CHEMICAL PROPERTIES

Inclusion of reactant molecules into CB[*n*] nanoreactors can have profound effects on the guest's solubility, chemical reactivity, chemical stability, reaction pathway, and other chemical properties [3,7,10,29]. A number of

TABLE 3.2 General Types of Reactions That Have Been Carried Out in Cucurbit[*n*]uril Nanoreactors

Type of reaction	CB[*n*] used	References
Acid–base hydrolysis	CB[6]	[112,117,159]
Charge transfer	CB[7], CB[8]	[139–141]
1.3-Dipolar cycloaddition	CB[6]	[142–147]
Desilylation	CB[6], CB[7], CB[8]	[158]
Electron transfer	CB[7]	[76,77]
Oxidation	CB[8]	[152–154]
Ozonolysis	CB[6]	[116]
Peptide bond formation	CB[8]	[138]
Photoinduced electron transfer	CB[7], CB[8]	[115]
Photocycloaddition	CB[7], CB[8]	[121–124,127–137,148–151]
Photolysis	CB[7]	[118,119]
Polymerization	CB[7]	[125,137]

previous review articles have included the effects of CB[*n*] on guest reactivity [1,3,5–7,10,17,28,29,54]. Table 3.2 provides an overview of the types of reactions which have been carried out in and/or catalyzed by CB[*n*] nanoreactors.

The presence of a CB[*n*] nanoreactor may simply increase the guest solubility or thermal stability in the solvent of interest. In some cases, inclusion of a single reactant can result in a different conformation (higher energy not typically seen in free solution) which results in a specific reaction pathway or product. In other cases, co-inclusion of two reactants (homogeneous dimerization or heterogeneous reaction) can provide a template for specific alignment which can lead to a specific reaction pathway and product. In addition, the presence of a CB[*n*] host may provide catalysis of a reaction by various mechanisms, that is, the same reaction occurs as in free solution but at an increased rate. These types of effects and applications of CB[*n*] nanoreactors will be described in detail in the following sections.

5.1 CB[*n*] Nanoreactor Control of Guest Acidity

The structures of guests **22–24** discussed in this section are shown in Fig. 3.12. Nau and coworkers have done extensive studies of the effects of inclusion into CB[*n*] cavities on guest acid–base properties and equilibria [110–113]. In particular, they have well differentiated the effect of inclusion into CB[*n*] cavities

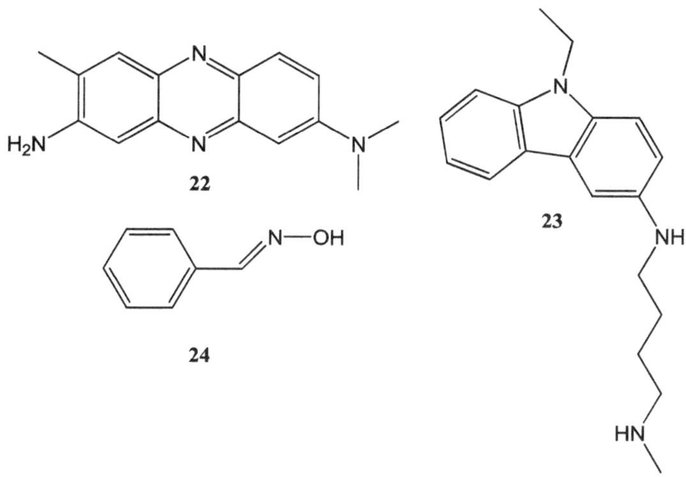

FIGURE 3.12 **Acid–base fluorescent guests 22–24.**

as compared to cyclodextrin cavities on the pK_a of fluorescent guests, such as Neutral Red **22** [110]. They showed that unlike the case of cyclodextrins, in which only the neutral form of Neutral Red formed inclusion complexes, both the neutral and cationic forms were bound by CB[7] [110]. Furthermore, they showed that the pK_a of this guest was *increased* significantly to around 8.8, a shift of nearly 2 pK_a units, whereas in cyclodextrin it was *decreased* to *ca.* 6.1. This increase in pK_a reflected the increased affinity of the CB[7] cavity for the protonated form of Neutral Red, and demonstrated the profound shifts in pK_a that can result from CB[*n*] binding.

Nau et al. exploited this large change in pK_a upon binding into CB[*n*] hosts to design a derivatized dye **23** which can be utilized as a competitive binding indicator [111]. As a result of the large change in pKa upon binding into CB[6], a concomitant large change in fluorescence signal is observed, most importantly a ninefold decrease in fluorescence intensity upon binding with CB[6]. This means that release of the dye **23** upon competitive binding by another guest will result in a large fluorescence increase, yielding a useful "switch-on" fluorescence sensor. The authors demonstrated the utility of this system by monitoring the enzymatic activity of lysine decarboxylase.

Nau et al. also used this CB[*n*]-induced effect on guest pKa to catalyze the acid hydrolysis of a set of three related guests, including benzaldoxime **24** [112]. Significant enhancements in the reaction rates were observed, for example by a factor of 300 in the case of **24**, thus demonstrating the utility of CB[*n*] nanoreactors in promoting base hydrolysis of guest molecules. In this case, the CB[6] nanoreactor facilitated protonation of the guest through stabilization of the resulting cationic from via ion–dipole interactions with the carbonyl portal oxygens. Bhasikuttan et al. have recently summarized the work in this area of CB[*n*]-induced pKa shifts in fluorescent guests, providing excellent detail and insight into its prospective applications in acid–base reactions [113].

5.2 CB[*n*] Nanoreactor Control of Guest Electrochemical Properties

In addition to acid–base properties, inclusion into CB[*n*] cavities can have profound effects on the electrochemical properties of guest, such as redox potential, and therefore have significant effects on electrochemical reactions of guests. Kaifer et al. have been leaders in this area, and have published two excellent reviews on the CB[*n*] complexes of redox-active guests [26,27]. A few representative examples will be discussed here. As mentioned in Section 3.1, the methylviologen dication **4** is bound extremely strongly inside the CB[7] nanoreactor [76,77]. This form of methylviologen is therefore by far the preferred form in CB[7]. Kim et al. showed that inclusion into CB[7] affects the value of the reduction potentials of the methylviologen system, and also the mechanism of the reduction reaction, that is, indirect versus direct electron transfer [76,114]. Kaifer et al. also studied the reduction of viologen dications included within CB[7] cavities [77]. They showed that one-electron reduction of 4,4′-bipyridinium viologen results in a decrease of the binding constant by a factor of two, and discussed the impact of CB[7] on the use of viologens in electron transfer reactions. As a final example in this section, Sun et al. investigated the effect of CB[7] and CB[8] on photoinduced electron transfer reactions in a series of [Ru(bpy)$_3$]-viologen-[Ru(bpy)$_3$] complexes [115]. They found that binding by either of these CB[*n*] nanoreactors lowered the oxidation potential of the ruthenium metal center and promoted the formation of a radical dimer. It is of interest to note that these reactions were carried out in acetonitrile, demonstrating the potential usefulness of CB[*n*] nanoreactors in nonaqueous solution.

CB[*n*] nanoreactors have also been used to template the formation of specific heterogeneous charge-dimer pairs within their cavities, with resulting controlled reactivity. This will be discussed in Section 5.5.

5.3 CB[*n*] Nanoreactors for Enhanced Reactant Solubility and Stability

As mentioned in Section 4.1, CB[*n*] hosts have been used to increase the solubility of organic compounds in water, in order to break them down through chemical reaction [55,94–96]. Karcher et al. studied the potential applicability of CB[6] nanoreactors to extract reactive dyes from wastewater [55], particularly in the presence of alkali and alkaline-earth metal cations, and then destroy them using ozonolysis [116]. They found that while oxidation through reaction with ozone was enhanced by the CB[6] nanoreactors, the dissolved organic content (DOC) in the CB[6]-added wastewater was actually higher than in the original water, so this was not deemed to be a useful approach.

A number of research groups have demonstrated the utility of CB[*n*] nanoreactors to increase guest dye stability and thereby enhance the dye's usefulness and the length of time the dye can be used [117–119]. This stability enhancement typically occurs through prevention of attack by hydrolysis and through oxidation of other reactants, physical confinement of the dye in the cavity, and the microenvironment experienced. Buschmann and Schollmeyer showed that several fluorescent dyes, including phenol blue **25** (Fig. 3.13), could be stabilized against acid hydrolysis by encapsulation in a variety of hosts, including CB[6], which in fact showed the highest degree of protection [117]. Nau et al. showed that Rhodamine 6G **26**, a commonly used fluorescent laser dye, can be made highly photostable by inclusion into CB[7] [118]. In this case, the CB[7] nanoreactor greatly inhibits photobleaching of the dye, which is a major issue in its use in laser applications. This process significantly increases the useful lifetime of a Rhodamine 6G solution. Wang et al. demonstrated significantly enhanced photostability of the linear cyanine dye **27** [119]. In the case of this long dye molecule, a 2:1 host:guest complex was formed with CB[7], and this complexation resulted in distinct promotion of the dye against photolysis. On the other hand, 1:1 complexes also formed at lower CB[7] concentration, but these afforded no significant stabilization of the dye. As a final example of stabilization of an included guest, Day, Collins et al. showed that encapsulation of dinuclear platinum complex drugs into CB[7] and especially CB[8] afforded greatly enhanced drug stability, for example, showing significantly reduced degradation by reaction with cysteine [120]. This has obvious positive implications for the use of CB[*n*] hosts as drug delivery agents.

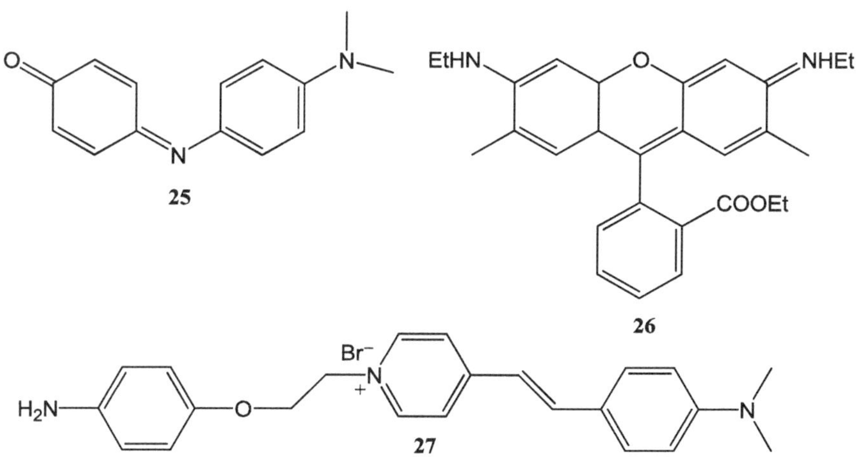

FIGURE 3.13 Fluorescent dyes 25–27 that have been stabilized in CB[*n*] nanoreactors.

Four examples were described of the use of CB[*n*] nanoreactors to enhance guest stability, by inhibiting various chemical reactions including hydrolysis and photolysis. In the remainder of this section, only examples in which chemical reaction of guests is *enhanced* will be described; the focus will be on the use of nanoreactors to improve, not inhibit, chemical reactivity.

5.4 CB[*n*] Nanoreactors for Reactant Geometry and Stereochemistry Control

A number of examples in which CB[*n*] nanoreactors have been used to control reactant geometry or stereochemistry prior to reaction will be described. A few more will be discussed in the next section, 5.5, which is focused on templating by CB[*n*] photoreactors; several of the examples in this section also fall under that related category. One of the earliest reports of the use of a CB[*n*] nanoreactor to perform a stereoselective reaction was that of Kim et al. in 2001, in which a [2 + 2] photoreaction was mediated by CB[8] [121]. Note that this report came out shortly after the publication of the synthesis for CB[*n*] cavities with $n > 6$. In this work, only the (*E*) stereoisomer **28** of diaminostilbene dihydrogen chloride was found to be encapsulated within CB[8] as a 1:2 host:guest complex. This resulted in a perfectly aligned pair of stilbene reactants and hence production of mainly the *syn*-4 **29** product via base-catalyzed UV photodimerization, as illustrated in Fig. 3.14. No products from the (*Z*) isomer were observed, even though UV exposure in solution leads to rapid isomerization. In this case, the CB[8] photoreactor served to ensure that only the (*E*) stereoisomer was present as reactant, providing a reaction with high stereoselectivity.

Rammamurthy et al. also used CB[*n*] nanoreactors to preselect specific orientations to make photodimerization reactions regioselective [122,123]. In the

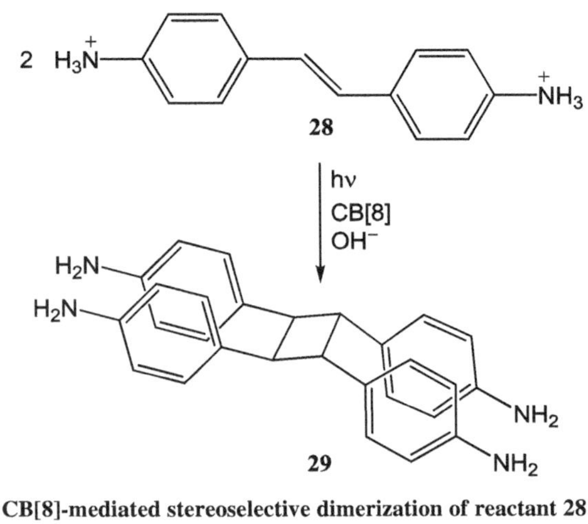

FIGURE 3.14 CB[8]-mediated stereoselective dimerization of reactant 28 to yield predominantly product 29. *(Adapted from Ref. [121], with permission of the Royal Society of Chemistry.)*

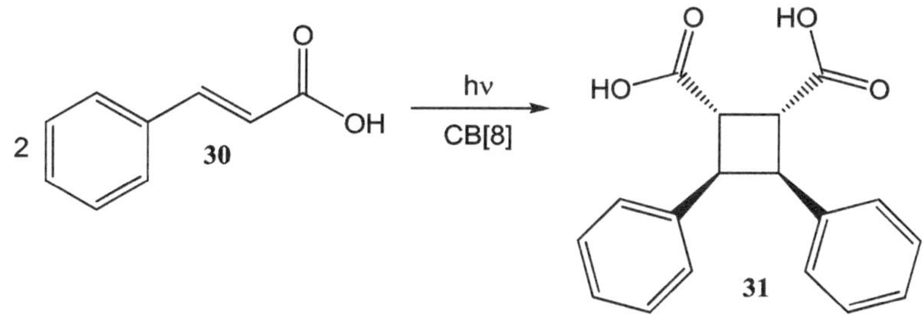

FIGURE 3.15 CB[8]-mediated stereoselective dimerization of reactant cinnamic acid 30 to yield predominantly dimer product 31. *(Adapted from Ref. [122] with permission of American Chemical Society. Copyright 2006.)*

case of cinnamic acid **30**, the formation of 1:2 host:guest inclusion complexes with the water-soluble guest actually helped to solubilize the CB[8], and the only the *syn* H–H dimer **31** resulted as product, as illustrated in Fig. 3.15. This result was clearly due to the confined orientation of the pair of reactants in the cavity [122]. These authors then did extensive further testing of the utility of CB[8] to cause stereoselective dimerization of a series of eight azastilbenes [123]. In each case, instead of a mixture of dimeric reaction products seen in free solution, only one reaction product was observed.

As a final specific example (in this section) of stereospecific control via a CB[*n*] nanoreactor. Kim, Inoue et al. described a highly stereoselective [4 + 4] photocyclodimerization of an α-cyclodextrin-appended 2-anthracene carboxylate guest, using both γ-cyclodextrin and CB[8] [124]. Both γ-cyclodextrin and CB[8] resulted in the formation of one major cycloaddition product, but while γ-cyclodextrin gave predominantly the HT dimer (98%), CB[8] yielded mainly the HH dimer (99%). The structures of the α-cyclodextrin-appended guest, as well as the various resulting dimers, can be found in Ref. [123]. This difference was explained as resulting from the very different shape and depth of the two host cavities. The presence of the α-cyclodextrin on the carboxylate group played a large role in this stereoselective process, demonstrating the potential effect of supramolecular interactions *outside* the main host cavity.

5.5 CB[*n*] Nanoreactors for Reaction Templating

In the last section, several examples of the use of CB[*n*] nanoreactors were described in which only specific stereoisomers of the products were observed, due to selective encapsulation of specific reactant stereoisomers and/or templating of specific reactant stereoisomers to yield specific products [121–124]. In this section, further and more general examples of CB[*n*]-mediated reaction templating will be described, to illustrate the wide range of template roles that these important nanoreactors can play.

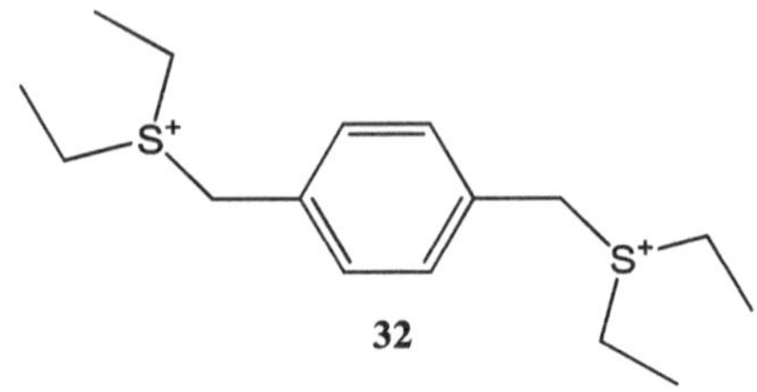
32

FIGURE 3.16 Structure of monomer 32, the elimination polymerization of which was templated by CB[7] as described in Ref. [125].

CB[*n*] nanoreactors can be used to template monomers to yield improved polymerization reactions. For example, Ling and Kaifer used CB[7] to help template the polymerization of two dicationic *p*-xylene guests, including bis(diethylsulfonium)-*p*-xylene **32** (Fig. 3.16), through an elimination polymerization reaction which released diethyl sulfide [125]. They showed that although the CB[7]inclusion complexes did not undergo the desired Wessling polymerization, the dicationic monomers could be prepolymerized to yield polyelectrolytes. When CB[7] was added to these polyelectrolytes, they developed conjugation more readily and at lower temperatures than in the case of the free polyelectrolytes to yield poly(phenylenevinylene) (PPV) polymer. The CB[7] also favored cleavage of side branches, and in addition complexed the diethyl sulfide elimination product. The authors also looked at CB[6] as a nanoreactor for this polymerization, but found that it had no effect, presumably due to its small cavity size.

One of the most common types of reactions to be template using CB[*n*] nanoreactors is photochemical cycloadditions; a few examples were discussed in Section 5.4 [120–124]. A number of other examples will be discussed here [126–137]. Svoboda and Köning reviewed template photochemical reactions back in 2006, and had a small section on such reactions in "molecular flasks" [126]. Although most examples at that time involved cyclodextrins as the molecular flask (a nice descriptive name for a nanoreactor), there were a few examples at that time involving CB[7] and CB[8]. Those examples and some more recent ones will be discussed here [127–137].

The photodimerization of *trans*-cinnamic acid **30** in the presence of CB[8] studied by Ramamurthy et al. was discussed in Section 5.4 [121]. Prior to that report, they specifically examined the templating ability of CB[8] and γ-cyclodextrin for this same dimeric photocycloaddition [127]. They also studied the templating action of CB[8] on the dimeric photocycloaddition of *trans*-1,2-bis(*n*-pyridyl)-ethylene **33** and other related olefins [128]. In the absence of CB[8], photoreaction of **33** produced only the monomer hydration product **34**. However, in the presence of CB[8], 1:2 host:guest inclusion complexes were formed, and upon irradiation gave predominantly (90%) the photodimerization product **35**. These reactions are illustrated in Fig. 3.17. Thus, in this case,

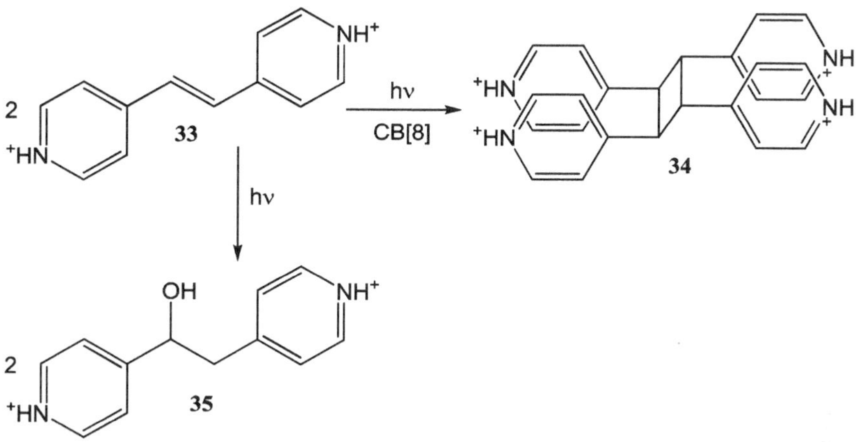

FIGURE 3.17 CB[8]-templated stereoselective photodimerization of reactant 33 to yield predominantly dimer product 34, but hydration product 35 in the absence of CB[8]. *(Adapted from Ref. [129] with permission of American Chemical Society. Copyright 2006.)*

templating by the CB[8] nanoreactor was required not just for regioselectivity, but for the photocycloaddition reaction to occur at all.

Macartney et al. used CB[7] to mediate and template the stereoselective [4 + 4] photodimerization of 2-aminopyridine hydrochloride **36** [129]. In the presence of CB[7], 1:2 host:guest inclusion complexes were formed, and photolysis gave exclusively the anti-trans addition photoproduct **37**. In the absence of CB[7], both the anti-trans and the syn-trans photoproduct are formed. Furthermore, they found that the formation of the photoproduct **37** inside the CB[7] nanoreactor stabilizes it significantly, preventing rearomatization back to the monomer reactants, which occurs thermally at room temperature in the absence of CB[7]. This system is illustrated in Fig. 3.18.

A number of other research groups have also used CB[n] nanoreactors to template photocycloaddition reactions. The structures of guests **38–42** discussed in

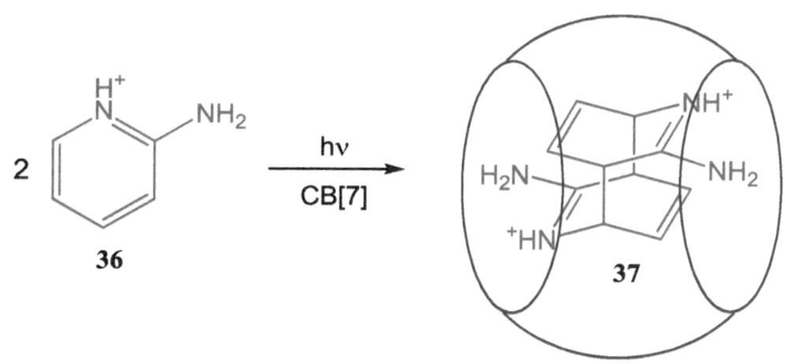

FIGURE 3.18 CB[7]-templated stereoselective photodimerization of reactant 34 to yield predominantly dimer product 35. *(Adapted with permission from Ref. [129]. Copyright 2006 American Chemical Society.)*

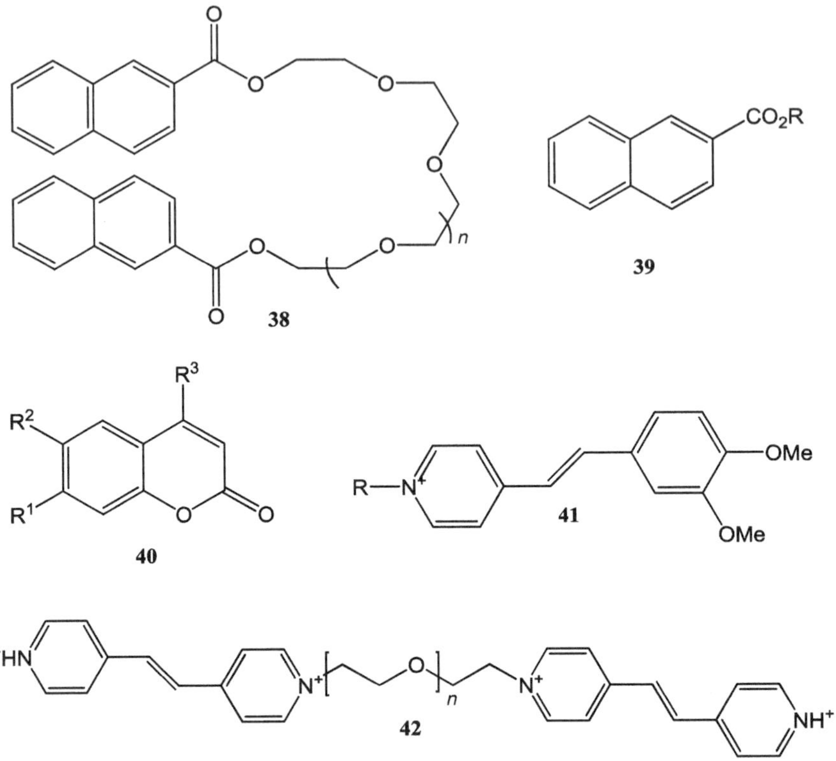

FIGURE 3.19 Some other reactants 38–42 which have been photochemically reacted within CB[*n*] nanoreactors.

this section are shown in Fig. 3.19. Wu et al. used CB[8] to template a highly efficient *intramolecular* photocycloaddition of the 2-naphthyl labeled poly(ethyl glycol) **38** [130]. They also used CB[8] to template the photodimerization of alkyl 2-naphthoate **39** [131]. Sivagaru et al. used CB[8] to template photochemical reactions of a series of coumarin derivatives **40** [132,133]. Gromov et al. used CB[8] as a template to create a photocontrolled molecular assembler based on autophotocycloaddition of styryl dyes such as **41** in the solid state as well as in aqueous solution [134]. Zhang et al. used CB[8] nanoreactors to control the photochemical reactions of azastilbene derivatives such as **42**, and thereby rationally adjust the emissions of the supramolecular photochemical products [135]. For example, products emitting purple, blue, and green light were generated using CB[8] (blue) or CB[7] (purple before UV irradiation, green after). This example well illustrates both the utility of using CB[*n*] nanoreactors, and the differences that can be obtained from different size CB[*n*]s. Da Silva et al. studied the photohydrolysis of pairs of benzimidazole (or thiabendazole) aligned in CB[8] cavities, and showed that the reaction dimerization products are different than from the free solution experiments [136]. As a final example of photochemical templating by a CB[*n*] nanoreactor, Zhang et al. prepared covalently attached hyperbranched polymers through an interesting combination of supramolecular and photochemistry [137]. CB[8] was used to bind with three-arm photosensitive monomers, forming supramolecular polymers (templating). UV photolysis then resulted in the covalent bonding of these to form hyperbranched covalent polymers.

CB[*n*] nanoreactors have also been used to template the dimerization of peptides. Urbach et al. showed that CB[8] selectively recognizes N-terminal aromatic peptides, and that through the formation of 1:2 host–guest inclusion complexes, it templates the dimerization of these peptides [138]. This templating was shown via X-ray crystallography to occur via inclusion of the hydrophobic aromatic side chains inside the CB[8] cavity, while the proximal N-terminal ammonium groups are chelated by the portal carbonyls.

As a final example of the use of CB[*n*] nanoreactors for templating chemical reactions and processes, Kim et al. have done extensive investigations of the formation of hetero-guest charge transfer pairs within CB[8] [139–141]. This work takes advantage of the capacity and tendency of the large CB[8] host cavity to bind two cationic guests, and builds on the previous work involving homo-guest 1:1 host–guest complexes of methyl viologen in CB[7][76,77,114], discussed in Sections 3.1 and 5.2. They showed for example that stable hetero-charge transfer complexes of the methyl viologen dication **4** and 2,6-dihydroxynaphthalene **43** form within the CB[8] nanoreactor cavity, as depicted in Fig. 3.20 [139]. The templating and microenvironment within the CB[8] cavity promote the charge transfer process.

These researchers subsequently showed that these CB[8]-encapsulated charge transfer pairs can act as supramolecular amphiphiles, and spontaneously form vesicle in solution [140]. They were able to investigate the structure and physical properties of these vesicles, which had a typical diameter of around 400–950 nm as indicated by SEM images. Furthermore, TEM images

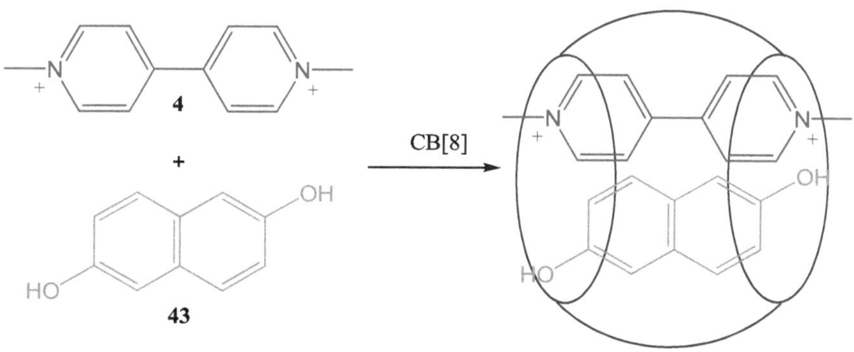

FIGURE 3.20 CB[8]-templated formation of the hetero-guest charge–transfer complex of acceptor 4 and donor 43. *(Adapted from Ref. [128], with permission of John Wiley & Sons.)*

revealed that these vesicles had hollow cavities, with the potential applications. These authors also looked at other CB[8]-stabilized included charge-transfer donor–acceptor pairs, with dipyridyliumylethylene-based electron acceptor units and (as previously) 2,6-dihydroxynaphthalene as the electron donor units, including compounds with these groups linked, forming *intramolecular* charge-transfer complexes within CB[8] [141].

5.6 CB[*n*] Nanoreactors for Reaction Catalysis

This final section will survey the use of CB[*n*] nanoreactors as catalysts in various types of reactions. In a sense, almost all applications of CB[*n*] hosts as nanoreactors are catalytic in nature, as the CB[*n*] itself is typically not used up or altered during the reaction. Catalysis of guest reactions by CB[*n*] hosts has been covered to some degree by most reviews of CB[*n*] chemistry [1,2,3,7,12,28,29,54], including two very recent reviews focused on CB[*n*] catalysis [28,29], so this section will just present a representative survey. The last two reviews in particular provide detailed discussion of the mechanisms for CB[*n*]-induced catalysis; readers who are interested are encouraged to go there for the details. Furthermore, this section will primarily focus on applications in which the presence of the CB[*n*] increases the efficiency of the reaction, and which did not fit one of the prior categories in this Section 5.

One of the first reports of the effect of a CB[*n*] on guest reactivity was that of Mock and Shih, published shortly after their seminal paper reporting the cucurbituril synthesis and structure, which showed that cucurbituril (CB[6]) can catalyze the 1,3-dipolar cycloaddition reaction of an alkyne with an azide [142]. They showed that CB[6] acts as a catalyst exhibiting enzyme-type kinetics, showing an increased rate of reaction with increased CB[6] concentration, with saturation behavior at high CB[6] concentration. They proposed a number of factors for the increased reaction rate in the presence of CB[6], including destabilization of the bound substrate, elimination of entropic constraints in the cycloaddition mechanism, and the binding (and holding for a period of time) of the two reactants promoting reaction to occur. They later studied the kinetics of the catalytic enhancement by CB[6] of the 1,3-cycloaddition reaction between another alkyne–azide pair [143]. Mock and coworkers then showed that the reaction is substantially accelerated (by a factor of 10^5!) and clearly demonstrated the significance of bound-substrate destabilization in the catalytic process in this case. In the past few years, alkyne–azide cycloaddition reactions have been recognized as having tremendous utility due to their ease of reaction and gentle reaction conditions, and are a prime example of the relatively new field of "click chemistry." Tuncel et al. have provided a recent review of the use of cucurbituril-catalyzed 1,3-dipolar cycloadditions as click reactions, which provides an excellent discussion and summary of this area [144]. Tuncel and Steinke used this click chemistry approach with CB[6] to catalytically produce polyrotaxanes, through an inclusion-induced self-threading mechanism [145].

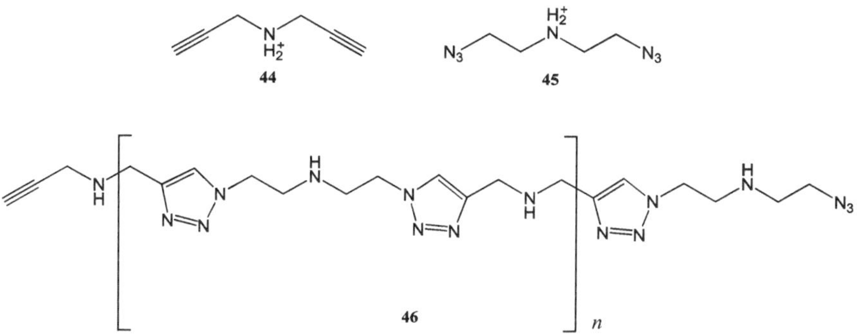

FIGURE 3.21 **Reactants 44 and 45 and product 46 in the CB[6]-catalyzed formation of oligotriazoles from a 1,3-dipolar azide–alkyne cycloaddition, as described in Ref.** [146].

Krasia and Steinke also used this click chemistry approach with catalytic CB[6] to form oligotriazoles **46**, from the reaction of bis(propargyl)amine HCl **44** with bis(2-azidoethyl)amine **45** [146] (Fig. 3.21).

As discussed in Section 5.3, CB[*n*] hosts can serve as template to position a pair of monomers to be perfectly aligned to undergo photocycloaddition [126–137]. CB[8] has proven to be especially useful for this, due to the optimum size of the cavity relative to most common reactants for this process. In many cases, formation of a 1:2 CB[8]:guest inclusion complex not only provides a template, but also catalyzes the reaction, yielding much faster kinetics than in water (for those systems which still occur in the absence of the CB[8] template) [147–151]. Sivagaru et al. clearly demonstrated this CB[8]-induced catalysis [147,148] in a continuation of their studies of CB[8]-templated photocycloadditions described previously [132,133]. They showed that CB[8] concentrations as low as 10% gave excellent catalytic results, and the system exhibited enzymatic-like saturation kinetics. The kinetic analysis revealed a cooperative mechanism in the catalytic process [147]. They presented an extensive experimental investigation into the mechanism of the catalysis, and provided a stepwise mechanism [148]. A number of other research groups have also demonstrated the high degree of stereoselectivity and catalysis that can be obtained by carrying out photocycloaddition reactions in CB[7] and CB[8] nanoreactors, for the monomeric guests *N*-allyl-2-aminopyridinium [149], 1,ω-bi(2-aminopyridinium) alkylates [150], and anthracene-labeled macrocycles [151].

A number of groups have used CB[*n*] photoreactors to catalyze the oxidation of guest molecules [152–154]. Cong and coworkers [152] showed that dehydration of veratry alcohol **47** to veratryaldehyde **48** by the oxidizing agent *o*-iodoxybenzoic acid **49** (Fig. 3.22) was enhanced by the presence of CB[8], both in terms of reaction time and in terms of % conversion. The same trends were observed for other benzylic alcohols as well, including benzyl alcohol and 3-methoxybenzyl alcohol. Furthermore, CB[8] was shown to provide

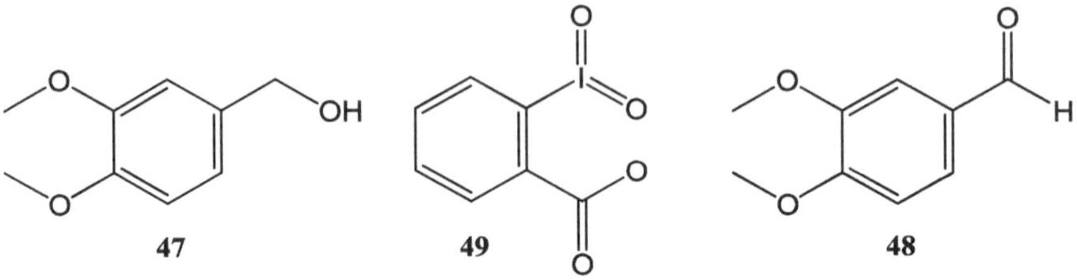

FIGURE 3.22 Reactants 47 and 49 and product 48 in the CB[8]-catalyzed oxidation reaction described in Ref. [152].

much better catalytic action than cyclodextrins. This same group also showed similar CB[8]-catalyzed enhancement of a wide range of alcohols, including alkyl, although the best catalytic enhancement was obtained for allyl and aryl alcohols [153,154].

Enhanced oxidations have also been observed using specific cucurbituril host–guest inclusion complexes as catalysts, as opposed to the hosts themselves [155–157]. For example, Demets et al. developed a novel oxovanadium(IV)-CB[6] complex **50**, in which the oxovanadium ion is coordinated to four of the six CB[6] portal carbonyl oxygens, as shown in Fig. 3.23 [155]. This novel complex was found to catalyze the oxidation of a range of substrates to their corresponding alcohols, using a variety of oxidizing agents, typically H_2O_2. Alkenes were not found to undergo reaction, which indicated that the catalysis did not involve the oxygen atom of the oxovanadium ion. However, alkanes such as *n*-pentane were catalytically converted to alcohols. The authors proposed, based

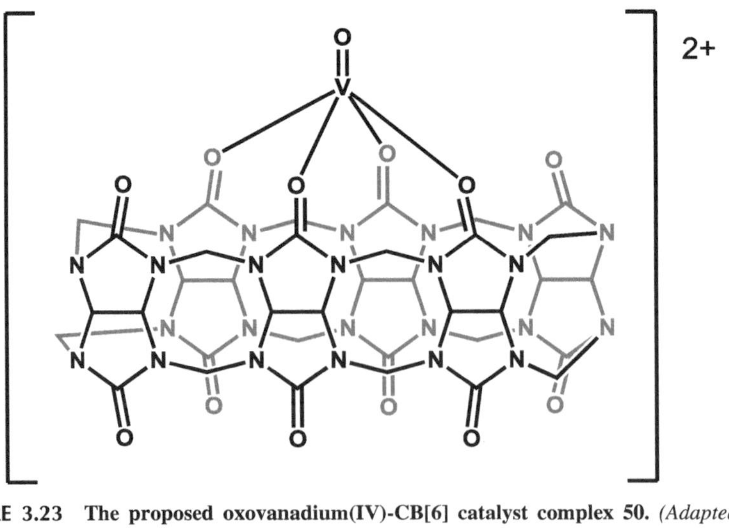

FIGURE 3.23 The proposed oxovanadium(IV)-CB[6] catalyst complex 50. *(Adapted from Ref. [155] with permission of Elsevier.)*

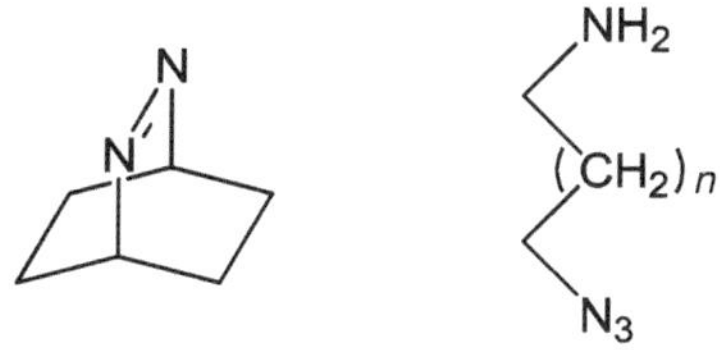

FIGURE 3.24 CB[*n*] guest reactants 51 and 52.

on this and other information, that catalysis required inclusion of the substrate into the cavity, where interaction with the vanadium atom could take place.

Nau et al. showed that CB[7] complexed at one of the rims by a transition metal cation (such as Ag^+) provides an effect catalytic nanoreactor [156]. As in the catalyst complex described previously, the substrate enters the CB[7] cavity, wherein it can interact with the transition metal catalyst. They were able to catalyze photochemical nitrogen elimination from the molecule DBO **51** (Fig. 3.24) using this Ag^+–CB[7] complex (Fig. 3.24).

Interestingly, Demets et al. prepared bromine- and iodine-CB[6] complexes for use as nanoreactors [157]. These complexes are quite different from the two transition-metal complexes described previously. Instead of a metal cation coordinated to one of the portals, in this case a Br_2 or I_2 molecule is included within the cavity, and is in place to catalyze or react with other molecules. The authors chose reactions that were known to be catalyzed by or involve as a reagent one of these halogen molecules. For example, Br_2-CB[6] was effectively applied to the electrophilic bromination of benzene. Such a nanoreactor which is "preloaded" with a catalyst or reagent could have significant and interesting potential applications.

Two other examples of the use of CB[*n*] nanoreactors will be mentioned. Lu and Masson showed that CB[6], CB[7], and CB[8] can catalyze the Ag^+-promoted desilylation of silicon containing pseidorotazanes [158]. This occurred via the stabilization of a key intermediate in the reaction within the CB[*n*] cavity. Finally, Brinker et al. showed that CB[6] itself can be used to catalyze the acidic decomposition of azidoaminoalkanes **52** (Fig. 3.24) [159]. These compounds were shown to decompose rapidly in acidic aqueous solution when complexed with CB[*n*], whereas in water they were stable for days. This provides an example of a guest that is actually *less stable* when included within a CB[*n*] cavity, and illustrates another aspect of the utility of CB[*n*] cavities as nanoreactors.

6 CONCLUSIONS

CB[*n*] are arguably the best of the families of molecular hosts (cyclodextrins, calixarenes, cavitands...) for utilization as nanoreactors, due to their favorable properties. They contain rigid (for $n \leq 8$), spherical-shaped cavities, with extremely low polarizability inside the cavities. The cavities have opposing restrictive carbonyl-lined portals, which serve to help keep guests included, and to help bind guests strongly through dipole–dipole and charge–dipole interactions with guests. They come in a wide range of sizes (from $n = 5–14!$), which means that a good size match can be found for a specific reactant guest. Although one of their limitations is their aqueous solubility (especially for $n = 6$ and 8), solubility can be increased by acidic solution, metal cations, and (in some cases) by inclusion of polar guests. Inclusion of reactant molecules into CB[*n*] hosts can have profound and sometimes useful effects on the guest physical properties (solubility, spectroscopic properties, and structure) and chemical properties (acidity, redox potential, reactivity). As demonstrated in the survey of reactants/reactions which have been carried out in CB[*n*] nanoreactors presented in this chapter, inclusion of reactants into CB[*n*] cavities can: (1) serve to template reactions, yielding region- and/or stereospecific products; (2) allow for the existence of a form of a reactant not usually seen in free solution, thus yielding unique products; (3) serve to catalyze reactions, increasing the and/or decreasing reaction time; and (4) have other miscellaneous and unique effects on reactions.

It is clear from the large numbers of studies that have been performed that CB[*n*] hosts are highly useful as nanoreactors in synthetic chemistry. It is also clear that research into the behavior and utilization of these "molecular beakers" will continue to grow in the future.

REFERENCES

[1] Cintas P. Cucurbituril: supramolecular perspectives for an old ligand. J Incl Phenom Molec Rec Chem 1994;17:205–20.
[2] Mock WL. Cucurbituril. Top Curr Chem 1995;175:1–24.
[3] Mock WL. Cucurbituril. Comp Supramol Chem 1996;2:477–93.
[4] Gerasko OA, Samsonenko DG, Fedin VP. Supramolecular chemistry of cucurbiturils. Russ Chem Rev 2002;71:840–61.
[5] Lee JW, Samal S, Selvapalam N, Kim H-J, Kim K. Cucurbituril homologues and derivatives: new opportunities in supramolecular chemistry. Acc Chem Res 2003;36:621–30.
[6] Kim K, Selvapalam N, Oh DH. Cucurbituril – a new family of hosts. J Inclus Phenom Macro Chem 2004;50:31–6.
[7] Lagona J, Mukhopadhyay P, Chakrabarti S, Isaacs L. The cucurbit[n]uril family. Angew Chem Int Ed 2005;44:4844–70.
[8] Das D, Scherman OA Isr. Cucurbituril: at the interface of small molecule host-guest chemistry and dynamic aggregates. J Chem 2011;51:537–50.
[9] Nau WM, Scherman OA Isr. The world of cucurbiturils – from peculiarity to commodity. J Chem 2011;51:492–4.
[10] Masson E, Ling X, Joseph R, Kyeremeh-Menash L, Lu X. Cucurbituril chemistry: a tale of supramolecular success. RSC Adv 2012;2:1213–47.
[11] Schalley CA, Angew. Of molecular gyroscopes, Matroshka dolls, and other "Nano"-toys. Chem Int Ed 2002;41:1513–5.
[12] Kottas GS, Clarke LI, Horienk D, Michl J. Artificial molecular rotors. Chem Rev 2005;105:1281–376.

[13] Wagner BD. In: Ariga K, Nalwa HS, editors. Bottom-up nanofabrication. Stevenson Ranch, CA: American Scientific Publishers; 2009. p. 1–33.
[14] Macartney DH Isr. Encapsulation of drug molecules by cucurbiturils: effects on their chemical properties in aqueous solution. J Chem 2011;51:600–15.
[15] Mandadapu V, Day AI, Ghanem A. Cucurbituril: chiral applications. Chirality 2014;26: 712–23.
[16] Turro NJ. Molecular structure as a blueprint for supramolecular structure chemistry in confined places. Proc Nat Acad Sci 2005;102:10766–70.
[17] Vriezema DM, Comellas Aragonès M, Elemans AAW, Cornelissen JJLM, Rowan AE, Nolte RJM. Self-assembled nanoreactors. Chem Rev 2005;105:1445–89.
[18] Koblenz TS, Wassenaar J, Reek JNH. Reactivity within a confined self-assembled nanospace. Chem Soc Rev 2008;37:247–62.
[19] Nardin C, Theoni S, Widmer J, Winterhalter M, Meier S. Nanoreactors based on (Polymerized) ABA-triblock copolymer vesicles. Chem Commun 2000;1433–4.
[20] Peters RJRW, Marguet M, Marais S, Fraaije MW, van Hest JCM, Lecommandoux S. Cascade reactions in multicompartmentalized polymersomes. Angew. Chem Int Ed 2014;53:146–50.
[21] Breslow R. Biomimetic chemistry and artificial enzymes: catalysis by design. Acc Chem Res 1995;28:146–53.
[22] Ikeda A, Hinkei S. Novel cavity design using calix[n]arene sckeletons: toward molecular recognition and metal binding. Chem Rev 1997;97:1713–34.
[23] Purse BW, Rebek J Jr. Functional cavitands: chemical reactivity in structured environments. Proc Nat Acad Sci 2005;102:10777–82.
[24] Marquez C, Huang F, Nau WM. Cucurbiturils: molecular nanocapsules for time-resolved fluorescence assays. IEEE Trans Nanobiosci 2004;3:39–45.
[25] Ayitou A, Pemberton BC, Kumarasamy E, Vallavoju N, Sivaguru J. Fun with photons: selective light induced reactions in solution and in water soluble nano-containers. Chimia 2011;65:202–9.
[26] Gadde S, Kaifer AE. Cucurbituril complexes of redox active guests. Curr Org Chem 2011;15:27–38.
[27] Kaifer AE, Li W, Yi S Isr. Cucurbiturils as versatile receptors for redox active substrates. J Chem 2011;51:496–505.
[28] Pemberton BC, Raghunathan R, Volla S, Sivaguru J. From containers to catalysis: supramolecular catalysis within cucurbiturils. Chem Eur J 2012;18:12178–90.
[29] Assaf KI, Nau WM. Cucurbiturils: from synthesis to high-affinity binding and catalysis. Chem Soc Rev 2015;44:394–418.
[30] Behrend R, Meyer E, Rusche F. Uber condensations producte ans glycoluril und formaldehyd. Liebigs Ann Chem 1905;339:1–36.
[31] Freeman WA, Mock WL, Shih N-Y. Cucurbituril. J Am Chem Soc 1981;103:7367–8.
[32] Flynn A, Hough GC, Stoddart JF, Williams DJ. Decamethylcucurbit[5]uril. Angew Chem Int Ed Engl 1992;31:1475–7.
[33] Kim J, Jung I-S, Kim S-Y, Lee E, Kang J-K, Sakamoto S, Yamaguchi K, Kim K. New cucurbituril homologues: synthesis, isolation, characterization, and X-ray crystal structures of cucurbi[n]turil (n = 5, 7, and 8). J Am Chem Soc 2000;122:540–1.
[34] Day AI, Arnold AP, Blanch RJ, Snushall B. Controlling factors in the synthesis of cucurbituril and its homologues. J Org Chem 2001;66:8094–100.
[35] Jansen K, Buschmann H-J, Wego A, Döpp D, Mayer C, Drexler H-J, Holdt H-J, Schollmeyer E. Cucurbit[5]uril, decamethylcucurbit[5]uril and cucurbit[6]uril. Synthesis, solubility and amine complex formation. J Inclus Phenom Macro Chem 2001;39:357–63.

[36] Liu S, Zavalij PY, Isaacs L. Cucurbit[10]uril. J Am Chem Soc 2005;127:16798–9.
[37] Isaacs L Isr. The mechanism of cucurbituril formation. J Chem 2011;51:578–91.
[38] Zhao J, Kim H-J, Oh J, Kim S-Y, Lee W, Sakamoto S, Yamaguchi K, Kim K. Cucurbit[n]uril derivatives soluble in water and organic solvents. Angew Chem Int Ed 2001;40:4233–5.
[39] Day AI, Arnold AP, Blanch RJ. A method for synthesizing partially substituted cucurbit[n] uril. Molecules 2003;8:74–84.
[40] Lagona J, Fettinger JC, Isaacs L. Cucurbit[n]uril analogues. Org Lett 2003;5:3745–7.
[41] Lagona J, Fettinger JC, Isaacs L. Cucurbit[n]uril analogues: synthetic and mechanistic studies. J Org Chem 2005;70:10381–92.
[42] Kim K, Slevapalam N, Ko YH, Park KM, Kim D, Kim J. Functionalized cucurbiturils and their applications. Chem Soc Rev 2007;36:267–79.
[43] Cheng X-J, Liang L-L, Chen K, ji N-N, Xiao X, Zhang J-X, Zhang Y-Q, Xue S-F, Zhu Q-J, Ni X-L, Tao Z. Twisted cucurbit[14]uril. Angew Chem 2013;125:7393–6.
[44] Mock WL, Pierpont J. A cucurbituril-based molecular switch. J Chem Soc Chem Commun 1990;1509–11.
[45] Blanch RJ, Sleeman AJ, White TJ, Arnold AP, Day AI. Cucurbit[7]uril and *o*-carborane self-assemble to from a molecular ball bearing. Nano Lett 2002;41:147–9.
[46] Day AI, Blanch RJ, Arnold AP, Lorenzo S, Lewis GR, Dance I. A cucurbituril-based gyroscane: a new supramolecular form. Angew Chem Int Ed 2002;2:275–7.
[47] Whang D, Heo J, Kim C-A, Kim K. Helical polyrotaxane: cucurbituril 'beads' threaded onto a helical one-dimensional coordination polymer. Chem Commun 1997;2361–2.
[48] Kim K. Mechanically interlocked molecules incorporating cucurbituril and their supramolecular assemblies. Chem Soc Rev 2002;31:96–107.
[49] Ko YH, Kim K, Kang J-K, Chun H, Lee JW, Sakamoto S, Yamaguchi K, Fettinger JC, Kim K. Designed self-assembly of molecular necklaces using host-stabilized charge-transfer interactions. J Am Chem Soc 2004;126:1932–3.
[50] Jeon WS, Kim E, Ko YH, Hwang I, Lee JW, Kim S-Y, Kim H-J, Kim K. Molecular loop lock: a redox-driven molecular machine based on a host-stabilized charge-transfer complex. Angew Chem Int Ed 2005;44:87–91.
[51] Liu S, Ruspic C, Mukhopadhyay P, Chakrabarti S, Zavalij PY, Isaacs L. The cucurbit[n]uril family: prime components for self-sorting systems. J Am Chem Soc 2005;127:15959–67.
[52] Saleh N, Koner AL, Nau WM. Activation and stabilization of drugs by supramolecular pK_a shifts: drug-delivery applications tailored for cucurbiturils. Angew Chem Int Ed 2008; 47:5398–401.
[53] Nau WM, Florea M, Assaf KI Isr. Deep inside cucurbiturils: physical properties and volumes of their inner cavity determine the hydrophobic driving force for host-guest complexation. J Chem 2011;51:559–77.
[54] Parvari O, Reany E, Keinan Isr. Applicable properties of cucurbiturils. J Chem 2011;51:646–63.
[55] Karcher S, Kornmüller A, Jekel M. Cucurbituril for water treatment. Part I: solubility of cucurbituril and sorption of reactive dyes. Wat Res 2001;35:3309–16.
[56] Buschmann H-J, Cleve E, Schollmeyer E. Cucurbituril as a ligand for the complexation of cations in aqueous solutions. Inorg Chim Acta 1992;193:93–7.
[57] Jeon Y-M, Kim J, Whang D, Kim K. Molecular container assembly capable of controlling binding and release of its guest molecules: reversible encapsulation of organic molecules in sodium ion complexed cucurbituril. J Am Chem Soc 1996;118:9790–1.
[58] Buschmann H-J, Jansen K, Meschke C, Schollmeyer E. Thermodynamic data for complex formation between cucurbituril and alkali and alkaline earth cations in aqueous formic acid solution. J Solut Chem 1998;27:135–40.

[59] Germain P, Létoffé JM, Merlin MP, Buschmann H-J. Thermal behaviour of hydrated and anhydrous cucurbituril A DSC, T.G. and calorimetric study in temperature range from 100 to 800 K. Thermochim Acta 1998;315:87–92.
[60] Marquez C, Nau WM. Polarities inside molecular containers. Angew Chem Int Ed 2001;40:4387–90.
[61] Koner AL, Nau WM. Cucurbituril encapsulation of fluorescent dyes. Supramol Chem 2007;19:55–66.
[62] Buschmann H-J, Wego A, Zielesny A, Schollmeyer E. Structure, electronic properties and NMR-shielding of cucurbit[n]urils. J Inclus Phenom Macro Chem 2006;54:85–8.
[63] Lehn JM. Supramolecular chemistry – scope and perspectives molecules, supermolecules, and molecular devices (Nobel lecture). Angew Chem Int Ed Engl 1988;27:89–112.
[64] Wagner BD. In: Nalwa HS, editor. Handbook of photochemistry and photobiology, 3. Stevenson Ranch, CA: American Scientific Publishers; 2003. p. 1–52.
[65] Wagner BD. Hydrogen bonding of excited states in supramolecular host-guest inclusion complexes. Phys Chem Chem Phys 2012;14:8825–35.
[66] Márquez C, Hudgins RH, Nau WM. Mechanism of host-guest complexation by cucurbituril. J Am Chem Soc 2004;126:5806–16.
[67] Mock WL, Shih N-Y. Host-guest binding capacity of cucurbituril. J Org Chem 1983;48: 3618–9.
[68] Mock WL, Shih N-Y. Organic ligand-receptor interactions between cucurbituril and alkylammonium ions. J Am Chem Soc 1988;110:4706–10.
[69] Meschke C, Buschmann H-J, Schollmeyer E. Complexes of cucurbituril with alkyl mono- and diammonium ions in aqueous formic acid studied by calorimetric titrations. Thermochim Acta 1997;297:43–8.
[70] Whang D, Park K-M, Heo J, Ashton P, Kim K. Molecular necklace: quantitative self-assembly of a cyclic oligorotaxane from nine molecules. J Am Chem Soc 1998;120: 4899–900.
[71] Buschmann H-J, Mutihac L, Jensen K, Schollmeyer E. Cucurbit[6]uril as ligand for the complexation of diamines, diazacrown ethers and cryptands in aqueous formic acid solution. J Inclus Phenom Macro Chem 2005;53:281–4.
[72] Hoffmann R, Knocke W, Fenn C. Host-guest complexes of cucurbituril with the 4-methylbenzylammonium ion, alkali-metal cations and NH_4^+. J Chem Soc Faraday Trans 1994;90:1507–11.
[73] Neugebauer R, Knoche W. Host-guest complexes of cucurbituril with 4-amino-4′-nitroazobenzene and 4,4′-diaminobenzene in acidic aqueous solutions. J Chem Soc Perkin Trans 1998;2:529–34.
[74] Buschmann H-J, Jansen K, Schollmeyer E. The formation of cucurbituril complexes with amino acids and amino alcohols in aqueous formic acid studied by calorimetric titrations. Thermochim Acta 1998;317:95–8.
[75] Fu H-Y, Xue S-F, Zhu Q-J, Tao Z, Zhang H-X, Day AI. Investigations of host-guest compounds of cucurbit[*n*=5]uril with some ortho aminopyridines and bispryridine. J Inclus Phenom Macro Chem 2005;53:101–7.
[76] Kim H-J, Jeon WS, Ko YH, Kim K. Inclusion of methylviologen in cucurbit[7]uril. PNAS 2002;99:5007–11.
[77] Ong W, Gómez-Kaifer M, Kaifer AE. Cucurbit[7]uril: a very effective host for viologens and their radical cations. Org Lett 2002;4:1791–4.
[78] Buschmann H-J, Jensen K, Schollmeyer E. Cucurbit[6]uril as ligand for the complexation of lanthanide cations in aqueous solution. Inorg Chem Commun 2003;6:531–4.

[79] Buschmann H-J, Schollmeyer E. Cucurbituril and β-cyclodextrin as hosts for the complexation of organic dyes. J Inclus Phenom Molec Recog Chem 1997;29:167–74.
[80] Wagner BD, Fitzpatrick SJ, Gill MA, MacRae AI, Stojanovic N. A fluorescent host-guest complex of cucurbituril in solution: a molecular jack o'lantern. Can J Chem 2001;79:1101–4.
[81] Wagner BD, Stojanovic N, Day AI, Blanch RJ. Host properties of cucurbit[7]uril: fluorescence enhancement of anilinonaphthalene sulfonates. J Phys Chem B 2003;107:10741–6.
[82] Dantz DA, Meschke C, Buschmann H-J, Schollmeyer E. Complexation of volatile organic molecules from the gas phase with cucurbituril and β-cyclodextrin. Supramol Chem 1998;9:79–83.
[83] Rankin MA, Wagner BD. Fluorescence enhancement of curcumin upon inclusion into cucurbituril. Supramol Chem 2004;16:513–9.
[84] Buschmann H-J, Jansen K, Schollmeyer E. Cucurbituril as host molecule for the complexation of aliphatic alcohols, acids and nitriles in aqueous solution. Thermochim Acta 2000;346:33–6.
[85] Whang D, Heo J, Park JH, Kim K. A molecular bowl with metal ion as bottom: reversible inclusion of organic molecules in cesium ion complexed cucurbituril. Angew Chem Int Ed 1998;37:78–80.
[86] El Haouaj M, Ko YH, Luhmer M, Kim K, Bartik K. NMR investigation of the complexation of neutral guests by cucurbituril. J Chem Soc Perkin Trans 2 2001;2104–7.
[87] Buschmann H-J, Schollmeyer E, Mutihac L. The formation of amino acid and dipeptide complexes with α-cyclodextrin and cucurbit[6]uril in aqueous solutions studied by titration calorimetry. Thermochim Acta 2003;399:203–8.
[88] Buschmann H-J, Mutihac L, Mutihac R-C, Schollmeyer E. Complexation behaviour of cucurbit[6]uril with short peptides. Thermochim Acta 2005;430:79–82.
[89] El Haouaj M, Luhmer YH, Ko K, Kim K, Bartik. NMR study of the reversible complexation of xenon by cucurbituril. J Chem Soc Perkin Trans 2 2001;804–7.
[90] Jian G, Li G. Preparation and biological activity of novel cucurbit[8]uril-fullerene complex. J Photochem Photobiol B Biol 2006;85:223–7.
[91] Kim S-Y, Jung I-S, Lee E, Kim J, Sakamoto S, Yamaguchi K, Kim K. Macrocycles within macrocycles: cyclen, cyclam, and their transition metal complexes encapsulated in cucurbit[8]uril. Angew Chem Int Ed 2001;40:2119–21.
[92] Jeon WS, Moon K, Park SH, Chun H, Ko YH, Lee JY, Lee ES, Samai S, Selvapalam N, Rekharsky MK, Sindelar V, Sobrasingh S, Inoue Y, Kaifer AE, Kim K. Complexation of ferrocene derivatives by cucurbit[7]uril host: a comparative study of the cucurbituril and cyclodextrin host families. J Am Chem Soc 2005;127:12984–9.
[93] Liu S, Shukla AD, Gadde S, Wagner BD, Kaifer AE, Isaacs L. Ternary complexes comprising cucurbit[10]uril, porphyrins, and guests. Angew Chem Int Ed 2008;47:2657–60.
[94] Karcher S, Kornmüller A, Jekel M. Effects of alkali and alkaline-earth cations on the removal of reactive dyes with cucurbituril. Acta Hydrochim Hydrobiol 1999;27:38–42.
[95] Buschmann H-J, Jonas C, Schollmeyer E. The selective removal of dyes from waste water. Eur Water Pollut Contr 1997;6:21–4.
[96] Dantz DA, Otyakmaz Ő, Buschmann H-J, Schollmeyer E. Removal of aromatic substances from water by the supramolecular recyclable complexing agent cucurbituril. Vom Wasser 1998;91:305–14.
[97] Huang Y, Xue S-F, Tao Z, Zhu Q-J, Zhang H, Lin J-X, Yu D-H. Solubility enhancement of kinetin through host-guest interactions with cucurbiturils. J Inclus Phenom Macro Chem 2008;61:171–7.

[98] Koner AL, Ghosh I, Saleh N, Nau WM. Supramolecular encapsulation of benzimidazole-derived drugs by cucurbit[7]uril. Can J Chem 2011;89:139–47.

[99] Mock WL, Shih N-Y. Structure and selectivity in host-guest complexes of cucurbituril. J Org Chem 1986;51:4440–6.

[100] Pemberton BC, Mumarasamy E, Jockusch S, Srivastava DK, Sivaguru J. Photophysical aspects of 6-methoxycoumarin-cucurbit[8]uril host-guest complexes. Can J Chem 2011; 89:310–6.

[101] Choi S, Park SH, Ziganshina AY, Ko YH, Lee JW, Kim K. A stable *cis*-stilbene derivative encapsulated in cucurbit[7]uril. Chem Commun 2003;2176–7.

[102] Bernal I, Mukhopadhyay U, Vivrovets AV, Fedin VP, Clegg W. A previously unrecognised hydronium di-cation in the crystal structure of a cucurbituril derivative. Chem Commun 2005;3791–2.

[103] Mitkina TV, Sokolov MN, Naumoc DM, Kuratieva NV, Gerasko OA, Fedin VP. Jørgensen complex within a molecular container: selective encapsulation of *trans*-$[CO(en)_2Cl_2]$ into cucurbit[8]uril and influence of inclusion on guest's properties. Inorg Chem 2006;45: 6950–5.

[104] Wang R, Yuan L, Macartney DH. Stabilization of the (*E*)-1-ferrocenyl-2-(1-methyl-4-pyridinium)ethylene cation by inclusion in cucurbit[8]uril. Organomet 2006;25:1820–3.

[105] Liu S, Zavalij PY, Lam Y-F, Isaacs L. Refolding foldamers: triazene-arylene oligomers that change shape with chemical stimuli. J Am Chem Soc 2007;129:11232–41.

[106] Rawashdeh AMM, Thangavel A, Sotiriou-Leventis C, Leventis N. Control of ketone to *gem*-diol equilibrium by host-guest interactions. Org Lett 2008;10:1131–4.

[107] Hart SL, Haines RI, Decken A, Wagner BD. Isolation of the *trans*-I and *trans*-II isomers of Cu^{II}(cyclam) via complexation with the macrocyclic host cucurbit[8]uril. Inorg Chim Acta 2009;362:4145–51.

[108] Gadde S, Batchelor EK, Weiss JP, Ling Y, Kaifer AE. Control of H- and J-aggregate formation via host-guest complexation using cucurbituril hosts. J Am Chem Soc 2008;130: 17114–9.

[109] Nally R, Scherman OA, Isaacs L. Polymer deaggregation and assembly controlled by a double cavity cucurbituril. Supramol Chem 2010;22:683–90.

[110] Mohanty J, Bhasikuttan AC, Nau WM, Pal H. Host-guest complexation of neutral red with macrocyclic host molecules: contrasting pK_a shifts and binding affinities for cucurbit[7]uril and β-cyclodextrin. J Phys Chem B 2006;110:5132–8.

[111] Praetorius A, Bailey DM, Schwarzlose T, Nau WM. Design of a fluorescent dye for indicator displacement form cucurbiturils: a macrocycle-responsive fluorescent switch operating through a pK_a shift. Org Lett 2008;10:4089–92.

[112] Klöck C, Dsouza RN, Nau WM. Cucurbituril-mediated supramolecular acid catalysis. Org Lett 2009;11:2595–8.

[113] Barooah N, Mohanty J, Pal H, Bhasikuttan AC. Cucurbituril-induced supramolecular pK_a shift in fluorescent dyes and its prospective applications. Proc Natl Acad Sci India Sect A Phys Sci 2014;84:1–17.

[114] Kim H-J, Jeon WS, Ko YH, Kim K. Inclusion of methylviologen in cucurbit[7]uril. Proc Nat Acad Sci 2002;99:5007–11.

[115] Monhaphol TK, Andersson S, Sun L. Isolated supramolecular $[Ru(bpy)_3]$-viologen-$[Ru(bpy)_3]$ complexes with trapped CB[7,8] and photoinduced electron-transfer study in nonaqueous solution. Chem Eur J 2011;17:11604–12.

[116] Kornmüller A, Karcher S, Jekel M. Cucurbiturilfor water treatment. Part II: ozonation and oxidative regeneration of cucurbituril. Wat Res 2001;35:3317–24.

[117] Buschmann H-J, Schollmeyer E. Stabilization of dyes against hydrolytic decomposition by the formation of inclusion compounds. J Inclus Phenom Molec Recog Chem 1992;14:91–9.
[118] Mohanty J, Nau WM. Ultrastable rhodamine with cucurbituril. Angew Chem Int Ed 2005;44:3750–4.
[119] Zhang H, Liu L, Gao C, Sun R, Wang Q. Enhancing photostability of cyanine dye by cucurbituril encapsulation. Dyes and Pigments 2012;94:266–70.
[120] Bali MS, Buck DP, Coe AJ, Day AI, Collins JG. Cucurbituril binding of *trans*-$[\{PtCl(NH_3)_2\}_2(\mu\text{-}NH_2(CH_2)_8NH_2)]^{2+}$ and the effect on the reaction with cysteine. Dalton Trans 2006;5337–44.
[121] Jon SY, Ko YH, Park SH, Kim H-J, Kim K. A facile, stereoselective [2 + 2] photoreaction mediated by cucurbit[8]uril. Chem Commun 2001;1938–9.
[122] Pattabiraman M, Kaanumalle LS, Natarajan A, Ramamurthy V. Regioselective photodimerization of cinnamic acids in water: templation with cucurbiturils. Langmuir 2006;22: 7605–9.
[123] Maddipatla MVSN, Kaanumalle LS, Natarajan A, Pattabiraman M, Ramamurthy V. Preorientation of olefins toward a single photodimer: cucurbituril-mediated photodimerization of protonated azastilbenes in water. Langmuir 2007;23:7545–54.
[124] Yang C, Mori T, Origane Y, Ko YH, Selvapalam N, Kim K, Inoue Y. Highly stereoselective photocylodimerization of α-cyclodextrin-appended anthracene mediated by γ-cyclodextrin and cucurbit[8]uril: a dramatic steric effect operating outside the binding site. J Am Chem Soc 2008;130:8574–5.
[125] Ling Y, Kaifer AE. Complexation of poly(phenylenevinylene) precursors and monomers by cucurbituril hosts. Chem Mater 2006;18:5944–9.
[126] Svoboda J, König B. Templated photochemistry: toward catalysts enhancing the efficiency and selectivity of photoreactions in homogeneous solutions. Chem Rev 2006;106:5413–30.
[127] Pattabiraman M, Natarajan A, Kaanumalle LS, Ramamurthy V. Templating photodimerization of *trans*-cinnamic acids with cucurbit[8]uril and γ-cyclodextrin. Org Lett 2005;7: 529–32.
[128] Pattabiraman M, Natarajan A, Kaliappan R, Mague JT, Ramamurthy V. Template directed photodimerization of *trans*-1,2-bis(*n*-pyridyl)-ethylenes and stilbazoles in water. Chem Commun 2005;4542–4.
[129] Wang R, Yuan L, Macartney DH. Cucurbit[7]uril mediates the stereoselective [4 + 4] photodeimerization of 2-aminopyridine hydrochloride in aqueous solution. J Org Chem 2006;71:1237–9.
[130] Wu X-L, Luo L, Lei L, Liao G-H, Wu L-Z, Tung C-H. Highly efficient cucurbit[8]uril-templated intramolecular photocycloaddition of 2-naphthalene-labeled poly(ethylene glycol) in aqueous solution. J Org Chem 2008;73:491–4.
[131] Lei L, Luo L, Wu X-L, Liao G-H, Wu L-Z, Tung C-H. Cucurbit[8]uril-mediated photodimerization of alkyl 2-naphthoate in aqueous solution. Tetrahedron Lett 2008;49:1502–5.
[132] Barooah N, Pemberton BC, Sivagaru J. Manipulating photochemical reactivity of coumarins within cucurbituril nanocavities. Org Lett 2008;10:3339–42.
[133] Barooah N, Pemberton BC, Johnston AC, Sivagaru J. Photodimerization and complexation dynamics of coumarins in the presence of cucurbit[8]urils. Photochem Photobiol Sci 2008;7:1473–9.
[134] Gromov SP, Yedernikov AI, Kuz'mina LG, Kondratuk DV, Sazonov SK, Strelemko YA, Alfimov MV, Howard JAK. Photocontrolled molecular assembler based on cucurbit[8]uril: [2 + 2]-autophotocycloaddition of styryl dyes in the solid state and in water. Eur. J Org Chem 2010;2010:2587–99.

[135] Yang H, Liu Y, Liu K, Yang L, Wang Z, Xhang X. Rational adjustment of multicolor emissions by cucurbiturils-based host-guest chemistry and photochemistry. Langmuir 2013;29: 12909–14.

[136] Smitka J, Lemos A, Porel M, Jockusch S, Belderrain TR, Tesařova E, Da Silva JP. Phototransformation of benzimidazole and thiabendazole inside cucurbit[8]uril. Photochem Photobiol Sci 2014;13:310–5.

[137] Yang H, Wang ZMZ, Zhang X. Fabricating covalently attached hyperbranched polymers by combining photochemistry with supramolecular polymerization. Polym Chem 2014;5: 1471–6.

[138] Heitmann LM, Taylor AB, Hart PJ, Urbach AR. Sequence-specific recognition and cooperative dimerization of N-terminal aromatic peptides in aqueous solution by a synthetic host. J Am Chem Soc 2006;128:12574–81.

[139] Kim H-J, Heo J, Jeon WS, Lee E, Kim J, Sakamoto S, Yamaguchi K, Kim K. Selective inclusion of a hetero-guest pair in a molecular host: formation of stable charge-transfer complexes of cucurbit[8]uril. Angew Chem Int Ed 2001;40:1526–9.

[140] Jeon YJ, Bharadwaj PK, Choi S-W, Lee JW, Kim K. Supramolecular amphiphiles: spontaneous formation of vesicles triggered by formation of a charge-transfer complex in a host. Angew Chem Int Ed 2002;41:4474–6.

[141] Ko YH, Kim K, Kim E, Kim K. Exclusive formation of 1:1 and 2:2 complexes between cucurbit[8]uril and electron donor-acceptor molecules induced by host-stabilized charge-transfer interactions. Supramol Chem 2007;19:287–93.

[142] Mock WL, Irra TA, Wepsiec JP, Manimaran TL. Cycloaddition induced by cucurbituril. A case of pauling principle catalysis. J Org Chem 1983;48:3619–20.

[143] Mock WL, Irra TA, Wepsiec JP, Adhya M. Catalysis by cucurbituril. The significance of bound-substrate destabilization for induced triazole formation. J Org Chem 1989;54: 5302–8.

[144] Tuncel D, Ŭnal O, Artar M Isr. Supramolecular assemblies constructed by cucurbituril-catalyzed click reaction. J Chem 2011;51:525–32.

[145] Tuncel D, Steinke JHG. Catalytically self-threading polyrotaxanes. Chem Commun 1999;1509–10.

[146] Krasia TC, Steinke JHG. Formation of oligotriazoles catalyzed by cucurbituril. Chem Commun 2002;22–3.

[147] Pemberton BC, Barooah N, Srivatsava DK, Sivagaru J. Supramolecular photocatalysis by confinement–photodimerization of coumarins within cucurbit[8]urils. Chem Commun 2010;46:225–7.

[148] Pemberton BC, Singh RK, Johnston AC, Jockusch S, Da Silva JP, Ugrinov A, Turro NJ, Srivatsava DK, Sivagaru J. Supramolecular photocatalysis: insights into cucurbit[8]uril catalyzed photodimerization of 6-methylcoumarin. Chem Commun 2011;47:6323–5.

[149] Wang Q, Cong H, Zhang J, Tao Z, Xue S. Stereoselective catalysis of cucurbit[n]uril (n=7, 8) on photocycloaddition of *N*-allyl-2-aminopyridinium. Chin J Org Chem 2011;31:1049–55.

[150] Cong H, Wang Q, Zhang J, Tao Z, Xue S. Supramolecular catalysis of cucurbit[8]uril on photocycloaddition of 1,ω-bi(2-aminopyridinium)alkylates. Chin J Org Chem 2012;32: 1093–100.

[151] Biesdermann F, Ross I, Scherman OA. Host-guest accelerated photodimerization of anthracene-labeled macromolecules in water. Polym Chem 2014;5:5375–82.

[152] Cong H, Chao FF, Zhang J-X, Zeng X, Tao Z, Xue S-F, Zhu Q-J. Rapid transformation of benzylic alcohols to aldehyde in the presence of cucurbit[8]uril. Catal Commun 2009;11: 167–70.

[153] Wang Y-H, Cong H, Zhao FF, Xue S-F, Tao Z, Zhu A-J, Wei G. Selective catalysis for the oxidation of alcohols to aldehydes in the presence of cucurbit[8]uril. Catal Commun 2011;12:1127–30.

[154] Cong H, Li Z-J, Wang Y-H, Tao Z, Yamoto T, Xue S-F, Wei G. Substituent effect of substrates on cucurbit[8]uril-catalytic oxidation of aryl alcohols. J Mol Catal A Chem 2013;374:32–8.

[155] de Lima SM, Gómez JA, Barros VP, de G, Vertuan S, das Dores Assis M, de Oliviera Graeff CF, Demets GJ-F. A new oxovanadium(IV)-cucurbit[6]uril complex: properties and potential for confined heterogeneous catalytic oxidation reactions. Polyhedron 2010;29:3008–13.

[156] Koner AL, Márquez C, Dickman MH, Nau WM. Transition-metal-promoted chemoselective photoreactions at the cucurbituril rim. Angew Chem Int Ed 2011;50:545–8.

[157] Reddy KRKK, Cavilini TS, Demets GJF, Sliva LF Jr. Bromine and iodine-cucurbit[6]uril complexes: preparation and applications in synthetic organic chemistry. New J Chem 2014;38:2262–4.

[158] Lu X, Masson E. Silver-promoted desilylation catalyzed by ortho- and allosteric cucurbiturils. Org Lett 2010;12:2310–3.

[159] Wieland M, Mieusset J-L, Brinker UH. Cucurbit[6]uril as a potential catalyst for the acidic decomposition of azidoaminoalkanes. Tetrahed Lett 2012;52:4351–3.

Chapter 4

Systems Based on Calixarenes as the Basis for the Creation of Catalysts and Nanocontainers

Ivan I. Stoikov, Luidmila S. Yakimova, Joshua B. Puplampu, Alena A. Vavilova

Organic Chemistry Department, A.M. Butlerov Chemical Institute, Kazan (Volga Region) Federal University, Kazan, Russian Federation

1 INTRODUCTION

An increase in the rate of chemical reactions has recently become a topic of research activity. In this direction, nanoreactors in which reactants are concentrated in small volumes of about several nanometers to be involved in appropriate reactions are considered as promising solutions. Micellar systems and emulsions can be mentioned as examples of such nanoreactors.

2 SYNTHESIS AND STRUCTURE OF CALIXARENES

Among the large number of macrocyclic compounds related to supramolecular chemistry, calixarenes offer unique possibilities for selectively forming, transporting, and catalyzing complexes, and also for serving as biomimetic and building blocks for the synthesis of supramolecular architectures. The cup-shaped structure of calixarenes ("calix" in Greek means "cup, vase") is a convenient platform for molecular design of the three-dimensional structures with a wide range of cavity sizes, different numbers and types of binding sites, and diverse spatial arrangement of binding groups, as well as the possibility of asymmetric cavity formation and the control of a balance between rigidity and flexibility of a receptor.

On this basis, calix[4]arene **1**, thiacalix[4]arene **2**, calix[4]resorcinarene **3**, and calix[4]pirogallolarene **4** (Scheme 4.1), which are cyclocondensation products of *p*-alkylphenols with formaldehyde or sulfur in alkaline medium and of acid-catalyzed reaction of resorcinol or pyrogallol with aldehydes, are used as basic blocks for the construction of host molecules.

Organic Nanoreactors. http://dx.doi.org/10.1016/B978-0-12-801713-5.00004-5

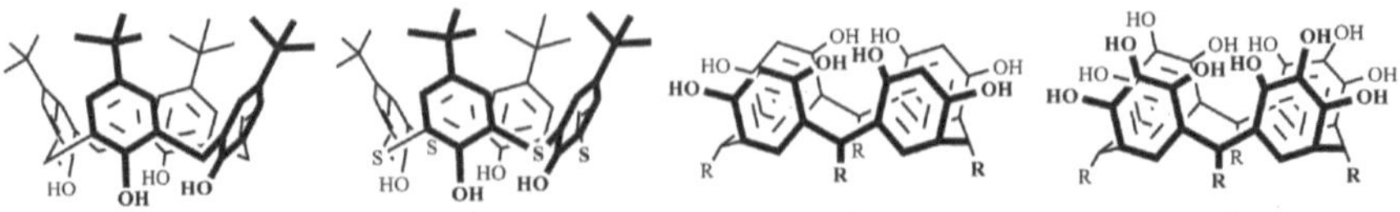

SCHEME 4.1 Structures of the calix[4]arene 1, thiacalix[4]arene 2, calix[4]resorcinarene 3, and calix[4]pirogallolarene 4.

2.1 Calixarenes and Thiacalixarenes

Phenol–formaldehyde condensation, conducted under different conditions, typically yields mixtures of oligomeric products. However, some phenols under certain conditions form macrocyclic products. Phenols having alkyl groups in the para-position form mainly tetra (I, $n = 4$), hexa (I, $n = 6$), and the octameric (I, $n = 8$) macrocycles in a base-catalyzed condensation with formaldehyde (Scheme 4.2).

Considering a molecular model of the cyclic tetramers, Gutsche introduced the term "calix[4]arenes" [1,2] ("calix"— vase, bowl, cup), a term that emphasizes the cup-like structure of these macrocycles, which have all four aromatic rings oriented in one direction (a so-called *cone* conformation). The word "arene" indicates the presence of aromatic rings in the macrocyclic ensembles. Later the term was extended to include large macrocyclic compounds ([1_n]metacyclophanes) and macrocycles, in which the methylene bridge is replaced by a heteroatom. The size of the cycle is indicated by a number placed in square brackets between the roots "calix" and "arene"; nature and position

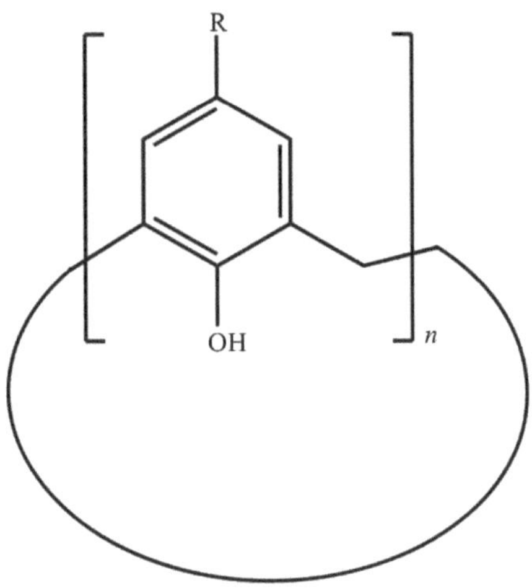

I

SCHEME 4.2 Schematic structure of calix[*n*]arenes.

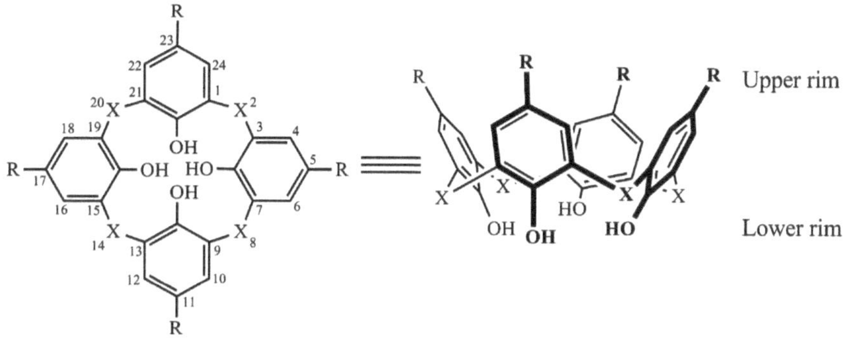

SCHEME 4.3 **The various structural presentation of calix[4]arene molecule.**

of the substituents on the aromatic ring is indicated by respective numbers and descriptors.

The phenolic hydroxyl group region of functionalized calix[*n*]arenes is termed the "lower rim," while the para-position of phenolic groups as the "upper rim" (Scheme 4.3). Neighboring aromatic rings in the calix[4]arene are called as adjacent (proximal), 1,2- or AB-opposing rings. Diametrically opposite (distal) rings occupy 1,3- or AC-positions.

The calix[4]arenes and thiacalix[4]arenes can exist in four possible conformations having a different arrangement of aromatic rings (upward, "u") and (downward, "d") on the molecule plane defined by the methylene (or heteroatom) bridges. They were named by Gutsche as *cone* (u, u, u, u), *partial cone* (u, u, u, d), *1,3-alternate* (u, d, u, d), and *1,2-alternate* (u, u, d, d) (Scheme 4.4) [2].

In most cases the signals of bridging methylene protons of calix[4]arenes **1** together with the signals of the lower or upper rim substituents in ^{1}H-NMR spectra clearly identify the macrocycle conformation as well as the spatial arrangement of the substituents [2]. Furthermore, Mendoza et al [3] introduced an exclusively useful rule that correlates ^{13}C-NMR spectra of calix[*n*]arenes ($n = 4 - 6$) with their conformations. Chemical shifts of bridging carbon atoms are equal to δ 31 if two adjacent aromatic groups are in the *syn*-orientation (u, u or d, d) and to δ 37, if they have anti-orientation (u, d).

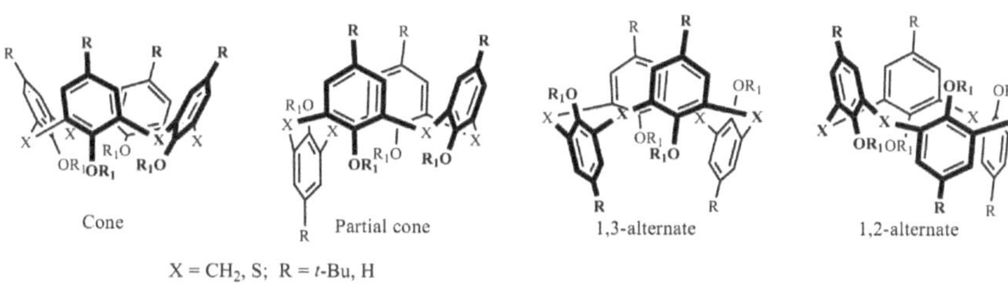

SCHEME 4.4 **The conformation of (thia)calix[4]arenes.**

Replacing of methylene bridges in calix[4]arenes **1** with sulfur atoms in thiacalix[4]arene **2** and its derivatives greatly complicates the establishment of the macrocycle conformation. This is due to the absence of bridging methylene protons, which otherwise allow determination of calix[4]arene **1** conformation by their NMR spectra. Two-dimensional NMR techniques are employed in the establishment of the conformation of the thiacalix[4]arene **2** molecules and its derivatives. Thiacalixarenes have larger cavity size, and their bridging sulfur atoms can be easily oxidized [4–6] to show new receptor properties [7–19].

2.2 Calix[4]resorcinarenes and Pyrogallolarenes

Calix[4]resorcinarenes (CRA) are synthetically available amphiphilic cyclic tetramers (cyclophanes). They are condensation products of aliphatic aldehydes and resorcinol (Scheme 4.5) [20–22]. Calix[4]resorcinarene molecules exist as three-dimensional structures with complex organization that can form several conformations, namely, *cone*, *partial cone*, *1,3-alternate*, and *1,2-alternate* conformation. The first of them have hydrophobic cavity. Due to structural factors, that is, presence of four aryl fragments serving as "rigid segment," hydrophilic hydroxyl groups, and hydrophobic hydrocarbon radicals present in the same molecule, calix[4]resorcinarenes are considered as striking examples of amphiphilic compounds with distinct hydrophilic and hydrophobic fragments. The amphiphilic nature of calix[4]resorcinarenes confers them with the ability to self-assemble and also to embed into organized structures (mono-, bi-, multilayers, or vesicles), thus making them promising in the control of the catalytic activity of supramolecular aggregates. In addition, the calix[4]resorcinarene platform provides a unique opportunity for targeted functionalization of the macrocycle, which enables the formation of various types of aggregates with forecasted nature and number of reaction centers and sites for the substrates attachment and binding. The resorcin[*n*]arene nomenclature is analogous to that of calix[*n*]arenes, in which *n* represents the number of repeating units in the ring. Pyrogallolarenes are related macrocycles derived from the condensation of pyrogallol (1,2,3-trihydroxybenzene) with an aldehyde.

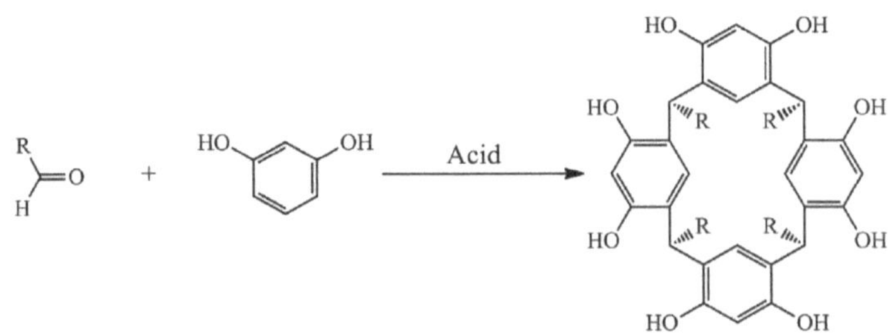

SCHEME 4.5 **The condensation of aliphatic aldehydes and resorcinol.**

3 MACROMOLECULAR CATALYSTS BASED ON MACROCYCLIC RECEPTORS

The design of the compounds capable of molecular recognition of cations, anions, and neutral molecules is one of the main achievements of the chemistry of molecular ensembles and intermolecular bonds. When such compounds are used as catalysts, the conversion of the reacting particles into products is preceded by the formation of supermolecules, due to the selective binding of the substrate molecule by the receptor molecule (Fig. 4.1) [23,24]. This approach enables the development of catalysts with the activity and selectivity determined by intermolecular interactions between the substrate and the catalyst.

Among molecules capable of molecular recognition, macrocyclic receptors are of great interest. They can form the host–guest complexes between the catalyst and substrate by noncovalent intermolecular interactions, such as Van der Waals interactions and hydrogen bonds (crown-ethers, cyclodextrins, calixarenes, pillar[5]arenes, etc.) [25–35]. Hydrophobic cavities of different shape in these compounds allow them binding substrates that correspond to the size and shape of the receptor cavity, and thus demonstrating molecular recognition.

The effect of macrocyclic receptors as components of catalytic systems based on metal complex on catalytic processes depends on the composition and structure of the resulting host–guest complexes. The macrocycle can interact either with the substrate or with the metal complex, which subsequently determines the characteristics of the catalyzed reaction.

In the first case (Fig. 4.2), the inclusion complex with the host molecule—but not with the substrate itself—is involved in the catalytic cycle. As a result, the role of a macrocycle in the catalyzed system can be described as follows [36,37]:

1. The size and structure of the host molecule determine the stability of the host–guest complex and affect the substrate selectivity of the reaction.
2. The host molecule, as a kind of nanoreactor, can stabilize the transition state of the reaction and increase its rate.

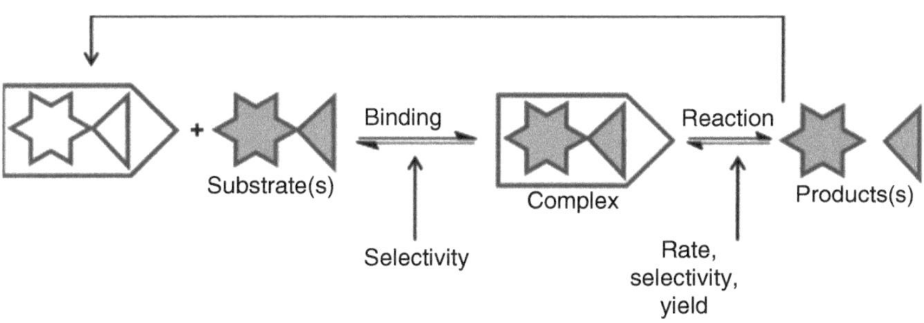

FIGURE 4.1 Schematic representation of the process involved in the use of a nanoreactor to catalyze a chemical reaction. *(Adapted from Ref. [23,24].)*

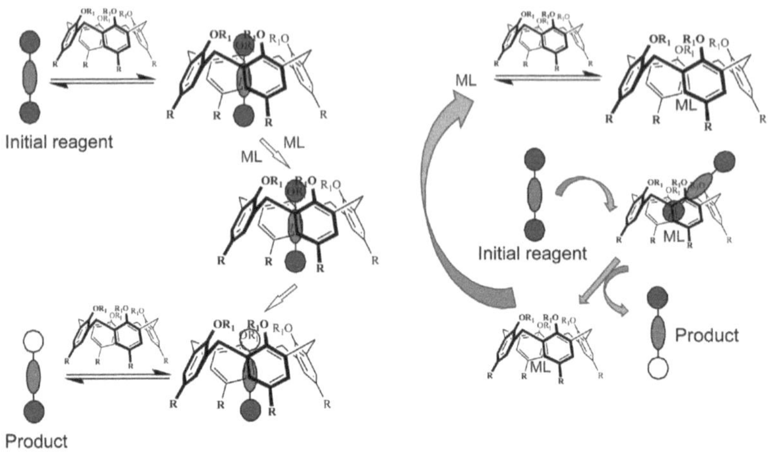

FIGURE 4.2 The effect of macrocyclic receptor on the mechanism of the catalytic process—possible routes. (a) Formation of the host–guest complex with the substrate. (b) Formation of the host–guest complex with the metal complex. *(Adapted from Ref. [36,37].)*

3. The formation of the host–guest complex can significantly affect the regio- and stereoselectivity of the process, due to the specific orientation of the substrate. Thus, individual fragments of the substrate molecules can be shielded and prevented from taking part in the reaction.
4. The possibility to bind the product to a host–guest complex reduces the concentration of receptor molecules and hence the rate of reaction.

In the second case, with a ligand or metal complex being guests, the role of the macroligand—host is limited to the stabilization of the catalytically active particles and the corresponding activated complex, as well as to the changes in the ligand environment.

The stability of the host–guest complex is significantly affected by the nature and composition of the solvent in which the processes occur. This is an important factor when the reaction is carried out in aqueous medium, where hydrophobic interactions mainly contribute to the energy of complex formation due to an increase in its stability. In this regard, the use of water-soluble macrocycles with hydrophobic cavities as components of water-soluble metal complexes in two-phase catalytic systems is of particular interest [38,39]. The catalyst, soluble in the aqueous phase, can be easily separated from the water-insoluble reaction products and reused. It should be emphasized that the activity of conventional catalysts is very low for the reactions involving substrates poorly soluble in water. Due to the formation of water-soluble inclusion host–guest complexes, the macrocyclic receptors not only influence the activity and selectivity of the reaction, but also perform the function of interfacial substrate carrier in aqueous phase.

4 SUPRAMOLECULAR CATALYSIS BY CALIXARENES

Construction of supramolecular associates is an important and promising trend in the field of catalysis. Catalytic systems based on calixarenes exhibit high selectivity toward most organic reactions. The activity of these catalysts depends on the conformity of the size of the host molecule cavity and that of the guest molecule so that the substrate selectivity is observed. The size of these cavities and consequently the selective conversion of certain substrates can be adjusted by the receptor composition.

There are two directions in the development of supramolecular catalytic compositions, that is, (1) creation of systems based on macrocyclic compounds as host molecules that bind substrates with their hydrophobic cavity and (2) development of the systems that bind substrates using aggregates formed by amphiphilic compounds. Compounds that form host–guest complexes like modified calixarenes are able to aid transport of substrates into the aqueous phase. This approach has been implemented in the Wacker oxidation [40,41], oxidation of alkylaromatic compounds [42], hydroxylation of aromatic compounds [43], hydrogenation [44,45], hydroformylation [45–48], and carbonylation [49]. In this case, the substrate is transported into the aqueous phase in the form of the corresponding inclusion complex. This not only affects the activity of the catalyst, but also provides selectivity of the process. Thus, in the Wacker oxidation of 1-alkenes the maximum yield of methyl ketone was achieved when 1-hexene is used, and for systems based on calix[6]arene with 1-octene among catalytic systems with modified calix[4]arenes [50].

In [51], Wacker oxidation of olefins was studied in the presence of catalytic systems comprising water-soluble calixarenes (sulfonated and glycydylated derivatives), palladium salt, and copper salt. The presence of nonpolar cavities in these molecules enables binding nonpolar substrates and their transfer into the aqueous phase where the reaction takes place. The activity of these catalysts depends on the complementarity between the cavity size of the host molecule and the size of the guest molecule. Therefore, substrate selectivity was exhibited. For example, the addition of calixarene increased the reaction rate for linear 1-alkenes which size corresponded to the size of the calixarene cavity (1-hexene for calix[4]arene and 1-octene for calix[6]arene). The activity of catalytic system applied for the oxidation of substituted styrenes also depended on the ratio of the size of the substrate molecule and that of the calixarene cavity.

Water-soluble calix[*n*]arenes ($n = 4$, 6, and 8) with ammonium groups were used as high-performance inverse phase-transfer catalysts in the nucleophilic substitution reaction in water, (Scheme 4.6) [52]. The efficiency of the calix[*n*]arenes as catalysts varied with the size and/or shape of the substrate molecules.

Calix[4]arenes **5–7** with one or several imidazole fragments were used as catalysts of the hydrolysis reaction of 4-nitrophenylbenzoate and acetate in buffer solution (Scheme 4.7). The authors showed that compounds containing benzyl radicals R exhibited the highest catalytic activity [53].

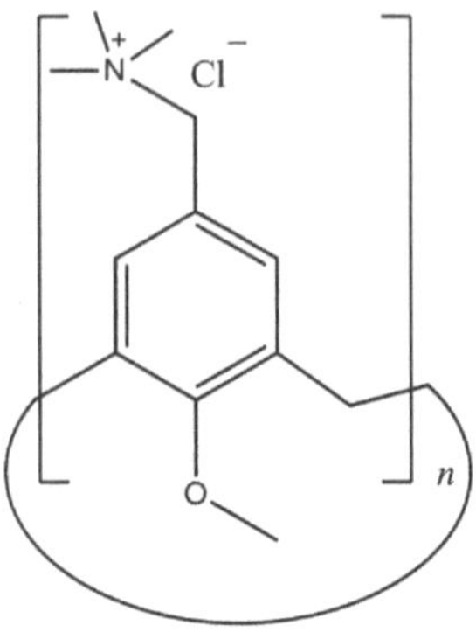

Catalyst, n=4,6,8

$$RX + MY \xrightarrow[t^\circ]{\text{Catalyst, water}} RY + MX$$

Nucleophiles: MX – NaCN, KSCN, KI
Substrates: RX – $C_8H_{17}Br$, $C_8H_{17}I$, $PhCH_2CH_2Br$, BnBr, BnCl, 2-naphthylCH_2Br

SCHEME 4.6 Calix[*n*]arenes as catalysts in the nucleophilic substitution reaction. *(Adapted from Ref. [52] with permission of Royal Society of Chemistry. Copyright 1997.)*

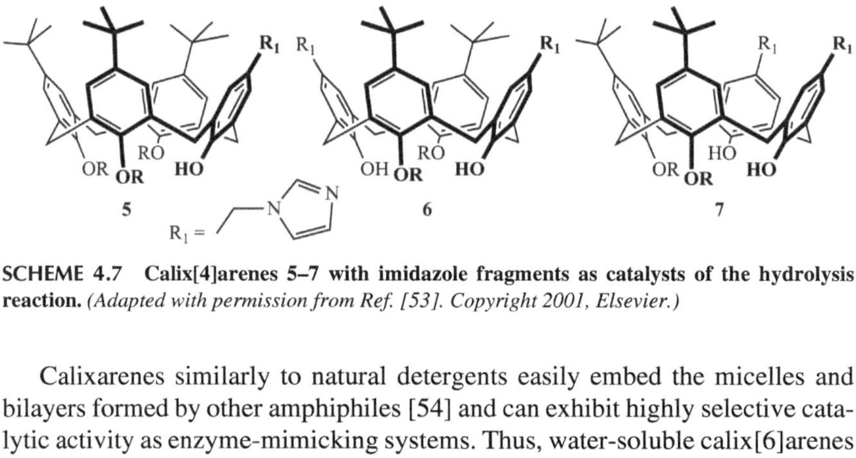

SCHEME 4.7 Calix[4]arenes 5–7 with imidazole fragments as catalysts of the hydrolysis reaction. *(Adapted with permission from Ref. [53]. Copyright 2001, Elsevier.)*

Calixarenes similarly to natural detergents easily embed the micelles and bilayers formed by other amphiphiles [54] and can exhibit highly selective catalytic activity as enzyme-mimicking systems. Thus, water-soluble calix[6]arenes **8–11** catalyzed the hydration of 1,4-dihydronicotinamide derivatives, important model compound used to study the properties of the reduced form of the coenzyme nicotinamide adenine dinucleotide (NADH) (Scheme 4.8) [55]. The reaction proceeded according to the Michaelis-Menten kinetics.

Calixarene **8** acted as a weak catalyst, whereas the compound **9** catalyzed fivefold more effectively than sodium dodecylsulfate. The compounds **10** and **11** with acidic protons were found to be extremely effective catalysts. The dependence of the observed reaction rate constants on concentration had a plateau typical for catalytic reactions in micellar media. Thus, *p*-sulfonate calix[6]arenes mimic the function of glyceraldehyde-3-phosphate dehydrogenase [55].

A series of water-soluble calix[*n*]arenes ($n = 4 - 8$) with different conformational abilities were investigated as artificial ribonucleases that hydrolyze cytidine-2′,3′-cyclophosphate at pH 2 and 30°C [56]. Best results were obtained with calix[4]arene **12**, which increased the rate by 16 times (Scheme 4.9).

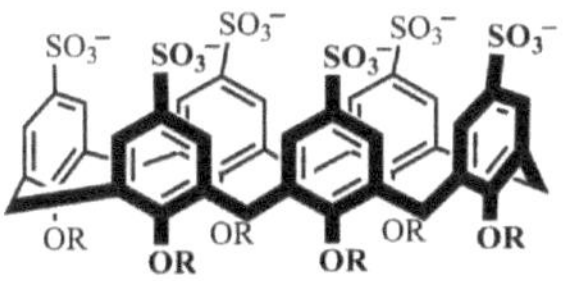

8, R = CH_3
9, R = n-$C_{12}H_{25}$
10, R = H
11, R = CH_2COOH

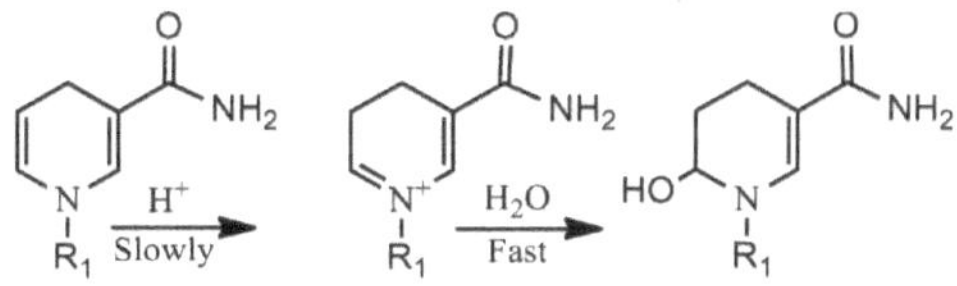

Synthetic model compounds (1,4-dihydronicotinamides) undergo an acid-catalyzed hydration reaction in aqueous solution

SCHEME 4.8 The water-soluble calix[6]arenes 8–11 as catalysts in the hydration of 1,4-dihydronicotinamide derivatives. *(Adapted from Ref. [55] with permission of American Chemical Society. Copyright 1986.)*

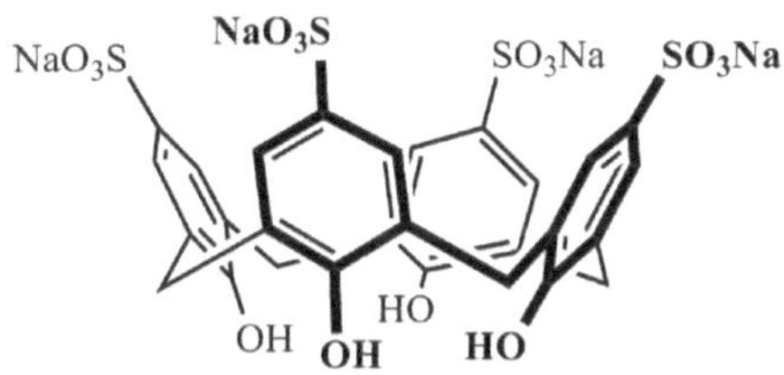

12

SCHEME 4.9 Structure of the water-soluble calix[4]arene 12. *(Adapted from Ref. [56].)*

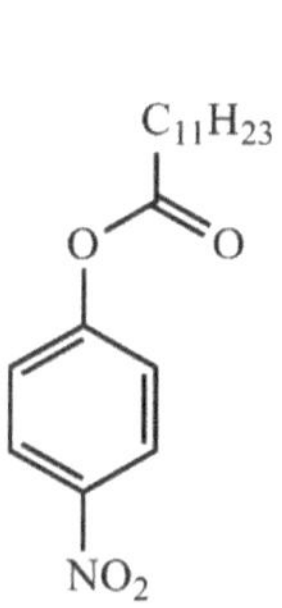

13

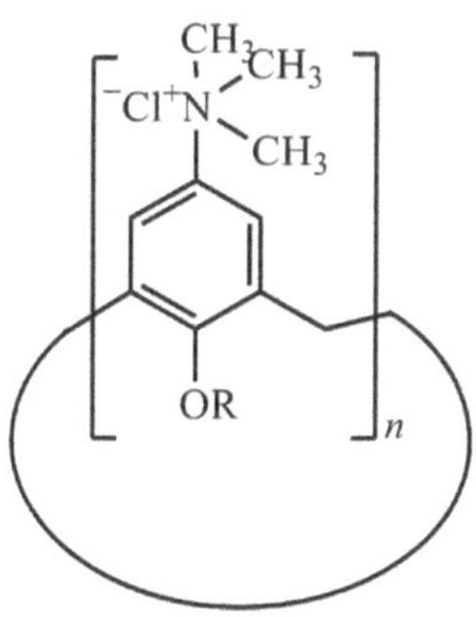

14, $n = 4$, R = CH_3
15, $n = 6$, R = CH_3
16, $n = 6$, R = C_8H_{17}

SCHEME 4.10 Structures of the 4-nitrophenyldodecanoate 13 and water-soluble salts of the calix[*n*]arenes. *(Adapted from Ref. [57] with permission of American Chemical Society. Copyright 1989.)*

Water-soluble calixarenes bearing positive charge at the upper rim can be efficient catalysts for the hydrolysis of 4-nitrophenyldodecanoate **13** (Scheme 4.10) [57]. Calix[4]arene **14** with the methyl fragment increased the rate of substrate hydrolysis, but to a lesser extent than the calix[6]arenes. Increase in the size of the calixarene molecular cavities (calixarene **15**) resulted in a faster reaction rate. Increasing hydrophobicity (calixarene **16**) led to further increase in the hydrolysis rate constants observed.

The calixarene **12** increased the rate of oxidation of 4-chloroaniline with hydrogen peroxide catalyzed by iron (II) salts [58]. Unfortunately, the composition of the products is not discussed in this case.

5 SUPRAMOLECULAR CATALYSIS BY METAL COMPLEXES BASED ON CALIXARENES

Ligands combining the properties of the metal complex with the molecular-recognition ability and phase transfer of substrate within one molecule are of particular interest in phase transfer catalysis. The amalgamation of a host site with metal center in a single molecule does not only preserve the characteristics of the catalytic systems, but also results in new properties for the catalysts. First, binding the substrate by ligand-receptor cavity determines the substrate orientation against that of the metal center. This can greatly affect the products distribution and substrate selectivity of the reaction. Enzyme models with a number of host molecule fragments in the ligand are most interesting examples. It is important to note that the target reaction involves the substrate coordination in a host–guest complex with the metal centers followed by its stabilization, including that of the reaction transition state, increased up to several orders of magnitude. This effect is associated with the cooperative binding of the metal center and that of the cavity of the receptor molecule (so-called "dual recognition" [59,60]). As a result of the coordinated substrate binding, there is no additional step involving the interaction of the inclusion complex with a metal ion, as is typical for catalytic systems described in the previous section. Thus, the energy and entropy of activation of the reaction decrease for these catalysts and the reaction rate significantly increases [61].

Calixarene **17** is one of such water-soluble calixarene ligands. The main advantages of catalysts based on such calixarenes include the substantial increase in activity compared with other catalysts and high selectivity of the processes (Scheme 4.11) [41,62].

In homogeneous catalysis it is possible to use complexes based on sulfur-containing calixarenes with gold in the catalytic oxidation of various harmful sulfides, such as mustard gas, under mild conditions [63]. Calixarene complexes

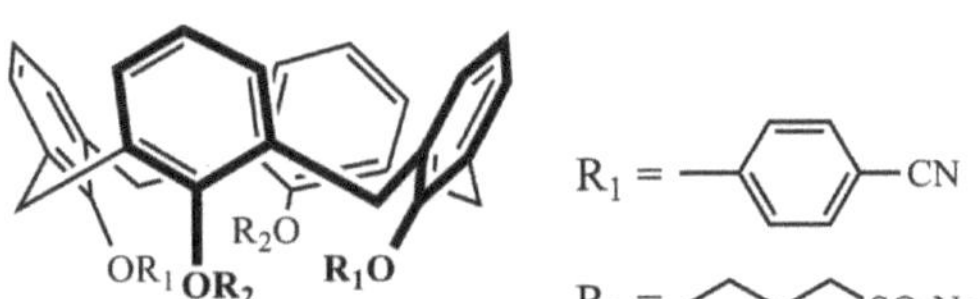

17

SCHEME 4.11 Structure of the water-soluble calix[4]arene 17. *(Adapted from Ref. [42,62] with permission of American Chemical Society and Elsevier.)*

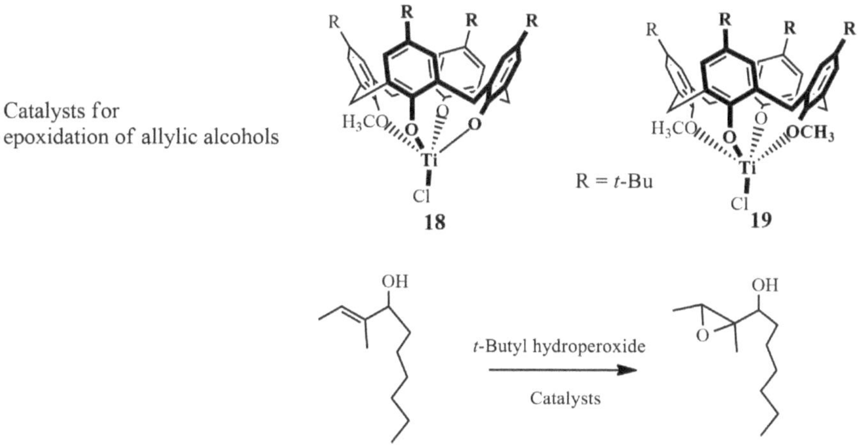

SCHEME 4.12 The catalysts 18 and 19 for epoxidation of allylic alcohols. *(Adapted with permission from Ref. [64]. Copyright 2001 Elsevier.)*

with Ti(IV) **18**, **19** were synthesized and used as catalysts in the reaction with allyl alcohol and butyl hydroperoxide in CH_2Cl_2 (Scheme 4.12) [64].

Water-soluble calix[*n*]arenes **20–24** with sulfonated fragments 4–6 aromatic rings) and aminomethylated calix[4]arenes were used as components in the catalytic system for the cross-coupling reactions of 1-iodo-4-phenylbenzene and phenylboronic acid (Suzuki reaction) catalyzed by palladium complexes with sulfonated triphenylphosphine (Scheme 4.13) [65]. Addition of water-soluble calixarenes increased the yield of reaction products more than twofold. The maximum activity was exhibited by calixarenes containing amino groups **22–24**. The appropriate constants of the formation of respective inclusion complexes were significantly higher.

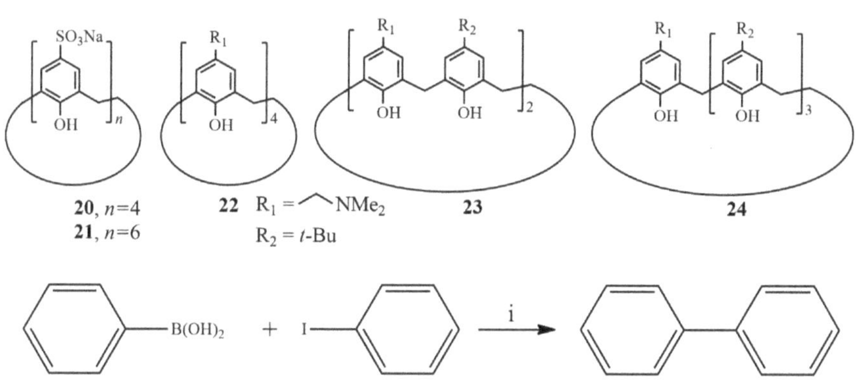

SCHEME 4.13 The catalysts 20–24 for Suzuki reaction. *(Adapted with permission from Ref. [65]. Copyright 2001 Elsevier.)*

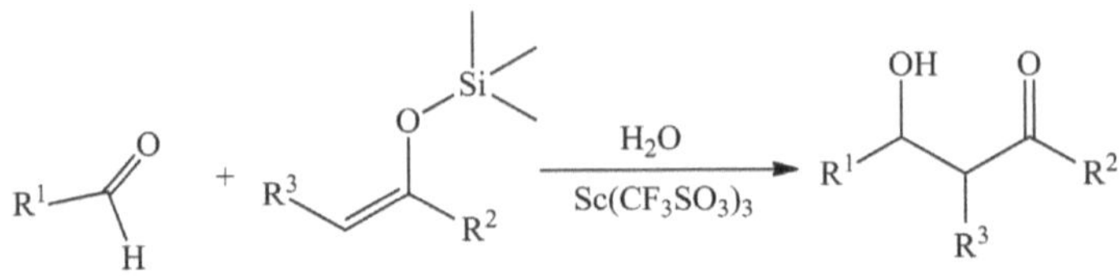

SCHEME 4.14 **Mukaiyama aldol addition.** *(Adapted with permission from Ref. [66]. Copyright 2000 Elsevier.)*

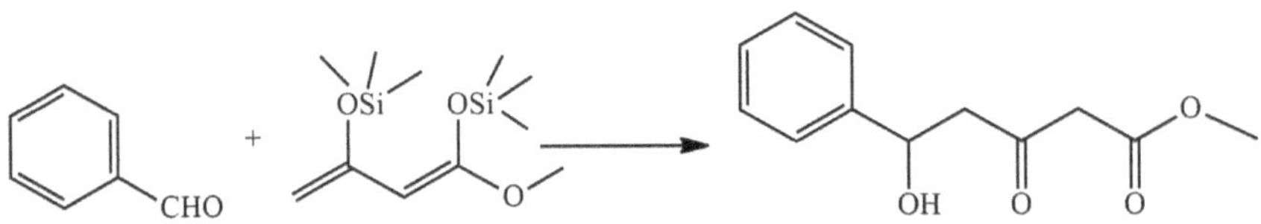

SCHEME 4.15 **The condensation reaction of dienes with benzaldehydes in the presence of calix[*n*]arene/Ti(IV) complex.** *(Adapted with permission from Ref. [67]. Copyright 2003 Elsevier.)*

Catalytic systems comprising of sulfonated calix[*n*]arenes with alkyl group at the lower rim ($n = 4, 6$) and scandium triflate were used for the Mukaiyama aldol addition (Scheme 4.14). It was established that calixarenes stabilized labile silyl enol ester and allowed reaction to be carried out in water [66].

The condensation reaction of dienes with benzaldehydes in the presence of calix[*n*]arene/Ti(IV) complexes formed in situ by treating appropriate calixarenes with $Ti(O\text{-}i\text{-}Pr)_4$ was studied [67], and the catalytic effects of the complexes confirmed (Scheme 4.15).

In the presence of barium salts, derivative of the *tert*-butylcalix[5]arene with crown-ether catalyzed methanolysis reaction of 4-nitrophenyl acetate (Scheme 4.16) [68].

Copper complex with calix[6]arene modified with pyridine was active in the oxidation of benzene to benzoquinone by hydrogen peroxide via the formation of phenol and hydroquinone [69,70]. Interestingly, catechol was not formed in this case. This same complex was also active in the oxidation of ethanol to aldehydes and acids [69]. Analogous complex containing copper and calix[4]arene modified with imidazole groups catalyzed the oxidation of 3,5-di-*tert*-butylpyrocatechol to corresponding quinone, benzene to benzoquinone, and ethyl alcohol to acetaldehyde by hydrogen peroxide [70].

Macromolecular complexes of rhodium containing the modified calixarene **26** as ligand were studied as catalysts for hydroformylation in two-phase media (Scheme 4.17) [71]. It should be noted that the catalysts were active not only in the hydroformylation of 1-alkenes, but to a less extent of 2-octene and 4-octene. Complexes with calixarenes modified with phosphine and phosphite groups were used as catalysts under homogeneous conditions in reactions involving the synthesis gas [72–75]. It is worth noting that the receptor molecule fragment

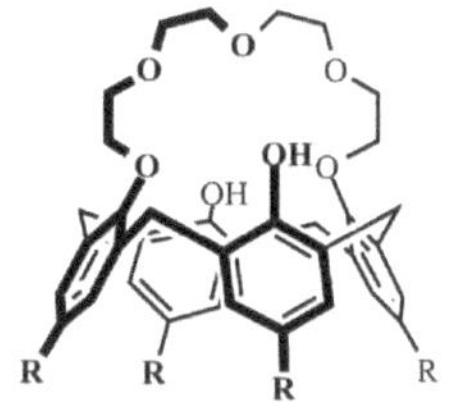

25

Catalyst – barium salt of *p-tert*-butylcalix[4]-arene-crown-5

Catalyst $\xrightarrow[-p\text{-Nitrophenol}]{k_1[p\text{-nitrophenyl acetate}]}$ Acylated intermediate catAc $\xrightarrow[-\text{AcOMe}]{\text{MeO}^-/\text{MeOH},\ k_2}$ Catalyst

SCHEME 4.16 **Methanolysis reaction of 4-nitrophenyl acetate in presence of the derivative of the *tert*-butylcalix[5]arene with crown-ether barium salts.** *(Adapted from Ref. [68] with permission of Royal Society of Chemistry. Copyright 1998.)*

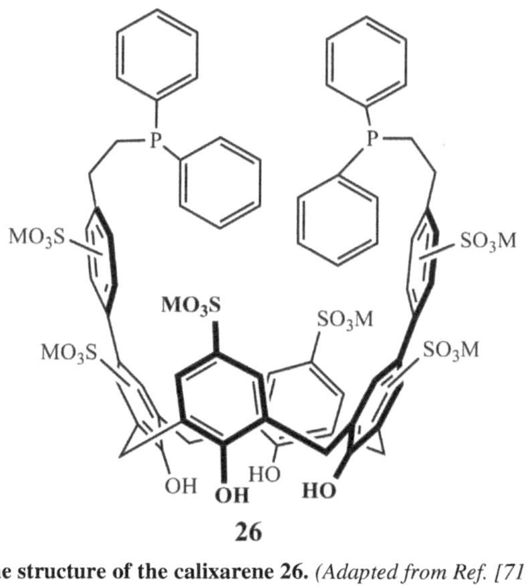

26

SCHEME 4.17 **The structure of the calixarene 26.** *(Adapted from Ref. [71] with permission of Wiley-VCH.)*

was mainly used in this case as a platform to create sterically hindered phosphine or phosphite ligands.

The hydroformylation of styrene and 1-octene with rhodium complexes containing diphenylphosphinated calixarenes **27, 28** was described in [74,76]. Loeber and coauthors [77] employed the complex **29** in the same reaction (Scheme 4.18). The quantitative conversion of styrene was observed within 48 h with a high selectivity for 2-phenylpropanal (95%).

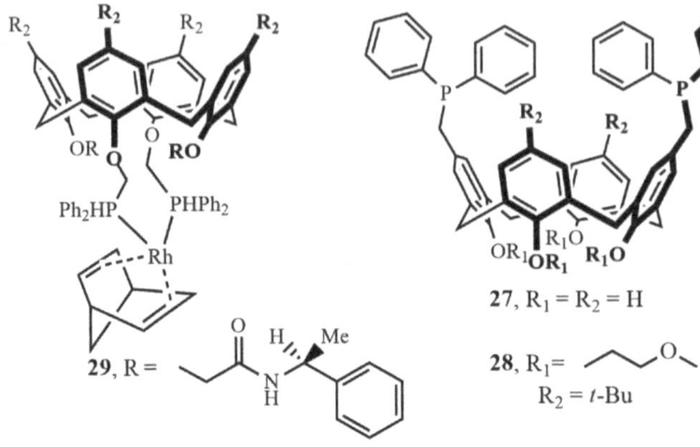

SCHEME 4.18 The structures of the calixarenes 27–29. *(Adapted with permission from Ref. [76] with permission of American Chemical Society. Copyright 1999.)*

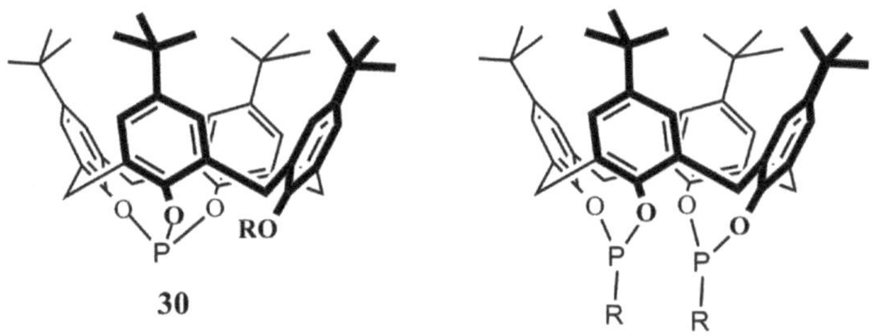

SCHEME 4.19 The structures of the calixarenes phosphite 30, 31. *(Adapted from Ref. [78,79] with permission of Wiley-VCH and Royal Society of Chemistry.)*

Calixarene phosphite **30** with unsubstituted hydroxyl group was used as a ligand in the rhodium complex–catalyzed hydroformylation of 1-hexene (Scheme 4.19) [78]. The almost quantitative conversion of initial alkene was observed in 3 hours with aldehyde yields of 80–85%. Unfortunately, the selectivity was low because the aldehydes with normal and isomerized chains were formed in similar amounts. A number of structurally similar phosphites **31** as components of catalytic systems for the hydroformylation of 1-octene are described (Scheme 4.19) [79]. It was shown that the nature and steric volume of R did not significantly affect main processes. Thus, the depth of conversion was 80–90% with about 60% selectivity in all cases.

Due to the large size of the calixarene molecules, they were also used for the synthesis of stereoregular polymers. Thus, neodymium complex containing the calixarene **32** was active in the formation of isotactic polypropylene oxide (Scheme 4.20) [80].

Palladium complexes with the calixarenes **33–35** were investigated in the alkylation of 1,3-diphenylprop-2-enyl acetate with dimethylmalonate and in

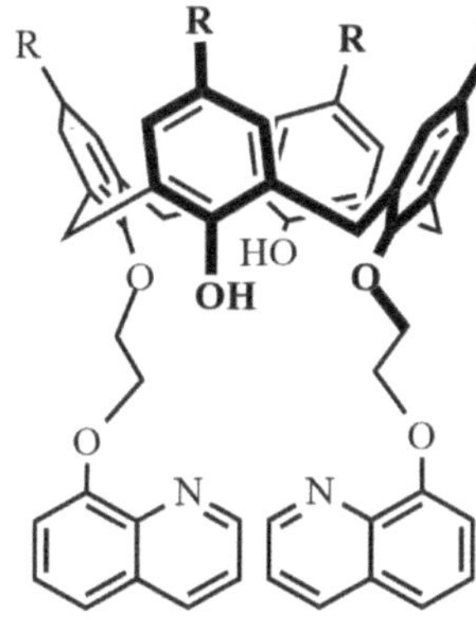

32, R = *t*-Bu

SCHEME 4.20 **The structure of the calixarene 32.** *(Adapted from Ref. [80] with permission of Elsevier.)*

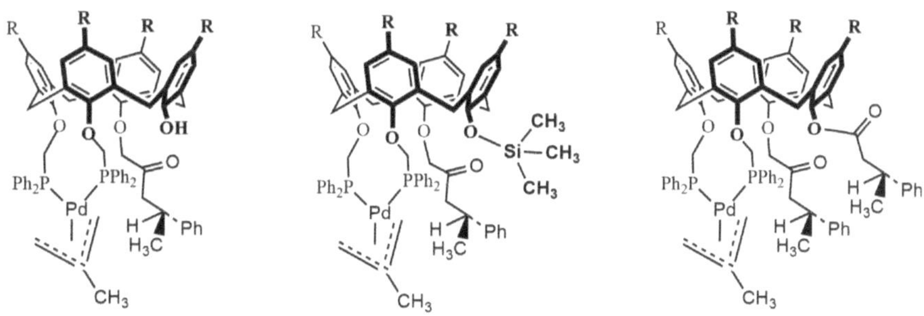

33 R = *t*-Bu **34** **35**

SCHEME 4.21 **The structures of the palladium complexes with the calixarenes 33–35.** *(Adapted from Ref. [81] with permission of Royal Society of Chemistry. Copyright 2001.)*

the presence of silane ($Me_3SiOC(NSiMe_3)CH_3$) (Scheme 4.21). The influence of calixarenes resulted in enantioselective addition with an enantiomeric excess value of 45 and 67%, which can be attributed to the optically active substituent near the active site. No asymmetric induction was observed with the complex **35** bearing two identical chiral substituents [81].

Dinuclear rhodium complex with **36** was active in the addition reactions of carbenes (Scheme 4.22). Moreover, the yield decreased for the sterically hindered olefins compared with rhodium acetate (II) due to steric factors. Furthermore, this complex preferably catalyzed the inclusion of carbene to the CH bond of aromatic fragment [82].

Complexes consisting of functionalized calix[4]arenes with zinc (II) (Scheme 4.23), barium (II), copper (II), and lanthanum (III) were proved as most interesting and efficient catalytic systems [83–85]. In these studies, bi- and trinuclear metal complexes consisting of Zn (II), Cu (II), and calixarene containing

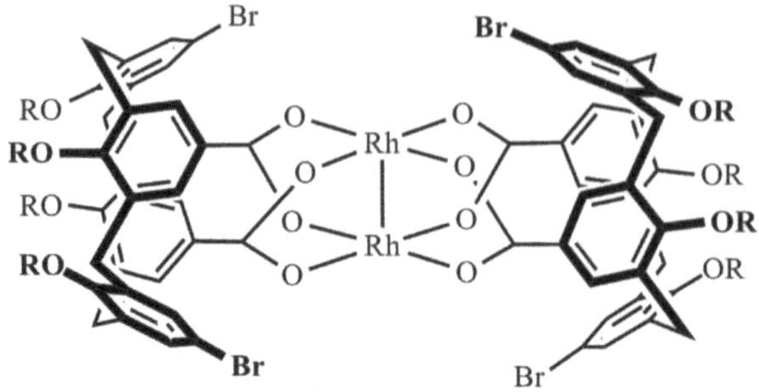

36, R = C_3H_7

SCHEME 4.22 Dinuclear rhodium complex with 36. *(Adapted from Ref. [82] with permission of Royal Society of Chemistry. Copyright 2002.)*

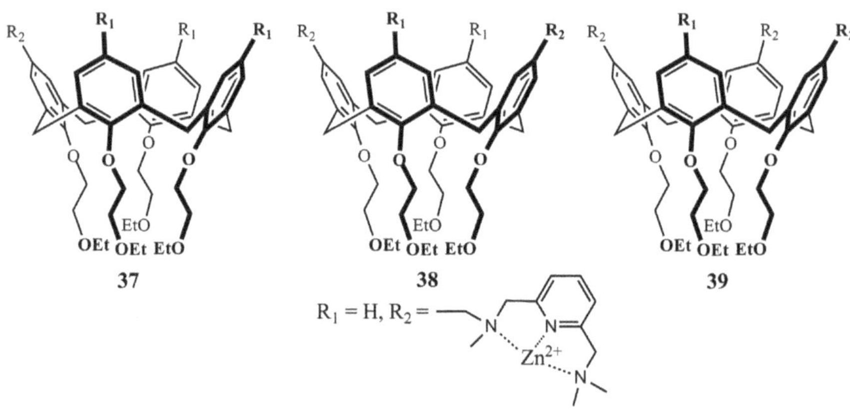

SCHEME 4.23 Complexes consisting of functionalized calix[4]arenes with zinc (II). *(Adapted from Ref. [83] with permission of American Chemical Society. Copyright 1999.)*

pyridine and imidazole fragments were synthesized. The rate of reaction of the hydrolysis of 2-(hydroxypropyl)-*p*-nitrophenylphosphate in the presence of the binuclear zinc complex **38** was increased by 23,000 times. The mononuclear complex **37** appeared to be 50 times less active. More impressive results were obtained with the zinc trinuclear complex **39**, able to catalyze cleavage of 2-(hydroxypropyl)-*p*-nitrophenyl phosphate (0.48×10^{-3} M, pH 7, 25°C) by 32,000 times faster than spontaneous hydrolysis of this substrate.

The trinuclear complex **39** was investigated in the cleavage of the RNA nucleotides [84]. The nucleotides investigated contain an acidic amide fragment, -NH-CO-, which coordinated to one zinc ion via a negatively charged nitrogen atom formed after deprotonation. Two remaining zinc ions activate a phosphoryl group. The cleavage of the leaving group can also facilitate the protonation of water molecules associated with zinc ions. The catalytic function of octa(dimethylaminopropyl)calix[4]resorcinol with a methyl group at the methine carbon atom was studied [86] in the hydrolysis of a number of 4-nitrophenyl esters of carboxylic acids (pH 9.6, 25°C). The catalytic activity of the calixarene studied was 180 times higher than that of dimethylamine in relation to 4-nitrophenylacetate.

Inhibition of alkaline hydrolysis of acetylcholine (60°C, pH = 10) caused by complexation of the positively charged fragment of acetylcholine with calix[4]resorcinol tetraanion was shown [87]. The tetradentate complexes of this calix[4]resorcinarene with copper (II) and zinc (II) ions exerted similar reactivity as original ligand [88]. The substrate binding constant depended on the structure and lipophilicity of calix[4]resorcinarene and was equal to (4–36) $\times 10^3$ mol/L.

The authors [89] have established that the reaction of 4-nitrophenyl esters of carboxylic acids with calixarene anions was related to nucleophilic substitution at the carbonyl group accompanied with the release of 4-nitrophenyl anion (stage I). The trans-esterification products formed in aqueous solution were hydrolyzed to corresponding carboxylic acid and initial calixarene (stage II). The CMC for the tetra- and octaanionic calixarene (R = C_9H_{19}) were found to be 8×10^{-4} and 6×10^{-5} mol/L, respectively, by tensiometric method. The kinetic dependencies of the rate constants observed for the cleavage of carboxylic acid esters against the concentration of the long chain calixarenes showed shape typical for micellar-catalyzed processes. The reactivity of calixarenes depended on the number of anionic centers. The highest reaction rate (170 times) was obtained for calix[4]resorcinarene octaanion (R = C_9H_{19}) in the reaction with 4-nirophenyl acetate [89].

In addition, bridged calix[4]resorcinols were used as templates in asymmetric catalysis (Scheme 4.24). The asymmetrically functionalized calix[4]resorcinol increased the rate of enantioselective addition of diethyl zinc to the benzaldehyde [88,90].

Water-soluble calix[4]resorcinols with sulfur group substituents function as catalysts in the amination (dehydration) of alcohols in aqueous medium (Scheme 4.25) [91].

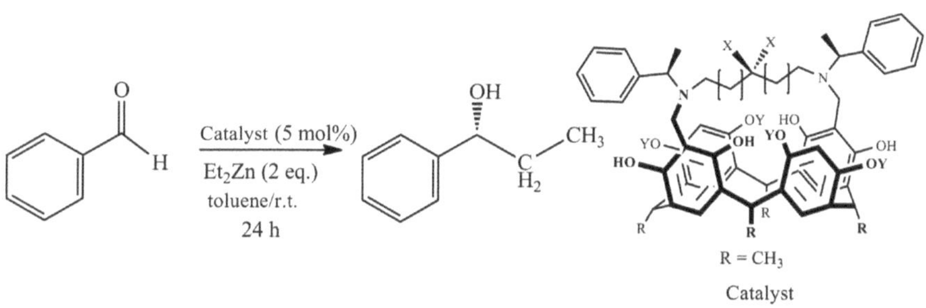

SCHEME 4.24 Bridged calix[4]resorcinols as templates in asymmetric catalysis. *(Adapted with permission from Ref. [90]. Copyright 2006 Elsevier.)*

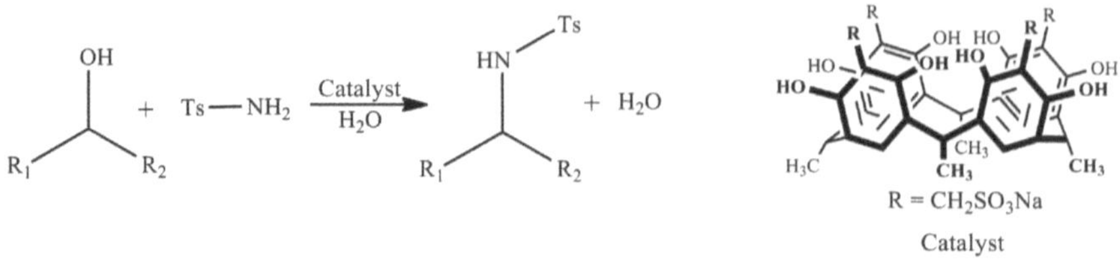

SCHEME 4.25 The water-soluble calix[4]resorcinols as catalysts in the amination of alcohols. *(Adapted from Ref. [91] with permission of Thieme. Copyright 2008.)*

6 SUPRAMOLECULAR SYSTEMS FOR CONTROLLED BINDING/ISOLATION OF ORGANIC MOLECULES AND BIOSUBSTRATES

Supramolecular architectures capable of self-inclusion of compounds through supramolecular "gates" are very attractive research objects. In such systems, the "gates" can be either open or closed, providing control over the guest release or binding. Analysis of literature data shows that controlled release/binding properties can be realized on surfaces [92], in the membranes [93], and in porous systems [94].

The development of systems with functions of controlled drug binding is one of the most promising areas in medical science. Recently, some works on systems that function in accordance with such "commands" by responding to external factors, for example, electric current [95–98], pH [99–101], enzyme action [102,103], and photoirradiation [104,105] have been performed. Despite the enormous progress in this area of research, many problems to be solved still remain. From a practical point of view, the strategy employed in this area is aimed at the development of new nontoxic controlled systems with improved biocompatibility and substrate selectivity. Most of such constructs are formed in organic solvents [106]. However, their applications for therapeutic purposes calls for their functioning in physiological conditions. Therefore, in order to maintain their viability in biological fluids, it is important to develop controlled systems in an aqueous medium.

Moreover, the process of controlled binding/release is widely used in the catalysis of reactions in molecular nanocapsules. Self-assembled capsules can be defined as receptors with closed cavities formed by reversible noncovalent interactions of two or more subunits [107–110]. Such aggregates in solution should have a rigid structure and can exhibit the ability to bind, which the individual components cannot do. The course of reactions taking place can be regulated by capsule functionalization [111]. The reaction product, in contrast to the original substrate, may have higher affinity to the capsule. Hence its release from the nanoreactor becomes a limiting stage of the reaction.

Various (thia)calix[4]arenes and calix[4]resorcinols can be used as building blocks for the construction of nanocapsules. The process of binding/release is even more important in gene therapy and engineering. Verrall's research group proposed a delivery mechanism where cationic surfactants were used as effective DNA transporters [112–114]. The concept of gene therapy is to replace defective or missing gene insertions by DNA delivery into healthy cells. This process is called transfection, and the gene carrier is called a vector. Most vectors have positive charge to bind to negatively charged DNA and facilitate their penetration through the hydrophobic cell membrane.

The potential to use calixarenes as DNA vectors has been demonstrated [115,116]. Sansone et al. [115] reported that a composition comprising calix[4]arene macrocycle derivative was more effective in delivering DNA than available commercial reagents for transfection. Previously, researchers modified the aromatic fragments of calix[4]arene with guanidine fragments for possible transport of molecules into the cells. Although these macrocycles could bind DNA, they showed low transfection efficiency and were toxic to cells. In new calix[4]arene, hydroxyl groups were introduced into the guanidine fragments to increase transfection efficiency with simultaneous significant decrease of the toxicity of these new transfection agents. However, despite these promising examples, this area of research is still at early stage and needs more active search for comprehensive and effective systems as well as further advancement in fundamental studies in this area at the crossroad between chemistry and biology.

7 CONCLUSIONS

Thus, the data discussed in this chapter indicate broad possibilities for the use of macrocycles in catalysis and as nanoreactors. The use of macrocyclic receptors on calixarene platform as components of catalytic systems results in a significant increase of catalytic activity, as well as the attainment of substrate regioselectivity.

ACKNOWLEDGMENTS

This work was funded by a subsidy allocated to Kazan Federal University for the project part of the state assignment in the sphere of scientific activities.

REFERENCES

[1] Gutsche CD. Calixarenes revisited. In: Stoddart JF, editor. Monographs in supramolecular chemistry. London: Royal Society of Chemistry; 1998.

[2] Gutsche CD. Synthesis of calixarenes and thiacalixarenes. In: Asfari Z, Böhmer V, Harrowfield J, Vicens J, Saadioui M, editors. Calixarenes 2001. Dordrecht: Kluwer Academic Publishers; 2001. p. 1–25.

[3] Van Duynhoven JPM, Janssen RG, Verboom W, Franken SM, Casnati A, Pochini A, Ungaro R, De Mendoza J, Nieto PM, Prados P, Reinhoudt DN. Control of calix[6]arene conformations by self-inclusion of 1,3,5-tri-O-alkyl substituents: synthesis and NMR studies. J Am Chem Soc 1994;116(13):5814–22.

[4] Iki N, Kumagai H, Morohashi N, Ejima K, Hasegawa M, Miyanari S, Miyano S. Selective oxidation of thiacalix[4]arenes to the sulfinyl- and sulfonylcalix[4]arenes and their coordination ability to metal ions. Tetrahedron Lett 1998;39:7559–62.

[5] Iki N, Horiuchi T, Oka H, Koyama K, Morohashi N, Kabuto C, Miyano S. Energy transfer luminescence of Tb^{3+} ion complexed with calix[4]arenetetrasulfonate and the thia and sulfonyl analogue. the effect of bridging groups. J Chem Soc Perkin Trans 2001;2:2219–25.

[6] Lhoták P, Morávek J, Smejkal T, Stibor I, Sykora J. Stereoselective oxidation of thiacalix[4]arenes with the $NaNO_3/CF_3COOH$ system. Tetrahedron Lett 2003;44:7333–6.

[7] Morohashi N, Iki N, Sugawara A, Miyano S. Selective oxidation of thiacalix[4]arenes to the sulfinyl and sulfonyl counterparts and their complexation abilities toward metal ions as studied by solvent extraction. Tetrahedron 2001;57:5557–63.

[8] Solovieva SE, Kleshnina SR, Kozlova MN, Galiullina LF, Gubaidullin AT, Latypov SK, Antipin IS, Konovalov AI. Synthesis and complexation properties of carbonyl-containing thiacalix[4]arenes. Russ Chem Bull 2009;57:1477–85.

[9] Stoikov II, Yushkova EA, Bukharaev AA, Biziaev DA, Ziganshina SA, Zharov I. Self-assembly of stereoisomers of *p-tert*-butyl thiacalix[4]arenes tetrasubstituted at the lower rim by a tertiary amide group with cations of p- and d-elements in the organic phase. J Phys Chem C 2009;2:15838–44.

[10] Amirov RR, Ziyatdinova AB, Burilova EA, Zhukov AY, Antipin IS, Stoikov II. Complexation of Gd^{III} with tetra-*p-tert*-butylthiacalix[4]arenoic acid in micellar media. Russ Chem Bull 2010;58:1400–7.

[11] Zheng X, Zhang W, Mu L, Zeng X, Xue S, Tao Z, Yamatob T. A novel rhodamine-based thiacalix[4]arene fluorescent sensor for Fe^{3+} and Cr^{3+}. J Incl Phenom Macrocycl Chem 2010;68:139–46.

[12] Mollard A, Ibragimova D, Antipin IS, Konovalov AI, Stoikov I, Zharov I. Molecular transport in thiacalix[4]arene-modified nanoporous colloidal films. Microporous Mesoporous Mater 2010;131:378–84.

[13] Sansone F, Baldini L, Casnati A, Ungaro R. Calixarenes: from biomimetic receptors to multivalent ligands for biomolecular recognition. New J Chem 2010;34:2715.

[14] Chen M, Zhang W, Jiang R, Diao G. Development of glucose amperometric biosensor based on a novel attractive enzyme immobilization matrix: amino derivative of thiacalix[4]arene. Anal Chim Acta 2011;687:177–83.

[15] Yushkova EA, Stoikov II, Puplampu JB, Antipin IS, Konovalov AI. Cascade and commutative self-assembles of nanoscale three-component systems controlled by the conformation of thiacalix[4]arene. Langmuir 2011;27:14053–64.

[16] Evtugyn GA, Kostyleva VB, Porfireva AV, Savelieva MA, Evtugyn VG, Sitdikov RR, Stoikov II, Antipin IS, Hianik T. Label-free aptasensor for thrombin determination based on the nanostructured phenazine mediator. Talanta 2012;102:156–63.

[17] Evtugyn G, Kostyleva V, Sitdikov R, Porfireva A, Savelieva M, Stoikov I, Antipin I, Hianik T. Electrochemical aptasensor based on a macrocyclic ligand bearing neutral red. Electroanalysis 2012;24:91–100.

[18] Kumar M, Kumar R, Bhalla V, Sharma PR, Kaur T, Qurishi Y. Thiacalix[4]arene based fluorescent probe for sensing and imaging of Fe^{3+} ions. Dalton Trans 2012;41:408–12.

[19] Mel'nikova NB, Kochetkov EN, Solov'eva SE, Popova EV, Antipin IS, Bol'shakova AE, Zhil'tsova OE, Konovalov AI. *tert*-Butylthiacalix[4]arene monolayers as a biomimetic model for the oxidation of antioxidants with cytochrome c. Russ Chem Bull 2012;60:1948–55.

[20] Hayashida O, Mizuki K, Akagi K, Matsuo A, Kanamori T, Nakai T, Sando S, Aoyama Y. Macrocyclic glycoclusters: self-aggregation and phosphate-induced agglutination behaviors

of calix[4]resorcarene-based quadruple-chain amphiphiles with a huge oligosaccharide pool. J Am Chem Soc 2003;125(2):594–601.
[21] Cevasco G, Thea S, Vigo D, Williams A, Zaman F. Catalysis and inhibition of ester hydrolysis in the presence of resorcinarene hosts functionalized with dimethylamino groups. J Phys Org Chem 2006;19(10):630–6.
[22] O'Farrell CM, Hagan KA, Wenzel TJ. Water-soluble calix[4]resorcinarenes as chiral NMR solvating agents for bicyclic aromatic compounds. Chirality 2009;21(10):911–21.
[23] Schneider H-J, Yatsimirsky A. Principles and methods in supramolecular chemistry. New York: Wiley; 1999. 362 p.
[24] Steed JW, Atwood JL. Supramolecular chemistry. New York: Wiley; 2009. 1002 p.
[25] Puplampu JB, Yakimova LS, Vavilova AA, Fayzullin DA, Zuev YF, Stoikov II. Synthesis of *p-tert*-butylthiacalix[4]arenes functionalized with tris(2-aminoethyl)amine fragments at the lower rim and their interaction with model lipid membranes. Macroheterocycles 2014;7:337–44.
[26] Puplampu JB, Yakimova LS, Vavilova AA, Rizvanov IK, Stoikov II. *p-tert*-Butyl thiacalix[4]arene derivatives functionalized in the lower rim with bis(3-aminopropyl)amine: synthesis and interaction with DNA. Macroheterocycles 2015;8(1):75–80.
[27] Shurpik DN, Yakimova LS, Makhmutova LI, Makhmutova AR, Rizvanov IK, Plemenkov VV, Stoikov II. Pillar[5]arenes with morpholide and pyrrolidide substituents: synthesis and complex formation with alkali metal ions. Macroheterocycles 2014;7(4):351–7.
[28] Gorbatchuk VV, Gatiatulin AK, Ziganshin MA, Gubaidullin AT, Yakimova LS. Unusually high efficiency of β-cyclodextrin clathrate preparation by water-free solid-phase guest exchange. J Phys Chem B 2013;117(46):14544–56.
[29] Stoikov II, Agafonova MN, Yakimova LS, Antipin IS, Konovalov AI. Molecular recognition of carboxylic acids and carboxylate anions by synthetic receptor. Molecular recognition: biotechnology, chemical engineering and materials applications. New York: Nova Publisher; 2011. pp. 1–44.
[30] Vavilova AA, Nosov RV, Yakimova LS, Antipin IS, Stoikov II. Synthesis of photo-switchable derivatives of *p-tert*-butyl thiacalix[4]arenes containing ethoxycarbonyl and 4-amidoazobenzene fragments in the lower rim substituents. Macroheterocycles 2013;6(3):219–26.
[31] Yakimova LS, Ziganshin MA, Sidorov VA, Kovalev VV, Shokova EA, Tafeenko VA, Gorbatchuk VV. Molecular recognition of organic vapors by adamantylcalix[4]arene in QCM sensor using partial binding reversibility. J Phys Chem B 2008;112(49):15569–75.
[32] Ziganshin MA, Yakimova LS, Khayarov KR, Gorbatchuk VV, Vysotsky MO, Böhmer V. Guest exchange in dimeric capsules of a tetraurea calix[4]arene in the solid state. Chem Commun 2006;37:3897–9.
[33] Smolko VA, Shurpik DN, Shamagsumova RV, Porfireva AV, Evtugyn VG, Yakimova LS, Stoikov II, Evtugyn GA. Electrochemical behavior of pillar[5]arene on glassy carbon electrode and its interaction with Cu^{2+} and Ag^{+} ions. Electrochim Acta 2014;147:726–34.
[34] Shen J, Yu X, Ye Y, Chen R, Jiang H, Zhou Q. Supramolecular catenane chemistry based on crown ether derivatives. Chin J Org Chem 2012;32(12):2265–75.
[35] Tero T-R, Nissinen MA. Perspective to resorcinarene crowns. Tetrahedron 2014;70(6):1111–23.
[36] Lehn J-M. Supramolecular chemistry: concepts and perspectives. New York: Wiley; 1995. 271 p.
[37] Weber E. Supramolecular chemistry II - Host design and molecular recognition. In: Top Curr Chem., 175. Berlin: Springer; 1995. 164 p.
[38] Reetz MT, List B, Jaroch S, Weinmann H. Organocatalysis. Ernst Schering Foundation Symposium Proceedings. Berlin, Heidelberg: Springer; 2008: 341 p.
[39] Shirakawa S, Shimizu S. Dehydrative amination of alcohols in water using a water-soluble calix[4]resorcinarene sulfonic acid. Synlett 2008;10:1539–42.

[40] Zahalka HA, Januszkiewicz K, Alper H. Olefin oxidation catalyzed by palladium chloride using cyclodextrins as phase transfer agents. J Mol Cat 1986;35(2):249–53.

[41] Maksimov AL, Sakharov DA, Filippova TYu, Zhuchkova AYa, Karakhanov EA. Supramolecular catalysts on the basis of molecules-receptors. Ind Eng Chem Res 2005;44(23):8644–53.

[42] Karakhanov EA, Maksimov AL, Ivanova EA. Supramolecular catalytic systems in biomimetic oxidation. Russ Chem Bull 2007;56(4):621–30.

[43] Karakhanov EA, Filippova TYu, Martynova SA, Maximov AL, Predeina VV, Topchieva IN. New catalytic systems for selective oxidation of aromatic compounds by hydrogen peroxide. Cat Today 1998;44(1-4):189–98.

[44] Chau NTT, Guégan J-P, Menuel S, Guerrero M, Hapiot F, Monflier E, Philippot K, Denicourt-Nowicki A, Roucoux A. β-Cyclodextrins grafted with chiral amino acids: a promising supramolecular stabilizer of nanoparticles for asymmetric hydrogenation? Appl Cat A 2013;467:497–503.

[45] Li Z-Y, Chen J-W, Liu Y, Xia W, Wang L. The use of calixarenes in asymmetric catalysis. Curr Org Chem 2011;15(1):39–61.

[46] Hapiot F, Bricout H, Tilloy S, Monflier E. Hydroformylation in aqueous biphasic media assisted by molecular receptors. Top Curr Chem 2013;342:49–78.

[47] Nguyen DH, Urrutigoïty M, Kalck P. Water-soluble hydroformylation catalysis. Metal-catalyzed reactions in water. Weinheim: Wiley-VCH; 2013. pp. 109–137.

[48] Jouffroy M, Sémeril D, Armspach D, Matt D. Phosphane-phosphite chelators built on a α-cyclodextrin scaffold: application in Rh-catalysed asymmetric hydrogenation and hydroformylation. Eur J Org Chem 2013;27:6069–77.

[49] Monflier E, Tilloy S, Bertoux F, Castanet Y, Mortreux A. New prospects for the palladium-catalyzed hydrocarboxylation of higher α-olefins in a two-phase system: the use of chemically modified β-cyclodextrin. New J Chem 1997;21(8):857–9.

[50] Karakhanov E, Buchneva T, Maximov A, Zavertyaeva M. Substrate selectivity in byphasic Wacker-oxidation of alkenes in the presence of water-soluble calixarenes. J Mol Cat A Chem 2002;184(1–2):11–7.

[51] Karakhanov EA, Buchneva TS, Maksimov AL, Runova EA. Calixarene-based catalyst systems in biphasic Wacker oxidation of olefins. Neftekhimiya 2003;43(1):42–9.

[52] Shimizu S, Kito K, Sasaki Y, Hirai C. Water-soluble calixarenes as new inverse phase-transfer catalysts: nucleophilic substitution of alkyl and arylalkyl halides in aqueous media. Chem Commun 1997;17:1629–30.

[53] Dospil G, Schatz J. Synthesis and characterization of imidazole-substituted calix[4]arenes as simple enzyme-mimics with acyltransferase activity. Tetrahedron Lett 2001;42(44):7837–40.

[54] Kimizuka N, Wakiyama T, Yanagi A, Shinkai S, Kunitake T. Calix[4]arene-mediated transport of alkali ions across synthetic black lipid membranes (BLM). Bull Chem Soc Jap 1996;69(12):3681–4.

[55] Shinkai S, Mori S, Koreishi H, Tsubaki T, Manabe O. Hexasulfonated calix[6]arene derivatives: a new class of catalysts, surfactants, and host molecules. J Am Chem Soc 1986;108(9):2409–16.

[56] Komiyama M, Isaka K, Shinkai S. Water-soluble calixarene as the first man-made catalyst for regioselective cleavage of ribonucleoside 2′,3′-cyclic phosphate. Chem Lett 1991;937–40.

[57] Shinkai S, Shirahama Y, Tsubaki T, Manabe O. Calixarene-catalyzed basic hydrolysis of *p*-nitrophenyl dodecanoate: a possible change in the mechanism front "deshielding" to "host–guest". J Am Chem Soc 1989;111(14):5477–8.

[58] Hykrdová L, Jirkovský J, Grabner G, Mailhot G, Bolte M. Degradation of supramolecular complexes photoinduced by iron (III): the influence of the host on the reactivity of the guest. Photochem Photobiol Sci 2003;2(2):163–70.

[59] Tabushi I. Cyclodextrin catalysis as a model for enzyme action. Chem Res 1982; 15(3):66–72.

[60] Agrawal YK, Bhatt H. Calixarenes and their biomimetic applications. Bioinorg Chem Appl 2004;2(3-4):237–74.

[61] Reetz MT. Supramolecular transition metal catalysts in two-phase systems. Catal Today 1998;42(4):399–411.

[62] Maksimov AL, Buchneva TS, Karakhanov EA. Supramolecular calixarene-based catalytic systems in the Wacker-oxidation of higher alkenes. J Mol Cat A Chem 2004;217(1-2):59–67.

[63] Boring E, Geletii YV, Hill CL. Catalytic aerobic oxidation of 2-chloroethyl ethylsulfide, a mustard simulant, under ambient conditions: effect of solvents, ligands, and transition metals on reactivity. J Mol Cat A Chem 2001;176(1–2):49–63.

[64] Massa A, D'Ambrosi A, Proto A, Scettri A. Critical importance of molecular sieves in titanium(IV)-calix[4]arene catalyzed epoxidation of allylic alcohols. Tetrahedron Lett 2001;42(10):1995–8.

[65] Baur M, Frank M, Schatz J, Schildbach F. Water-soluble calix[*n*]arenes as receptor molecules for non-polar substrates and inverse phase transfer catalysts. Tetrahedron 2001;57(32):6985–91.

[66] Tian H-Y, Chen Y-J, Wang D, Zeng C-C, Li C-J. Calix[6]arene derivatives bearing sulfonate and alkyl groups as surfactants in $Sc(OTf)_3$-catalyzed Mukaiyama aldol reactions in water. Tetrahedron Lett 2000;41(15):2529–32.

[67] Soriente A, Fruilo M, Gregoli L, Neri P. Calix[*n*]arene/Ti(IV) complexes as active catalysts in aldol reaction of Chan's diene. Tetrahedron Lett 2003;44(33):6195–8.

[68] Cacciapaglia R, Mandolini L, Arnecke R, Böhmer V, Vogt W. Barium(II) complexes of calixcrowns derived from *p-tert*-butylcalix[5]arene as potential transacylation catalysts. Regio- and stereo-selective monoacylation of the calixcrown. J Chem Soc Perkin Trans 1998;2(2):419–23.

[69] Le Clainche L, Rondelez Y, Sénèque O, Blanchard S, Campion M, Giorgi M, Duprat F, Le Mest Y, Reinaud O. Calix[6]arene-based models for mono-copper enzymes: a promising supramolecular system for oxidation catalysis. CR Acad Sci 2000;3(10):811–9.

[70] Sénèque O, Campion M, Douziech B, Giorgi M, Riviére E, Journaux Y, Le Mest Y, Reinaud O. Supramolecular assembly with calix[6]arene and copper ions: formation of a novel tetranuclear core exhibiting unusual redox properties and catecholase activity. Eur J Inorg Chem 2002;8:2007–14.

[71] Shimizu S, Shirakawa S, Sasaki Y, Hirai C. Novel water-soluble calix[4]arene ligands with phosphane-containing groups for dual functional metal-complex catalysts: the biphasic hydroformylation of water-insoluble olefins. Angew Chem Int Ed 2000;39(7):1256–9.

[72] Shirakawa S, Shimizu S, Sasaki Y. Rhodium-catalyzed biphasic hydroformylation of 4-octene using water-soluble calix[4]arene-phosphine ligands. New J Chem 2001;25(6):777–9.

[73] Breit B, Winde R, Mackewitz T, Paciello R, Harms K. Phosphabenzenes as monodentate π-acceptor ligands for rhodium-catalyzed hydroformylation. Chem A Eur J 2001;7(14): 3106–21.

[74] Csók Z, Szalontai G, Czira G, Kollár L. Carbonylation (hydroformylation and hydrocarbalkoxylation) reactions in the presence of transition metal: *p-tert*-butyl-calix[4]arene-based phosphine and phosphinite systems. J Organomet Chem 1998;570(1):23–9.

[75] Parlevliet FJ, Kiener C, Fraanje J, Goubitz K, Lutz M, Spek AL, Kamer PCJ, Van Leeuwen PWNM. Calix[4]arene based monophosphites, identification of three conformations and their use in the rhodium-catalysed hydroformylation of 1-octene. J Chem Soc Dalton Trans 2000;7:1113–22.

[76] Bagatin IA, Matt D, Thönnessen H, Jones PG. Calix[4]arene ligands with phosphorus-containing groups tethered at the upper rim. Inorg Chem 1999;38(7):1585–91.

[77] Loeber C, Wieser C, Matt D, De Cian A, Fischer J, Toupet L. Cavity-shaped phosphane ligands: phosphane-amide hybrids based on a calix[4]arene matrix and their chelating behavior towards platinum and rhodium. Bull Soc Chim Fr 1995;132(2): 166–77.
[78] Cobley CJ, Ellis DD, Guy Orpen A, Pringle PG. Rhodium(I) complexes of robust phosphites derived from calix[4]arenes and their application in the hydroformylation of 1-hexene. J Chem Soc Dalton Trans 2000;7:1109–12.
[79] Paciello R, Siggel L, Röper M. Chelated bisphosphites with a calix[4]arene backbone: new ligands for rhodium-catalyzed low-pressure hydroformylation with controlled regioselectivity. Angew Chem Int Ed 1999;38(13-14):1920–3.
[80] Zheng Y-S, Ying L-Q, Shen Z-Q. Polymerization of propylene oxide by a new neodymium complex of calixarene derivative. Polymer 2000;41(4):1641–3.
[81] Dieleman C, Steyer S, Jeunesse C, Matt D. Diphosphines based on an inherently chiral calix[4]arene scaffold: synthesis and use in enantioselective catalysis. J Chem Soc Dalton Trans 2001;17:2508–17.
[82] Seitz J, Maas G. Calixarenes as ligands for transition-metal catalysts: a bis(calix[4]arene-11,23-dicarboxylato) dirhodium complex. Chem Commun 2002;4:338–9.
[83] Molenveld P, Stikvoort WMG, Kooijman H, Spek AL, Engbersen JFJ, Reinhoudt DN. Dinuclear and trinuclear Zn(II) calix[4]arene complexes as models for hydrolytic metalloenzymes: synthesis and catalytic activity in phosphate diester transesterification. J Org Chem 1999;64(11):3896–906.
[84] Molenveld P, Engbersen JFJ, Reinhoudt DN. Specific RNA dinucleotide cleavage by a synthetic calix[4]arene-based trinuclear metallo(II)-phosphodiesterase. Angew Chem Int Ed 1999;38(21):3189–92.
[85] Baldini L, Bracchini C, Cacciapaglia R, Casnati A, Mandolini L, Ungaro R. Catalysis of acyl group transfer by a double-displacement mechanism: the cleavage of aryl esters catalyzed by calixcrown – Ba^{2+} complexes. Chem Eur J 2000;6(8):1322–30.
[86] Pirrincioglu N, Zaman F, Williams A. A calixresorcinarene provides the framework for an artificial esterase. J Chem Soc Perkin Trans 1996;2(12):2561–2.
[87] Schneider H-J, Schneider U. Inhibition of choline acetate hydrolysis in the presence of a macrocyclic polyphenolate. J Org Chem 1987;52(8):1613–5.
[88] Arnott G, Hunter R, Su H. Synthesis and characterization of chiral, bridged resorcinarenes as templates for asymmetric catalysis. Tetrahedron 2006;62(5):977–91.
[89] Mirgorodskaya AB, Kudryavtseva LA, Kazakova EKh, Konovalov AI. Reactions of calix[4] resorcinolarene anions with esters of carboxylic acids in H_2O-DMF solvent. Russ Chem Bull 2000;49(2):261–4.
[90] Arnott G, Hunter R. Enantioselective addition of diethylzinc to benzaldehyde catalysed by chiral, bridged resorcinarenes: a stereoselectivity model based on chirality transfer. Tetrahedron 2006;62(5):992–1000.
[91] Shirakawa S, Shimizu S. Dehydrative amination of alcohols in water using a water-soluble calix[4]resorcinarene sulfonic acid. Synlett 2008;10:1539–42.
[92] Wen L, Hou X, Tian Y, Nie F-Q, Song Y, Zhai J, Jiang L. Bioinspired smart gating of nanochannels toward photoelectric-conversion systems. Adv Mat 2010;22(9):1021–4.
[93] Karavasili C, Spanakis M, Papagiannopoulou D, Vizirianakis IS, Fatouros DG, Koutsopoulos S. Bioactive self-assembling lipid-like peptides as permeation enhancers for oral drug delivery. J Pharm Sci 2015;104(7):2304–11.
[94] Noël T, Hessel V. Membrane microreactors: gas-liquid reactions made easy. ChemSusChem 2013;6(3):405–7.

[95] Sun Y-L, Yang Y-W, Wu W, Zhang SX-A. Supramolecular nanovalve systems based on macrocyclic synthetic receptors. Gaodeng Xuexiao Huaxue Xuebao/Chem J Chin Univ 2012;33(8):1635–42.

[96] Yang N, Chen T, Fu J. Controlled release of corrosion inhibitor from intelligent nanocontainers based on acid and base dual-responsive. Gaodeng Xuexiao Huaxue Xuebao/Chem J Chin Univ 2014;35(5):971–5.

[97] Sun Y-L, Yang B-J, Zhang SX-A, Yang Y-W. Cucurbit[7]uril pseudorotaxane-based photoresponsive supramolecular nanovalve. Chem A Eur J 2012;18(30):9212–6.

[98] Du L, Liao S, Khatib HA, Stoddart JF, Zink JI. Controlled-access hollow mechanized silica nanocontainers. J Am Chem Soc 2009;131(42):15136–42.

[99] Liu J, Du X. pH- and competitor-driven nanovalves of cucurbit[7]uril pseudorotaxanes based on mesoporous silica supports for controlled release. J Mat Chem 2010;20(18):3642–9.

[100] Guo W, Wang J, Lee S-J, Dong F, Park SS, Ha C-S. A general pH-responsive supramolecular nanovalve based on mesoporous organosilica hollow nanospheres. Chem A Eur J 2010;16(29):8641–6.

[101] Lee S-F, Zhu X-M, Wang Y-XJ, Xuan S-H, You Q, Chan W-H, Wong C-H, Wang F, Yu JC, Cheng CHK, Leung KC-F. Ultrasound, pH, and magnetically responsive crown-ether-coated core/shell nanoparticles as drug encapsulation and release systems. ACS Appl Mat Int., 2013;5(5):1566–74.

[102] Palanikumar L, Choi ES, Cheon JY, Joo SH, Ryu J-H. Noncovalent Polymer-Gatekeeper in Mesoporous Silica Nanoparticles as a Targeted Drug Delivery Platform. Adv Funct Mat 2015;25(6):957–65.

[103] Harnoy AJ, Rosenbaum I, Tirosh E, Ebenstein Y, Shaharabani R, Beck R, Amir RJ. Enzyme-responsive amphiphilic PEG-dendron hybrids and their assembly into smart micellar nanocarriers. J Am Chem Soc 2014;136(21):7531–4.

[104] Wang X, Li Z, Yang Y, Gong X, Liao Y, Xie X. Photomechanically controlled encapsulation and release from pH-responsive and photoresponsive microcapsules. Langmuir 2015;31(19):5456–63.

[105] Zhang X, Zhao H, Tian D, Deng H, Li H. A photoresponsive wettability switch based on a dimethylamino calix[4]arene. ChemA Eur J 2014;20(30):9367–71.

[106] Park C, Lee K, Kim C. Photoresponsive cyclodextrin-covered nanocontainers and their sol-gel transition induced by molecular recognition. Angew Chem Int Ed 2009;48:1275–8.

[107] Bassyouni F, Elhalwany N, Abdel Rehim M, Neyfeh M. Advances and new technologies applied in controlled drug delivery system. Res Chem Int 2015;41(4):2165–200.

[108] Bansode SS, Banarjee SK, Gaikwad DD, Jadhav SL, Thorat RM. Microencapsulation: a review. Int J Pharm Sci Rev Res 2010;1(2):38–43.

[109] Shekhar K, Naga Madhu M, Pradeep B, Banji D. A review on microencapsulation. Int J Pharm Sci Rev Res 2010;5(2):58–62.

[110] Laughrey Z, Gibb BC. Water-soluble, self-assembling container molecules: an update. Chem Soc Rev 2011;40(1):363–86.

[111] Buxboim A, Daube SS, Bar-Ziv R. Synthetic gene brushes: a structure–function relationship. Mol Syst Bio 2008;4:181–9.

[112] Singh J, Michel D, Getson HM, Chitanda JM, Verrall RE, Badea I. Development of amino acid substituted gemini surfactant-based mucoadhesive gene delivery systems for potential use as noninvasive vaginal genetic vaccination. Nanomedicine 2015;10(3):405–17.

[113] Singh J, Yang P, Michel D, Verrall RE, Foldvari M, Badea I. Amino acid-substituted gemini surfactant-based nanoparticles as safe and versatile gene delivery agents. Curr Drug Deliv 2011;8(3):299–306.

[114] Wang C, Li X, Wettig SD, Badea I, Foldvari M, Verrall RE. Investigation of complexes formed by interaction of cationic gemini surfactants with deoxyribonucleic acid. Phys Chem Chem Phys 2007;9(13):1616–28.
[115] Bagnacani V, Franceschi V, Fantuzzi L, Casnati A, Donofrio G, Sansone F, Ungaro R. Lower rim guanidinocalix[4]arenes: macrocyclic nonviral vectors for cell transfection. Bioconj Chem 2012;23(5):993–1002.
[116] Goldring WPD, Jubeli E, Downs RA, Johnston AJS, Abdul Khalique N, Raju L, Wafadari D, Pungente MD. Novel macrocyclic and acyclic cationic lipids for gene transfer: synthesis and in vitro evaluation. Bioorg Med Chem Lett 2012;22(14):4686–92.

Chapter 5

Carbon Nanotube Nanoreactors for Chemical Transformations

M. Rosa Axet, Philippe Serp
Coordination Chemistry Laboratory, University of Toulouse, Emile Monso, Toulouse Cedex, France

1 INTRODUCTION

Carbon nanotubes (CNTs) constitute a class of nanostructured materials that possess outstanding mechanical, electrical, and optical properties, which can be exploited for various applications [1]. Besides these properties, a specific feature of CNTs is their well-defined one-dimensional (1D) cavity, which makes of them a desirable material for (1) encapsulation of various molecules or materials, and (2) for generation of confinement effects. According to CNT type (single- or multiwalled CNT; SWCNTs or MWCNTs), the cavity diameter may change from 0.5–2 nm for SWCNTs, which correspond to microporosity, to 2–100 nm for MWCNTs, which correspond to meso- (2–50 nm) and even macroporosity (>50 nm) [2]. This is an important point, since the extent of confinement phenomena should be largely dominated by CNT diameter, even if we cannot exclude an effect from tube helicity. The insertion of foreign species into these 1-D nanocavities has considerable impact on the host molecules/materials, including changes in their structure, size distribution, surface area, and dynamics. This leads to many interesting and striking properties that are quite different from those in bulk, and opens the way to exciting applications, such as quantum molecular sieving [3,4], nanofiltration [5,6], drug delivery [7,8], and catalysis [9–11]. For nanofiltration, extremely high aspect ratios, molecularly smooth hydrophobic graphitic walls, and nanoscale inner diameters of CNTs give rise to the unique phenomenon of ultra-efficient transport of water and gas through these narrow molecular pipes. Water and gas molecules move through nanotube pores orders of magnitude faster than through other pores of comparable size [6]. The frictionless transport of water molecules makes CNTs suitable for the development of high fluxing separation techniques, particularly for water purification/desalination [12]. Appropriate pore diameters can constitute energy barriers at the channel entries, rejecting salt ions and permitting water through the nanotube hollows. It is also possible to modify CNT pores to selectively sense

Organic Nanoreactors. http://dx.doi.org/10.1016/B978-0-12-801713-5.00005-7

and reject ions. Thus, CNT membrane can be used as a "gatekeeper" for size-controlled separation of multiple pollutants. For drug delivery, hydrophobic CNTs can penetrate through cell membranes and reach locations that are generally inaccessible to polar drug molecules. Hence, one can encapsulate drugs inside CNTs and deliver them to desired parts of the living body. In the field of catalysis, it is generally admitted that antagonistic effects at the mesoscale result in the existence of an optimal catalytic efficiency depending on pore size [13–15]. More generally, to perform confined chemical transformations [16], the inner cavity of CNTs constitutes a real nanoreactor, in which a variety of effects such as diffusion, concentration, electronic, or entropic effects can affect the course of a specific reaction. Thus, reaction energetics, mechanisms, and dynamics could be significantly altered inside CNTs due to their large intrinsic polarizabilities and the severely decreased reaction volume. In this chapter we will mainly focus on catalytic chemical transformations, since they have been by far the most studied. We will describe the various routes proposed to prepare confined catalysts and will discuss their characterization. Confinement effects inside CNTs will be discussed in detail, and an exhaustive review of chemical transformations performed inside CNTs will be presented.

2 CONFINEMENT EFFECTS INSIDE CARBON NANOTUBES

The confinement of metal (or metal oxide) nanoparticles (NPs) inside CNTs often leads to significantly enhanced catalytic activity/selectivity/stability with respect to the same bare metal NPs or those deposited on the outer walls of CNTs. Such a different behavior originates from both the physical (spatial restriction of the channels) and chemical factors (electronic interaction of confined species with the curved graphene walls) inside CNTs. In this sense, SWCNTs are very interesting because they have a higher degree of uniformity and a smaller channel compared than MWCNTs. More interestingly, they exhibit intrinsically either metallic or semiconducting properties with different electronic structures around the Fermi level [17].

2.1 Electronic Effects

The electronic configuration of the carbon atom is $1s^2 2s^2 2p^2$. The 2s and 2p atomic orbitals can readily mix to form sp^2 and sp^3 hybrids. For graphite and CNTs, the hybridization is sp^2 type. In graphite, the π orbitals are perpendicular to the graphene plane, while in CNTs they are oriented along the normal to nanotube surface and thus are not parallel with each other, due to the surface bending. Thus, at a local scale, the σ and π orbitals are no longer perpendicular to each other. Overlap of the π orbitals is introduced. As a consequence, the parts of the π orbitals inside and outside of a nanotube rearrange themselves in such a way that the outer contribution is larger than the inner one (see Fig. 5.1). The curvature induces a mixed state of σ and π orbitals, called rehybridization. Consequently, the π-orbital is always inclined by the hybridization angle $\delta = a/(2\sqrt{3}d)$ (a the

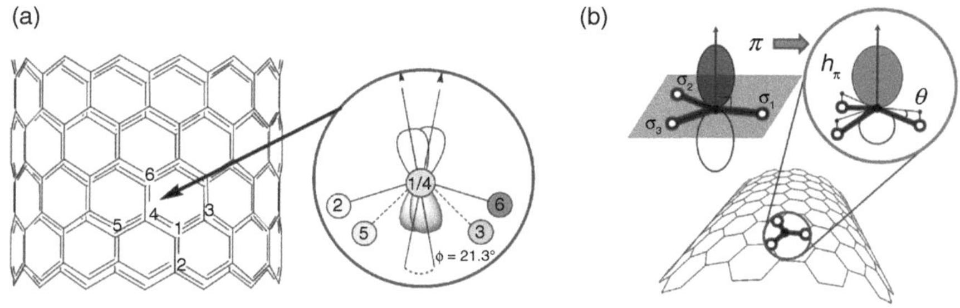

FIGURE 5.1 (a) The π-orbital misalignment angles ϕ along the C1–C4. (b) Diagram showing the π orbital in planar graphene and its change into h_π under bending together with the pyramidalization angle θ. *Reprinted from Ref. [20] with permission of Royal Society of Chemistry.*

length of the lattice unit vector and d the CNT diameter) relative to the normal direction to the tube's surface [18]. This tilting angle is strongly dependent on CNT diameter and helicity [19]. Curvature in the CNT introduces misalignment of π-orbitals within the graphene sheet (Fig. 5.1a). The π-orbitals of a CNT are not pointed directly toward the central axis of the nanotube, and some adjacent carbon π-orbitals have a misalignment angle, ϕ, between them. The rehybridization can also be explained by a mixture of sp^2 and sp^3 orbitals. While single sp^2 and sp^3 orbitals are saturated, the mixed state contains unsaturated orbitals. The mixing of a tetravalent sp^3 orbital to a trivalent sp^2 orbital leaves one hybrid orbital free for binding [20].

The higher the ratio of the sp^3 contribution, the more free bonds exist and the higher the reactivity. As an index for local reactivity, the pyramidalization angle θ is used (see Fig. 5.1b), which is sufficient for describing the curvature-induced shift in sp^2 hybridization. θ is the angle between the σ and π orbitals minus 90 degrees ($\theta = 0$ degree for the sp^2 hybridization). The tetrahedral sp^3 orbital has a pyramidalization angle of 19.5 degrees. This angle changes depending on the local mixture of sp^2 and sp^3 orbitals. The degree of hybridization $sp^{2+\eta}$ could be obtained, η being a number between 0 and 1. The comparison between different pyramidalization angles allows the comparison of reactivity, where a higher angle results in a larger reactivity. Thus, the degree of hybridization is highly curvature dependent, so that it is even possible to increase the reactivity alone by bending a CNT.

At a macroscopic scale, the σ and π orbitals also mix with orbitals of the neighboring sites, and such mixing affects the band structure of CNTs. Of course, distortions of CNTs will accentuate this phenomenon, suggesting that local reactivity can be significantly enhanced. Optical absorption spectroscopy and scanning tunnel microscopy have shown the electronic structure changes near the Fermi level along with the curvature for a single nanotube [21]. Thus, the π electron density of the graphene sheet shifts from the concave (inner) to the convex (outer) surface because of the tube curvature. This results in the occurrence of an electron potential difference across the CNT walls (Fig. 5.2).

For CNTs with a diameter of 1.36 nm, a strong electric field exists in the cavity due to a potential difference as high as 0.20 eV. Such a potential difference

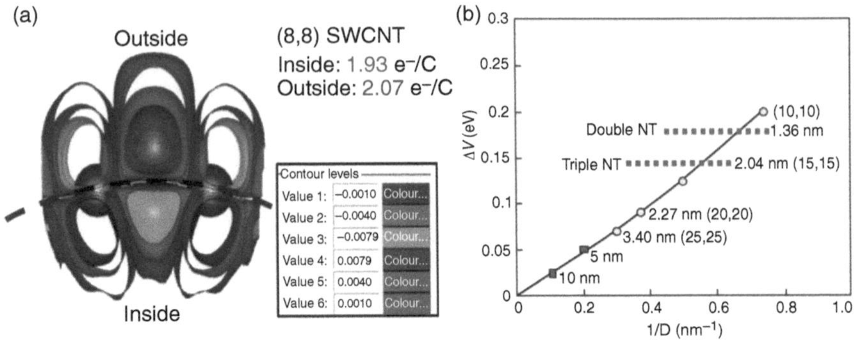

FIGURE 5.2 (a) Number of electrons on carbon atoms on the concave and convex surface of a (8,8) SWCNT. (b) Electron static potential difference inside and outside CNTs as a function of tube diameter. *(Adapted from Ref. [11] with permission of Oxford Journals.)*

decreases to 0.05 eV for MWCNTs presenting an internal diameter of 5 nm [11]. The curvature effect opens a small band gap for metallic nanotubes. A semiconductor tube has a band gap inversely proportional to its diameter d, while the chiral metallic and zigzag metallic tubes are predicted to have narrow-gap scaling by $1/d^2$. The chemical reactivity showed dependence on the tube diameter or curvature. Although the electronic structure near the Fermi level has been widely investigated, much less is known concerning the whole band structure and especially the σ bands. The quantum confinement of SWCNTs on bonding and rehydridization of π and σ orbitals and band gap change has been investigated by using resonant X-ray emission spectroscopy [22]. At the excitation energy of 285.5 eV (π*), the emission spectral features change drastically due to the surface bending of SWCNTs. The changes can be related to the π and σ band structure evolution near K point. The observed quantum confinement of SWCNTs affects both π and σ bands and band gap (Fig. 5.3) through the rehybridization of π and σ orbitals.

The energy difference between LUMO and HOMO reflects the band gap (Eg) of SWCNTs, and thus Eg (LUMO-HOMO) versus d (diameter) can reflect the change for SWCNTs with different diameters.

A universal structure-reactivity relationship was found for reactions occurring both on the inner and outer surface of CNTs with different diameters and different N and O concentrations [23]. No matter whether the reactions occur inside or outside of the channel, the C—H activation barrier can be predicted by C—O bond length and the spin density and charge carried by the active oxygen. They provide a clear connection between reactions in and outside of nanotubes.

If we now consider the interaction of a metallic NP with CNTs, we will find it perturbs the CNT electronic structure. For example, calculations of the density of states (DoS) for a (12,12)-SWCNT show that implantation of Os_{55} clusters into the nanotube sidewall perturbs the electronic structure of the SWCNT quite significantly, due to a strong hybridization between the d orbitals of the Os cluster and the π-electronic system of the host-nanotube (Fig. 5.4) [24].

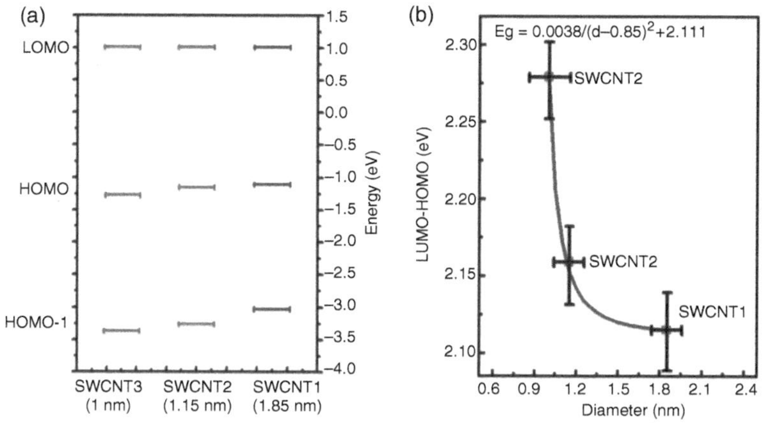

FIGURE 5.3 (a) The energy levels of LUMO, HOMO, and HOMO-1 for SWCNTs of different diameters. (b) The energy difference between LUMO and HOMO and the fitting curve. *(Reprinted from Ref. [22] with permission of AIP Publishing LLC. Copyright 2008.)*

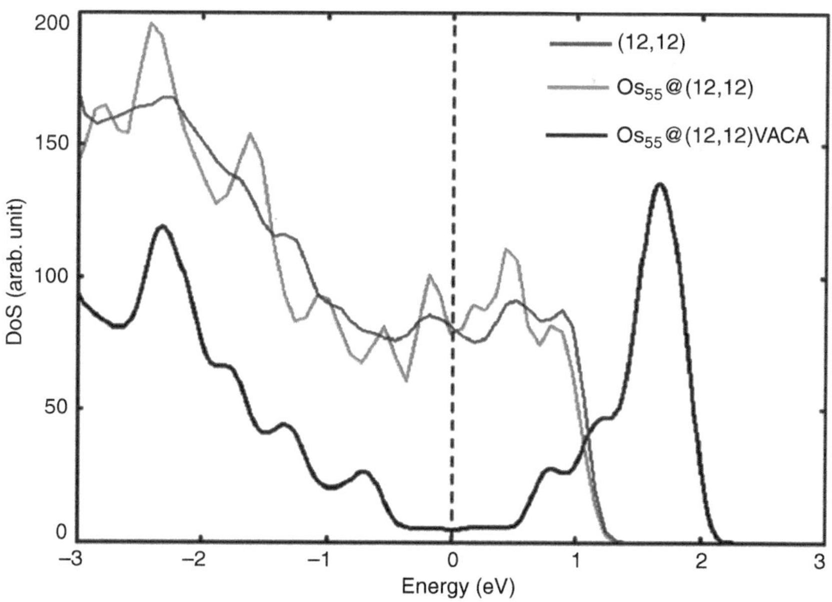

FIGURE 5.4 Density of states plots for solid line (black) a pristine (12,12)-SWCNT, [gray (red in the web version) line] Os_{55} inside (12,12)-SWCNT, and [dark gray (blue in the web version) line] Os_{55} inside (12,12)-SWCNT with vacancies (12 carbon atoms missing). The zero corresponds to the Fermi energy. *(Reprinted from Ref. [24] with permission of American Chemical Society. Copyright 2012.)*

The effect of confinement was also studied using DFT calculations on Al_n cluster geometries (for $n = 1–13$) inside armchair SWCNTs (m,m) of varying diameter, which shows linear structures for $n \leq 5$ in SWCNTS (m,m) $[m < 7]$ and 3D structures thereafter for $n > 5$ and $m > 6$ [25]. In strong confinement, Al–C interaction becomes more dominant, and if the effect of confinement is relaxed, the Al–Al interactions become relatively more important. It was observed that the clusters have a tendency to form nanowires in strong confinement and retain their free space geometries under less confinement.

First-principles calculations were also performed to investigate the binding energies, geometric structures, and electronic properties of 4d transition metal (TM) (particularly Ru), atoms, and clusters adsorbed outside/inside the single-walled or double-walled CNTs. The equilibrium adsorption structures of the TM atoms depend on the valence electron configuration of the metal atoms. Due to curvature effect, all TM atoms adsorbed inside and outside (6,6) CNTs and donate different amounts of electrons to the nanotube, with a nearly constant difference of about 0.5 electrons/TM atom (Fig. 5.5a). The analysis of electronic DoS (Fig. 5.5b) revealed hybridization between the p electrons from C and the d electrons from Ru, a factor that results in charge transfer from metal to carbon. The similarity between the DoS of Ru^+ ion and the DoS on the Ru atom within the Ru–CNT hybrid system clearly shows that there are electrons transferred from Ru atom to CNT. The analysis of the partial DoS of the Ru-CNT hybrid systems demonstrates that the Fermi level locates at the d states of Ru atom and the s, p states of CNT, implying that the Ru–CNT interaction mainly comes from the coupling of d electrons of ruthenium atom and sp electrons of carbon atoms. Similar phenomena were observed for Ru adsorption on other CNTs of different sizes and chiralities. The amount of charge transfer shows systematic trend with the electronegativity of 4d TMs. When Ru atom or cluster adsorbs on double-walled CNTs, the effect of charge transfer is slightly enhanced with regard to the SWCNTs [26]. Finally, it was demonstrated that reactions confined within SWCNT channels are modulated by the metallic and semiconducting character of the hosts [17]. For rhenium NPs, in situ Raman and XANES spectroscopies provide information about the electronic state of CNTs and the confined Re species, which reveal electronic interactions between encapsulated species and nanotubes. More electrons are transferred from metallic tubes (m-SWCNTs) to oxidic rhenium clusters, leading to a lower-valence state rhenium oxide than that in semiconducting tubes (s-SWCNTs). The high-valence

▶ **FIGURE 5.5** (a) Charge transfer as a function of the inverse of nanotube diameter (positive values mean electrons transferred from Ru to CNT). Dark gray (blue in the web version) squares stand for interior adsorption and gray (red in the web version) circles for exterior adsorption. (b) Local density of states DoS on Ru atom for four different systems: free Ru atom, Ru^+, and Ru inside and outside (6,6) SWCNT. Black dashed line stands for Fermi level; middle solid (black), right gray (red in the web version), and left dark gray (blue in the web version) lines represent the 5s, 5p, and 4d states, respectively. *(Reprinted from Ref. [26] with permission of AIP Publishing LLC. Copyright 2010.)*

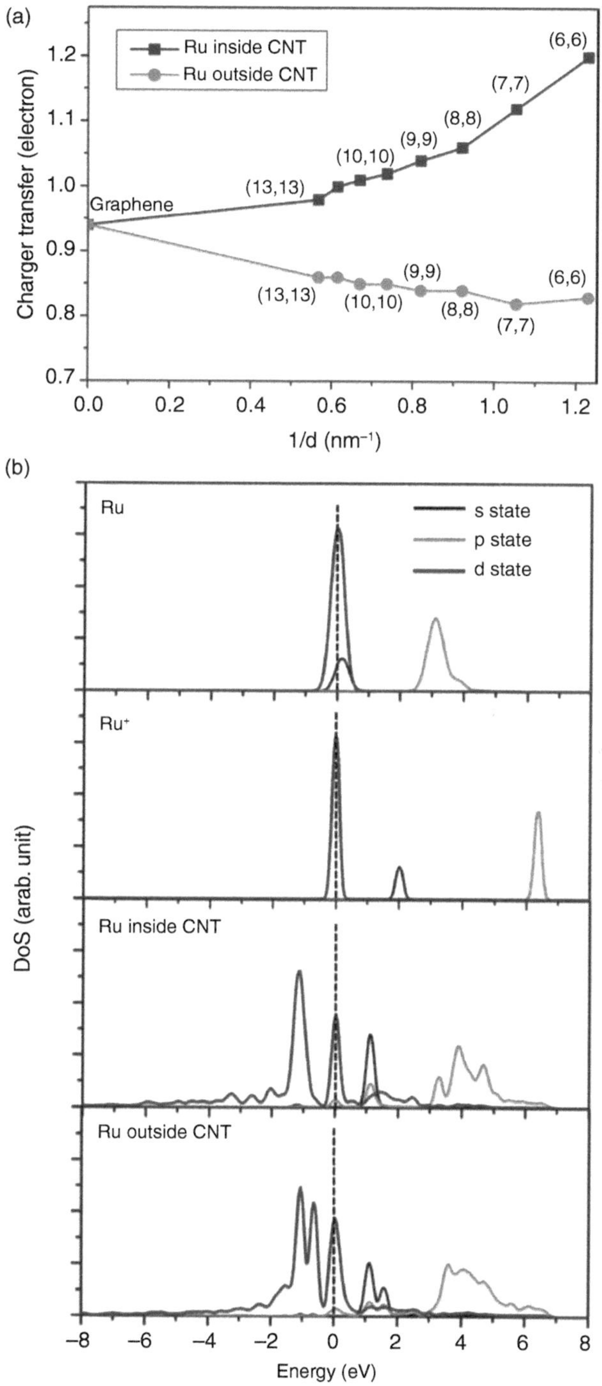
(a)
Ru inside CNT
Ru outside CNT
Charger transfer (electron)
Graphene
(13,13)
(10,10)
(9,9)
(8,8)
(7,7)
(6,6)
1/d (nm⁻¹)
(b)
Ru
Ru⁺
Ru inside CNT
Ru outside CNT
s state
p state
d state
DoS (arab. unit)
Energy (eV)

state Re within s-SWCNTs is more readily reduced when raising the temperature, whereas only a sluggish change is observed for Re within m-SWCNTs.

The theory indicates that the strength of the adsorption and bonding of reactants on catalyst surfaces is closely related to the symmetry and spin state of reactants' molecular or atomic orbitals, and depends on matching the energy levels of the reactants to those of the catalyst surface as well. Norskov and coworkers invoked a so-called d-band center theory based on a large body of experimental and theoretical results, which emphasizes that the density of d-band valence electrons near the Fermi level is an important factor affecting catalytic reactions [27]. It is thus particularly attractive to be able to modulate metallic NP reactivity by controlling: (1) their location (inside *vs.* outside CNTs), (2) CNT diameter, and (3) CNT helicity (metallic vs. semiconductor).

2.2 Surface and Cavity Effects

The internal cavity of CNTs, which presents a peculiar electronic structure, can be considered as a nanoreactor providing confinement effects for catalysis. This can be rationalized by the electronic effects just discussed, but also by additional confinement effects such as a facilitated diffusion, an enrichment of molecules (concentration gradients), and the spatial restriction.

2.2.1 Diffusion and Concentration Gradients

The atomically smooth surface and the rigidity of CNTs suggest an ideal medium for rapid nanoscale mass transport. Both experimental and modeling studies have established a diffusion driven mass transport of light gases and liquids inside CNTs exceeding those of other porous materials [28,29].

2.2.1.1 Diffusion of Liquids

From the fundamental point of view, understanding liquid behavior in nanoscale spaces is of primary importance, since the resulting properties may be very different from those in bulk liquid, solid, and gaseous phases of the same substance. The transport of liquids along their central pores is now of considerable interest both for testing classical theories of fluid flow at the nanoscale and for potential nanofluidic device applications [30]. The topic of the static and dynamic behavior of liquids inside CNTs is relatively broad, comprising the investigation of liquid entering inside the tubes, their subsequent filling, the overall flow through tubes, and the wetting of the nanotube walls. Flow of liquids through porous membranes can be predicted using the Hagen-Poiseuille equation [31]. The basic assumptions of this equation are laminar flow and no-slip at the boundary layer; that is, the velocity of the fluid at the CNT wall is zero. Water appears to wet CNTs with flow governed by the Washburn equation, but not for subnanosecond timescales [30]. Another important variable is represented by CNT diameters. On one hand, much interest has been placed in liquids inside CNTs having a diameter of a few nanometers with the promise of observing noncontinuum fluid dynamics effects due to confinement. On the other hand, numerous researchers have focused on larger nanotubes (diameters between 10 and 100 nm, or above) for applications where handling of a larger volume of liquid is required, with the liquid not necessarily experiencing non-continuum behavior. The continuum approximation seems to break down below 10 nm in the case of water in CNTs. Smooth liquid-gas interface (meniscus) disappears in tubes with the diameter less than 8–10 nm, and an anomalous behavior of water is observed in 1–7 nm CNTs. It was shown that the water structure, such as the -OH bond orientations and the hydrogen bonding in the depletion region where the water concentration is less than 5% of the bulk value, has a direct effect on the enhancement of flow rates [32]. A unique combination of features such as hexagonal structure, interatomic distances of 0.142 nm, and water orientations with "free" OH bonds pointing to the wall and the decreased hydrogen bonding in the depletion region contribute to large flow rates in CNTs, in contrast to other materials such as polymer membranes with significant wall roughness. The average number of hydrogen bonds HB(*n*) was computed from the center to the surface of (30,30) and (20,20) CNTs, and these results are shown in Fig. 5.6 [33]. In the center of the tube, HB(*n*) is 3.7, which is equal to HB(*n*) for bulk water. As we approach the nanotube wall, HB(*n*) is reduced to about 2.1 in both (20,20) and (30,30) CNTs. Depletion of hydrogen bonds near the surface is due to the geometrical configuration, which prevents the hydrogen

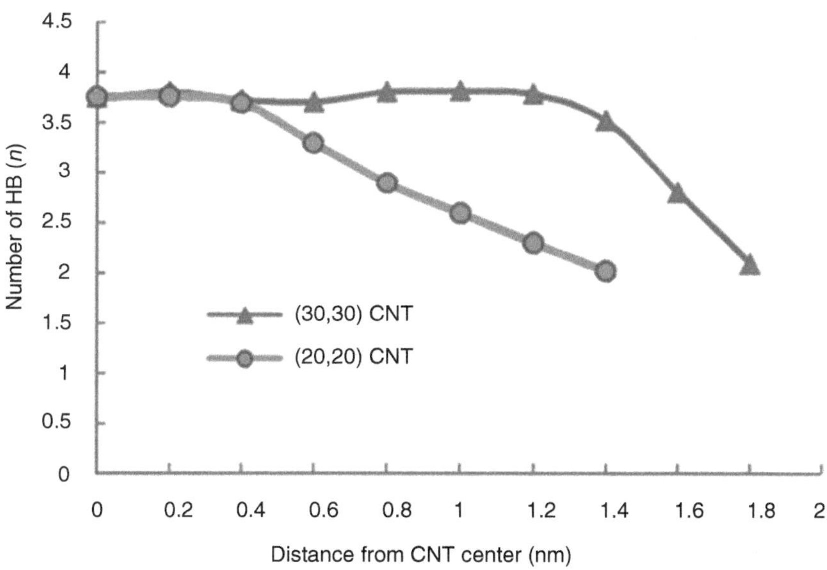

FIGURE 5.6 Spatial variation of the average number of hydrogen bonds [HB(*n*)] from the center to the surface of the CNT. *(Adapted from Ref. [33] with permission of American Chemical Society. Copyright 2011.)*

bonds from forming. For example, free OH bonds are observed near the surface (because of geometrical constraints), and these bonds are not involved in hydrogen bonding. These dangling OH bonds change the dynamics of molecules near the surface. The reduced number of hydrogen bonds provides more degrees of freedom for the water molecules, which results in enhanced diffusion near the surface of the CNT. The average diffusion coefficient in the nanotube as a function of the CNT diameter indicates that, as the diameter increases, the diffusion mechanism changes from "single-file" to "transition-to-normal" to "normal" to "transition-to-ballistic" to "normal" diffusion [33].

A dual-control-volume grand canonical molecular dynamics simulation study of the transport of a water and methanol mixture under a fixed concentration gradient through CNTs of various diameters and surface chemistries was also performed [34]. Carboxyl acid (—COOH) groups were anchored onto the inner wall of a CNT to alter the hydrophobic surface into a hydrophilic one. Results show that the transport of the mixture through hydrophilic tubes is faster than through hydrophobic nanotubes, although the diffusion of the mixture is slower inside hydrophilic than hydrophobic pores due to a hydrogen network. Thus, the transport of the liquid mixture through the nanotubes is controlled by the pore entrance effect, for which hydrogen bonding plays an important role. In contrast to water, which shows inhomogeneous variation of both translational and rotational diffusion with CNT diameter [35], the diffusion coefficient of acetonitrile changes uniformly and can be described by a simple analytic model [36].

Molecular dynamics simulations have suggested that CNTs have very low surface friction (almost frictionless interface) with respect to fluid flow, resulting in high fluid velocity [37]. This has also been confirmed by direct experiment on timescales, showing fast flow corresponding to large slip lengths. Generally speaking, liquid flow through the noninteracting hydrophobic CNT cores is found to be dramatically enhanced, with flow velocities ~1,000–10,000 times faster than predictions of liquid transport from conventional no-slip hydrodynamic predictions [30,31]. The extent of slip decreases with increasing hydrophobicity of the liquid, which is consistent with stronger interactions with the CNT surface reducing the flow enhancement. Finally, even if ionic liquids (ILs) and apolar CNTs are disparate objects; their interaction leads to spontaneous CNT filling with ILs. Moreover, ionic diffusion of highly viscous ILs can increase fivefold inside CNTs, approaching that of molecular liquids, even though the confined IL phase still contains exclusively ions.

It is worth noting that the surface chemistry and structure of nanotubes must be controlled with high precision to control flow rate and interaction of fluid components with tube walls. There are two main obstacles that researchers must overcome in order to achieve all these results. The first is understanding the physical origin (mechanism) of the nonclassical behavior of liquid under confinement. First, experimental results have confirmed some of the molecular dynamic conclusions, but a general theoretical framework is still missing. The second one is more technical and relates to improving the manufacturing techniques of nanofluidic devices based on CNTs, both regarding the uniformity of CNT functionalization and device integration.

2.2.1.2 Diffusion of Gases

Gas transport through mesoporous materials is generally a combination of (1) viscous or Poiseuille flow, (2) Knudsen diffusion, and (3) surface diffusion. Due to the intrinsic smoothness of CNT surface, the absolute gas flow through membranes of aligned double walled CNTs with pore size less than 2 nanometers largely exceeds (one or two order of magnitude) predictions of the Knudsen diffusion model [38]. Compared to commercial polycarbonate membranes presenting pore sizes of 15 nm, the gas and also water permeabilities in these CNT-based membranes are several orders of magnitude higher. It was also found that transport rates in CNTs are orders of magnitude faster than in zeolites or in any microporous material for which experimental data are available [29]. The exceptionally high transport rates in CNTs are shown to be a result of the inherent smoothness of the nanotubes [6]. Gases are transported by a Knudsen-like transport mechanism but are ~15–30-fold faster than Knudsen diffusion predictions, due to highly specular reflection by the smooth CNT walls retaining the forward momentum of the gas. The dynamic of molecules in a confined system is often affected by the interaction between gas molecules and the confining walls. Generally, the wall of these porous mediums can be distinguished into two extremes: that is, a perfectly smooth wall or a very rough wall. When the gas molecules are transported through a perfectly smooth wall, the collisions of the particles are completely elastic and the component of the velocity parallel to the wall remains constant. In such conditions, the diffusive behavior can arise only from collisions with other gas particles. On the other hand, when gas molecules passed through a rough wall, as in a zeolite confined system, the collision of the particles would be sufficiently random and hence allows the observation of diffusion behavior without collisions with other adsorbed gas particles. Despite some thermal fluctuations of the wall atoms that interact with the adsorbed molecules, the wall surface of CNTs is considered almost defect-free and close to a perfectly smooth wall. Therefore, the interaction of the gas with the inner walls of the CNTs can be described by a smooth potential energy surface. If many studies have indicated that the diffusion of light gases adsorbed as single components in CNTs is extremely rapid, an extremely rapid diffusion can also occur for mixtures of adsorbed molecules (CH_4/H_2 mixtures) [39].

By simulating the CNT as a flexible framework it has been demonstrated that the flexibility has a crucial influence on self-diffusion at low loadings [40]. The inclusion of nanotube flexibility reduces the transport diffusion relative to the rigid nanotube by roughly an order of magnitude close to zero pressure, but at pressures above about 1 bar the transport diffusivities for flexible and rigid nanotubes are very similar, differing by less than a factor or two on average [41]. Hence, the transport diffusivities are still extremely large compared to other known materials when flexibility is taken into account. The rapid diffusion of adsorbed molecules and the strong coupling between the fluxes of the adsorbed species in a mixture have interesting implications for uses of CNTs in membrane-based applications.

2.2.1.3 Concentration Gradients

In addition to an electron-poor environment, the internal cavity of CNTs composed of $sp^{2+\eta}$-hybridized carbon atoms presents a hydrophobic character. In that context, it can be anticipated that the confinement of hydrophobic and/or nucleophilic species inside the CNT cavity should be possible. The adsorption selectivity of molecules inside the tube is greatly driven by the different interaction potentials of the molecules with the graphitic carbon. Using molecular dynamics experiments, the analysis of equimolar mixtures of water and acetonitrile confined within CNTs was reported [42]. Two different CNT sizes were investigated: (8,8) tubes, with a radius of 0.55 nm, and (16,16) ones, with a radius of 1.1 nm. In the narrowest tubes, the cylindrical cavity is filled exclusively by acetonitrile; as the radius of the tube reaches ~1 nm, water begins to get incorporated within the inner cavities. In (16,16) tubes, the analysis of global and local concentration fluctuations shows a net increment of the global acetonitrile concentration; in addition, the aprotic solvent is also the prevailing species at the vicinity of the tube walls.

Adsorption properties of nitrogen and oxygen gases in SWCNT bundles were investigated by isotherm adsorption and XRD [43]. The results of this study indicate that the inside of nanotube has a stronger affinity for gas adsorption than the interstitial channels of bundles. This was confirmed in several experimental and modeling studies [44]. Grand canonical Monte Carlo simulations were performed to investigate the adsorption behavior of an equimolar CO_2/CH_4 mixture in CNTs [45]. The results show that the CNTs have a preferential adsorption of CO_2 in the binary CO_2/CH_4 mixture. For pore size effect on the adsorption behavior, it was found that the adsorption of CO_2 is much larger than that of CH_4 in the same CNT, and that the CO_2 adsorption in the CNTs increases dramatically with an increase of CNT's diameter, whereas the absolute amount of adsorbed CH_4 changes little with the CNT's pore size. The existence of a concentration gradient between the outside and the inside of the CNT has opened the way to applications as adsorbent and to hydrogen storage [46], but this molecule enrichment can also be exploited for chemical reactivity. Indeed, this finding might make possible to conduct reactions that conventionally require high pressures under much milder conditions. It has been shown that not only H_2 but also N_2 and CO can be enriched within CNT channels [47]. For example, at 300 K and 1 MPa, CO is enriched by 20 times within a (10,10) CNT: this means that the CO pressure within the CNT cavity is enhanced by a factor of 20 in an external CO atmosphere of 1 MPa. CO and O_2 enrichment has also been proposed to explain the high activity of confined Ru catalysts for CO oxidation in H_2-rich streams [48]. The selective enrichment of molecules with different properties in solution has also been reported [49]. Taking the hydroxylation of benzene to phenol as an example, the authors observed by means of solid-state NMR studies that benzene is enriched inside CNT channels while the product phenol was discriminatively expelled out of the channels. The benzene diffusion in the inner cavity of CNTs induces a local high concentration of this molecule. Comparing the reactivity of two rhenium catalysts (confined Re and Re dispersed on the external surface), the authors have observed a four-fold enhancement of the benzene conversion rate into phenol when using the rhenium-confined catalyst. This is the result, in part, of the local high concentration of benzene inside the CNT cavity, but also of the selective desorption of the hydrophilic phenol molecules. CNT-based catalysts may thus find applications in high-pressure reactions such as Fischer–Tropsch synthesis (FTS) [50–52], hydrogenation reactions [53,54], and some selective oxidations [55,56].

2.2.2 Spatial Restriction and Stabilization

It is well known that materials confined in nanoscale dimensions have properties that strongly differ from the properties of bulk systems. This is due to the reducing the dimensionality of the system and interface effects. Confining boundaries bias the spatial distribution of the constituent molecules and the ways by which those molecules can dynamically rearrange. These effects play important roles in the thermodynamics of the confined systems [57,58].

A simulation study of benzene molecules confined into SWCNT was reported [59]. It was found that the molecules of benzene mostly group in the middle distance from the axis of the tube to the wall. The molecules located in the vicinity of the wall demonstrate some deviation from planar shape. In the vicinity of the tube wall the benzene molecules have orientations, which are almost parallel to the tube axis. It was also reported that the confined space of CNTs and its interactions with peptides stabilizes the structure of biomacromolecules [60].

The relatively rigid channels of CNTs formed by the curved graphene walls can also affect the size and shape and properties of confined NPs. As a result, NPs can be dispersed there more homogeneously, with a uniform size. For example, it was reported in a study on confined Co, Ru, and Pd NPs that the confined NPs systematically show a smaller particle size than those supported on the external surface of the CNTs [61]. The configuration space of Pt NPs inside and outside CNTs shows distinct features [62]. Compared to the NPs outside the CNT, Pt NPs inside the CNT are structurally more flexible at low temperatures, but have much less liquid-like structure configurations at high temperature. These structural differences are reflected by the fact that the peak of the heat capacity curve for the NP inside the CNT shifts toward the low temperature region.

CNTs can be used as templates for the growth of low-dimensional inorganic materials whose structures and properties often differ greatly from those of the bulk. Indeed, it has been shown that the accommodation of nanosize metal species inside CNTs may lead to essential changes of the structural, magnetic, and electronic properties of the CNTs and the metal species [63]. For example, a cobalt diiodide nanostructure encapsulated within a SWCNT was reported [64]. This material has an unprecedented twisted double tetrahedral chain structure arising from a rotation of Co_2I_4 units along its length. The complete nanostructure comprises two distinct regions with oppositely handed helices separated by a short disordered region. Atomic chains and crystal of iodine were successfully generated in a controlled manner inside SWCNTs [65]. The structure is strongly dependent on the diameter of SWCNTs; the single, double, and triple helical structures becomes quite stable when the diameter of SWCNTs matches a specific size. More than three chains of iodine are not very stable, and they often crystallize inside the CNT when the diameter is larger than 1.45 nm. Covalently bonded selenium double helices were obtained within the cavity of double-walled CNTs [66]. The double-helix structure is completely different from the bulk atomic arrangement and may be considered a new structural phase of Se. Besides helicoidal nanostructures, nanowires (sometime atomic chains) were also produced. Nanowires of MoO_2 [67], Bi (1 nm inner diameter and 30 nm in length) [68], FeCo (2–5 nm, single crystals) [69], $Fe_{65}Ni_{35}$ (lengths <500 nm and widths <40 nm) [70], Gd [71], Au [72], Ag [72,73], Pt [72], and Pd (1.0–1.4 nm diameter) [72] have been produced inside CNTs. Ultrathin europium nanowires with diameters of a single atom were selectively produced within SWCNTs [74]. Because the nanowires are confined within and protected by the walls of the CNTs, the nanowires resist oxidation and structural disintegration even in ambient conditions. Individual atomic Mo chains were obtained inside double-walled CNTs, exhibiting inner diameters ranging from 0.6 to 0.8 nm [75]. These individual atomic chains form spontaneously within the hollow core of tubes in the absence of any reducing agent. However, quantum mechanical calculations have shown that Mo does not organize into linear chains but rather prefers to form four atom per unit cell nanowires that consist of a subunit of a Mo body-centered cubic crystal [76]. The model explains the 0.3 nm separation between features measured by high-resolution transmission electron microscopy, and why the nanotube diameter must be in the 0.70–1.0 nm range to accommodate the smallest stable one-dimensional wire.

2.3 Combination of Effects

Obviously, if the effects previously cited may operate independently, it is also possible that a combination of these primary effects can lead to a specific secondary confinement effect. For example, previous studies revealed that confinement inside CNTs leads to facilitated reduction of a series of metal oxides, such as iron [51,77,78], ruthenium [79], and cobalt [80], and retarded oxidation of metallic iron [78] with respect to their counterparts located on the outer walls of CNTs. This was attributed to the electronic interactions of the confined materials with the interior surface and the spatial restriction on the particle aggregation under reaction conditions. Iron nitride NPs were synthesized by confinement inside CNTs, which facilitated the formation of cubic iron nitride with a high

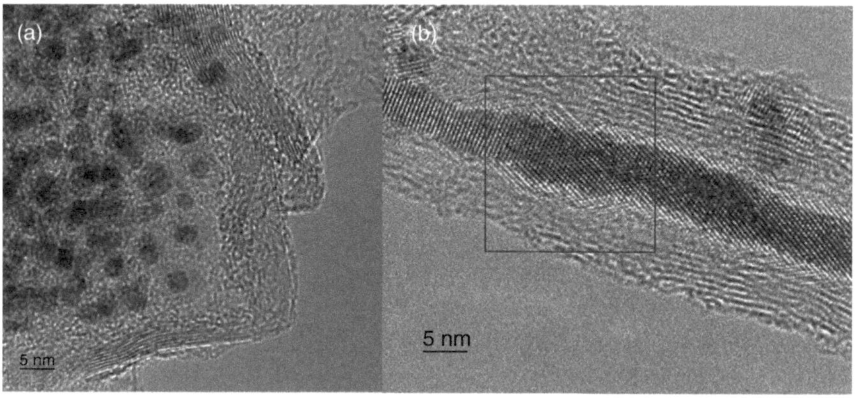

FIGURE 5.7 Confined (a) PtCo NPs in large diameter CNTs, and (b) PtCo (Pt rich) PtCo nanowires in small diameter CNTs.

N content and exhibited better resistance to oxidation and hence a much higher activity than reduced iron catalyst [81]. PtCo and PtFe NPs and nanowires have been selectively confined in CNTs (Fig. 5.7). A marked confinement effect has been evidenced, both on bimetallic nano-object shape and composition. Large diameter nanotubes favor the confinement of small Co- or Fe-rich NPs (Fig. 5.7a), while in small-diameter CNTs Pt-rich nanowires are formed (Fig. 5.7b).

3 CHARACTERIZATION OF CONFINED SPECIES IN CARBON NANOTUBES

Determining the selectivity toward particles inside CNTs is not an easy task. Conventional 2D TEM is normally not applicable for a quantitative determination of inside/outside selectivity. The accurate identification of a particle's relative location inside or outside the MWCNT usually requires 3D TEM (TEM tomography) [53, 82–85], while sometimes a 2D TEM with tilting and rotational sample stages may also provide an estimate of the distribution between inside-particles and outside-particles [86,87]. Since most researchers have limited access to a 3D TEM, a simple analysis method is desired to extract selectivity information from normal 2D TEM images. Two methods were developed for the analysis of microscopic images from TEM so that the fraction of particles inside MWCNTs can be accurately characterized without 3D TEM [61,88]. These methods are based on the measurement of r/R, the ratio of the inside to outside diameters of the MWCNTs. Indeed, since the surface chemistry of MWCNTs is homogeneous around the outer circular perimeter of the tubes, the probability of locating a NP is constant around the outer circular perimeter.

With XPS analysis, metal NPs located in MWCNT cavity give no contribution in signal, due to the short mean free path of excited electrons in graphite (2.6 nm) and the relatively large thickness of the MWCNT (10–100 nm)

[53,87]. This loss in signal gives valuable information about the percentage of metal particles situated at the inner surface of the tube.

DFT calculations have shown that the NMR spectroscopy of molecules encapsulated within SWCNTs show a characteristic dependence on the chirality of the surrounding CNTs, with a weaker dependence on the nanotube diameter [89]. The primary difficultly in the measurement of the NMR spectra of CNTs is the large magnetic inhomogeneity arising from the residual metal catalyst in the sample. The measured spectra consist of a broadband centered on about 125 ppm, although more recent work has achieved narrower bands. Ring currents in the nanotube induce shifts in the chemical shift of the nuclei comprising the encapsulated molecule. The calculations suggest that the change in chemical shift (Table 5.1) of the encapsulated molecules is dependent on the surrounding nanotube, and relatively independent of the molecule itself.

The most striking feature of the results is a significant difference in the predicted shifts for armchair and zigzag nanotubes. The change in chemical shift on encapsulation in the zigzag nanotubes is typically 10–20 ppm greater than that for the armchair nanotubes. This effect is generally larger than the variations arising from the diameter of the tubes.

Other theoretical and experimental works have confirmed that NMR responses of molecules confined in CNTs are different from those of the molecules in other confining environments [49,90–92]. For water confined in (14,0)

TABLE 5.1 Computed Change in Chemical Shift Upon Encapsulation in SWCNTs (in ppm)

Molecule		(5,5)/6.9 Å	(8,8)/11 Å	(10,0)/8 Å	(12,0)/9.6 Å
He	^{3}He	−4.6	−3.7	−26.2	−18.1
HF	^{19}F	−3.0	−3.8	−19.6	−15.6
	^{1}H	−4.8	−3.7	−19.8	−15.0
CO	^{13}C	−0.4	−3.7	−26.3	−18.4
	^{17}O	−1.5	−3.6	−27.9	−17.7
CH_4	^{13}C	+0.3	−4.2	−19.9	−18.7
	^{1}H	−4.5	−3.8	−20.9	−18.3
C_2H_2	^{13}C	−1.4	−3.7	−25.7	−18.1
	^{1}H	−5.2	−3.6	−25.6	−17.7
C_2H_4	^{13}C	−6.6	−3.8	−25.5	−18.3
	^{1}H	−5.0	−3.7	−26.0	−18.0
H_2CO	^{13}C	−1.0	−3.6	−17.4	−15.6
	^{1}H	−5.4	−6.7	−19.8	−16.2

Reprinted from Ref. [89] with permission of AIP Publishing LLC. Copyright 2008.

and (19,0) CNTs with diameters $d = 11$ and 14.9 Å, respectively, the confinement leads to a large (c. −23 ppm) upfield shift relative to bulk liquid water. This shift is a consequence of strongly anisotropic magnetic fields induced in the CNT by an applied magnetic field. A study by Bao and coworkers suggested that the chemical shift can be used as a probe to distinguish the molecules inside and outside of the CNT channels, as well as the type of CNTs (such as metallic and semiconducting) [90].

It was also demonstrated that confinement has a profound effect on the IR spectra of the internally adsorbed species. The spectral changes relate to the enhanced binding of the adsorbates to the nanotube interior walls and to the spatial limitations that prohibit formation of bulk-like structures. An experimental vibrational spectroscopy study has evidenced the existence of a water phase inside SWCNTs that exhibits an unusual form of hydrogen-bonding due to confinement [93]. Water adopts a stacked-ring structure inside CNTs, forming intra- and interring hydrogen bonds. The intraring hydrogen bonds are bulk-like while the interring hydrogen bonds are relatively weak, having a distorted geometry that gives rise to a distinct OH stretching mode. The experimentally observed infrared mode at 3507 cm^{-1} is assigned to vibrations of the interring OH-groups based on detailed atomic-level modeling (Fig. 5.8).

CF_4 adsorbs on closed SWCNTs, exhibiting its ν_3 asymmetric stretching mode at 1267 cm^{-1} (red shift relative to the gas phase, 15 cm^{-1}) [94]. When adsorption occurs in the interior of opened SWCNTs, internally bound CF_4 exhibits its ν_3 mode at 1247 cm^{-1} (red shift relative to the gas phase, 35 cm^{-1}). Qualitatively similar red shifts were seen in the case of adsorption of NO on SWCNTs [94]. Upon NO adsorption, the molecules dimerized, forming $(NO)_2$, with the asymmetric stretch vibration for the interior-adsorbed dimer occurring at 1257 cm^{-1}, while the gas phase frequency for this vibration is known to be 1289 cm^{-1}. The vibrational frequency of the Si—H bond in octasiloxane, $Si_8H_8O_{12}$ (a cube-shaped molecule), normally appears as a sharp band at 2277 cm^{-1} in solution. Upon insertion in CNTs, the Si—H absorption band is

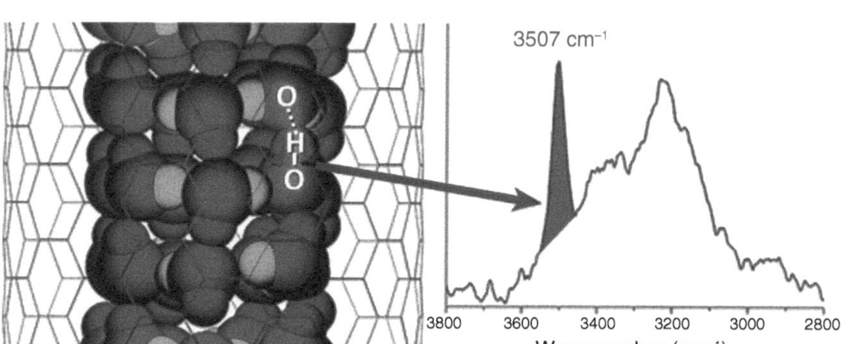

FIGURE 5.8 Infrared spectrum of confined water molecules. *(Adapted from Ref. [93] with permission of American Chemical Society. Copyright 2006.)*

substantially broadened for $Si_8H_8O_{12}$@SWCNT and its center is red-shifted by c. $-15\ cm^{-1}$ [95]. The C—C vibrations in ethane and propane and the symmetric stretch in SF_6 were also seen to produce red-shifted features in Raman spectra following adsorption on MWCNTs [96]. The energy of the interaction of $CoCp_2$ ($Cp = \eta^5 - C_5H_5$) with a CNT was predicted to be only about 0.1 eV [97]. However, the nature of the $CoCp_2$/CNT interaction appears to be more complex than pure van der Waals forces. $CoCp_2$ and $Co(CpEt)_2$ in SWCNTs were probed by UV-vis absorption spectroscopy, revealing that the oxidation state of Co is changed from +2 to +3 upon insertion of these metallocenes into SWCNTs. Normally, p-doped SWCNTs under the ambient conditions withdraw a valence electron from Co ion, creating positive charge on the molecule and negative charge on the nanotube.

It was also shown that thermal desorption kinetics clearly resolve adsorbates bound in different sites on SWCNT bundles [98]. The molecules $n\text{-}C_9H_{20}$ and CCl_4 were compared and it was found that the CNT internal sites exhibited the highest desorption temperature, whereas external groove sites exhibited the next highest desorption temperature for both molecules. A similar tendency was observed by temperature-programmed desorption measurements for several linear and branched alkanes [98]. When $n\text{-}C_9H_{20}$ and CCl_4 coadsorb, the more strongly bound $n\text{-}C_9H_{20}$ quantitatively displaces CCl_4 from internal sites to groove sites.

Finally, adsorption kinetics of various species confined in CNTs have been reported [99–104]. Formation of two condensed phases of Ar in the interior of SWCNTs has been observed at 77 K [101]. The similarity between the hexane adsorption isotherms and those of other organic molecules demonstrates that the adsorption mechanisms are not specific to hexane, and that this methodology can be potentially applicable to other sorbates [99].

4 SYNTHESIS OF CONFINED METAL NANOPARTICLES IN CARBON NANOTUBES

A few years after the observation of CNTs by Ijima [105], some studies of the opening, wetting, capillarity, and filling of CNTs appeared in the literature [106–109], following that, several strategies to confine NPs inside CNTs have been developed; and nowadays the relatively selective confinement of discrete NPs inside the CNT cavity is achieved in a straightforward manner. The methods to fill CNTs include wet chemistry, simultaneous growth of CNTs and filling of the inner cavity by arc-discharge, chemical vapor deposition (CVD) or pyrolysis, filling of volatile or melted compounds, and the use of a supercritical medium. Even if early studies on filling CNTs led to the formation of metallic nanowires [107,109,110], the use of wet chemistry methods rapidly allowed obtaining the confinement of discrete NPs inside CNTs [111]. The simplicity, versatility and up-scalability of this method has made it the most widely used to synthesize this kind of material [112].

4.1 Wet Chemistry Method

4.1.1 Wet Impregnation From a Metallic Precursor Solution

The wet chemistry method consists in the filling of CNTs by capillary forces of a solution containing a metallic precursor, followed by a thermal and/or a hydrogenation treatment to yield the NPs. Ebbesen [113] established that low surface tension solvents ($<$100–200 mNm^{-1}) can fill the channels at atmospheric pressure by capillarity. However, the capillarity effect depends on the surface functionalization and CNT diameters. SWCNTs, DWCNTs, and MWCNTs [83,114] with small internal diameters ($<$10 nm) are difficult to fill through this method, and filling with volatile precursors in vapor phase is more suitable. However, SWCNTs [115], DWCNTs, [116] and small-diameter MWCNTs [51,77–79, 86,117–121] can be filled by wet impregnation assisted by ultrasonic treatment and stirring.

Generally, CNTs used in wet chemistry are purified and opened by HNO_3 oxidation and used without any further treatment. CNTs oxidized and cut via oxidation catalyzed by Ag also provide good confinement of Ru NPs [117,122]. CNTs first purified by HNO_3 are impregnated with $AgNO_3$ and thermal treated under Ar, to decompose the Ag precursor to Ag NPs. Next Ag/CNTs are cut by oxidation using an O_2/Ar flow. This treatment permits the modulation of the CNT length varying the Ag loading.

Recent studies point out that removing the oxygenated functional groups on the CNT surface before performing the filling favors the synthesis of NPs inside CNTs by giving more hydrophobic surfaces and avoiding the attachment of the metal species through the oxygen functionalities at the outside walls of the nanotubes [123–125]. The oxygen groups are easily removed after the acid-purification step by thermal treatment at high temperatures under an inert gas flow. The effect of the thermal treatment temperature of CNTs support on FTS was investigated in cobalt-based catalysts [124]. CNTs treated at 650°C showed better catalytic activity, lower CH_4 selectivity, and higher C_{5+} hydrocarbonselectivity compared with other cobalt-supported CNTs catalysts prepared by employing thermal treatment at other temperatures, or the untreated ones containing oxygenated groups. On the other hand, it has been also reported that the confinement of Ru, Co, and Pd NPs was more effective in an oxidized CNT support than in CNTs treated at 900°C [61].

Confined metal NPs in CNTs synthesized with this method are most often used as catalysts. In order to better understand the effect of confining nanocatalysts inside the CNTs channels, it is essential to synthesize a catalyst for comparison with metal NPs dispersed exclusively on the outer surface of CNTs. The first reported studies achieved the synthesis of NPs entirely supported outside the CNT cavity by keeping the CNTs' ends closed [48,51,77,78,86,119,121,126]. CNTs with closed tips are also obtained by HNO_3 oxidation under mild oxidation conditions (diluted HNO_3, short reaction times) in order to remove exclusively the catalysts and the amorphous carbon coming from the CNT synthesis.

Nevertheless, CNTs produced in this way may exhibit different characteristics from the opened CNTs due to the different treatment, such as purity, surface area, and number and type of surface oxygen groups. These differences can affect the dispersion and properties of metal catalysts, and therefore the catalytic performance. Temporarily blocking the CNTs' channels with a solvent can overcome this issue of standard wet chemistry procedure and exclusively disperse the NPs outside the open CNTs. This strategy, first described by Bao [117], made it possible to obtain only the exterior surfaces of opened CNTs decorated with Ru NPs. The CNTs were filled with xylene, which is immiscible with the reaction solvent (water), preventing the aqueous $RuCl_3$ solution from infiltrating the channels. In addition, the higher boiling point of xylene with respect to water allows the preferential evaporation of water on the outside and deposition of $RuCl_3$ on the exterior surface of CNTs. The two-step biphasic impregnation has been widely used to produce selectively decorated NPs on the outer surface due to its simplicity, as it avoids any additional step of selected functionalization, which is difficult, time consuming, and altering the surface chemistry of the CNT walls. In that way FeCo [127], Ni, [128] Fe [129], Ru [48,79,117,118,122,130], Pt [54,131], and Pd [132] NPs have been selectively decorated on the inner and outer surface of CNTs. Xylene is the most used solvent to block the CNT channel; however, other solvents such as acetone [130], water [129] and benzene [127] have been successfully used. Indeed, it has been shown that parameters such as boiling point and miscibility in water have no crucial effect on the selectivity of the deposition, and it is the difference in the liquid/solid interface energies of both solvents with the CNT surface which is the decisive parameter [128].

Another strategy to fill the inner cavity of CNTs is the use of anodic aluminum oxide (AAO) template. First, an AAO template is synthesized and the CNTs selectively grown in the channels by CVD, permitting CNTs with uniform diameter and length. Next, the metallic precursor solution is introduced in the CNTs, followed by a calcination/reduction process to produce the NPs [133–136]. Finally, the AAO template is dissolved in NaOH and the NPs confined inside CNTs can be recovered. Even if this method produces uniform CNTs it suffers of scale-up difficulties.

4.1.2 Wet Impregnation From Preformed NPs Suspensions

The filling of CNTs with preformed NPs has also been achieved. This wet chemistry method makes it possible to confine preformed NPs with controlled surface properties inside the CNTs. Early studies demonstrated that preformed NPs can fill the CNTs; for instance, commercial ferrofluids can wet the inside walls of CNTs synthesized in AAO templates [137], or fluorescent polystyrene can be selectively filled by wet impregnation [138,139]. Oxidized CNTs were also filled with a commercial ferrofluid solvent; however, the deposition was not selective, and an extra washing step was needed to remove NPs from the outer walls of the CNTs [140]. Preformed gold NPs were deposited on several

CNTs presenting different surface groups due to different surface treatments. However, the deposition inside the CNTs was not completely selective [141].

In order to avoid the unselective deposition on the outside wall of the CNTs, the interaction of the preformed NPs with the inner surface and the repulsion with the outer wall has to be favored. In that sense, functionalization of CNT external surface by introducing a long alkyl chain amide group blocks the outer surface for NP deposition [53]. In addition, PtRu preformed NPs synthesized by an organometallic route [142] in the presence of a ligand possessing two functionalities—one has an affinity for the surface of the NPs, and the other for the inner surface—have been selectively confined in CNTs (Fig. 5.9).

This simple and selective procedure, based on molecular recognition, permitted to build a nanoreactor consisting of approximately 2 nm PtRu NPs selectively confined in segmented CNTs of 40 nm internal diameter (Fig. 5.9). Preformed gold NPs stabilized with isocyanide ligands [143] and PtRu NPs synthesized in supercritical carbon dioxide ($scCO_2$) [144] have also been successfully confined using long alkyl chain functionalized CNTs.

In summary, the wet chemistry method has proven extremely versatile in confining a plethora of different NPs inside the CNT channels. This strategy allows selective introduction of metallic precursor solutions or preformed NP suspensions in CNTs through playing with experimental parameters such as the nature of the solvent, ultrasonic and extended stirring, introduction of selectively washing

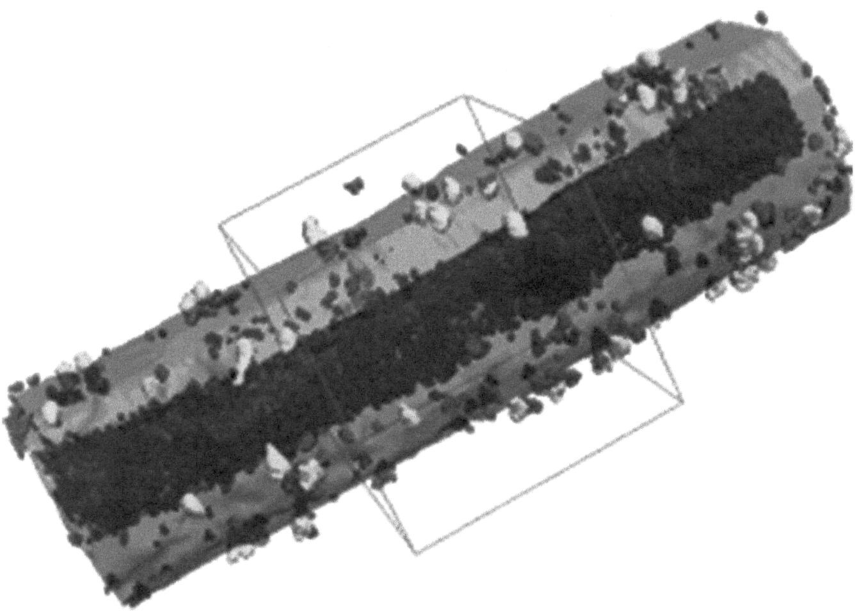

FIGURE 5.9 Modeling of the reconstructed nanoreactor volume by 3D TEM. The white NPs are Au NPs deposited on the CNT surface for image reconstruction. *(Adapted from Ref. [53] with permission of Wiley.)*

step, and controlling the surface chemistry of the CNTs. Usually, MWCNTs with diameters greater than 10 nm are filled by this procedure, and even if SWCNTs and small MWCNTS can be successfully filled using wet chemistry, filling of small tubes by gas phase techniques seems to be more appropriate.

4.2 Use of Melted Compounds

Capillary forces also drive melted compounds inside the CNTs. This procedure consists in the melting of metals or metal salts at high temperature (above the melting point) under high vacuum in the presence of opened CNTs. This simple method allows production of nanowires inside the CNTs, and usually SWCNTs are filled in this manner. The first observation of this capillarity filling was made with lead [107]. Pb NPs were first deposited on the outer surface of MWCNTs. The grid containing the sample was annealed at 400°C (above the melting point of Pb), and the examination of the sample by TEM showed that nanowires were formed in the interior of the CNTs. Opened CNTs could be filled with melted $AgNO_3$ by capillary forces by heating both compounds at 230°C for 30 min ($AgNO_3$ melting point 212°C). The filling efficiency was low, and only tubes with inner diameters greater than 4 nm were filled. The subsequent decomposition of the salt in situ in the TEM produced Ag nanowires [145]. The low filling of the nanotubes is overcome by applying high vacuum and high temperature to the CNT-metal mixture. $ErCl_3$ has been driven inside SWCNTs using the following procedure: open-ended SWCNTs vacuum-sealed (10^{-7} Torr) in a quartz tube with anhydrous $ErCl_3$ were heated at 800°C for 72 h. The compound was then washed with MeOH in order to remove any $ErCl_3$ adsorbed on the outer surface of the SWCNTs and then dried at 80°C for 12 h. Stable and crystalline $ErCl_3$ nanowires were obtained with high filling ratio (90%) [146]. The confined $ErCl_3$ nanowires can be treated at high temperature under vacuum in order to obtain Er nanowires confined inside the SWCNTs, although a decrease of filling yields was detected [147]. Ultrathin Eu nanowires of a single atom encapsulated in SWCNTs in a very high filling were also produced using this procedure [74]. The CNT walls protect the nanowires against oxidation and structural disintegration, allowing investigation in detail of the structure and properties of the nanowires, and making them excellent candidates for use in electronic devices. Nanowires of GeTe [148] were also confined inside SWCNTs, as well as GdI_3 [149] inside MWCNTs.

4.3 Use of Volatile Compounds

C_{60} peapods, C_{60} encapsulated inside a CNT, were first produced and observed during CNT synthesis by pulsed laser vaporization [150]. Nowadays the synthesis of encapsulated fullerenes [151], endohedral fullerenes [71,151–156] and volatile organometallic complexes of fullerenes [157] within the CNTs is mainly achieved by vapor phase filling technique, consisting in heating at the

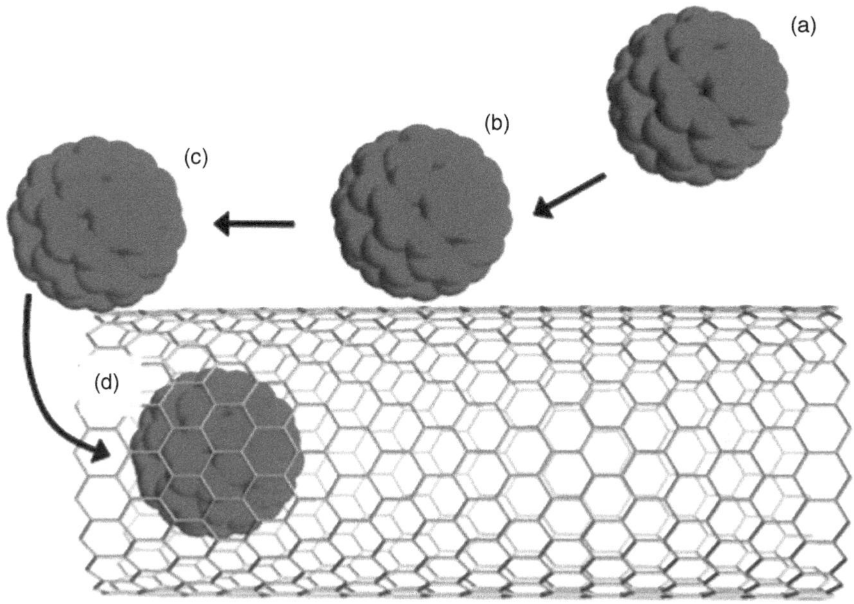

FIGURE 5.10 Stages of CNT filling with fullerene C_{60}*via* the vapor phase mechanism. (a) Fullerene in the gas phase, (b,c) fullerene adsorbed on the nanotube surface diffusing toward the open terminus, (d) fullerene encapsulated inside the nanotube. *(Reprinted (adapted) from Ref. [158] with permission of American Chemical Society. Copyright 2010.)*

desired temperature, usually under static vacuum, the fullerene and the CNT (Fig. 5.10). Typically, vapor phase filling with C_{60} leads to high filling yields of up to 100%. However, recent publications have also described the successful synthesis of this kind of material using wet chemistry methods [158–160] or $scCO_2$ [161]. The vapor phase filling technique has been proven to be suitable to introduce small molecules such as adamantane [162] or octasilasesquioxane [163] ($H_8Si_8O_{12}$) inside small CNTs, and also volatile organometallic complex and clusters. First examples of CNT filling with metallic complexes by gas phase processes were described for metallocenes. Cobaltocene ($CoCp_2$) have been introduced in SWCNTs [97], and ferrocene ($FeCp_2$) in DWCNTs [164] and SWCNTs [165], using this procedure. Also, cobaltphthalocyanine complex has been introduced within 2 nm diameter MWCNTs [166]. All these complexes have been observed inside the CNTs by HRTEM techniques.

Discrete NPs can be confined in small CNT channels (SWCNTs, DWCNTs, or small-diameter MWCNTs) by gas phase deposition. First, volatile metallic complexes are introduced in the CNTs followed by a decomposition step (annealing, oxidation, or reduction) to give ultrasmall metallic (or metal oxide) NPs due to the restricted space.

The introduction of volatile Re_2O_7 clusters inside SWCNTs made it possible to obtain, first, isolated rhenium oxide NPs of 1 nm size, and then, by H_2

reduction, small Re NPs that sometimes aggregated to give small nanowires [167]. Similarly, TiO_2 NPs were synthesized inside DWCNTs by vaporization of $TiCl_4$ in the presence of the CNTs, followed by a hydrolysis step with EtOH [168]. Carbonyl complexes of transition metals ($[M_x(CO)_y]$, where $x = 1, 2,$ or 3 and $y = 6, 10,$ or 12 for M = W, Re, Os [169] or Ru [170], respectively) inserted into SWCNTs (diameter 1.5 nm) by gas phase transformed into small metallic NPs under heat treatment or electron beam irradiation. The authors pointed out that the synthesis of discrete NPs inside the CNTs was possible due to the release of CO during the decomposition step, as they suggest that the gas creates pockets of high pressure between NPs, preventing their collision and coalescence into larger structures. $[Re(CH_3)O_3]$ give also discrete Re NPs inside SWCNTs [17] and DWCNTs [49] after vaporization. The Re NPs inside MWCNTs [171] aggregated gradually by successive steps of annealing to give ultrathin nanowires within the confined space of CNTs with a diameter less than 1.5 nm with an unusual face-centered cubic (fcc) structure along the CNTs (usually Re presents a hexagonal close-packed (hcp) structure). Discrete small Pt NPs within small diameter CNTs were also obtained by hydrogenation of the $[Pt(Cp)(Me)_3]$ complex introduced previously inside the CNT channels [172].

4.4 Use of Supercritical Medium

Filling CNTs with functionalized fullerenes cannot be achieved by gas phase synthesis, since their functional groups decompose at the temperatures generally used for filling empty fullerenes (400–600°C). The use of $scCO_2$ allowed introducing in CNTs this kind of compound at low temperature with high yields. $C_{61}(COOEt)_2$ and $C_{61}(COOH)_2$ were first introduced in SWCNTs by immersing the fullerene-SWCNTs mixture in $scCO_2$ at 50°C under 150 bar and keeping the supercritical conditions several days [173]. The same procedure has been successfully used to fill SWCNTs with other fullerenes, such as C_{60} [161], $C_{60}O$ [174], and long alkyl chain functionalized C_{60} [175].

The use of supercritical fluids has also allowed synthesizing discrete metal NPs inside the CNTs. Ag_2O was thermally decomposed and mixed to MWCNTs. The mixture was then placed under supercritical water (scW), allowing Ag NPs to open the CNTs and fill the cavity of the tubes [176]. Recently, $RuCl_3$ was introduced inside MWCNTs by $scCO_2$ using MeOH as assisting solvent. Once the Ru precursor was inside the channels, it was reduced under a H_2 flow to obtain small Ru NPs (2–4 nm) confined inside the MWCNTs [177].

4.5 Simultaneous Growth of CNTs and Filling of the Inner Cavity

With this method, the tubes are generated along with simultaneous filling by the doped element. The major drawbacks are the inability to control the size of the filled material, which generally presents nanowire morphology, and the fact that the resulting metallic NPs are often encapsulated inside the CNTs and will not be accessible to reactants for catalytic reactions.

Arc-discharge synthesis of CNTs in the presence of several metals has been discussed for filling CNTs [110]. Mainly nanowires were obtained using this procedure and some of them do not completely fill the CNT.

The pyrolysis or chemical vapor deposition of a metal precursor, a volatile complex such as $FeCp_2$, $CoCp_2$, or $NiCp_2$, and a carbon source, such as acetylene, toluene, dichlorobenzene, or C_{60}, also permits the simultaneous synthesis of CNTs and confined metal nanowires. Encapsulated nanowires of Fe [56,178–181], FeCo [69], and FeNi [182] within CNTs were synthesized using elevated temperatures. When $[Co(CO)_3NO]$ was used both as catalyst and the source for CNT growth, the nanowires and the filling yields were poor [183].

A recent work showed that Fe NPs can be encapsulated into the compartments of pea-pod–like CNTs through one-step synthesis. $FeCp_2$ and NaN_3were heated together at 350°C for 6 h in a stainless steel autoclave under nitrogen. By this simply method well-defined nanocatalysts were produced and applied in the oxygen reduction reaction (ORR) [184].

In conclusion, several methods are used nowadays to introduce metals inside CNTs. Among them, the wet chemical method is the most versatile and simplest for production of discrete metallic NPs inside the CNT cavity. However, small CNTs (<10 nm) are usually difficult to fill by wet impregnation. The gas phase synthesis or the use of supercritical fluids can overcome this issue, although the metallic precursors have to be either volatile or soluble in the supercritical medium. The fact that discrete NPs can be produced inside the tubes using these methods explains why materials synthesized in these ways have found applications in catalysis. On the other hand, the introduction of metals inside the CNTs by melting metallic precursors, giving nanowires, or simultaneously synthesizing CNTs and metallic NPs, thereby producing encapsulated and inaccessible NPs, explains why their application in catalysis is limited.

5 CHEMICAL TRANSFORMATIONS INSIDE CARBON NANOTUBES

The control of chemical reaction pathways at the molecular level presents one of the most important scientific challenges to achieve fully sustainable, thermodynamically efficient chemical processes. Molecular simulation studies of chemical equilibrium for several reactions in pores of slit-like and cylindrical geometry have shown a significant effect of the confinement on the equilibrium compositions, with differences of several orders of magnitude with respect to the bulk fluid phase in some cases [185]. There are many examples in host–guest chemistry where the inner cavity of supramolecular cavitands have been used to emulate enzymatic pockets to accelerate reactions or gain unusual regioselectivity. Another space that could be utilized for controlling stereochemistry is the hollow region inside a CNT, which has already been demonstrated to be an effective vessel for filling with molecular nanomaterials. Owing to the spatial confinement imparted by the CNTs, we have shown that novel and distinct physical and chemical properties are found for the encapsulated materials. Thus, chemical reactions inside CNTs can yield unusual outcomes. Despite the plethora of research on nanomaterials encapsulated within CNTs, there have been only a few studies illustrating the use of the inner space of CNTs to control noncatalytic chemical reactions. Catalytic transformation, however, have been much more studied.

5.1 Noncatalytic Chemical Reactions

Metal-free reactions represent a direct way to study the confinement effects induced by the nanotube itself. Few such reactions have been carried out within CNTs, but notable examples include the formation of linear structures, such as fullerene [151] and fullerene epoxide oligomers [174], formation of DWCNTs from endohedral fullerenes encapsulated in SWCNTs [71,155,156], or from encapsulated ferrocene [165], graphene nanoribbons (GNR) [186,187], the transformation of [Fe($C_{60}Me_5$)Cp] into encapsulated C_{70} [157], or the transformation of adamantane [188] and functionalized diamantine [189] in carbon chains.

Small molecules encapsulated within CNTs can undergo reactions in only one direction by applying heat or electron beam irradiation, the latter making it possible to follow the reaction in situ by HRTEM techniques. Fullerene epoxide, $C_{60}O$, encapsulated within SWCNTs, forms strictly linear polymer chains within the nanotube, whereas the same reaction taking place outside the nanotube leads to the formation of branched, highly convoluted polymers [173]. Dimerization of fullerene C_{60} was visualized by means of TEM, with images of the motion of fullerene during the dimerization process revealing the specific orientations of the interaction and the bond reorganization after the initial contact [151]. Annealing or e-beam exposure of coronene-filled SWCNTs with diameters close to the size of coronene results in the formation of internal nanotubes through complicated polymerization steps. These might involve dehydrogenated nanoribbon formation as intermediate structures; the same treatment in MWCNTs yields polymerized coronene products and no internal nanotubes [190].

Ferrocene encapsulated within SWCNTs was annealed at different temperatures; the study reveals that annealing at different temperatures leads to the selective growth of iron doped and nondoped DWCNTs [165]. Similarly, Gd nanowires inside DWCNTs can be obtained from heat treatment of encapsulated Gd@C_{82} endofullerene within SWCNTs [71]. Controlling the SWCNTs diameter, the authors were able to produce single Gd atomic chain, one-dimensional alignment of Gd squares, and Gd nanowires that correspond to a one-dimensional segment of the bulk close-packed structure. Reactions of endohedral fullerenes [155,156] and [Fe($C_{60}Me_5$)Cp] [157] encapsulated inside SWCNTs promoted by electron beam irradiation were imaged in situ by electron microscopy. La@C_{82} peapods formed a LaC_2 cluster attached to the CNT inside the CNT host. The interaction of the cluster with the CNT walls was also examined, revealing that the cluster can open the CNT walls, exit, and seal the hole in the CNT wall [155]. $Pr_2@C_{72}$ peapods inside thin-walled CNTs also formed a new shell of carbon together with the formation of small Pr nanoclusters [156]. The process tracked in situ by TEM revealed the mechanism of formation of the nanoclusters: the nanocrystals were formed through aggregation of several small clusters instead of by the addition of single atoms to one pre-existing cluster. [Fe($C_{60}Me_5$)Cp] transformed into fullerene C_{70} under electron beam, and the C—C bond reorganization has also been followed in situ under TEM conditions [157].

Synthesis of graphene nanoribbons [187] and sulfur-terminated GNR [186] was possible through the use of SWCNTs as nanoreactors. The polycyclic aromatic hydrocarbons coronene and pyrelene were encapsulated inside SWCNTs by gas phase deposition and then annealed. TEM analyses revealed the formation of GNR inside the SWCNTs displaying different shapes depending on the precursor used. Similarly, sulfur-terminated nanoribbons can also be formed inside the SWCNTs from encapsulate sulfur-containing organic molecules by annealing at 1000°C or under electron beam irradiation.

C—C bond reorganization in confined space is also possible with sp^3-hybridized carbon-based molecules. Adamantane was confined in DWCNTs forming linear adamantine assemblies [188]. The annealing of the sp^3-hybridized nanostructures under vacuum at temperatures higher than 500°C convert the adamantine into carbon chains inside the CNTs; however, the introduction of H_2 during the annealing step suppressed their formation. Similarly, diamond nanowires formed inside DWCNTs from encapsulated diamantane dicarboxylic acid [189].

A quantum mechanical study of the D + H_2 hydrogen exchange reaction in a CNT has shown that the chemical reactivity can be enhanced considerably compared to the gas phase for the smaller CNTs studied [191]. The origin of this enhancement is mostly steric in nature and has to do with alignment of the H_2 along the long tube axis. It has been demonstrated that the confinement of physisorbed 1-heptene molecules inside of SWCNTs results in lowering of their reactivity toward atomic hydrogen compared to 1-heptene molecules adsorbed on external SWCNT sites [192]. It was found that CNT walls prevent atomic H species from entering and undergoing reaction with the double bonds of confined alkene molecules.

In an effort to gain insight into the effect of CNT confinement on reaction enthalpies and activation energies, calculations using hybrid-density functional theory have been carried out for the Menshutkin S_N2 reaction in gas phase and inside CNTs. The polarizability of the CNTs provides an interaction mechanism, which leads to the stabilization of confined dipolar species. Inside hydrogen-terminated (8,0) and (9,0) CNTs, the activation energy and reaction endothermicity for the Menshutkin S_N2 reaction are significantly reduced, compared to those in the gas phase [193]. The CNT-confined chloride exchange S_N2 reaction for methyl chloride has been examined using DFT. It was found that inside the CNT the reaction is disfavored (E_a = 17.9 kcal mol^{-1}) with respect to the gas phase (E_a = 13.2 kcal mol^{-1} in the gas phase). The augmented barrier, with respect to the gas phase, can be ascribed to a complex interplay between Cl···π and C—H···π interactions; that is, interactions of the two Cl atoms and the C—H bonds of the substrate with the carbon electron cloud of the tube wall [194].

Molecular dynamic simulations show that the confined space of SWCNTs may influence both the kinetics and thermodynamics of the Diels–Alder reactions between anthracenes and N-substituted maleimide guests, favoring an unusual regioselectivity. Thus, in the confined space of CNTs, the 1,4-exo adduct of a Diels–Alder cycloaddition may be produced instead of the 9,10-adduct, which is favored in bulk. It is possible that the nanotube chirality could influence the outcome of the reaction, as this could alter the preferred orientation of the reactants inside the SWCNT. However, it was suspected that the diameter of the SWCNT is likely to have by far the largest influence on the reaction outcome, as this will have the biggest influence on how close the reactants can approach each other [195].

5.2 Catalytic Chemical Reaction

Catalysis, as a key and enabling technology, plays a pivotal role in fields from energy and the environment to fine chemistry. Rational design and synthesis of highly active, selective, and stable catalysts has become the ultimate goal of catalysis research. Reactions or molecules confined in nanochannels are quite different from those outside of channels. Such confinement effects widely exist in CNTs, zeolites, and other mesoporous materials. The confinement provides a potential method to control the reactivity of catalysts dramatically. If catalytic NPs are placed within such an environment, the nanospace not only exerts a physical restriction on them (and on the reactions they facilitate), but also provides a unique electronic microenvironment that modulates electron transfer processes. An increasing number of studies reveal that confinement of metal or metal oxide NPs inside CNTs often leads to significantly enhanced catalytic activity with respect to the same bare metal NPs or those deposited on the outer walls of CNTs.

5.2.1 Gas-Phase Reactions

5.2.1.1 Fischer–Tropsch Synthesis

Cobalt- and iron-based catalysts have been both employed in industry for FTS. Iron catalysts can operate under a wide range of temperatures and H_2/CO ratios with low CH_4 selectivity, and exhibit higher catalytic activity for the water gas shift (WGS) reaction than cobalt catalysts. However, the iron catalysts deactivate rapidly. Cobalt catalysts are preferred due to their high activity, high selectivity to long-chain paraffin, low WGS activity, and high resistance to deactivation by water [196]. Confinement effects of cobalt- and iron-based catalysts supported on CNTs have been investigated for FTS.

Bao and coworkers have found that Fe_2O_3 redox properties are modified when the NPs are confined in CNTs. The reduction of Fe_2O_3 NPs is facilitated when supported inside of the CNT channels, compared to the ones supported on the outer walls [77], and the temperature of reduction decreases with the inner diameter of the CNTs [78]. A similar effect was observed for Fe_3O_4 nanowires confined inside the CNTs [197]. Keeping in mind that the redox properties can significantly affect the catalytic performances, the authors studied the effect of confinement of Fe_2O_3 in CNTs on the FTS [51]. An in situ XRD investigation showed that the distribution of iron carbide and oxide phases is modulated in the CNT confined system, and the iron species encapsulated inside CNTs exist in a more reduced state, tending to form more iron carbides under the reaction conditions, which have been recognized to be essential to obtaining high FTS activity. Indeed, FTS activity is increased when iron is confined within the CNTs and the selectivity toward C_{5+} hydrocarbons is higher. The authors point out that the modified redox properties play a role in the increased catalytic performances of the confined catalysts. Iron particles confined in CNT channels have been used as a catalyst for the direct conversion of syngas to light olefins [198]. Compared with iron catalysts supported on other materials, such as Silica-1, SBA-15 and carbon black, the CNT-confined catalyst exhibits a higher CO conversion and selectivity to the light olefins.

Similarly, iron oxide NPs confined inside the CNTs' channels revealed easier reduction of the oxide at a lower temperature than those supported on the outer wall [129]. Even if both in and out NPs exhibit similar conversion, confined NPs displayed higher selectivity to heavier hydrocarbons and were more stable. Cubic iron nitride FeN NPs were synthesized inside or outside CNTs [81]. The FeN confined within the CNTs exhibits higher activity than the nitride iron particles located on the CNT exterior walls in CO hydrogenation. The authors correlated this effect to the smaller particle size and the stronger retention of nitrogen atoms in the cubic nitride lattice, resulting in formation of more carbides in than out.

Effects of confinement have also been studied for Co based catalysts [50,52,124,199]. Co NPs supported on CNTs with several filling yields and modified with Ru and K were tested as catalysts in the FTS [199]. Increasing the Co loading increased the amount of Co on the outer surface of the CNTs, increased the cobalt cluster sizes, and decreased the reduction temperature and dispersion. This affected the conversion and selectivity of the reaction. Furthermore, Ru additive enhanced the reducibility of oxidized Co species, increased the dispersion and decreased the average cobalt cluster size, and enhanced the selectivity of FTS toward the higher molecular-weight hydrocarbons. Deactivation mechanism of Co catalysts supported and confined in CNTs was also investigated [52]. Two different kinds of deactivation were found: cobalt oxidation and sintering. NPs located inside CNTs displayed lower rates of sintering as compared with the particles located on the outer layers, as found by TEM investigations. The effect of the confinement and the size of the inner diameter of CNTs were investigated for Co nanocatalysts [50]. As in the case of iron, the TPR results indicated that it was easier to reduce the catalyst with Co particles inside CNT than the one with

Co particles outside CNTs, and the reducibility of cobalt oxide particles inside the CNT decreased as the cobalt oxide particle size increased. The activity of the nanocatalysts was enhanced when the Co NPs were confined inside of the CNTs. The diameter of the inner channel also affected the activity; smaller pore size increased the activity. The effect of the surface chemistry of the CNTs on FTS was also investigated with Co based nanocatalysts [124]. Oxygen groups present on the CNTs were removed by thermal treatment under an inert atmosphere. CNTs treated at 650°C showed better catalytic activity, lower CH_4 selectivity and higher C_{5+} hydrocarbon selectivity compared with other cobalt supported CNTs catalysts prepared by employing thermal treatment at other temperatures or the untreated ones containing oxygenated groups.

5.2.1.2 CO Preferential Oxidation

CO preferential oxidation (CO-PROX) has been investigated in order to remove the trace amounts of CO in H_2 rich stream, as remaining traces of CO in H_2 can easily poison the Pt based anode electrocatalyst of polymer electrolyte membrane fuel cells (PEMFC).

Preformed gold NPs deposited on several kinds of CNTs were tested as catalysts for the CO-PROX. No significant confinement effects were detected in these catalysts; however, promising results were obtained with Au NPs deposited over nitrogen-doped CNTs [143]. Confinement effects on Ru based nanocatalysts were also investigated [48,119]. Both studies reveal that Ru NPs confined in the inner cavity of the CNTs show better activities in the CO-PROX. The authors conclude that the channels of CNTs can selectively increase the density of reactants, and the encapsulation of Ru NPs in the channels of CNTs can provide a microenvironment to reinforce the reactivity of catalytic sites.

5.2.1.3 Other Reactions Involving CO

Confinement effects were also studied in the syngas conversion to ethanol, acetaldehyde, acetic acid, and C_2 oxygenates [86]. When an Rh–Mn catalyst was confined into the CNT channels, much higher conversion was reported than that for Rh–Mn dispersed on the outer CNT surface. TEM characterizations showed that the confinement prevented metal sintering, and that the Rh-Mn catalysts likely existed in a more reduced state than those on the exterior of the CNTs. In addition, TPD experiments showed that the H_2 desorption peak observed for Rh-Mn inside the CNTs was of higher intensity than that for Rh–Mn outside the CNTs, indicating a higher concentration of active hydrogen species inside the CNTs. This will further supplement the hydrogenation rate, leading to a significant increase in the overall yield of C_2 oxygenates.

5.2.1.4 Oxygen Reduction Reaction

Fe NPs encapsulated into the compartments of peapod-like CNTs synthesized at 350°C through a one-step procedure using ferrocene and sodium azide as the precursors have been used as catalysts in the ORR [184]. The encapsulation of the iron NPs inside the closed compartments of the peapod structure avoided the direct contact of metal particles with harsh environments including acid medium, oxygen, and sulfur contaminations. However, O_2 was activated efficiently, showing high activity and stability in PEMFC. DFT calculations show that catalytic activity could emerge from the electron transfer from Fe particles to the CNTs.

A strategy to effectively confine iron carbide nanocrystals in N-doped carbon coated on CNT (CNx@CNT) was reported [200]. This strategy prevented the agglomeration of iron carbide during pyrolysis and thus provided sufficient highly active catalytic sites. Together with the benefit from a three-dimensional conductive network of CNT-based core-shell structures for fast electron transfer and rapid mass transfer, the developed nanocatalyst exhibited a significantly enhanced electrocatalytic activity for ORR, as well as high durability and methanol tolerance. Moreover, the types of the confined iron compounds appreciably affected the performance of the catalysts, and Fe_3C might be most effective on improving ORR activity in this case.

5.2.1.5 Other Gas-Phase Reactions

Ni was filled in the channel or deposited on the outer surface of CNTs by capillarity or deposition [201]. The catalytic performance was evaluated by the hydrogenation of benzene to cyclohexane in the gas phase at 160°C and under atmospheric pressure. The conversion of benzene on the Ni-in-CNTs catalyst was 4.6 times that of Ni-out-CNTs. These results indicate that the enhanced catalytic activity can be attributed to the confinement in CNTs with more defects, which provides facile reduction, reinforced reactivity, and increased reactants concentrations, due to a larger charge transfer and deficient electrons in tubular nanoreactor. FeCo NPs confined within CNTs, with different Fe/Co ratios, are active as Co-based catalysts for the NH_3 decomposition to produce H_2, but the stability of Co was significantly improved by alloying with Fe. Confinement inside CNTs was found to improve the thermal stability rather than the turnover rate of the NPs [127]. The effect of the electron structure of CNTs on the confined metal particles and their catalytic activity was investigated using Ru NPs supported inside and outside CNTs. The synthesis of ammonia was chosen as probe reaction [79]. In this study Ru NPs supported on the outer wall of the CNTs were more active than the confined ones. Several experimental techniques together with DFT calculations suggest that the Ru NPs located outside exhibit a higher electron density than the ones confined within the CNTs. The rate-determining step of ammonia synthesis is the dissociation of N_2, which is an electrophilic process, being favored with the Ru NPs with higher electron density supported on the outer walls of CNTs.

CNTs filled with CeO_2 particles were prepared by wet impregnation assisted by capillary force [202]. Compared to CeO_2 outside CNTs, these composites show superior catalytic performance of oxidative dehydrogenation of ethylbenzene to styrene in the presence of CO_2. The results of this study indicate that CNT tubular structure results in strengthened interaction between CeO_2 and inner wall, which induces distortion and reducibility of CeO_2 lattices to promote the activation of surface lattice oxygen and the formation of oxygen vacancy. The activated surface oxygen and oxygen vacancy from CeO_2@CNT catalysts play an important role in two-step oxidative dehydrogenation reaction by promoting reverse water–gas shift reaction. In addition, CeO_2 filled into shorter CNTs exhibits higher catalytic activities due to decreasing the diffusion resistance of reactants and products in CNT channels.

5.2.2 Liquid-Phase Reactions

5.2.2.1 Hydrogenation

The confinement effects of CNTs have been thoroughly studied using liquid phase hydrogenation reactions. Selective hydrogenation of *trans*-cinnamaldehyde (CAL) has become a model reaction for testing metal catalysts confined within CNTs, as it can evaluate the structural properties of the catalyst [53]. Albeit other substrates have been studied, including prochiral ones [54,131,132].

Pd NPs confined inside MWCNTs, with several metal loadings ranging from 2 to 12%, were compared to Pd/AC- and Pd/zeolite-supported catalysts for the hydrogenation of benzene [203]. The confined Pd NPs exhibited higher activity and higher TOF value than Pd supported on conventional activated carbon and Y-type zeolites. Pd NPs confined within MWCNTs were also tested in the CAL hydrogenation and compared to commercial Pd/AC, showing high selectivity for the C=C hydrogenation [204]. Generally, confined metal catalysts in CNTs give better selectivity toward the hydrogenation of the C=O. Pt NPs were supported on the exterior wall of small diameter tubes (< 10 nm) and on the inner walls of larger diameter CNTs (60–100 nm) [114]. Both nanocatalysts showed high selectivity in the hydrogenation of the C=O bond, but Pt/CNT supported inside the channels showed almost 90% of selectivity against 60% reached with Pt/CNT attached outside the CNT. Confined PtRu catalysts synthesized using molecular recognition in MWCNTs with a diameter of 40 nm, and showing 80% of NPs located inside MWCNTs, reached 95% of selectivity toward cynnamyl alcohol (C=O hydrogenation) [53]. The C=O hydrogenation selectivity was correlated to the percentage of NPs located inside the CNTs. Activity was also enhanced with the confinement; unsupported PtRu NPs showed a TOF of 30 h^{-1}, while confined bimetallic NPs attained a TOF of 85 h^{-1}. Similar behavior was observed for preformed PtRu NPs synthesized in *sc*CO_2 and deposited over functionalized CNTs [144]. High C=O hydrogenation selectivity and high activity were attained with confined catalysts. The same authors studied the effects of the CNTs' pretreatment and diameter, and the nature of the metal (Co, Ru, and Pd) on the filling yield of the metallic NPs in the MWCNTs [61]. It was possible to modulate the filling yield (from 10 to 80%), controlling experimental parameters; furthermore, a correlation between the filling yields and the catalytic performances in CAL hydrogenation was established: hydrogenation of C═O bond and TOFs increased with the number of NPs located inside the CNT channels. Gold NPs supported on different treated CNTs showed different filling ratios, depending on the surface chemistry of the CNTs [141]. The different Au/CNTs, together with Au/AC catalysts, were tested in the CAL hydrogenation reaction. In all cases, the hydrogenation of the C═C bond was more favorable and became more selective when the percentage of gold NPs inside the CNTs was higher. In addition, better TOFs were also observed with the confined nanocatalysts.

Wet impregnation and double impregnation method made it possible to selectively obtain Ru NPs supported inside and outside the same type of CNTs [130]. The nanocatalysts were used in the hydrogenation of several substrates to investigate the effect of the electron density on the catalysis. For that reason they also prepared thermally treated supported catalysts, the heat treatment making Ru NPs electron rich. The authors can correlate the catalytic performances to the electron density of the metal NPs. The electron-rich Ru would be favorable for *p*-chloronitrobenzene hydrogenation but unfavorable for benzene hydrogenation. For cinnamaldehyde hydrogenation, the electron-rich Ru would be favorable for adsorption of C═O bond rather than C═C bond, and thus promote the selectivity to cinnamyl alcohol. A confinement effect induced by the electronic effect has different influences on these substrates, and it can be enhanced by heat treating for all the reactions.

The hydrogenation of other substrates has been also tested with confined nanocatalysts in CNTs. Hydrogenation of methyl acetate with copper NPs [121], cyclooctene and butyraldehyde hydrogenation in $scCO_2$ with Ru NPs [170], and D-glucose hydrogenation to sorbitol with Ru NPs [177] showed that confinement has a positive effect on the catalytic performances.

Asymmetric hydrogenation using Pt [54,131] and Pd [132] nanocatalysts has been reported by Li and coworkers. Improved wet chemistry was used to synthesize Pt and Pd NPs inside CNTs, and by-phasic impregnation to deposit selectively the NPs on the outer walls. The Pt nanocatalysts modified with cinchonidine (chiral modifier) were very active in the hydrogenation of α-ketosters. The activity was highly enhanced by confining the catalyst inside the CNTs, as compared to the NPs supported outside the CNTS or to Pt/AC and Pt/Al_2O_3 catalysts. The high enantioselectivity reached with the Pt nanocatalysts confined inside the CNTs (up to 96% ee) was attributed to the enrichment of both cinchonidine and the reactants inside the nanochannels [54]. The effect of water as a cosolvent was also investigated [131]. Adsorption experiments confirmed that the water can enrich the nanochannels with the chiral modifier and enhance its interaction with the inside walls of the CNTs, making it possible to obtain better enantioselectivity. Pd nanocatalysts synthesized in the same manner and also modified with cinchonidine were tested in the asymmetric hydrogenation of α,β-unsaturated carboxylic acids. The Pd nanocatalyst inside the channels of CNTs also showed higher activity and enantioselectivity than that of the Pd nanocatalyst located outside the channels [132].

5.2.2.2 Oxidation

We have seen that the charge transfer between encapsulated metals and the CNTs can result in an increase of the electron density of the outside walls of the CNTs. Furthermore, the catalytically active surface area does not decrease because the surface area that contacts the reactants is not occupied, and thus the oxidative activity may be enhanced. Fe nanowires encapsulated inside CNTs were tested as catalysts in the oxidation of cyclohexane in order to prove this concept [56]. Fe nanowires encapsulated in CNTs were synthesized by simultaneous growth of CNTs and filling of the inner cavity by CVD using ferrocene as catalysts and xylene and dichlorobenzene as carbon sources. The Fe-based catalysts was efficient in the aerobic oxidation of cyclohexane. The best results were obtained with a catalyst containing 20% of Fe. In addition, the encapsulated Fe allows the magnetic recovery of the catalyst. Selective oxidation of cyclohexane was also investigated with Au NPs confined inside CNTs [205]. The Au NPs were photocatalytically active with 14.6% conversion of cyclohexane and a high selectivity of 86.9% to cyclohexanol using air and visible light at room temperature. The catalysts were stable and showed excellent repeatability. Small Pt clusters (1 nm) confined inside thin-walled CNTs with an inner diameter of 1.0–1.5 nm were synthesized by vaporization of a volatile Pt precursor and used as catalysts in the toluene oxidation reaction [172]. The encapsulation within the small-diameter CNTs restricted the size of Pt nanoclusters around 1 nm and enabled modulating of the Pt species at the active reduced states. The Pt NPs were protected from oxygen under reaction conditions, leading to higher activity and stability than for the NPs located on the outer walls of the CNTs and in AC support. Pd-V NPs confined inside DWCNTs and MWCNTs with small inner diameters were utilized as catalyst in the hydroxylation reaction of benzene [206]. The catalytic activity was enhanced by confinement of Pd-V species inside the DWCNTs (1–1.5 nm inner diameter) with respect to Pd–V catalysts confined inside MWCNTs (4–8 nm inner diameter) or supported on the outer walls of DWCNTs.

Sub-nanometer Ti clusters have been homogeneously dispersed within DWCNTs presenting an inner diameter ranging from 1.0 to 1.5 nm. The confined titania exhibits a much higher activity than the titania particles attached on the outside walls of the DWNTs in the epoxidation of propylene by H_2O_2 [207]. XPS, XANES, and Raman spectroscopy data suggest electron transfer from titanium to the inner surfaces of the DWCNTs. In contrast, no electron transfer has been observed for the outside titania. The enhanced activity of the confined titania clusters is likely attributed to their small size and interaction with the DWCNT surface.

5.2.2.3 Other Reactions

It was shown that CNTs provide an excellent environment for Cu-catalyzed cycloaddition reactions [208]. The kinetics of reactions are accelerated and conversion rates improved in CNTs relative to catalysts deposited on the outer surfaces. Most significantly, catalytic centers embedded in the CNT cavity are stabilized by interactions with the nanoscale graphitic step-edges, which prevent the loss of catalyst during reactions, thus allowing efficient recycling of Cu@ graphite nanofibers, with retention of high catalytic activity cycle after cycle.

Palladium NPs confined in hollow graphite nanofibers exhibit attractive catalytic properties in Suzuki–Miyaura cross-coupling reactions [209]. There also, confinement of NPs at the step-edges facilitates retention of catalytic centers and recycling without any significant loss of activity or selectivity over multiple catalytic cycles. Furthermore, careful comparison of the catalytic properties of Pd NPs either on or in graphite nanofibers reveals that nanoscale confinement of catalysts fundamentally affects the pathways of the Suzuki-Miyaura reaction. The yield and selectivity for the cross-coupled product are critically dependent on the steric properties of the aryl iodide reactant, whereas no effects of confinement are observed for aryl boronic acid reactants possessing substituents in different positions.

6 SUMMARY

The tubular morphology of CNTs, as well as their nanometric diameter, offers great opportunities for studying chemical reactions in confined spaces. Indeed, the thin graphitic walls provide a protection against the environmental degradation, and the tubes exhibit electronic, surface, and cavity confinement effects. The confinement of molecules or materials may generate unique one-dimensional supramolecular nanostructures that may bring forth novel electronic, magnetic, and chemical properties. Several methodologies, mainly based on wet chemistry, have been developed to confine nanomaterials inside CNTs; and nowadays the relatively selective confinement of discrete nano-objects inside the CNT cavity can be achieved in a straightforward manner. Numerous studies have shown that confinement could allow the tuning of catalytic performances, including activity, selectivity and stability, without changing the catalyst compositions. As far as the nature of the confinement effects is concerned, it can be either physical or chemical. The physical nature is related to the geometry of the CNT channels, that is, to the space restriction, which can affect the orientation, conformation, and diffusion of confined molecules and nanomaterials. The chemical factors are related to the chemical nature of the pores, which influence the interaction of the confined phase with the pore wall. This can further affect the adsorption, diffusion, and selective stabilization of reactants, intermediates, or products, and thus have a pronounced effect on the equilibria and the reaction rates. The interaction can also modify the catalyst structures, which may alter and completely change the reaction mechanisms.

It appears that all the aforementioned effects can influence chemical reactions to different extents depending on the selected metals (if a catalyst is present), the diameter of CNTs, and the specific reactions carried out. It is frequently a collective result of more than just one of these effects. Although wide efforts have been devoted to assess the relative importance of each factor in order to gain fundamental insights, such insights are still, in most cases, rather difficult to distinguish. It would certainly require CNTs with better-defined size, structure, and properties, and efficient techniques for filling of small nanotubes at a bulk scale, as well as advanced characterization techniques to examine the catalytic roles of local structures under working conditions. At the same time, theoretical simulations on models closer to realistic systems could offer further insights into confinement effects.

REFERENCES

[1] Monthioux M, Serp P, Flahaut E, Razafinimanana M, Laurent C, Peigney A, Bacsa W, Broto J-M. Introduction to carbon nanotubes. In: Bhushan B, editor. Springer handbook of nanotechnology. Berlin Heidelberg: Springer; 2007. p. 43–112.

[2] Rouquerol J, Avnir D, Fairbridge CW, Everett DH, Haynes JM, Pernicone N, Ramsay JDF, Sing KSW, Unger KK. Recommendations for the characterization of porous solids (Technical Report). Pure Appl Chem 1994;66(8):1739.

[3] Wang Q, Challa SR, Sholl DS, Johnson JK. Quantum sieving in carbon nanotubes and zeolites. Phys Rev Lett 1999;82:956–9.

[4] Challa SR, Sholl DS, Johnson JK. Light isotope separation in carbon nanotubes through quantum molecular sieving. Phys Rev B 2001;63:245419.

[5] Youn S, Buchheim J, Park H. Confined water in carbon nanotubes and its applications. In: Mercury L, Tas N, Zilberbrand M, editors. Transport and reactivity of solutions in confined hydrosystems. Netherlands: Springer; 2014. p. 19–27.

[6] Noy A, Park HG, Fornasiero F, Holt JK, Grigoropoulos CP, Bakajin O. Nanofluidics in carbon nanotubes. Nano Today 2007;2:22–9.

[7] Chaban VV, Prezhdo OV. Water boiling inside carbon nanotubes: toward efficient drug release. ACS Nano 2011;5:5647–55.

[8] Cruz FJAL, dePablo JJ, Mota JPB. Endohedral confinement of a DNA dodecamer onto pristine carbon nanotubes and the stability of the canonical B form. J Chem Phys 2014;140:225103.

[9] Serp P, Castillejos E. Catalysis in carbon nanotubes. Chem Cat Chem 2010;2:41–7.

[10] Pan X, Bao X. The effects of confinement inside carbon nanotubes on catalysis. Acc Chem Res 2011;44:553–62.

[11] Yang F, Deng D, Pan X, Fu Q, Bao X. Understanding nano effects in catalysis. Nat Sci Rev 2015;2:183–201.

[12] Das R, Ali ME, Hamid SBA, Ramakrishna S, Chowdhury ZZ. Carbon nanotube membranes for water purification: a bright future in water desalination. Desalination 2014;336:97–109.

[13] Goettmann F, Sanchez C. How does confinement affect the catalytic activity of mesoporous materials? J Mater Chem 2007;17:24–30.

[14] Yu C, He J. Synergic catalytic effects in confined spaces. Chem Commun 2012;48:4933–40.

[15] Thomas JM, Raja R. Exploiting nanospace for asymmetric catalysis: confinement of immobilized, single-site chiral catalysts enhances enantioselectivity. Acc Cem Res 2008;41:708–20.

[16] Santiso EE, Kostov MK, George AM, Nardelli MB, Gubbins KE. Confinement effects on chemical reactions: toward an integrated rational catalyst design. Appl Surf Sci 2007;253:5570–9.

[17] Zhang F, Pan X, Hu Y, Yu L, Chen X, Jiang P, Zhang H, Deng S, Zhang J, Bolin TB, Zhang S, Huang Y, Bao X. Tuning the redox activity of encapsulated metal clusters via the metallic and semiconducting character of carbon nanotubes. Proc Natl Acad Sci 2013;110:14861–6.
[18] Kleiner A, Eggert S. Curvature, hybridization, and STM images of carbon nanotubes. Phys Rev B 2001;64:113402.
[19] Ouyang Y, Peng J-C, Wang H, Peng Z-H. The rehybridization of electronic orbitals in carbon nanotubes. Chinese Phys B 2008;17:3123–7.
[20] Serp P, Machado B, editors. Nanostructured carbon materials for catalysis. Cambridge: Royal Society of Chemistry; 2015.
[21] Odom TW, Huang J-L, Kim P, Lieber CM. Atomic structure and electronic properties of single-walled carbon nanotubes. Nature 1998;391:62–4.
[22] Zhong J, Chiou J, Dong C, Song L, Liu C, Xie S, Cheng H, Pong W-F, Chang C, Chen Y, Wu Z, Guo J. Probing quantum confinement of single-walled carbon nanotubes by resonant soft-x-ray emission spectroscopy. Appl Phys Lett 2008;93:023107.
[23] Shao J, Yuan L, Hu X, Wu Y, Zhang Z. The effect of nano confinement on the C–H activation and its corresponding structure-activity relationship. Sci Rep 2014;4:7225.
[24] Zoberbier T, Chamberlain TW, Biskupek J, Kuganathan N, Eyhusen S, Bichoutskaia E, Kaiser U, Khlobystov AN. Interactions and reactions of transition metal clusters with the interior of single-walled carbon nanotubes imaged at the atomic scale. J Am Chem Soc 2012;134:3073–9.
[25] Garg I, Sharma H, Dharamvir K, Jindal VK, Kanhere DG. DFT study of Aln (1–13) clusters encapsulated inside single walled carbon nanotubes. J Phys Chem C 2010;114:18762–72.
[26] Gao H, Zhao J. First-principles study of Ru atoms and clusters adsorbed outside and inside carbon nanotubes. J Chem Phys 2010;132. 234704.
[27] Greeley J, Nørskov JK, Mavrikakis M. Electronic structure and catalysis on metal surfaces. Annu Rev Phys Chem 2002;53:319–48.
[28] Mattia D, Gogotsi Y. Review: static and dynamic behavior of liquids inside carbon nanotubes. Microfluid Nanofluid 2008;5:289–305.
[29] Skoulidas AI, Ackerman DM, Johnson JK, Sholl DS. Rapid transport of gases in carbon nanotubes. Phys Rev Lett 2002;89:185901.
[30] Whitby M, Quirke N. Fluid flow in carbon nanotubes and nanopipes. Nat Nano 2007;2:87–94.
[31] Majumder M, Chopra N, Hinds BJ. Mass transport through carbon nanotube membranes in three different regimes: ionic diffusion and gas and liquid flow. ACS Nano 2011;5:3867–77.
[32] Joseph S, Aluru NR. Why are carbon nanotubes fast transporters of water? Nano Lett 2008;8:452–8.
[33] Barati Farimani A, Aluru NR. Spatial diffusion of water in carbon nanotubes: from Fickian to ballistic motion. J Phys Chem B 2011;115:12145–9.
[34] Zheng J, Lennon EM, Tsao H-K, Sheng Y-J, Jiang S. Transport of a liquid water and methanol mixture through carbon nanotubes under a chemical potential gradient. J Chem Phys 2005;122:214702.
[35] Mashl RJ, Joseph S, Aluru NR, Jakobsson E. Anomalously immobilized water: a new water phase induced by confinement in nanotubes. Nano Lett 2003;3:589–92.
[36] Kalugin ON, Chaban VV, Loskutov VV, Prezhdo OV. Uniform diffusion of acetonitrile inside carbon nanotubes favors supercapacitor performance. Nano Lett 2008;8:2126–30.
[37] Majumder M, Chopra N, Andrews R, Hinds BJ. Nanoscale hydrodynamics: enhanced flow in carbon nanotubes. Nature 2005;438:44–144.
[38] Holt JK, Park HG, Wang Y, Stadermann M, Artyukhin AB, Grigoropoulos CP, Noy A, Bakajin O. Fast mass transport through sub-2-nanometer carbon nanotubes. Science 2006;312:1034–7.

[39] Chen H, Sholl DS. Rapid diffusion of CH_4/H_2 mixtures in single-walled carbon nanotubes. J Am Chem Soc 2004;126:7778–9.

[40] Jakobtorweihen S, Verbeek MG, Lowe CP, Keil FJ, Smit B. Understanding the loading dependence of self-diffusion in carbon nanotubes. Phys Rev Lett 2005;95:044501.

[41] Chen H, Johnson JK, Sholl DS. Transport diffusion of gases is rapid in flexible carbon nanotubes. J Phys Chem B 2006;110:1971–5.

[42] Rodriguez J, Elola MD, Laria D. Confined polar mixtures within cylindrical nanocavities. J Phys Chem B 2010;114:7900–8.

[43] Fujiwara A, Ishii K, Suematsu H, Kataura H, Maniwa Y, Suzuki S, Achiba Y. Gas adsorption in the inside and outside of single-walled carbon nanotubes. Chem Phys Lett 2001;336:205–11.

[44] Kondratyuk P, Yates JT. Molecular views of physical adsorption inside and outside of single-wall carbon nanotubes. Acc Chem Res 2007;40:995–1004.

[45] Huang L, Zhang L, Shao Q, Lu L, Lu X, Jiang S, Shen W. Simulations of binary mixture adsorption of carbon dioxide and methane in carbon nanotubes: temperature, pressure, and pore size effects. J Phys Chem C 2007;111:11912–20.

[46] Darkrim FL, Malbrunot P, Tartaglia GP. Review of hydrogen storage by adsorption in carbon nanotubes. Int J Hydrogen Energ 2002;27:193–202.

[47] Guan J, Pan X, Liu X, Bao X. Syngas segregation induced by confinement in carbon nanotubes: a combined first-principles and Monte Carlo study. J Phys Chem C 2009;113:21687–92.

[48] Gao Y, Xie K, Mi S, Liu N, Wang W, Huang W. Preferential oxidation of CO in a H2-rich stream over multi-walled carbon nanotubes confined Ru catalysts. Int J Hydrogen Energ 2013;38:16665–76.

[49] Zhang H, Pan X, Han X, Liu X, Wang X, Shen W, Bao X. Enhancing chemical reactions in a confined hydrophobic environment: an NMR study of benzene hydroxylation in carbon nanotubes. Chem Sci 2013;4:1075–8.

[50] Xie W, Zhang Y, Liew K, Li J. Effect of catalyst confinement and pore size on Fischer-Tropsch synthesis over cobalt supported on carbon nanotubes. Sci China Chem 2012;55:1811–8.

[51] Chen W, Fan Z, Pan X, Bao X. Effect of confinement in carbon nanotubes on the activity of Fischer−Tropsch iron catalyst. J Am Chem Soc 2008;130:9414–9.

[52] Tavasoli A, Trépanier M, Dalai AK, Abatzoglou N. Effects of confinement in carbon nanotubes on the activity, selectivity, and lifetime of Fischer−Tropsch Co/carbon nanotube catalysts. J Chem Eng Data 2010;55:2757–63.

[53] Castillejos E, Debouttière P-J, Roiban L, Solhy A, Martinez V, Kihn Y, Ersen O, Philippot K, Chaudret B, Serp P. An efficient strategy to drive nanoparticles into carbon nanotubes and the remarkable effect of confinement on their catalytic performance. Angew Chem Int Ed 2009;48:2529–33.

[54] Chen Z, Guan Z, Li M, Yang Q, Li C. Enhancement of the performance of a platinum nanocatalyst confined within carbon nanotubes for asymmetric hydrogenation. Agnew Chem Int Ed 2011;50:4913–7.

[55] Yang M, Ling Q, Yang H, Li C, Zhang A. Enhanced catalytic activity of K-birnessite MnO_2 confined in carbon nanotubes for selective oxidation of benzyl alcohol. Catal Commun 2014;46:238–41.

[56] Yang X, Yu H, Peng F, Wang H. Confined iron nanowires enhance the catalytic activity of carbon nanotubes in the aerobic oxidation of cyclohexane. Chem Sus Chem 2012;5:1213–7.

[57] Alba-Simionesco C, Coasne B, Dosseh G, Dudziak G, Gubbins KE, Radhakrishnan R, Sliwinska-Bartkowiak M. Effects of confinement on freezing and melting. J Phys Condens Matter 2006;18:R15.

[58] Jiang J, Sandler SI, Smit B. Capillary phase transitions of n-alkanes in a carbon nanotube. Nano Lett 2004;4:241–4.

[59] Fomin YD, Tsiok EN, Ryzhov VN. The behavior of benzene confined in a single wall carbon nanotube. J Comput Chem 2015;36:901–6.

[60] Trzaskowski B, Jalbout AF, Adamowicz L. Molecular dynamics studies of protein-fragment models encapsulated into carbon nanotubes. Chem Phys Lett 2006;430:97–100.

[61] Trang Nguyen T, Serp P. Confinement of metal nanoparticles in carbon nanotubes. Chem Cat Chem 2013;5:3595–603.

[62] Wei G-F, Shang C, Liu Z-P. Confined platinum nanoparticle in carbon nanotube: structure and oxidation. Phys Chem Chem Phys 2015;17:2078–87.

[63] Ivanovskaya VV, Köhler C, Seifert G. 3d metal nanowires and clusters inside carbon nanotubes: structural, electronic, and magnetic properties. Phys Rev B Condens Matter Mater Phys 2007;75:075410.

[64] Philp E, Sloan J, Kirkland AI, Meyer RR, Friedrichs S, Hutchison JL, Green MLH. An encapsulated helical one-dimensional cobalt iodide nanostructure. Nat Mater 2003;2: 788–91.

[65] Guan L, Suenaga K, Shi Z, Gu Z, Iijima S. Polymorphic structures of iodine and their phase transition in confined nanospace. Nano Lett 2007;7:1532–5.

[66] Fujimori T, dos Santos RB, Hayashi T, Endo M, Kaneko K, Tománek D. Formation and properties of selenium double-helices inside double-wall carbon nanotubes: experiment and theory. ACS Nano 2013;7:5607–13.

[67] Chen YK, Green MLH, Tsang SC. Synthesis of carbon nanotubes filled with long continuous crystals of molybdenum oxides. Chem Commun 1996;2489–90.

[68] Kiang C-H, Choi J-S, Tran TT, Bacher AD. Molecular nanowires of 1 nm diameter from capillary filling of single-walled carbon nanotubes. J Phys Chem B 1999;103:7449–51.

[69] Elías AL, Rodríguez-Manzo JA, McCartney MR, Golberg D, Zamudio A, Baltazar SE, López-Urías F, Muñoz-Sandoval E, Gu L, Tang CC, Smith DJ, Bando Y, Terrones H, Terrones M. Production and characterization of single-crystal FeCo nanowires inside carbon nanotubes. Nano Lett 2005;5:467–72.

[70] Grobert N, Mayne M, Terrones M, Sloan J, Dunin-Borkowski RE, Kamalakaran R, Seeger T, Terrones H, Ruhle M, Walton DRM, Kroto HW, Hutchison JL. Alloy nanowires: invar inside carbon nanotubes. Chem Commun 2001;471–2.

[71] Kitaura R, Imazu N, Kobayashi K, Shinohara H. Fabrication of metal nanowires in carbon nanotubes via versatile nano-template reaction. Nano Lett 2008;8:693–9.

[72] Govindaraj A, Satishkumar BC, Nath M, Rao CNR. Metal nanowires and intercalated metal layers in single-walled carbon nanotube bundles. Chem Mater 2000;12:202–5.

[73] Sloan J, Wright DM, Bailey S, Brown G, York APE, Coleman KS, Green MLH, Hutchison JL, Woo H-G. Capillarity and silver nanowire formation observed in single walled carbon nanotubes. Chem Commun 1999;699–700.

[74] Kitaura R, Nakanishi R, Saito T, Yoshikawa H, Awaga K, Shinohara H. High-yield synthesis of ultrathin metal nanowires in carbon nanotubes. Agnew Chem Int Ed 2009;48:8298–302.

[75] Muramatsu H, Hayashi T, Kim YA, Shimamoto D, Endo M, Terrones M, Dresselhaus MS. Synthesis and isolation of molybdenum atomic wires. Nano Lett 2008;8:237–40.

[76] Meunier V, Muramatsu H, Hayashi T, Kim YA, Shimamoto D, Terrones H, Dresselhaus MS, Terrones M, Endo M, Sumpter BG. Properties of one-dimensional molybdenum nanowires in a confined environment. Nano Lett 2009;9:1487–92.

[77] Chen W, Pan X, Willinger M-G, Su DS, Bao X. Facile autoreduction of iron oxide/carbon nanotube encapsulates. J Am Chem Soc 2006;128:3136–7.

[78] Chen W, Pan X, Bao X. Tuning of redox properties of iron and iron oxides via encapsulation within carbon nanotubes. J Am Chem Soc 2007;129:7421–6.

[79] Guo S, Pan X, Gao H, Yang Z, Zhao J, Bao X. Probing the electronic effect of carbon nanotubes in catalysis: NH_3 synthesis with Ru nanoparticles. Chem Eur J 2010;16:5379–84.

[80] Trépanier M, Dalai AK, Abatzoglou N. Synthesis of CNT-supported cobalt nanoparticle catalysts using a microemulsion technique: role of nanoparticle size on reducibility, activity and selectivity in Fischer–Tropsch reactions. Appl Catal A 2010;374:79–86.

[81] Yang Z, Guo S, Pan X, Wang J, Bao X. FeN nanoparticles confined in carbon nanotubes for CO hydrogenation. Energy Environ Sci 2011;4:4500–3.

[82] Friedrich H, Guo S, de Jongh PE, Pan X, Bao X, de Jong KP. A quantitative electron tomography study of Ruthenium particles on the interior and exterior surfaces of carbon nanotubes. Chem Sus Chem 2011;4:957–63.

[83] Ersen O, Werckmann J, Houllé M, Ledoux M-J, Pham-Huu C. 3D Electron microscopy study of metal particles inside multiwalled carbon nanotubes. Nano Lett 2007;7:1898–907.

[84] Ziese U, de Jong KP, Koster AJ. Electron tomography: a tool for 3D structural probing of heterogeneous catalysts at the nanometer scale. Appl Catal A 2004;260:71–4.

[85] Ersen O, Florea I, Hirlimann C, Pham-Huu C. Exploring nanomaterials with 3D electron microscopy. Mater Today 2012;42:33–58.

[86] Pan X, Fan Z, Chen W, Ding Y, Luo H, Bao X. Enhanced ethanol production inside carbon-nanotube reactors containing catalytic particles. Nat Mater 2007;6:507–11.

[87] Winter F, Leendert Bezemer G, van der Spek C, Meeldijk JD, Jos van Dillen A, Geus JW, de Jong KP. TEM and XPS studies to reveal the presence of cobalt and palladium particles in the inner core of carbon nanofibers. Carbon 2005;43:327–32.

[88] Wang X, Li N, Pfefferle LD, Haller GL. Metal nanoparticles inside multi-walled carbon nanotubes: a simple method of preparation and of microscopic image analysis. Micropor Mesopor Mat 2013;176:139–44.

[89] Besley NA, Noble A. NMR chemical shifts of molecules encapsulated in single walled carbon nanotubes. J Chem Phys 2008;128:101102.

[90] Ren P, Zheng A, Pan X, Han X, Bao X. DFT study on the NMR chemical shifts of molecules confined in carbon nanotubes. J Phys Chem C 2013;117:23418–24.

[91] Liu X, Pan X, Shen W, Ren P, Han X, Bao X. NMR study of preferential endohedral adsorption of methanol in multiwalled carbon nanotubes. J Phys Chem C 2012;116:7803–9.

[92] Huang P, Schwegler E, Galli G. Water confined in carbon nanotubes: magnetic response and proton chemical shieldings. J Phys Chem C 2009;113:8696–700.

[93] Byl O, Liu J-C, Wang Y, Yim W-L, Johnson JK, Yates JT. Unusual hydrogen bonding in water-filled carbon nanotubes. J Am Chem Soc 2006;128:12090–7.

[94] Byl O, Kondratyuk P, Forth ST, FitzGerald SA, Chen L, Johnson JK, Yates JT. Adsorption of CF4 on the internal and external surfaces of opened single-walled carbon nanotubes: a vibrational spectroscopy study. J Am Chem Soc 2003;125:5889–96.

[95] Khlobystov AN, Britz DA, Briggs GAD. Molecules in carbon nanotubes. Acc Chem Res 2005;38:901–9.

[96] Matranga C, Bockrath B, Chopra N, Hinds BJ, Andrews R. Raman spectroscopic investigation of gas interactions with an aligned multiwalled carbon nanotube membrane. Langmuir 2006;22:1235–40.

[97] Li L-J, Khlobystov AN, Wiltshire JG, Briggs GAD, Nicholas RJ. Diameter-selective encapsulation of metallocenes in single-walled carbon nanotubes. Nat Mater 2005;4:481–5.

[98] Kondratyuk P, Yates JT Jr. Desorption kinetic detection of different adsorption sites on opened carbon single-walled nanotubes: the adsorption of *n*-nonane and CCl_4. Chem Phys Lett 2005;410:324–9.
[99] Agnihotri S, Mota JPB, Rostam-Abadi M, Rood MJ. Theoretical and experimental investigation of morphology and temperature effects on adsorption of organic vapors in single-walled carbon nanotubes. J Phys Chem B 2006;110:7640–7.
[100] Rawat DS, Calbi MM, Migone AD. Equilibration time: kinetics of gas adsorption on closed- and open-ended single-walled carbon nanotubes. J Phys Chem C 2007;111:12980–6.
[101] Jakubek ZJ, Simard B. Two confined phases of argon adsorbed inside open single-walled carbon nanotubes. Langmuir 2004;20:5940–5.
[102] Striolo A, Chialvo AA, Gubbins KE, Cummings PT. Water in carbon nanotubes: adsorption isotherms and thermodynamic properties from molecular simulation. J Chem Phys 2005;122:234712.
[103] Kuznetsova A, Yates JT, Liu J, Smalley RE. Physical adsorption of xenon in open single walled carbon nanotubes: observation of a quasi-one-dimensional confined Xe phase. J Chem Phys 2000;112:9590–8.
[104] Jiang J, Sandler SI, Schenk M, Smit B. Adsorption and separation of linear and branched alkanes on carbon nanotube bundles from configurational-bias Monte Carlo simulation. Phys Rev B 2005;72:045447.
[105] Iijima S. Helical microtubules of graphitic carbon. Nature 1991;354:56–8.
[106] Ajayan PM, Ebbesen TW, Ichihashi T, Iijima S, Tanigaki K, Hiura H. Opening carbon nanotubes with oxygen and implications for filling. Nature 1993;362:522–5.
[107] Ajayan PM, Iijima S. Capillarity-induced filling of carbon nanotubes. Nature 1993;361:333–4.
[108] Dujardin E, Ebbesen TW, Hiura H, Tanigaki K. Capillarity and wetting of carbon nanotubes. Science 1994;265:1850–2.
[109] Tsang SC, Chen YK, Harris PJF, Green MLH. A simple chemical method of opening and filling carbon nanotubes. Nature 1994;372:159–62.
[110] Guerret-Piecourt C, Le Bouar Y, Loiseau A, Pascard H. Relation between metal electronic structure and morphology of metal compounds inside carbon nanotubes. Nature 1994;372:761–5.
[111] Wu H-Q, Wei X-W, Shao M-W, Gu J-S, Qu M-Z. Preparation of Fe-Ni alloy nanoparticles inside carbon nanotubes via wet chemistry. J Mater Chem 2002;12:1919–21.
[112] Serp P, Philippot K, editors. Nanomaterials in catalysis. Germany: Wiley-VCH Verlag GmbH & Co. KGaA; 2013.
[113] Ebbesen TW. Wetting, filling and decorating carbon nanotubes. J Phys Chem Solids 1996;57:951–5.
[114] Ma H, Wang L, Chen L, Dong C, Yu W, Huang T, Qian Y. Pt nanoparticles deposited over carbon nanotubes for selective hydrogenation of cinnamaldehyde. Catal Commun 2007;8:452–6.
[115] Borowiak-Palen E, Mendoza E, Bachmatiuk A, Rummeli MH, Gemming T, Nogues J, Skumryev V, Kalenczuk RJ, Pichler T, Silva SRP. Iron filled single-wall carbon nanotubes: a novel ferromagnetic medium. Chem Phys Lett 2006;421:129–33.
[116] Jorge J, Flahaut E, Gonzalez-Jimenez F, Gonzalez G, Gonzalez J, Belandria E, Broto JM, Raquet B. Preparation and characterization of α-Fe nanowires located inside double wall carbon nanotubes. Chem Phys Lett 2008;457:347–51.
[117] Wang C, Guo S, Pan X, Chen W, Bao X. Tailored cutting of carbon nanotubes and controlled dispersion of metal nanoparticles inside their channels. J Mater Chem 2008;18:5782–6.

[118] Zheng W, Zhang J, Zhu B, Blume R, Zhang Y, Schlichte K, Schloegl R, Schueth F, Su DS. Structure-function correlations for Ru/CNT in the catalytic decomposition of ammonia. Chem Sus Chem 2010;3:226–30.
[119] Li B, Wang C, Yi G, Lin H, Yuan Y. Enhanced performance of Ru nanoparticles confined in carbon nanotubes for CO preferential oxidation in a H_2-rich stream. Catal Today 2011;164:74–9.
[120] Chen W, Fan Z, Zhang B, Ma G, Takanabe K, Zhang X, Lai Z. Enhanced visible-light activity of titania via confinement inside carbon nanotubes. J Am Chem Soc 2011;133:14896–9.
[121] Wang D, Yang G, Ma Q, Wu M, Tan Y, Yoneyama Y, Tsubaki N. Confinement effect of carbon nanotubes: copper nanoparticles filled carbon nanotubes for hydrogenation of methyl acetate. ACS Catal 2012;2:1958–66.
[122] Wang L, Ge L, Rufford TE, Chen J, Zhou W, Zhu Z, Rudolph V. A comparison study of catalytic oxidation and acid oxidation to prepare carbon nanotubes for filling with Ru nanoparticles. Carbon 2011;49:2022–32.
[123] Baaziz W, Begin-Colin S, Pichon BP, Florea I, Ersen O, Zafeiratos S, Barbosa R, Begin D, Pham-Huu C. High-density monodispersed cobalt nanoparticles filled into multiwalled carbon nanotubes. Chem Mater 2012;24:1549–51.
[124] Xing C, Yang G, Wang D, Zeng C, Jin Y, Yang R, Suehiro Y, Tsubaki N. Controllable encapsulation of cobalt clusters inside carbon nanotubes as effective catalysts for Fischer–Tropsch synthesis. Catal Today 2013;215:24–8.
[125] Liu X, Marangon I, Melinte G, Wilhelm C, Menard-Moyon C, Pichon BP, Ersen O, Aubertin K, Baaziz W, Pham-Huu C, Begin-Colin S, Bianco A, Gazeau F, Begin D. Design of covalently functionalized carbon nanotubes filled with metal oxide nanoparticles for imaging, therapy, and magnetic manipulation. ACS Nano 2014;8:11290–304.
[126] Shi Z-Q, Dong Z-P, Sun J, Zhang F-W, Yang H-L, Zhou J-H, Zhu X-H, Li R. Filled cobalt nanoparticles into carbon nanotubes as a rapid and high-efficiency catalyst for selective epoxidation of styrene with molecular oxygen. Chem Eng J 2014;237:81–7.
[127] Zhang J, Muller J-O, Zheng W, Wang D, Su D, Schlogl R. Individual Fe-Co alloy nanoparticles on carbon nanotubes: structural and catalytic properties. Nano Lett 2008;8:2738–43.
[128] Tessonnier J-P, Ersen O, Weinberg G, Pham-Huu C, Su DS, Schlogl R. Selective deposition of metal nanoparticles inside or outside multiwalled carbon nanotubes. ACS Nano 2009;3:2081–9.
[129] Abbaslou RMM, Tavassoli A, Soltan J, Dalai AK. Iron catalysts supported on carbon nanotubes for Fischer–Tropsch synthesis: effect of catalytic site position. Appl Catal A 2009;367:47–52.
[130] Wang Y, Rong Z, Wang Y, Zhang P, Wang Y, Qu J. Ruthenium nanoparticles loaded on multiwalled carbon nanotubes for liquid-phase hydrogenation of fine chemicals: an exploration of confinement effect. J Catal 2015;329:95–106.
[131] Guan Z, Lu S, Chen Z, Li C. An unexpected effect of water on the asymmetric hydrogenation of α-ketoesters on platinum nanoparticles confined in carbon nanotubes. J Catal 2013;305:19–26.
[132] Guan Z, Lu S, Li C. Enantioselective hydrogenation of α,β-unsaturated carboxylic acid over cinchonidine-modified Pd nanoparticles confined in carbon nanotubes. J Catal 2014;311:1–5.
[133] Wang X-H, Orikasa H, Inokuma N, Yang Q-H, Hou P-X, Oshima H, Itoh K, Kyotani T. Controlled filling of Permalloy into one-end-opened carbon nanotubes. J Mater Chem 2007;17:986–91.

[134] Yu W-J, Hou P-X, Li F, Liu C. Improved electrochemical performance of Fe_2O_3 nanoparticles confined in carbon nanotubes. J Mater Chem 2012;22:13756–63.

[135] Hevia S, Homm P, Cortes A, Nunez V, Contreras C, Vera J, Segura R. Selective growth of palladium and titanium dioxide nanostructures inside carbon nanotube membranes. Nanoscale Res Lett 2012;7:348. 342.

[136] Segura RA, Contreras C, Henriquez R, Haberle P, Acuna JJS, Adrian A, Alvarez P, Hevia SA. Gold nanoparticles grown inside carbon nanotubes: synthesis and electrical transport measurements. Nanoscale Res Lett 2014;9:207.

[137] Korneva G, Ye H, Gogotsi Y, Halverson D, Friedman G, Bradley J-C, Kornev KG. Carbon nanotubes loaded with magnetic particles. Nano Lett 2005;5:879–84.

[138] Kim BM, Qian S, Bau HH. Filling carbon nanotubes with particles. Nano Lett 2005;5:873–8.

[139] Bazilevsky AV, Sun K, Yarin AL, Megaridis CM. Room-temperature, open-air, wet intercalation of liquids, surfactants, polymers and nanoparticles within nanotubes and microchannels. J Mater Chem 2008;18:696–702.

[140] Jain D, Wilhelm R. An easy way to produce α-iron filled multiwalled carbon nanotubes. Carbon 2007;45:602–6.

[141] Zhang X, Guo YC, Zhi CZ, Gao JS, Xu CM. High performance of carbon nanotubes confining gold nanoparticles for selective hydrogenation of 1,3-butadiene and cinnamaldehyde. J Catal 2012;292:213–26.

[142] Amiens C, Chaudret B, Ciuculescu-Pradines D, Colliere V, Fajerwerg K, Fau P, Kahn M, Maisonnat A, Soulantica K, Philippot K. Organometallic approach for the synthesis of nanostructures. New J Chem 2013;37:3374–401.

[143] Castillejos E, Chico R, Bacsa R, Coco S, Espinet P, Perez-Cadenas M, Guerrero-Ruiz A, Rodriguez-Ramos I, Serp P. Selective deposition of gold nanoparticles on or inside carbon nanotubes and their catalytic activity for preferential oxidation of CO. Eur J Inorg Chem 2010;2010:5096–102.

[144] Castillejos E, Jahjah M, Favier I, Orejon A, Pradel C, Teuma E, Masdeu-Bulto AM, Serp P, Gomez M. Synthesis of platinum-Ruthenium nanoparticles under supercritical CO_2 and their confinement in carbon nanotubes: hydrogenation applications. Chem Cat Chem 2012;4:118–22.

[145] Ugarte D, Chatelain A, de Heer WA. Nanocapillarity and chemistry in carbon nanotubes. Science 1996;274:1897–9.

[146] Kitaura R, Ogawa D, Kobayashi K, Saito T, Ohshima S, Nakamura T, Yoshikawa H, Awaga K, Shinohara H. High-yield synthesis and characterization of the structural and magnetic properties of crystalline $ErCl_3$ nanowires in single-walled carbon nanotube templates. Nano Res 2008;1:152–7.

[147] Ayala P, Kitaura R, Nakanishi R, Shiozawa H, Ogawa D, Hoffmann P, Shinohara H, Pichler T. Templating rare-earth hybridization via ultrahigh vacuum annealing of ErCl3 nanowires inside carbon nanotubes. Phys Rev B Condens Matter Mater Phys 2011;83. 085407.

[148] Giusca CE, Stolojan V, Sloan J, Borrnert F, Shiozawa H, Sader K, Rummeli MH, Buchner B, Silva SRP. Confined crystals of the smallest phase-change material. Nano Lett 2013;13:4020–7.

[149] Fidiani E, Costa PMFJ, Wolter AUB, Maier D, Buechner B, Hampel S. Magnetically active and coated gadolinium-filled carbon nanotubes. J Phys Chem C 2013;117:16725–33.

[150] Smith BW, Monthioux M, Luzzi DE. Encapsulated C60 in carbon nanotubes. Nature 1998;396:323–4.

[151] Koshino M, Niimi Y, Nakamura E, Kataura H, Okazaki T, Suenaga K, Iijima S. Analysis of the reactivity and selectivity of fullerene dimerization reactions at the atomic level. Nat Chem 2010;2:117–24.

[152] Hirahara K, Suenaga K, Bandow S, Kato H, Okazaki T, Shinohara H, Iijima S. One-dimensional metallofullerene crystal generated inside single-walled carbon nanotubes. Phys Rev Lett 2000;85:5384–7.

[153] Suenaga K, Okazaki T, Wang CR, Bandow S, Shinohara H, Iijima S. Direct imaging of Sc_2@C_{84} molecules encapsulated inside single-wall carbon nanotubes by high-resolution electron microscopy with atomic sensitivity. Phys Rev Lett 2003;90. 055506/055501-055506/055504.

[154] Sato Y, Yumura T, Suenaga K, Moribe H, Nishide D, Ishida M, Shinohara H, Iijima S. Direct imaging of intracage structure in titanium-carbide endohedral metallofullerene. Phys Rev B Condens Matter Mater Phys 2006;73. 193401/193401-193401/193404.

[155] Warner JH, Ito Y, Rummeli MH, Buchner B, Shinohara H, Briggs GAD. Capturing the motion of molecular nanomaterials encapsulated within carbon nanotubes with ultrahigh temporal resolution. ACS Nano 2009;3:3037–44.

[156] Warner JH, Plant SR, Young NP, Porfyrakis K, Kirkland AI, Briggs GAD. Atomic scale growth dynamics of nanocrystals within carbon nanotubes. ACS Nano 2011;5:1410–7.

[157] Nakamura E, Koshino M, Saito T, Niimi Y, Suenaga K, Matsuo Y. Electron microscopic imaging of a single group 8 metal atom catalyzing C-C bond reorganization of fullerenes. J Am Chem Soc 2011;133:14151–3.

[158] Chamberlain TW, Popov AM, Knizhnik AA, Samoilov GE, Khlobystov AN. The role of molecular clusters in the filling of carbon nanotubes. ACS Nano 2010;4:5203–10.

[159] Chamberlain TW, Meyer JC, Biskupek J, Leschner J, Santana A, Besley NA, Bichoutskaia E, Kaiser U, Khlobystov AN. Reactions of the inner surface of carbon nanotubes and nanoprotrusion processes imaged at the atomic scale. Nat Chem 2011;3:732–7.

[160] Chamberlain TW, Champness NR, Schroeder M, Khlobystov AN. A piggyback ride for transition metals: encapsulation of exohedral metallofullerenes in carbon nanotubes. Chem Eur J 2011;17:668–74. S668/661-S668/665.

[161] Khlobystov AN, Britz DA, Wang J, O'Neil SA, Poliakoff M, Briggs GAD. Low temperature assembly of fullerene arrays in single-walled carbon nanotubes using supercritical fluids. J Mater Chem 2004;14:2852–7.

[162] Yao M, Stenmark P, Abou-Hamad E, Nitze F, Qin J, Goze-Bac C, Wagberg T. Confined adamantane molecules assembled to one dimension in carbon nanotubes. Carbon 2011;49:1159–66.

[163] Wang J, Kuimova MK, Poliakoff M, Briggs GAD, Khlobystov AN. Encapsulation and IR probing of cube-shaped octasilasesquioxane $H_8Si_8O_{12}$ in carbon nanotubes. Agnew Chem Int Ed 2006;45:5188–91.

[164] Li YF, Hatakeyama R, Kaneko T, Izumida T, Okada T, Kato T. Synthesis and electronic properties of ferrocene-filled double-walled carbon nanotubes. Nanotechnology 2006;17:4143–7.

[165] Shiozawa H, Pichler T, Grueneis A, Pfeiffer R, Kuzmany H, Liu Z, Suenaga K, Kataura H. A catalytic reaction inside a single-walled carbon nanotube. Adv Mater 2008;20:1443–9.

[166] Schulte K, Swarbrick JC, Smith NA, Bondino F, Magnano E, Khlobystov AN. Assembly of cobalt phthalocyanine stacks inside carbon nanotubes. Adv Mater 2007;19:3312–6.

[167] Costa PMFJ, Sloan J, Rutherford T, Green MLH. Encapsulation of RexOy clusters within single-walled carbon nanotubes and their in tubulo reduction and sintering to Re metal. Chem Mater 2005;17:6579–82.

[168] Zhang H, Pan X, Liu J, Qian W, Wei F, Huang Y, Bao X. Enhanced catalytic activity of sub-nanometer titania clusters confined inside double-wall carbon nanotubes. Chem Sus Chem 2011;4:975–80.

[169] Chamberlain TW, Zoberbier T, Biskupek J, Botos A, Kaiser U, Khlobystov AN. Formation of uncapped nanometer-sized metal particles by decomposition of metal carbonyls in carbon nanotubes. Chem Sci 2012;3:1919–24.

[170] Chamberlain TW, Earley JH, Anderson DP, Khlobystov AN, Bourne RA. Catalytic nanoreactors in continuous flow: hydrogenation inside single-walled carbon nanotubes using supercritical CO_2. Chem Commun 2014;50:5200–2.

[171] Zhang F, Ren P, Pan X, Liu J, Li M, Bao X. Self-assembly of atomically thin and unusual face-centered cubic Re nanowires within carbon nanotubes. Chem Mater 2015;27:1569–73.

[172] Zhang F, Jiao F, Pan X, Gao K, Xiao J, Zhang S, Bao X. Tailoring the oxidation activity of Pt nanoclusters via encapsulation. ACS Catal 2015;5:1381–5.

[173] Britz DA, Khlobystov AN, Wang J, O'Neil AS, Poliakoff M, Ardavan A, Briggs GAD. Selective host-guest interaction of single-walled carbon nanotubes with functionalized fullerenes. Chem Commun 2004;176–7.

[174] Britz DA, Khlobystov AN, Porfyrakis K, Ardavan A, Briggs GAD. Chemical reactions inside single-walled carbon nano test-tubes. Chem Commun 2005;37–9.

[175] Chamberlain TW, Camenisch A, Champness NR, Briggs GAD, Benjamin SC, Ardavan A, Khlobystov AN. Toward controlled spacing in one-dimensional molecular chains: alkyl-chain-functionalized fullerenes in carbon nanotubes. J Am Chem Soc 2007;129:8609–14.

[176] Chang J-Y, Mai F-D, Lo B, Chang J-J, Tzing S-H, Ghule A, Ling Y-C. Transportation of silver nanoparticles in nanochannels of carbon nanotubes with supercritical water. Chem Commun 2003;2362–3.

[177] Xu S, Zhang P, Li H, Wei H, Li L, Li B, Wang X. Ru nanoparticles confined in carbon nanotubes: supercritical CO2 assisted preparation and improved catalytic performances in hydrogenation of d-glucose. RSC Adv 2014;4:7079–83.

[178] Han W, Kohler-Redlich P, Scheu C, Ernst F, Ruhle M, Grobert N, Terrones M, Kroto HW, Walton DRM. Carbon nanotubes as nanoreactors for boriding iron nanowires. Adv Mater 2000;12:1356–9.

[179] Satishkumar BC, Govindaraj A, Vanitha PV, Raychaudhuri AK, Rao CNR. Barkhausen jumps and related magnetic properties of iron nanowires encapsulated in aligned carbon nanotube bundles. Chem Phys Lett 2002;362:301–6.

[180] Kim H, Sigmund W. Iron nanoparticles in carbon nanotubes at various temperatures. J Cryst Growth 2005;276:594–605.

[181] Wang W, Wang K, Lv R, Wei J, Zhang X, Kang F, Chang J, Shu Q, Wang Y, Wu D. Synthesis of Fe-filled thin-walled carbon nanotubes with high filling ratio by using dichlorobenzene as precursor. Carbon 2007;45:1127–9.

[182] Lv R, Kang F, Wang W, Wei J, Gu J, Wang K, Wu D. Effect of using chlorine-containing precursors in the synthesis of FeNi-filled carbon nanotubes. Carbon 2007;45:1433–8.

[183] Liu S, Zhu J, Mastai Y, Felner I, Gedanken A. Preparation and characteristics of carbon nanotubes filled with cobalt. Chem Mater 2000;12:2205–11.

[184] Deng D, Yu L, Chen X, Wang G, Jin L, Pan X, Deng J, Sun G, Bao X. Iron encapsulated within pod-like carbon nanotubes for oxygen reduction reaction. Agnew Chem Int Ed 2013;52:371–5.

[185] Santiso E, George A, Sliwinska-bartkowiak M, Nardelli M, Gubbins K. Effect of confinement on chemical reactions. Adsorption 2005;11:349–54.

[186] Chuvilin A, Bichoutskaia E, Gimenez-Lopez MC, Chamberlain TW, Rance GA, Kuganathan N, Biskupek J, Kaiser U, Khlobystov AN. Self-assembly of a sulphur-terminated graphene nanoribbon within a single-walled carbon nanotube. Nat Mater 2011;10:687–92.

[187] Talyzin AV, Anoshkin IV, Krasheninnikov AV, Nieminen RM, Nasibulin AG, Jiang H, Kauppinen EI. Synthesis of graphene nanoribbons encapsulated in single-walled carbon nanotubes. Nano Lett 2011;11:4352–6.

[188] Zhang J, Feng Y, Ishiwata H, Miyata Y, Kitaura R, Dahl JEP, Carlson RMK, Shinohara H, Tomanek D. Synthesis and transformation of linear adamantane assemblies inside carbon nanotubes. ACS Nano 2012;6:8674–83.

[189] Zhang J, Zhu Z, Feng Y, Ishiwata H, Miyata Y, Kitaura R, Dahl JEP, Carlson RMK, Fokina NA, Schreiner PR, Tomanek D, Shinohara H. Evidence of diamond nanowires formed inside carbon nanotubes from diamantane dicarboxylic acid. Agnew Chem Int Ed 2013;52:3717–21.

[190] Botka B, Füstös ME, Tóháti HM, Németh K, Klupp G, Szekrényes Z, Kocsis D, Utczás M, Székely E, Váczi T, Tarczay G, Hackl R, Chamberlain TW, Khlobystov AN, Kamarás K. Interactions and chemical transformations of coronene inside and outside carbon nanotubes. Small 2014;10:1369–78.

[191] Lu T, Goldfield EM, Gray SK. Chemical reactivity within carbon nanotubes: a quantum mechanical study of the $D + H_2 \rightarrow HD + H$ reaction. J Phys Chem C 2008;112:2654–9.

[192] Kondratyuk P, Yates JT. Effects of molecular confinement inside single-walled carbon nanotubes on chemical reactivity − Atomic H + 1-heptene. J Am Chem Soc 2007;129:8736–9.

[193] Halls MD, Schlegel HB. Chemistry inside carbon nanotubes: the Menshutkin SN2 reaction. J Phys Chem B 2002;106:1921–5.

[194] Giacinto P, Bottoni A, Calvaresi M, Zerbetto F. Cl(−) Exchange SN2 reaction inside carbon nanotubes: C–H···π and Cl···π interactions govern the course of the reaction. J Phys Chem C 2014;118:5032–40.

[195] Smith NM, Swaminathan Iyer K, Corry B. The confined space inside carbon nanotubes can dictate the stereo- and regioselectivity of Diels-Alder reactions. Phys Chem Chem Phys 2014;16:6986–9.

[196] Fu T, Li Z. Review of recent development in Co-based catalysts supported on carbon materials for Fischer–Tropsch synthesis. Chem Eng Sci 2015;135:3–20.

[197] Cao F, Zhong K, Gao A, Chen C, Li Q, Chen Q. Reducing reaction of Fe_3O_4 in nanoscopic reactors of a-CNTs. J Phys Chem B 2007;111:1724–8.

[198] Wang C, Pan X, Bao X. Direct production of light olefins from syngas over a carbon nanotube confined iron catalyst. Chin Sci Bull 2010;55:1117–9.

[199] Trepanier M, Tavasoli A, Dalai AK, Abatzoglou N. Co, Ru and K loadings effects on the activity and selectivity of carbon nanotubes supported cobalt catalyst in Fischer–Tropsch synthesis. Appl Catal A 2009;353:193–202.

[200] Zhang Y, Jiang W-J, Guo L, Zhang X, Hu J-S, Wei Z, Wan L-J. Confining iron carbide nanocrystals inside CN_x@CNT toward an efficient electrocatalyst for oxygen reduction reaction. ACS Appl Mater Interfaces 2015;7:11508–15.

[201] Yang H, Song S, Rao R, Wang X, Yu Q, Zhang A. Enhanced catalytic activity of benzene hydrogenation over nickel confined in carbon nanotubes. J Mol Catal A-Chem 2010;323:33–9.

[202] Rao R, Zhang Q, Liu H, Yang H, Ling Q, Yang M, Zhang A, Chen W. Enhanced catalytic performance of CeO_2 confined inside carbon nanotubes for dehydrogenation of ethylbenzene in the presence of CO_2. J Mol Catal A-Chem 2012;363-364:283–90.

[203] Zhang AM, Dong JL, Xu QH, Rhee HK, Li XL. Palladium cluster filled in inner of carbon nanotubes and their catalytic properties in liquid phase benzene hydrogenation. Catal Today 2004;93-95:347–52.

[204] Tessonnier J-P, Pesant L, Ehret G, Ledoux MJ, Pham-Huu C. Pd nanoparticles introduced inside multi-walled carbon nanotubes for selective hydrogenation of cinnamaldehyde into hydrocinnamaldehyde. Appl Catal A 2005;288:203–10.

[205] Liu J, Liu R, Li H, Kong W, Huang H, Liu Y, Kang Z. Au nanoparticles in carbon nanotubes with high photocatalytic activity for hydrocarbon selective oxidation. Dalton Trans 2014;43:12982–8.
[206] Zhang H, Pan X, Bao X. Facile filling of metal particles in small carbon nanotubes for catalysis. J Energy Chem 2013;22:251–6.
[207] Zhang H, Pan X, Liu J, Qian W, Wei F, Huang Y, Bao X. Enhanced catalytic activity of subnanometer titania clusters confined inside double-wall carbon nanotubes. Chem Sus Chem 2011;4:975–80.
[208] Rance GA, Solomonsz WA, Khlobystov AN. Click chemistry in carbon nanoreactors. Chem Commun 2013;49:1067–9.
[209] Cornelio B, Saunders AR, Solomonsz WA, Laronze-Cochard M, Fontana A, Sapi J, Khlobystov AN, Rance GA. Palladium nanoparticles in catalytic carbon nanoreactors: the effect of confinement on Suzuki-Miyaura reactions. J Mater Chem A 2015;3:3918–27.

Chapter 6

Dendrimers as Nanoreactors

Samahe Sadjadi
Gas Conversion Department, Iran Polymer and Petrochemical Institute, Faculty of Petrochemicals, Tehran, Iran

1 INTRODUCTION TO DENDRIMERS

1.1 Dendrimers

Dendrimers are highly ordered, hyperbranched, three-dimensional macromolecules with three distinguished architectural regions: (1) the core, (2) layers of branched repeat units emanating from this core, and (3) end groups on the outer layer of repeat units [1].

Dendrimers are different from hyperbranched polymers. They can be produced using a bottom-up approach and iterative sequences of reaction steps, while hyperbranched polymers are the products of noniterative polymerization [2].

At each repeatable reaction step an additional layer of branches, called generation, will be constructed [3]. Two major approaches, divergent and convergent, have been widely used for synthesis of dendrimers (Fig. 6.1) [1].

In a divergent approach pioneered by Tomalia and Newkome et al., dendrimers grow from the core to the periphery [4].

A divergent approach is an ideal choice for large-scale manufacturing of dendrimers of high generation due to its ease of procedure. Various commercial dendrimers, such as poly(propyleneimine) (PPI) and poly(amidoamine) (PAMAM), have been synthesized through this synthetic approach. However, the quality of synthesized dendrimers cannot be controlled in this method. This fact is attributed to the exponential increase of terminal groups with each generation that can result in various side reactions or incomplete reactions [1] and structural defects.

Fréchet et al. introduced the convergent method in 1989–90. In this approach, the growth will be initiated from the periphery fragments and continue inward by coupling end groups to each branch of the monomer [1,4]. A representative example of dendrimers obtained by convergent approach is polyether dendrimers (also known as Fréchet dendrimers).

Limited reactive terminal groups provide better structural control and prevent from structural defects. However, steric hindrance in this approach prohibits

Organic Nanoreactors. http://dx.doi.org/10.1016/B978-0-12-801713-5.00006-9

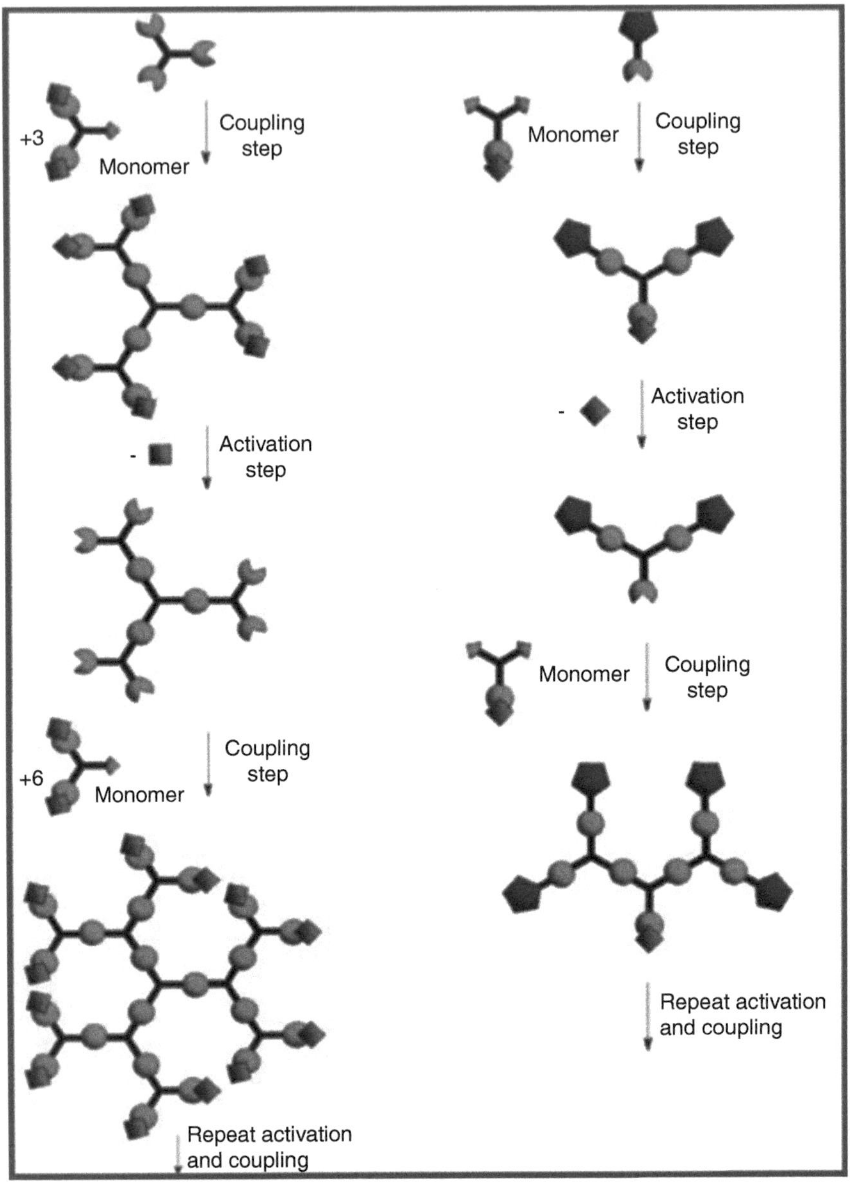

FIGURE 6.1 Schematic representation of divergent (left) and convergent (right) methods of dendrimer synthesis. *(Reprinted from Ref. [2] with permission of American Chemical Society. Copyright © 2011.)*

synthesis of dendrimers of high generation. Therefore, this method is useful for the synthesis of dendrimers of low generations [5].

Combination of divergent and convergent methods can be exploited to improve the synthesis and purity of dendrimers [4].

Recently, new methods such as orthogonal synthesis, double-stage convergent method, double-exponential method, hypermonomer method, and click chemistry have been disclosed for synthesis of dendrimers [5].

The cavity and well-defined shape of dendrimers, as well as their high chemical activity and solubility and low viscosity, result in various applications across a range of disciplines. [3]. There have been several recent review articles on specific applications of dendrimers, including dendritic catalysts [6], dendrimers as encapsulation, stabilizing, or directing agents for inorganic nanoparticles [2], dendrimer-based nanocarriers and nanovectors for tumor-targeted drug and gene delivery [7,8], dendrimer-based molecular imaging contrasting agents [9], and dendrimer-encapsulated and cored metal nanoparticles for electrochemical nanobiosensing [10].

1.2 Metallodendrimers

Metallodendrimers are a mature area of nanoscience with versatile potential applications. These compounds have been subjected to numerous studies. Several review articles summarized the recent advances and applications of metallodendrimers[11,12]. Metallodendrimers can be classified as monometallic and bimetallic dendrimers.

Monometallic dendrimers can be synthesized by chemical reduction of metal salts or complexes as nanoparticle precursors with a reducing agent such as $NaBH_4$ or H_2 (Fig. 6.2) [13].

Bimetallic dendrimers can be obtained by three prominent methods, including cocomplexation of metal salts, partial displacement, and sequential reduction. These methods were comprehensively studied and reviewed before [4,14]. The schematic representation of these methods is outlined in Fig. 6.3. The cocomplexation method results in alloy-type bimetallics, while the other two methods can lead to alloy or core/shell materials depending on the identity of the two metals and the details of how the synthesis is carried out [15].

2 NANOREACTORS

Nanoreactors are nanoscale reaction vessels that are able to encapsulate guests. It is proved that a confined nanoenvironment within a nanoreactor may influence the reaction in terms of reactivity, selectivity, the reaction pathway, yield, kinetics, or mechanism and conversion. These features of nanoreactors are reminiscent of hydrophobic pockets in enzymes, which are capable of molecular recognition and selective encapsulation of guests. Enzymes can reduce the energy of the transition states of key reaction intermediates selectively. Furthermore, they accelerate reaction rates by increasing the local concentration.

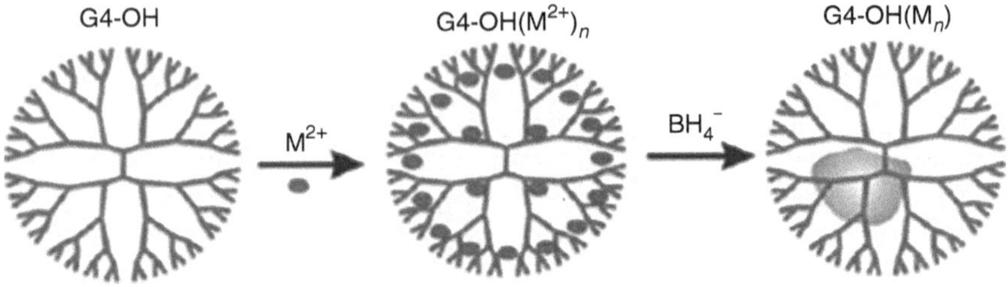

FIGURE 6.2 Schematic representation of the synthesis of dendrimer-encapsulated metal nanoparticles (DEMNs). *(Reprinted from Ref. [13] with permission of Elsevier. Copyright © 2003.)*

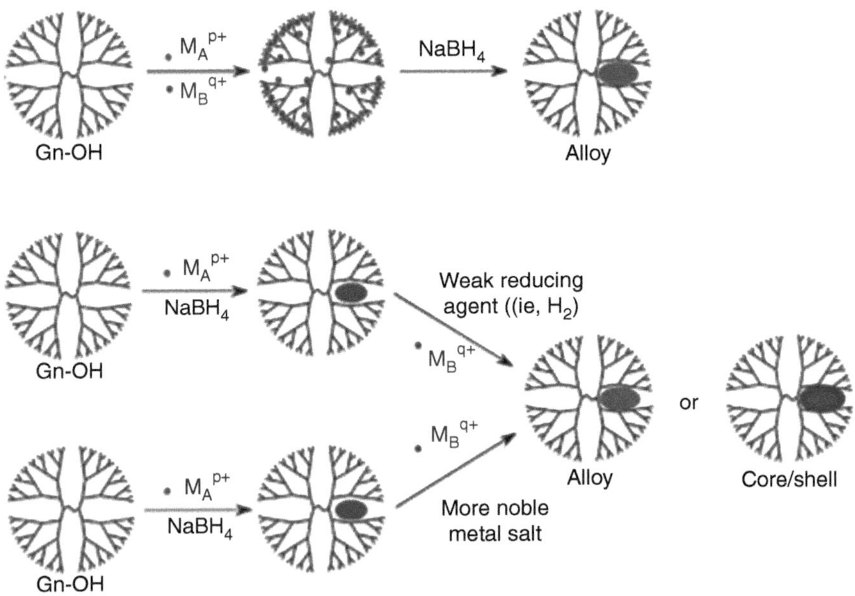

FIGURE 6.3 The schematic representation of cocomplexation, partial displacement, and sequential reduction methods. *(Reprinted from Ref. [15] with permission of American Chemical Society. Copyright 2005.)*

Nanoreactors ranging from polymeric to supramoleular have been designed and synthesized [16].

3 DENDRIMERS AS NANOREACTORS

3.1 Dendrimer Cavity

A dendrimer has a better-defined shape than ordinary polymers or surfactants and possesses dynamic inner cavities. That makes them capable of encapsulating of guests and acting as "nanoreactors" and nanoporous stabilizers [2,17].

The existence of a hollow core as well as a densely packed exterior layer in PAMAM dendrimers was proven by studying their conformational behavior [3]. Meijer et al. introduced the concept of "dendritic boxes" by synthesis and characterization of dendrimer with flexible core and solid shell structure. The flexible core was based on poly(propyleneimine) dendrimer. The rigid shell was obtained by modification of terminal groups with bulky amino acid derivatives, that is (*t*-Boc)-protected L-phenylalanine. Dendrimer structure was fully studied by ^{1}H and ^{13}CNMR, IR and UV techniques. Additionally, solid-phase behavior of the shell was confirmed by spin-lattice (T_1) and spin-spin (T_2) relaxation measurements and molecular mechanics calculations [3,18,19].

The evolution of such cavities in dendrimers is seen more profoundly in the higher-generation dendrimers, due to the increased number of branch points and the attendant increase in the number of peripheral groups, leading to more dense peripheries and less-dense interior cavities [20]. Furthermore, the existence of the interior cavity in dendrimers is related to conformational behavior and the degree of backfolding of terminal branches into the interior of the dendrimer [21].

The structural and chemical properties and polarity of the nanoenvironment within a dendrimer are greatly influenced by the choice of internal branching units, core, and rigidity of the building blocks. It is worth nothing that external parameters, such as pH and solvent can also affect the conformational features of dendrimers [3]. For example, it was established that at basic pH, poly(propylene imine) dendrimers tolerate backfolding, which results in changing conformation from a "soft-core, dense-shell" dendrimer to a "dense-core" structure (Fig. 6.4) [21,22].

The internal environment of the dendrimer can be designed in a way that mimics enzyme pockets and facilitates substrate preconcentration. For example, a nonpolar substrate placed in an aqueous environment containing a nonpolar dendrimer will tend to concentrate within the dendrimer, thus affecting subsequent reaction kinetics. Additionally, the interior cavity of dendrimers can be modulated to stabilize charged intermediates or transition states, thereby enhancing reaction rates. While an interior environment capable of reducing transition state energy can be designed, poor activity may still be observed due to steric issues involving branches (Fig. 6.5)[23].

3.2 Periphery of Dendrimers

The dendrimer periphery can have a filtering role and control the penetration of reagents to the interior of the dendrimer. In this context, dendrimers can be considered as size-selective nanofilters, ones able to distinguish between substrates that differ slightly in chemical structures [24]. The utility of the filtering role of the dendrimer periphery in catalysis was proven by Crooks et al. The authors performed hydrogenation of olefins using palladium nanoparticles encapsulated

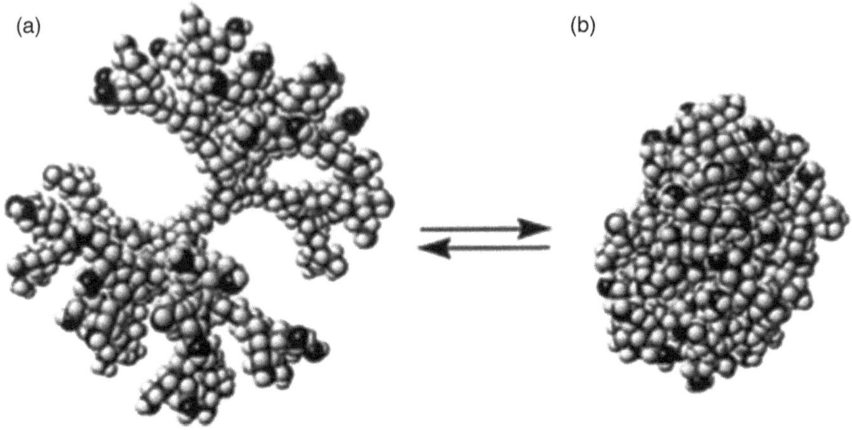

FIGURE 6.4 Gas-phase molecular modeling pictures of a fifth generation dendrimer (a) fully protonated, (b) nonprotonated. *(Reprinted from Ref. [22] with permission of American Chemical Society. Copyright 1999.)*

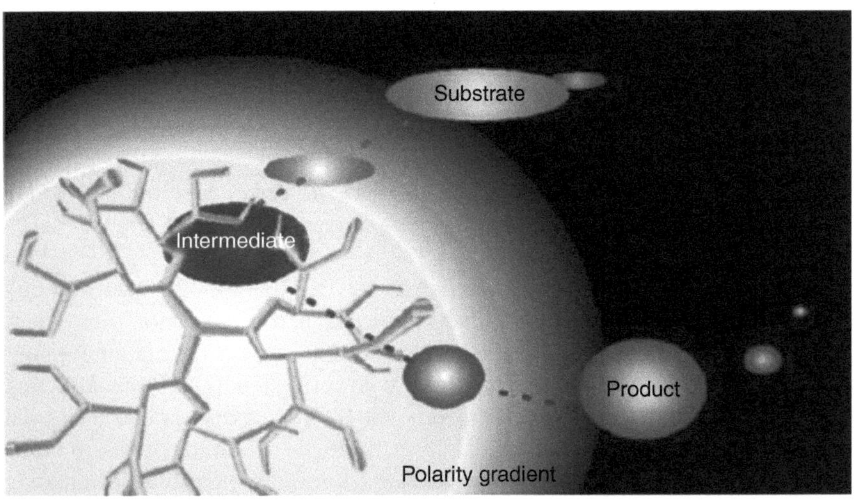

FIGURE 6.5 Illustration of substrate migration into the dendrimer interior. The nanoenvironment is greatly influenced by the branching units and creates a gradient of polarity with respect to the distinct interior and exterior solvent. These favorable conditions encourage substrate concentration and transition-state stabilization. The resulting product is released from the nanoreactor into the solvent. *(Reprinted from Ref. [23] with permission of Elsevier. Copyright 2005.)*

in three generations of PAMAM dendrimers, G_n-OH ($n = 4, 6, 8$). Keeping the size of Pd nanoparticles constant, 40 Pd atoms, a series of catalysts which differed only in porosity and diameter of dendrimers were obtained. To verify the filtering and size-selectivity of the dendrimer periphery (Fig. 6.6), the hydrogenation rates of allylic alcohols of different size and shapes were investigated

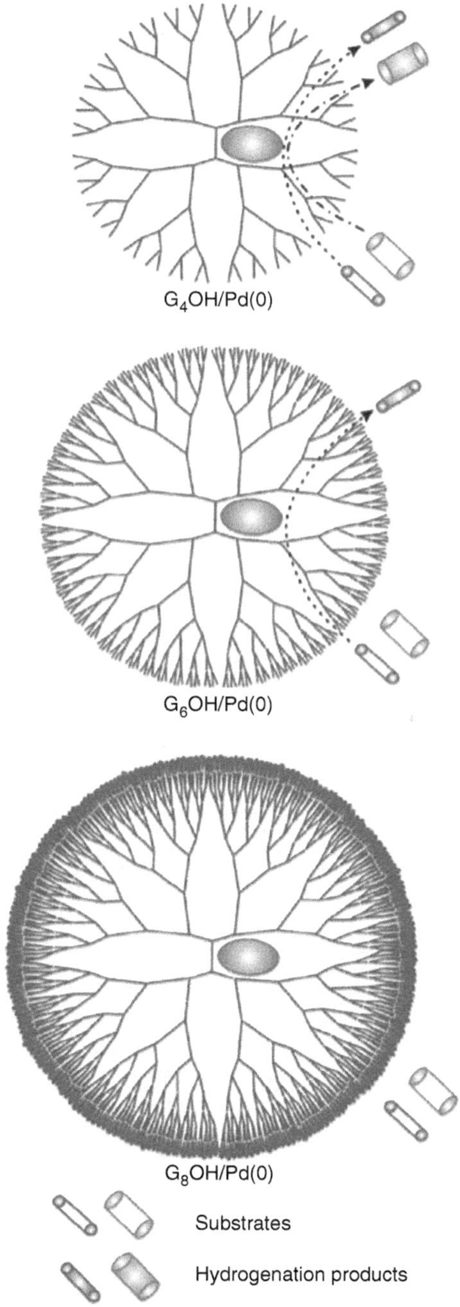

FIGURE 6.6 Schematic representation of size-selective hydrogenation by DEMNs prepared within different generation dendrimers. *(Reprinted from Ref. [13] with permission of Elsevier. Copyright 2003.)*

TABLE 6.1 Hydrogenation Reaction Rates Using GnOH/Pd(0)40 Catalysts for Structurally Related Allylic Alcohols[a]

Substrate	G_4-OH/Pd(0)40	TOF[mol H_2(mol Pd)$^{-1}$ h^{-1}] G_6-OH/Pd(0)40	G_8-OH/Pd(0)40
(1) 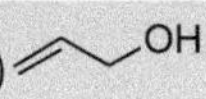	480/470[b]	450/460[b]	120
(1) 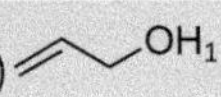	220[c]	200[c]	130[c]
(2) 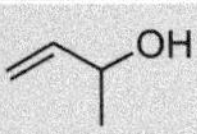	450/460[b]	380	93
(3) 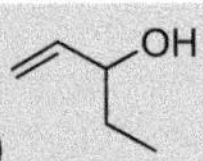	260	280	68
(4) 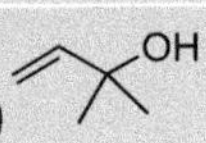	150	75	62
(5) 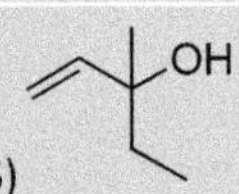	100	40	50

[a]*Hydrogenation reactions were carried out at 25 ± 2°C using 2 × 10^{-4} M Pd(0) composite catalysts in MeOH–H_2O (4:1 v/v) mixtures. The turnover frequency (TOF) was calculated based on H_2 uptake (mol of H_2 per mol of Pd(0) per h).*
[b]*Duplicate measurements were performed to illustrate the level of run-to-run reproducibility.*
[c]*Hydrogenation reactions were carried out at 25 ± 2°C using 2 × 10^{-4} M Pd(0) composite catalysts in water only (no methanol).*
Reprinted from Ref. [24] with permission of American Chemical Society. Copyright 2001.

(Table 6.1). It was established that turnover frequencies (TOFs) depend on substrate and dendrimer generation.

The nanofiltering role of the dendrimer was also confirmed by molecular modeling. The distance between adjacent terminal groups for the different-generation dendrimers was calculated using appropriate molecular models and reasonable assumptions. The calculated average edge-to-edge van der Waals distance between two terminal hydroxyl groups in G_4-OH, G_6-OH, and G_8-OH were 8.2, 5.4, and 3.2 Å, respectively [13,24]. The largest linear dimension perpendicular to the O—H bond direction of the allylic alcohols (a good measure of substrate size and free volumes of substrates) was also calculated (Table 6.2).

Comparison of experimental data, Table 6.1, and parameters obtained from molecular modeling revealed that, except substrate 4, for all dendrimer generations there was an inverse relationship between TOF and substrate size:

TABLE 6.2 The Calculated Free Volume and Substrate Size for Substrates 1–5. Adapted from Ref. [24]

Substrate	Free volume (Å)	Substrate size (Å)
(1) 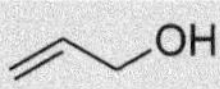	60.7	5.5
(2) 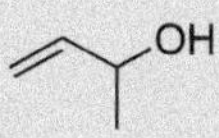	75.4	7.0
(3) 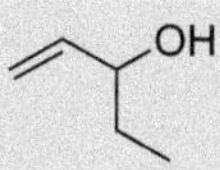	96.1	7.5
(4) 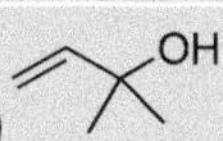	89.7	7.0
(5) 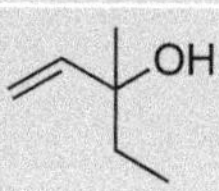	103	8.0

the larger the substrate size, the smaller the TOF. Furthermore, the TOF for a specific substrate decreased as the dendrimer size increased. These observations confirmed the filtering role of dendrimer. However, the results for substrate 4 indicated that other parameters, such as chemical and more complex structural aspects of the substrate, can also have a role in this issue.

4 DENDRITIC HOSTS

The presence of interior functional groups and cavities in dendrimers allows them to act as hosts and encapsulate guest molecules. In 1982 Maciejewskiin suggested the possibilities of guest encapsulation within dendrimers for the first time [25]. Thereafter, Tomalia provided evidence for unimolecular encapsulation of guests in dendritic hosts [21]. The chemical and structural features of guest and dendrimer, such as cavity size, and the structure and chemical composition of the terminal groups can influence the host–guest chemistry of dendrimers [19].

Various interactions—such as hydrogen-bond, van der Waals, hydrophobic force, electrostatic and metal–ligand interactions—can be involved in the guest encapsulation within the dendrimer [19,21].

Up to now, encapsulation of various guests such as anticancer drugs [26], photosensitizers [27], fullerenes [28], dyes [21], and solvents such as water, carbon tetrachloride, and 1,4-dioxane have been reported [3].

The host–guest properties of dendritic molecules and interactions involved in the encapsulation of the guest were studied comprehensively [21,29].

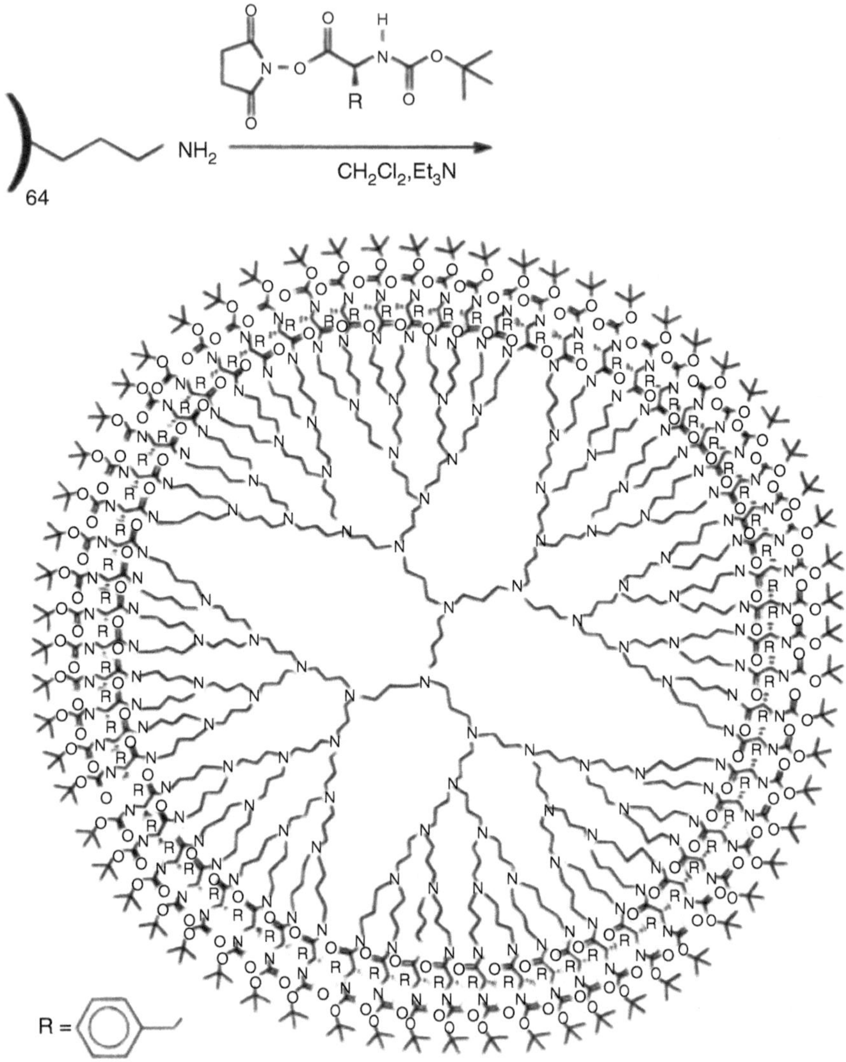

FIGURE 6.7 Dendritic box. *(Reprinted from Ref. [25] with permission of American Chemical Society. Copyright 1995.)*

Recently, the utility of NMR techniques for studying the host behavior of various dendrimers has been reviewed by Cheng [30].

Using a dendritic box which was constructed from a flexible poly(propyleneimine) dendrimer with 64 amine end groups and an L-phenylalanine derivative (Fig. 6.7) as host and Rose Bengal and *p*-nitrobenzoic acid as two guests with different sizes, Meijer et al. demonstrated shape-selective liberation of guests from dendritic boxes (Fig. 6.8) [25]. Four molecules of

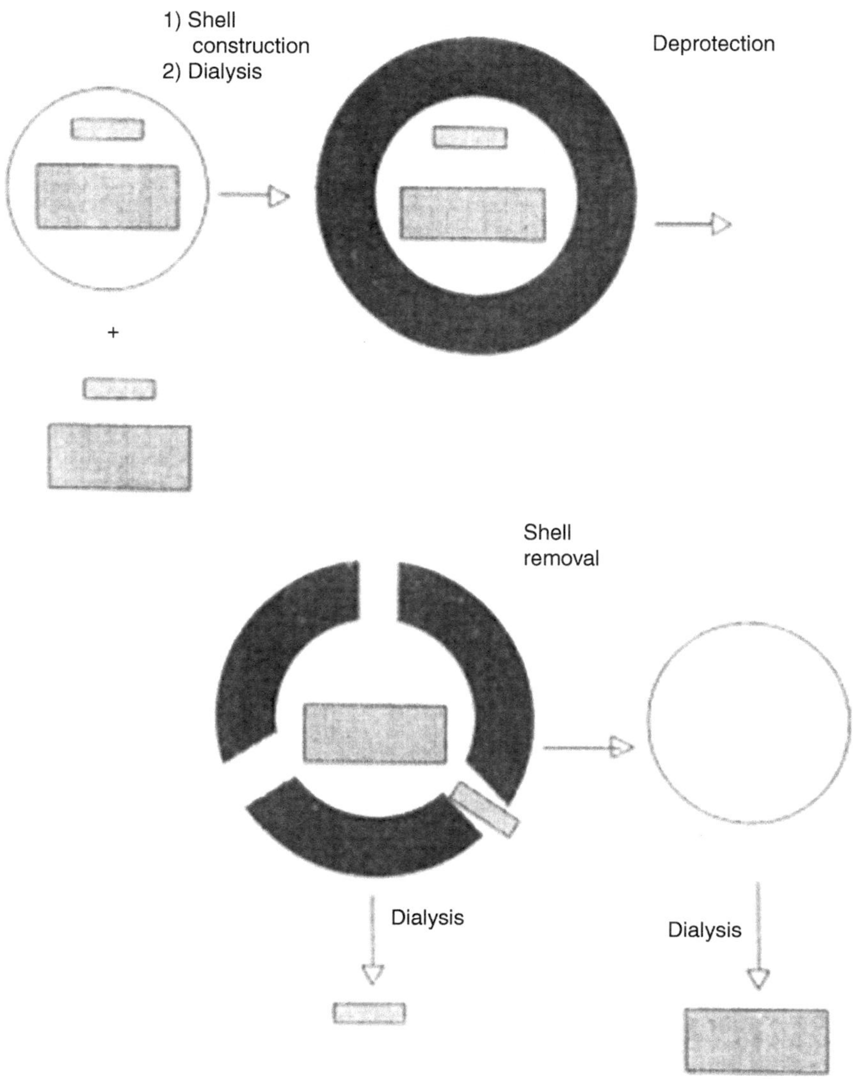

FIGURE 6.8 Schematic presentation of the principle of encapsulation and shape-selective liberation. First, two guests that differ in size are encapsulated in the box, and dialysis is used to remove adhered and excess guest. Next, the shell is partially perforated, yielding a modified dendritic box in which only the larger guest is entrapped and from which the smaller guest is liberated. Subsequent removal of the shell liberates the larger guests, and the starting poly(propylene imine) dendrimer is recovered. *(Reprinted from Ref. [25] with permission of American Chemical Society. Copyright 1995.)*

Rose Bengal and 8–10 molecules of *p*-nitrobenzoic acid could be encapsulated in each dendritic host. To achieve selective liberation of *p*-nitrobenzoic acid, the dendritic box was penetrated by hydrolysis of the *t*BOC and dialysis of the reaction mixture. In this way all *p*-nitrobenzoic acids were dissolved in the mixture while guests Rose Bengal remained encapsulated. Hydrolysis of the outer shell released Rose Bengal after dialysis [25].

5 DENDRITIC NANOREACTOR EFFECTS ON GUEST(S)

5.1 Dendritic Nanoreactor Effect on Solubility of Guest(s)

The interior cavity of most of dendrimers is hydrophobic and can encapsulate insoluble guests such as drugs. Encapsulation within a dendrimer can enhance aqueous solubility of hydrophobic guests. Dendrimer-assisted solubilization depends on various factors such as dendrimer concentration, generation size, pH, dendritic core, and so on [31]. Up to now, improvement of aqueous solubility of various drugs and biological molecules, including antitumor compound methyl (5-[propylthio]-1*H*-benzimidazol-2-yl) carbamate, Furosemide, 5-(aminosulfonyl)-4-chloro-2-((2-furanylmethyl) amino) benzoic acid [32], Nicotinic acid, 3-carboxylic pyridine [33], Nifedipine (4-(2-nitrophenyl)-2,6 dimethyl-3,5-dicarbomethoxy-1,4-dihydropyridine) [34], and Aceclofenac [35] have been reported. Jian et al. reviewed the applications of dendrimers in enhancement of drug solubility and the parameters influencing the solubilization through dendrimers [31]. In most cases, the increase in dendrimer concentration and generation leads to an increase in solubility of hydrophobes. The molecular mass of guest is also an effective parameter on solubility enhancement. Some studies showed that the increase in solubility of smaller guests was better than that of larger ones [31]. For example, dendrimer-induced aqueous solubility enhancement of small drugs such as ibuprofen and flurbiprofen was > 500-fold, while few fold enhancement of solubility of larger drugs, such as amphotericin B were obtained. Investigating dendrimer-assisted aqueous solubility of guest, one must keep in mind other factors, such as pH and protonation status of dendrimer and dendrimer terminal groups.

Dendrimers have also been used for improving biomedical applications of some carbon nanomaterials through increasing their poor water solubility.

Kono et al. reported improvement of water-solubility of fullerene, C_{60} by using PAMAM dendrimer modified with both poly(ethylene glycol) (PEG) and β-cyclodextrin (CD) [28]. The authors believed that the presence of CD at the periphery of dendrimer can increase the formation of complex with fullerene, while PEG moiety can promote water-solubility and biocompatibility. The selection of β-cyclodextrin was based on its low affinity for interaction with PEG. The synthetic procedure for preparation of PEG/CD-modified dendrimer and encapsulation of fullerene is depicted in Fig. 6.9.

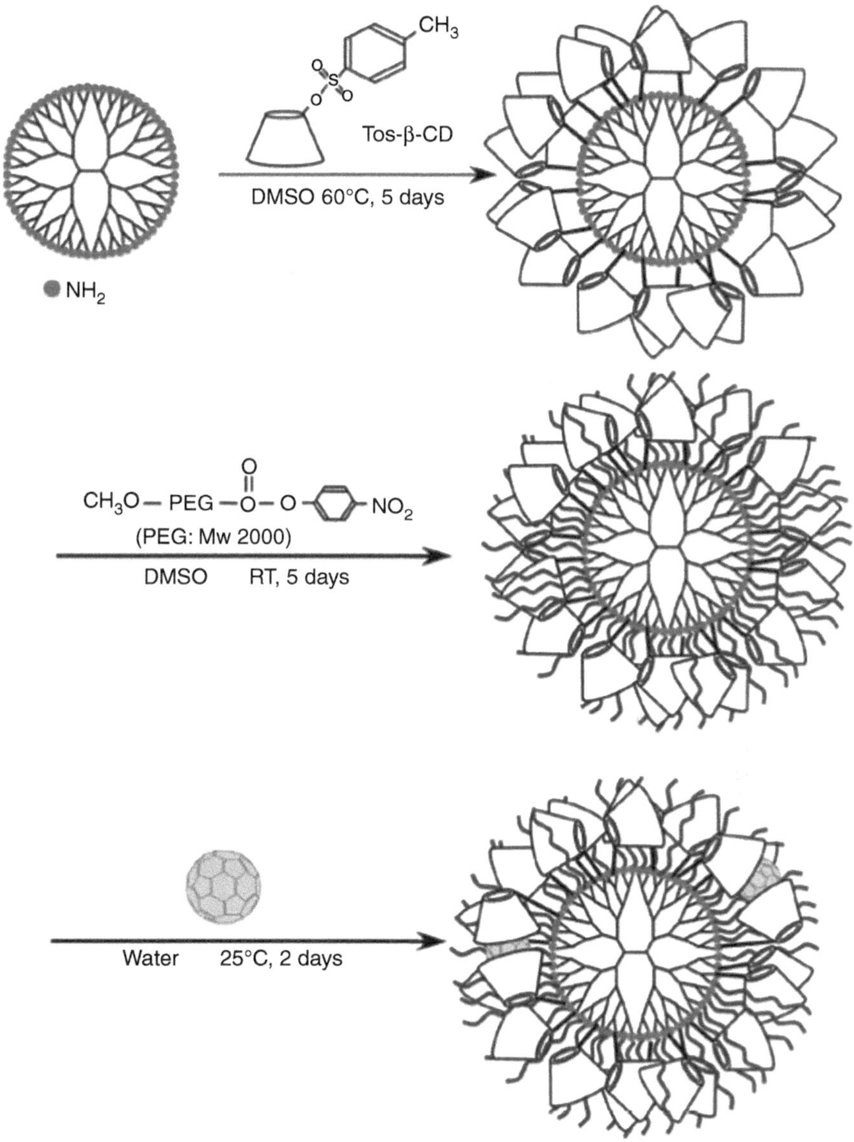

FIGURE 6.9 Preparation method of the PEG/CD-modified dendrimer including C_{60}. *(Reprinted from Ref. [28] with permission of American Chemical Society. Copyright 2008.)*

To elucidate the effect of PEG/CD modified dendrimer on C_{60} water-solubility, the UV-vis spectra of C_{60} aqueous solutions of the PEG/CD modified dendrimers, the CD-modified dendrimer, the PEG dendrimer, free CD, and free PEG were recorded (Fig. 6.10). Comparing UV-vis spectra of free CD and CD-modified dendrimers implies that attachment of CD to a modified dendrimer

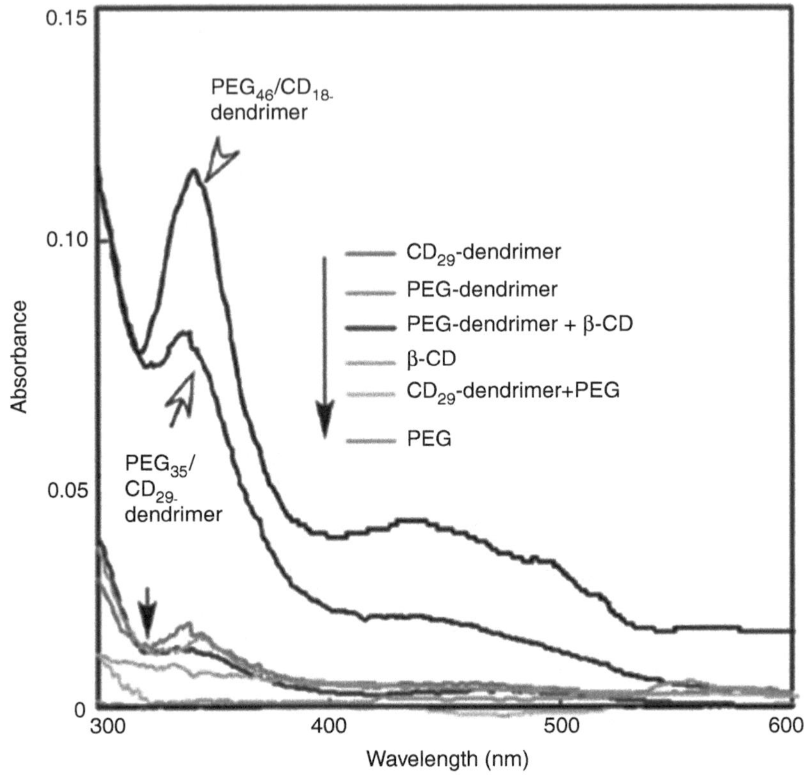

FIGURE 6.10 UV-vis spectra of C_{60} aqueous solutions containing the PEG_{46}/CD_{18}-modified dendrimer (2.0 μM), the PEG_{35}/CD_{29}-modified dendrimer (2.0 μM), the CD29-modified dendrimer (2.0 μM), and the PEG_{64}-modfied dendrimer (2.0 μM), free CD (58 μM), free PEG (70 μM), or the mixtures. These spectra were obtained by subtracting the spectra of these compounds without C_{60} from those of the complex. *(Reprinted from Ref. [28] with permission of American Chemical Society. Copyright 2008.)*

can increase the ability to complex with C_{60}. It was also observed that neither free PEG nor PEG-modified dendrimer dissolved C_{60} effectively. The solution containing both PEG- and CD-modified dendrimers had stronger adsorption, suggesting that the presence of both PEG and CD of the dendrimer were crucial for the efficient complex formation with C_{60}.

The authors assumed that the effect of CD and PEG resulted from clustering at the surface of the dendrimer. Comparing the PEG_{46}/CD_{18}-modified dendrimer with the PEG_{24}/CD_{29}-modified dendrimer established that the former is more effective in solubilizing C_{60}. It was concluded that in the case of PEG_{24}/CD_{29}-modified dendrimers, CD molecules might be too crowded to interact with C_{60} at the surface of the dendrimer. It is also possible that the sufficient number of PEG chains at the surface of the dendrimer need to disperse the complex with C_{60} [28].

5.2 Dendritic Nanoreactor Effect on Biocompatibility of Guest(s)

The cytotoxicity of dendrimers has been primarily studied in vitro, however, a few in vivo studies have been published. Dendrimers with positively charged surface groups are prone to destabilize cell membranes and cause cell lysis. One way to reduce the cytotoxicity of cationic dendrimers may reside in partial surface derivatization with chemically inert functionalities such as PEG or fatty acids [36]. The modified dendrimers can be used for synthesis of dendrimer-encapsulated nanoparticles with improved biocompability. As an example, modified amine-terminated poly(amidoamine) dendrimers of generation 5, $G_5.NH_2$, were used to develop biocompatible and stable computed tomography (CT) imaging contrast agents. Initially, dendrimers were modified with polyethylene glycol (PEG). This can render dendrimer biocompatibile. PEGylated dendrimers were subsequently used as a template for synthesis of dendrimer-encapsulated gold nanoparticles ($[(Au^0)_n$-$G_5.NH_2$-$m$$PEG_{20}]$ DENPs, n denotes the molar ratio between dendrimers and Au salt). Finally, the remaining dendrimer terminal amines were neutralized by acetylation and $[(Au^0)_n$-$G_5.NHAc$-$m$$PEG_{20}]$ DENPs with a size range of 2–4 nm were obtained (Fig. 6.11) [37].

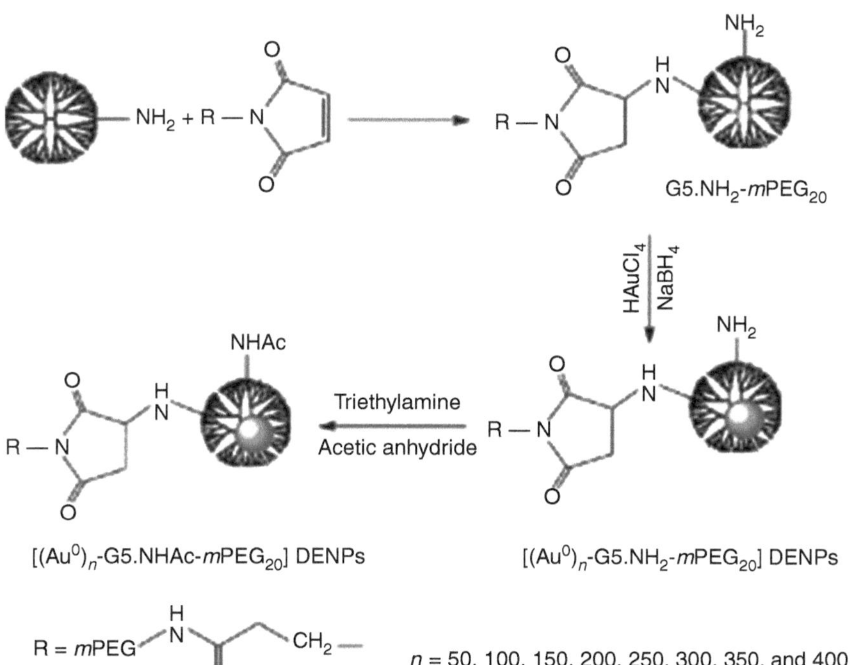

FIGURE 6.11 Schematic illustration of the preparation of the $[(Au^0)_n$-$G_5.NHAc$-$m$$PEG_{20}]$ DENPs. *(Reprinted from Ref. [37] with permission of Elsevier. Copyright 2012.)*

The formed PEGylated Au DENPs were water-dispersible, stable in a pH range of 5–8 and a temperature range of 0–50°C, and noncytotoxic at a concentration as high as 100 μM. They enabled not only X-ray CT blood pool imaging of mice and rats after intravenous injection of the particles, but also effective CT imaging of a xenograft tumor model in nude mice [37].

5.3 Dendritic Nanoreactor Effect on Fluoresces of Guest(s)

Balzani and Vogtle et al. investigated the encapsulation of eosin Y (E^{2-}), Fig. 6.12, in fluorescent dansyl functionalized poly(propylene amine) dendrimer of various generations, *n*D, $n = 1–5$, and other fluorescent dyes, Fig. 6.12, in 4D [38]. It was found that dendrimer generation and electronic properties affect the formation of host–guest species. The higher the dendrimer generation, the more dye molecules can be encapsulated. Using generation 5 dendrimer, 5D, up to 12 eosin molecules could be hosted. 4D could maximally encapsulate 25 relatively large Rose Bengal molecules and 1 smaller fluorescein molecule. This indicates that the size of dye does not have a crucial role in encapsulation.

Furthermore, upon encapsulation and formation of dendrimer-encapsulated dyes, new fluorescence properties which were different from each constituent were observed and assigned to quenching/sensitization processes. The fluorescence of peripheral dansyl groups of the dendrimer was totally quenched through energy transfer by the encapsulated dyes while hosted dyes experienced partial quenching.

Excitation of dendrimer-encapsulated eosin results in sensitization of the dioxygen emission via eosin triplet excited state. The behavior of fluorescein and Rose Bengal was qualitatively similar to that of eosin, whereas 5-(and-6)-carboxynaphthofluorescin was not encapsulated.

It's worth mentioning that, encapsulation of dyes in dendrimers can be used for dye extraction. The used dendrimers were soluble in dichloromethane while insoluble in water. As eosin is colored in aqueous solution at $pH > 5$, mixing the colored aqueous solution of eosin with dendrimer (2D–5D) containing dichloromethane results in phase transfer of eosin from aqueous to dichloromethane phase through dye encapsulation in dendrimer. This can be detected visually and quantitatively by measuring the absorbance changes at the wavelength of maximum of eosin absorption spectrum in water ($\lambda_{max} = 515$ nm).

5.4 Dendritic Nanoreactor and Guest Isomerization

Ramamurthy and Jayaraman employed the cavity of poly(alkyl aryl ether) dendrimers of the third generation with various *n*-alkyl linker (ethyl to *n*-pentyl) connecting the branch points (C_2G_3–C_5G_3) for photochemical isomerizations of azobenzene and stilbene. Isomerization was carried out in aqueous dendrimer solutions. Irradiation of solutions for 6 h furnished products. The results of isomerization of stilbene in the presence of various dendrimer solutions and in

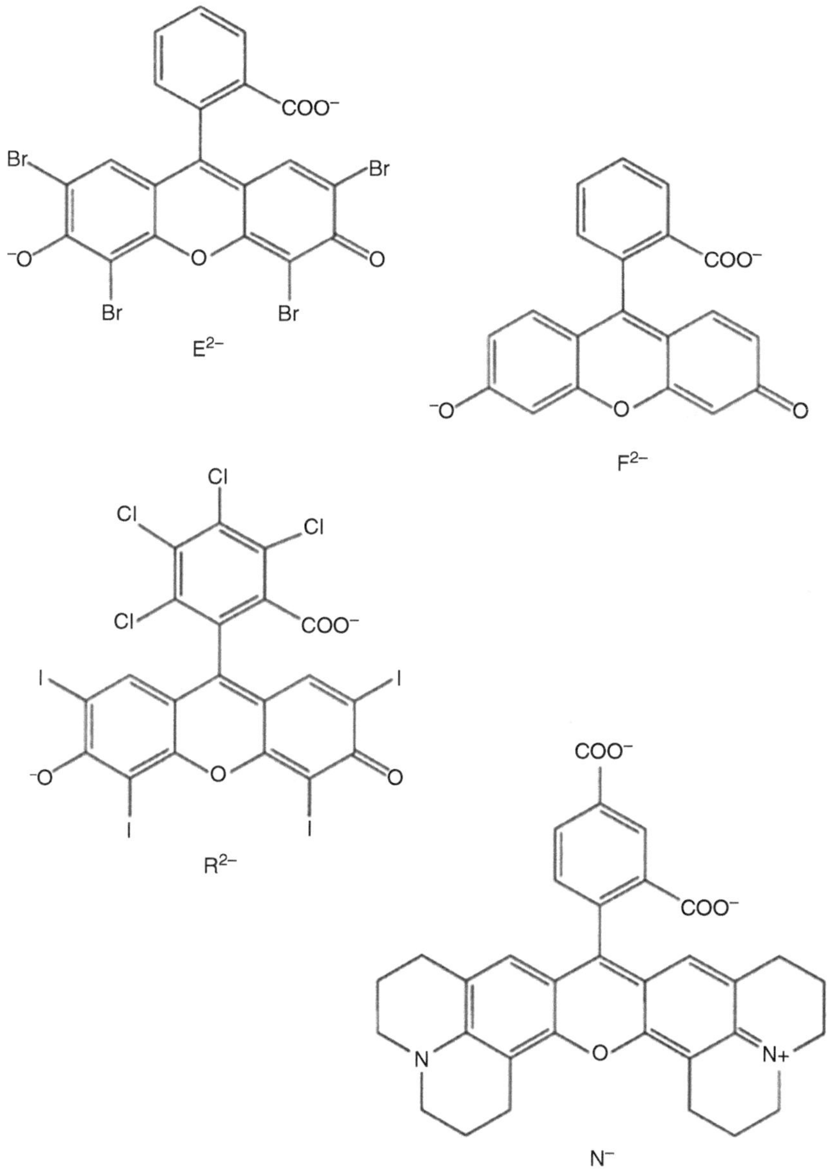

FIGURE 6.12 Structure formulas of the dyes (guests) [38]. Eosin Y (E^{2-}), Fluorescein (F^{2-}), 5-(and-6)-carboxynaphthofluorescin (N^{-}) and Rose Bengal (R^{2-}). *(Reprinted from Ref. [38] with permission of Elsevier. Copyright 2002.)*

TABLE 6.3 Isomerization of *trans*-Stilbene in Aqueous Dendrimer Solutions and in EtOH

	trans-Stilbene (%)	*cis*-Stilbene (%)	Phenanthrene (%)
EtOH	15	56	29
C_5G_3		72	28
C_4G_3		91	9
C_3G_3		99	trace
C_2G_3		99	trace

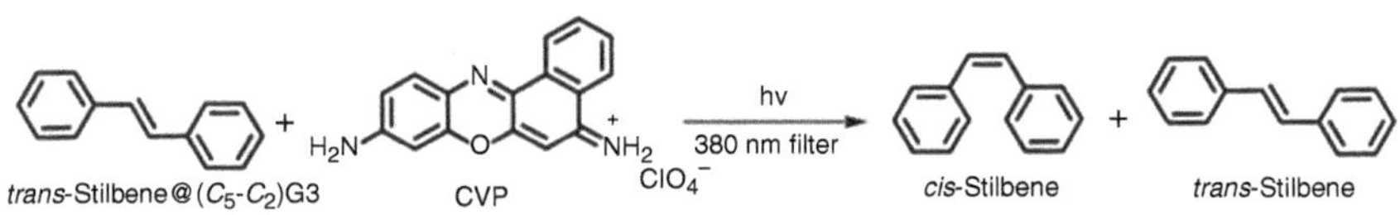

FIGURE 6.13 Photoisomerization of trans-stilbene@(C_5—C_2)G_3 dendrimers in the presence of CVP as electron-transfer sensitizer. *(Reprinted from Ref. [20] with permission of American Chemical Society. Copyright 2012.)*

the absence of dendrimer in ethanol are summarized in Table 6.3. The formation of phenanthrene in addition to isomerization product, *cis*-stilbene, in ethanol solution and in samples with C_5G_3 and C_4G_3 dendrimers implied that long linker in dendrimer facilitates the formation of cyclization product. Furthermore, the absence of formation of tetraphenylcyclobutane which is [2 + 2] cycloaddition product was attributed to confinement of stilbene within the cavity of the dendrimer. Comparison of results of various dendrimers suggested that dendrimers with smaller inner cavities can stabilize the *cis*-isomer favorably [20].

Using cresyl violet perchlorate (CVP) as the electron-transfer sensitizer, photo-induced electron transfer of stilbene was also studied (Fig. 6.13). As expected, isomerization of *trans*-stilbene in ethanol resulted in *trans*-isomer as main product (Table 6.4). Interestingly, in aqueous media containing dendrimers, *cis*-isomers were obtained as main products. Additionally, like direct irradiation, dendrimers with smaller cavity size favored *cis*-stilbene formation. This observation can confirm the role of dendrimer cavity size.

6 DENDRITIC NANOREACTORS AS TEMPLATING AND STABILIZING AGENT

The cavity of a dendrimer can be exploited for templating and stabilizing of nanoparticles. In this way a single nanoparticle will be stabilized within the interior space of a single dendrimer. The obtained dendrimer-encapsulated

TABLE 6.4 Photo-Isomerization of *trans*-Stilbene Encapsulated Within C_2G_3–C_5G_3 Dendrimers, in the Presence of CVP as Electron-Transfer Sensitizer Dye, Using a 380 nm Filter and Irradiation for 2 h

Dendrimer	*trans*-Stilbene (%)	*cis*-Stilbene (%)
EtOH	98	2
C_5G_3	16	84
C_4G_3	5	95
C_3G_3	1	99
C_2G_3	2	98

nanoparticles, DENs, have wide range of applications especially in catalysis. In this case, dendrimer cavity can control the size of nanoparticles and prevent the agglomeration while providing sufficient space for substrates molecules to enter and react in the presence of metallic catalyst [21].

Poly(amidoamine) dendrimers (PAMAM) and Poly(propyleneimine) dendrimers (PPI) and their derivatives have been widely used as a template for preparation of DENs.

There are some requisites to achieve effective encapsulation and avoid the exchange of metal ions: (1) the appropriate dendrimer generation (size) has to be chosen for encapsulation purposes, (2) proper reaction conditions should be used, and (3) interdendrimer interaction should be prevented [2]. This can be achieved by using inert or charged terminal groups or inert tails.

Another approach for nanoparticle stabilization is using dendrimers as surfactants for the formation of nanoparticles between dendrimers, that is, interdendrimer stabilization [2]. In this case, dendrimers play the role of surfactants and will be adsorbed on the growing nanoparticles. Although larger nanoparticles with broader size distribution are expected in this approach, its versatility makes it attractive for stabilizing nanoparticles. One of the features of this method is that dendrimer stabilization can be accomplished after the formation of nanoparticles by replacing other surfactants with dendrimers. The driving force for this replacement can be stronger chelating groups and /or increase of entropy [2].

Coating nanoparticles with dendron and formation of nanoparticle-core dendrimer is the third approach for dendrimer-assisted stabilization of nanoparticles. Three strategies for preparing dendron-coated nanoparticles are ligand exchange for dendron, dendron growth from the nanoparticle core, and nanoparticle formation in the presence of dendrons [2].

Dendrimer-encapsulated nanoparticles can be characterized using various techniques, including UV-vis and ATR-IR spectroscopies, X-ray methods

(EXAFS, EDS, XPS), EPR, and NMR, and electron and atomic force microscopy techniques [39].

6.1 Stabilization and Templating of Metallic Nanoparticles

Up to now templating and stabilizing of various mono and bimetallic dendrimer-encapsulated nanoparticles, such as Ag [40], Au [41], Co [42], Pd [43], Ni [44], Cu [45], Ag–Pd [46], Au–Ag [47], Pd–Au [48], and so on, have been reported. There are also some review articles and book chapter covering this topic [14]. In this section some recent examples of dendrimer- stabilized nanoparticles are presented.

6.1.1 *Stablizing and Templating of Monometallic Dendrimer-encapsulated Nanoparticles*

Miyahara et al. employed hydroxyl-terminated PAMAM dendrimer of generation 4, G_4-OH, and carboxyl-terminated PAMAM dendrimer of generation 4.5, G_4.5-COO^-, as a template for synthesis of dendrimer-encapsulated Pt nanoparticles. Using TEM and UV-vis techniques, authors studied the effects of reaction conditions, such as temperature, pH, coordination time, and surface functional groups of the dendrimer on Pt^{2+} coordination (Fig. 6.14). As shown in Fig. 6.14, the coordination proceeds at neutral and acidic pH for G_4-OH and G_4.5-COO^- solutions, respectively. The decreased absorbance after 10 days in basic condition was attributed to replacement of Cl^- ligands with OH^- ligands. The unchanged absorbance in acidic condition was attributed to absence of ligand replacement.

It was demonstrated that high Pt^{2+} coordination ratios (α) can result in dendrimer-encapsulated Pt nanoparticles with narrow size distribution, while low values of α lead to nonencapsulated nanoparticles. Furthermore, the coordination can be accelerated by increasing temperature to 50°C [49].

Amiridis et al. reported the application of fourth-generation hydroxyl-terminated poly(amidoamine) (PAMAM) dendrimers, G_4OH-PAMAM, as a template for synthesis of ZrO_2-supported Rh subnanometer particles. Using EXAFS and HRTEM techniques, it was established that partially hydrolyzed Rh^{3+} species in an aqueous solution can interact with amide/amine groups in the interior of G_4OH dendrimers to form $(Rh^{3+})_x$–G_4OH complexes where x is the Rh/dendrimer molar ratio. Room temperature reduction with $NaBH_4$ results in the formation of Rh_3 clusters.

The resulting Rh-dendrimer nanocomposites can be deposited onto a high surface area ZrO_2 support. Removal of dendrimer and the formation of stable Rh nanoparticles was obtained by thermal treatments in O_2 and H_2. Very limited sintering of Rh was observed during these thermal treatments, leading to the formation of Rh particles with subnanometer dimensions [50].

Using alkyl-functionalized PAMAM dendrimers, Crooks et al. synthesized dendrimer-encapsulated Ni nanoparticles containing 55 and 147 metal atoms. The hydrophobic dendrimer which is soluble in organic solvents prevents Ni

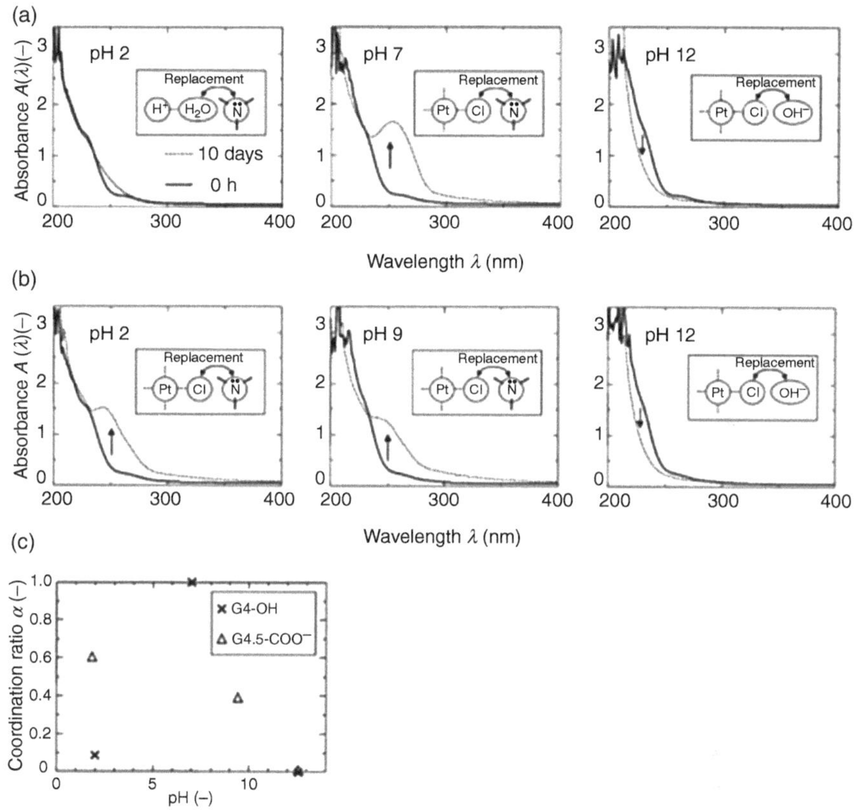

FIGURE 6.14 UV-vis spectroscopic data for Pt^{2+} coordination with (a) G_4-OH and (b) G_4.5-COO^- at various pH after 0 h (solid curves) and 10 days (dotted curves). (c) Relation between pH and R after 10 days. *(Reprinted from Ref. [49] with permission of American Chemical Society. Copyright 2010.)*

from oxidation by water and oxygen [44]. Magnetic analysis indicated that Ni DENs were ferromagnetic at 5 K and exhibited low hysteresis and suppressed magnetic saturation, both of which were directly correlated to the small size of the particles.

Synthesis of dendrimer-encapsulated coppers as a novel electrochemical label for sensitive electrochemical stripping detection of DNA hybridization has been reported by Zhu et al. Cu^{2+}/dendrimer nanocomposite can be considered as a new and flexible electrochemical label for biomolecules due to its good reproducibility and detection sensitivity [45].

6.1.2 Stabilizing and Templating of Bimetallic Dendrimer-Encapsulated Nanoparticles

Using amine-terminated PAMAM dendrimers of generation 5 as templates or stabilizers and adjusting the gold atom/silver atom/dendrimer molar ratios, a series of Au–Ag alloy DENPs were synthesized in aqueous solution. To improve the cytocompatibility of the obtained DENPs, the surface charge of particles was neutralized by acetylation of dendrimer terminal amine groups.

The synthesized Au–Ag alloys were fully characterized before and after acetylation. It was demonstrated that both Au atom/Ag atom/dendrimer molar ratio and the acetylation modification influence the size of binary nanoparticles. With constant total metal atom/dendrimer molar ratio, the size of alloy nanoparticles decreases with gold composition. For particles with similar metal composition, the size of the particles becomes slightly larger after acetylation [47].

Evaluation of colloidal stability of synthesized binary nanoparticles at different pH and temperature conditions indicated their stability at different pH (pH = 5–8) and temperature (4–50°C) conditions. To assess the potential application of Au–Ag alloy as CT imaging agents, their X-ray attenuation characteristics were also studied. It was demonstrated that the attenuation of the binary nanoparticles is dependent on both gold content and the particle surface characteristics [47].

Using silver(I)–bis(oxalato)palladate(II) complex, Rhee et al. reported synthesis of ultrafine Ag-Pd bimetallic nanoparticles. The advantage of this complex, which contains both metal ions, is prevention of formation of silver halides. Moreover, the oxalate ligand rapidly decomposes by light irradiation or chemical reduction. It was demonstrated that the particle size distribution is dependent on the Ag/Pd ratio [46].

The UV-vis spectra of Ag–Pd bimetallic nanoparticles with various Ag/Pd ratios are displayed in Fig. 6.15. As the Ag/Pd ratio increases, the λ_{max} shifts to a longer wavelength. Moreover, it should be noted that the spectra of the resulting nanoparticles are different not only from those of the monometallic Ag or Pd nanoparticles but also from those of their physical mixtures. Therefore, the characteristic absorption spectra strongly suggest the formation of Ag–Pd bimetallic nanoparticles. The change in the absorption spectra of the bimetallic nanoparticles from those of monometallic ones can be primarily attributed to the change in dielectric function caused by mixing the two different metal atoms.

The variations of particle size with the change of Ag/Pd ratio are presented in Fig. 6.16. It is observed that the average particle size increases with the increased Ag content.

Crooks et al. reported oxidation-driven structural conversion of quasi-alloy PdAu dendrimer-encapsulated nanoparticles to a Au-core/Pd-shell configuration. The six generation of poly(amidoamine) dendrimer was used as a template for synthesis of the initial quasi-alloy by cocomplexation of correspondent metal ions, $PdCl_4^{2-}$ and $AuCl_4^-$. Exposure to air resulted in selective reoxidation of the Pd atoms, and subsequent rereduction led to deposition of a Pd-rich shell on the surface of the remaining Au core (Fig. 6.17).

Dodecanethiol extraction of the materials into hexanes led to formation of Au-core/Pd-shell monolayer-protected clusters [48].

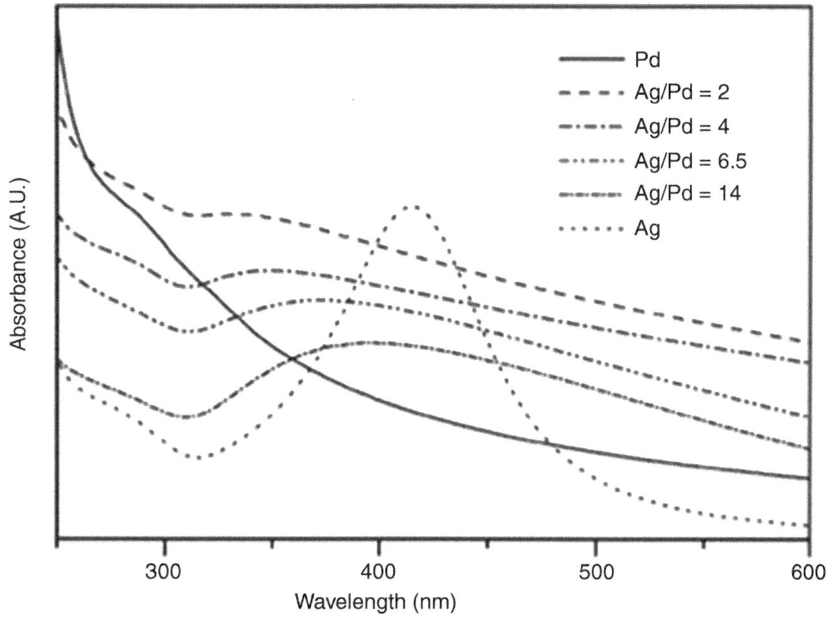

FIGURE 6.15 UV-vis spectra of Ag–Pd bimetallic nanoparticles with various Ag/Pd ratios. *(Reprinted from Ref. [46] with permission of Elsevier. Copyright 2004.)*

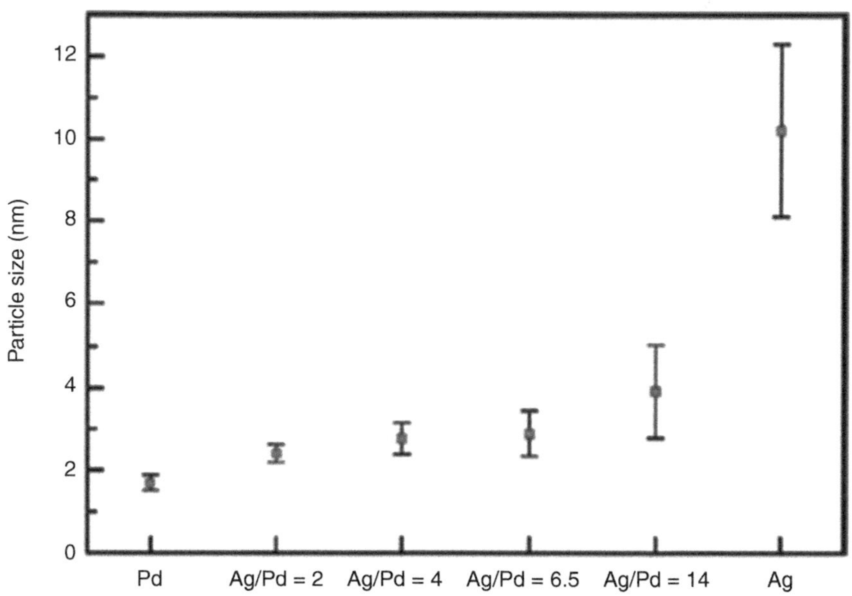

FIGURE 6.16 Variations of particle size with the change of Ag/Pd ratio. *(Reprinted from Ref. [46] with permission of Elsevier. Copyright 2004.)*

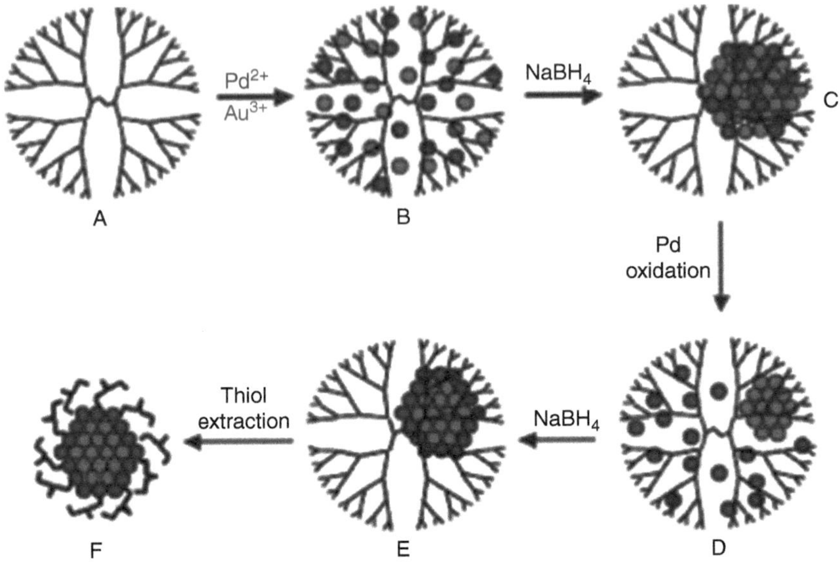

FIGURE 6.17 Oxidation-driven structural conversion of quasi-alloy PdAu dendrimer-encapsulated nanoparticles to a Au-core/Pd-shell configuration. *(Reprinted from Ref. [48] with permission of American Chemical Society. Copyright 2008.)*

6.2 Stabilization and Templating of Quantum Dots

Semiconductor nanocrystals and quantum dots (QDs) have been introduced as promising labeling materials for biological and biomedical applications. In comparison with organic flurophoroes which have narrow excitation bands, broad emission spectra with a long tail at red wavelength, low photochemical stability, and sophisticated synthetic chemistry, the band-edge photoluminescence (PL) emissions of high-quality semiconductor nanocrystals are narrow, symmetric, and strongly size-dependent due to quantum confinement effects [51].

The prerequisite for biomedical applications of QDs is availability of photostable, compatible, and water-soluble nanocrystals [52].

A monolayer of organic ligand on the surface of nanocrystals can stabilize them and provides binding sites for the outside world [51].

Most of the reported synthesis methods result in QDs coated with hydrophobic molecules, which make them insoluble in water. For using these QDs as biolables, the hydrophobic outer surface has to become hydrophilic to make them water-soluble and biocompatible.

Three main strategies exist for converting hydrophobic QDs into hydrophilic ones. The first approach is based on either micelle or polymer wrapping, with either synthetic or natural polymers. The supramolecular assembly is, however, large and mainly maintained by local hydrophobic interactions. Consequently, the optical properties of the nanocrystals can be largely retained. The thick polymer-wrapping layer provides a sufficient steric protection for the nanocrystals, but it also makes the complex relatively large in size and limited in physical permeability. The second approach is to coat a hydrophilic shell, such as a silica coating, on the surface of nanocrystals. This method results in low yield, and the sensitivity of the silica shell to pH may cause precipitation and gel formation. In addition, the silica shell is typically relatively thick and as irregular as the polymer wrapping layers. The third approach consists of exchanging of the original organic ligands from the synthesis with hydrophilic ligands. This strategy has successfully been applied using functional groups which strongly adhere to the QD surface, such as phosphines, amines, and thiols. Among the various ligands used, hydrophilic dendron ligands [53] and dendron-box [51,54] ligands are promising because of their exceptional stability in comparison to that of other hydrophilic ligands and relatively thin ligand layer [52].

Li et al. reported synthesis of dendrimer-stabilized CdSe/ZnS core–shell semiconductor nanocrystals. A ligand exchange process—with a dendron ligand with two primary amine-anchoring groups as its focal point, and hydroxyl groups as its terminal groups—has been exploited for making nanocrystals water soluble. In this dendron ligand, two primary amine-anchoring groups serve as bidentate ligands for the QD, while the protruding hydroxyl groups enhance the water-solubility of the dendron nanocrystals (Fig. 6.18).

The stable and bright dendron nanocrystals were soluble in various aqueous media, including all common biological buffer solutions tested. However, coating with dendron led to an increase of nanocrystal diameters [52]. In this example, dendritic nanoreactor influenced the solubility of QDs and improved its stability.

In an attempt to develop synthesis of QDs using dendrimers as template, amino-polyester dendrons were used for templating and stabilizing of CdS nanoparticles on the surface of single-wall carbon nanotubes (SWNTs) (Fig. 6.19). Attachment to SWNTs can be achieved by focal group while the interior space of the dendron serves for nanoparticle formation. The obtained CdS nanoparticles with diameter of 1.4 nm were stable for more than 90 days at 25°C. It has been suggested that these characteristics may be useful for fabricating molecular electronic devices [55].

In another example, the third generation of isocyanate-based dendrons was used to regioselectively react with primary alcohols of cellulose and form dendrimerized cellulose polycarbamate (Fig. 6.20). To obtain modified cellulose/CdS composites, the terminal ester groups of the dendronized cellulose were cleaved using formic acid to generate the {[$(HO_2C)_{27}$-Den]-cellulose}.

Formation of the CdS with {[$(HO_2C)_{27}$-Den]-cellulose} was performed similar to that of dendrimer-templated nanoparticles. The authors believed that the size of QDs was in the quantum-confined regime with luminescence properties [56].

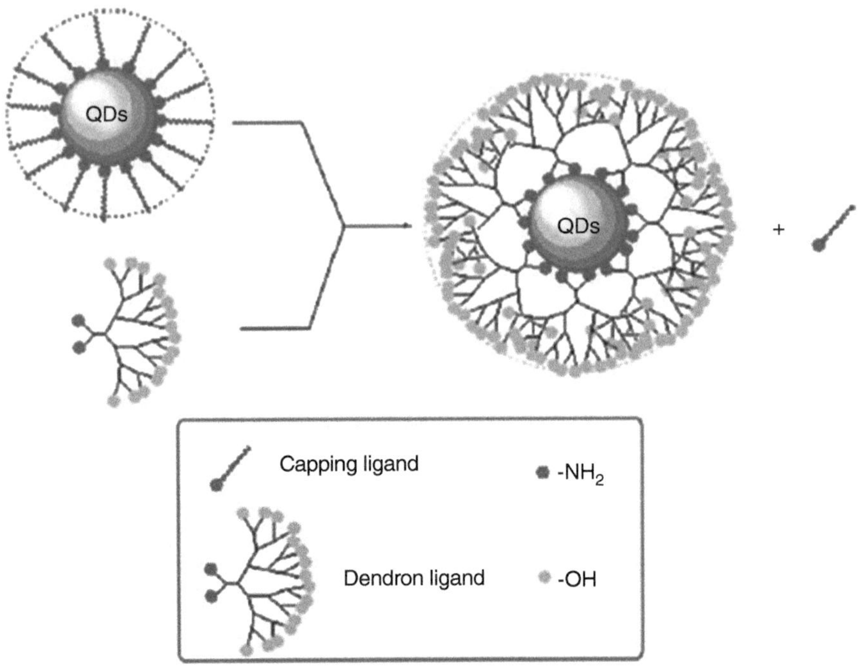

FIGURE 6.18 Schematic illustration of the formation of the core-shell CdSe/ZnS QDs by using bidentate dendron ligands. *(Reprinted from Ref [52] with permission of Elsevier. Copyright 2009.)*

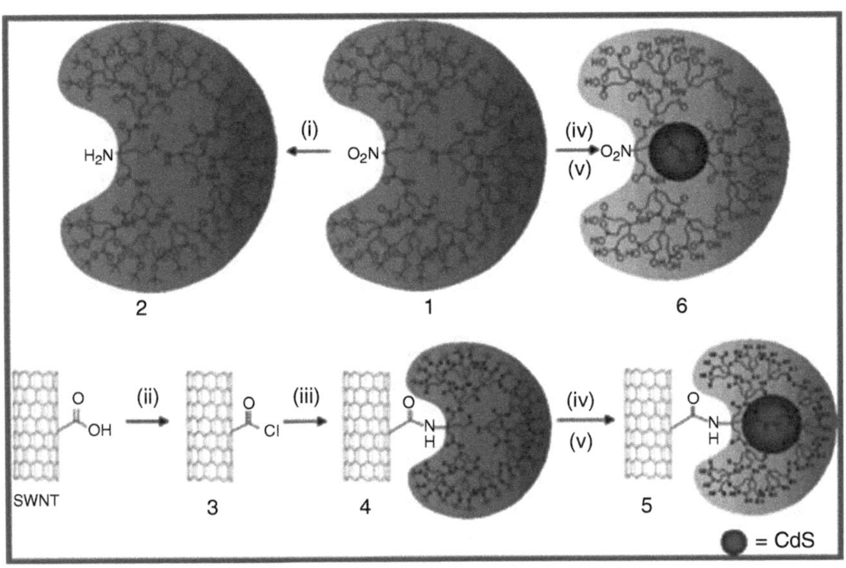

FIGURE 6.19 Synthesis of the tethered (5) and untethered (6) CdS QDs. *(Reprinted from Ref. [55] with permission of American Chemical Society. Copyright 2006.)*

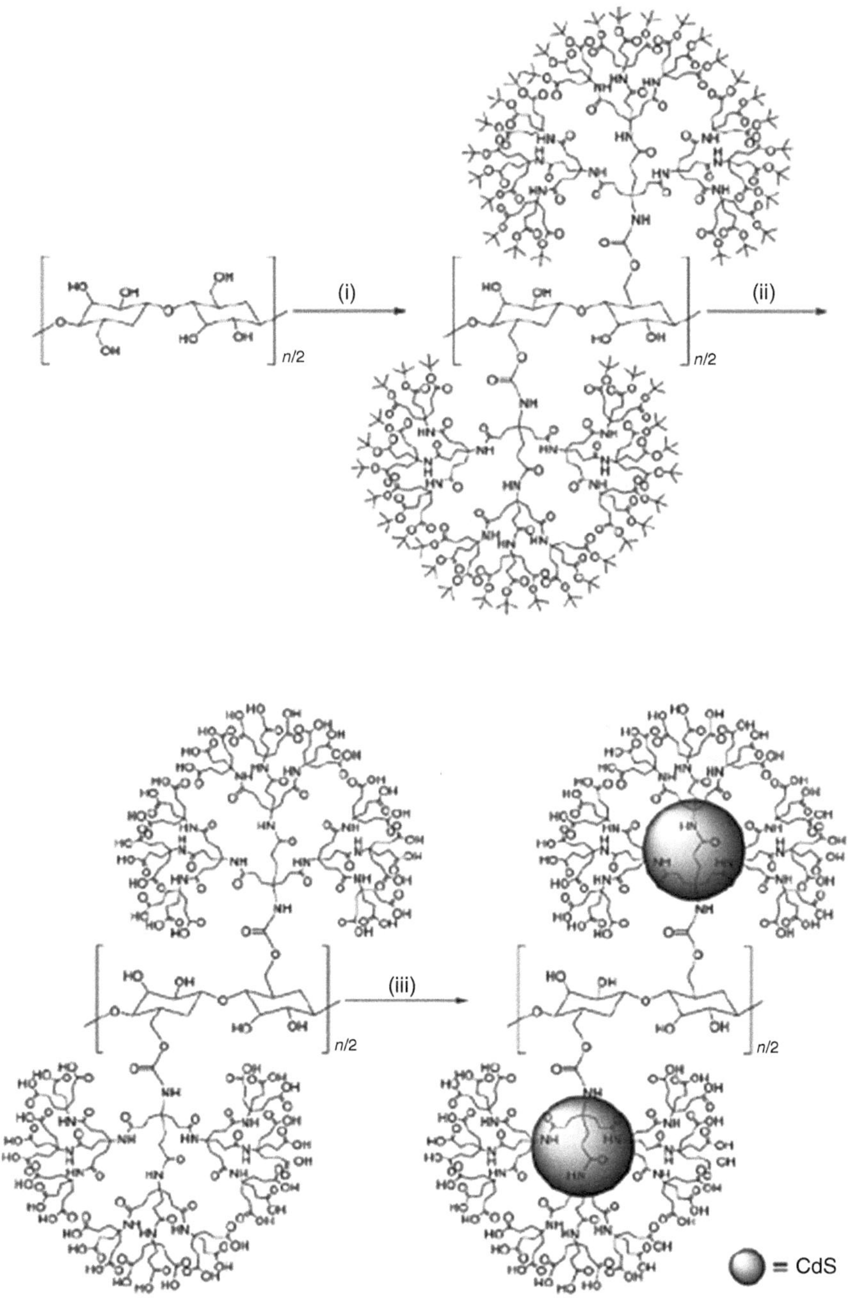

FIGURE 6.20 Synthesis of cellulose/CdS composites: (1) G_3-NCO dendron, dibutyltin dilaurate, LiCl, DMAc, 60°C; (2) HCO_2H, 25°C; (3) $Cd(NO_3)_2.4H_2O$, Na_2S, MeOH. *(Reprinted from Ref. [56] with permission of Royal Society of Chemistry. Copyright 2006.)*

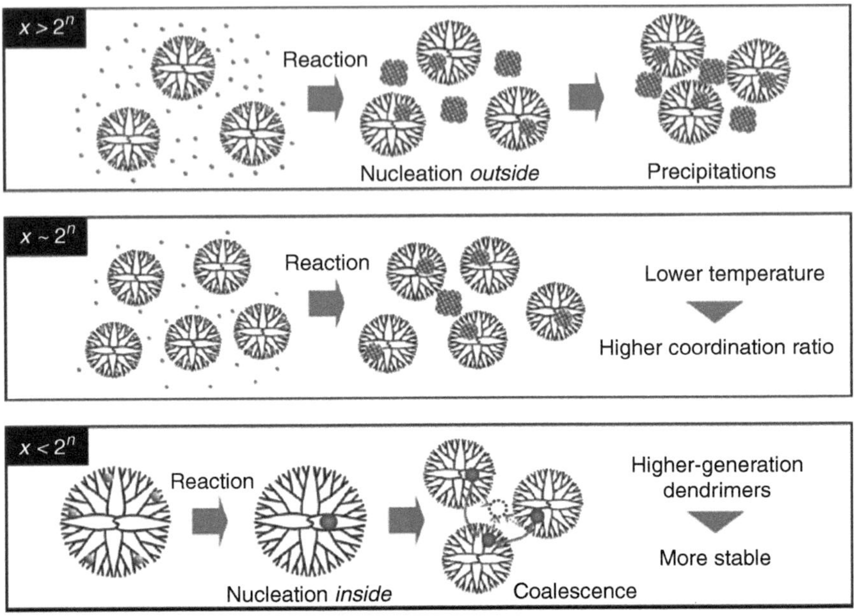

FIGURE 6.21 **Synthetic mechanism for CdS QDs encapsulated within G_n-OH dendrimers.** *(Reprinted from Ref. [57] with permission of Elsevier. Copyright 2012.)*

The last example of this section is related to the application of hydroxyl-terminated PAMAM dendrimers (G_n-OH) for synthesis of CdS QDs. The authors investigated the effects of Cd^{2+}ion/dendrimer ratio (x), dendrimer generations, and temperatures on the resultant CdS QDs. It was demonstrated that adjusting x to be less than $2n$ enables most of the nucleation to occur inside the dendrimers, which leads to stable dendrimer-encapsulated CdS QDs. It was established that higher-generation dendrimers can confine smaller QDs due to their denser surfaces. Reducing temperature proved simple and effective in enhancing the coordination ratio at equilibrium and decreasing nucleation outside the dendrimers, although it slows the coordination rate (Fig. 6.21) [57].

7 DENDRITIC NANOREACTOR IN CATALYSIS

Dendrimers can be used to address the major problems in the field of catalysis, including catalyst recovery, selectivity, and efficiency.

Dendrimers have been proposed to bridge the gap between homogeneous and heterogeneous catalysis. They have the potential to approach catalysts benefiting from high activity, high selectivity, high stability, and easy separation [6]. In these systems, catalytic moiety can be located in the core, periphery, and building blocks of the dendrimer.

Dendritic catalysis have been used in various chemical reactions, including the Suzuki–Miyaura reaction, Mizoroki–Heck reaction, hydrogenation reaction, carbonylation and hydroformylation reactions, oxidation reaction, polymerization and oligomerization reactions, arylation reaction, alkylation reaction, and asymmetric synthesis [6]. Recently, dendritic catalysts have been reviewed by Astruc et al. [6]. In another review article, Reek et al. reviewed the applications of dendrimers as support for recoverable catalysts and reagents [58]. The authors believed that catalytic performance in these systems depends on used dendritic architecture.

The effects of dendrimers on catalytic performance of metallodendrimers can be discussed in term of templating and stabilizing of metallic nanoparticles and size-selectivity and filtering role of periphery of dendrimer.

7.1 Representative Examples of Dendritic Catalysts

Amphiphilic dendritic nanoreactors containing triethylene glycol (TEG) termini and intradendritic triazole rings, 1 (G_0) and 2 (G_1) (Fig. 6.22), were proposed by Astruc et al. for stabilizing very small, 1–2 nm Pd nanoparticles. The TEG termini renders dendrimer water solubility, while a hydrophobic interior with 1,2,3-triazole ligands can stabilize Pd nanoparticles. K_2PdCl_4 and $NaBH_4$ were used as Pd precursor and reducing agent, respectively.

The Pd(II) complexation sites in dendrimer 1 were studied by Uv-vis spectroscopy. Upon addition of dendrimer 1 to Pd salt, a new band at 217 nm appeared and was assigned to a ligand-to-metal charge-transfer transition of Pd(II). This observation established the intradendritic complexation of Pd(II) at

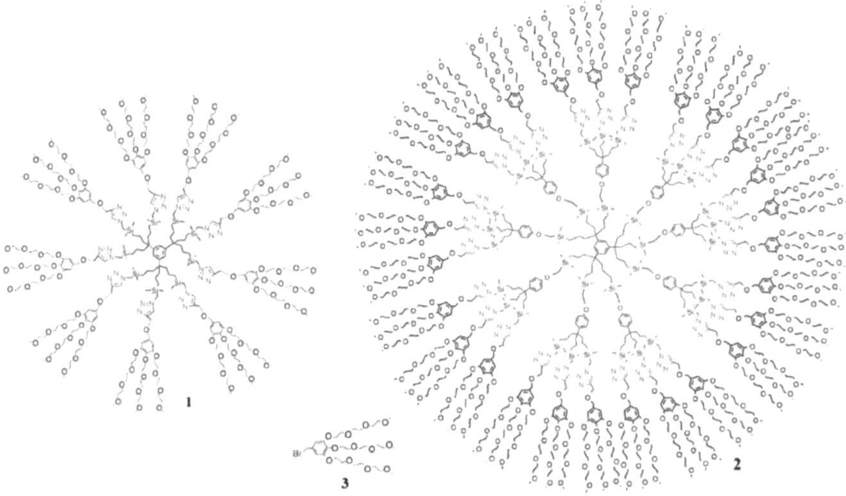

1

3

2

FIGURE 6.22 **G_0-27 TEG 1 dendrimer, G_1-81 TEG 2 dendrimer, and dendron TEG 3.** *(Reprinted from Ref. [59] with permission of Wiley-VCH. Copyright 2014.)*

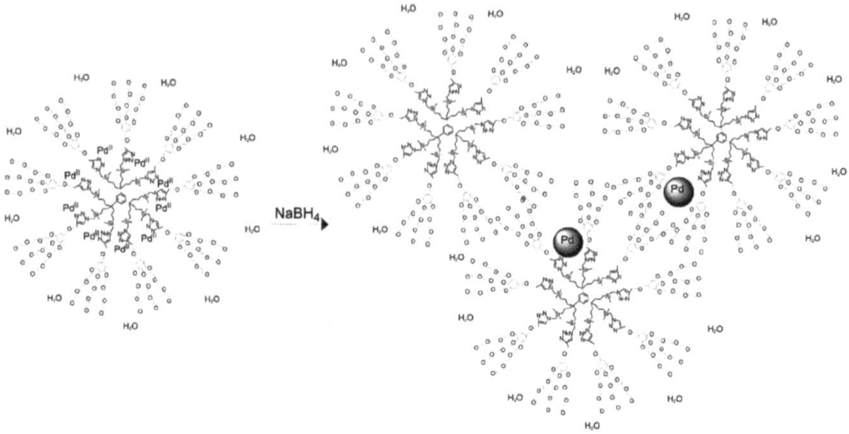

FIGURE 6.23 The Pd(II) complexation sites in dendrimer 1. *(Reprinted from Ref. [59] with permission of Wiley-VCH. Copyright 2014.)*

the triazole sites of dendrimer 1 and ruled out the strong Pd(II) complexation of the terminal TEG groups (Fig. 6.23). It's worth mentioning that mixing Pd salt with dendron 3 did not result in the distinguishing band at 217 nm [59].

DOSY NMR and DLS analysis were performed to determine the hydrodynamic diameters of dendrimers 1 and 2. Interestingly, very large values of DLS were obtained, which indicates the aggregation of a number of dendrimers and the formation of supramolecular assemblies of dendrimers. This observation was attributed to interdendritical interactions of TEG-termini.

The authors reported the catalytic performance of this dendrimer-stabilized nanoparticle in C–C crosscoupling reactions, including the Suzuki–Miyaura, Sonogashira and Heck, and reduction of 4-nitrophenol.

Suzuki–Miyaura reactions of iodoarene and bromoarene derivatives were remarkably catalyzed by this dendrimer-stabilized Pd nanoparticle, with the resulting TONs equal to or larger than 10^6. The catalytic activity for Sonogashira coupling was also impressive, and only 0.01% mol Pd was required.

In the case of Heck coupling, promising results were obtained. However, 0.1 mol% catalyst was used, due to the catalyst instability at high temperature (>100°C).

The catalytic activity of this catalyst for reduction of 4-nitrophenol was exceptional. The amount of Pd was quite low (down to 0.02 mol%) and the TOFs were very high.

Dendritic nanoreactor 1 was also employed as a recyclable micellar nanoreactor, stabilizer, and activator of Cu^I-catalyzed alkyne−azide cycloaddition click reaction, CuAAC reaction, in the presence of a very low amount (only 0.1%) of Cu (hexabenzyltren) Br (tren = triaminoethylamine) catalyst, 4, in aqueous media (Fig. 6.24) [60].

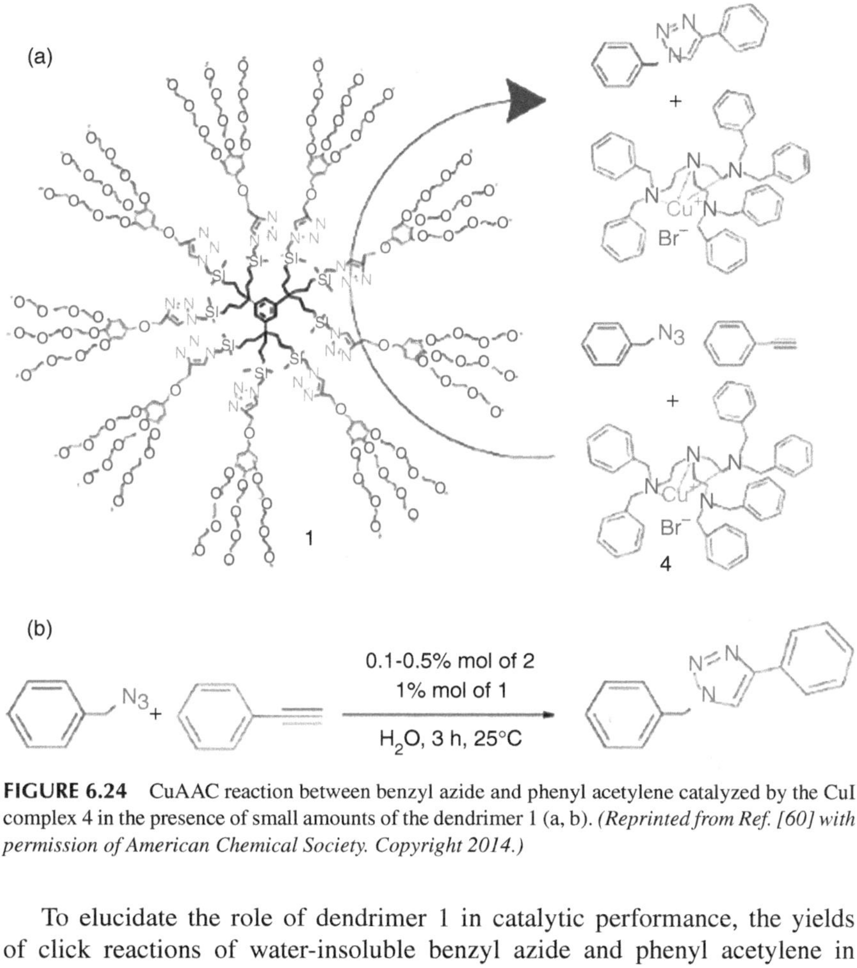

FIGURE 6.24 CuAAC reaction between benzyl azide and phenyl acetylene catalyzed by the CuI complex 4 in the presence of small amounts of the dendrimer 1 (a, b). *(Reprinted from Ref. [60] with permission of American Chemical Society. Copyright 2014.)*

To elucidate the role of dendrimer 1 in catalytic performance, the yields of click reactions of water-insoluble benzyl azide and phenyl acetylene in the presence of various amounts of catalyst 2, 0.1–0.5% mol, and 1% mol dendrimer 1 were investigated. Furthermore, the yields of reaction using 0.1% mol catalyst 4 in the presence and absence of dendrimer 1 were compared. Excellent yields (92–98%) were obtained for various amounts of catalyst in the presence of dendrimer 1. However, in the absence of dendrimer, only 2% product was furnished, indicating the essential role of dendrimer in catalysis. The authors studied the encapsulation and water-solubility of catalyst 4 by ^{1}HNMR technique and believed that the substrate can approach the catalyst easier in the hydrophobic core of the dendrimer.

To obviate the need for synthesizing the catalyst with nitrogen ligand, intradendritic triazole ligands were used as an activator of Cu^{I} catalyst. In this case, $CuSO_4.5H_2O$ and sodium ascorbate (NaAsc) can be used as precatalyst and reducing agent, respectively. Coordination of Cu^{II} to intradendritic triazole ligands and subsequent reduction to Cu^{I} can lead to1- Cu^{I} catalyst (Fig. 6.25).

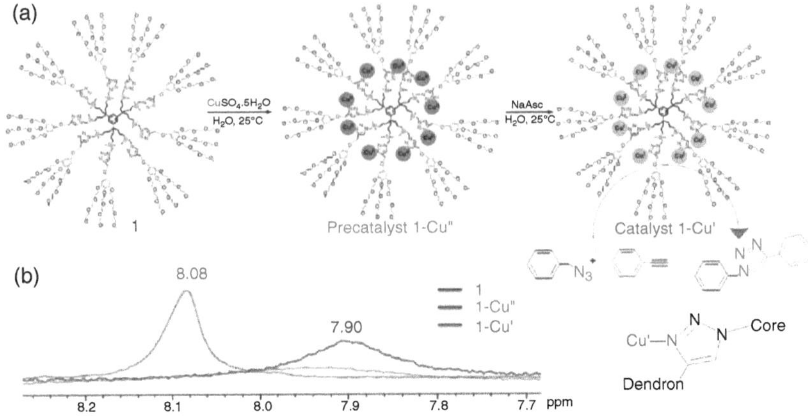

FIGURE 6.25 (a) Dendrimer 1 as nanoreactor/ligand for CuAAC catalysis. (b) Comparison of the NMR signals of the triazole proton of 1 alone (7.90 ppm), with Cu^{II} (very broad due to the paramagnetism) and Cu^{I} (shift to 8.08 ppm) showing the coordination of the intradendritic triazoles of 1 to the copper ions. *(Reprinted from Ref. [60] with permission of American Chemical Society. Copyright 2014.)*

The catalytic performance of this catalyst in CuAAC reaction was investigated for various alkynes and azides. High yields (81–99%) were obtained in the presence of very low amounts of catalyst.

Chemoselective hydrogenation of carbonyl and olefin functional groups by using dendrimer-encapsulated copper cluster, Cu_{30}@PAMAM—OH(G_6), in aqueous media was reported by Tsukuda et al. [61]. The catalytic activity of this catalyst for hydrogenation of various groups, including C═C, C═O and NO_2, was investigated (Table 6.5). The results in Table 6.5 indicate that the Cu_{30}@ PAMAM—OH(G_6) can catalyze hydrogenation of both carbonyl and olefins to corresponding alcohols and saturated alkanes. However, the catalyst is more efficient in the case of carbonyl hydrogenation. Furthermore, nitrobenzene, entry 12, could not be hydrogenated under this condition.

Taking different hydrogenation efficiency of C═C, C═O, and NO_2 functional groups into account, authors developed chemoselective hydrogenation of substrates with various functional groups, entries 13, 14, and 15. In the case of cinnamaldehyde (CAL), cinnamyl alcohol (COL), and 3-phenyl propanol (HCOL) were obtained as the main product and byproduct, respectively. The selectivity can be increased to 93% by decreasing the reaction temperature from 25 to 10°C.

The effects of cluster size and dendrimer generation were also investigated. Measuring the catalytic activities of a series of Cu_n@ PAMAM—OH(G_6), n = 10, 20, 30, 40, 50, 60, in model reaction, hydrogenation of 2-hexanone, an inverse relationship between conversion and the molar ratio [Cu^{2+}]/[PAMAM—OH(G_6)] was established (Fig. 6.26a).

Previous reports indicated that using dendrimer with higher generation and steric crowding on the periphery would result in the hydrogenation rate

TABLE 6.5 Catalytic Hydrogenation by Cu_{30}@PAMAM—OH(G_6)[a]

Entry	Substrate	Conversion (%)	Product	Selectivity
1	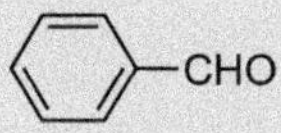	95	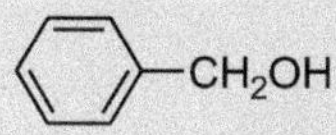	100
2	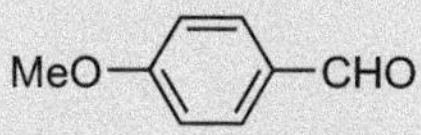	88	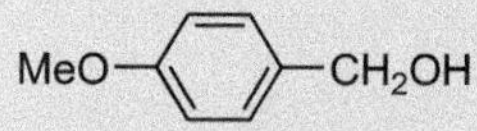	100
3	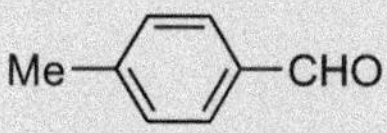	90	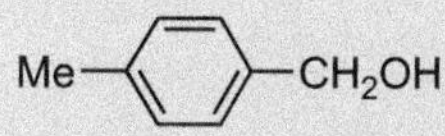	100
4	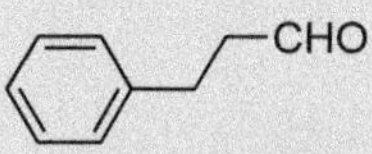	99	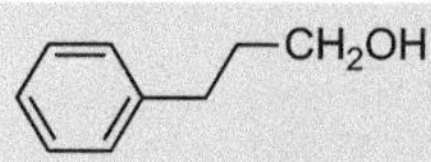	100
5	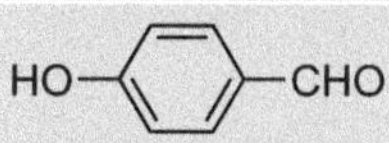	99	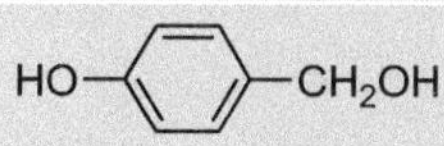	100
6	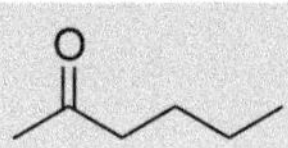	99 0[b] 0[c]	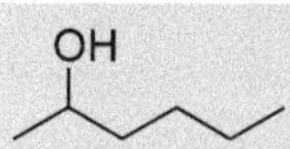	100 — —
7	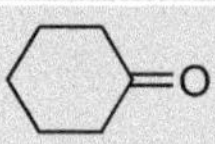	99	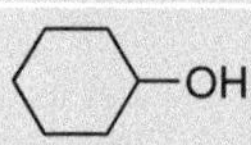	100
8	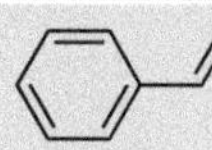	22	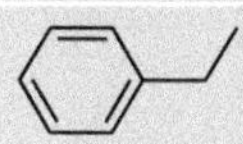	100
9	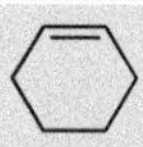	20	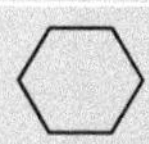	100
10	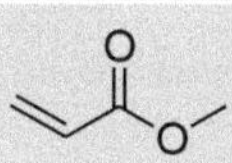	15	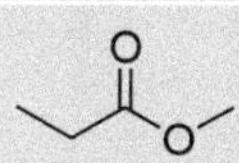	100
11	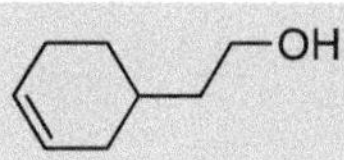	18	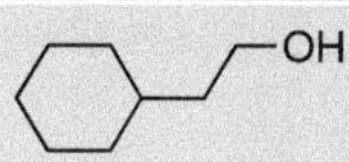	100
12	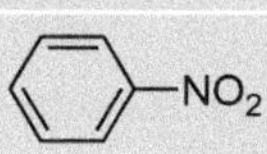	0	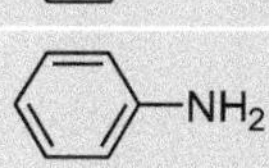	—
13	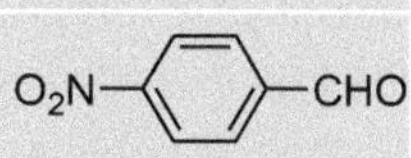	88	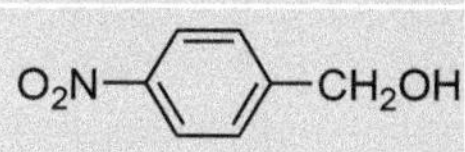	100
14	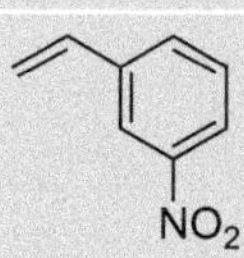	12	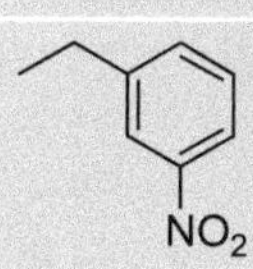	100

(Continued)

TABLE 6.5 Catalytic Hydrogenation by Cu_{30}@PAMAM—OH$(G_6)^a$ *(Cont.)*

Entry	Substrate	Conversion (%)	Product	Selectivity
15	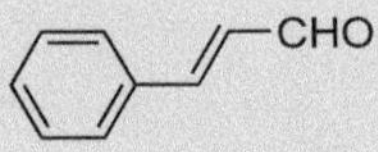 CAL	99 67^d	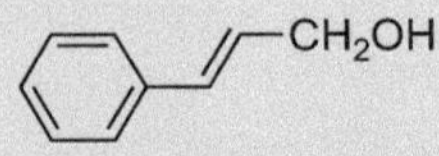 COL 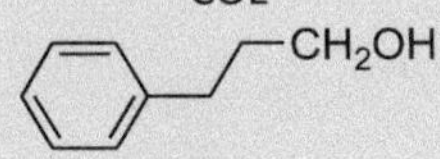 HCOL	86 93^d 13 7^d

[a]Reaction conditions: amount of Cu in catalyst, 4.6 μmol; amount of substrate, 0.23 mmol; volume of water, 5 mL; pressure of H_2, 3 MPa; temperature, 25°C; stirring rate, 900 rpm; reaction time, 6 h for entries 1–7, 14, 15, and 12 h for entries 8–13.
[b]In the absence of H_2, but with 5 equiv of $NaBH_4$.
[c]$(Cu^{2+})_{30}$@PAMAM—OH(G_6) as catalyst.
[d]Temperature, 10°C; pressure of H_2, 1 MPa; reaction time, 30 min.
Reprinted from Ref. [61] with permission of American Chemical Society. Copyright 2013.

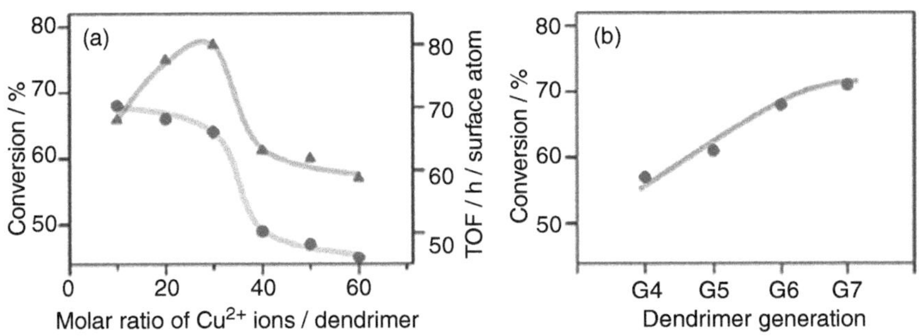

FIGURE 6.26 (a) Conversions (red) and TOF values (green) for hydrogenation of 2-hexanone as a function of the molar ratio of $[Cu^{2+}]/[PAMAM - OH(G_6)]$. (b) Conversion for hydrogenation of 2- hexanone as a function of Cu_{10}@PAMAM – OH generation. The curves are guides for the eye. In both cases, the experimental conditions are the same as in Table 6.5, with a reaction time of 30 min. *(Reprinted from Ref. [61] with permission of American Chemical Society. Copyright 2013.)*

decreasing because of inaccessibility of the substrate to the encapsulated catalyst. However, quite opposite results were obtained in hydrogenation of 2-hexanone with Cu_{10}@PAMAM—OH(G_n) catalysts (Fig. 6.26b). The authors assigned this unusual observation to more efficient condensation of organic substrates into the dendrimer cavity in water due to increased hydrophobicity with generation.

Using peripherally modified PAMAM dendrimer-supported chiral salen Mn(III) complex, asymmetric epoxidation of unfunctionalized olefins was performed. Modification of dendrimer was achieved through reaction of the terminal primary amion group of three generations of PAMAM dendrimer, $(G_n, n = 1, 2, 3)$

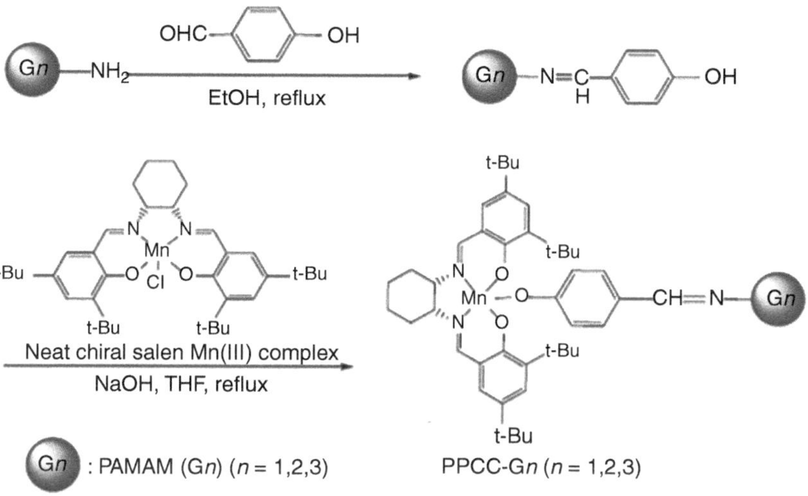

FIGURE 6.27 Synthesis of the PPCC-G_n (n = 1, 2, 3). *(Reprinted from Ref. [62] with permission of Royal Society of Chemistry. Copyright 2013.)*

with aldehyde group of *p*–hydroxybenzaldehyde. The hydroxyl group at the termini of dendrimer was subsequently transformed to phenolate ion under basic condition and exchanged with chlorine ion in chiral salen Mn(III) to furnish the supported catalyst, PPCC-G_n (n =1, 2, 3) (Fig. 6.27) [62].

The results of catalytic activity of these catalysts in enantioselective epoxidation of various unfunctionalized olefins indicated that low amount of PPCC-G_1 catalyst, 0.25 mol%, can catalyze the epoxidation of styrene with yield for more than 99 and 38% ee, while these values for neat complex were 97 and 20%, respectively.

This superior ee was assigned to the positive "dendrimer effect" of the PAMAM dendrimer support. The high local concentrations of the active sites were confined to the surface of the PAMAM dendrimer through a phenoxy group, which resulted in the cooperative interaction between neighboring catalytic sites.

Additionally, investigation of the dendrimer generation effect established that lower dendrimer generation results in higher ee values. This was ascribed to the increased dense packing of the chiral end groups at the periphery, which resulted in increasing difficulty for all end groups in adopting their preferred conformation in the catalytic reaction.

Dendrimers were also used for synthesis and improvement of performance of photocatalysts. As an example, synthesis of dendrimer-protected TiO_2 nanoparticles was accomplished by hydrolysis of $TiCl_4$ in a cold solution of dendrimers. It was reported that the size of the obtained nanoparticles depends on terminal groups of dendrimers and the pH of the aqueous solution before addition of

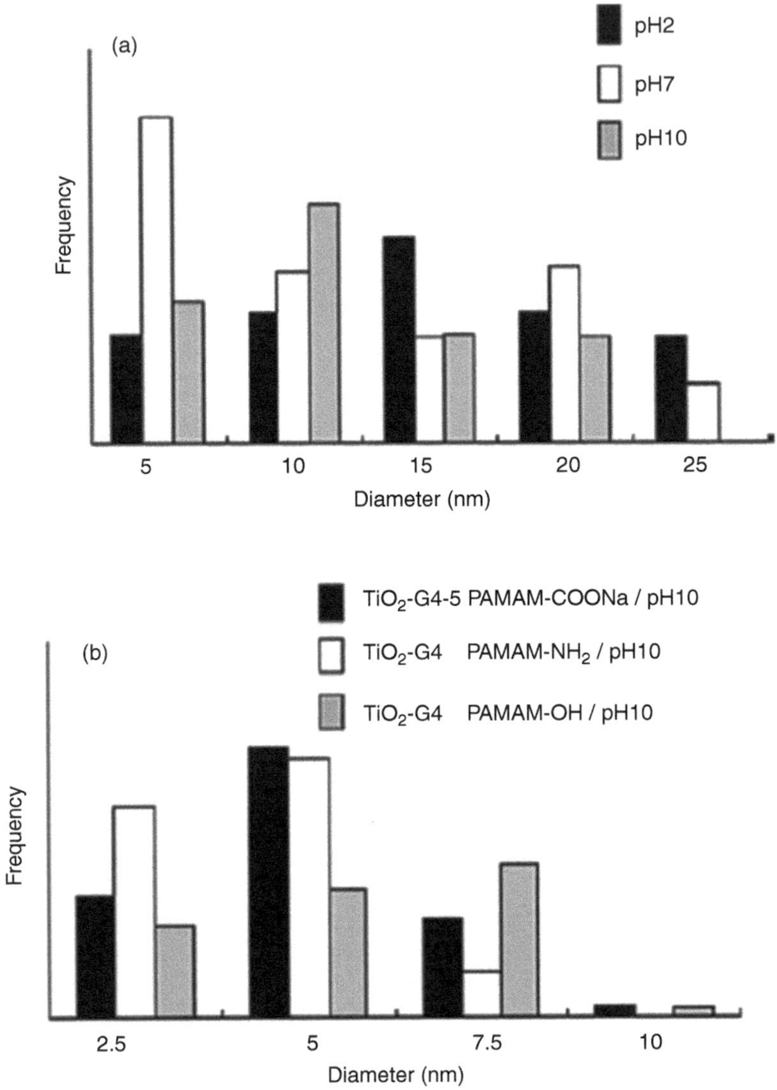

FIGURE 6.28 Size histograms of TiO_2 nanoparticles without (a) and with (b) dendrimer. *(Reprinted from Ref. [63] with permission of Elsevier. Copyright 2005.)*

$TiCl_4$ (Fig. 6.28). Comparison of photocatalytic activity of dendrimer-stabilized TiO_2 and TiO_2 nanoparticles in photodegradation of 2,4-dichlorophenoxyacetic acid indicated the superior activity of dendritic photocatalysts. The effects of the dendrimer were attributed to stabilization of nanoparticles and reinforcement of photocatalytic activity [63].

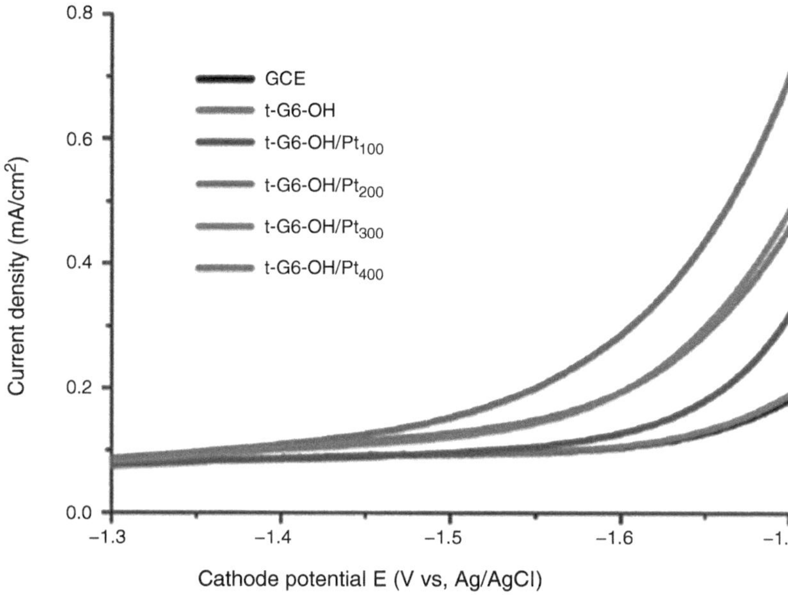

FIGURE 6.29 **Cathodic polarization curves of GCE, t-G_6-OH, and t-G- OH/Pt*x* (*x* = 100, 200, 300, 400)-modified GCE electrodes in 0.1 M KCl aqueous solution (scan rate, 5 mV/s; electrode area, 0.071 cm^2; reference electrode, Ag/AgCl).** *(Reprinted from Ref. [65] with permission of American Chemical Society. Copyright 2012.)*

8 DENDRITIC NANOREACTOR IN ENERGY SECTOR

Dendrimer-encapsulated nanoparticles have potential applications in the energy sector. For example, polyamidoamine (PAMAM) dendrimer-encapsulated platinum nanoparticles (Pt-DENs) have been introduced as a promising cathode catalyst for air-cathode single-chamber microbial fuel cells (SCMFCs) [64]. This novel catalyst led to higher power production with 129.1%, as compared to cathodes with electrodeposited Pt.

Dendrimer-encapsulated Pt nanoparticles, DENPt, were used for designing artificial hydrogenases and producing hydrogen. The large cavity of sixth generation PAMAM dendrimers, terminated with triple-hydroxyl groups (t-G_6-OH) or single-hydroxyl groups (s-G_6-OH), was used as a template and stabilizer for synthesis of a series of DENPt with various sizes (t-G_6-OH/Pt*x* and s-G_6-OH/Pt*x*) (*x* = 100, 200, 300, 400, 500, 600).

To investigate the catalytic activity of DENPt, cathodic polarization curves of the t-G_6-OH/Pt*x* (*x* = 100, 200, 300, 400)-modified working electrodes were measured (Fig. 6.29). The t-G_6-OH/Pt_{200} had the maximum catalytic activity, followed by t-G_6-OH/Pt_{300} and t-G_6-OH/Pt_{400}. The low catalytic activity of t-G_6-OH/Pt_{100} was attributed to the small size of the Pt nanoparticle, which was below some threshold limit required for full catalytic activity, and the weak

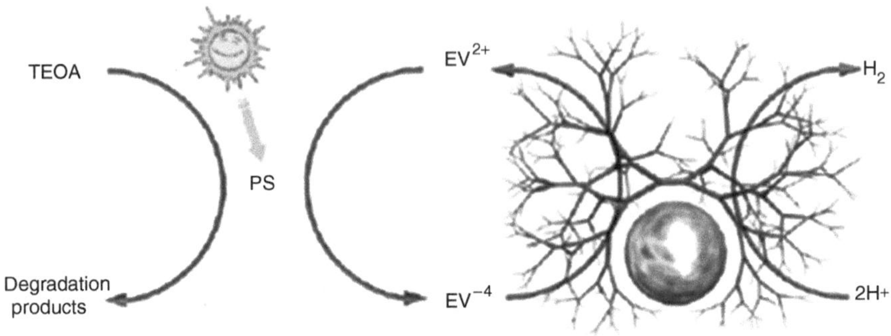

FIGURE 6.30 Schematic diagram of light-driven hydrogen production catalyzed by DENPt. *(Reprinted from Ref. [65] with permission of American Chemical Society. Copyright 2012.)*

contact of Pt nanoparticles with the glassy carbon electrode surface caused by the sizable surrounding t-G_6-OH molecule. A similar trend of catalytic activity for s-G_6-OH/Pt_x (x = 100, 200, 300, 400) was observed [65].

Sensitizer Pt(II) terpyridyl phenylacetylide perchlorate [(Pt-tppa) ClO_4] was prepared and used for investigating light-induced hydrogen production. For this propose, Pt-$tppa^+$ and ethyl viologen (EV^{2+}) were applied as the photosensitizer and the electron relay, respectively, and TEOA as the sacrificial electron donor (Fig. 6.30). It was established that the size of Pt nanoparticles has an essential role in catalytic efficiency. The t-G_6-OH/ Pt_{200} and s-G_6-OH/Pt_{200} exhibit the highest catalytic activity. The lower catalytic performance of other catalytic systems was ascribed to the smaller surface area of the Pt clusters, which decreases the virtual amount of catalyst due to the inactive interior Pt atoms. Similar catalytic activity for t-G_6- OH/Pt_x and s-G_6-OH/Pt_x indicates no passivation effect caused by the periphery hydroxyl groups.

An important challenge in hydrogen energy is hydrogen storage. Up to now various candidates such as metal organic frameworks (MOFs), carbon nanotubes, metal hydrides, and ammonia borane have been studied for chemical hydrogen storage.

Ammonia borane with high hydrogen content, 19.6 wt%, can be considered as an interesting candidate for hydrogen storage. Hydrolysis of ammonia borane and hydrogen production can be performed through thermal decomposition or catalytic hydrolysis under milder conditions.

In this regard, developing an efficient, economical, and durable catalyst to improve the kinetic properties under moderate reaction conditions has remained a challenge.

To address this problem, Xu et al. synthesized dendrimer-encapsulated Co nanoparticles, (G_6-OH(Co_{60})) with narrow size distribution (1.6 ± 0.4 nm), by using hydroxyl-terminated poly (amidoamine) dendrimer of generation six [42], and they investigated its catalytic activity for hydrolysis of ammonia borane. The results indicated that hydrogen generation from room temperature

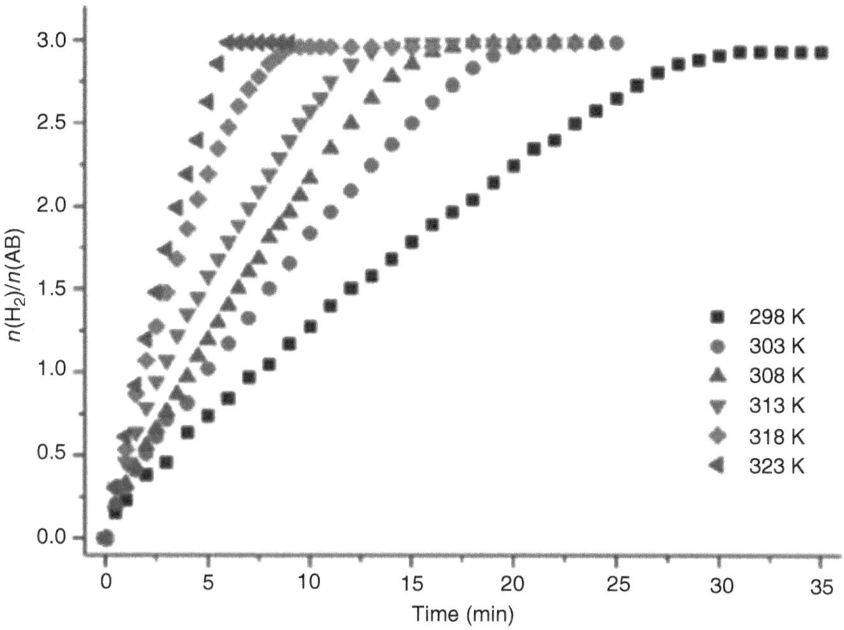

FIGURE 6.31 Temperature effects on the rate of H_2 generation from the hydrolysis of AB catalyzed by G_6-OH(Co_{60}) (0.8 m, 1.0 mL, catalyst/AB = 0.013) at different reaction temperatures. *(Reprinted from Ref. [42] with permission of WILEY-VCH Verlag GmbH & Co. Copyright 2014.)*

(G_6-OH(Co_{60})) catalyzed hydrolysis of ammonia borane was 3 equiv (Fig. 6.31). This is the complete hydrogen generation from hydrosis of ammonia borane according to Eq. 6.1.

$$NH_3BH_{3(s)} + 2H_2O_{(l)} \rightarrow NH^+_{4(aq)} + BO^-_2(aq) + 3H_{2(g)}$$

Equation 6.1. Hydrolysis of ammonia borane. Reprinted with permission from Ref. [42]. Copyright 2014 WILEY-VCH Verlag GmbH & Co.

Another example is related to fabrication of NiSn nanomaterial with potential application as anodes of Li ion batteries. In this context, Crooks et al. reported synthesis of nearly monodispersed NiSn dendrimer-encapsulated nanoparticles with diameter of ~1.2 nm using hydrophobic, sixth-generation poly(amidoamine) dendrimers in toluene [66]. The interesting fact is that this dendrimer-encapsulated bimetallic alloy consists of two reactive metals, which typically degrade in the presence of O_2 and H_2O [2].

9 CONCLUSIONS

Dendrimers are a class of host molecules with a high level of structural regularity and molecular monodispersity. Their unusual chemical and physical properties as well as well-defined shape make them ideal candidates for utilization as nanoreactors. Both cavity and periphery of dendrimers play roles in their utility as nanoreactor.

Dendritic nanoreactors can be used as a template and/or stabilizer for synthesis of nanoparticles with definite size and disparity. Various kinds of mono and bimetallic nanoparticles, as well as QDs, have been prepared using dendrimers.

The role of dendritic nanoreactors is not limited to templating and stabilizing. They are capable of molecular recognition through the filtering role of their periphery.

Dendritic shells can be used to create a nanoenvironment favorable for catalysis, or to provide shielding for functional groups at the dendritic core. From various examples provided in this chapter, the utility of dendritic catalysis in various chemical reactions can be concluded. They have the potential to approach catalysts benefiting from high activity, high selectivity, high stability, and easy separation.

Applications of dendritic nanoreactors are beyond the issues discussed in this chapter, and that readers who are interested are encouraged to study the references for more details.

REFERENCES

[1] Grayson SM, Frechet JMJ. Convergent dendrons and dendrimers: from synthesis to applications. Chem Rev 2001;101:3819–67.

[2] Bronstein LM, Shifrina ZB. Dendrimers as encapsulating, stabilizing, or directing agents for inorganic nanoparticles. Chem Rev 2011;111:5301–44.

[3] Vögtle F. Dendrimers II, architecture, nanostructure and supramolecular chemistry. Springer: Berlin, Heidelberg, and New York; 2000.

[4] Gade LH. Dendrimer catalysis. Springer: Germany; 2006.

[5] Vögtle F, Richardt G, Werner N. Dendrimer chemistry. WILEY-VCH: Germany; 2009.

[6] Wang D, Astruc D. Dendritic catalysis: basic concepts and recent trends. Coord Chem Rev 2013;257:2317–34.

[7] Kesharwani P, Iyer AK. Recent advances in dendrimer-based nanovectors for tumor-targeted drug and gene delivery. Drug Discov Today 2015;20:536–47.

[8] Kesharwani P, Jain K, Jain NK. Dendrimer as nanocarrier for drug delivery. Prog Polym Sci 2014;39:268–307.

[9] Qiao Z, Shi X. Dendrimer-based molecular imaging contrast agents. Prog Polym Sci 2015;44:1–27.

[10] Hasanzadeh M, et al. Dendrimer-encapsulated and cored metal nanoparticles for electrochemical nanobiosensing. Trends Anal Chem 2014;53:137–49.

[11] Hwang S-H, et al. Recent progress and applications for metallodendrimers. New J Chem 2007;31:1192–217.

[12] Andres R, de Jesus E, Flores JC. Catalysts based on palladium dendrimers. New J Chem 2007;31:1161–91.

[13] Niu Y, Crooks RM. Dendrimer-encapsulated metal nanoparticles and their applications to catalysis. C R Chimie 2003;6:1049–59.

[14] Peng X, Pan Q, Rempel GL. Bimetallic dendrimer-encapsulated nanoparticles as catalysts: a review of the research advances. Chem Soc Rev 2008;37:1619–28.
[15] Scott RWJ, Wilson OM, Crooks RM. Synthesis, characterization, and applications of dendrimer-encapsulated nanoparticles. J Phys Chem B 2005;109:692–704.
[16] Vriezema DM, et al. Self-assembled nanoreactors. Chem Rev 2005;105:1445–89.
[17] Astruc D, Chardac F. Dendritic catalysts and dendrimers in catalysis. Chem Rev 2001;101:2991–3023.
[18] Jansen JFGA, de Brabander-van den Berg EMM, Meijer EW. Encapsulation of guest molecules into a dendritic box. Science 1994;266:1226–9.
[19] Wiesler UM, et al. Dendrimers III: design, dimension, function, vol. 3. Berlin: Springer-Verlag; 2001.
[20] Natarajan B, et al. Dynamic internal cavities of dendrimers as constrained media: a study of photochemical isomerizations of stilbene and azobenzene using poly(alkyl aryl ether) dendrimers. J Org Chem 2012;77:2219–24.
[21] Frechet JMJ, Tomalia DA. Dendrimers and other dendritic polymers. New York: John Wiley & Sons; 2001.
[22] Weener J-W, van Dongen JLJ, Meijer EW. Electrospray mass spectrometry studies of poly(propylene imine) dendrimers: probing reactivity in the gas phase. J Am Chem Soc 1999;121:10346–55.
[23] Liang C, Frechet JMJ. Applying key concepts from nature: transition state stabilization, pre-concentration and cooperativity effects in dendritic biomimetics. Prog Polym Sci 2005;30:385–402.
[24] Niu Y, Yeung LK, Crooks RM. Size-selective hydrogenation of olefins by dendrimer-encapsulated palladium nanoparticles. J Am Chem Soc 2001;123:6840–6.
[25] Jansen JFGA, Meijer EW. The dendritic box: shape-selective liberation of shell deprotection encapsulated guests. J Am Chem Soc 1995;117:4417–8.
[26] Kojima C, et al. Synthesis of polyamidoamine dendrimers having poly(ethylene glycol) grafts and their ability to encapsulate anticancer drugs. Bioconjugate Chem 2000;11:910–7.
[27] Kojima C, et al. Preparation of polyethylene glycol-attached dendrimers encapsulating photosensitizers for application to photodynamic therapy. Bioconjugate Chem 2007;18:663–70.
[28] Kojima C, et al. Aqueous solubilization of fullerenes using poly(amidoamine) dendrimers bearing cyclodextrin and poly(ethylene glycol). Bioconjugate Chem 2008;19:2280–4.
[29] Reinhoudt DN. Supramolecular materials and technologies: perspectives in supramolecular chemistry, vol. 4. New York: John Wiley & Sons; 1999.
[30] Hu J, Xu T, Cheng Y. NMR insights into dendrimer-based host−guest systems. Chem Rev 2012;112:3856–91.
[31] Jain NK, Gupta U. Application of dendrimer–drug complexation in the enhancement of drug solubility and bioavailability. Expert Opin Drug Metab Toxicol 2008;4:1035–52.
[32] Devarakonda B, et al. Effect of pH on the solubility and release of furosemide from polyamidoamine (PAMAM) dendrimer complexes. Int J Pharm 2007;345:142–53.
[33] Yiyun C, Tongwen X. Solubility of nicotinic acid in polyamidoamine dendrimer solutions. Eur J Med Chem 2005;40:1384–9.
[34] Devarakonda B, Hill RA, de Villiers MM. The effect of PAMAM dendrimer generation size and surface functional group on the aqueous solubility of nifedipine. Int J Pharm 2004;284:133–40.
[35] Patel J, et al. Solubility of aceclofenac in polyamidoamine dendrimer solutions. Int J Pharm Investig 2011;1:135–8.

[36] Svenson S, Tomalia DA. Dendrimers in biomedical applications—reflections on the field. Adv Drug Deliv Rev 2005;57:2106–29.
[37] Peng C, et al. PEGylated dendrimer-entrapped gold nanoparticles for in vivo blood pool and tumor imaging by computed tomography. Biomaterials 2012;33:1107–19.
[38] Balzani V, et al. Fluorescent guests hosted in fluorescent dendrimers. Tetrahedron 2002;58:629–37.
[39] Caminade A-M, et al. Dendrimers towards catalytic, material and biomedical uses. Hoboken: John Wiley & Sons Ltd; 2011.
[40] Tao L, et al. Modification of multi-wall carbon nanotube surfaces with poly(amidoamine) dendrons: synthesis and metal templating. Chem Commun 2006;4949–51.
[41] D'Aleo A, et al. Oligothia dendrimers for the formation of gold nanoparticles. Adv Funct Mater 2004;14:1167–77.
[42] Aranishi K, Zhu QL, Xu Q. Dendrimer-encapsulated cobalt nanoparticles as high- performance catalysts for the hydrolysis of ammonia borane. Chem Cat Chem 2014;6:1375–9.
[43] Mizugaki T, et al. PAMAM dendron-stabilised palladium nanoparticles: effect of generation and peripheral groups on particle size and hydrogenation activity. Chem Commun 2008; 241–3.
[44] Knecht MR, Garcia-Martinez JC, Crooks RM. Synthesis, characterization, and magnetic properties of dendrimer-encapsulated nickel nanoparticles containing <150 atoms. Chem Mater 2006;18:5039–44.
[45] Gao H, et al. Dendrimer-encapsulated copper as an oveloligonucleotides label for sensitive electrochemical stripping detection of DNA hybridization. Biosens Bioelectron 2013;48:210–5.
[46] Chung Y-M, Rhee H-K. Dendrimer-templated Ag–Pd bimetallic nanoparticles. J Colloid Interface Sci 2004;271:131–5.
[47] Liu H, et al. Tunable synthesis and acetylation of dendrimer-entrapped or dendrimer-stabilized gold–silver alloy nanoparticles. Colloids Surf B 2012;94:58–67.
[48] Knecht MR, et al. Structural rearrangement of bimetallic alloy Pd–Au nanoparticles within dendrimer templates to yield core/shell configurations. Chem Mater 2008;20:1019–28.
[49] Yamamoto D, Watanabe S, Miyahara MT. Coordination and reduction processes in the synthesis of dendrimer-encapsulated Pt nanoparticles. Langmuir 2010;26:2339–45.
[50] Siani A, et al. Dendrimer-mediated synthesis of subnanometer-sized Rh particles supported on ZrO_2. J Catal 2009;266:331–42.
[51] Guo W, et al. Conjugation chemistry and bioapplications of semiconductor box nanocrystals prepared via dendrimer bridging. Chem Mater 2003;15:3125–33.
[52] Zhao Y, et al. Architecture of stable and water-soluble CdSe/ZnS core–shell dendron nanocrystals via ligand exchange. J Colloid Interf Sci 2009;339:336–43.
[53] Wang YA, et al. Stabilization of inorganic nanocrystals by organic dendrons. J Am Chem Soc 2002;124:2293–8.
[54] Guo W, et al. Luminescent CdSe/CdS core/shell nanocrystals in dendron boxes: superior chemical, photochemical and thermal stability. J Am Chem Soc 2003;125:3901–9.
[55] Hwang S-H, et al. Dendron-tethered and templated CdS quantum dots on single-walled carbon nanotubes. J Am Chem Soc 2006;128:7505–9.
[56] Hwang S-H, et al. Construction of CdS quantum dots via a regioselective dendritic functionalized cellulose template. Chem Commun 2006;3495–7.
[57] Yamamoto D, et al. Synthesis and photoluminescence characterization of dendrimer-encapsulated CdS quantum dots. Colloids Surf A 2012;411:12–7.
[58] van Heerbeek R, et al. Dendrimers as support for recoverable catalysts and reagents. Chem Rev 2002;102:3717–56.

[59] Deraedt C, Salmon L, Astruc D. "Click" dendrimer-stabilized palladium nanoparticles as a green catalyst down to parts per million for efficient C-C cross-coupling reactions and reduction of 4-nitrophenol. Adv Synth Catal 2014;356:2525–38.

[60] Deraedt C, Pinaud N, Astruc D. Recyclable catalytic dendrimer nanoreactor for part-per-million CuI catalysis of "click" chemistry in water. J Am Chem Soc 2014;136:12092–8.

[61] Maity P, Yamazoe S, Tsukuda T. Dendrimer-encapsulated copper cluster as a chemoselective and regenerable hydrogenation catalyst. ACS Catal 2013;3:182–5.

[62] Peng M, et al. A highly efficient and recyclable catalyst: dendrimer supported chiral salen Mn(III) complex for asymmetric epoxidation. RSC Adv 2013;3:20684–92.

[63] Nakanishi Y, Imae T. Synthesis of dendrimer-protected TiO_2 nanoparticles and photodegradation of organic molecules in an aqueous nanoparticle suspension. J Colloid Interface Sci 2005;285:158–62.

[64] Yang X, et al. Microbial fuel cell cathode with dendrimer encapsulated Pt nanoparticles as catalyst. J Power Sources 2011;196:10611–5.

[65] Yu T, et al. Dendrimer-encapsulated Pt nanoparticles: an artificial enzyme for hydrogen production. J Phys Chem C 2012;116:10516–21.

[66] Gates AT, et al. Synthesis and characterization of NiSn dendrimer-encapsulated nanoparticles. Langmuir 2010;26:12994–9.

Chapter 7

Catalysis Within the Self-Assembled Resorcin[4]arene Hexamer

Giuseppe Borsato, Alessandro Scarso
Dipartimento di Scienze Molecolari e Nanosistemi, Università Ca' Foscari di Venezia, Mestre (VE), Italy

1 CATALYSIS WITHIN CAVITIES

Since most chemical processes require the presence of a catalyst in order to be synthetically effective, the study of catalysis at all levels is one of the fundamental topics in chemistry. The development of new systems able to mimic some of the astonishing features typical of enzymes is a strategic issue for the scientific community and one of the main endeavors of research in the field of catalysis. Enzymes are polypeptidic molecules that, thanks to their folding in water and sometimes aided by the presence of specific molecular species as coenzymes, are able to impart accelerations of several order of magnitude to organic transformations, showing also high chemo-, regio-, and/or enantioselectivity and specificity [1]. These properties are basically due to recognition events between enzyme and substrate, that is, selected from the medium, but even more importantly by recognition and stabilization of the reaction transition state hosted by the active site. In this specific cavity, substrate molecules can be bound and selectively transformed into products through a perfectly juxtaposed combination of supramolecular weak interactions as well as reversible covalent interactions. The affinity between the active site and the substrate has been extensively investigated to explain the remarkable catalytic activity of enzymes. At first sight the high size and shape complementarities between the two are quite evident, especially considering the adaptation undergone by the enzyme during the process. In fact, a sort of continuous variation in the cavity shape ensures efficient substrate recognition, desolvation, preorganization, and the stabilization of the transition state of the reaction [2], eventually leading to substrate selective conversion [3].

The fascinating properties characterizing enzymes are still unrivaled by artificial catalysts. For this reason, an increasing number of research groups

Organic Nanoreactors. http://dx.doi.org/10.1016/B978-0-12-801713-5.00007-0

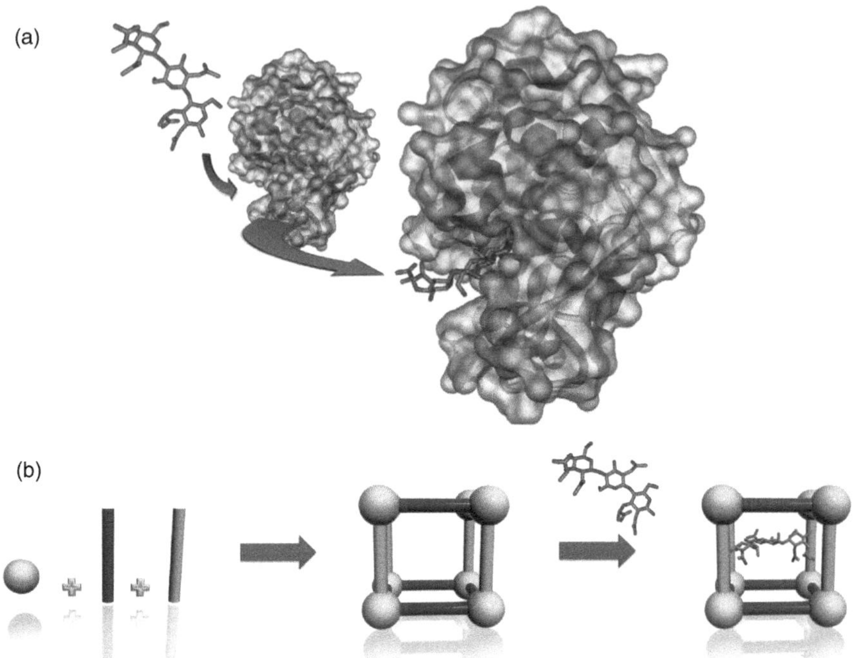

FIGURE 7.1 Comparison between (a) the binding of the substrate by the enzyme and (b) the construction of supramolecular assemblies exploiting weak intermolecular interactions to obtain supramolecular self-assembled catalysts.

are focusing their activities toward the development of artificial homogeneous catalysts capable of mimicking some of the properties of natural enzymes. Even though molecular recognition in artificial host–guest systems has reached extremely high levels of sophistication, the application of these principles to the stabilization of transition states is still an uphill struggle. As one of the most obvious and peculiar properties of enzymes is the presence of a very large surface area compared to the substrate size, the creation of de novo–designed catalysts has to face the difficulty of synthesizing elaborated tridimensional structures large enough to completely surround the substrates in order to provide a large surface for contact. Examples of templated polymeric catalysts and dendrimers have appeared in recent years, but supramolecular chemistry and self-assembly processes provide in some cases much simpler solutions.

The idea consists in spontaneously creating the catalyst in solution through the self-assembly of simple units, thus creating higher molecular weight structures by highly directive and predesigned interactions [4] with proper cavities or environments where substrates can be bound and transformed (Fig. 7.1) [5], with the ultimate goal of enzyme mimic [6].

Depending on the chemical nature of the solvent, examples of self-assembling supramolecular catalysts [7,8] have been found in water using surfactants forming micelles characterized by catalytic properties [9] or with metal–ligand-based capsules [10,11] and hydrophobic capsules [12], all showing excellent examples of artificial enzymes (Fig. 7.2a–d) driven by a combination of hydrophobic effect, ion pairing, and other weak interactions.

As far as artificial catalysts in organic solvents are concerned, the formation of large assembled structures able to provide cavities to bind substrates is a rather difficult task and, compared to supramolecular catalysis in water, hydrogen-bonded capsules are the most rewarding structures for the development of self-assembled supramolecular catalysts, thanks to the reversible formation of the assemblies that ensures easy substrate ingress and product egress (Fig. 7.2e). Nevertheless, because of the necessity of excluding substrates and experimental conditions that compete with H-bond formation, the use of hydrogen bond-based assemblies in catalysis is less developed compared to supramolecular catalysis in water, as previously mentioned.

It is worth noting that the first landmark example of supramolecular catalysis within a self-assembled capsule was the so-called "soft ball," a larger version of the original "tennis ball" [13] reported by Rebek and coworkers operating in aromatic solvents. The softball capsule is a reversible supramolecular dimer obtained through hydrogen bonding between two subunits in which two glycoluril moieties are linked through rigid spacers. The cavity within this capsule has a volume of about 240–320 Å^{3,} that is capable of accommodating up to two guests slightly larger than benzene. This enabled the use of the "softball" as supramolecular catalyst for the bimolecular Diels–Alder reaction between benzoquinone and cyclohexadiene (Fig. 7.3a). The fast encapsulation of the two substrates within the "softball" promoted the formation of the Diels–Alder adduct with an acceleration of about two orders of magnitude with respect to the same reaction carried out in *p*-xylene in the absence of the supramolecular vessel. Unfortunately, the product proved to be an excellent guest for the capsule, and because of that the system was unable to perform more than one catalytic cycle [14]. To enable this supramolecular catalytic system to perform catalytic cycles, Rebek and coworkers simply optimized the size and shape of the two reagents with the aim of forming cycloaddition products characterized by low affinity for the capsule cavity. To this end, cyclohexadiene was replaced by 2,5-dimethylthiophene dioxide, and the Diels–Alder reaction was carried out in the presence of a catalytic amount of "softball" (10% with respect to each substrate) (Fig. 7.3b). Interestingly, the reaction rate was enhanced by the "softball" observing a 75% yield after 4 days at 40°C, while under the same condition only 17% yield was achieved in the absence of the supramolecular vessel [15]. Similarly to enzymes, the supramolecular catalyst showed a marked decrease of catalytic activity in the presence of *p*-cyclophane as competitive inhibitor that occupied the cavity, thus disabling access for the substrates that did not get converted into products.

Since this pioneering contribution, the development of artificial supramolecular catalysis within self-assembled capsules in organic solvents passed through the development of several new capsules characterized by different sizes, shapes,

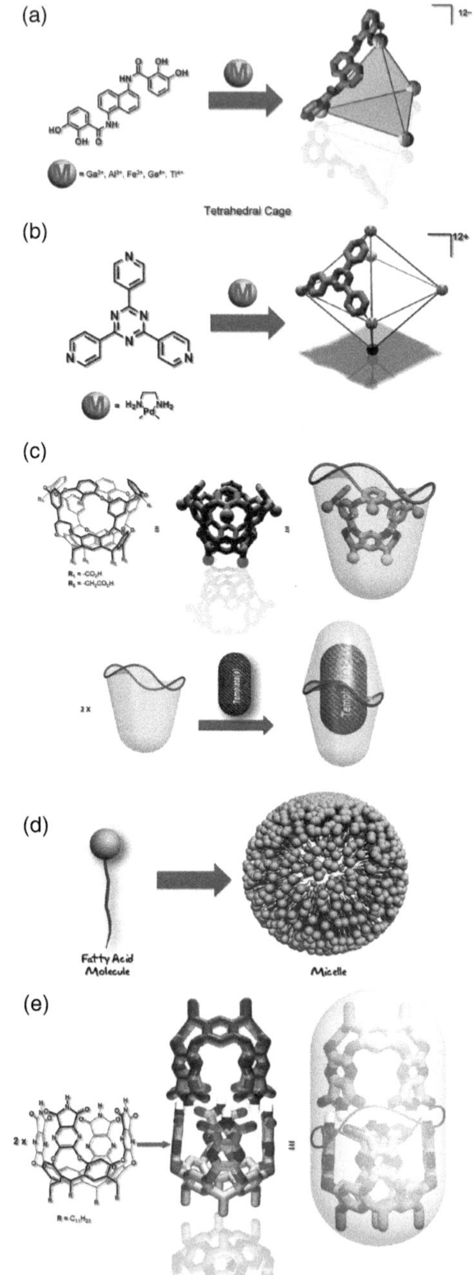

FIGURE 7.2 Examples of supramolecular capsular catalyst in water: (a) tetrahedral capsule, (b) octahedral capsule, (c) hydrophobic dimeric capsule, (d) micelles from surfactants. Example of supramolecular capsular catalyst in organic solvent (e) H-bonded cylindrical capsule. *(Adapted from Zecchina A, Bordiga S, Groppo E, editors. Selective nanocatalysts and nanoscience. Weinheim: Wiley-VCH Verlag GmbH; 2011.)*

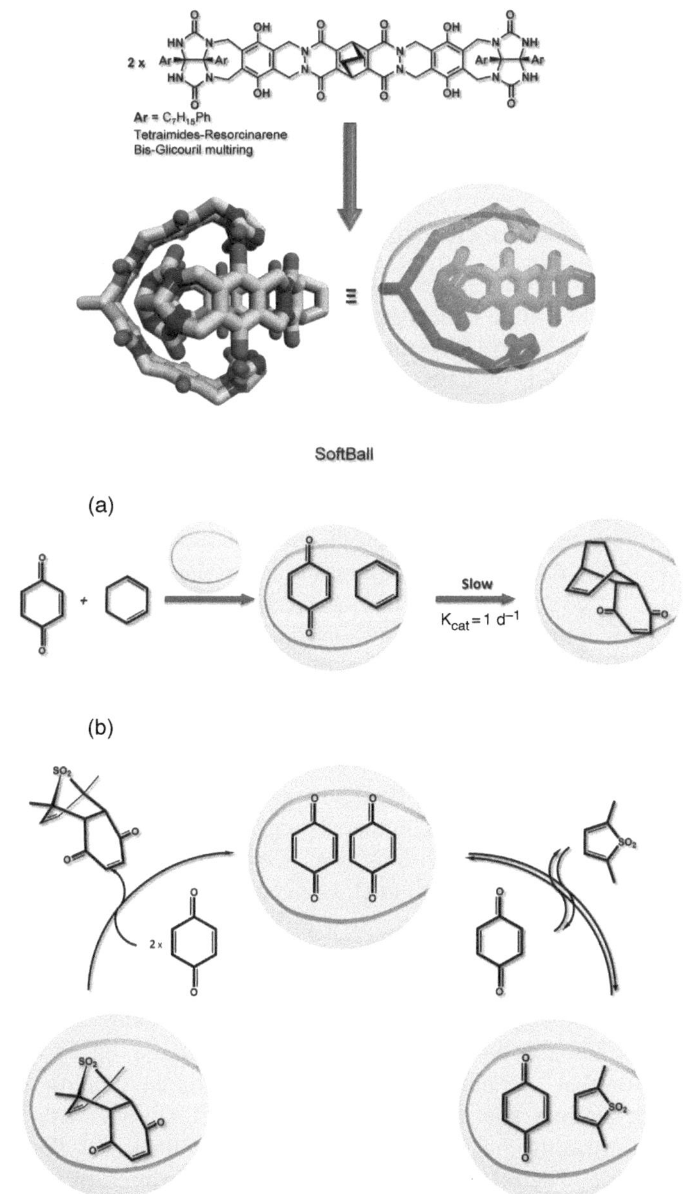

FIGURE 7.3 Molecular structure of the "softball" (a) Diels–Alder reaction between benzoquinone and cyclohexadiene mediated by the "softball" in stoichiometric amount and (b) the catalytic cycle of the "softball" in the Diels–Alder reaction between benzoquinone and 2,5-dimethylthiophene dioxide. *(Adapted from Zecchina A, Bordiga S, Groppo E, editors. Selective nanocatalysts and nanoscience. Weinheim: Wiley-VCH Verlag GmbH; 2011.)*

and properties. Recently, the attention has been focused on the employment of one of the largest known self-assembling capsules in organic solvents based on the interaction between monomeric resorcin[4]arene structures, greatly favored by its simple and straightforward synthesis in multigram scale and its large cavity that allows multiple encapsulation of substrates and catalysts.

1.1 Hexamer of Resorcin[4]arene: Synthesis and Properties

Resorcin[4]arene is an easy to prepare molecule directly available by condensation between one to one stoichiometric amount of resorcinol and aliphatic aldehydes under acid catalysis in ethanol [16]. The calyx-like monomeric unit is obtained with all the OH moieties on one rim of the calyx structure; when long aldehydes are employed, nonflipping of the aromatic units is possible and the molecule is conformationally stable.

In 1997 MacGillivray and Atwood demonstrated the peculiar behavior in the crystal state of six resorcin[4]arene units and eight water molecules to self-assemble forming a spherical hexamer [17] through the simultaneous formation of 60 O—H...O hydrogen bonds in appropriate apolar organic solvents such as deuterated chloroform or benzene. Each water molecule develops three hydrogen bonds with the OH groups of three neighboring resorcin[4]arene molecules, while each monomer interacts through four intramolecular hydrogen bonds with other neighboring resorcin[4]arene causing a pseudospherical chiral racemic structure with a cavity of about 1375 $Å^3$. The retention in solution of the same hexameric structure observed in the crystal state was further demonstrated a few years later through NMR diffusion experiments by Cohen and coworkers that disclosed that C-undecyl-resorcin[4]arene self-assembles in water-saturated chloroform-d without any guest showing that eight solvent molecules fill the cavity (Fig. 7.4) [18]. Concomitantly, Rebek and coworkers showed that a positively charged guest with appropriate shape and size, such as quaternary ammonium compounds, could be encapsulated [19] thanks to extended cation-π interactions [20,21] with the internal surface of the capsule due to the high number of electron-rich aromatic rings present.

Moreover Tiefenbacher and coworkers reported the encapsulation of tertiary amines that are neutral guests incapable of developing hydrogen bonds. These authors studied the encapsulation of amines with different basicity demonstrating that the protonation of the amines was the cause of the encapsulation. The results showed that the delocalization of the negative charge along the entire structure of the self-assembled hexamer causes an acidity of the hexameric capsule considered as a sole entity much higher than the single molecules of resorcin[4]arene, estimating a pK_a of about 5.5–6 that could be further increased through encapsulation of ammonium salts [22]. Apart from cationic guests, the capsule was shown to efficiently bind neutral species through H-bonding like carboxylic acids, aminoacids [23], and alcohols [24], present in large excess, even though in these cases it was not

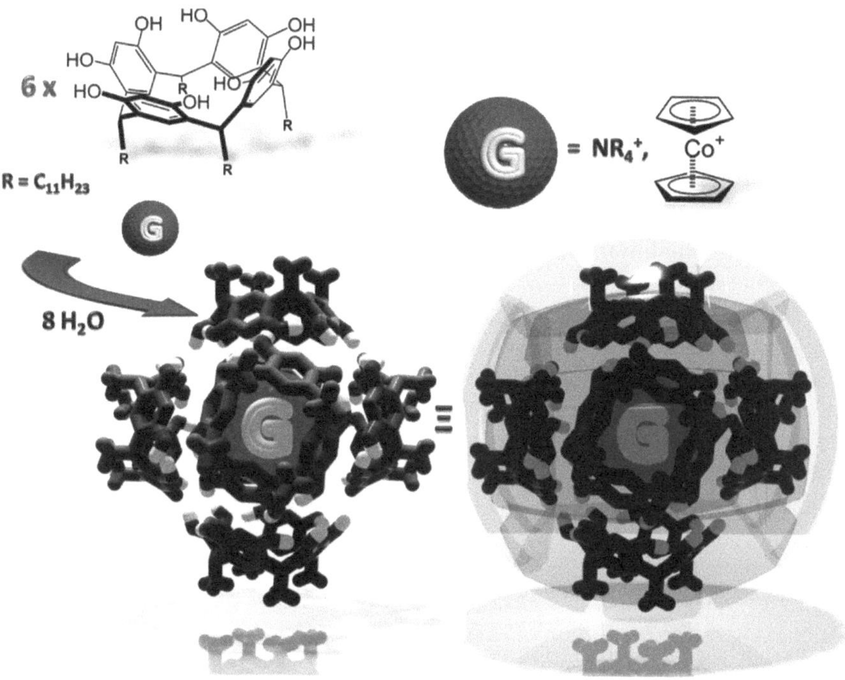

FIGURE 7.4 **Resorcin[4]arene in a wet organic solvent self-assembles forming the hexameric capsule that can host cationic guests.** *(Adapted from Scarso A, Borsato G. Self-assembly of organic supramolecular capsules. In: Gale PA, Steed JW, editors. Supramolecular chemistry: from molecules to nanomaterials, vol. 5. Chichester, UK: Wiley; 2012, p. 2085–2114.)*

clear whether the binding was between one resorcin[4]arene and one guest or between the self-assembled hexamer and six guests.

Soon after the discovery of the traditional host–guest properties of the resorcin[4]arene capsule, it became evident that this capsule represents an ideal supramolecular structure for several applications in homogeneous catalysis. Its use in catalysis is spurred by the simple and accessible synthesis of the monomer combined with the formation of a very large cavity that is a feature extremely difficult to find in other structures.

2 HEXAMERIC CAPSULE AS AN INHIBITOR

The first example published in the literature concerning the use of the hexameric capsule in catalysis was actually an example of catalytic modulation through encapsulation. Scarso and coworkers selected the very well known photocatalyst $[Ru(bpy)_3]^{2+}$ (bpy, 2,2′-*bis*-pyridine) that, thanks to its pseudospherical shape and its cationic character, perfectly fits the cavity of the capsule with the aim of investigating the effect of reversible binding on the photocatalytic activity of the metal catalyst [25].

It is in fact known that $[Ru(bpy)_3]^{2+}$ under visible light irradiation and in the presence of oxygen provides reactive species like singlet oxygen [26] or hydrogen peroxide [27]. The latter terminal oxidants can efficiently transform electron-rich substrates like sulfides into the corresponding sulfoxides. In their work, Scarso et al. investigated the possibility of modulating the catalytic activity of the $[Ru(bpy)_3]^{2+}$ by changing the solvation sphere of the catalyst through encapsulation and release in solution, while maintaining all the other experimental conditions such as temperature, presence of oxygen, and light irradiation unaltered.

The encapsulation of the Ru(II) complex in the presence of one equivalent of a capsule was confirmed by ^1H-NMR analysis, demonstrating the complete disappearance of the four resonances of the original metal species in the aromatic region of the spectrum while new, extremely weak, broad resonances emerged upfield at 6.6 ppm. The complex could be released in solution with reappearance of the corresponding resonances by addition of an excess of a competitive cationic guest [28] such as $(NEt_4)(OTf)$. The reversible binding within the cavity was confirmed by UV-vis analyses demonstrating that, upon encapsulation, the maximum of absorbance for the complex at 455 nm slightly shifted about 8 nm to shorter wavelengths, and that in the presence of an excess of competitive cationic guest, the maximum of absorption in the UV-vis spectrum was restored to 453 nm. This showed that no alteration of the absorption properties of the complex was caused by encapsulation.

In the presence of visible light irradiation provided by a 120 W lamp and with 1 atm of O_2, the Ru(II) complex in solution promoted the oxidation of dibutyl sulfide to the corresponding dibutyl sulfoxide obtained in 32% yield. Under identical experimental conditions but in the presence of the capsule, the reaction did not occur as a consequence of the encapsulation of the Ru(II) photocatalyst. The catalytic activity was restored when repeating the same experiment with $[Ru(bpy)_3]^{2+}$ in the presence of both capsule and tetraethylammonium competitive guest, due to the displacement of the metal catalyst from the cavity of the capsule. Similar results were observed with other thioethers showing conversions to the corresponding sulfoxides that were dependent on the electron density on the sulfur atom; in all cases, no conversion was observed with the encapsulated Ru(II) catalyst. Since it was demonstrated that the absorption properties of the Ru(II) metal center were substantially not influenced by the capsule, it is likely that the inactivation provided by the capsule could be due to interrupted energy transfer from the Ru(II) center to O_2.

Operating a series of alternating additions of capsule to stop the oxidation reaction by encapsulation of $[Ru(bpy)_3]^{2+}$ (OFF state) followed by additions of excess of tetraethylammonium competitive guest with concomitant release in solution of $[Ru(bpy)_3]^{2+}$ (ON state), it was possible to operate a reversible switching of the photocatalytic activity of the complex, simply changing its solvation sphere without changing either the coordination on the metal or the experimental conditions (Fig. 7.5). Overall a stepwise kinetic profile for the

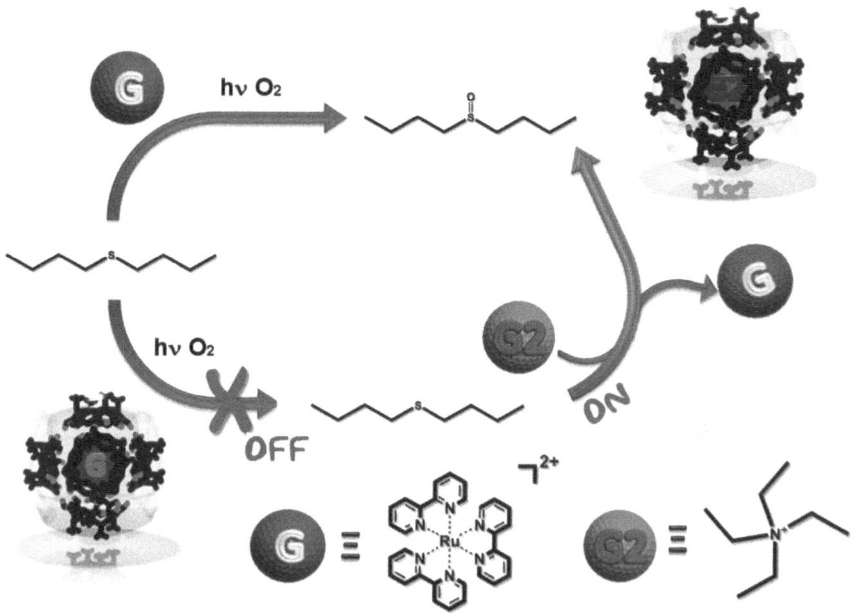

FIGURE 7.5 **Reversible sulfoxidation reaction in the presence of visible light and 1 atm. of O_2 mediated by encapsulation (OFF state) of $[Ru(bpy)_3]^{2+}$ and release in the presence of tetraethylammonium competitive guest (ON state).** *(Adapted from Ref. [25].)*

formation of the sulfoxide was obtained, representing the first example of this ON/OFF modulation. Instead of operating in a covalent manner on the catalyst, it worked at a supramolecular level on its solvation sphere.

3 HEXAMERIC CAPSULE AS A SUPRAMOLECULAR NANOREACTOR

The size of the cavity of the hexamer is sufficiently large to accommodate several solvent molecules, and the encapsulation of metal catalysts can be exploited to steer the reaction pathway of a catalytic reaction toward unexpected products due to the steric hindrance experienced by the substrate within the cavity on the way to the catalyst. It is known that, for several encapsulation phenomena, the space occupied by the guest molecules is slightly larger than the total volume available; in other words, the packing coefficient given as the ratio between the total volume of guests and the volume of the cavity is usually in the range 0.45–0.55, which is the same observed for most liquids [29].

Therefore, choosing transition metal catalysts characterized by the positive charge (a crucial need for encapsulation) and with reduced volume to allow the coencapsulation of substrates, it is possible to design reactions where the capsule dramatically changes the fate of the chemical transformation. This is a typical property of enzymes where substrate binding, preorganization, and folding are pivotal for high levels of chemo, regio, and stereoselectivities.

In the following sections, examples are presented where the synergic effect between transition metal catalysts and the hexamer of resorcin[4]arene imparts unique product and substrate selectivities. In all cases, coencapsulation of the substrate and the catalyst is crucial.

3.1 Effects on Product Selectivity

While the study and the application of transition metal catalysis within water-soluble supramolecular hosts have been reported in recent years, the extension of the same approach to H-bonded capsules in organic solvent is still a rather uninvestigated field. The first example was reported by Scarso and coworkers using a cationic elongated Au(I) organometallic complex bearing a NHC ligand (NHC=the N-heterocyclic carbene 1,3-*bis*(2,6-diisopropylphenyl)imidazol-2-ylidene) that was suitable for encapsulation within the cavity of the resorcin[4] arene capsule. These complexes are well known for their catalytic activity in the alkyne hydration reaction; since the hexameric capsule requires the presence of water for self-assembling, the complex was investigated to compare the effects on its activity and selectivity of being free in solution or within the cavity of the capsule [30]. 4-Phenyl-1-butyne was selected as substrate since it could provide two hydration products, 4-phenyl-2-butanone or 4-phenyl-butanal, respectively obtained by attack of the water molecule either on the C2 of the alkyne moiety or on C1, together with a cyclic 1,2-dihydronaphthalene product derived by an intramolecular ring-closing reaction when activated by the Au(I) catalyst in the absence of water (Fig. 7.6).

Initially, the quantitative encapsulation of the cationic Au(I) complex was investigated, showing that in the presence of the equivalent of one capsule, all the resonances of the free complex disappeared and the ^{1}H-NMR spectrum clearly revealed the presence of new upfield shifted resonances at −0.26 and 0.37 ppm attributed to the *i*-Pr residues of the ligand, while the vinyl resonances of the encapsulated complex appeared between 5 and 7 ppm. It is worth noting that the cationic nature of the complex is an extremely important prerequisite, as observed when comparing the affinity of the neutral (NHC)AuCl complex with one equivalent of capsule that showed only partial encapsulation and much larger amounts of free complex in solution. Confirmation of the quantitative encapsulation when using the cationic NHC–Au complex was obtained by 2D NOESY experiments that showed close contacts between the *i*-Pr residues of the Au complex and both the OH residues and the aryl proton in between the OH groups of the capsule, while DOSY experiments showed similar diffusion coefficients for the resonances of the capsule and those attributed to the encapsulated complex. Similarly to what was previously observed with the Ru(II) photocatalyst, the encapsulated NHC–Au(I) catalyst could be released in solution by addition of an excess of cationic tetraethyl ammonium tetrafluoroborate as competitive guest.

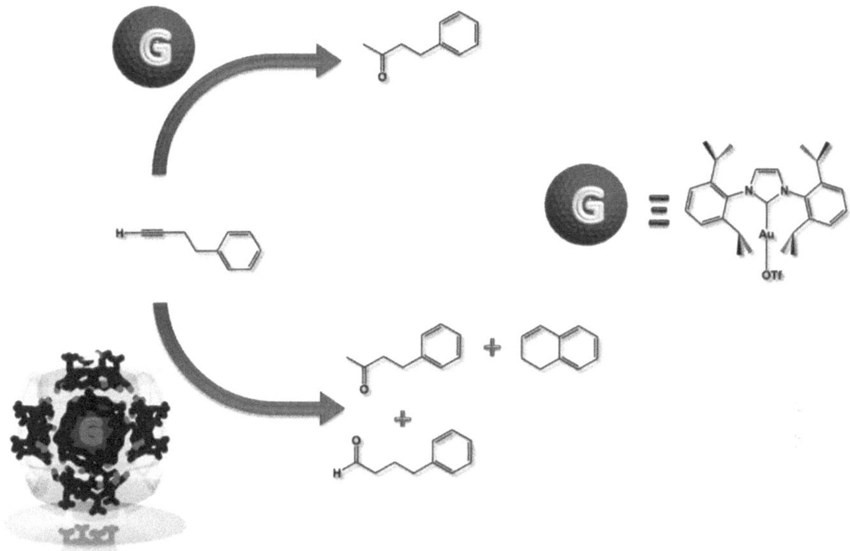

FIGURE 7.6 The encapsulation of the NHC–Au(I) catalyst within the hexamer of resorcin[4] arene drastically changes the product distribution in the terminal alkyne hydration of 4-phenyl-1-butyne. *(Adapted from Ref. [30].)*

When the Au(I) catalyst was tested at 70°C in the hydration reaction of 4-phenyl-1-butyne, two different scenarios were observed for the free complex in solution and for the encapsulated complex. In the first scenario, the reaction led exclusively and quantitatively within 30 min to the formation of the methyl ketone hydration product. In the second, the reaction was slower, showing after 400 min only 28% of conversion of the substrate; more interestingly, apart from the expected methyl ketone with 12% yield, significant amounts (4%) of the aldehyde were obtained, a novelty for Au catalysts. This demonstrates that the sterically constrained environment within the cavity was responsible for the steering of the regioselectivity of the reaction. Even more unexpected were both the presence of 1,2-dihydronaphthalene (12%), whose formation is likely to be favored by unusual folding of the substrate, and the difficult encapsulation of water, making the intermolecular hydration reaction comparable in rate with the intramolecular one.

As a further confirmation for the effect of the simple change of the solvation sphere of the catalyst, the hydration reaction with the encapsulated NHC–Au(I) complex was repeated, and after 400 min a large excess of tetraethylammonium competitive guest was added with concomitant release in the solution of the catalyst, changing the catalyst's selectivity and forming exclusively the methyl ketone as hydration product.

Overall, these experiments demonstrated that simply by changing the solvation sphere of the NHC–Au(I) catalyst, it is possible to drastically alter the product distribution of the reaction, showing that yields and in particular chemo- and regioselectivities can be strongly affected by the capsule, leading to the formation of unusual products. In fact, the Au(I) catalyst has a volume of about 400 Å^3 and its packing coefficient is roughly 30%, leaving sufficient space for coencapsulating from two to four solvent molecules that can be replaced by the substrate that however, experiences more stringent steric requirements with respect to the reaction in the bulk solvent. The unusual regioselectivity observed in the hydration products could be due to the forced folding of the substrate, while the partial formation of the cyclic product could be due to difficult access for water into the cavity.

3.2 Effects on Substrate Selectivity

Selectivity is an extremely important property of chemical transformations, but usually we consider only the right side of a chemical equation. For catalytic reactions, the presence of several competitive substrates introduces another kind of selectivity, this one related to the ability of the catalyst to pick preferentially one substrate from the mixture and to convert the substrate into the corresponding products while leaving the other substrates behind. Artificial catalysts developed to display substrate selectivity are rare; in particular, heterogeneous catalysts such as zeolites due to their porosity can display this kind of selectivity due to their porosity, and selected examples are also known with catalytic polymers [31,32] and imprinted polymers [33]. As far as homogeneous catalysts are concerned, examples are limited to the kinetic resolution of racemates in which one of the enantiomers reacts more rapidly than the other [34] through covalent interaction of the substrate with the catalyst, while other examples deal with mixtures of substrates that differ quite remarkably in their structure away from the reacting functional group. Far less frequent are examples related to "homologous" series of substrates tested together toward a certain catalyst [35].

Conversely, enzymes, as hypothesized for the first time by Arrhenius in 1888 [36], are known for also being extraordinarily substrate–selective catalysts [37–40], and at first sight their ability to select substrates is related to their mechanism of action, which involves a recognition event forming a substrate–enzyme adduct before the true catalytic event (Fig. 7.7). To observe this property, a three-dimensional structure and a large interacting surface are required. To implement such concepts into traditional homogenous catalysts, it is necessary to increase the contact surface between catalyst and substrate, and this is possible through the use of encapsulated catalysts where the capsule impart the steric and electronic rules for the selection of the substrate. The substrate that is preferentially bound has access to the catalyst and is therefore converted. This approach has been observed more frequently for encapsulated metal catalysis within water-soluble capsules [41,42], while examples in organic solvents are very limited [43–45].

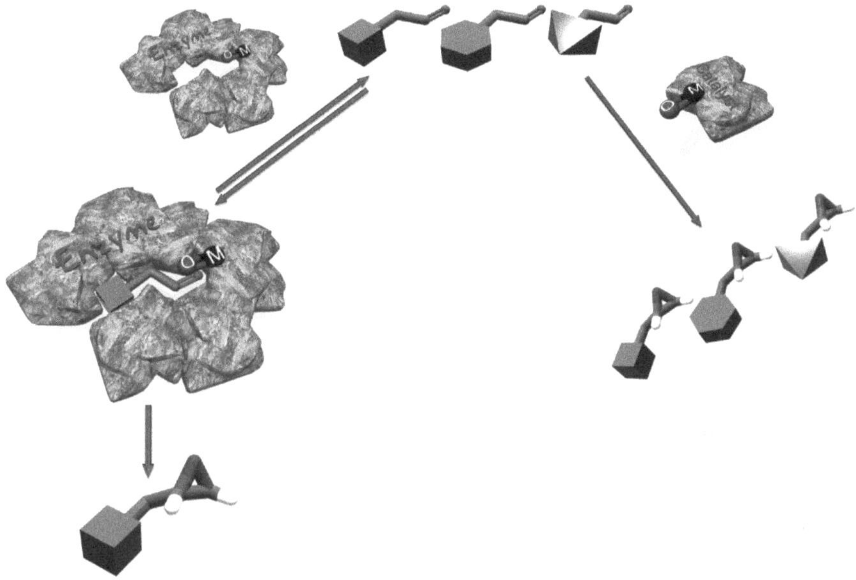

FIGURE 7.7 Comparison between typical poor substrate selectivity displayed by a traditional homogeneous catalyst and the extremely high substrate selectivity typical of enzymes. *(Adapted from Ref. [46].)*

On the basis of the effects observed on product selectivity in the alkyne hydration reaction mediated by the encapsulated NHC–Au(I) catalyst within the hexamer of resorcin[4]arene, Scarso and coworkers investigated the ability of the same supramolecular catalytic system in the selection of homologous series of substrates [46].

The competitive experiments were carried out with a series of terminal alkynes characterized by similar electronic properties but differing in portions of the molecule remote from the reaction site. For aliphatic alkyne, the three substrates ethynyl cyclohexane, 1-octyne (which represents an acyclic isomer of the former substrate), and 1-dodecyne were tested, with the free catalyst compared to the encapsulated catalyst. In the former case, after a short induction time, the initial reaction rate for the three substrates showed similar behavior for 1-dodecyne and 1-octyne, while the cyclic isomer, being slightly more electron rich, reacted 1.5 times faster (Fig. 7.8a). Encapsulation of the catalyst led to a magnification of the favorable hydration of the cyclic substrate that reacted more than twice as rapidly as the longer substrate. It is likely that extended linear substrates, that in their extended conformation are approximately 1.4 and 2.1 times longer than ethynyl cyclohexane, have to fold to better complement the residual space left available within the cavity occupied by the catalyst.

In order to exclude folding effects, three rigid aromatic terminal alkynes like phenylacetylene, 4-methyl-phenylacetylene, and 4-*t*-butyl-phenylacetylene, characterized by the same general shape but different lengths and electronic properties, were tested with the free and the encapsulated catalyst. In the bulk

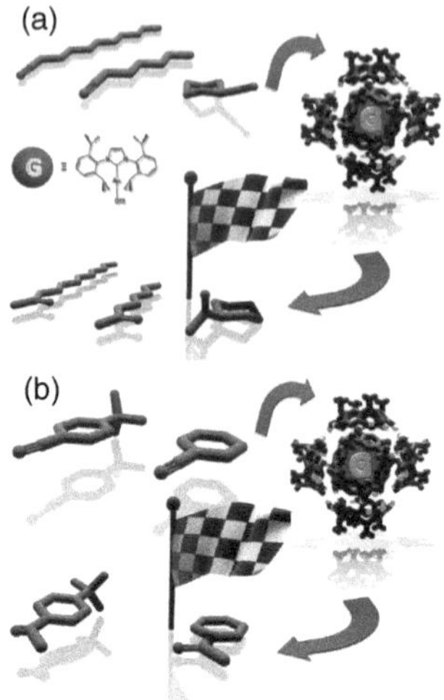

FIGURE 7.8 Comparison of substrate selectivity displayed by the NHC–Au(I) catalyst in the hydration reaction of alkynes free in solution and encapsulated in the hexamer: (a) preferential hydration of cyclic ethynyl cyclohexane rather than acyclic aliphatic terminal alkynes and of (b) shorter rather than longer aromatic rigid alkynes.

solvent, the reaction was faster with the more electron-rich substituted alkynes, with relative initial rates of reaction equal to 1 to 1.4 to 1.5 for phenylacetylene, 4-methyl-phenylacetylene and 4-*t*-butyl-phenylacetylene, respectively. It is worth noting that, upon encapsulation, a complete inversion of the substrate selectivity was observed, with the smaller substrate that is more easily hosted in the residual space within the cavity, reacting faster than the 4-substituted phenylacetylenes (phenylacetylene:4-methyl-phenylacetylene:4-*t*-butyl-phenylacetylene 1.6:1.3:1.0) (Fig. 7.8b).

The previously described results represent another example of how the hexameric capsule can be efficiently employed as a nanometric reactor that provides unexpected selectivities by simply changing the second coordination sphere around a given homogeneous catalyst; especially because of the restricted space, selection of the substrates is possible, based on steric ground. The differences observed are quite significant if one considers that, apart from catalyst and substrates, space is also available for solvent molecules in order to reach the optimal packing coefficient range [29].

Aiming at demonstrating the potentialities of the hexameric capsule as a substrate selective nanoreactor, Scarso and coworkers extended the concept to a stoichiometric reaction. In particular, the steric hindrance due to the confined space within the cavity of the capsule was exploited in the substrate selective amide synthesis mediated by the cationic condensing agent 1-ethyl-3-(-3-dimethylaminopropyl) carbodiimide hydrochloride [47]. Similarly to the encapsulation of other cationic species, the cationic condensing agent that in

TABLE 7.1 Catalytic Tests for the Competitive Coupling of Pairs of Amines with one Acid and Pairs of Acids with one Amine Mediated by the Carbodiimide Cationic Condensing Agent in the Presence or Absence of the Hexameric Capsule as Supramolecular Nanoreactor at 60°C, Time 18 h. +: Presence; −: Absence

#	Acid	Amine	Capsule	Short/long amide ratio
1	C_4	C_4 C_8	−	1.7
2	C_4	C_4 C_8	+	2.2
3	C_{12}	C_4 C_8	−	1.2
4	C_{12}	C_4 C_8	+	2.0
5	C_6 C_{12}	C_4	−	0.5
6	C_6 C_{12}	C_4	+	8.1
7	C_6 C_{12}	C_{16}	−	1.2
8	C_6 C_{12}	C_{16}	+	28

Adapted from Ref. [47], Royal Society of Chemistry.

chloroform exists as a mixture of two isomers (an an acyclic and a cyclic one) showed complete disappearance of all the resonances at the ^{1}H-NMR upon addition of the equivalent of one capsule. The condensing agent did not suffer the presence of water when encapsulated, since the carbodiimide structure remained stable for several days without formation of the corresponding urea derivative. Conversely, addition of 10 equivalents of a competitive tetraethylammonium guest was sufficient to induce the release of the carbodiimide cation from the capsule.

Initially, solutions of the free and encapsulated condensing agent were tested for amide synthesis using combinations of two aliphatic amines characterized by different length of the aliphatic chain and one aliphatic carboxylic acid monitoring the formation of the two amides (Table 7.1). Under these conditions the substrate selectivity imparted by the capsule was not so high and the presence of the supramolecular system caused only a slight improvement in the short versus long amide products. The reason for the low substrate selectivity was ascribed

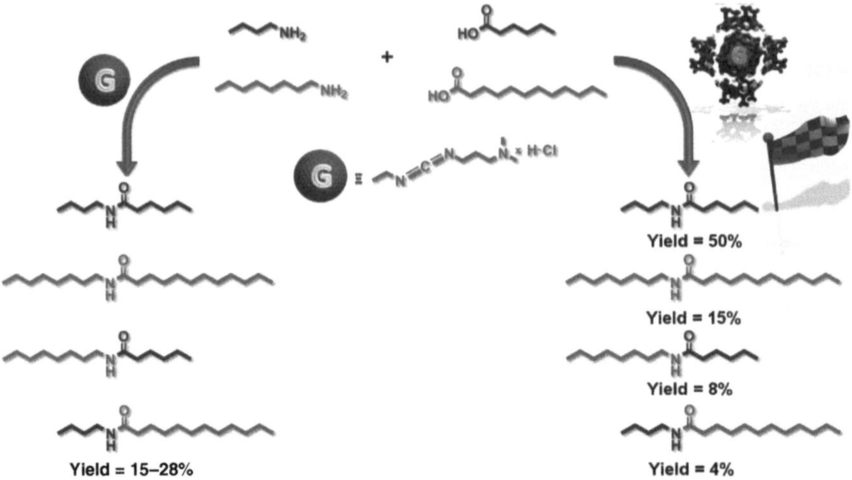

FIGURE 7.9 Competitive coupling of a pair of carboxylic acids with a pair of amines mediated by the carbodiimide cationic coupling agent either in the presence or in the absence of the resorcinarene capsule.

to the mechanism of the carbodiimide coupling, where the rate-determining step of the reaction is the attack of the carboxylic acid on the condensing agent forming an activated O-acylisourea followed by fast amine attack [48]. Better results were observed testing pairs of acids with one amine.

Much better substrate selectivity was observed in this second case, as the carbodiimide species reacted much faster with the shorter acids that are more easily accommodated within the cavity than the longer ones. Values for the shorter/longer amide ratios were observed between 8 up to 28 in the case of hexadecylamine (Table 7.1). No selectivity was observed with the condensing agent free in solution.

A further series of competitive experiments comprising two amines (butylamine and octylamine) and two (hexanoic and dodecanoic acid) acids were carried out, with the carbodiimide in pure solvent showing no selectivity and the four possible amide products being obtained in similar amounts (yields 28–12%). Conversely, when using the encapsulated condensing agent, the capsule showed a marked preference for selecting the combination of the shorter acid and the shorter amine; this led to the corresponding amide in 50% yield, while the longer amide was formed with only 4% yield (Fig. 7.9).

Overall, the hexameric resorcin[4]arene capsule showed very interesting selective properties for the condensation reaction between carboxylic acids and amines mediated by a cationic carbodiimide-based condensing agent; similarly to what was observed in the previous case for the alkyne hydration reaction, the capsule prefers the encapsulation of smaller and more compact substrates that react faster than longer ones.

4 HEXAMERIC CAPSULE AS A CATALYST

Examples of the application of the resorcin[4]arene capsule as the true organocatalyst are the majority. This is a consequence of the great opportunities offered by the capsule for the stabilization of reactions involving cationic intermediate species.

4.1 Hydrolysis and Hydration Reactions

Recent examples of promotion by the resorcin[4]arene hexameric capsule are the hydrolysis of acetals[22], exploiting the presence of water molecules as hydrogen-bonding units present in the supramolecular assembly. As an example, 1,1-diethoxyethane was efficiently converted into the corresponding carbonyl compound acetaldehyde in 85% yield after 1 h at room temperature in the presence of only 10 mol% of capsule, while negligible conversion was observed with the concomitant presence of tetrabutylammonium as a competitive guest for the cavity. The reaction proved to be sensitive to the substrate size, showing a gradual decrease of activity with the increase in the length of the acetal, with butanal formed in 62% yield, pentanal in 44%, octanal 13% and dodecanal in less than 4% yield. This observation was probably due to more efficient binding of smaller substrates within the cavity in combination with some solvent molecules to fit the proper packing coefficient [29].

As a simple example of competitive substrate selection, the reactivity of two very different substrates was considered in a competitive experiment. In a control experiment, the reaction of 1,1-diethoxyethane and the much longer 1,1-diethoxydodecane in the presence of trifluoroacetic acid led to the formation of both aldehydes in 24 and 41% yield, respectively. When the capsule was used as a supramolecular catalyst, the reaction was much more selective, with 83 and 2% yield for the shorter and longer aldehyde, respectively (Fig. 7.10).

The high electron density present in the cavity of the hexamer spurred the investigation of other classes of molecules that, apart from cationic species and H-bonding molecules, could be encapsulated and eventually act as substrates.

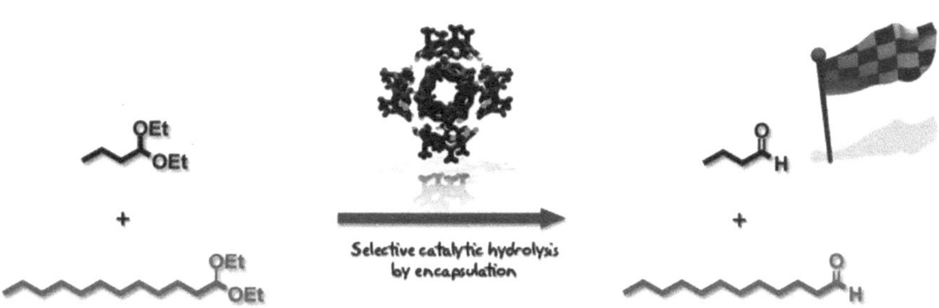

FIGURE 7.10 Supramolecular catalysis in the hydrolysis of acetals imparted by the hexameric capsule that showed to be sensitive to the size of the substrate. *(Adapted from Ref. [22] with permission of American Chemical Society.)*

Isonitriles are a special class of molecules [49], characterized by an extremely repellent odor, where the R—N≡C functional group can be described with either the zwitterionic form with positive N and negative C, or the predominantly carbenic electron-poor structure [50] that overall confer a linear geometry to the NC bond. Because of its electronic structure, isonitriles are largely investigated in multicomponent reactions, since the C atom of the isonitrile moiety reacts either as an electrophile [51] or as a nucleophile [52].

Scarso and coworkers disclosed the unique affinity of this class of neutral molecules for the cavity of the hexamer, observing that several isonitrile species, either with aliphatic or aromatic residues, were encapsulated in the presence of the resorcin[4]arene capsule [53], with the usual disappearance of the ^{1}H NMR resonances for the free isonitrile and appearance of new broad upfield shifted resonances. For instance, addition of cyclohexyl isonitrile led to the formation of new resonances in the NMR spectrum in the range 0.1 to −0.7 ppm. In the cases of isopropyl-isonitrile, several new resonances were observed in the range −0.3 to −0.7 ppm and with *t*-butyl isonitrile a major single sharp resonance at −0.62 ppm (Δδ −2.07 ppm) was present. Similar results were observed for benzyl isonitrile and for an aromatic derivative. It was proposed that encapsulation of isonitriles was driven by the stabilization imparted by the electron-rich cavity to the electron-poor carbenic partial character of this class of molecules. Even more interestingly, it was observed that, upon gentle heating the system at 60°C, the complete conversion of the cyclohexyl isonitrile took place, with concomitant formation of the corresponding *N*-formylamide due to water addition to the R-NC moiety [54], while in the absence of capsule hydration of the substrate was negligible (Fig. 7.11).

After a series of control experiments carried out excluding the effect of the Brønsted acidity of the capsule and the involvement of the OH functional groups of the resorcin[4]arene, the effect of the cavity on the catalytic supramolecular hydration reaction was confirmed. In order to further substantiate this hypothesis, experiments were carried out repeating the hydration reactions of several isonitriles in the presence of the capsule and with an excess of tetraethylammonium as competitive guest, observing in all cases a marked decrease of the formation of the corresponding formamides.

Considering all the isonitriles tested, the catalytic activity of the supramolecular capsule was highly dependent on the size, shape, and electronic properties of the substrates, once again underlining the importance of proper packing coefficient of the coencapsulated species in the cavity. The catalytic activity of the capsule was due both to its mild acidity [22] that favors the protonation of the C atom of the isonitrile unit, and the stabilizing effects of the electron-rich cavity through cation-π interactions, which facilitated the hydration reaction.

4.2 C—C Bond-Forming Reactions

Diazoacetates represent a class of organic compounds bearing a nitrogen molecule as a terminal functional group (general formula $R_2C{=}N_2$) characterized

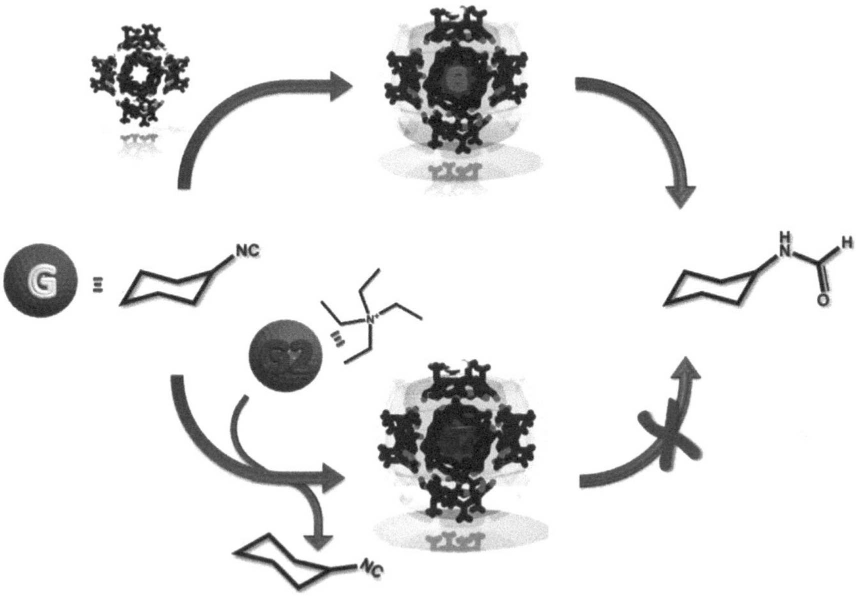

FIGURE 7.11 Hydration reaction of cyclohexyl isonitrile catalyzed by the hexameric capsule and inhibited by the presence of tetraethylammonium cation as a competitive guest for the cavity. *(Adapted from Ref. [53].)*

by an electronic structure with a positively charged central nitrogen atom and a negative charge delocalized between the terminal nitrogen and the carbon [55]. Beside this, these compounds are well-known carbene-precursors simply through the loss of a neutral di-nitrogen molecule [56]. Their carbene-like character prompted their employment as potential substrates for the hexamer capsule. Scarso and coworkers investigated their encapsulation in the resorcin[4] arene assembly and found that the efficient capsule catalyzed 1,3-dipolar cycloaddition between diazoacetate esters and electron-poor alkenes, leading to 4,5-dihydro-1*H*-pyrazole derivatives [57]. It is worth noting that Diels–Alder reactions are among the most investigated chemical transformations, displaying the effects of supramolecular unimolecular, or self-assembled capsules in water [58,59], while examples in organic solvents are rare apart from the seminal work of Rebek [15].

By means of ^{1}H-NMR investigation, it was possible to observe the affinity of diazoacetate esters for the resorcin[4]arene capsule. In fact, in the presence of 10 equivalents of ethyl, *t*-butyl, or benzyl diazoacetate with respect to the capsule, in all cases new upfield shifted resonances appeared related to the encapsulated substrate in slow exchange on the NMR timescale (at $\delta = -0.86$ to -1.06 ppm for ethyl diazoacetate, $\delta = -0.50$ and -0.52 ppm for *t*-butyl diazoacetate, at $\delta \approx 6.4$ ppm and 3.8 ppm for benzyl diazoacetate, respectively). Simple integration of the resonances attributed to the encapsulated substrates revealed a slightly higher affinity for *t*-butyl diazoacetate with respect to ethyl diazoacetate, while for benzyl diazoacetate the determination was not possible due to partial signal overlap. Especially with *t*-butyl diazoacetate, further 2D NMR experiments supported the encapsulation of the substrate with the NOESY spectrum that showed cross-peaks between the *tert*-butyl moiety of the diazoacetate compound and the aromatic CH group between the hydroxyl groups of resorcin[4]arene and the DOSY spectrum, showing same diffusion coefficients for the resonances of the capsule and those at $\delta \approx -0.50$ ppm for the encapsulated *t*-butyl moiety.

The interaction of the capsule with the diazoacetate esters did not lead to their decomposition. In fact, all substrates remained unaltered both in the presence and in the absence of the capsule, even at 50°C for 20 h in water saturated chloroform-d. More interestingly, since this class of molecules are good partners for the 1,3-dipolar cycloaddition reactions with a large series of dipolarophiles [60], their interaction with electron-poor alkenes was investigated in the presence and in the absence of the capsule in order to ascertain its supramolecular catalytic effects.

The cycloaddition reaction usually occurs under high concentration conditions, or it can be promoted by the presence of Lewis acids, leading initially to the formation of 4,5-dihydro-3*H*-pyrazoles that are unstable intermediates and that later convert into the corresponding 4,5-dihydro-1*H*-pyrazole isomers [61]. To explore the reactivity of diazoacetate esters in the cycloaddition reaction, initial experiments were carried out with acrolein. These showed that, in the absence of the capsule, the spontaneous formation of the corresponding 4,5-dihydro-1*H*-pyrazole was low (12% yield), while in the presence of the capsule it increased up to 47% in yield. Usual experiments with tetraethylammonium as competitive guest for the cavity added in large excess led to the formation of only 8% yield of the cycloaddition product, indicating a drop in the reaction rate when the cavity was occupied. In the presence of reduced amounts of competitive ammonium cation, an increase of product formation was observed as a consequence of the more accessible cavity of the capsule. In order to exclude a possible catalytic effect due to interactions between the hydroxyl group of the capsule and the substrate molecules, the reaction with 24 equivalents of resorcinol as a capsule mimic was carried out without observing substantial catalytic effects imparted by the hydroxyl groups of resorcinol.

The same catalytic effect imparted by the empty capsule was observed changing the diazoacetate ester in the reaction with acrolein. In fact, with the *t*-butyl derivative the reaction with the capsule led to 97% yield, and with benzyl diazoacetate 54% yield of the corresponding cycloaddition product was observed. In both cases, the presence of the capsule occupied by the ammonium inhibitor led to a marked decrease of product formation. The higher conversions obtained in the presence of *t*-butyl diazoacetate as substrate were probably due to the good affinity with the cavity by means of CH-π interactions.

The catalytic effect of the capsule was extended also to the use of *t*-butyl diazoacetate, with other less reactive electron-poor alkenes following the reactions

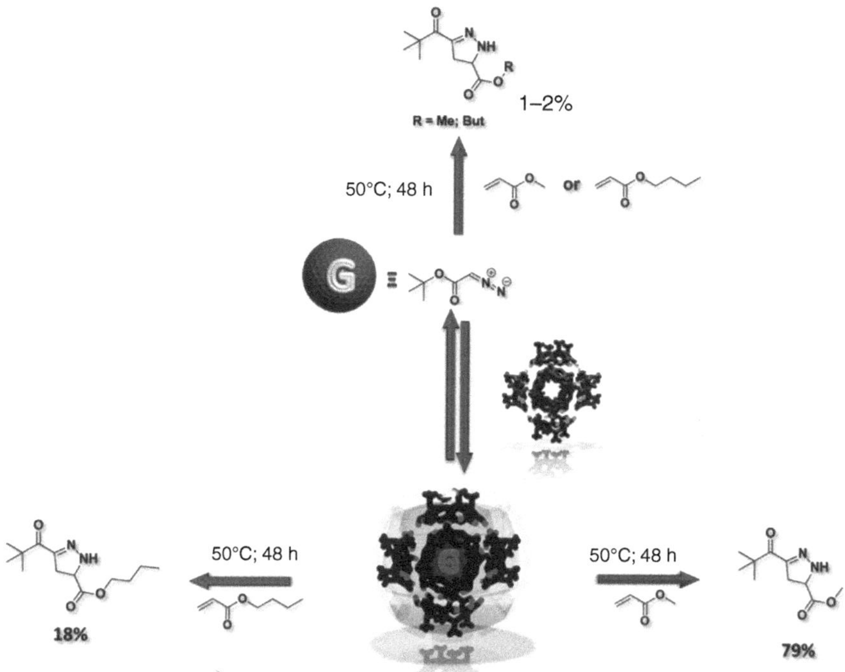

FIGURE 7.12 **1,3-Dipolar cycloaddition reaction between diazoacetate esters and electron-poor alkenes like acrylate esters mediated by the hexameric capsule.** *(Adapted from Ref. [57].)*

at 50°C because of the intrinsic lower reactivity of these alkenes. In particular, the reaction with α,β-unsaturated aldehydes crotonaldehyde and *trans*-2-hexenal did not lead to the formation of the corresponding 4,5-dihydro-1*H*-pyrazoles when the reaction was carried out in the absence of the capsule or with the capsule occupied by the ammonium competitive guest, while with the free capsule yield of 95 and 79% for the corresponding cycloaddition products were observed after 48 h at 50°C.

The reaction, as further extended to acrylate esters such as methyl-acrylate and butyl-acrylate, displayed 79 and 18% yield in the reaction with *t*-butyl diazoacetate catalyzed by the resorcin[4]arene capsule, while with the occupied cavity yields were both lower than 5% and in the absence of the capsule decomposition products were detected (Fig. 7.12).

The results demonstrated the versatility of the resorcin[4]arene capsule in closely favoring the encapsulation of neutral diazoacetate esters and promoting through encapsulation the reaction with electron-poor alkenes with, once again, a supramolecular catalytic effect imparted by the electron rich cavity of the capsule.

Another example of the use of a modified hexameric resorcin[4]arene capsule for the activation of a cycloaddition reaction was reported by Shimizu and

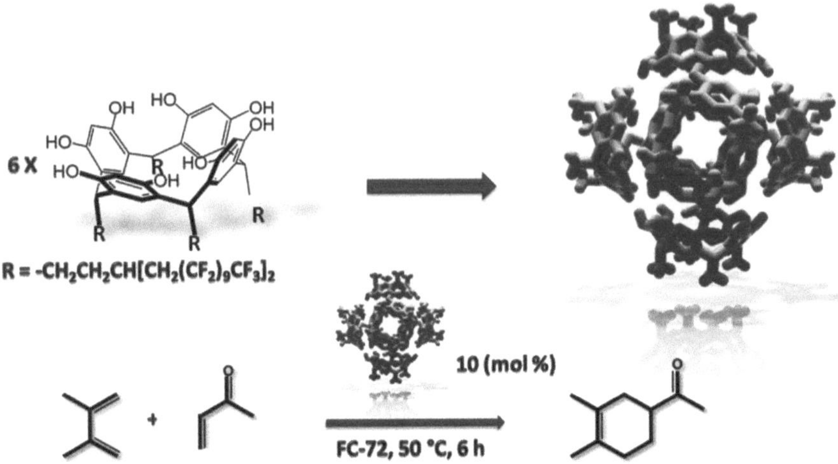

FIGURE 7.13 Diels–Alder cycloaddition mediated by a hexameric capsule bearing fluorous feet operating in a fluorinated solvent and recycle of the supramolecular catalytic system. *(Adapted from Ref. [62].)*

coworkers based on the fluorophobic effect exploiting the use of a hexameric capsule bearing fluorinated ponytails and operating in a biphasic fluorous/organic phase [62], with the latter composed of the mixture of diene and dienophile. It was demonstrated that the modified partially fluorinated resorcin[4]arene self-assembles in fluorinated solvents with small amounts of water, forming the hexameric capsule exactly as in traditional organic solvents. Considering the rather low solubility of traditional organic dienes and dienophiles for the fluorinated liquids, the idea was to force the reagents through the fluorophobic effect to enter the cavity of the capsule, thus favoring their cycloaddition reaction as a consequence of their high local concentration and proximity. The authors investigated initially the reaction between methyl vinyl ketone with 2,3-dimethyl-1,3-butadiene in perfluorohexane FC-72 at 50°C for 6 h observing only 10% yield in the absence of capsule, while with the capsule the yield of the adduct increased to 74% (Fig. 7.13). In order to prove that the reaction takes place within the cavity, a control experiment with a tetra-acetylated version of the resorcin[4]arene that cannot form the capsule was used, showing less than 18% yield, an indication that in this case too the reaction occurs predominantly within the cavity and that easy egress of the products and ingress of fresh substrates are possible, leading to interesting turnover numbers.

The effect of the composition of the fluorous solvent (from pure perfluorobenzene to pure perfluorohexane) was investigated observing higher yields in the cycloaddition products with higher fluorous content of the solvent, with a plateau region at approximately 50:50 perfluorobenzene:perfluorohexane; this is because of increased association constant of the capsule. The study was further extended to other combinations of dienes and dienophiles, showing in all cases increases from 2.3- to 21-fold in the experiments carried out in the presence of the capsule, with interesting changes in the regioselectivity of the reaction showing higher *endo/exo* ratios than those obtained in the control reactions without a capsule, as in the case of the reaction between cyclopentadiene and methyl acrylate, with an increase of the *endo/exo* ratio from 2.9:1 to 5.5:1. The biphasic nature of the reaction spurred the investigation of the recyclability of the catalytic system. In fact, in the reaction between 2,3-dimethyl-1,3-butadiene and 3-buten-2-one in perfluorohexane with the capsule it was possible to recycle the supramolecular catalyst for overall five consecutive runs observing yields in the range 50-61% were observed for overall five consecutive runs. This represented the first example of the application of the resorcin[4]arene capsule in a catalytic reaction displaying efficient recycling of the supramolecular catalyst.

4.3 C-heteroatom Bond-Forming Reactions

The good affinity of isonitrile substrates for the cavity and the wide range of possible reactions offered by this class of molecules spurred the research of other chemical transformations where the capsule could act as a supramolecular catalyst. It is important to note that the combination with another substrate should be much faster than the reaction with water that is always present in the reaction medium because it is important for the self-assembling of the capsule. After seeking for various nucleophiles, Scarso and coworkers disclosed that the reaction between isonitriles and trimethylsilyl azide could be promoted by the capsule, leading to the rapid formation of the corresponding 1*H*-tetrazoles [63] without observing the formation of formylamide products typical of the hydration reaction.

In particular, the reaction between cyclohexyl isonitrile and the $TMSN_3$ was investigated both in the presence and in the absence of the hexameric capsule in water-saturated chloroform-d. While the reaction did not occur spontaneously, in the presence of a catalytic amount of capsule (10 mol%), the reaction was complete in slightly more than 6 h (Fig. 7.14). Also in this case, rigid control experiments like the use of a weak Brønsted acid of comparable acidity of the capsule and the use of resorcinol as a H-bond activating species similar to resorcin[4]arene did not justify the catalytic activity observed. Experiments in the presence of the capsule occupied by a competitive tetraethylammonium guest showed a marked decrease of product formation after the same reaction time (33% yield), confirming the crucial role of the encapsulation in the catalytic activity of the supramolecular capsule.

It also ruled out the formation of the tetrazole from the prior reaction of the corresponding formylamide with trimethylsilyl azide, which is a viable alternative reaction pathway [64]. In fact, the reaction of only the isonitrile with the capsule led to the formation of the corresponding formylamide and subsequent addition of trimethylsilyl azide to the this solution did not lead to the formation of the corresponding tetrazole.

The inhibiting effect imparted to the reaction by the presence of the ammonium species was even more evident with isonitrile substrates bearing

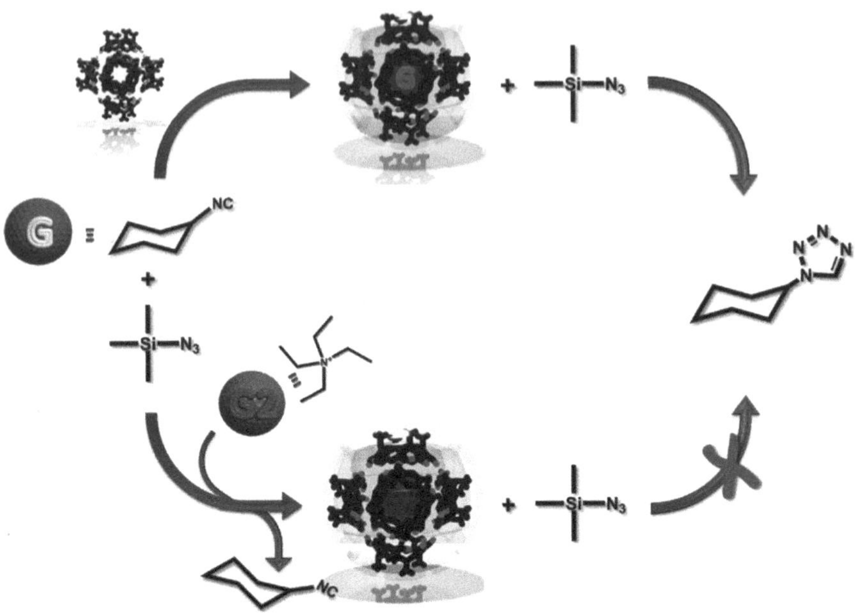

FIGURE 7.14 Cycloaddition reaction between cyclohexyl isonitrile and trimethylsilyl azide catalyzed by the hexameric capsule and hampered by the rpesence of tetraethylammonium as a competitive guest for the cavity of the capsule. *(Adapted from Ref. [63].)*

aromatic moieties in their molecular structure, as with benzyl isonitrile (free capsule 93% yield, occupied capsule 6% yield) and naphthyl isonitrile (free capsule 81% yield, occupied capsule 4% yield). The catalytic activity observed in this cycloaddition reaction was a further confirmation of the ability of the hexameric capsule to favor the stabilization of isonitrile addition intermediates.

4.4 Cyclization Reactions

As mentioned in the introduction, exploiting the known weak Brønsted acidity of the capsule (pKa 5.5–6) in combination with its affinity to uptake alcohols as neutral guests through H-bonding, Tiefenbacher and coworkers investigated the effect of the capsule in the uptake and intramolecular hydroalkoxylation of unactivated hydroxy olefins catalyzed inside a self-assembled cavity of the hexamer behaving as an enzyme-like host structure [65]. This reaction can be promoted by a series of strong Brønsted acids like triflic acid, leading initially to the protonation of the C=C double bond forming cationic intermediate species, with subsequent intramolecular attack of the hydroxyl alcoholic moiety. Usually this reaction is carried out in rather harsh conditions that often lead to the possible formation of acid-promoted side reactions also due to poor functional group compatibility that overall limits the applicability of this chemical transformation.

The addition of 2,6-dimethylhept-5-en-2-ol to a solution of the resorcin[4]arene led firstly to the encapsulation of the substrate as evidenced by the formation of new upfield shifted resonances in the region of 0.5 to −0.6 pm in the ^{1}H-NMR spectrum, while within a few days at 30°C the formation of the corresponding cyclic 2,2,6,6-tetramethyltetrahydro-2*H*-pyran product was observed in 96% yield; the control experiment without capsule showed only 7% yield of the product. The reaction turned out to be sensitive to the content of water present in solution under better conditions, with regular chloroform-d rather than with water-saturated chloroform-d. As observed in several formed examples, the reaction with the capsule occupied by the tetrabutyl ammonium as a competitive guest under identical experimental conditions showed only weak background conversion of 7%, confirming that the reaction occurs within the cavity through stabilization imparted by the aromatic electron-rich internal surfaces on cationic intermediates and transition states, with the eventual release of the product ether product in favor of new incoming alcoholic substrate.

The catalytic method was extended to a series of different unsaturated alcohols in order to investigate the scope of the reaction. More specifically, only substrates leading upon protonation to the formation of tertiary carbocations turned out to be suitable substrates for the reaction. In all cases, the synthesis of a wide range of differently substituted tetrahydropyran and oxepane derivatives was achieved under mild experimental conditions within up to 6 days, with only 10 mol% of the supramolecular capsule as catalyst, while in the presence of the ammonium competitive guest the reactions were sluggish. For all substrates, the hydroalkoxylation always showed the formation of the cyclic product corresponding to the Markovnikov addition to the unsaturated double bond.

The substrate selectivity of the capsule on this reaction was investigated for the substrates reported in Fig. 7.15. The two alcohols reacted in the presence of the capsule, leading to formation of the cyclic product derived from the smaller substrate with respect to the longer substrate in a 98:2 ratio, while the same pair of substrates with just acid catalysis provided by triflic acid reacted almost with comparable rates, leading to the corresponding cyclic products in a 46:54 ratio. This selectivity arises by the preferential binding of the capsule for smaller, more globular substrates with respect to longer and more elongated ones as observed also in other catalytic reactions [65].

Another striking example of the combination of the weak acidity of the capsule and its ability to stabilize cationic intermediates and related transition states is the reaction of acyclic terpene derivatives, which was found by Tiefenbacher and coworkers [66]. In particular, it was observed that the capsule could efficiently activate substrates like geraniol, nerol, linalool, and related acetate esters because of the presence in all of them of allylic moieties with leaving groups sensitive to the presence of acids like hydroxyl moiety or acetate. All these substrates could be converted into complicated mixtures of cyclic products deriving from the formation of intermediate allylic cationic species that eventually undergo intramolecular cyclization reactions or that

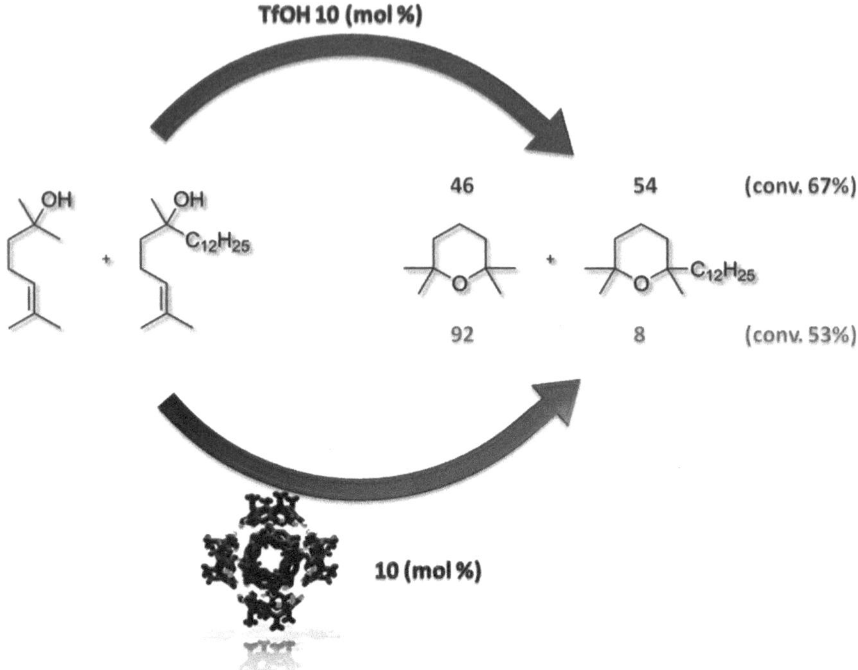

FIGURE 7.15 Intramolecular hydroalkoxylation of unactivated hydroxy olefins promoted by the hexameric capsule of resorcin[4]arene through encapsulation compared to triflic acid catalysis. *(Adapted from Ref. [65].)*

isomerize and subsequently react with nucleophiles. Looking at this extremely complicated scenario composed of several intermediate species and even more possible products, it is really impressive that enzymes like *cyclases* and *terpene synthases* operate on the similar substrates bearing phosphate-leaving groups, but in this case the reactions are much more selective. The key to the success for the enzymes consists in the sequestration of the substrate-limiting side reactions due to the attachment of different nucleophiles, the stabilization of the cationic allylic intermediates, and their preorganization and folding, thus promoting the formation of only certain cyclic products.

The resorcin[4]arene capsule was shown to efficiently mimic enzymes when it came to these aspects, thanks to its ability to stabilize the cationic allylic intermediate, limiting the attack from undesired nucleophiles and favoring the selective cyclization to certain products. In fact, it was observed that geranyl alcohol could be a suitable guest for the capsule that, in the presence of 10 mol% of capsule at 30°C, could be completely converted in 28 h with the initial formation of terpinene, linalool, and α-terpineol, all species derived from the selective formation of the intermediate carbocation. The two intermediate alcohols formed were further converted over time, giving rise to eucalyptol (Fig. 7.16a). Confirmation of the peculiar role played by the cavity of the

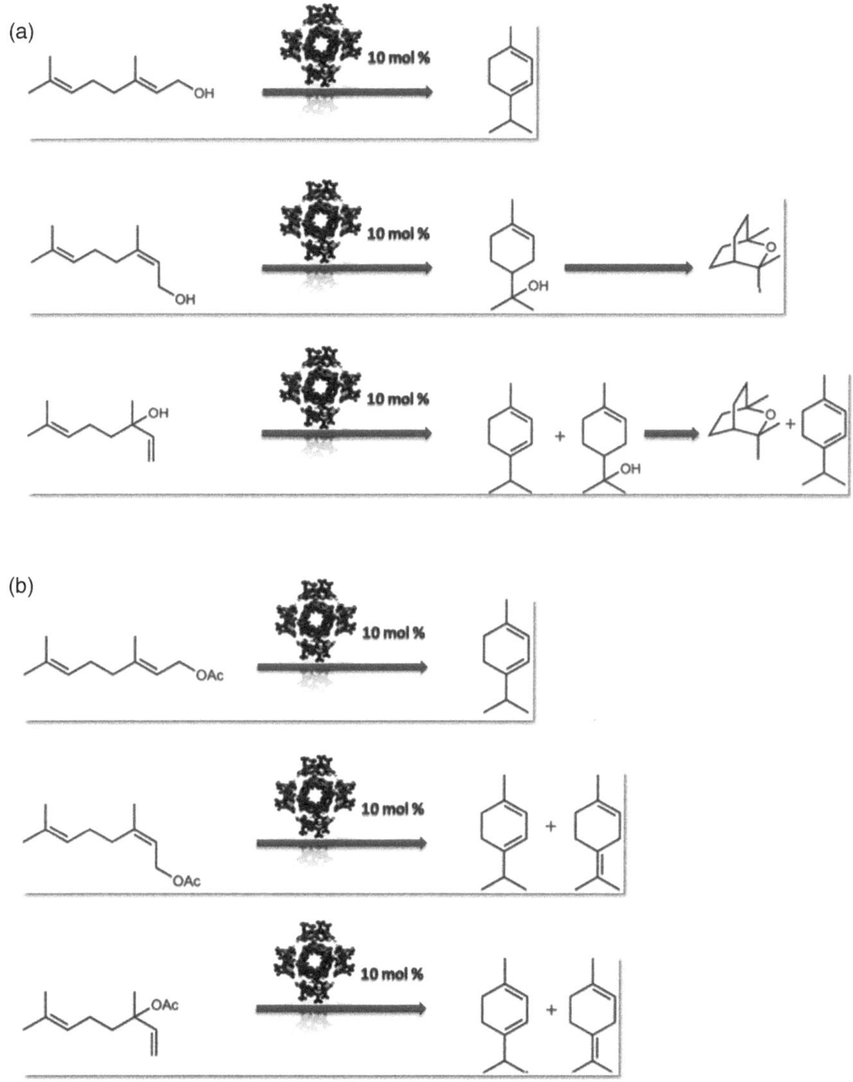

FIGURE 7.16 Intramolecular acyclic terpene derivatives isomerization through the formation of allylic cationic species and the effect of the resorcin[4]arene capsule on the product distribution of the reaction for (a) allyl alcohols and corresponding (b) acetate esters as substrates. *(Adapted from Ref. [66] with permission of Nature Publishing.)*

capsule in the reaction was demonstrated using tetrabutylammonium as competitive guest, showing no conversion of the substrate, thus excluding simple Brønsted acid catalysis effects of the capsule.

When nerol as a similar substrate was investigated as substrate in the presence of the capsule in catalytic amounts, α-terpineol was observed as a dominant product that further converted into eucalyptol together with other products but with an overall higher selectivity. The formation of eucalyptol from nerol and geraniol as acyclic terpenes was an unprecedented result that underlined the product selectivity features possible when using the resorcin[4]arene capsule.

Using acetate esters of geraniol and nerol, it was possible to greatly favor the formation of a different product distribution (Fig. 7.16b), limiting the amounts of linalool and α-terpineol and favoring the 1,2-hydride shift to form the intermediate cationic species, eventually leading to α-terpinene, which is an impressive result difficult to achieve in the absence of well-defined nanoenvironments. The capsule demonstrated great acceleration ofthe cyclization reaction with k_{cat} only two orders of magnitude lower than *cyclase* enzymes. A deep investigation of the possible mechanism of action of the capsule to explain the formation of the different possible products led to the conclusion that a possible transoid–cisoid isomerization of the intermediate cation should be possible and the same could be possible also in natural *cyclase* enzymes.

Overall, the earlier described reactions of terpene derivatives mediated by the capsule really represents important examples of how, with simple self-assembled nanometric containers endowed with a few functionalities, it is possible to mimic some of the fundamental properties of enzymes. This demonstrates that the way opened by the development of supramolecular catalysis will soon reward the scientists with better and better catalytic systems.

5 CONCLUSIONS AND FUTURE PERSPECTIVES

Supramolecular catalysis is now enriched by a new principal actor: the resorcin[4]arene hexamer as an example of a very large, substrate-selective nanohost able to greatly stabilize reactions involving electron-poor and cationic intermediates. This simple and pseudospherical aggregate that spontaneously forms in organic solvents from a vase-shape molecule obtained in a single synthetic step, has been exploited as a well-defined nanoreactor for a wide range of chemical transformations. All the features observed are consequences of the encapsulation of the substrates or of specific cationic catalysts. These species experience unique solvation effects imparted by the electron-rich aromatic surfaces of the assembly. Thanks to its very large cavity, this assembly has attracted the interest of several research groups looking to use it as a simple but efficient example of artificial nanoenzyme.

REFERENCES

[1] Wong C-H, Whitesides GM. Enzymes in synthetic organic chemistry. Oxford, UK: Pergamon Press; 1994.

[2] (a) Gao J, Ma S, Major DT, Nam K, Pu J, Truhlar DG. Mechanisms and free energies of enzymatic reactions. Chem Rev 2006;106:3188–209. (b) Tehila S, Koblenz, Jeroen Wassenaar, Reek Joost NH. Reactivity within a confined self-assembled nanospace. Chem Soc Rev 2008;37:247–62.

[3] Koshland DE Jr, Neet KE. The catalytic and regulatory properties of enzymes. Annu Rev Biochem 1968;37:359–411.

[4] Vriezema DM, Aragonès MC, Elemans JAAW, Cornelissen JJLM, Rowan AE, Nolte RJM. Self-assembled nanoreactors. Chem Rev 2005;105:1445–89.

[5] Dydio P, Reek JNH. Supramolecular control of selectivity in transition-metal catalysis through substrate preorganization. Chem Sci 2014;5:2135–45.

[6] (a) Wei H, Wang E. Nanomaterials with enzyme-like characteristics (nanozymes): next-generation artificial enzymes. Chem Soc Rev 2013;42:6060–93. (b) Lin Y, Ren J, Qu X. Catalytically active nanomaterials: a promising candidate for artificial enzymes. Acc Chem Res 2014;47:1097–105.

[7] van Leeuwen PWNM. Supramolecular catalysis. Wiley-VCH Weinheim, Germany; 2008.

[8] (a) Raynal M, Ballester P, Vidal-Ferran A, van Leeuwen PWNM. Supramolecular catalysis. Part 1: non-covalent interactions as a tool for building and modifying homogeneous catalysts. Chem Soc Rev 2014;43:1660–733. (b) Raynal M, Ballester P, Vidal-Ferran A, van Leeuwen PWNM. Supramolecular catalysis. Part 2: artificial enzyme mimics. Chem Soc Rev 2014;43:1734–87.

[9] La Sorella G, Strukul G, Scarso A. Recent advances in catalysis in micellar media. Green Chem 2015;17:644–83.

[10] Brown CJ, Toste FD, Bergman RG, Raymond KN. Supramolecular catalysis in metal-ligand cluster hosts. Chem Rev 2015;115:3012–35.

[11] (a) Leenders SHAM, Gramage-Doria R, de Bruin B, Reek JNH. Transition metal catalysis in confined spaces. Chem Soc Rev 2015;44:433–48. (b) Rebilly J-N, Colasson B, Bistri O, Over D, Reinaud O. Biomimetic cavity-based metal complexes. Chem Soc Rev 2015;44:467–89.

[12] Jordan JH, Gibb BC. Molecular containers assembled through the hydrophobic effect. Chem Soc Rev 2015;44:547–85.

[13] de Mendoza J. Self-Assembling Cavities: Present and Future. Chem Eur J 1998;4:1373–7.

[14] Kang J, Rebek J Jr. Acceleration of a Diels-Alder reaction by a self-assembled molecular capsule. Nature 1997;385:50–2.

[15] (a) Kang J, Hilmersson G, Santamaría J, Rebek J Jr. Diels–Alder reactions through reversible encapsulation. J Am Chem Soc 1998;120:3650–6. (b) Kang J, Santamaría J, Hilmersson G, Rebek J Jr. Self-assembled molecular capsule catalyzes a Diels–Alder reaction. J Am Chem Soc 1998;120:7389–90.

[16] (a) Sverker Hoegberg AG. Two stereoisomeric macrocyclic resorcinol-acetaldehyde condensation products. J Org Chem 1980;45:4498–500. (b) Sverker AG, Hoegberg. Cyclooligomeric phenol-aldehyde condensation products. 2. Stereoselective synthesis and DNMR study of two 1,8,15,22-tetraphenyl[14]metacyclophan-3,5,10,12,17,19,24,26-octols. J Am Chem Soc 1980;102:6046–50. (c) Antesberger J, Cave GWV, Ferrarelli MC, Heaven MW, Raston CL, Atwood JL. Solvent-free, direct synthesis of supramolecular nano-capsules. Chem Commun 2005;892–4.

[17] MacGillivray LR, Atwood JL. A chiral spherical molecular assembly held together by 60 hydrogen bonds. Nature 1997;389:469–71.

[18] Avram L, Cohen Y. Spontaneous formation of hexameric resorcinarene capsule in chloroform solution as detected by diffusion NMR. J Am Chem Soc 2002;124:15148–9.

[19] Shivanyuk A, Rebek J Jr. Reversible encapsulation by self-assembling resorcinarene subunits. Proc Natl Acad Sci USA 2001;98:7662–5.

[20] Dougherty DA. The cation-π interaction. Acc Chem Res 2013;46:885–93.

[21] (a) Barrett ES, Dale TJ, Rebek J Jr. Stability, dynamics, and selectivity in the assembly of hydrogen-bonded hexameric capsules. J Am Chem Soc 2008;130:2344–50. (b) Avram L,

Cohen Y. The role of water molecules in a resorcinarene capsule as probed by NMR diffusion measurements. Org Lett 2002;4:4365–8. (c) Avram L, Cohen Y, Rebek J. Recent advances in hydrogen-bonded hexameric encapsulation complexes. Chem Commun 2011;47:5368–75. (d) Purse BW, Rebek J. Self-fulfilling cavitands: packing alkyl chains into small spaces. Proc Natl Acad Sci USA 2006;103:2530–4. (e) Shivanyuk A, Rebek J. Assembly of resorcinarene capsules in wet solvents. J Am Chem Soc 2003;125:3432–3. (f) Yamanaka M, Shivanyuk A, Rebek J. Kinetics and thermodynamics of hexameric capsule formation. J Am Chem Soc 2004;126:2939–43.

[22] Zhang Q, Tiefenbacher K. Hexameric resorcinarene capsule is a Brønsted acid: investigation and application to synthesis and catalysis. J Am Chem Soc 2013;135:16213–9.

[23] (a) Aoyama Y, Tanaka Y, Toi YH, Ogoshi H. Polar host-guest interaction. Binding of nonionic polar compounds with a resorcinol-aldehyde cyclooligomer as a lipophilic polar host. J Am Chem Soc 1988;110:634–5. (b) Aoyama Y, Nonaka Y, Tanaka Y, Toi H, Ogoshi H. Polar host–guest interactions. Solubilization of some polar compounds with lipophilic calix[6]arenes containing polar groups in apolar media. J Chem Soc Perkin Trans 1989;2:1025–9.

[24] (a) Slovak S, Cohen Y. The effect of alcohol structures on the interaction mode with the hexameric capsule of resorcin[4]arene. Chem Eur J 2012;18:8515–20. (b) Slovak S, Avram L, Cohen Y. Encapsulated or not encapsulated? Mapping alcohol sites in hexameric capsules of resorcin[4]arenes in solution by diffusion NMR spectroscopy. Angew. Chem Int Ed 2010;49:428–31. (c) Schnatwinkel B, Stoll I, Mix A, Rekharsky MV, Borovkov VV, Inoue Y, Mattay J. Monomeric, dimeric and hexameric resorcin[4]arene assemblies with alcohols in apolar solvents. Chem Commun 2008;3873–5. (d) Ugono O, Travis Holman K. An achiral form of the hexameric resorcin[4]arene capsule sustained by hydrogen bonding with alcohols. Chem Commun 2006;2144–6.

[25] Bianchini G, Scarso A, La Sorella G, Strukul G. Switching the activity of a photoredox catalyst through reversible encapsulation and release. Chem Commun 2012;48:12082–4.

[26] DeRosa MC, Crutchley RJ. Photosensitized singlet oxygen and its applications. Coord Chem Rev 2002;233, 234:233–4.

[27] Zen J-M, Liou S-L, Senthil Kumar A, Hsia M-S. An efficient and selective photocatalytic system for the oxidation of sulfides to sulfoxides. Angew Chem Int Ed 2003;42:577–9.

[28] Palmer LC, Shivanyuk A, Yamanaka M, Rebek J Jr. Resorcinarene assemblies as synthetic receptors. Chem Commun 2005;857–8.

[29] Mecozzi S, Rebek J Jr. The 55 % Solution: A Formula for Molecular Recognition in the Liquid State. Chem Eur J 1998;4:1016–22.

[30] Cavarzan A, Scarso A, Sgarbossa P, Strukul G, Reek JNH. Supramolecular control on chemo- and regioselectivity via encapsulation of (NHC)-Au catalyst within a hexameric self-assembled host. J Am Chem Soc 2011;133:2848–51.

[31] Skouta R, Wei S, Breslow R. High rates and substrate selectivities in water by polyvinylimidazoles as transaminase enzyme mimics with hydrophobically bound pyridoxamine derivatives as coenzyme mimics. J Am Chem Soc 2009;131:15604–5.

[32] Liu L, Rozenman M, Breslow R. Hydrophobic effects on rates and substrate selectivities in polymeric transaminase mimics. J Am Chem Soc 2002;124:12660–1.

[33] Matsui J, Nicholls IA, Karube I, Mosbach K. Carbon−Carbon bond formation using substrate selective catalytic polymers prepared by molecular imprinting: An artificial Class II Aldolase. J Org Chem 1996;61:5414–7.

[34] Tokunaga M, Larrow JF, Kakiuchi F, Jacobsen EN. Asymmetric catalysis with water: efficient kinetic resolution of terminal epoxides by means of catalytic hydrolysis. Science 1997;277:936–8.

[35] Lindbäck E, Dawaigher S, Wärnmark K. Substrate-selective catalysis. Chem Eur J 2014;20:13432–81.
[36] Purich DL, Allison RD. Handbook of biochemical kinetics: A guide to dynamic processes in the molecular life sciences. Waltham, Massachusetts, USA: Academic Press; 1999.
[37] Case A, Stein RL. Mechanistic origins of the substrate selectivity of serine proteases. Biochemistry 2003;42:3335–48.
[38] Zou Y, Li C, Brunzelle JS, Nair SK. Molecular basis for substrate selectivity and specificity by an LPS biosynthetic enzyme. Biochemistry 2007;46:4294–304.
[39] Nau WM, Ghale G, Hennig A, Bakirci H, Bailey DM. Substrate-selective supramolecular tandem assays: monitoring enzyme inhibition of arginase and diamine oxidase by fluorescent dye displacement from calixarene and cucurbituril macrocycles. J Am Chem Soc 2009;131:11558–70.
[40] Thulasiram HV, Dale Poulter C. Farnesyl diphosphate synthase: the art of compromise between substrate selectivity and stereoselectivity. J Am Chem Soc 2006;128:15819–23.
[41] Leung DH, Bergman RG, Raymond KN. Highly selective supramolecular catalyzed allylic alcohol isomerization. J Am Chem Soc 2007;129:2746–7.
[42] Pluth MD, Bergman RG, Raymond KN. Catalytic deprotection of acetals in basic solution with a self-assembled supramolecular "nanozyme". Angew Chem Int Ed 2007;46:8587–9.
[43] Chow H-F, Ching Mak C. Dendritic Bis(oxazoline)copper(II) Catalysts. 2.1 Synthesis, Reactivity, and Substrate Selectivity. J Org Chem 1997;62:5116–27.
[44] Bhyrappa P, Young JK, Moore JS, Suslick KS. Dendrimer-Metalloporphyrins: Synthesis and catalysis. J Am Chem Soc 1996;118:5708–11.
[45] Lee SJ, Cho S-H, Mulfort KL, Tiede DM, Hupp JT, Nguyen ST. Cavity-tailored, self-sorting supramolecular catalytic boxes for selective oxidation. J Am Chem Soc 2008;130:16828–9.
[46] Cavarzan A, Reek JNH, Trentin F, Scarso A, Strukul G. Substrate selectivity in the alkyne hydration mediated by NHC–Au(I) controlled by encapsulation of the catalyst within a hydrogen bonded hexameric host. Catal Sci Technol 2013;3:2898–901.
[47] Giust S, La Sorella G, Sperni L, Strukul G, Scarso A. Substrate selective amide coupling driven by encapsulation of a coupling agent within a self-assembled hexameric capsule. Chem Commun 2015;51:1658–61.
[48] (a) Williams A, Ibrahim IT. Carbodiimide chemistry: recent advances. Chem Rev 1981;81: 589–636. (b) Mironova DF, Dvorko GF. Chem Abstr 1971;75. 62706. (c) Joullié MM, Lassen KM. Evolution of amide bond formation. ARKIVOC 2010;8:189–250.
[49] Ugi I, Werner B, Dömling A. The Chemistry of Isocyanides, their MultiComponent Reactions and their Libraries. Molecules 2003;8:53–66.
[50] Ramozzi R, Chéron N, Braïda B, Hiberty PC, FleuratLessard P. A valence bond view of isocyanides' electronic structure. New J Chem 2012;36:1137–40.
[51] Li Y, Xu X, Xia C, Zhang L, Pana L, Liu Q. Double nucleophilic attack on isocyanide carbon: a synthetic strategy for 7-aza-tetrahydroindoles. Chem Commun 2012;48:12228–30.
[52] Paquette LA, Allen GR Jr, Broadhurst MJ. Addition reactions of the uniparticulate electrophile chlorosulfonyl isocyanate to highly strained bicyclic hydrocarbons. J Am Chem Soc 1971;93:4503–8.
[53] Bianchini G, La Sorella G, Canever N, Scarso A, Strukul G. Efficient isonitrile hydration through encapsulation within a hexameric self-assembled capsule and selective inhibition by a photo-controllable competitive guest. Chem Commun 2013;49:5322–4.
[54] (a) Lakshminarasimhan M, Madzelan P, Nan R, Milkovic NM, Wilson MA. Evolution of new enzymatic function by structural modulation of cysteine reactivity in Pseudomonas fluorescens isocyanide hydratase. J Biol Chem 2010;285:29651–61. (b) Hashimoto K, Suzuki H,

Taniguchi K, Noguchi T, Yohda M, Odaka M. Catalytic mechanism of nitrile hydratase proposed by time-resolved X-ray crystallography using a novel substrate, tert-butylisonitrile. J Biol Chem 2008;283:36617–23.

[55] Regitz M. Diazo compounds: properties and synthesis. Amsterdam, Netherlands: Academic Press Inc., Elsevier Inc.; 1986.

[56] (a) Padwa A, Krumpe KE. Application of Intramolecular Carbenoid Reactions in Organic Synthesis. Tetrahedron 1992;48:5385–453. (b) Maas G. New syntheses of diazo compounds. Angew Chem Int Ed 2009;48:8186–95.

[57] La Sorella G, Sperni L, Strukul G, Scarso A. Supramolecular Encapsulation of Neutral Diazoacetate Esters and Catalyzed 1,3-Dipolar Cycloaddition Reaction by a Self-Assembled Hexameric Capsule. ChemCatChem 2015;7:291–6.

[58] Yao J, Yan Z, Ji J, Wu W, Yang C, M- Nishijima G, Fukuhara T, Mori Y, Inoue. Ammonia-driven chirality inversion and enhancement in enantiodifferentiating photocyclodimerization of 2-anthracenecarboxylate mediated by diguanidino-γ-cyclodextrin. J Am Chem Soc 2014;136:6916–9.

[59] (a) Inokuma Y, Yoshioka S, Fujita M. A molecular capsule network: guest encapsulation and control of Diels-Alder reactivity. Angew Chem Int Ed 2010;49:8912–4. (b) Smulders MMJ, Nitschke JR. Supramolecular control over Diels–Alder reactivity by encapsulation and competitive displacement. Chem Sci 2012;3:785–8.

[60] (a) Bojilova A, Videnova I, Ivanov C. Regio- and stereo-selective 1,3-dipolar cycloaddition reactions of ethyl diazoacetate to 3-substituted 2H-1-benzopyran-2-ones. Tetrahedron 1994;50:13023–36. (b) Tuktarova AR, Akhmetova AR, Kamalova RF, Khalilova LM, Pudasb M, Ibragimova AG, Dzhemileva UM. Catalytic [2+1]-cycloaddition of ethyl diazoacetate to fullerene. Russ J Org Chem 2009;45:1168–2117.

[61] (a) Noels AF, Btaham JN, Hubert AJ, Teyssié PH. Cycloadditions of diazoesters to α,β-unsaturated aldehydes. Tetrahedron 1978;34:3495–7. (b) Lévai A. Synthesis of pyrazolines by the reactions ofα,β-enones with diazomethane and hydrazines. Chem Heterocycl Comp 1997;33:647–59.

[62] Shimizu S, Usui A, Sugai M, Suematsu Y, Shirakawa S, Ichikawa H. Hexameric capsule of a resorcinarene bearing fluorous feet as a self-assembled nanoreactor: A Diels–Alder reaction in a fluorous biphasic system. Eur J Org Chem 2013;4734–7.

[63] Giust S, La Sorella G, Sperni L, Fabris F, Strukul G, Scarso A. Supramolecular catalysis in the synthesis of substituted 1 H-tetrazoles from isonitriles by a self-assembled hexameric capsule. Asian J Org Chem 2015;4:217–20.

[64] Pinto DJP, Pruitt JR, Cacciola J, Favig JM, Han Q, Orwat MJ, Quan ML, Rossi KA. PCT Int Appl 1998, WO 9828,269 (Cl. C07D207/34), CA (1998) 129:109090n.

[65] Catti L, Tiefenbacher K. Intramolecular hydroalkoxylation catalyzed inside a self-assembled cavity of an enzyme-like host structure. Chem Commun 2015;51:892–4.

[66] Zhang Q, Tiefenbacher K. Terpene cyclization catalysed inside a self-assembled cavity. Nature Chem 2015;7:197–202.

Chapter 8

The Varied Supramolecular Chemistry of Pyrogallol[4]arenes

Saeedeh Negin*, George W. Gokel**
**Departments of Chemistry & Biochemistry, Center for Nanoscience, University of Missouri, St. Louis, MO, United States of America*
***Department of Biology, Center for Nanoscience, University of Missouri, St. Louis, MO, United States of America*

1 INTRODUCTION

The origins of supramolecular chemistry are generally traced to two pioneering discoveries that were made in the 1960s. The first of these was the discovery of crown ethers by Charles Pedersen [1]. The other was the invention of cryptands by Lehn and Sauvage [2]. Pedersen was searching for a divalent cation binder based on the catechol molecule and he inadvertently prepared a small amount of a macrocycle. He found that these macrocycles, which he named crown ethers, complexed alkali metal cations such as sodium and potassium. This was a long sought and previously unrealized goal in chemistry.

Lehn and Sauvage deliberately designed compounds that they expected to be three-dimensional alkali metal binders. They succeeded in this with compounds that were similar to those Pedersen had isolated but bicyclic rather than monocyclic [3]. The development of these two classes of related compounds caused an explosion of activity both on the synthetic [4] and the analytical fronts [5]. As structures became more elaborate, new binding selectivities emerged, as did previously undocumented binding interactions (Fig. 8.1).

Gradually, the binding of metallic cations gave way to the complexation of organic cations and then to salts, dications, some anions, and even molecules. Lehn offered the term "supramolecular chemistry" to describe chemistry beyond the molecule, which is to say assemblies organized by noncovalent interactions [6].

Although the development of crown ethers and cryptands gave intense impetus to supramolecular chemistry, macrocycle chemistry can be traced to a much earlier origin. As early as the late 19th century, von Baeyer [7] condensed acetone with pyrrole to give a macrocyclic tetramer. A hundred years later, these compounds would be recognized as anion binders and called "calixpyrroles."

Organic Nanoreactors. http://dx.doi.org/10.1016/B978-0-12-801713-5.00008-2

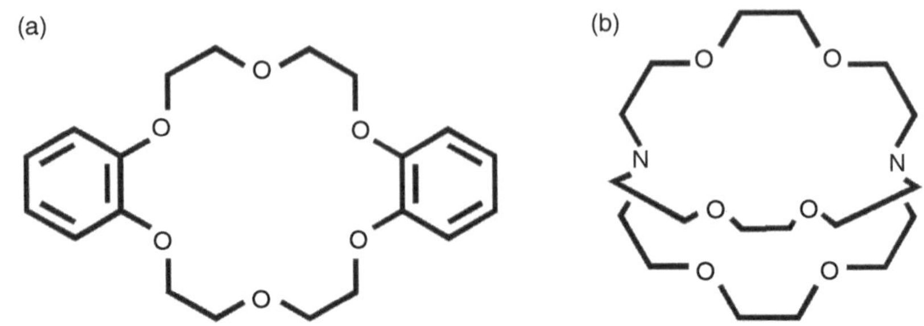

FIGURE 8.1 Structures of dibenzo-18-crown-6 (a) and [2.2.2] cryptand (b).

An analogous tetramer was reported in 1955 that resulted from the condensation of furan with acetone [8]. This process is reminiscent of that reported by von Baeyer. In both cases a cyclic tetramer forms that has heteroatoms directed to the center of a macroring.

In 1909, Baekelund patented [9] a condensation between phenol and formaldehyde that gave a plastic product he called Bakelite. By the middle of the 20th century, relatives of the macrocycles that formed from phenols, rather than pyrroles, had been prepared and studied by Zinke and Ziegler [10], Hayes and Hunter [11], Cornforth et al. [12], and others. The compounds of greatest interest were the known family of macrocycles to which Gutsche gave the cognomen "calixarenes" [13] (Fig. 8.2).

Calixarenes at this early time were being prepared in a stepwise fashion. Although many pioneers studied phenol-formaldehyde macrocycles, an important advance occurred in the 1980s. It was at that time that Gutsche and his coworkers developed a simple procedure to prepare calixarenes in collaboration with the Petrolite company. This procedure made calixarenes more readily available [14], and the family was elaborated in many laboratories [15] with the same enthusiasm that had been applied to crown ethers after their discovery.

Resorcinol and pyrogallol are benzene derivatives that have two and three hydroxyl groups, respectively. They react readily with aldehydes to form macrocycles that are similar in some aspects to the calixarenes [16]. The formation of calixarenes involves reaction of phenol at the 2,6-carbons, through which the methylene link forms. Resorcinol is 1,3-dihydroxybenzene, and as with pyrogallol, 1,2,3-trihydroxybenzene, the ring-forming links occur at the 4- and 6-positions. Exemplary structures are shown in Fig. 8.3.

The pyrogallol[4]arenes have proved to be an intriguing group of molecules that are generally easy to synthesize. While the calixarenes often require the phenol's 4-position to be protected during cyclization, no such precaution is required with pyrogallol[4]arenes. The synthesis of pyrogallol[4]arenes appears to be a messy affair, but the product is readily obtained. Typically, one simply heats an aldehyde with pyrogallol in the presence of ethanolic-aqueous HCl. After several hours, the very dark reaction mixture is allowed to cool. It is not uncommon for relatively pure product to crystallize from this mixture. Although

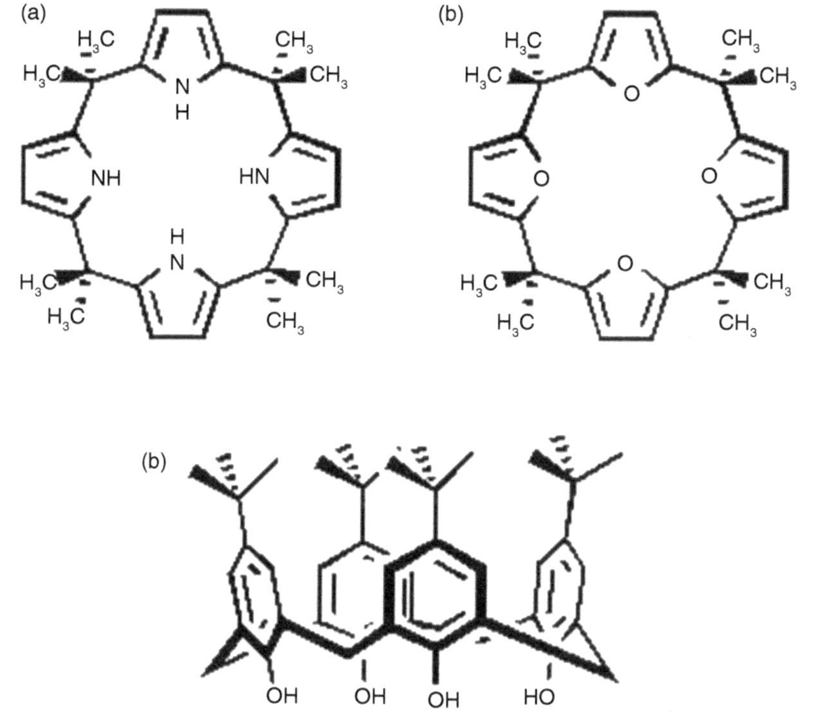

FIGURE 8.2 **Structures of calixpyrrole.** (a) The furan analog of a calixpyrrole, (b) a furanocycle, and (c) a calix[4]arene.

yields are typically only in the 20–30% range, the starting materials are usually simple and the reaction is straightforward to perform. The procedure requires only reflux, and the solid product is often easy to isolate. No doubt other materials are formed in the reaction mixture, but it is typically the tetrameric pyrogallol[4]arene that crystallizes.

The selective crystallization of relatively pure pyrogallol[4]arenes from the complex reaction mixture is somewhat surprising. It likely occurs in consequence of dynamics within the acidic solution leading to crystallization of the cycles having aligned alkyl chains. In pyrogallol[4]arenes, the fully

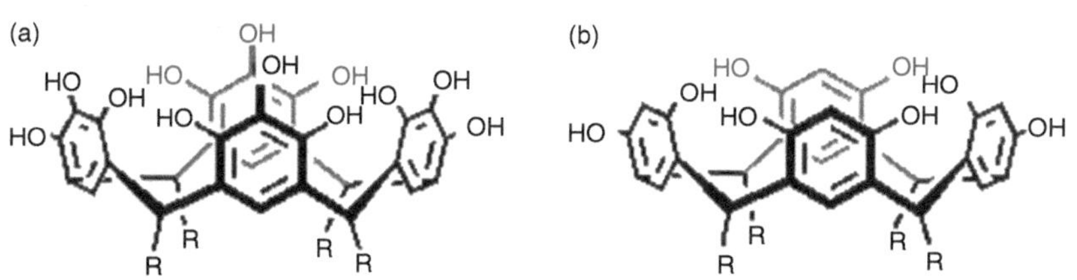

FIGURE 8.3 **Structures of a generalized pyrogallol[4]arene (a) and a resorcin[4]arene (b).**

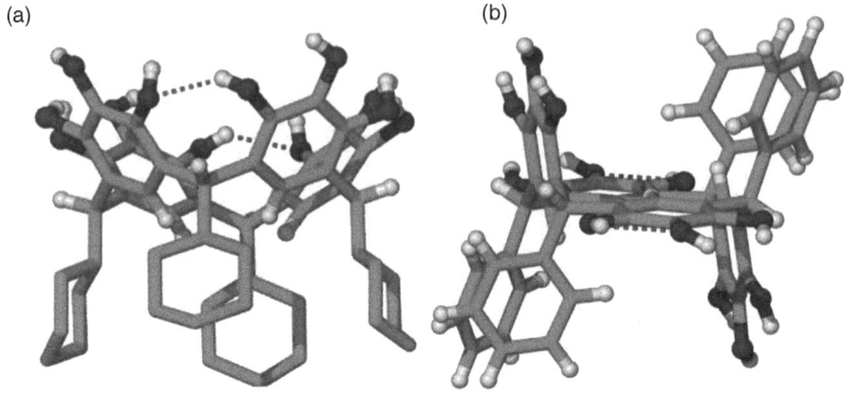

FIGURE 8.4 Comparison of the solid-state structures of tetracyclohexylpyrogallol[4]arene (a) and tetraphenylpyrogallol[4]arene (b).

aligned side-chain condition is referred to as "*rccc*." The letter "*r*" stands for reference, and "*c*" or "*t*" indicates whether another chain is either *cis* or *trans* to it. In virtually all of the studies reported here, the chains are in the aligned, *rccc* arrangement. A notable exception is the report of tetraphenyl-pyrogallol[4]arene, which has the *rctt* side chain disposition. The structure of the tetraphenyl compound [17] along with the solid-state structure of tetracyclohexylpyrogallol[4]arene is shown in Fig. 8.4. The latter exhibits the *rccc* arrangement of side chains.

2 PYROGALLOL[4]ARENE CAPSULES

Atwood and coworkers have extensively explored the interactions of pyrogallol[4]arene tetramers with each other in the self-assembly mode. They have shown that the hydroxyl groups along the aromatic perimeter of the six pyrogallol[4]arene tetramers can interact to form essentially octahedral capsules, as illustrated in Fig. 8.5 [18]. These compounds form in a variety of solvents and they can enclose significant amounts of space. They can encapsulate guests such as the solvents from which they form [19]. These remarkable capsules comprise an early example of molecules that enclose specific volumes and in principle could be used as nanoreactors [20].

Although most substituted pyrogallol[4]arene capsules have quite reasonable stability based on their 72 hydrogen bond network, Atwood showed that they could be further stabilized by replacing the network with various metals. Atwood, Dalgarno, and coworkers explored these compounds extensively and found that only certain elements were appropriate to the formation of these metal organic nanocapsules, or "MONCs" [21]. One such metal was copper, which we used in developing a membrane-resident ion conductor as described later in this chapter.

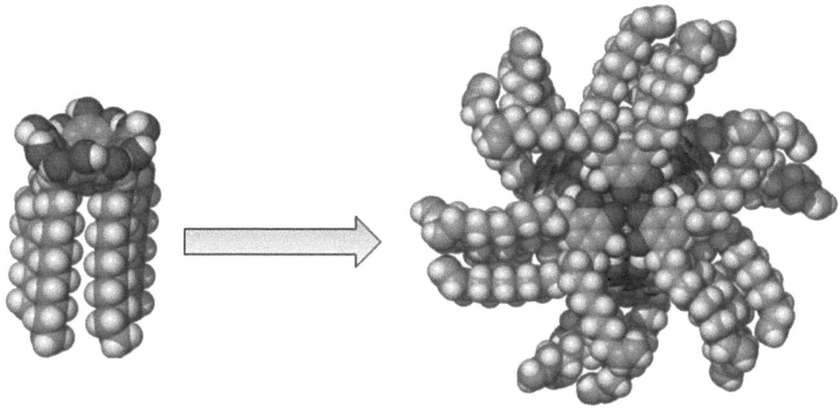

FIGURE 8.5 **The solid-state structures of tetra-*n*-undecylpyrogallol[4]arene "monomer" that results from hexamerization into a capsule.**

Early examples of both capsules and the metal organic nanocapsules were prepared from pyrogallol[4]arene that had relatively short side chains. A typical side chain is *n*-butyl or isobutyl, and solid-state structures have been reported for these capsules by several groups. Work in this area continues until today in several labs, but our interest in pyrogallol[4]arenes diverged from the capsule approach.

3 PYROGALLOL[4]ARENE CAPSULE–MEMBRANE INTERACTIONS

Our original interest in pyrogallol[4]arene-derived capsules was to prepare molecules that had relatively long alkyl side chains. There were two initial notions associated with this research direction. The first was to see if the capsule would comprise a stable scaffold upon which a *pseudo*-bilayer type membrane could be constructed. The octahedral capsule would have 24 alkyl chains radiating from it in all directions. For example, exposure of the capsule to phospholipids might create a *pseudo*-liposome. The concept of a phospholipid "encrusted" pyrogallol[4]arene capsule is illustrated schematically in Fig. 8.6.

The second notion was to see if the capsules surrounded with alkyl chains would themselves have a high affinity for bilayer membranes. It is well known, of course, that large protein molecules can insert into bilayers and reside in a stable environment therein. We were interested to see if capsules that were poly-alkylated would also show a membrane affinity based on their hydrocarbon chain perimeter, and if so, what would be their properties? The long-term goal of such an effort was to see if the capsules could encompass, surround, or contain, for example, a drug substance. If the capsule guest liposome could readily penetrate into the membrane or through it, the complex would comprise a novel drug-delivery vehicle. To the extent the complex deaggregated within the bilayer, it could deliver a guest such as an active pharmaceutical compound.

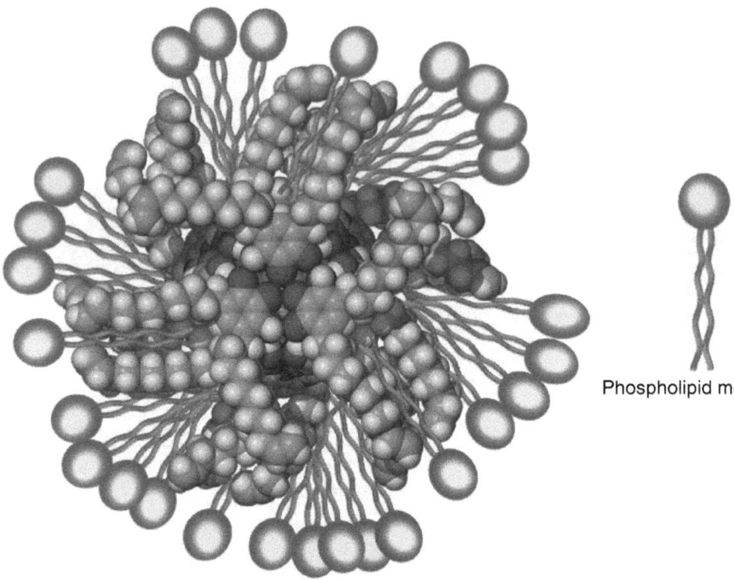

FIGURE 8.6 Schematic of a *pseudo*-liposome formed from pyrogallol[4]arene capsule alkyl chains and phospholipid hydrophobic chains.

4 PYROGALLOL[4]ARENE MEMBRANE AGGREGATION—PLANAR BILAYER STUDIES

The notion of developing capsules that could insert into bilayers and have rather sophisticated functions clearly required some control studies. The obvious first step was to determine if pyrogallol[4]arenes themselves would insert into bilayer membranes. Pyrogallol[4]arenes are obviously amphiphiles, but not all amphiphiles immediately insert into a membrane. Depending on the properties of the amphiphile, the membrane, and the bulk medium, the amphiphile may prefer to aggregate rather than insert into the bilayer. Alternately, it may form a layer on the surface of the membrane without actually penetrating it.

There is an incredibly complex relationship among amphiphiles and membranes. The compounds may aggregate in the bulk phase, and the aggregate may partition into the membrane. The aggregate may be in equilibrium with monomers in the bulk phase, and the monomers may partition into the membrane as well as into the aggregate(s). As noted earlier, monomers or aggregates may have an affinity for the membrane surface that is more energetically favorable than penetration. The same situation can obtain for monomers and aggregates within the bilayer. In fact, various aggregation states within the membrane may lead to different functions of the amphiphiles within the bilayer.

We used a planar bilayer voltage clamp conductance apparatus with an embedded asolectin membrane obtained from soybean. The aqueous buffer solutions used potassium chloride as the salt and the pH was kept at 7.0. A sample of the amphiphile was typically dissolved in dimethylsulfoxide (DMSO). The solution was then added to the so-called "cis" chamber of the planar bilayer apparatus so that the final concentration within the buffer solution was 7 μM.

Initial studies were conducted with pyrogallol[4]arenes having either 3- or 11-carbon side chains. The 11-carbon side chain pyrogallol[4]arenes, isolated both as capsules and as crystalline bilayers, were studied [22]. When pyrogallol[4]arenes having a 3-carbon side chain were added to the asolectin membranes in the planar bilayer conductance apparatus, only so-called "spiking behavior" was observed. Spiking behavior means that some current flow is detected through the membrane, but the flow is irregular and apparently results from local disruptions in the membrane.

When a stable conducting pore forms, one typically observes a square wave pattern. In a well-behaved system, the peak of each square wave will be of a consistent height. Some peaks may be observed at twice the more common height, suggesting that two pores, rather than one, have formed. The duration of the open states can vary substantially depending upon the compound, the compound concentration, the membrane in which the compounds are embedded, and the applied potential. Still, the observation of well-formed peaks and the characteristics of "open-close" behavior indicate the existence of a pore within the bilayer.

The difference in membrane behavior between the tetra-*n*-propylpyrogallol[4]arene and the tetra-*n*-undecylpyrogallol[4]arene was dramatic. Unlike the irregular behavior of tetra-*n*-propylpyrogallol[4]arene, fairly regular open-close behavior at applied voltages in the 10 to 50 mV range was observed for samples of the tetra-*n*-undecyl analog that had originally been isolated as a crystalline bilayer and then sonicated. More remarkable was the observation that this open-close behavior gradually led to stable pores that remained consistently open. At voltages that were either positive or negative and in the range of ±10 mV to ±50 mV, flat-topped open states were observed, as is apparent in the trace shown in Fig. 8.7. When the voltage was switched from positive to negative, an equivalent current flow was observed at either the positive or negative potential. This remarkable finding led us to assume that some type of stable aggregate was forming within the bilayer.

In order to examine this phenomenon more closely, we decided to alter the asolectin membrane. We were operating on the assumption that six tetrameric pyrogallol[4]arene molecules were inserted into the upper leaflet of the bilayer and that these were forming a self-assembled, internal pore. The conclusion that the aggregate was hexameric was based upon the conductance of the pore. Calculation of an estimated pore size gave a diameter of approximately 11 Å. The interior of a pyrogallol[4]arene macrocycle is barely 3 Å. Molecular models suggested that six pyrogallol[4]arene molecules aligned laterally would give a pore of approximately 15 Å in diameter, which corresponds reasonably well (13 ± 2 Å) with the calculated size.

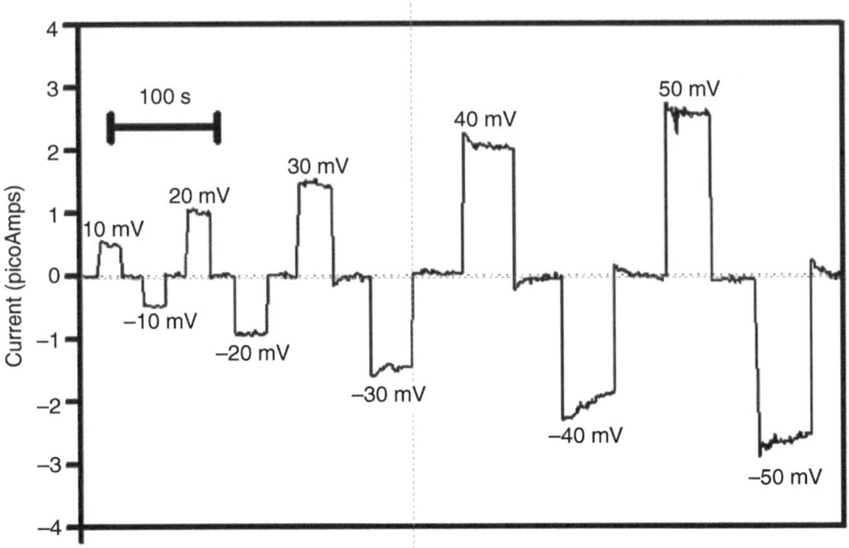

FIGURE 8.7 Voltage switching results for tetra-*n*-undecylpyrogallol[4]arene in a cholesterol/asolectin membrane and applied potential in the range of ±10 mV and ±50 mV.

We reasoned that if an aggregated pore formed in the asolectin membrane, the addition of cholesterol might alter the membrane dynamics. Addition of 20% by weight of cholesterol to the asolectin membrane was expected to make it more rigid and compact. It is this membrane composition that is reflected in the conductance trace shown in Fig. 8.7. If pore formation was the result of a dynamic aggregation phenomenon (shown schematically in Fig. 8.8), the more rigid membrane could potentially compress the aggregate and perhaps change its aggregation number. A change in aggregation number and pore size would certainly be reflected in the measured conductance. In the presence of cholesterol, the "switching behavior" persisted, but the conductance diminished. A calculation of the pore size from the conductance observed in the cholesterol-augmented membrane led us to conclude that the new pore diameter was only 5 Å.

An examination of existing molecular models revealed that four pyrogallol[4]arenes aligned laterally in the bilayer would have an internal opening of about 5 Å. Such a self-assembled pore, whether it is four or six monomers, has a stable opening within it that would explain the "always open" property. An aggregate pore as illustrated in Fig. 8.9 could only conduct ions or collapse. If it collapsed, there would be no conductance from it and activity would be lost unless or until another pore formed.

The model envisioned for these types of aggregates was an insertion of the alkyl chains into the upper leaflet of the bilayer from the bulk phase. This would be followed by a gradual association of monomers into a conducting pore. It is unclear what combination of forces would make a hexamer the favored aggregation

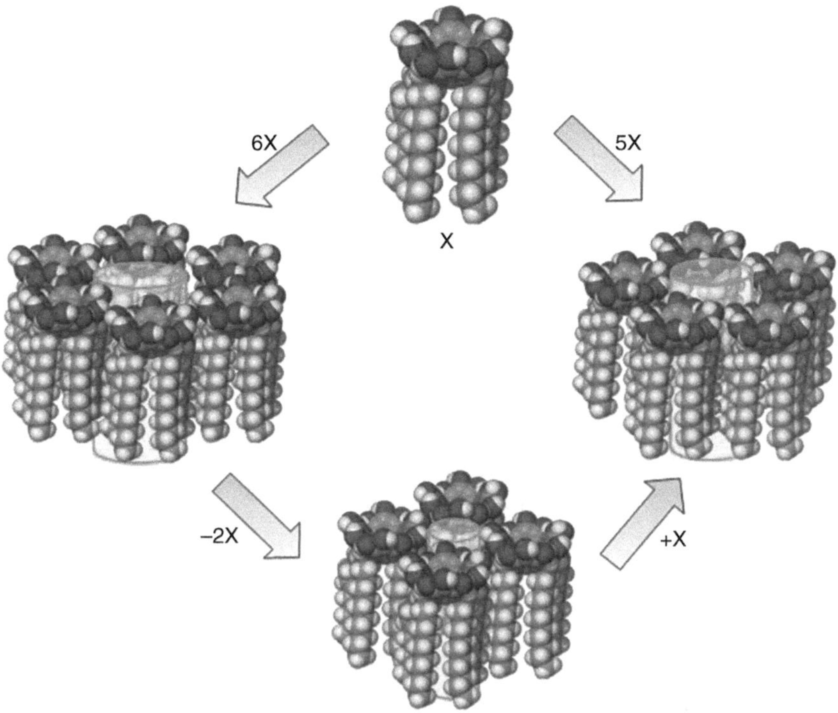

FIGURE 8.8 Schematic of possible aggregation states for pyrogallol[4]arenes embedded within a phospholipid bilayer membrane.

state in asolectin membranes. Such a pore would be rich in the polar hydroxyl residues facing the aqueous layer in the upper leaflet of the bilayer. It is well known with toroidal pores that reorganization of the lower leaflet can occur in order to facilitate the formation of a channel. We infer from the data that this is happening in the present case. It seems unlikely that compounds so hydroxyl-rich on one surface of the structure would be able to pass through a bilayer membrane into the lower leaflet and have the hydroxyl groups penetrating into the aqueous or cytosolic interior. Such a process would be equivalent to phospholipids undergoing transverse relaxation or "flip-flop." This is a very energetically costly process compared to movement of an amphiphile within a single leaflet of the bilayer.

5 PYROGALLOL[4]ARENE MEMBRANE AGGREGATION—LANGMUIR TROUGH STUDIES

In order to better understand the possible association of pyrogallol[4]arenes in a single bilayer leaflet, we availed ourselves of the Langmuir trough. This is a device that consists essentially of a pan of water and two movable barriers on opposite sides of the pan. A compound is applied to the surface of the water and allowed to disperse into a noncompressed monolayer. In the experiment, the

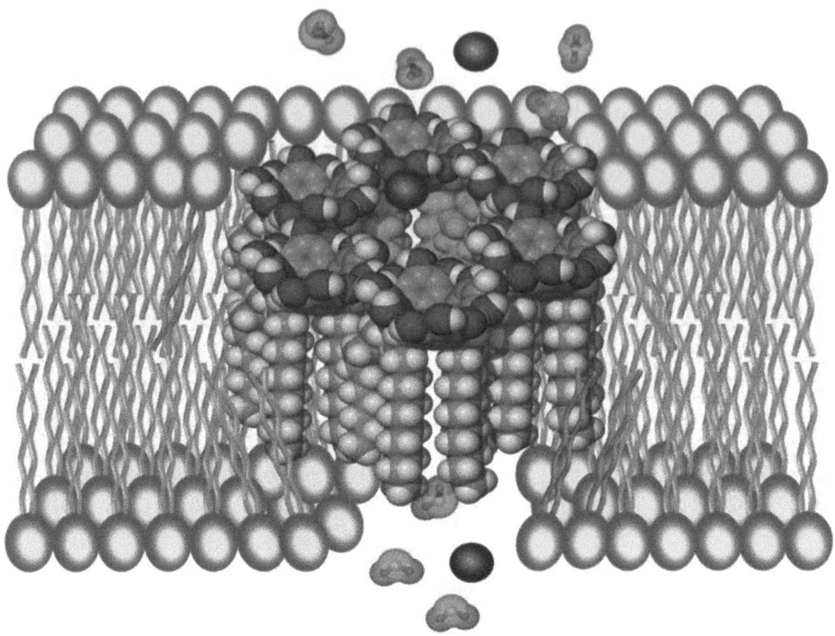

FIGURE 8.9 **Schematic showing possible hexameric pore formation in the membrane by tetra-*n*-undecylpyrogallol[4]arenes.**

barriers are moved inward applying physical pressure to the initially dispersed material. As the monolayer is compressed, the forces required to change the level of molecular organization are detected by a transducer (a so-called Wilhelmy plate) and reported to appropriate analytical software. A plot is obtained of surface pressure as a function of molecular area.

The Langmuir trough was used to study pyrogallol[4]arenes having side chains ranging in length from 1 to 12 carbons [23], both linear and branched. What we expected for the longer chain compounds was that the polyhydroxylated surface of the macrocycle would penetrate the aqueous phase. The side chains would be turned upward into the air. As the barriers were moved toward each other, we would expect the dispersed macrocycles to be compressed into a minimum area arrangement. Once the closest possible contact was reached, it would take significant energy for further collapse. Such a collapse means that the molecules would break from the aqueous surface and tumble over each other or perhaps form a new lamellar assembly. The point of maximum compression before collapse reveals the area occupied by each individual pyrogallol[4]arene. If the area calculated corresponds to the area estimated for the macrocycle, close contact of the rings can be presumed. In such a case, the hydrocarbon chains must be turned upward (perpendicular to the aqueous phase) into the air (hence, the air–water interface). A schematic of the Langmuir trough is illustrated in Fig. 8.10.

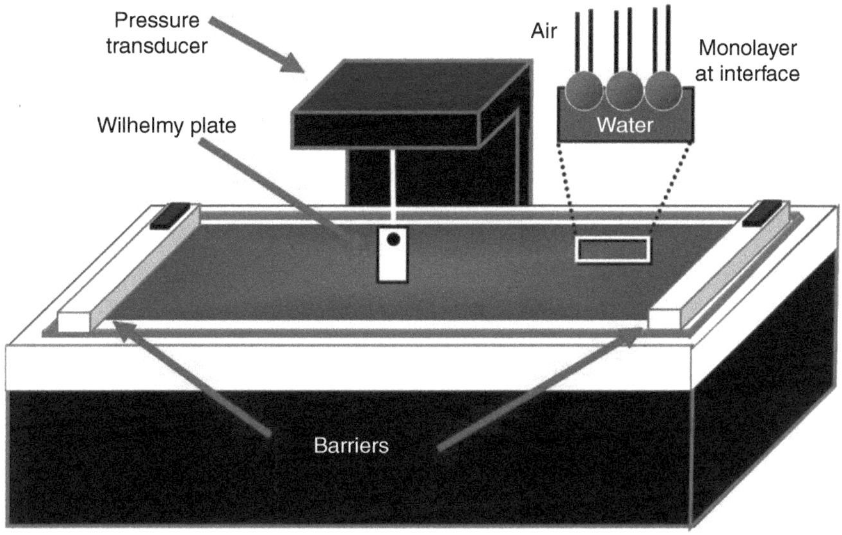

FIGURE 8.10 Schematic representation of the Langmuir trough instrument used to assess the surface behavior of pyrogallol[4]arenes.

The most interesting examples among the compounds studied were pyrogallol[4]arenes having methyl, 2-propyl (isopropyl), and 3-pentyl side chains [24]. If one considers the pyrogallol[4]arenes to be amphiphiles, the four methyl side chains make the compound perhaps the shortest amphiphile known. Because the cycle is so rigid and the hydroxyl groups are so intimately involved with the aqueous surface, these molecules form orderly arrays on the surface of an aqueous phase.

A surprising result was found with the simplest branched-chain pyrogallol[4]arene. In this case, 2-propylpyrogallol[4]arene formed stable arrays at the air–water interface. When compressed, however, it appeared that the tetramers were smaller in size than expected. A closer analysis suggested that as the compounds compressed on the aqueous surface, they tilted to an angle of about 15 degree downward into the aqueous phase, and the side chains interlocked along the assembly. This provided a stable and overlapped array on the aqueous surface.

An even more unusual situation was observed with 3-pentylpyrogallol[4]arene. These side chains interact in a gear-like fashion and form a surface that is highly ordered across the aqueous phase. The surface behavior of the 3-pentylpyrogallol[4]arenes exhibits both ion transport and a remarkable nanotube structure. These are discussed in Section 8.

6 A MONC-BASED ION CONDUCTING CHANNEL

Although the accumulating evidence showed that the pyrogallol[4]arenes functioned in a bilayer membrane to form self-assembled pores, the question of whether a capsule could also show transport activity intrigued us. Solid-state structure analysis of the octahedral pyrogallol[4]arene capsular assemblies suggested that ions could flow through the small orifices in the channel's exterior.

In order to explore the possibility of whether one of the longer chain compounds could function as a pore, we prepared *n*-undecyl pyrogallol[4]arene. The capsule formed readily from six monomers having 11-carbon side chains and was isolated. It was then converted into the copper metal organic nanocapsule (MONC) by using the techniques developed in the Atwood laboratory [21]. The MONC was then added to an asolectin membrane and studied by the planar bilayer conductance method. Classic open-close behavior was observed for this compound at voltages in the range of +30 to +60 mV. The conductance data were recorded for potassium cation, which was present in a concentration of 450 mM [25].

A somewhat surprising result was found when the current voltage plot was prepared. Typically, symmetrical materials that form pores conduct ions equally well in both directions, leading to a linear plot that passes through zero. The voltage current (I–V) plot did not pass through zero. Instead, an unexpected ion selectivity that favored K^+ over Cl^- by 2.4 was observed. A possible explanation for this is that K^+ is slightly smaller than Cl^- and the capsule's interior is electron rich, which would favor a cation over an anion.

It is interesting to consider what mechanism might allow for open-close behavior in this case. From the conductance data, we surmised that potassium ions were passing through the opening in the capsule. The conductance (in picoSiemens) was consistent with a relatively small opening. However, passing through the capsule does not tell whether only one or two stacked capsules were present in the bilayer. If a single capsule was present in the membrane, simply rocking from side to side could disrupt a channel formed within the bilayer. Although it might seem less likely that the smaller, single- capsule system would be viable, it is well known that when pores form in bilayers there is often a local contraction in thickness of the membrane.

A second possibility was that two capsules had inserted in the bilayer and that ions passed freely through either of them. The open-close gating in such a case would then result from movement of the two capsules with respect to each other. When they were aligned, flow would be rapid, but when they were offset from each other, ions would not pass through them effectively. Of course, the rocking of either capsule in such a configuration would also be problematic.

Definitive information was not obtained that would allow the distinction between one or two copper MONCs stacked within the membrane. Either situation could presumably afford a conducting channel. The two-stack pore notion seemed less likely because two MONCs — possibly at different levels in the bilayer — would have to organize vertically. A single capsule is about 16 Å across, and the insulator regime of a bilayer is about 30 Å. If some thinning of a membrane occurred, as is typical of pore formation, then a stack of two MONCs becomes even more unlikely.

7 SOLID-STATE STRUCTURES OF LINEAR AND BRANCHED PYROGALLOL[4]ARENES

As noted previously, our initial plan was to develop compounds that could form capsules, that have a hydrocarbon perimeter, and that could interact with bilayer membranes. We thus prepared and studied three pyrogallol[4]arenes. The tetramers were formed from three different aldehydes: $CH_3(CH_2)_7CH{=}O$, $CH_3(CH_2)_9CH{=}O$, and $CH_2{=}CH(CH_2)_8CH{=}O$. Reactions of pyrogallol with these aldehydes resulted in the formation of capsules having *n*-octyl (35% yield), *n*-decyl (41%), or 9-decenyl (40%) side chains. All three capsules were isolated by crystallization from a variety of solvents [26].

The difference in behavior of these capsules was surprising. The *n*-octyl substituted capsule crystallized from EtOAc. A solid-state structure showed that six molecules of EtOAc were enmeshed in the alkyl chains. In contrast, the *n*-decyl capsule crystallized with six molecules of EtOAc and six of MeCN, all within the capsule. None was detected in the pendant alkyl chains. The decenyl-substituted capsule enclosed six ordered EtOAc molecules. These differences in behavior were subtle and somewhat unexpected, considering the capsules themselves are the same and the difference lies only in how similar the side chains are in each.

The pyrogallol[4]arenes typically separate from complex reaction mixtures. As they are crystalline, a number of solid-state structures have been reported from a number of laboratories. The majority of these structures involve linear chains, although some are unsaturated or terminated by a halogen or alcohol function. Relatively little information was available concerning branched chain pyrogallol[4]arenes. We anticipated that, like the linear-chained counterparts, the pyrogallol[4]arene derivatives would crystallize either in the capsular or bilayer form.

The bilayer form is what one might expect from amphiphilic structures such as the pyrogallol[4]arenes: layers are obtained in which the head groups of one layer interact with those of the next. Further, the hydrocarbon chains of one layer interact with those of the next. Thus, the structure involves an infinite stack of opposing side chains alternating with opposing hydroxyl groups.

In our studies, we were unable to confidently predict whether the capsule or the bilayer form would be observed upon crystallization. Isopropylpyrogallol[4]arene formed both bilayers (from 1:1 EtOAc:EtOH) and capsules (from 2:1 $PhNO_2$:MeOH). When the side chains were 3-heptyl-, 4-heptyl-, 5-nonyl, or cyclohexyl, bilayers were formed in the solid state. In some cases, solvent molecules interceded between the layers to either fill void space or to interact directly with the various chains. For example, when isopropylpyrogallol[4]arene crystallized in the bilayer arrangement rather than in the capsular form, EtOAc molecules were linked to the hydroxyls within their same plane. Ethanol molecules were present in the interlamellar space.

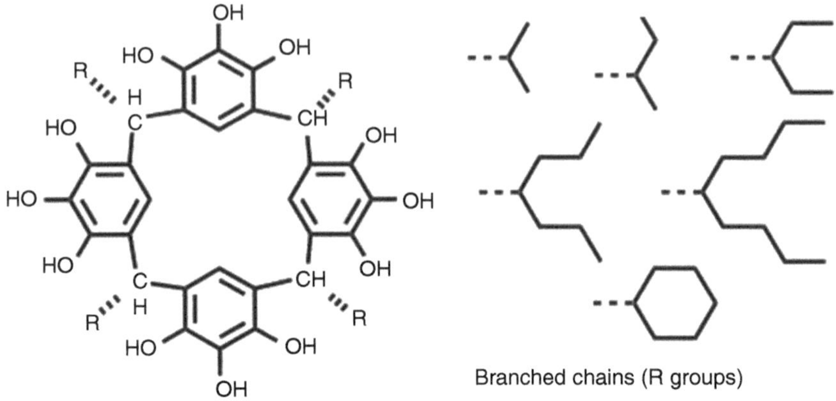

FIGURE 8.11 Branched-chain pyrogallol[4]arenes prepared and studied by X-ray analysis.

When the side chains were either 4-heptyl or 5-nonyl, an extensive hydrogen bonding network was noted, typically involving EtOH, EtOAc, and H_2O if any of these three solvents was present (Fig. 8.11) [27].

8 PYROGALLOL[4]ARENE-BASED NANOTUBES

When the pyrogallol[4]arene side chains were 3-pentyl (diethylmethyl), a unique nanotube structure was isolated [24]. The nanotubes that formed from tetra-3-pentylpyrogallol[4]arene were different from those isolated by Rissanen [28] and by Atwood [29] and their coworkers. The Rissanen nanotube was an assembly of resorcinarenes, and dimethylquinuclidinum dibromide was an integral component in the structure. The Atwood nanotube assembled from tetra-*n*-hexylpyrogallol[4]arene, but the void spaces within the structure were filled either by pyrene or bromopyrene. In short, in both cases, other molecules participated in a critical way in the formation of the nanotube. In our tetra-3-pentylpyrogallol[4]arene system, the nanotubes were formed by a combination of hydrogen bond and solvent interactions that afforded a cylinder.

The capsules that typically form from pyrogallol[4]arenes do so in an octahedral array. They enclose >1000 Å^3 of space and various molecules may be incorporated within the capsule or entwined externally in their side chains. The pyrogallol[4]arene nanotubes also comprise six of the tetramer macrocycles in each repeating unit, but they can be thought of as a doughnut-shaped hexamer instead of a capsule. These "doughnuts" organize into a cylindrical stack, although each individual hexamer is canted by about 30 degrees from a normal axis (Fig. 8.12).

Each of the individual doughnut units is slightly unsymmetrical, leading to an elliptical opening and overall aspect. The internal dimensions of this nanotube are about 18 × 20 Å [24]. The outer dimensions of the ellipse are approximately 26 × 33 Å. The thickness of each doughnut unit is approximately 10 Å. The individual doughnut units are each linked laterally by two hydrogen

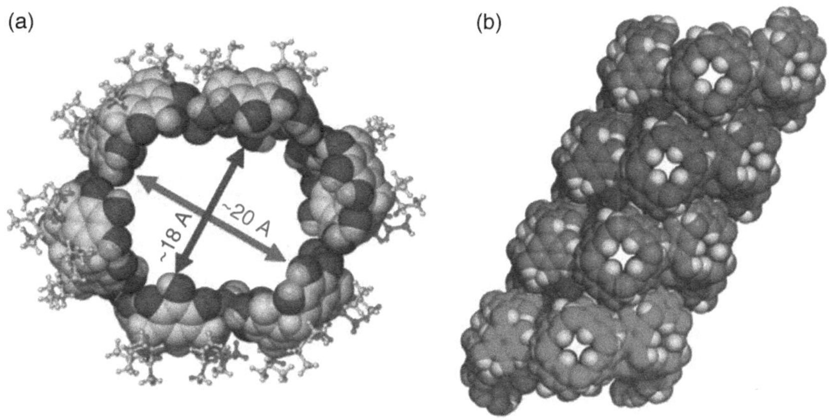

FIGURE 8.12 Renderings of the tetra-3-pentylpyrogallol[4]arene nanotube showing the hexameric doughnut and internal dimensions (a) and a stack of four hexamers (b) that leads to the extended nanotube.

bonds that sustain the overall structure. It is interesting to note that within each hexameric doughnut, the six monomers may be grouped into three distinctive dimer structures. These dimer structures are consistently distributed throughout the solid-state structure. Thus, the nanotubes have an overall consistency about them. This is apparent in the structure shown at the right of Fig. 8.12.

Perhaps the most interesting feature exhibited by these nanotubes is that the pendant alkyl groups interlock laterally with those of adjacent nanotubes. The regular positions of the 3-pentyl (or diethylmethyl) side chains form something that resembles a slightly elliptical gear. The ethyl groups act as the teeth of the gear and interlock the structures laterally throughout the solid state. Fig. 8.13 shows how this interaction occurs between two hexameric "doughnuts."

The nanotubes have been found to contain both ions and water. It was thought that these compounds might function as organic conductors because they appear to be water and ion-filled pipes. Another motivation for attempting to make conductors from them is that electron microscopic analysis revealed the presence of linear nanotube arrays of multiple conductor strands that extended to lengths of 5 or more microns [27]. An end-on view of three nanotubes and their internal ions and water is shown in Fig. 8.14.

We also envisioned the possible insertion of molecules within the nanotubes so that the overall structures might be used as a delivery mechanism. Work was conducted with a variety of fluorescent dyes. Attempts were made both to preform the tubes and then to expose them to dye solutions, and alternately, to cocrystallize various dyes and tetra-3-pentylpyrogallol[4]arene. In no case did X-ray analysis reveal dyes encapsulated within the nanotubes. Interestingly, many of the crystals did appear fluorescent, suggesting possible surface attachment of dyes at some level. Fig. 8.14 suggests the internal openings into which it was hoped that dyes would insinuate.

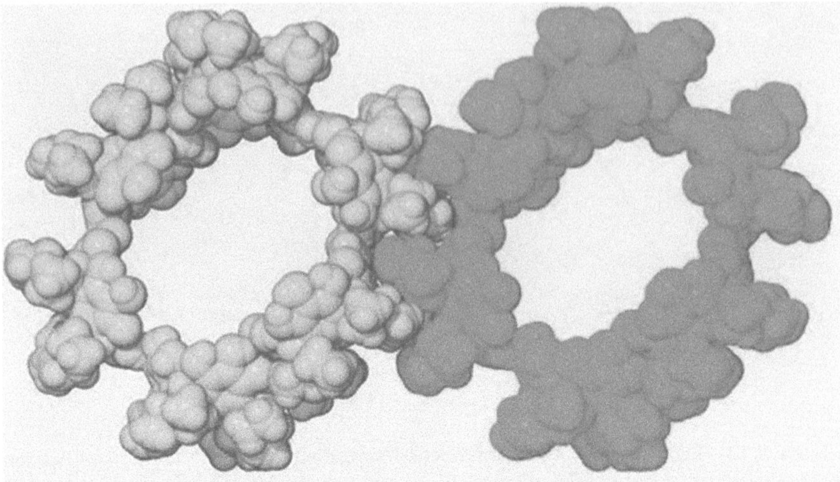

FIGURE 8.13 Image showing the "gearing" of the ethyl groups present in the nanotubes that form from tetra-3-pentylpyrogallol[4]arene.

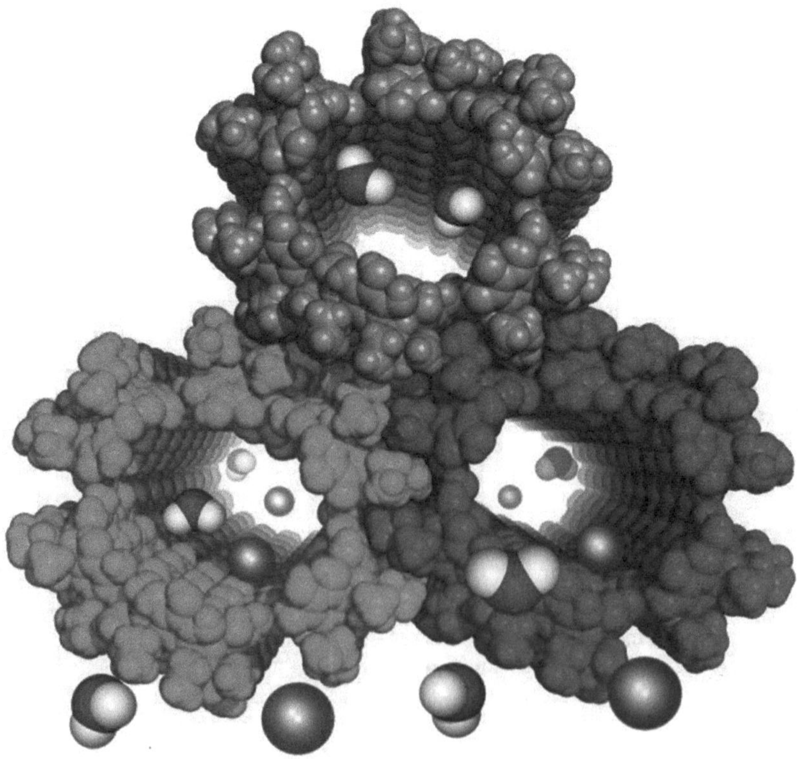

FIGURE 8.14 End-on view of three nanotube structures that form from tetra-3-pentylpyrogallol[4]arene. The image was produced from the solid-state structure and shows ions and water that are associated with the tube.

9 TETRA-3-PENTYLPYROGALLOL[4]ARENE-MEDIATED ION TRANSPORT

The behavior of alkyl-chained pyrogallol[4]arenes in membranes as assayed by the planar bilayer method is described in Section 4. The compounds to which those studies refer all had linear alkyl chains. The fact that branched chain compounds, especially tetra-3-pentylpyrogallol[4]arene, differed in some respects in their behavior from those having linear chains, suggested that different transport activity might be observed in the membrane. We determined to compare the membrane behavior of the 3-pentyl- and 4-heptylpyrogallol[4]arenes. As noted earlier, the former crystallizes as a unique nanotube and the latter forms bilayers in the solid state [30].

A particularly interesting question was how, if at all, the nanotube structure would insert into a bilayer membrane. If it did insert, the tube might transcend the membrane, perhaps forming a new type of conductor. Alternately, we could imagine that forces within the membrane might simply disperse the tubular pyrogallol[4]arene assembly into nonoriented monomers, leading to no conductance at all.

One reason that we felt it was possible that decomposition of the nanotube system would be observed was an experiment that we had done earlier with the *n*-decyl tetramer. The tetramer was crystallized into either bilayers or capsules by changing the crystallization solvent. We dissolved samples of each crystal in EtOAc in preparation for HPLC analysis [25]. When either sample was injected into the HPLC, the chromatogram observed was essentially identical. This suggested that the system was highly dynamic, despite the reported ability of pyrogallol[4]arene capsules to retain their integrity in aqueous solution [31].

The nanotubes formed from tetra-3-pentylpyrogallol[4]arene can be crystallized either from EtOAc/EtOH or from EtOAc/MeCN. The resulting nanotubes are quite similar, but the H-bond interactions with solvent vary. Attempts to crystallize other branched chain pyrogallol[4]arenes led either to bilayers or to capsules rather than nanotubes, at least in our hands. Our efforts included isopropyl, 3-pentyl-, 4-heptyl-, and 5-nonyl-substituted pyrogallol[4]arenes (Fig. 8.15).

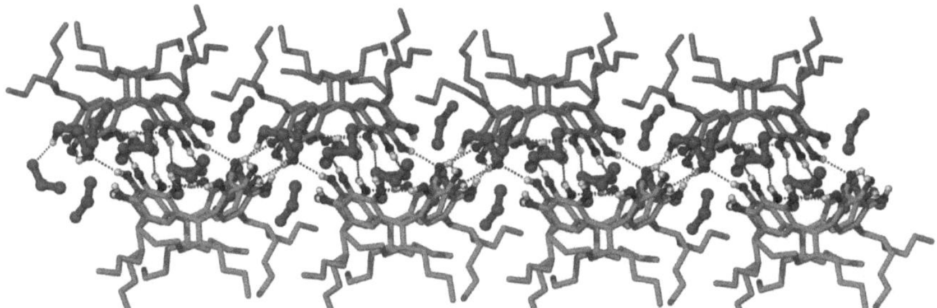

FIGURE 8.15 Crystal structure of 5-nonylpyrogallol[4]arene that forms a bilayer illustrating H-bond interactions involving the pyrogallol[4]arenes and ethanol solvent. *(Adapted from Ref. [27]. With permission from Taylor & Francis. Copyright 2014.)*

Channel behavior was assessed for tetra-3-pentylpyrogallol[4]arene and tetra-4-heptylpyrogallol[4]arene by using the planar bilayer conductance technique. This method allows observation of open-close behavior in a bilayer membrane, which connotes channel activity. In the studies discussed here, soybean asolectin was used as the bilayer membrane.

Tetra-4-heptylpyrogallol[4]arene showed at least seven distinctive open states (different channels), some of which were of significant duration. Overall, this behavior was not considered to reflect a single, stable organization of the monomers within the bilayer. In contrast, the planar bilayer conductance trace for tetra-3-pentylpyrogallol[4]arene showed a single open state that dominated the trace.

One of the most interesting and unusual feature of tetra-3-pyrogallol[4]arene behavior in the bilayer was a rather long opening that appeared at twice the conductance of typical openings. What made this remarkable was the fact that what appeared to be two channels opening and two channels closing occurred at exactly the same time. It is not uncommon to observe the opening of two distinct but otherwise identical channels at nearly the same time. Normally, though, one observes a single opening and then a step to a second opening (twice the conductance) and then a return to a single state followed eventually by a close state. The odds that two channels would open and close at identical times seemed small. A more reasonable interpretation was that the gearing shown in Fig. 8.13 persisted within the bilayer and two pairs of "doughnuts" formed two adjacent channels when appropriately aligned within the bilayer.

Our conclusion was that two of the geared, cyclic hexamers stacked on top of each other within the bilayer membrane. This presented a stack thickness within the membrane of about 24 Å, which is certainly feasible from a structural perspective. It appears that the two hexamers maintain their structural integrity despite the fact that the overall nanotube degrades and a channel is formed when two of these molecules slide into congruence with each other. Sliding past each other eliminates both open states simultaneously, leading to the sometimes rapid open-close behavior that is observed.

Because this mode of channel formation was surprising, we also examined the behavior of this compound on the Langmuir trough. This technique was discussed in Section 5. Typically, the area occupied by the tetramer head groups will be similar for both linear and branched pyrogallol[4]arenes assuming no unusual interactions either among the side chains or the head groups. Fig. 8.16 shows schematically how linear chained pyrogallol[4]arenes organize at the air–water interface.

If the nanotubes' hexameric doughnuts retained their structural integrity on the aqueous surface, their presence would leave void spaces on the aqueous surface at the close-packed state. In fact, this was observed and the phenomenon is illustrated schematically in Fig 8.17. The hexagons superimposed on the circles indicate the void spaces leading to larger-than-expected estimated areas. The presence of these void spaces contrasts with the surface behavior of other pyrogallol[4]arenes studied.

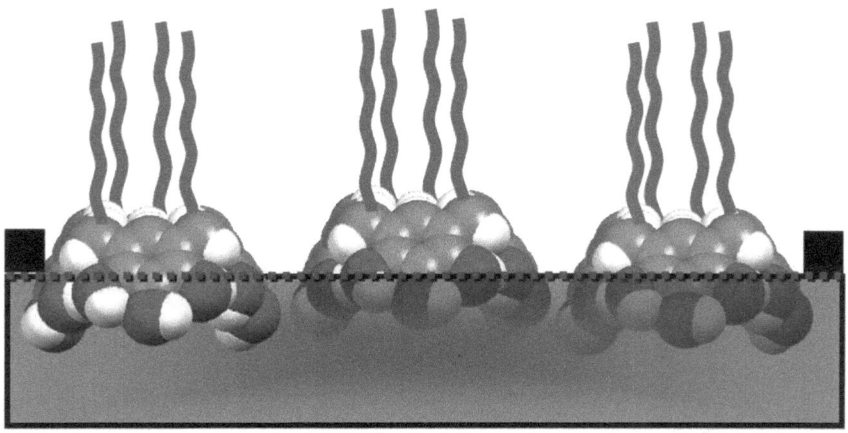

FIGURE 8.16 **Schematic representation of a pyrogallol[4]arene on the surface of a Langmuir trough at the air–water interface.**

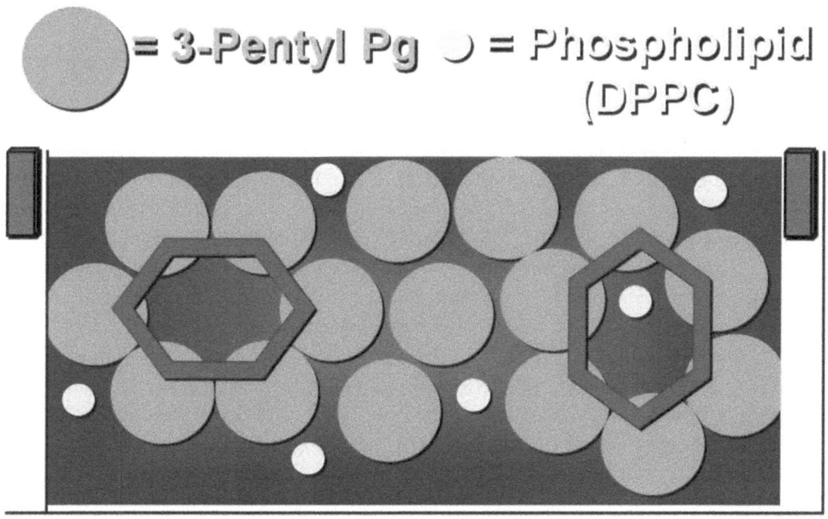

FIGURE 8.17 **Schematic representation of the organization of nanotube hexamers on the water surface of a Langmuir trough.**

10 CONCLUSIONS

The pyrogallol[4]arenes have shown remarkable properties. They are amphiphiles even when their alkyl side chains are short. When their alkyl side chains are longer (eg, in the C_{11} range), they self-assemble into ion-conducting pores within bilayer membranes. The alkyl side-chained pyrogallol[4]arenes can also form into previously known capsules, but these can entrap solvent molecules within their chains. When converted into a metal organic nanocapsule, ion conducting activity is observed in planar bilayers. When the side chains are branched, a new type of nanotube and a new type of membrane conductance activity is observed. These properties are remarkable and the compounds no doubt will be further explored for new applications that involve both encapsulation and mediators of reactions.

ACKNOWLEDGMENTS

We warmly thank the National Science Foundation for grants (CHE-0957535 and CHE-1307324) supporting much of the work that is described herein. We are also grateful to Dr. Oleg Kulikov for kindly providing images of compounds he studied.

REFERENCES

[1] Pedersen CJ. Cyclic polyethers and their complexes with metal salts. J Am Chem Soc 1967;89:7017–36.

[2] Dietrich B, Lehn JM, Sauvage JP. Diaza-polyoxa-macrocyclic and macrobicyclic compounds. Tetrahedron Lett 1969;34:2885–8.

[3] Dietrich B, Lehn JM, Sauvage JP, Blanzat J, Cryptates X. Syntheses and physical properties of diazapolyoxamacrobicyclic systems. Tetrahedron 1973;29:1629–45.

[4] Gokel GW, Korzeniowski SH. Macrocyclic polyether syntheses. Berlin: Springer-Verlag; 1982. pp. 410.

[5] (a) Frensdorff HK. Salt complexes of cyclic polyethers. Distribution equilibriums. J Am Chem Soc 1971;93:4684–8. (b) Pedersen CJ, Frensdorff HK. Macrocyclic polyethers and their complexes. Angew Chem, Int Ed Engl 1972;11:16–25. (c) Izatt RM, Bradshaw JS, Nielsen SA, Lamb JD, Christensen JJ, Sen D. Thermodynamic and kinetic data for cation-macrocycle interaction. Chem Rev 1985;85:271–339. (d) Izatt RM, Pawlak K, Bradshaw JS, Bruening RL. Thermodynamic and kinetic data for macrocycle interactions with cations and anions. Chem Rev 1991;91:1721–85. (e) Izatt RM, Bradshaw JS, Pawlak K, Bruening RL, Tarbet BJ. Thermodynamic and kinetic data for macrocycle interaction with neutral molecules. Chem Rev 1992;92:1261–354.

[6] Lehn JM. Supramolecular chemistry: receptors, catalysts, and carriers. Science 1985;227:849–56.

[7] Baeyer A. Ueber ein Condensationsproduct von Pyrrol mit Aceton. Ber Dtsch Chem Ges 1886;19:2184–5.

[8] Ackman RG, Brown WH, Wright GF. The condensation of methyl ketones with furan. J Org Chem 1955;20:1147–58.

[9] Baekeland LH. Condensation product and method of making same. US patent 942,809; 1909.

[10] Zinke A, Ziegler E. Zur Kenntnis des Härtungsprozesses von phenol-formaldehyd-harzen, X. Mitteilung. Ber Deutsch Chem Gesell 1944;77:264–72.

[11] Gutsche CD. Calixarenes revisited, vol. 6. Cambridge: Royal Society of Chemistry; 1998.

[12] Cornforth JW, Morgan ED, Potts KT, Rees JW. Preparation of antituberculous polyoxyethylene ethers of homogeneous structure. Tetrahedron 1973;29:1659.

[13] Gutsche CD. Calixarenes. Cambridge: Royal Society of Chemistry; 1989. pp. 223.

[14] Gutsche CD, Dhawan B, No KH, Muthukrishnan R. Calixarenes. 4. The synthesis, characterization, and properties of the calixarenes from *p-tert*-butylphenol. J Am Chem Soc 1981;103:3782–92.

[15] (a) Vicens J, Böhmer V, Calixarenes. A versatile class of macrocyclic compounds (topics in inclusion science). Berlin: Springer Verlag; 1990. pp. 280. (b) Gutsche CD. Calixarenes

revisited. Cambridge, UK: Roy Soc Ch; 1998. (c) Mandolini L, Ungaro R. Calixarenes in action. Hackensack, NJ: World Scientific Publishing Company; 2000. pp. 271. (d) Asfari M-Z, Böhmer V, Harrowfield J, Vicens J. Calixarenes. Weinheim: Springer Verlag; 2001. pp. 700.

[16] Sliwa W, Kozlowski C. Calixarenes and resorcinarenes: synthesis, properties, and applications. Weinheim: Wiley-VCH Verlag GmbH & Co; 2009. pp. 316.

[17] Kaas JP, Cesar HZ, Zeller M, Hunter AD, Dueno EE. 2,8,14,20-Tetraphenylpyrogallol[4]arene dimethylformamide octasolvate. Acta Cryst 2006;E 26:o3179–80. Section E Structure Reports Online.

[18] (a) Atwood JL, Barbour LJ, Jerga A. Hydrogen-bonded molecular capsules are stable in polar media. Chem Commun 2001;2376–7. (b) Atwood JL, Szumna A. Hydrogen bonds seal single-molecule capsules. J Am Chem Soc 2002;124:10646–7. (c) Atwood JL, Barbour LJ, Jerga A. Organization of the interior of molecular capsules by hydrogen bonding. Proc Natl Acad Sci USA 2002;99:4837–41. (d) Atwood JL, Barbour LJ, Jerga A. Very large supramolecular capsules based on hydrogen bonding. Perspectives in Supramol Chem 2003;7: 153–75. (e) Atwood JL, Szumna A. Anion-sealed single-molecule capsules. Chem Commun 2003;940–1.

[19] (a) Atwood JL, Barbour LJ, Dalgarno SJ, Hardie MJ, Raston CL, Webb HR. Toward mimicking viral geometry with metal-organic systems. J Am Chem Soc 2004;126:13170–1.

[20] (a) Atwood JL, Barbour LJ, Jerga A, Schottel BL. Guest transport in a nonporous organic solid via dynamic van der Waals cooperativity. Science 2002;298:1000–2. (b) Dalgarno SJ, Tucker SA, Bassil DB, Atwood JL. Fluorescent guest molecules report ordered inner phase of host capsules in solution. Science 2005;309:2037–9. (c) Jin P, Dalgarno SJ, Barnes C, Teat SJ, Atwood JL. Ion transport to the interior of metal-organic pyrogallol[4]arene nanocapsules. J Am Chem Soc 2008;130:17262–3.

[21] (a) McKinlay RM, Cave GW, Atwood JL. Supramolecular blueprint approach to metal-coordinated capsules. Proc Natl Acad Sci USA 2005;102:5944–8. (b) McKinlay RM, Thallapally PK, Atwood JL. Hexameric C-alkylpyrogallol[4]arene molecular capsules sustained by metal-ion coordination and hydrogen bonds. Chem Commun 2006;2956–8. (c) Power NP, Dalgarno SJ, Atwood JL. Guest and ligand behavior in zinc-seamed pyrogallol[4]arene molecular capsules. Angew Chem Int Ed 2007;46:8601–4. (d) Kumari H, Dennis CL, Mossine AV, Deakyne CA, Atwood JL. Exploring the magnetic behavior of nickel-coordinated pyrogallol[4]arene nanocapsules. ACS Nano 2012;6:272–5. (e) Kumari H, Kline SR, Wycoff WG, Paul RL, Mossine AV, Deakyne CA, Atwood JL. Solution-phase structures of gallium-containing pyrogallol[4]arene scaffolds. Angew Chem Int Ed Engl 2012;51:5086–91. (f) Kumari H, Kline SR, Schuster NJ, Barnes CL, Atwood JL. Exploring the ellipsoidal and core-shell geometries of copper-seamed C-Alkylpyrogallol[4]arene nanocapsules in solution. J Am Chem Soc 2011;133:18102–5.

[22] Li R, Kulikov OV, Gokel GW. Pyrogallarene-based ion-conducting pores that show reversible conductance properties. Chem Commun 2009;6092–4.

[23] Daschbach MM, Kulikov OV, Long EF, Gokel GW. Pyrogallol[4]arenes show highly variable amphiphilic behavior at the air-water interface dependent upon sidechain length and branching. Chem Eur J 2011;17:8913–21.

[24] Kulikov OV, Daschbach MM, Yamnitz CR, Rath N, Gokel GW. Self-assembled, cogged hexameric nanotubes formed from pyrogallol[4]arenes with a unique branched side chain. Chem Commun 2009;7497–9.

[25] Kulikov OV, Li R, Gokel GW. A synthetic ion channel derived from a metallogallarene capsule that functions in phospholipid bilayers. Angew Chem Int Ed Engl 2009;48:375–7.

[26] Kulikov OV, Rath NP, Zhou D, Carasel IA, Gokel GW. Guest molecule entrapment by both capsule and hydrocarbon sidechains in self-assembled pyrogallol[4]arenes. New J Chem 2009;33:1563–9.

[27] Kulikov OV, Negin S, Rath NP, Gokel GW. Morphologies of branched-chain pyrogallol[4]arenes in the solid state. Supramol Chem 2014;26:506–16.

[28] Mansikkamaki H, Nissinen M, Rissanen K. Noncovalent pi.pi-stacked exo-functional nanotubes: Subtle control of resorcinarene self-assembly. Angew Chem Int Ed Engl 2004;43: 1243–6.

[29] Dalgarno SJ, Cave GW, Atwood JL. Toward the isolation of functional organic nanotubes. Angew Chem Int Ed Engl 2006;45:570–4.

[30] Negin S, Daschbach MM, Kulikov OV, Rath N, Gokel GW. Pore formation in phospholipid bilayers by branched-chain pyrogallol[4]arenes. J Am Chem Soc 2011;133:3234–7.

[31] (a) Dalgarno SJ, Antesberger J, McKinlay RM, Atwood JL. Water as a building block in solid-state acetonitrile-pyrogallol[4]arene assemblies: Structural investigations. Chemistry 2007;13:8248–55. (b) Dalgarno SJ, Power NP, Atwood JL. Metallo-supramolecular capsules. Coord Chem Rev 2008;252:825–41.

Chapter 9

Supramolecular Coordination Cages as Nanoreactors

Samahe Sadjadi
Gas Conversion Department, Iran Polymer and Petrochemical Institute, Faculty of Petrochemicals, Tehran, Iran

1 INTRODUCTION

Nanoreactors are nanocontainers with an internal cavity and the ability to encapsulate one or more guests. Encapsulation can not only isolate the guest from the bulk, but also induce specific conformation of the guest(s). Furthermore, confined space within a nanoreactor increases the concentration of reactants and can effectively influence the reaction inside the nanoreactor through binding interactions. These features result in various applications for nanoreactors, especially in catalysis [1,2].

Among various types of nanoreactors, self-assembled capsules have proven their use as catalysts of wide range of reactions [3]. Self-assembled capsules can be obtained through hydrogen bonding or metal–ligand interactions [1].

In metal–ligand self-assembled capsules (also referred as self-assembled coordination cages), metal ions act as the vertices in the geometrical structure that connects organic ligands as edges or faces [4,5]. Like hydrogen-bond self-assembled capsules, the cavity in these structures can accommodate one or more guests. However, guest exchange can be achieved by expansion of these capsules apertures (*vide infra*) [6].

Correct choice of ligands, metal ions, and conditions is essential for assembly of these capsules [4,6]. Two kinds of ligands (ie, one-dimensional and two-dimensional ligands) can be used for assembly of metal–ligand self-assembled capsules [7].

One-dimensional ligands which form the edges of the coordination cage can bridge only two metal centers, while the faces of the final structure are formed by two-dimensional ligands which are cable of bridging between more than two metal atoms [4].

Saalfrank and coworkers [8] report on assembly of M_4L_6-type coordination tetrahedrons demonstrated the metal–ligand coordination as an organizing force for the formation of coordination cages.

Organic Nanoreactors. http://dx.doi.org/10.1016/B978-0-12-801713-5.00009-4

During the last decade, several research groups—mainly the Raymond and Bergman group and the Fujita group—developed supramolecular assemblies based on metal–ligand interactions, and reported their applications for molecular recognition, stabilization, and catalysis of various chemical reactions. There are several review articles and book chapters on metal–ligand self-assembled capsules and their applications [6,9–12].

In this chapter, we briefly review the use of metal–ligand self-assembled capsules as nanoreactors for encapsulation and stabilization of reactive compounds and the concept of cavity-derived synthesis of chemicals within the self-assembled coordination cage.

2 M_4L_6 TETRAHEDRAL SELF-ASSEMBLED CAPSULES

2.1 Guest Encapsulation in M_4L_6 Cages

Raymond et al. studied M_4L_6 cages comprehensively. In these cages, four metal atoms were located at the corners of tetrahedrons and bridged by six naphthalene-based bis-bidentate catechol amide. Systems of various metals such as Ga^{3+}, Al^{3+}, In^{3+}, and Fe^{3+} were reported by this research group (Fig. 9.1) [13]. The ligand that spans each edge provides strong mechanical coupling between the vertices, forcing the assemblies to be homochiral and adopt either the $\Delta,\Delta,\Delta,\Delta$- or $\Lambda,\Lambda,\Lambda,\Lambda$-configuration. The enantiomers are stable, noninterconverting, and resolvable [11].

These coordination cages have a hydrophobic interior environment and highly negative charge, $[M_4L_6]^{12-}$. It makes them soluble in polar solvents. Encapsulation of mono cationic guests NR_4^+ (R = Me, Et, Pr) is favored due to the anionic feature of the cage [6,13], while neutral guests and highly charged cationic guests are not suitable for encapsulation [11]. Other factors which affect the encapsulation are size and hydrophobicity of guests [13].

Raymond et al. encapsulated two cationic ruthenium sandwich complexes, $[CpRu(\eta^6\text{-}C_6H_6)]^+$ and $[CpRu(p\text{-cymene})]^+$ ($Cp = C_5H_5$) in M_4L_6 assembly by guest exchange with NMe_4^+ [14].

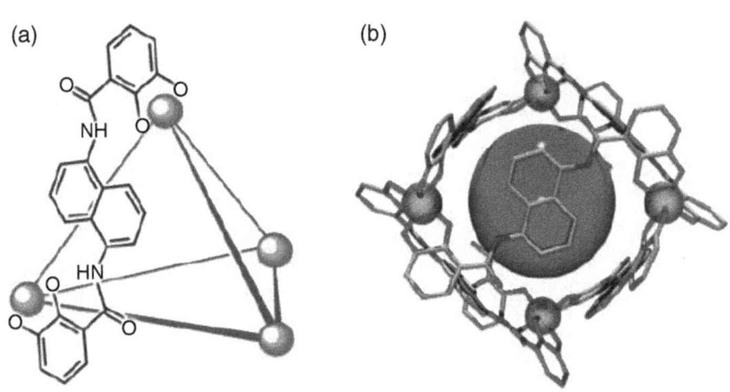

FIGURE 9.1 (a) The edges of the tetrahedron are spanned by six bis-bidentate catechol amide ligands. (b) Assembled tetrahedron with a generic spherical guest. *(Reprinted from Ref. [11] with permission of Royal Society of Chemistry. Copyright 2007.)*

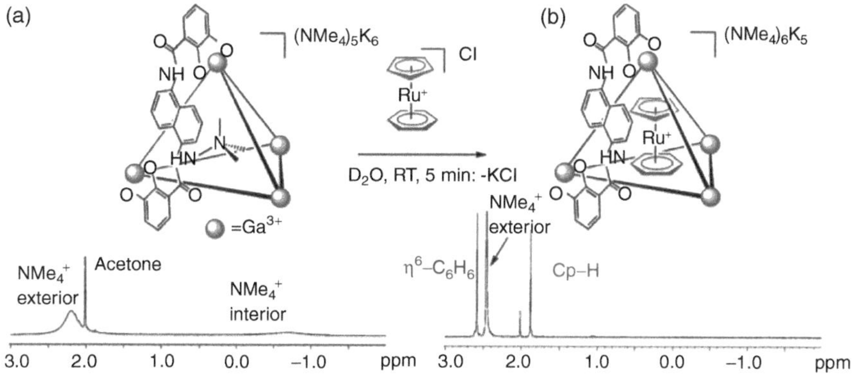

FIGURE 9.2 Guest exchange reaction of NMe_4^+ for $[CpRu(\eta^6\text{-}C_6H_6)]^+$. (a) $K_6(NMe_4)_5[NMe_4 \subset Ga_4L_6]$ before the addition of $[CpRu(\eta^6\text{-}C_6H_6)]^+$. (b) After 5 min of the addition of $[CpRu(\eta^6\text{-}C_6H_6)]^+$, the guest exchange is complete. *(Reprinted from Ref. [14] with permission of American Chemical Society. Copyright 2004.)*

The driving force for guest exchange is replacing NMe_4^+ with a lower binding constant with a guest that has more favored size and binding affinity. This phenomenon can be monitored by ^{1}HNMR spectroscopy (Fig. 9.2). As shown in Fig. 9.2, upon addition of $[CpRu(\eta^6\text{-}C_6H_6)]Cl$ to the solution of $K_6(NMe_4)_5[NMe_4 \subset Ga_4L_6]$ and subsequent guest exchange, the signals of interior NMe_4^+ (−0.70 ppm) disappeared and there appeared two new signals related to encapsulated ruthenium complex at 2.60 ppm ($\eta^6\text{-}C_6H_6$) and 1.89 ppm (Cp-H) [14].

Encapsulation of the prochiral complex $[CpRu(p\text{-cymene})]^+$ by the chiral host renders enantiotopic protons diastereotopic, as evidenced by ^{1}HNMR spectroscopy.

Encapsulation of chiral guests resulted in formation of diastereometric complexes. Taking two racemic pairs of ruthenium complex and M_4L_6 cage into account, one can expect four stereoisomers (Δ/R, Δ/S, Λ/R, Λ/S) (Fig. 9.3) [13].

The utility of chiral metal–ligand assemblies for an asymmetric recognition process was demonstrated by Raymond et al. [15]. In this study, a series of hydrophobic ruthenium half-sandwich complexes, CpRu(diene)Cl and Cp*Ru(diene)Cl [$Cp^* = \eta^6\text{-}C_5(CH_3)_5$], were encapsulated in M_4L_6 assembly. Upon encapsulation and formation of $[CpRu(\text{isoprene})(H_2O) \subset Ga_4L_6]^{11-}$ and $[CpRu(\text{2-ethylbutadiene})(H_2O) \subset Ga_4L_6]^{11-}$, two diastereomers with equal quantity were obtained. However, strically more demanding Cp* resulted in better recognition (Table 9.1) [15]. Comparison of diastereomeric excess of reported derivatives (Table 9.1) demonstrate that encapsulation is controlled by both shape and size of guests.

Recently, Würthner et al. have developed a giant electroactive M_4L_6 tetrahedral with an edge length of 3.9 nm and estimated internal volume greater

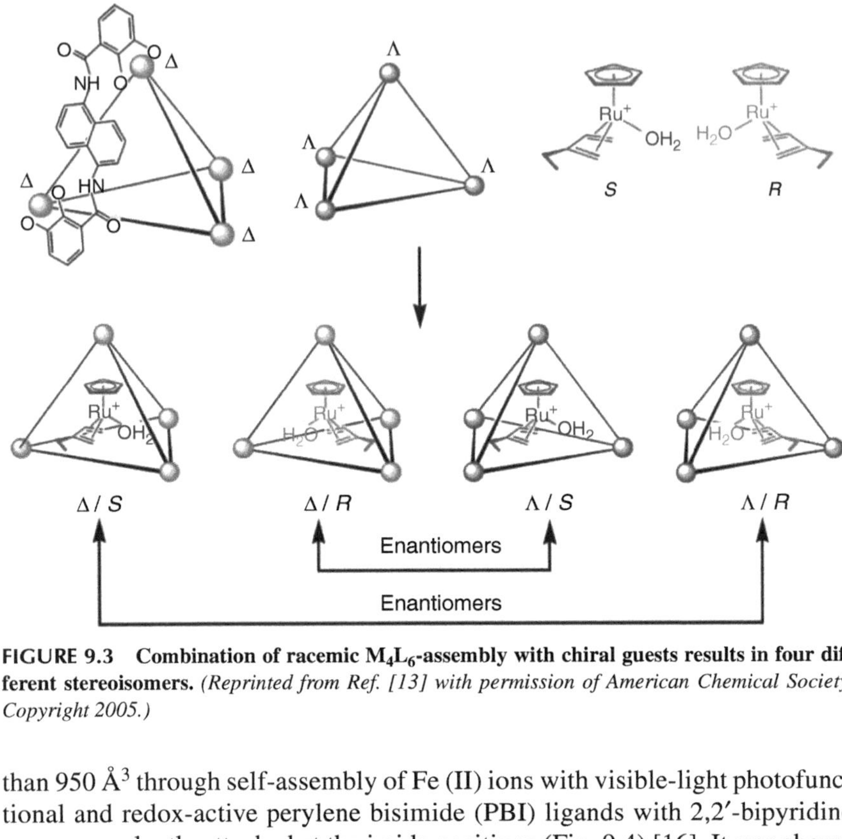

FIGURE 9.3 Combination of racemic M_4L_6-assembly with chiral guests results in four different stereoisomers. *(Reprinted from Ref. [13] with permission of American Chemical Society. Copyright 2005.)*

than 950 Å^3 through self-assembly of Fe (II) ions with visible-light photofunctional and redox-active perylene bisimide (PBI) ligands with 2,2′-bipyridine groups covalently attached at the imide positions (Fig. 9.4) [16]. It was shown that many of the desirable photo- and electro-active properties of the PBI

TABLE 9.1 Observed Diastereoselectivities in the Encapsulation Reactions[a]

$[Cp^* Ru($ (alkene with R_1, R_2) $)(H_2O) \subset Ga_4L_6]^{11-}$		de (%)
R^1 = H	R^2 = Me	4
R^1 = H	R^2 = Et	70
R^1 = H	R^2 = *i*-Pr	26
R^1 = H	R^2 = *n*-Pr	64
R^1 = Me	R^2 = H	15
R^1 = Et	R^2 = H	18
R^1 = *n*-Pr	R^2 = H	17

[a]*The de (diastereomeric excess) values were determined by integration of the ^{1}HNMR spectra and have an estimated ±3% error.*
Reprinted from Ref. [15] with permission of American Chemical Society. Copyright 2004.

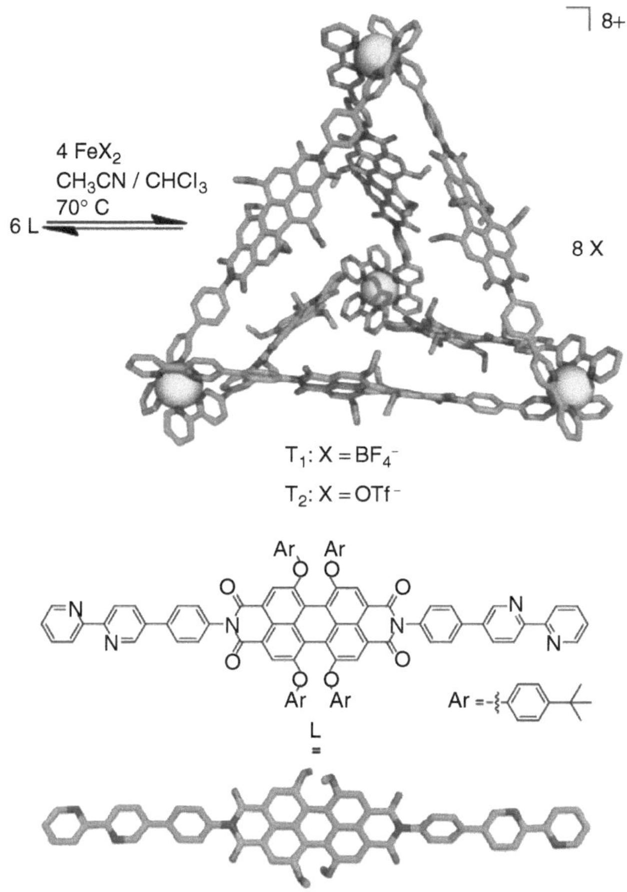

FIGURE 9.4 Self-assembly of T_1 ($[Fe_4L_6](BF_4)_8$) and T_2 ($[Fe_4L_6](OTf)_8$) from L. The tetrahedron model is depicted with all $Fe(bpy)_3$ vertices in the Δ conformation. Hydrogen atoms are omitted and OPhtBu substituents are represented as OMe for clarity. Fe, circle. *(Reprinted from Ref. [16] with permission of American Chemical Society. Copyright 2013.)*

ligands were transferred to the nanoscale metallosupramolecule. The tetrahedron absorbs strongly across the visible spectrum out to 650 nm and exhibits a total of 7 highly reversible electrochemical oxidation and reduction waves spanning a 3.0 V range. This facile cycling of 34 electrons between +18 and −16 charged species is probably made possible by the porous nature of the tetrahedron that allows the necessary counterions to freely flow in and out of the host [16].

To establish the utility of this giant host for encapsulation of large guests, the authors studied the encapsulation of C_{60} within this huge tetrahedron by using various techniques such as ^{13}C NMR spectroscopy, UV-vis spectroscopy, and ESI–MS. The results indicated that this cage can be a good candidate for accommodating large guest molecules.

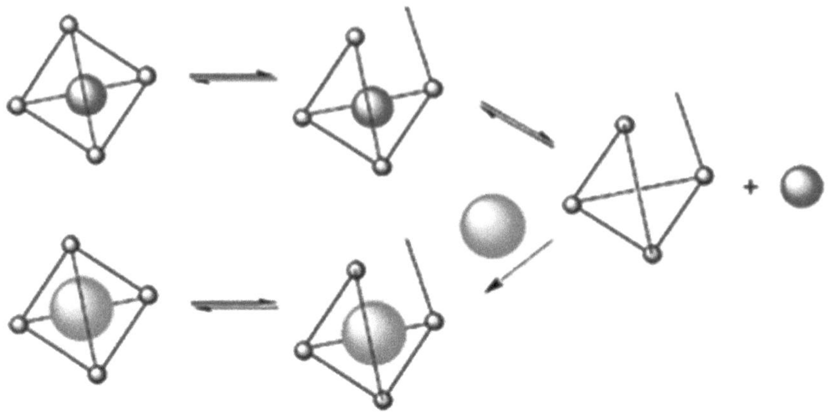

FIGURE 9.5 **Schematic representation of a putative guest exchange mechanism for the supramolecular M_4L_6 host based on the partial dissociation of one ligand.** In this mechanism, guest exchange would be mediated by metal–ligand lability. *(Reprinted from Ref. [17] with permission of American Chemical Society. Copyright 2002.)*

2.2 Guest Exchange Pathway in M_4L_6 Cages

Raymond et al. investigated the guest exchange pathway in M_4L_6 metal–ligand self-assembled cages to determine whether capsule rupture is required for guest ingress and egress (Figs. 9.5 and 9.6) [17]. In one pathway, partial dissociation of the cage provides a portal for guest passage; in the second pathway, deformation

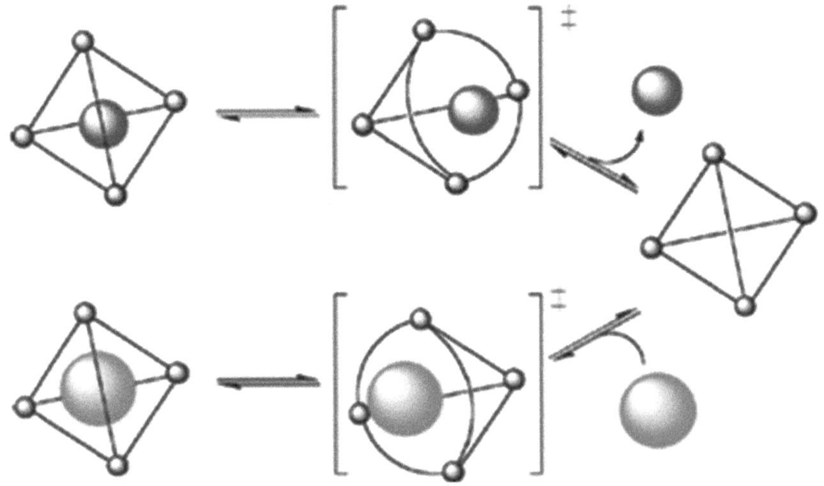

FIGURE 9.6 **Schematic representation of a nondissociative guest exchange mechanism for the M_4L_6 host.** It is likely that deformation of the host structure is necessary to accommodate the passage of guests through the assembly's C_3 symmetric apertures. *(Reprinted from Ref. [17] with permission of American Chemical Society. Copyright 2002)*

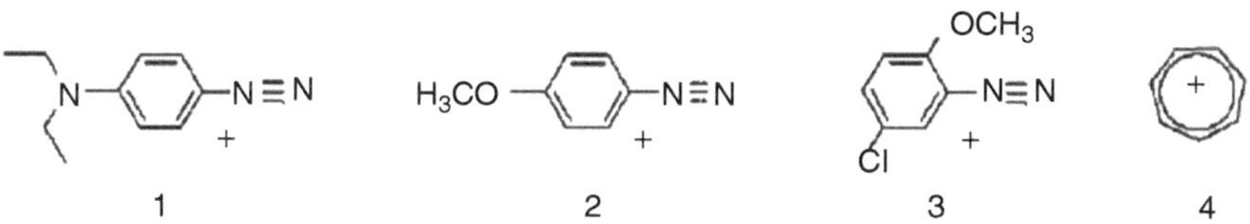

FIGURE 9.7 Structures of the test guest cations: 4-(diethylamino)benzenediazonium (1), 4-methoxybenzenediazonium (2), 5-chloro-2- methoxybenzenediazonium chloride (Diazo Red RC, 3), tropylium ion (4). *(Reprinted from Ref. [19] with permission of Wiley-VCH. Copyright 2004.)*

of the cage creates a dilated aperture for guest passage without cage rupture [17]. The results indicated that guest exchange did not require metal–ligand bond rupture of the cage. The rate of guest exchange in M_4L_6 metal–ligand self-assembled cages is not influenced by the nature of metal–ligand interactions of the cage.

2.3 Protection and Stabilization of Reactive Species

2.3.1 Stabilization of Diazonium and Tropylium Cations

The hydrophobic cavity of M_4L_6 cage and its strong affinity for cationic guests can be exploited for encapsulation and stabilization of reactive intermediates [18].

In this context, $[Ga_4L_6]^{12-}$ coordination cage was used for encapsulation and stabilization of aromatic diazonium cations and tropylium cations (Fig. 9.7) [19]. The results indicated that only 4-(diethylamino) benzenediazonium cation, 1, and aromatic tropylium cation, 4, could be encapsulated within a $[Ga_4L_6]^{12-}$ coordination cage. Encapsulation protected reactive cations and resulted in deceleration of their decomposition [19].

2.3.2 Protection and Stabilization of Iminium Cations

Raymond et al. reported encapsulation of various iminium cations in a tetrahedral $[Ga_4L_6]^{12-}$ coordination cage (Fig. 9.8). Molecular encapsulation could cause a remarkable shift in unfavorable equilibrium between keton and iminium. This can be considered a strategy to generate iminium ions under unconventional

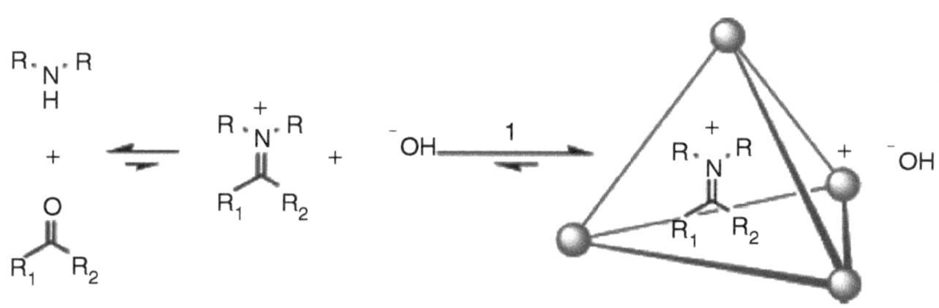

FIGURE 9.8 Stabilization of iminium ions in aqueous solution by molecular encapsulation. *(Reprinted from Ref. [18] with permission of American Chemical Society. Copyright 2006.)*

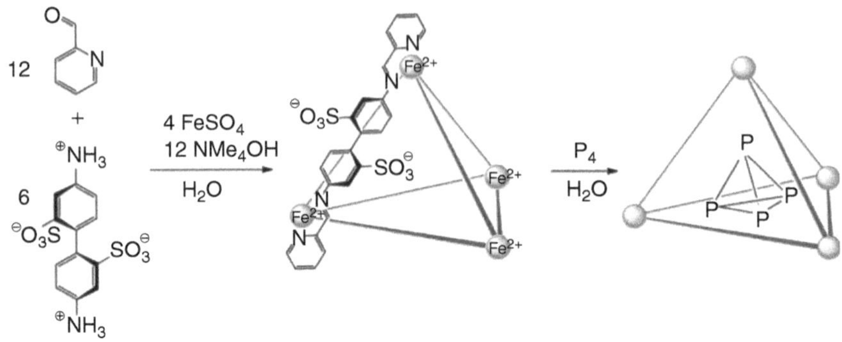

FIGURE 9.9 **Trapping of white phosphorus (P_4) in supramolecular coordination cage.** *(Reprinted from Ref. [10] with permission of American Chemical Society. Copyright 2015.)*

conditions. It is worth noting that encapsulated iminium ions were stable for months at room temperature. Investigation of various iminium cations indicated that molecular recognition depended on the charge, size, shape, and hydrophobicity of the guests [18].

2.3.3 Stabilization of P_4 Molecules

Nitscheke et al. disclosed the utility of tetrahedral anionic assembly for encapsulation and stabilization of reactive tetrahedral P_4 molecules (Fig. 9.9) [20]. Phosphorous was air-sensitive explosive upon interaction with atmospheric oxygen, while encapsulated phosphorous was air stable and water soluble [10]. This was attributed to the size of the oxidized species, which were too large to fit the cavity of the cage. The release of P_4 was simply obtained by guest exchange with benzene without rupturing the cage.

2.3.4 Protective Environment in Diels–Alder Reaction

In their next attempt to exploit a tetrahedral coordination cage as protective environment, Nitscheke et al. [21] reported a Diels–Alder reaction between furan and maleimide in the presence of a coordination cage (Fig. 9.10). The authors showed that the Diels–Alder reaction could be blocked by encapsulation of furan within the coordination cage. Upon addition of a competing guest with stronger binding affinity (benzene), furan could be released from the cage and undergo Diels–Alder cycloaddition.

2.3.5 Encapsulation and Storage of SF_6

The utility of the above mentioned tetrahedral coordination cage for reversible, selective trapping of inert, hydrophobic SF_6 was also reported [22]. The encapsulation of SF_6 within the interior of the coordination cage was confirmed by X-ray diffraction and NMR techniques. The results indicated that SF_6 occupies 53% of the void volume enclosed by the cage, above the reported optimal volume ratio of 40% for a gas encapsulated within an organic host.

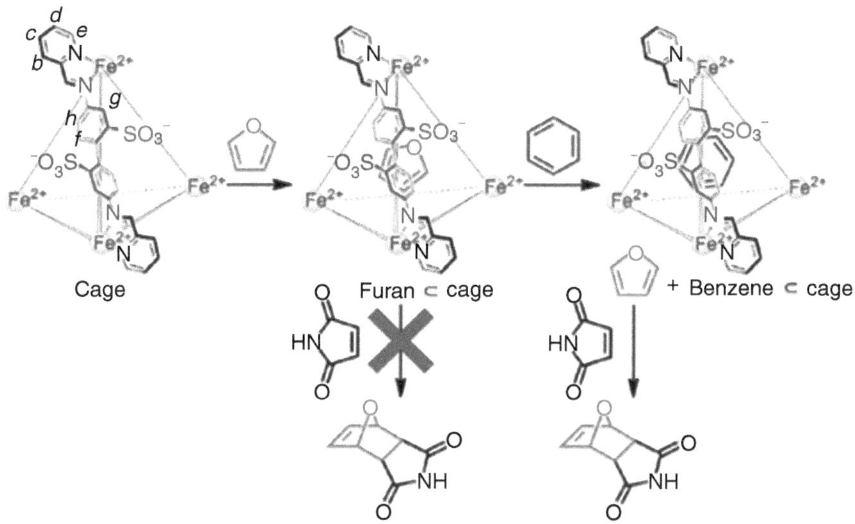

FIGURE 9.10 A schematic representation of the capsule-mediated Diels–Alder reaction: encapsulation of furan by the coordination cage prevented furan from reacting with maleimide. Upon the addition of the competing guest benzene, furan was released and the Diels–Alder reaction was initiated. *(Reprinted from Ref. [21] with permission of Royal Society of Chemistry. Copyright 2012.)*

Given that this cage has no affinity for Xe, Ar, N_2, O_2, C_2H_4, CO_2, or N_2O gases, and that it has a strong host–guest binding constant, this coordination cage is a potential candidate for removing SF_6 from gas mixtures [22]. Encapsulated SF_6 is significantly compressed and can be readily stored in solution. Increasing temperature and the addition of acid or tris (2-ethylamino) amine can be used as stimuli for releasing the encapsulated SF_6.

2.4 Catalysis by M_4L_6 Cages

2.4.1 Hydrolysis

A water-soluble tetrahedral M_4L_6 assembly was exploited to catalyze acidic hydrolysis of orthoformates $HC(OR)_3$, (R = alkyl or aryl) in basic solution [23]. It is worth nothing that orthoformates are stable in neutral or basic solutions, and their hydrolysis was reported in acidic solution. The merit of the M_4L_6 assembly-based approach is that the catalytic amount of metal–ligand cage can catalyze the so-called reaction in basic media in a short reaction time.

As mentioned before, M_4L_6 assembly has a high affinity for monocationic guests due to its negative charge. Therefore, the protonated form of neutral and even weakly basic guests (orthoformates, phosphines, amines) will be encapsulated in M_4L_6 assembly (Fig. 9.11). Hydrolysis of orthoformates to formate esters and then formats will proceed in the presence of a catalytic amount of M_4L_6 assembly, 1mol% [6]. Product inhibition was not observed due to the anionic characteristic of product, carboxylate [10]. To establish the catalytic role

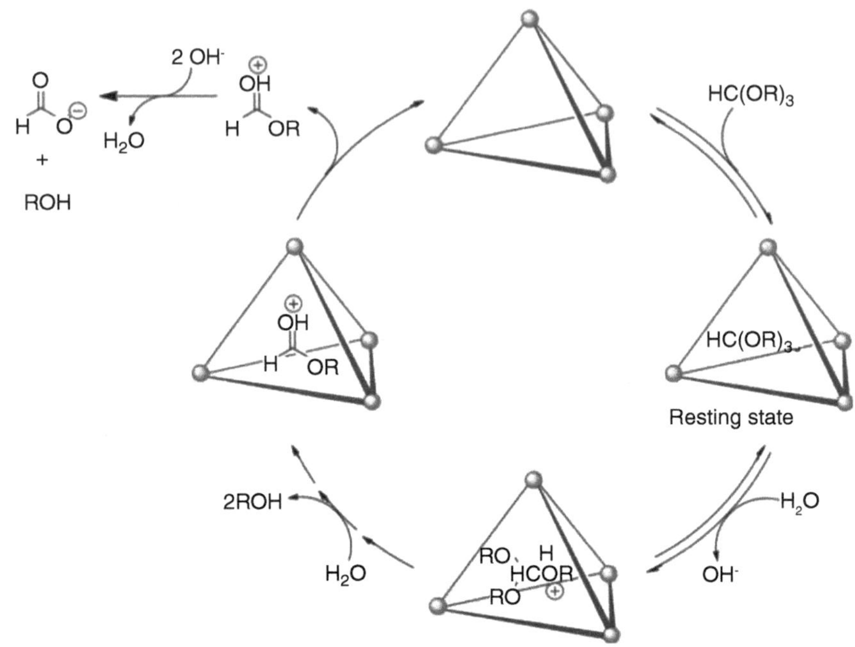

FIGURE 9.11 Mechanism for catalytic orthoformate hydrolysis in the presence of a catalytic cage. *(Reprinted from Ref. [10] with permission of American Chemical Society. Copyright 2015.)*

of the interior cavity of M_4L_6 assembly, a guest with a strong binding constant, NEt_4^+, was added to the reaction solution. The hydrolysis of orthoformate was blocked in the presence of NEt_4^+, confirming the essential role of the cavity of M_4L_6 assembly in catalysis [23].

2.4.2 Aza-Cope Rearrangement

The cationic 3-aza-Cope rearrangement was performed in the confined environment of M_4L_6 assembly [24]. Rearrangement of various enammonium substrates was reported for both free and encapsulated enammonium cations (Fig. 9.12). The results indicated that all encapsulated substrates underwent rearrangement faster than free ones (Table 9.2).

To determine whether the observed rate acceleration originated from the negative charge of the cage, KCl was added in the absence of the cage. This did not lead to significant change in rearrangement rate of free substrate, ruling out the role of the negative charge of the cage in rate enhancement.

Comparing the rate constant of various substrates, (Table 9.2) indicates that introduction of additional groups on the substrate can accelerate the rate of rearrangement of encapsulated substrate. However, this trend was not observed for free substrate. For example, the presence of isopropyl substituent at R^2 in substrate (entry 7 in Table 9.2) increases the rate of rearrangement of

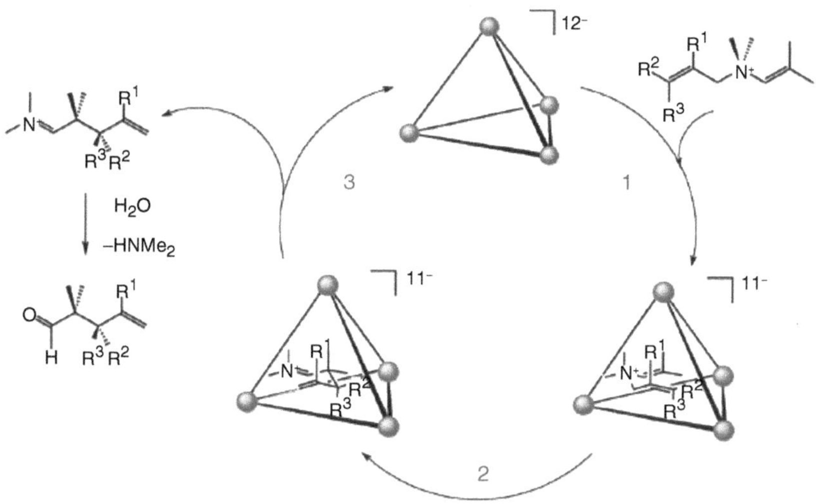

FIGURE 9.12 Proposed catalytic cycle for the cationic 3-aza-Cope rearrangement. *(Reprinted from Ref. [24] with permission of Wiley-VCH Verlag GmbH & Co. Copyright 2004.)*

encapsulated substrate while decreases that of free substrate. It was proposed that with free substrate, an isopropyl group can increase the steric repulsion between the ends of alkyl chain, resulting in decrease of the percentage of reactive conformations in free solution. However, this effect was not observed for the encapsulated counterpart (entry 7 in Table 9.2), and its reaction rate was comparable to those of the other encapsulated substrate. In this case, encapsulation within the cavity of M_4L_6 assembly brings two pendant alkyl chains in close proximity.

The authors investigated the activation parameters for the rearrangement of free and encapsulated substrates to clarify the origin of rate acceleration [24].

TABLE 9.2 Rate Constants for Free (K_{free}) and Encapsulated (K_{encaps}) Rearrangements (Measured at 50°C) and Their Acceleration Factors

Substrate	R^1	R^2	R^3	K_{free}[$\times10^{-5}S^{-1}$]	K_{encaps}[$\times10^{-5}S^{-1}$]	Acceleration
1	H	H	H	3.49	16.3	5
2	Me	H	H	7.61	198	26
3	H	Et	H	3.17	446	141
4	H	H	Et	1.50	135	90
5	H	*n*Pr	H	4.04	604	150
6	H	H	*n*Pr	1.69	74.2	44
7	H	*i*Pr	H	0.37	316	854

Reprinted from Ref. [24] with permission of Wiley-VCH Verlag GmbH & Co. Copyright 2004.

Interestingly, almost identical enthalpies of activation were obtained for both free and encapsulated substrates of entry 3, 23.1(±0.8) $kcalmol^{-1}$ and 23.0 (±0.9) $kcalmol^{-1}$, respectively. However, entropies of activation of free and encapsulated substrates were remarkably different, −8 ± 2 eu and +2 ± 3 eu, respectively.

Encapsulation results in loss of several rotational degrees of freedom of substrate and its preorganization into a reactive conformation. For larger substrates, which fit more tightly in the cavity of M_4L_6 assembly, the effect of preorganization is greater [13]. Product inhibition was not observed in this case. As the neutral product, aldehyde, has lower affinity to metal–ligand cage, it can be displaced by enammonium substrate [24].

2.4.3 Nazarov Cyclization

The Nazarov cyclization of pentamethylcyclopentadienol was catalyzed in the presence of a self-assembled metal–ligand supramolecular capsule $[Ga_4L_6]^{12-}$ at mild temperature [25]. Remarkable acceleration—2.1 million times comparable to that of enzymes—was observed.

The mechanism of catalyzed reaction was investigated comprehensively using ^{18}O-exchange experiments, kinetic analysis of the reaction, and computational studies.

To elucidate the role of encapsulation, the mechanisms of catalyzed and uncatalyzed reactions were compared. The results indicated that encapsulation, protonation, and water loss from the substrate are reversible, while electrocyclization is irreversible (Fig. 9.13).

In the uncatalyzed reaction, the rate-determining step was electrocyclization; in the catalyzed reaction, water loss and electrocyclization faced equal barriers. The observed rate enhancement was ascribed to the transition-state stabilization and enhanced basicity, which originated from encapsulation within the metal–ligand self-assembled capsule.

2.4.4 Hydroalkoxylation of Allene

Tetrahedral Ga_4L_6 (L=*N*,*N*′-bis(2,3-dihydroxybenzoyl)-1,5-diaminonaphthalene) cage was used by Raymond et al. to encapsulate Gold(I)-phosphine complexes, PR_3AuX, (X = halides) [26]. The encapsulated gold complex PMe_3AuX was subsequently used as a catalyst for intramolecular hydroxylation of allenes to produce allylic ether and diene (Fig. 9.14). It was observed that the catalytic activity of encapsulated catalyst, PMe_3AuBr, was improved eightfold. Furthermore, the reaction can be performed in aqueous media if the encapsulated gold catalyst is used. It is worth noting that this study was the first example in which encapsulation of gold complex in a supramolecular cage increased the rate of gold-catalyzed reaction.

In their subsequent report, Raymond et al. used the encapsulated gold complex, Au–Ga_4L_6, in combination with enzymes, esterases, or lipases in a one-pot tandem reaction sequence. Enzymatic ester/imide hydrolysis of allenic ester or amide results in an intermediate [27], which subsequently undergoes

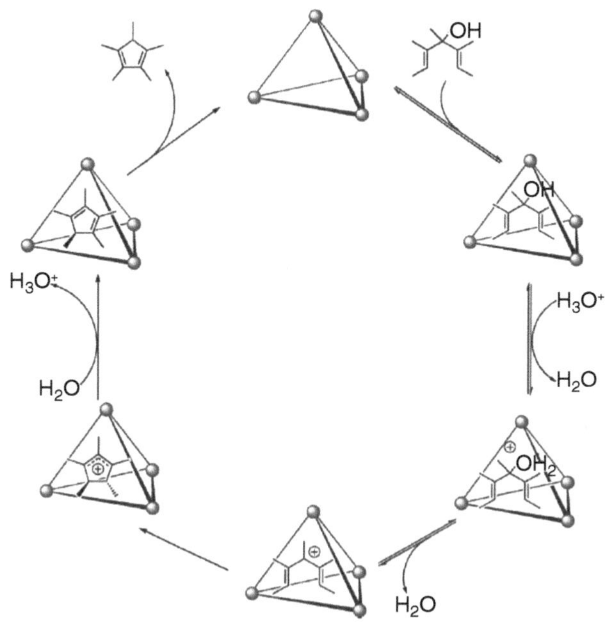

FIGURE 9.13 Proposed catalytic cycle for the coordination cage-catalyzed Nazarov cyclization of 1,4-pentadien-3-ols, with a representative substrate. *(Reprinted from Ref. [25] with permission of Wiley-VCH Verlag GmbH & Co. Copyright 2014.)*

$Me_3PAu^+ \subset 1$-catalyzed intramolecular hydroxyalkylation at the allene group to afford products (Table 9.3) [28].

It was demonstrated that addition of dimethyl sulfoxide (DMSO) or MeOH to the reaction mixture can improve the conversion of 5. Furthermore, increasing the temperature to 37°C allowed the reaction to proceed to complete conversion and excellent yield using only six units of enzyme [27]. Neither the enzyme nor Me_3PAu^+C1 could, in isolation, produce the tetrahydrofuran product; it arose as a product of the system rather than the individual components [28].

The authors have also developed a three-component multicatalytic system in which an alcohol dehydrogenase (ADH) was employed in a sequence with

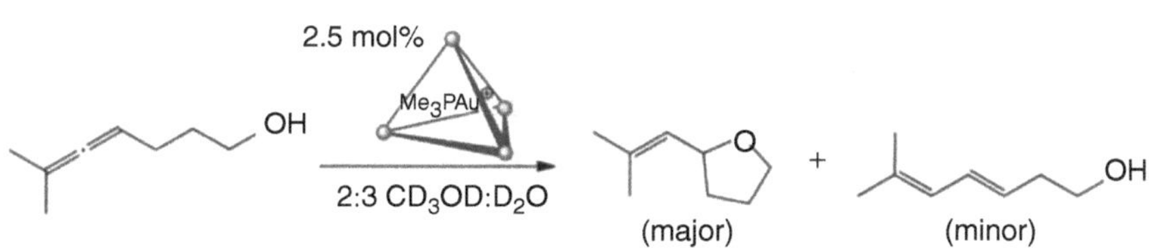

FIGURE 9.14 Hydroalkoxylation of allene 51 to yield allylic ether and diene. *(Reprinted from Ref. [26] with permission of American Chemical Society. Copyright 2011, and Ref. [10]. American Chemical Society. Copyright 2015.)*

TABLE 9.3 Tandem Esterase- or Lipase-Mediated Acetate Hydrolysis Followed by $Me_3PAu^+ \subset 1$- or Me_3PAuCl-Catalysed Hydroalkoxylation

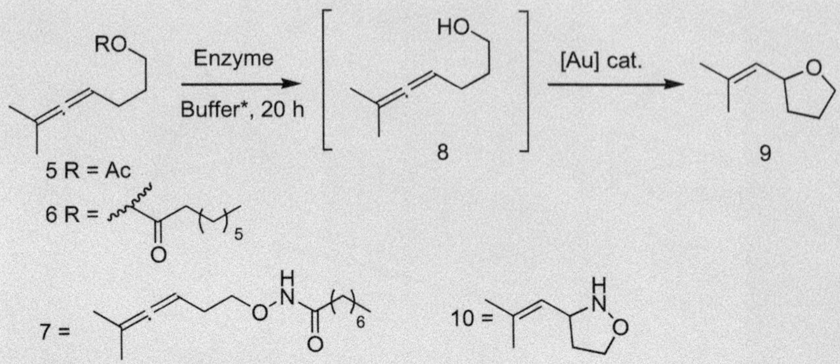

Substrate	Product	Enzyme	[Au] cat.	Product ratio (4:2:3)[a]
5	9	Rabbit liver esterase[b]	$Me_3PAu^+ \subset 1$ Me_3PAuCl	(0:0:100) (14:24:62)
5	9	Hog liver esterase[c]	$Me_3PAu^+ \subset 1$ Me_3PAuCl	(2:4:94) (35:22:44)
5	9	Horse liver esterase[c]	$Me_3PAu^+ \subset 1$ Me_3PAuCl	(4:43:53) (11:35:54)
6	9	Mucor miehei lipase[d]	$Me_3PAu^+ \subset 1$ Me_3PAuCl	(0:0:100) (15:0:65)
7	10	Mucor miehei lipase[d]	$Me_3PAu^+ \subset 1$ Me_3PAuCl	(38:2:60) (32:2:66)

Tris buffer pH 8 was used in the esterase reactions and phosphate buffer pH 7 was used in the lipase reactions. 5% DMSO or MeOH was also used in all the reactions to increase the solubility of 4 in aqueous buffer. The substrate concentration was 2.2 mM. Cat., catalyst.
[a]*Determined by proton NMR spectroscopy (see the supplementary information for NMR yields relative to internal standard).*
[b]*Six units of enzyme used at 37°C.*
[c]*One unit of esterase used at 23°C.*
[d]*Three units of enzyme used at 37°C.*

$(Me_3P)CpRu(NCMe)_2^+$ encapsulated in a coordination cage. Isomerization of allylic alcohols into the corresponding aldehydes was catalyzed by an encapsulated catalyst, and was followed by reduction to alcohol by ADH (Fig. 9.15) [27]. To avoid using a stoichiometric amount of nicotinamide adenine dinucleotide phosphate (NADPH) cofactor, which is required as a fuel in the ADH step, the authors employed coupling of reduction to oxidation of sodium formate by formate dehydrogenase (FDH), such that NADPH could be regenerated in situ. Like the aforementioned system, neither the enzyme couple nor the encapsulated ruthenium catalyst alone could carry out both reactions in the sequence.

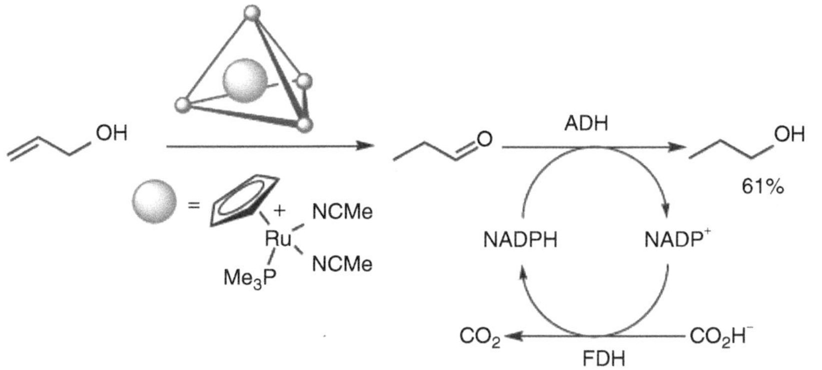

FIGURE 9.15 Ru(II)-mediated olefin isomerization of allyl alcohol to give propanal, followed by reduction to propanol via ADH. *(Reprinted from Ref. [27] with permission of Nature. Copyright 2013.)*

2.4.5 Monoterpene Cyclization

Using the hydrophobic cavity of a M_4L_6 coordination cage, catalytic cyclization of monoterpene (±)-citronellal and two homologues was performed at moderate temperatures and physiological pH.

According to previous reports, under Brønsted acid catalysis, three classes of products can be expected (Fig. 9.16). Four stereoisomeric products 13 and products 12a–d can be generated as major and minor products, respectively. Additionally, product 14 can be formed from condensation of 13 with 11. Using a coordination cage as a catalyst, 11 can be encapsulated within the hydrophobic interior of the cage, as confirmed by ^{1}HNMR technique. It was established that, in contrast to cyclization in acidic aqueous solution, the hydrophobic interior of the cage prevents the capture of reactive intermediates by water [29]. Cyclization followed by elimination resulted in diastereomeric products (Fig. 9.17) [10].

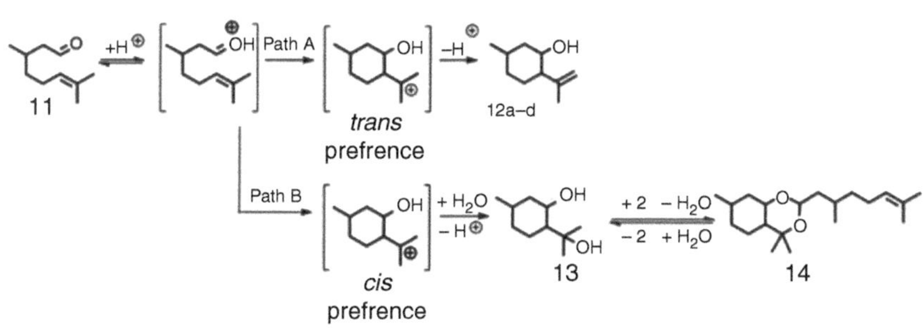

FIGURE 9.16 Proton-mediated cyclization of 2 to products 12–14. Under catalysis by acidic solution, 12–14 are observed with 4 as the major product by path B. Catalysis with 1 affords 12a–d as the major class of products, demonstrating that path A is instead favorable. *(Reprinted from Ref. [29] with permission of American Chemical Society. Copyright 2012.)*

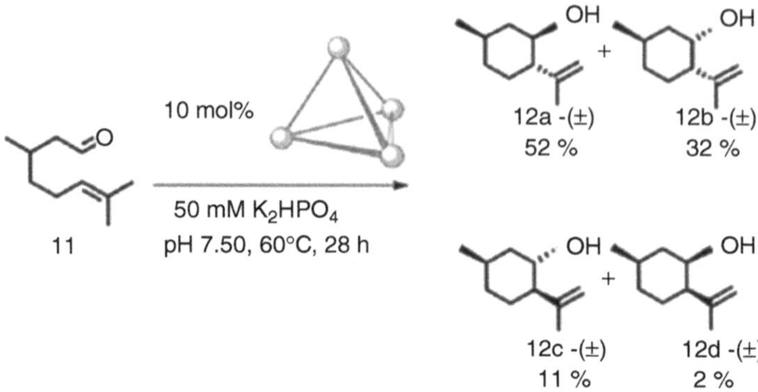

FIGURE 9.17 Selectivity of alkene products from the cyclization of 11 by the coordination cage. *(Reprinted from Ref. [29] with permission of American Chemical Society. Copyright 2012.)*

2.4.6 Self-Organizing Chemical Assembly Line

Nitschke et al. reported self-organization of simple chemical precursors to generate a functional "assembly line" of molecular actors to transform furan into 5-hydroxy-3- (nitromethyl) dihydrofuran-2(3*H*)-one, c, (Fig. 9.18) [30]. In this system, the coordination cage self-assembled in situ in the presence of all other reaction components [28]. In the first cycle, cycle A, furan tolerated the hetero-Diels–Alder cycloaddition, with singlet oxygen photogenerated by methylene blue to form high-energy intermediate endoperoxide. As can be seen in cycle B, the in situ self-assembled cage catalyzes the transformation of endoperoxide into fumaraldehydic acid. In the final cycle, an L-proline-catalyzed 1,4-addition of nitromethane to intermediate **b** and cyclization yield product **c** [30].

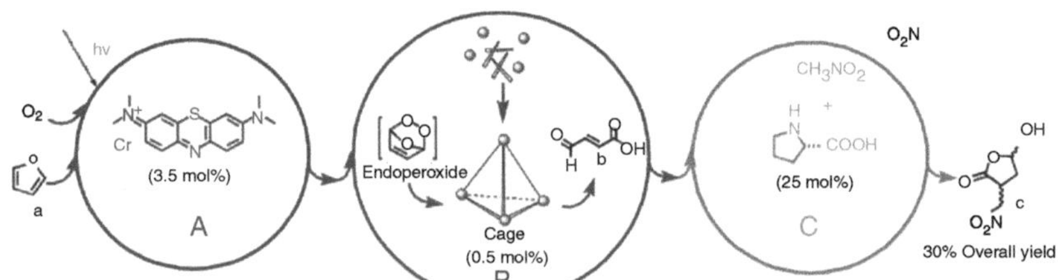

FIGURE 9.18 Relay multicatalytic system, in which all steps take place in water at room temperature at pH 4.0. In this one-pot sequential transformation, reaction of furan a with singlet oxygen (photogenerated by methylene blue (cycle A) gave the corresponding endoperoxide. This high-energy intermediate was transformed into fumaraldehydic acid b in the presence of a catalytic amount of cage (cycle B), which assembled in situ without interference from the other components within the mixture. The L-proline-catalyzed 1,4-addition of nitromethane to b (cycle C) afforded the final product c in 30% overall yield. The derivatization of c allowed the determination of its ratio of enantiomers (er = 84:16). *(Reprinted from Ref. [30] with permission of American Chemical Society. Copyright 2013.)*

3 SELF-ASSEMBLED CAPSULES WITH TWO-DIMENSIONAL LIGANDS

Fujita et al. developed self-assembly of predesigned organic ligands with transition metals having 90° coordination angles. They incorporated *cis*-protected square planar metals, $[(L)M]^{2+}$, into metal-organic frameworks such as cages, cones, capsules, and boxes (Fig. 9.19) [9].

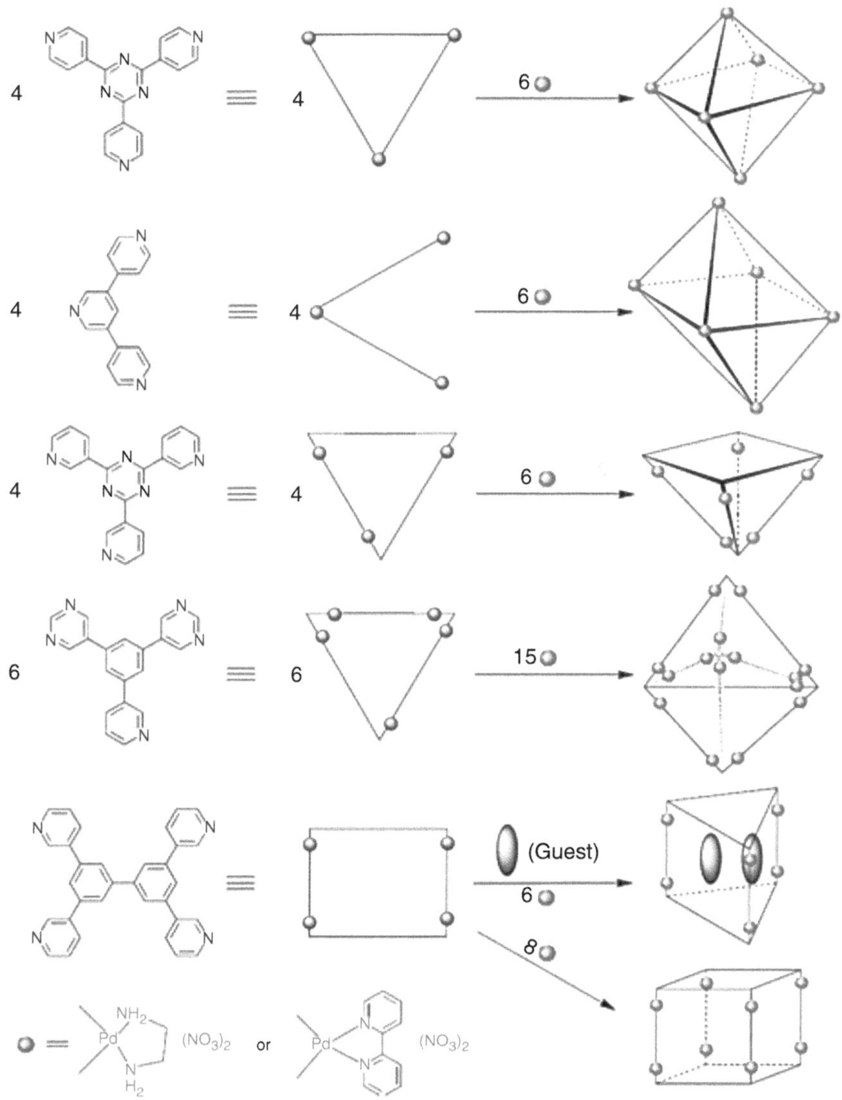

FIGURE 9.19 Examples of coordination cages developed by Fujita et al. *(Reprinted from Ref. [9] with permission of Elsevier. Copyright 2002.)*

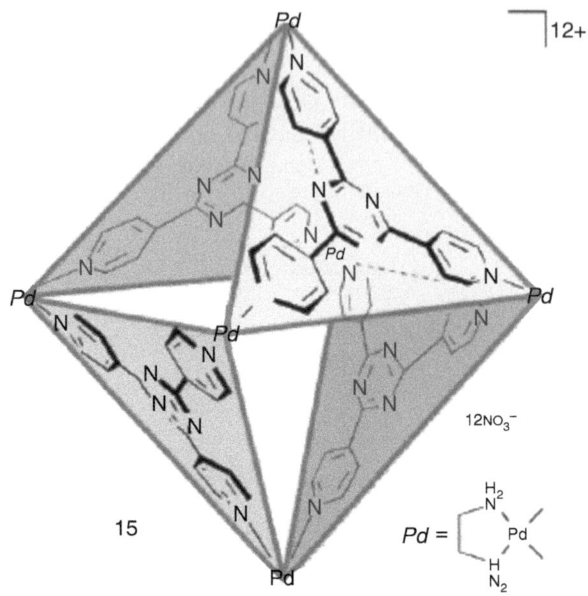

FIGURE 9.20 **Octahedral M_6L_4 coordination cage 15.** *(Reprinted from Ref. [33] with permission of Wiley-VCH Verlag GmbH & Co. Copyright 2012.)*

Using two-dimensional ligands in which panel ligands act as faces rather than the edges of polyhedrons, more enclosed coordination cages were obtained [5].

Water-soluble octahedral $[M_6L_4]^{12+}$ coordination cages (Fig. 9.20) with a large hydrophobic cavity, 15, was obtained from the self-assembly of triangular ligands, 2,4,6-tris(4-pyridyl)-1,3,5-triazine, and a 90° unit block, $[enPd(II)]^{2+}$ in a 6:4 stoichiometry [5]. This nanocage can encapsulate a variety of neutral substrates such as *o*-carborane, adamantine, 1 and 2-adamantol, and tri-*tert*-butylbenzene [31,32]. Encapsulation within the cavity of $[M_6L_4]^{12+}$ coordination cages can improve the water solubility of guests [32].

3.1 Protection and Stabilization

3.1.1 Stabilization of Dinuclear Ruthenium Complexes

Fujita et al. reported encapsulation of dinuclear ruthenium complex $[(\eta^5\text{-indenyl})Ru(CO)_2]_2$ within the cavity of a M_6L_4 coordination cage [34]. Four isomers of this complex are in rapid equilibrium (Fig. 9.21). Besides this, the complex is photosensitive, and CO dissociation and cleavage of labile metal-metal bonds in this complex can occur photochemically. Encapsulation can trap the CO-bridged *cis* configuration, 16a, by suppressing *cis-trans* isomerization and ligand exchange. Furthermore, encapsulation can improve the photostability of the entrapped isomer.

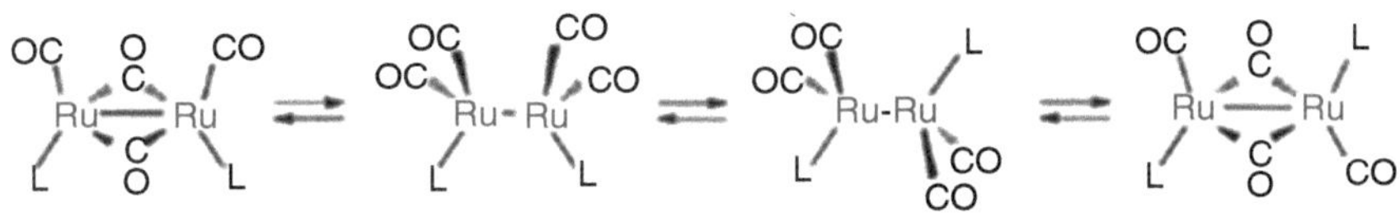

cis-bridged 16a | *cis*-nonbridged 16b | *trans*-nonbridged 16c | *trans*-bridged 16d

FIGURE 9.21 **Equilibrium among the four isomers 7a–d (L = η^5-indenyl).** *(Reprinted from Ref. [34] with permission of American Chemical Society. Copyright 2011.)*

3.1.2 Protection of a Specific Reaction Site

The M_6L_4 coordination cage was also used as a protective group for control of the regioselectivity of nucleophilic substitution of aryl-substituted allylic chlorides [35]. Normally, both α and γ carbons of allylic chlorides can be the sites of reaction and attacked by nucleophile (Fig. 9.22). Therefore, two regioisomeric products are expected in this reaction.

Using encapsulation of substrate within the cavity of a M_6L_4 coordination capsule, the internal reaction site of a substrate can be shielded, and nucleophilic attack will be directed to a less protected terminal site (Fig. 9.22). In this way, regioselectivity of the reaction can be controlled without changing steric and electronic effects of nucleophiles and substrates, or solvent polarity.

3.1.3 Dye Sequestering

Although encapsulation of organic dyes within the cavity of coordination cages improves their solubility and prevents dye aggregation, it also quenches

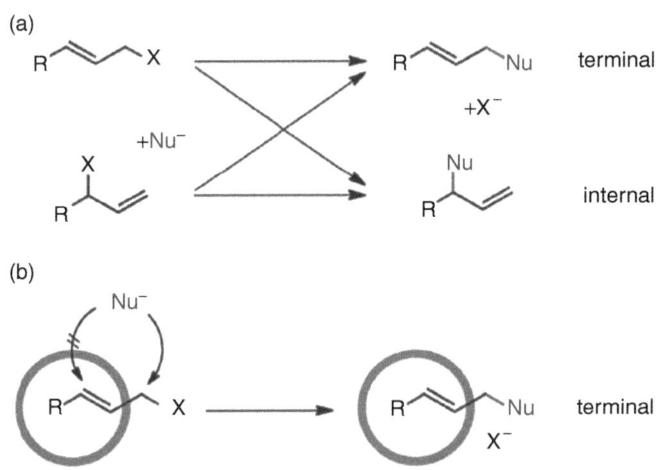

FIGURE 9.22 (a) Nucleophilic substitution of allylic halides. The crossed arrows mean that allylic rearrangement takes place. (b) Schematic representation of noncovalent protection of allylic halides by a host cage to promote the substitution at the terminal reaction site. *(Reprinted from Ref. [35] with permission of Royal Society of Chemistry. Copyright 2012.)*

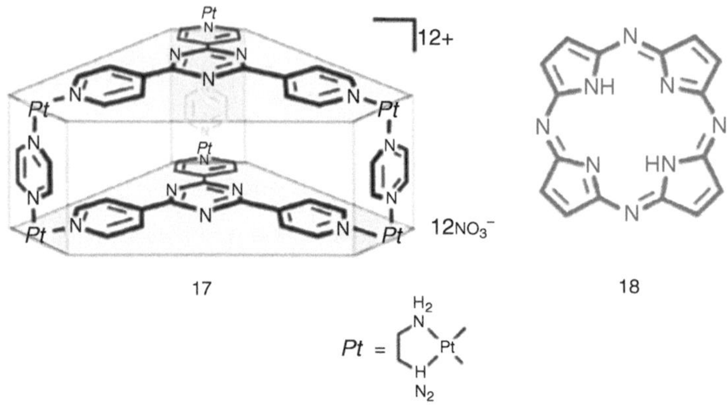

FIGURE 9.23 Structures of coordination cage 17 and tetraazaporphine 9. *(Reprinted from Ref. [36] with permission of American Chemical Society. Copyright 2009.)*

the guest emission. This observation was attributed to energy transfer from electron-rich dye to low-lying LUMO of ligand panels [36]. Using electron-deficient fluorophore, Tetraazaporphine (TAP),18, Fujita et al. demonstrated the utility of coordination cage 17, (Fig. 9.23) for sequestering dye. Interestingly, the charge–transfer interaction between host and guest was suppressed, and encapsulated TAP remained emissive. Furthermore, the water solubility of dye was improved and dye aggregation was prevented. The emission of TAP can be quenched by addition of NEt_3. However, quenching of encapsulated TAP by well-known quenchers such as DMSO and DMF was decreased.

3.1.4 Stabilization of Coordinatively Unsaturated Transition-Metal Complex

Encapsulation of coordinatively unsaturated manganese complex within the cavity of coordination cage of M_6L_4 provided an opportunity to stabilize and observe this labile species by crystallography [37]. Encapsulation of $Cp'Mn(CO)_3$ (Cp′ = methylcyclopentadienyl), 19, within the cavity of a coordination cage resulted in $15.(19)_4$ host–guest product. Photoirradiation of $15.(10)_4$ led to dissociation of one CO ligand from one of the four guests (Fig. 9.24). The measurement of the geometry of the unsaturated transition metal center by X-ray crystallography confirmed the pyramidal geometry of 16-electron unsaturated manganese complex [12].

3.2 Catalysis

3.2.1 Cavity-Directed Synthesis of Labile Silanol Oligomers

Performing selective oligomerization of trialkoxysilane, $RSi(OMe)_3$, using three coordination cages of the same stoichiometry but different cavity sizes

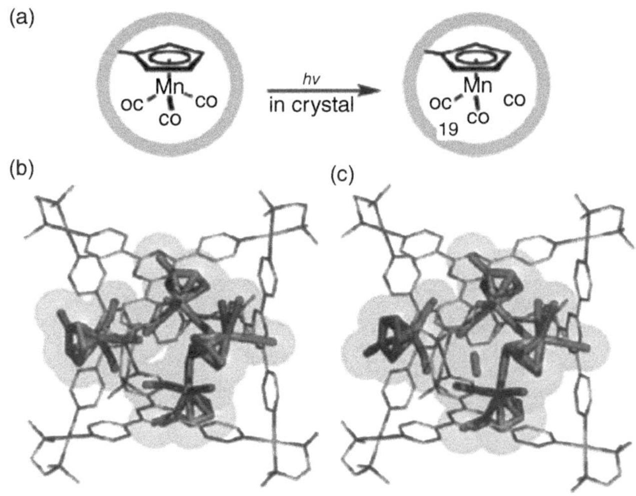

FIGURE 9.24 (a) Photochemical generation of the coordinatively unsaturated species 19 within cage in the crystal. (b) X-ray crystal structures of 15 ⊂ [Cp′Mn(CO)$_3$] before and (c) after photoirradiation; guest: green C, blue N, red CO, orange Mn; host: gray C, blue N, orange Pd. *(Reprinted from Ref. [12] with permission of Wiley-VCH Verlag GmbH & Co. Copyright 2009.)*

and shapes (20, 21, 15), Fujita et al. presented an example of cavity-directed synthesis [38].

It was established that in the presence of tubular coordination, cage 20 (which has the smallest cavity) trialkoxysilan can undergo hydrolysis but not polycondensation. In this strategy, the very labile silanetriol intermediate, 22, can be isolated (Fig. 9.25).

In the case of coordination cage 21, which has a larger cavity and can accommodate two guests, silanol dimer, 23, was obtained as the main product. However, the spherical cavity of cage 15 induced formation of silanol cyclic trimer, 24. This study established the utility of coordination cages' cavities for controlling oligomerization of reactive monomers at desired degrees.

3.2.2 *Photocleavage of α-Diketones*

Another example of cavity-directed synthesis is related to photocleavage of α-diketones in the cavity of a $[M_6L_4]^{12+}$ coordination cage, which results in unexpected cyclization products [39]. Under normal conditions, various degradation products are expected from hemolytic cleavage of α-diketones and formation of acyl radicals. In confined systems, however, kinetically unfavorable pathways without hemolytic cleavage became major pathways.

Fujita et al. encapsulated diphenylethanedione in the cavity of $[M_6L_4]^{12+}$ coordination cage. ^{1}HNMR and single-crystal X-ray diffraction studies established the encapsulation of two α-diketones per cage. After irradiation for 6 h, three main products 25–27 were isolated (Fig. 9.26). The authors

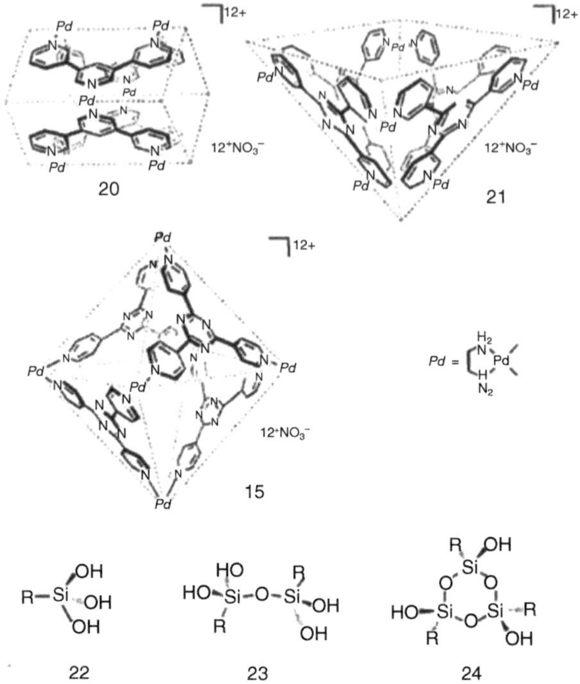

FIGURE 9.25 The structures of cages 20; 21; and 15; and silanetriol, 22; silanol dimer, 23; and silanol cyclic trimer, 24. *(Reprinted from Ref. [38] with permission of American Chemical Society. Copyright 2001.)*

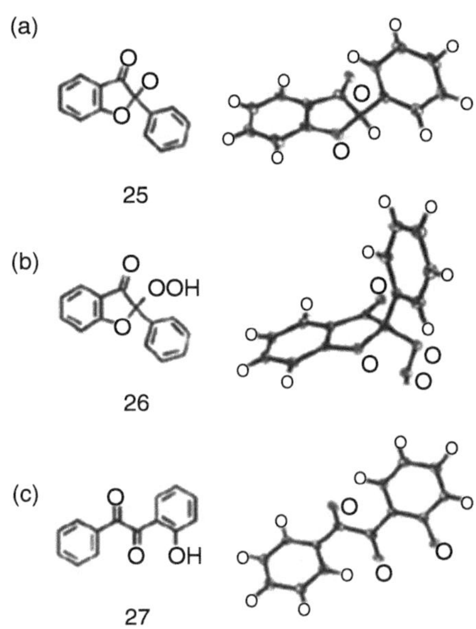

FIGURE 9.26 Thermal-ellipsoid plots (50% probability level) of the molecular structures of (a) 25, (b) 26, and (c) 27. *(Reprinted from Ref. [39] with permission of Wiley-VCH Verlag GmbH & Co. Copyright 2007.)*

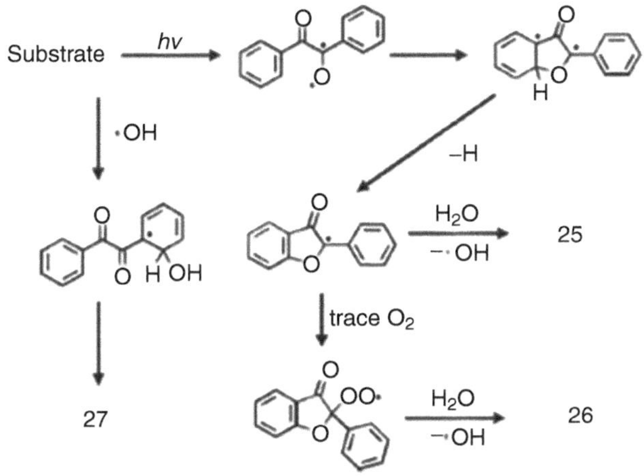

FIGURE 9.27 Proposed reaction mechanisms for the formation of 25, 26, and 27. *(Reprinted from Ref. [39] with permission of Wiley-VCH Verlag GmbH & Co. Copyright 2007.)*

proposed a mechanism based on the formation of a biradical by photoexcitation of the carbonyl n–π* transition (Fig. 9.27). Subsequently, the nucleophilic radical center on the oxygen atom attacked the terminal of the α,β-unsaturated carbonyl moiety of the adjacent benzoyl moiety, producing a cyclized structure through a radical addition–elimination reaction. The resulting carbon radical abstracted the hydrogen radical either directly or after being trapped by molecular oxygen to give the cyclized products 25 and 26, respectively [39].

3.2.3 *Knoevenagel Condensation*

Fujita et al. successfully used a M_6L_4 coordination cage for Knoevenagel condensation of a variety of aromatic aldehydes in aqueous media and under mild reaction conditions [40]. Product inhibition did not take place, due to poor fit of product to cage. Reaction of 2-naphtaldehyde and Meldrum's acid in water or methanol proceeded poorly. However, in the presence of the coordination cage, high yield of condensation product, 96%, was obtained. It was demonstrated that the role of M_6L_4 cage was more than dissolving reagents, and that it could promote the condensation reaction in water. The hydrophobic cavity of the M_6L_4 cage was essential for reaction, and the cage components did not promote the reaction.

The proposed mechanism of this reaction in the presence of a coordination cage is illustrated in Fig. 9.28. It was observed that the reaction of sterically demanding substrates such as 9-anthraldehyde, which furnishes low yields under normal conditions, proceeded efficiently in the presence of the M_6L_4 cage, while small aldehydes (such as benzaldehyde) resulted in low yields. The authors attributed this observation to low binding ability of benzaldehyde to the cage. Electron-rich guests were more preferable, due to cationic characteristics of the cage.

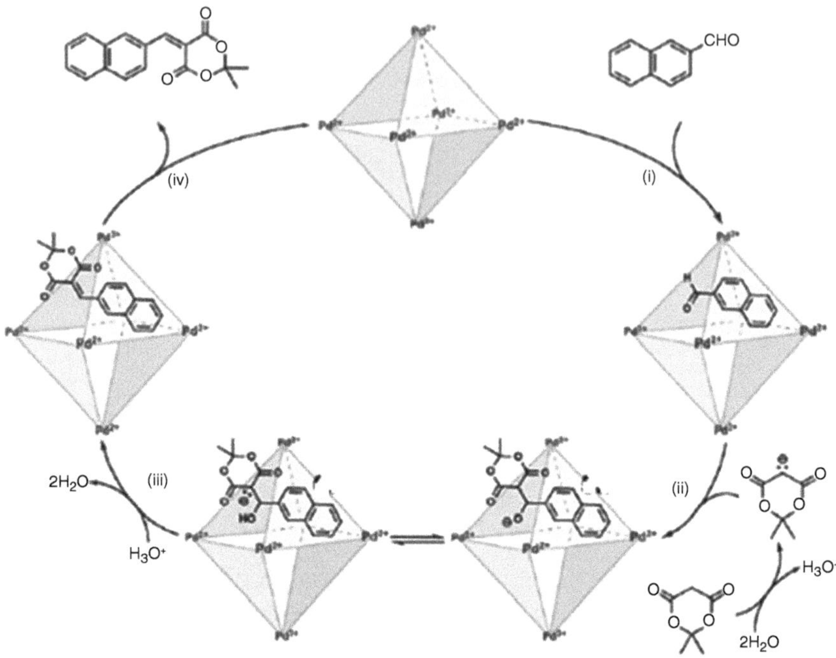

FIGURE 9.28 Proposed mechanism for the catalytic Knoevenagel condensation of aldehyde with Meldrum's acid in the presence of a coordination cage. For simplicity, the encapsulation number of guest molecules is not considered in this catalytic cycle. *(Reprinted from Ref. [40] with permission of American Chemical Society. Copyright 2012.)*

3.2.4 Highly Stereoselective [2 + 2] Photodimerization of Olefins

Fujita et al. exploited the M_6L_4 coordination cage as a molecular flask for performing [2 + 2] cross-photodimerization of olefins, acenaphthylene, and quinone derivatives, all in a very selective manner [41]. In this reaction, the M_6L_4 nanocage could recognize two different olefin molecules in a pairwise selective fashion, resulting in cross-photoisomerization. It is worth noting that under normal conditions (ie, in the absence of the nanocage), cross-photoisomerization had no preference over the homocoupling [6]. The authors believed that the key step in this reaction was the selective formation of a ternary complex of nanocage and two different olefins, governed by the size compatibility of the guests with the restricted space of the cavity. For example, replacing 5-ethoxynaphthoquinone with less sterically demanding 5-methoxynaphthoquinone resulted in formation of hetero [2 + 2] adduct as major product and homo [2 + 2] adduct as homo adduct in low yield.

3.2.5 Diels–Alder Reactions of Inert Aromatic Compounds

Using a coordination cage, Fujita et al. promoted regio- and stereocontrolled Diels–Alder reactions of inert aromatics (such as naphtalenes and triphenylene) with *N*-cyclohexylmaleimide [42]. The results can be explained by preorganization of substrate in the cavity of the cage (which decreases the entropy cost and switches the reaction profile from a bimolecular to a pseudo-intramolecular reaction) and an increase of local concentration within this nanoreactor [42].

Diels–Alder reactions within the cavity resulted in unusual products. In the case of reaction of *N*-cyclohexylmaleimide and anthracenes, the unusual *syn*-adduct was characterized by 2D NMR technique ,which was formed from a Diels–Alder reaction at an electronically unfavorable terminal benzene ring. Replacing bulky cyclohexyl group of *N*-cyclohexylmaleimide with small groups (such as methyl or ethyl) resulted in a reaction at the usual central benzene ring of anthracene, indicating the essential role of sterically demanding *N*-substituent on *N*-cyclohexylmaleimide for determining the reaction site on anthracene. A Diels–Alder reaction of a remarkably inert aromatic compound, naphthalene, was also studied. As entropic cost is the main cause of inertness of naphthalene, it was expected that this reaction proceeded in the cage of the M_6L_4 nanoreactor. It is worth nothing that unsubstituted naphthalene was not an ideal substrate, and it adapted the wrong geometry regarding which reaction sites were separated. However, substitution of proper positions resulted in products with unusual regio- and stereoselectivity. Increasing the bulkiness of substituent (replacing methyl with ethyl group) increased the yield of reaction remarkably. The authors believed that preorganization surpassed the electronic effects of substituent.

3.2.6 Alkane Oxidation

Photochemical excitation of a coordination cage was used for inducing oxidation of photochemically inactive alkane [43]. This process consists of three steps: (1) Encapsulation of an inert guest, (2) photochemical excitation of the cage, and (3) subsequent oxidation of the encapsulated guest (Fig. 9.29).

In an example, adamantane as a guest was encapsulated within the cavity of a $[M_6L_4]^{12+}$ coordination cage and irradiated under aerobic conditions. ^{1}HNMR studies showed the formation of 1-adamantylhydroperoxie and 1-adamantanol. The reaction did not proceed in the absence of the coordination cage or in the presence of only the components of the cage, demonstrating the essential role of the cavity. Using this procedure, photo-oxidation of cyclic alkanes (C_nH_{2n},

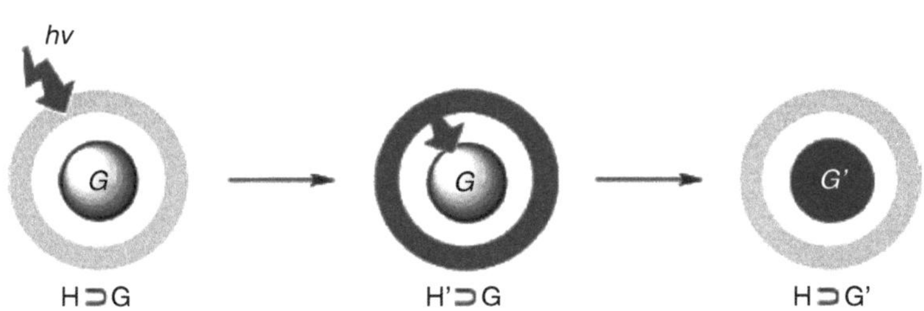

FIGURE 9.29 Photoinduced energy and electron transfer between self-assembled coordination cage and guest. *(Reprinted from Ref. [43] with permission of American Chemical Society. Copyright 2004.)*

$n = 6–8$) to corresponding hydroperoxides and ketones was also performed. The size and shape of alkanes were important, and linear or large cyclic alkanes, which could not be encapsulated, were not oxidized. Besides, this reaction was not mediated by a M_6L_4 bowl-shaped coordination cage, which indicated the importance of host–guest tight contact.

3.2.7 Photo-driven Anti-Markovnikov Alkyne Hydration

A $[M_6L_4]^{12+}$ coordination cage, which can act as electron acceptor, was used by Fujita et al. for anti-Markovnikov photohydration of arylalkynes [44].

The authors proposed a mechanism based on a cage-mediated guest-to-host electron transfer (Fig. 9.30) in which the cage acted as a photosensitizing molecular flask. Excitement of the coordination cage, followed by electron transfer from alkyne to an electron-deficient cage and the reaction of a molecule of water (solvent) with the obtained phenyl alkyne radical cation, results in benzylic radicals and subsequently the anti-Markovnikov product.

3.2.8 Photomediated 1,4-Radical Addition to o-quinones

The utility of the $[M_6L_4]^{12+}$ coordination cage for the control of regio- and stereoselectivity of radical reaction was established by Fujita et al. [45]. Preorganization of *o*-quinone, 28, and bulky toluene 29, in the coordination cage resulted in an unusual 1,4-adduct, 30. The authors proposed that photoexcitation of quinone, and subsequent hydrogen abstraction from the methyl group of toluene, led to

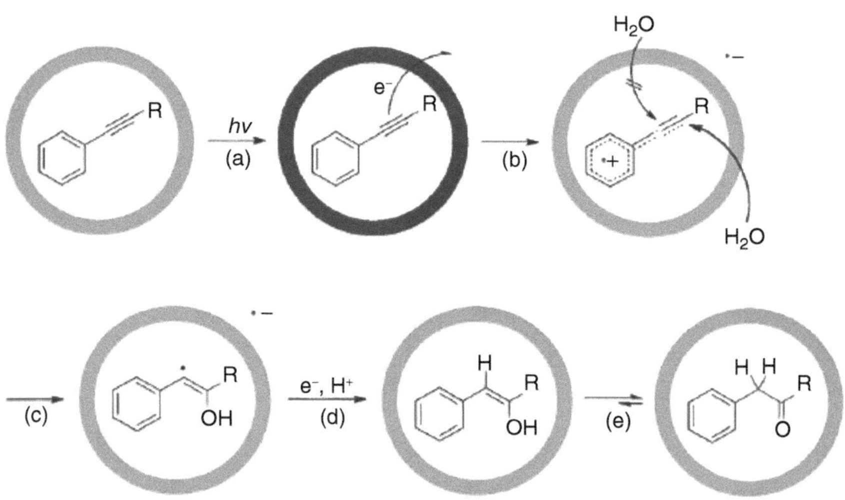

FIGURE 9.30 Plausible mechanism for the photohydration of arylalkynes within a self-assembled coordination cage. (a) Photoexcitation of the coordination cage; (b) guest-to-host electron transfer; (c) hydration of the alkyne (two possible pathways shown); (d) back electron transfer then protonation; (e) keto–enol tautomerization. *(Reprinted from Ref. [44] with permission of Royal Society of Chemistry. Copyright 2011.)*

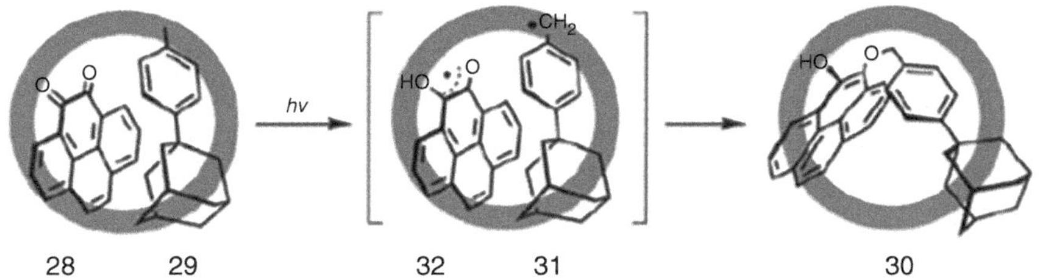

FIGURE 9.31 Photoreaction of quinone 28 with 29 in the cavity of a self-assembled coordination cage. *(Reprinted from Ref. [45] with permission of Wiley-VCH Verlag GmbH & Co. Copyright 2008.)*

formation of benzylic radical 31. Selective recombination of semiquinone, 32, and benzylic radicals resulted in the formation of 1,4 adduct, 30 (Fig. 9.31). The role of the cage in fixing substrate in a specific orientation and favoring a specific reaction pathway is reminiscent of enzymatic reaction control.

3.2.9 Unusual Photoreaction of Triquinacene

Fujita et al. studied photochemical behavior of the encapsulated triquinacene within the coordination cage $[M_6L_4]^{12+}$ [33]. It was demonstrated that the coordination cage can serve as a photochemically active molecular flask that induces unusual photo-oxidation of triquinacene at the peripheral allylic position. Addition of triquinacene to the aqueous solution of the coordination cage resulted in encapsulation of four guests in the cage. After photoirradiation, the product was extracted and characterized by ^{1}HNMR and GC–MS techniques. The results indicated the formation of product 34 (Fig. 9.32). The mechanistic studies were accomplished by performing the reaction under anaerobic conditions and ^{18}O-labling. It was confirmed that dissolved O_2 in aqueous solution was the source of oxygen in the

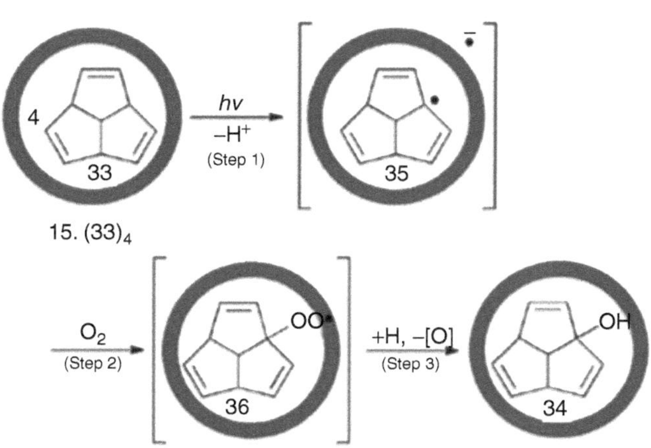

FIGURE 9.32 The proposed mechanism for the formation of 34 within cage 15 via guest-to-host photoinduced electron transfer. *(Reprinted from Ref. [33] with permission of Wiley-VCH Verlag GmbH & Co. Copyright 2012.)*

product. Furthermore, H_2O had no role in photo-oxidation. A plausible mechanism is depicted in Fig. 9.32. It was proposed that, in the first step, there took place an electron transfer from 33 to triazine ligand of the cage, and deprotonation from the allylic position of 33. In the second step, the trapping of 1-triquinacenyl radical, 35, with dissolved O_2 resulted in 1-triquinacenylperoxide radical, 36. The product 34 would be obtained from decomposition of 36 in final step.

Replacing the triazine panel ligand of the cage with benzene did not lead to selective photo-oxidation, indicating the essential role of electron-deficient triazine ligand.

3.2.10 Cavity-Directed Chromism of Dye

Fujita et al. exploited the cavity of the $[M_6L_4]^{+12}$ coordination cage to induce chromism of phthalein dyes 37 [46]. Phthalein dyes are in pink quinone dianion form at basic pH, and colorless lactone form at acidic pH (Fig. 9.33).

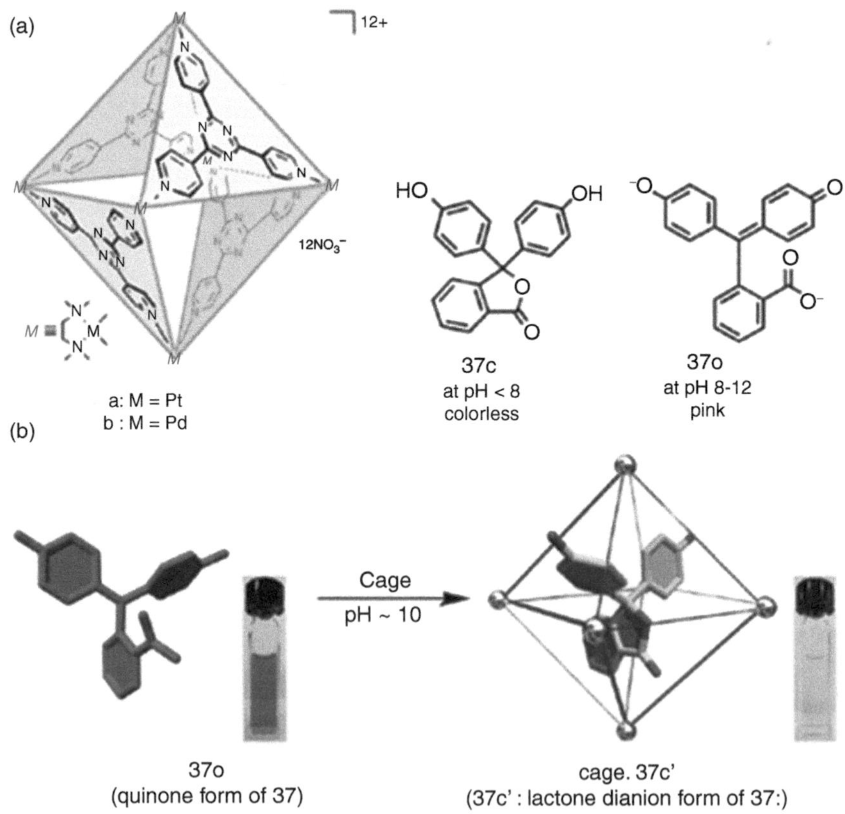

FIGURE 9.33 (a) Chemical structures of cage and phenolphthalein (37c, lactone form; 37o, quinone form). (b) Cartoon representation of the cavity-directed chromism of phenolphthalein. *(Reprinted from Ref. [46] with permission of American Chemical Society. Copyright 2015.)*

As encapsulation of the guest depends on shape and size, tetrahedral lacton, which can better fit the cavity of the coordination cage, encapsulated and stabilized more than quinone. This can shift the equilibrium of two forms toward lactonization and disappearance of pink color. In this way, cavity-directed (and not pH-dependent) chromism can be obtained. Therefore, phthalein dyes remain colorless, even in a basic solution. Upon release of phthalein dyes by the addition of a competitive guest, the pink color will be recovered. The authors believed that this procedure could be used to control the responsiveness of target compounds to external stimuli without any chemical modification [46].

4 GIANT SELF-ASSEMBLED M_NL_{2N} SPHERICAL COMPLEXES

The self-assembly of metal ions with square planar coordination geometry (M) and rigid bent organic ligands (L) can result in a family of roughly spherical polyhedra with a general formula of M_nL_{2n}. For entropically favored regular or semiregular polyhedra, n is limited by simple geometrical constraints to be 6, 12, 24, 30, or 60 (Fig. 9.34) [47,48].

Variation of the ligand bend angle enables the metal ions and the ligands to be mapped effectively onto the vertices and edges, respectively, of the series of polyhedra [48].

Using bent ligands 38a and 38b and Pd (II) ions, Fujita et al. developed 36-component $M_{12}L_{24}$ cuboctahedron 39a,b (Fig. 9.35) [49]. The self-assembled polyhedron structures were characterized by using cold-spray ionization mass

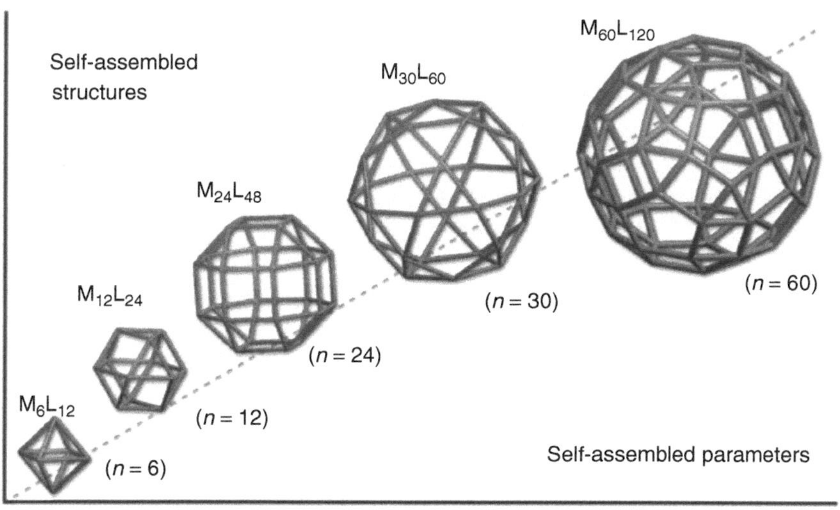

FIGURE 9.34 The family of roughly spherical coordination polyhedra with general formula M_nL_{2n}, where metals (M) and bridging bis(pyridine) ligands (L) are mapped onto the vertices and edges, respectively, of the polyhedra. *(Reprinted from Ref. [48] with permission of Royal Society of Chemistry. Copyright 2013.)*

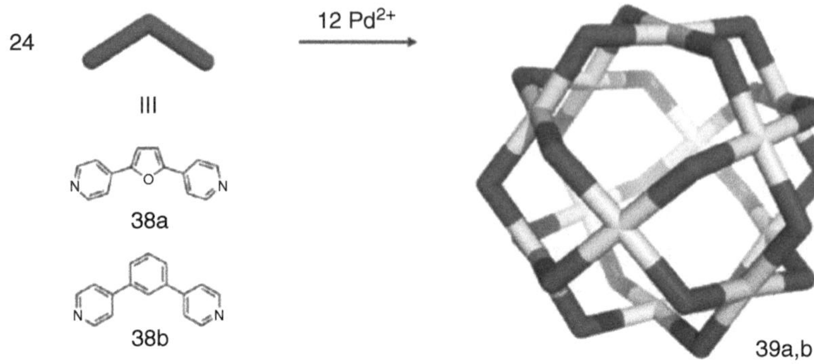

FIGURE 9.35 Self-assembly of $M_{12}L_{24}$ spheres. *(Reprinted from Ref. [48] with permission of Royal Society of Chemistry. Copyright 2013.)*

spectrometry (CSI–MS), Diffusion-ordered NMR spectroscopy (DOSY), and ^{1}HNMR and STM techniques. It was established that the structure formed from self-assembly of ligand 38b behaved as a group of "molecular particles," with precise chemical structure and a uniform dimension (height) of 3.5 nm [49]. Molecular modeling predicted that the diameter of $M_{12}L_{24}$ cuboctahedron formed from an expanded version of ligand of 38b, ligand 40 (Fig. 9.36), would be 5.2 nm. The authors disclosed the possibility of attaching a functional group such as porphyrin on each ligand and achieving 24 functional groups equivalently aligned at the periphery of the sphere.

In this regard, Fujita et al. reported clustering of 24 (oligo) saccharides at the periphery of the self-assembled $M_{12}L_{24}$ sphere. For this purpose, a series of mono-, di-, and trisaccharide-substituted bridging ligands, 41a–f, was synthesized and complexed with Pd(II) ions (Fig. 9.37). Furthermore, the interactions of (oligo) saccharides functionalized $M_{12}L_{24}$ sphere with lectines, concanavalin A (ConA), and peanut agglutinin (PNA) were studied.

ConA, which selectively recognizes α-mannopyranoside and α-glucopyranoside, forms an aggregate with 42a, which has 24 α-mannopyranoside and 42e and 42f containing α-glucopyranoside. In the case of PAN, a galactose-binding lectin from *Arachis hypogaea,* 42b and 42d bearing β-galactopyranoside form aggregates.

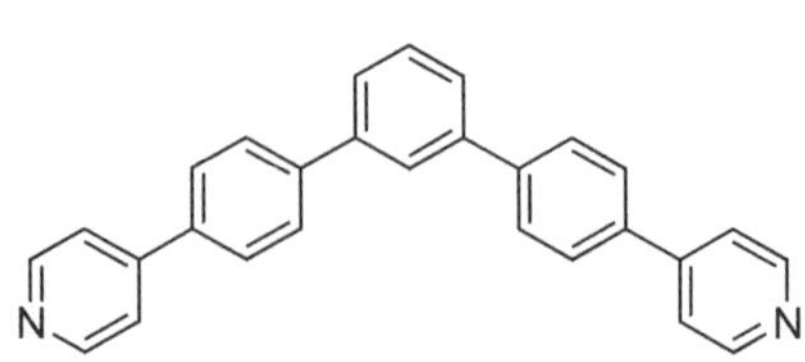

FIGURE 9.36 The structure of ligand 40.

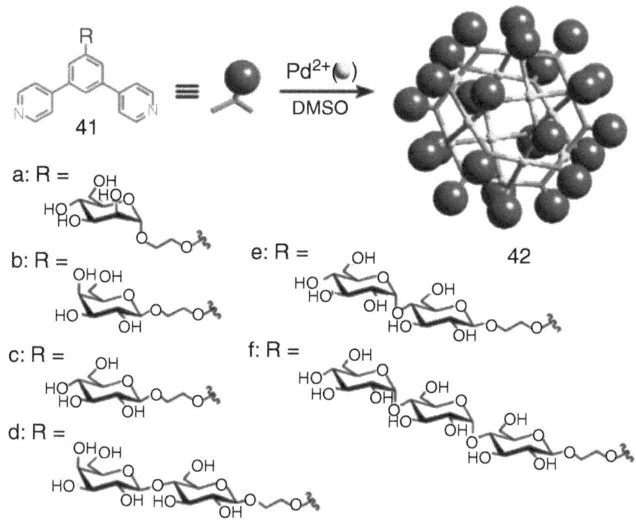

FIGURE 9.37 Self-assembly of $M_{12}L_{24}$ complexes with 24 saccharide moieties at the periphery. *(Reprinted from Ref. [50] with permission of American Chemical Society. Copyright 2007.)*

Recently, the geometrically restricted intermediates in self-assembly of $M_{12}L_{24}$ cuboctahedron have been studied by using time-dependent NMR spectroscopy and CSI–MS analyses [51]. For this purpose, a triazine-based ligand, 43, was synthesized and self-assembled into a $M_{12}L_{24}$ cuboctahedron 44 by using $Pd(NO_3)_2$. The formation of this polyhedron was confirmed by ^{1}HNMR, DOSY, and CSI-MS techniques. To study the intermediates that can exist prior to formation of $M_{12}L_{24}$, ^{1}HNMR spectrum of ligand and $Pd(NO_3)_2$ mixture after 3 days at 80°C, (Fig. 9.38b) was compared with that of the mixture of ligand and metal at room temperature after 30 min (Fig. 9.38c). The observed additional peaks in Fig. 9.38c were assigned to metastable intermediates. These studies, as well as CSI–MS analysis, confirmed two intermediates: first the short-lived M_8L_{16} and then the relatively long-lived M_9L_{18}. Heating intermediates for 1h at 60°C resulted in disappearance of M_8L_{16} selectively over M_9L_{18}. Additional heating for 6 h at 80°C caused the remaining M_9L_{18} to disappear as well. This indicated that these two species are intermediates but not side products. Finally, by using a ligand with smaller bend angle, 112°, ligand 45 (Fig. 9.39), M_8L_{16} and M_9L_{18} species were kinetically trapped because their framework distortions were considerably reduced.

Another member of this giant spherical complex is 72-component $M_{24}L_{48}$ rhombicuboctahedra, which can form from self-assembly of square planar Pd^{2+} ions and bent ligands with a bend angle of 135° (Fig. 9.40) [48,52]. It is noteworthy that switching between two spherical complexes of $M_{24}L_{48}$ and $M_{12}L_{24}$

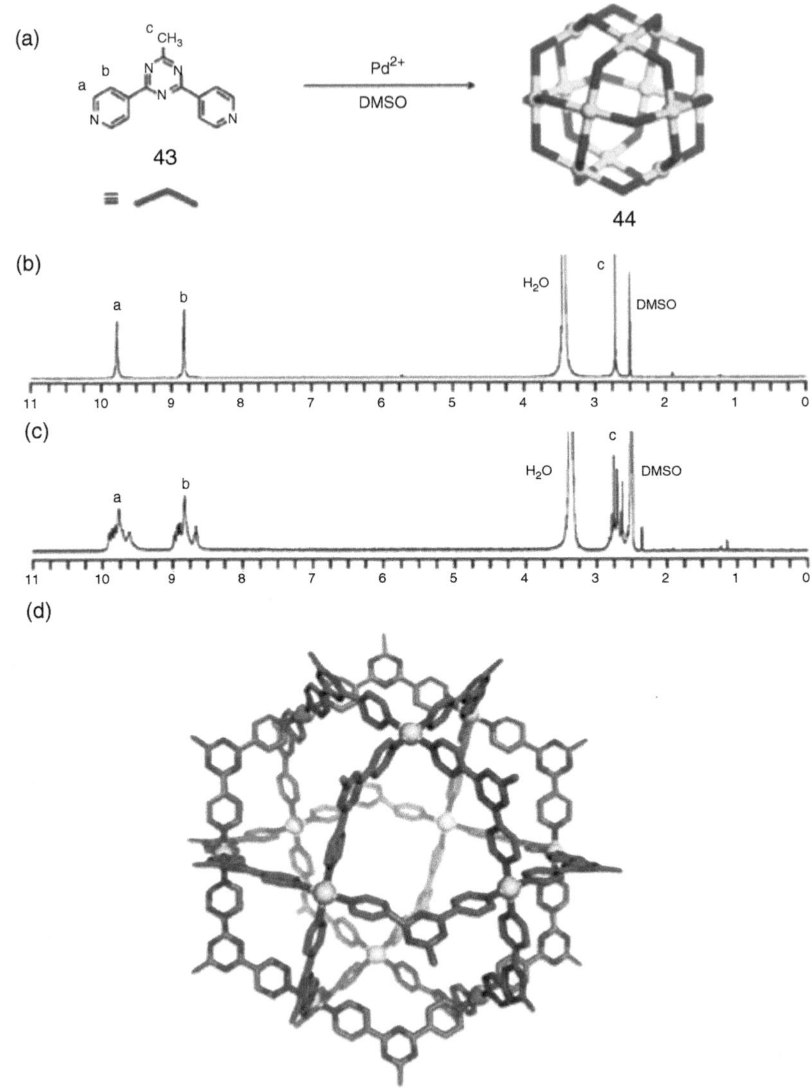

FIGURE 9.38 (a) Schematic representation of the self-assembly of $M_{12}L_{24}$ complex 44. (b,c) ^{1}HNMR spectra (500 MHz, [D$_6$]DMSO, 300 K) after mixing ligand 43 and Pd^{II} at (b) 80°C for 3 days, and (c) rt for 30 min. (d) X-ray crystal structure of 44 (BF_4^- salt). Hydrogen atoms, counter ions, and solvent molecules have been omitted for clarity. *(Reprinted from Ref. [51] with permission of Wiley-VCH Verlag GmbH & Co. Copyright 2014.)*

can be achieved by changing the ligand bend angle. Using a furan-based ligand with a bend angle of ~120°, a $M_{12}L_{24}$ sphere self-assembled, while applying thiophene-based ligand with a wider bend angle resulted in self-assembly of a $M_{24}L_{48}$ sphere.

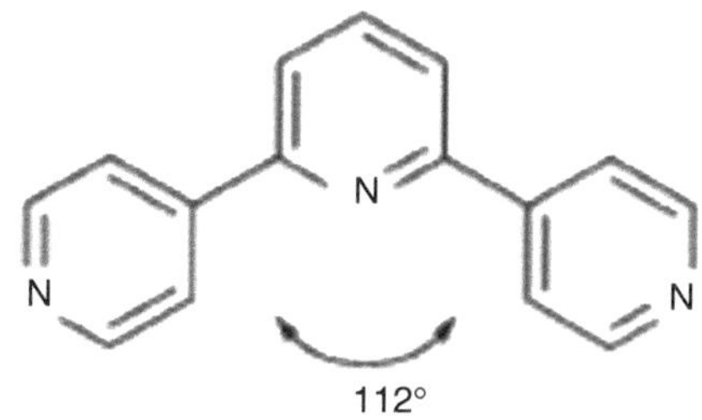

45

FIGURE 9.39 Structure of ligand 45. *(Reprinted from Ref. [51] with permission of Wiley-VCH Verlag GmbH & Co. Copyright 2014.)*

To elucidate which structure self-assembles in the mixture of ligands, Pd^{2+} was added to mixtures with various ratios of ligands 38 and 46 (Fig. 9.40). The results indicated that either $M_{12}L_{24}$ or $M_{24}L_{48}$ complex formed. A difference of only 3° in the mean ligand bend angle (between 131° and 134°) behaves as a "critical structural switch" determining the structure of the resulting assembly [48].

In this regard, the structures of self-assembled complexes formed from Pd^{2+} and three new ligands with angles between those of 38 and 46 were predicted empirically [53]. It was demonstrated that for all three new ligands, $M_{24}L_{48}$ complex was formed exclusively and no mixture of $M_{12}L_{24}$ and $M_{24}L_{48}$ was observed (Fig. 9.41).

Another member of this family is a M_6L_{12} coordination cube that can form from self-assembly of 6 metal ions and 12 bent ligands with a 90° bend angle. Fujita et al. reported the synthesis and characterization of a Pd_6L_{12} molecular

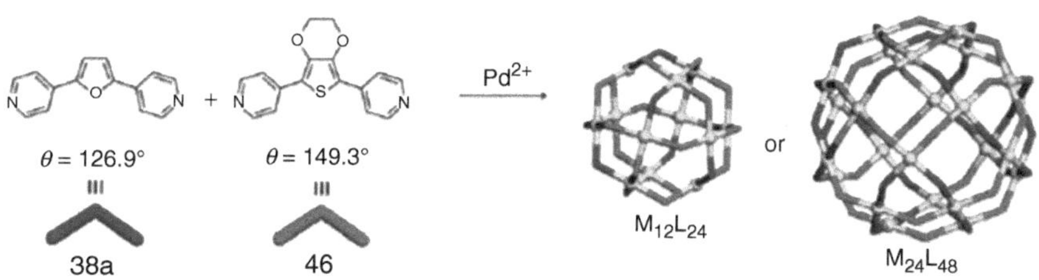

Mixing ratio (38a:46)	$\bar{\theta}$	Product
8 : 2	131.4	$M_{12}L_{24}$ only
7 : 3	133.6	$M_{24}L_{48}$ only

131° < threshold < 134°

FIGURE 9.40 Self-organization of mixtures of furan-cored ligand 38a and thiophene-based ligand 46 exclusively into either M_12L_{24} or $M_{24}L_{48}$ structures, with a critical structural switch occurring at an average bend angle between 131 and 134°. *(Reprinted from Ref. [48] with permission of Royal Society of Chemistry. Copyright 2013.)*

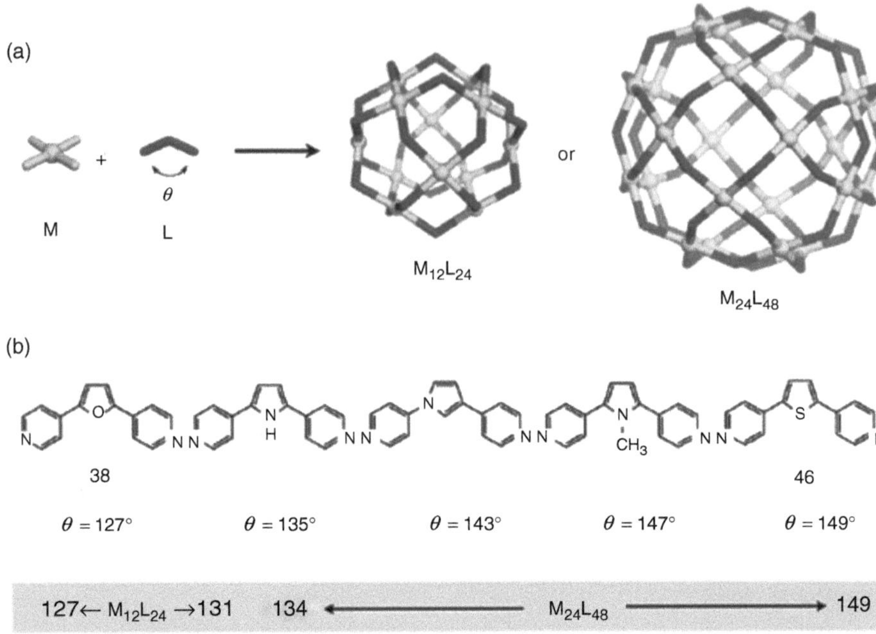

FIGURE 9.41 Self-assembly of $M_{12}L_{24}$ and $M_{24}L_{48}$ polyhedra in which the metal centers define a cuboctahedron and a rhombicuboctahedron, respectively. (a) Representation. (b) Structural formulae of ligands with their bend angles. Angle spans used to predict the formation of $M_{12}L_{24}$ and $M_{24}L_{48}$ are indicated below the formula. *(Reprinted from Ref. [53] with permission of Wiley-VCH Verlag GmbH & Co. Copyright 2012.)*

cube using 12 bidentate bent ligands 47 and 6 Pd (II) ions (Fig. 9.42) [54]. The large cavity and remarkable stability of this cube was promising for guest encapsulation, recognition, and reaction in the cavity.

Fujita et al. reported orthogonal self-assembly of 24 ligand 48, which consists of two effectively identical coordination sites, A and B, linked with triethylene glycol with 24 Pd^{II} ions into a giant sphere-in-sphere complex (Fig. 9.43) [55]. The structure and composition of this sphere-in-sphere complex were characterized by using several techniques, such as cold spray ionization time-of-flight mass spectrometry (CSI–TOFMS), DOSY, ^{1}HNMR, and single-crystal X-ray analysis. In the crystal state, the two independent spherical frameworks of 49, with diameters of 6.3 and 3.5 nm, were clearly visible. Similarly to double-shelled virus capsids, the formation of the inner and the outer shells was orthogonal and simultaneous [55].

In another attempt, Fujita et al. designed and synthesized a $M_{18}L_{24}$ stellated polyhedron with a concave surface. Complexation of Pd (II) with ligand 50, which possesses three coordination sites, results in $M_{12}L_{24}$ cuboctahedral core 51 with a M: L molar ratio of 1:2 in which two coordination sites which attached directly to benzene participate in self-assembly (Fig. 9.44). Upon addition of more

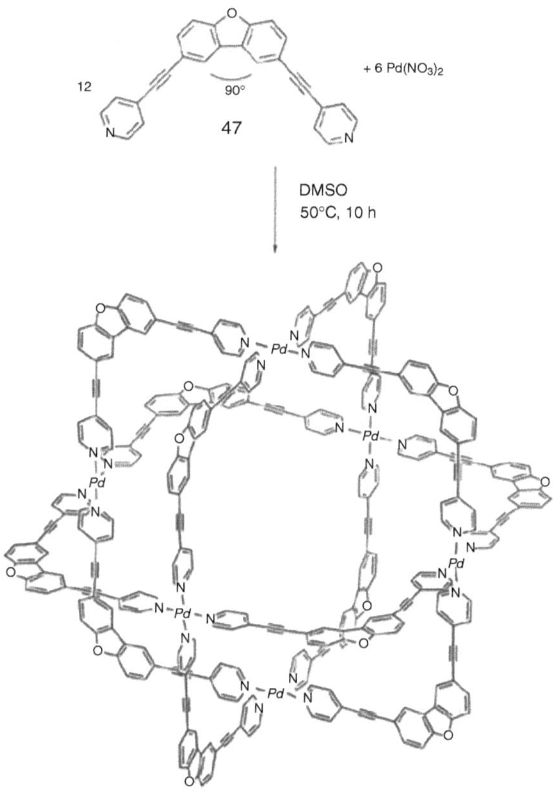

FIGURE 9.42 Self-assembly of M_6L_{12} cubic. *(Reprinted from Ref. [54] with permission of Royal Society of Chemistry. Copyright 2009.)*

Pd (II), the remaining red coordination sites can participate in the coordination and consequently convert the $M_{12}L_{24}$ cuboctahedral into $M_{18}L_{24}$ stellated cuboctahedron 52 (Fig. 9.45). This process can be reversed by removal of six Pd(II) ions from the vertices by the addition of *N,N,N′,N′*-tetramethylethylenediamine (TMEDA). This behavior is reminiscent of the analogous metal-triggered gate opening–closing switches found in spherical virus capsids [56].

4.1 Guest Encapsulation in Spherical Complexes

Using giant $M_{12}L_{24}$ spherical complex 55, Fujita et al. reported encapsulation of a small protein, ubiquitin [57]. For this purpose, protein was attached to one of 24 bidentate ligand 54 to form ligand 53. The self-assembly around the protein was achieved through addition of Pd(II) ions and additional bidentate ligands 54 (Fig. 9.46). The protein-encapsulated structure could be well analyzed by NMR spectroscopy, ultracentrifugation, and X-ray crystallography, coupled

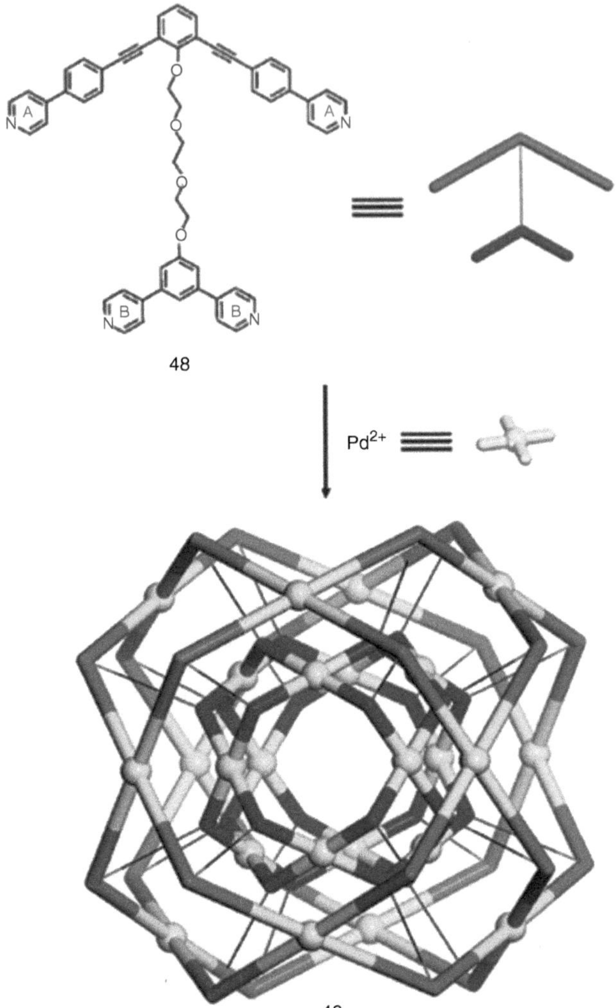

49

FIGURE 9.43 Self-assembly of sphere-in-sphere complex 49. *(Reprinted from Ref. [55] with permission of Wiley-VCH Verlag GmbH & Co. Copyright 2011.)*

with the maximum-entropy method that clearly mapped the electron density of the protein in the coordination cage [57].

4.2 Solubility Enhancement by Using Spherical Complexes

To enhance the solubility of C_{60}, Fujita et al. developed a nanoscopic coronene pseudosolvent phase within a $M_{12}L_{24}$ sphere. Using bidentate ligand 57a with pendant coronene, the team synthesized $M_{12}L_{24}$ sphere 58 with 24 large aromatic coronene molecules located inside the 4.6 nm-sized shell framework

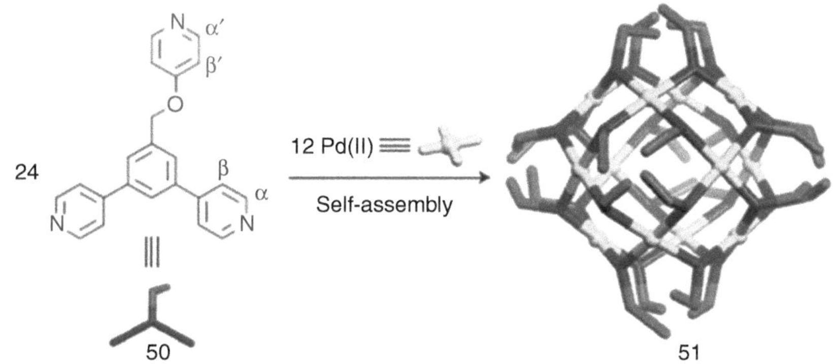

FIGURE 9.44 **Synthesis and structural analyses of cuboctahedron 51.** Cartoon illustration of the self-assembly of 51. *(Reprinted from Ref. [56] with permission of Nature. Copyright 2012.)*

(Fig. 9.47). The solubility of C_{60} within the coronene pseudosolvent phase was much greater (~30 times) than bulk toluene [58]. It's worth mentioning that the aromatic inner phase is essential for dissolving C_{60} and solubility was not achieved by using $M_{12}L_{24}$ sphere formed from ligands containing methyl groups instead of coronene molecules. Furthermore, a sphere with 24 chloronaphthalene molecules was not effective for dissolving $C_{60.}$ The authors suggested that the total aromatic surface area of the 24 chloronaphthalenes was not enough to efficiently dissolve C_{60} within the sphere [58].

4.3 Cavity-Templated Synthesis

4.3.1 Polymerization

A spherical self-assembled $M_{12}L_{24}$ complex bearing 24 methyl methacrylate (MMA) units in its interior was prepared from ligands 59a–d by Fujita et al.

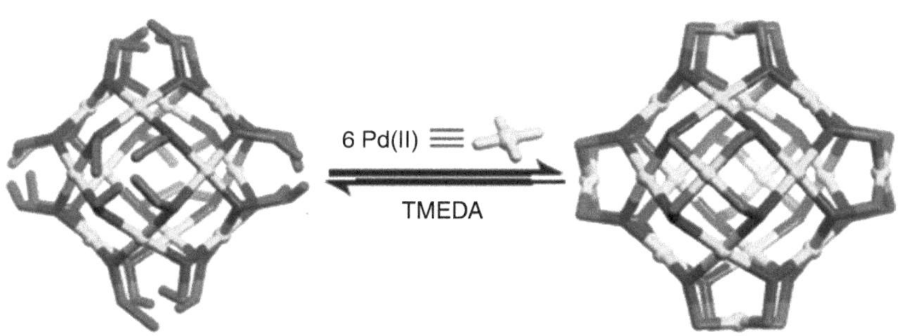

FIGURE 9.45 **Reversible conversion between an $M_{12}L_{24}$ cuboctahedron (51) and an $M_{18}L_{24}$ stellated cuboctahedron (52) through gate opening–closing.** *(Reprinted from Ref. [56] with permission of Nature. Copyright 2012.)*

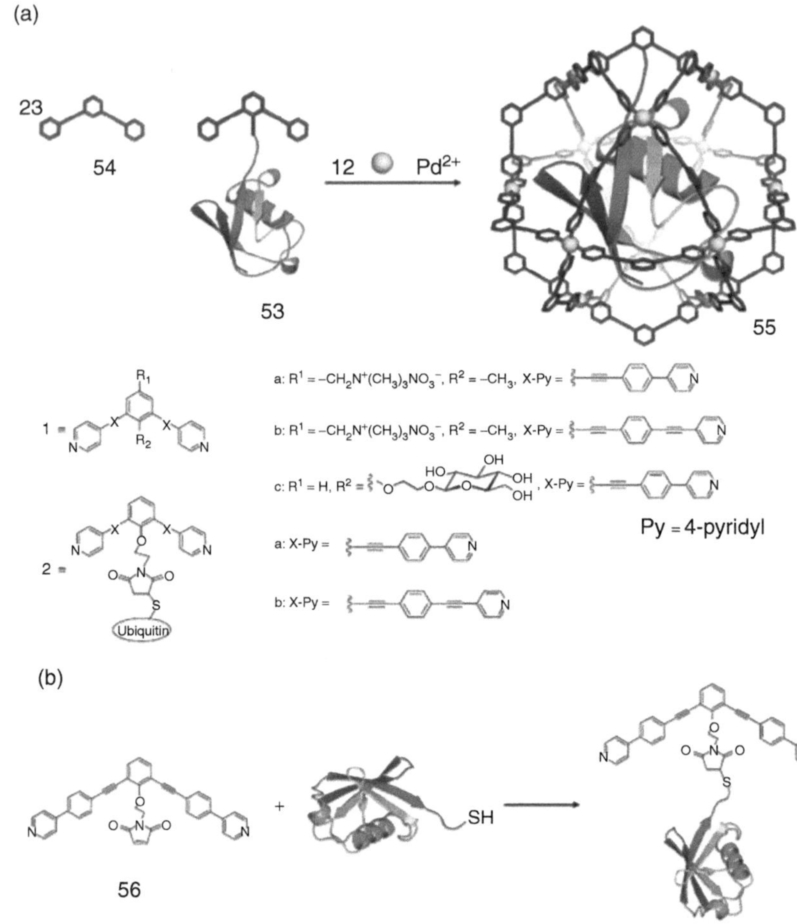

FIGURE 9.46 Schematic representation of the encapsulation of ubiquitin. (a) Self-assembly of ubiquitin-containing spheres 55. Ubiquitin was encapsulated within 55 by assembly of ligand 54 around ubiquitin ligand 53 upon addition of Pd(II) ions at 45°C for 3 h. (b) Preparation of ubiquitin-attached ligand 53a from 56a and Gly76Cys-mutated ubiquitin. *(Reprinted from Ref. [57] with permission of Nature. Copyright 2012.)*

(Fig. 9.48) [59]. Radical polymerization in the sphere was performed by using 2, 2′- azobis(isobutyronitrile) as radical initiator. It was demonstrated that endohedral radical polymerization can be promoted due to the increased concentration of monomer units at the core of the complex [48]. Radical polymerization in 60c resulted in conversion of 73% of MMA units into PMMA oligomers, while polymerization in spheres 60a, 60b, and 60d led to 22, 29, and 62% conversion, respectively. The superior results in the case of 60c (n = 3) were attributed to close packing of 24 MMA units at the core of the sphere. The long linker, n = 4, caused MMA units to repel each other and lie away from the center

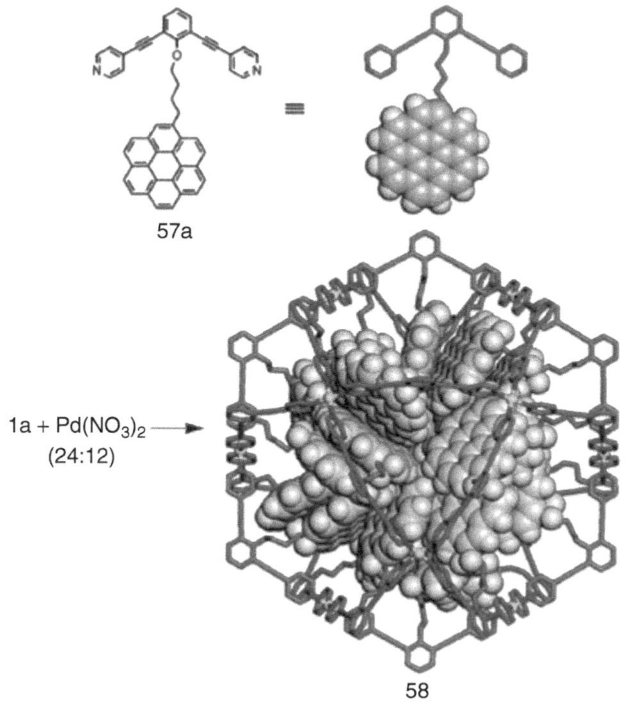

FIGURE 9.47 **Self-assembly of $M_{12}L_{24}$ coordination sphere 58 with 24 pendant coronene molecules.** *(Reprinted from Ref. [58] with permission of American Chemical Society. Copyright 2010.)*

of the complex, while short linkers (n = 1 and 2) prevented the frequent close approach of MMA units [48].

Another example of polymerization in self-assembled spherical complexes is related to polymerization of an anionic monomer within the nanocavity of a highly cationic $M_{12}L_{24}$ coordination sphere with 24 ammonium cations on its interior surface [60]. The highly cationic nanoenvironment of this nanocage was exploited for controlling radical polymerization of anionic sodium *p*-styrenesulphonates. It was demonstrated that a cationic sphere can have electrostatic interactions with anionic guests and encapsulate them within its cationic interior.

The valance of anionic guests influences the strength of cation–anion interaction remarkably.

Furthermore, performing radical polymerization inside the sphere increased the rate of polymerization, and the molecular weight of the resultant polymer was controlled, depending on the number of positive charges on the template [60].

4.3.2 Templating Synthesis of Nanoparticles

Using a well-defined framework of self-assembled hollow spherical compounds as an "endotemplate," Fujita et al. reported synthesis of highly monodispersed

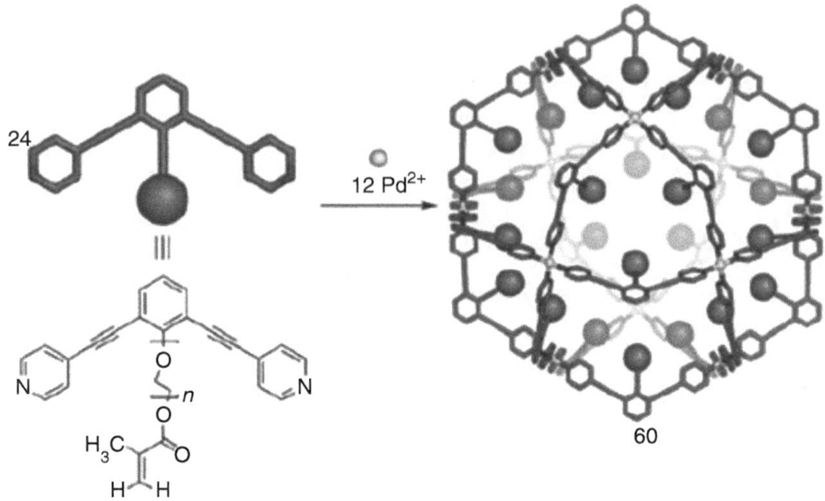

59 a(n = 1)
b(n = 2)
c(n = 3)
d(n = 4)

FIGURE 9.48 Self-assembly of $M_{12}L_{24}$ molecular nanoparticles 60 with 24 MMA units. *(Reprinted from Ref. [59] with permission of Wiley-VCH Verlag GmbH & Co. Copyright 2007.)*

silica nanoparticles with a diameter of 2–4 nm [61]. Upon mixing Pd $(NO_3)_2$ and bis(4-pyridyl)-substituated bent ligand 61, the $M_{12}L_{24}$ organometallic sphere 62 self-assembled (Fig. 9.49). To modify the periphery and the interior of the sphere, 24 pendant glucose units were attached to ligand before self-assembly. Tetramethoxysilane, TMOS, was used as a precursor of silica nanoparticles. It was confirmed that in the presence of the sphere with interior glucose, 62a and 62c, all TMOS react within the sphere. However, in 62b where glucose was replaced with methyl group, condensation proceeded in the bulk solvent. Furthermore, the size of the organometallic sphere controlled the size of silica nanoparticles. The more expanded the diameter of the sphere, the larger the nanoparticle that will be obtained. The authors believed that this method could be generalized for synthesis of various metal oxides, metal-clusters, and metal-organic frameworks with a diameter of less than 10 nm.

In another example, highly monodispersed (M_w/M_n <1.01) core-shell nanoparticles with monodispersed silica core nanoparticle, and a metal oxide (TiO_2 or ZrO_2) layer (SiO_2/TiO_2, SiO_2/ZrO_2) with a diameter of 3 nm, were synthesized by using a $M_{12}L_{24}$ self-assembled sphere as a template (Fig. 9.50) [62]. The protective framework of the template rendered core-shell nanoparticles stable in solution. Neither aggregate formation nor changes in molecular weight were observed, even after several months in solution [62].

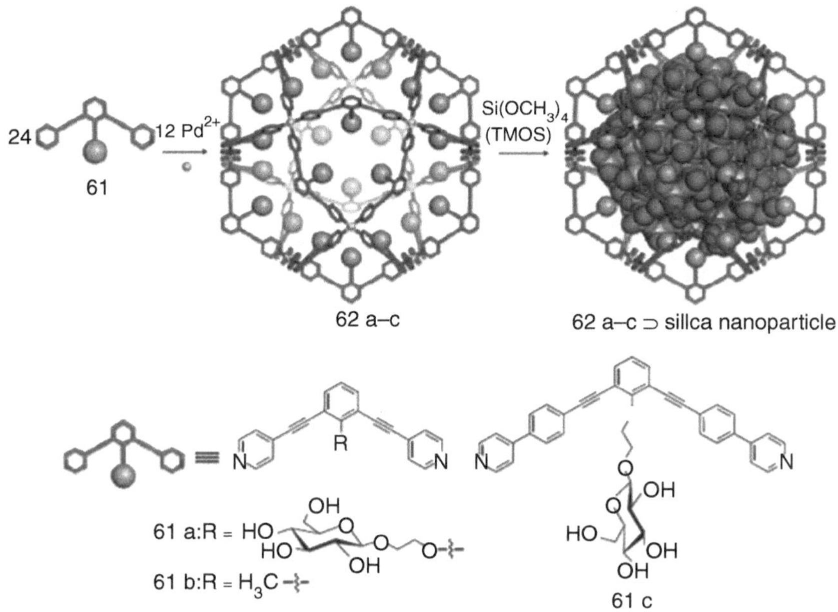

FIGURE 9.49 Synthesis of silica nanoparticles within spheres 62. Spheres 62a,b (4.6 nm) and 62c (6.3 nm) quantitatively self-assembled from ligands 61a–c and Pd(II) ions in DMSO. The sol–gel condensation of TMOS was carried out in the presence of 61a or 61c (DMSO–$CHCl_3$–H_2O, room temperature, four days) to form monodispersed silica nanoparticles within the spherical shells of 62a,c. *(Reprinted from Ref. [61] with permission of Nature. Copyright 2010.)*

5 MISCELLANEOUS COORDINATION CAGES

Recently, Yoshizawa et al. developed a new M_2L_4 molecular capsule with an aromatic shell and a cavity with a diameter of ~1 nm capable of encapsulation of medium and very large molecules such as C_{60}. The distinguishing features of this capsule, which make it different from previous M_2L_4 coordination cages with wire-like frameworks (Fig. 9.51) are (1) its controlled solubility and crystallinity, which can be achieved through attachment of exterior functional groups, (2) its aromatic shell, which is provided with extended aromatic panels, and (3) its rigid and curved ligand, which provides a large cavity [63].

Mimicking fullerenes as spherical molecules with aromatic shells, anthracene was used for aromatic panels of the ligands of the so-called capsule. The structure of rigid bent bridging bispyridine ligand 63, which consists of two anthracene rings connected by an *m*-phenylene spacer, is displayed in Fig. 9.52. The pendant alkoxy groups improve the solubility of the capsule in protic solvents. The self-assembly of Pd (II) ions and ligand 63 in 1:2 ratio results in formation of a M_2L_4 capsule, 64. Encapsulation can improve the solubility of guests in protic solvents dramatically.

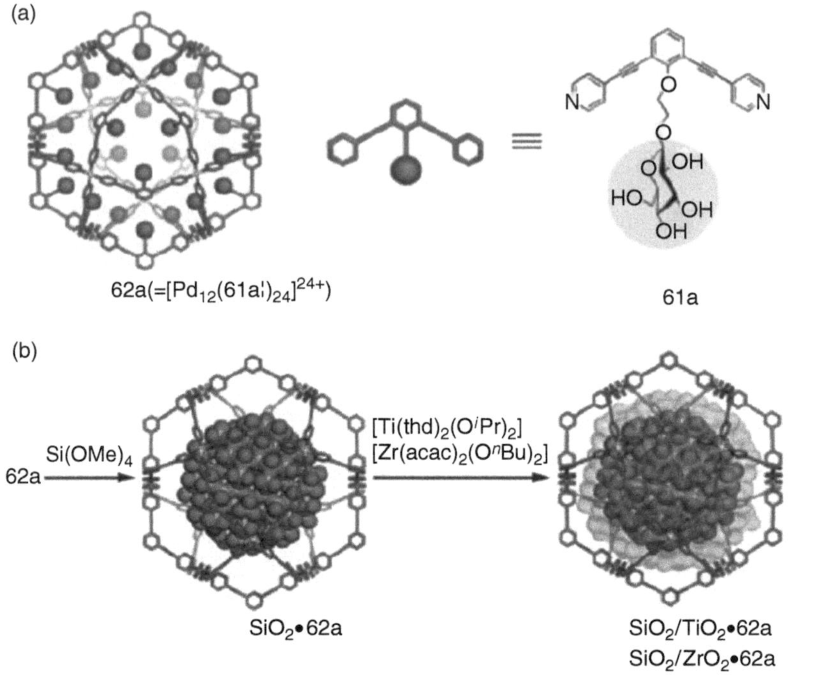

FIGURE 9.50 (a) Structures of 62a and 61a. Yellow spheres in compound 62a denote Pd^{2+} ions. (b) Schematic representation of the template-directed synthesis of core–shell nanoparticles within hollow sphere 62a. The $ROCH_2CH_2O$ pendant groups (R = β glucosyl) are omitted for clarity. *(Reprinted from Ref. [62] with permission of Wiley-VCH Verlag GmbH & Co. Copyright 2011.)*

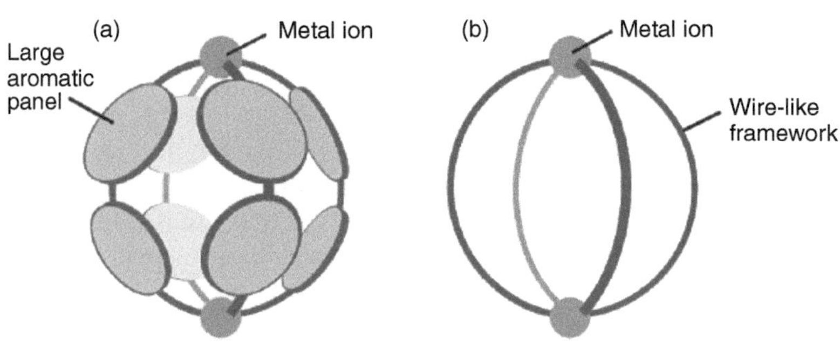

FIGURE 9.51 Cartoons of (a) the M_2L_4 capsule with large aromatic panels reported herein and (b) a previousM_2L_4 cage with a wirelike framework. *(Reprinted with permission from Ref. [63]. Copyright 2011, American Chemical Society.)*

Fujita et al. reported a new nanochanneled material through coordination-driven folding and self-assembly of a short peptide fragment, Gly–Pro–Pro (Gly: glycine, Pro: L-proline) (Fig. 9.53) [64]. Two types of one-dimensional chiral nanochannels were recognized by crystallographic studies. The larger channel, with a diameter of more than 2 nm, could encapsulate various guests

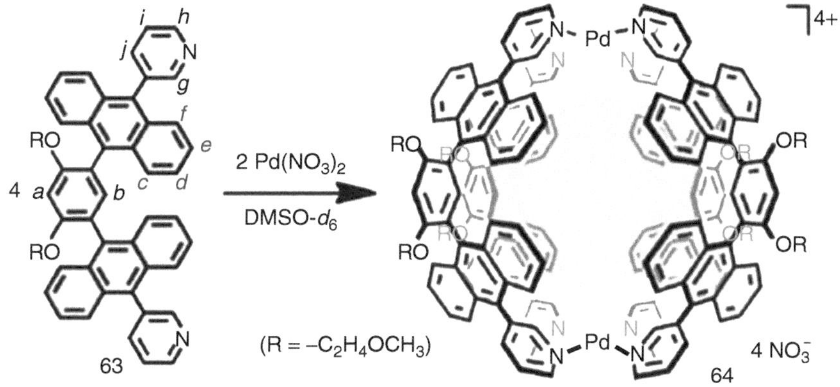

FIGURE 9.52 Schematic representation of the self-assembly of molecular capsule 64. *(Reprinted from Ref. [63] with permission of American Chemical Society. Copyright 2011.)*

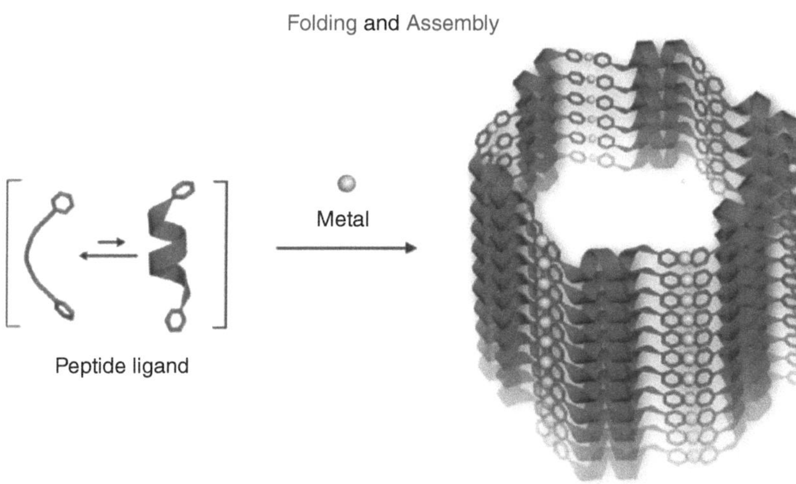

FIGURE 9.53 The concept of this work. Random-coil peptide ligands are concomitantly folded through metal-directed networking. Note that the illustration of the network does not represent the network topology of the complex in the present study. *(Reprinted from Ref. [64] with permission of Wiley-VCH Verlag GmbH & Co. Copyright 2014.)*

such as anions, organic chiral molecules and bio-oligomers with moderate to high chiral recognition for chiral guests.

Prior to assembly, coordinative ligand 65 was obtained through modification of the peptide Gly-Pro-Pro by introduction of 3-pyridyl groups (py) with amid bonds at both termini. Upon addition of an ethanol solution of ligand 65 to an aqueous solution of $AgBF_4$, block crystals of $[(AgBF_4).65]_n$ were furnished after one week (Fig. 9.54). It was established that peptid ligand 65 is fixed in P_{II} helix conformation. All of the peptide ligands 65 are linked by py-Ag^I-py

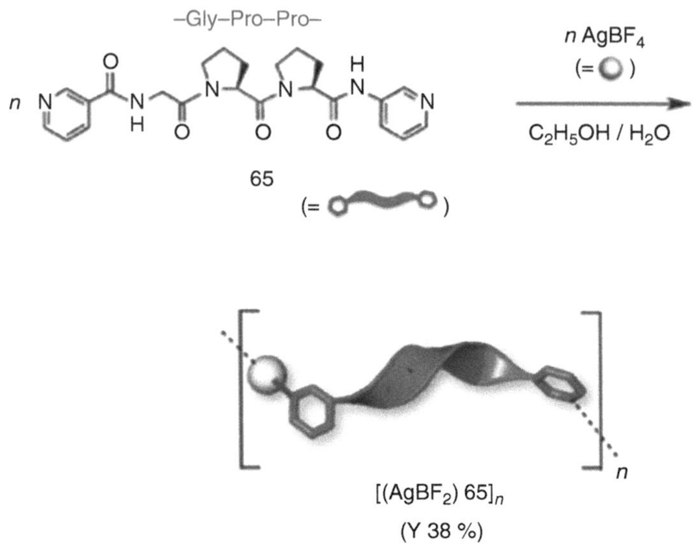

FIGURE 9.54 The crystallization of Gly–Pro–Pro ligand with $AgBF_4$. *(Reprinted from Ref. [64] with permission of Wiley-VCH Verlag GmbH & Co. Copyright 2014.)*

coordination in a directional N→C-terminal manner, similar to natural polypeptide chains. Furthermore, the linear helical strand of Ag^I-65 repeats shows a unique hexagonal entanglement in the crystal, which affords a three-dimensional honeycomb network.

6 CONCLUSIONS

Metal–ligand self-assembled capsules are one of the most attractive classes of self-assembled nanoreactors. Up to now, various kinds of coordination cages with different shape and cavity size have been developed. As many of the examples discussed in this chapter illustrate, this class of nanoreactors can effectively be used for encapsulation and stabilization of various guests, including reactive intermediates and species. The applications of these nanoreactors in catalysis of a wide range of organic reactions led to promising results, which indicated the utility of these nanoreactors for controlling regio- and stereoselectivity of reactions, altering the priority of reaction sites and increasing the yield of reactions. In other words, fixing substrate in a specific orientation in coordination cages and making a specific reaction pathway favored is reminiscent of enzymatic reaction control.

The interior cavity of coordination cages has also been exploited as endotemplate for the synthesis of highly monodispersed nanoparticles.

It is worth to mention that this research field is growing and the kinds and applications of coordination cage are beyond the discussed issues in this chapter. Therefore readers who are interested are encouraged to study the references for more details.

REFERENCES

[1] Vriezema DM, et al. Self-assembled nanoreactors. Chem Rev. 2005;105:1445–89.
[2] Llabres i Xamena FX. Metal organic frameworks as heterogeneous catalysts. Cambridge: RSC Catal Series No. 12; 2013.
[3] Koblenz TS, Wassenaar J, Reek JNH. Reactivity within a confined self-assembled nanospace. Chem Soc Rev 2008;37:247–62.
[4] Steed JW, Turner DR, Wallace KJ. Core concepts in supramolecular chemistry and nanochemistry. Hoboken, NJ: John Wiley & Sons; 2007.
[5] Sato S, Murase T, Fujita M. Self-assembly of coordination cages and ppheres. Supramolecular chemistry: From molecules to nanomaterials. Hoboken, NJ: John Wiley & Sons; 2012.
[6] Brinker UH. Molecular encapsulation. Weinheim: WILEY-VCH Verlag GmbH & Co; 2010.
[7] Zecchina A, Bordiga S, Groppo E. Selective nanocatalysts and nanoscience: concepts for heterogeneous and homogeneous catalysis. Wiley-VCH Germany; 2011.
[8] Saalfrank RW, et al. The First "adamantoid" alkaline earth metal chelate complex: synthesis, structure, and reactivity. Angew Chem Int Ed Engl 1988;27:851–3.
[9] Sun W-Y, et al. Multicomponent metal–ligand self-assembly. Curr Opin Chem Biol 2002;6: 757–64.
[10] Brown CJ, et al. Supramolecular catalysis in metal–ligand cluster hosts. Chem Rev 2015;115: 3012–35.
[11] Pluth MD, Raymond KN. Reversible guest exchange mechanisms in supramolecular host–guest assemblies. Chem Soc Rev 2007;36:161–71.
[12] Yoshizawa M, Klosterman JK, Fujita M. Functional molecular flasks: new properties and reactions within discrete, self-assembled hosts. Angew Chem Int Ed 2009;48:3418–38.
[13] Fiedler D, et al. Selective molecular recognition, C-H bond activation, and catalysis in nanoscale reaction vessels. Acc Chem Res 2005;38:351–60.
[14] Fiedler D, et al. Encapsulation of cationic ruthenium complexes into a chiral self-assembled cage. Inorg Chem 2004;43:846–8.
[15] Fiedler D, et al. Enantioselective guest binding and dynamic resolution of rationic ruthenium complexes by a chiral metal-ligand assembly. J Am Chem Soc 2004;126:3674–5.
[16] Mahata K, Frischmann PD, Würthner F. Giant electroactive M_4L_6 tetrahedral host self-assembled with Fe(II) vertices and perylene bsimide dye edges. J Am Chem Soc 2013;135:15656–61.
[17] Davis AV, Raymond KN. The big squeeze: guest exchange in an M_4L_6 supramolecular host. Proc Natl Acad Sci 2002;99:4793–6.
[18] Dong VM, et al. Molecular recognition and stabilization of iminium ions in water. J Am Chem Soc 2006;128:14464–5.
[19] Brumaghim JL, et al. Encapsulation and stabilization of reactive aromatic diazonium ions and the tropylium ion within a supramolecular host. Eur J Org Chem 2004;2004:5115–8.
[20] Mal P, et al. White phosphorus is air-stable within a self-assembled tetrahedral capsule. Science 2009;324:1697–9.
[21] Smulders MMJ, Nitschke JR. Supramolecular control over Diels–Alder reactivity by encapsulation and competitive displacement. Chem Sci 2012;3:785–8.
[22] Riddell IA, et al. Encapsulation, storage and controlled release of sulfur hexafluoride from a metal–organic capsule. Chem Commun 2011;47:457–9.
[23] Pluth MD, Bergman RG, Raymond KN. Acid catalysis in basic solution: a supramolecular host promotes orthoformate hydrolysis. Science 2007;316:85–8.
[24] Fiedler D, Bergman RG, Raymond KN. Supramolecular catalysis of a unimolecular transformation: aza-cope rearrangement within a self-assembled host. Angew Chem Int Ed 2004;43: 6748–51.

[25] Hastings CJ, Bergman RG, Raymond KN. Origins of large rate enhancements in the Nazarov cyclization catalyzed by supramolecular encapsulation. Chem Eur J 2014;20:3966–73.

[26] Wang ZJ, et al. Hydroalkoxylation catalyzed by a gold(I) complex encapsulated in a supramolecular host. J Am Chem Soc 2011;133:7358–60.

[27] Wang ZJ, et al. A supramolecular approach to combining enzymatic and transition metal catalysis. Nat Chem 2013;5:100–3.

[28] Zarra S, et al. Molecular containers in complex chemical systems. Chem Soc Rev 2015;44:419–32.

[29] Hart-Cooper WM, et al. Selective monoterpene-like cyclization reactions achieved by water exclusion from reactive intermediates in a supramolecular catalyst. J Am Chem Soc 2012;134: 17873–6.

[30] Salles AG Jr, et al. A self-organizing chemical assembly line. J Am Chem Soc 2013;135:19143–6.

[31] Kusukawa T, Yoshizawa M, Fujita M. Probing guest geometry and dynamics through host–guest interactions. Angew Chem Int Ed 2001;40:1879–84.

[32] Kusukawa T, Fujita M. Encapsulation of large, neutral molecules in a self-assembled nanocage incorporating six Palladium(ii) ions. Angew Chem Int Ed 1998;37:3142–4.

[33] Murase T, Nishijima Y, Fujita M. Unusual photoreaction of triquinacene within self-assembled hosts. Chem Asian J 2012;7:826–9.

[34] Horiuchi S, Murase T, Fujita M. Noncovalent trapping and stabilization of dinuclear ruthenium complexes within a coordination cage. J Am Chem Soc 2011;133:12445–7.

[35] Kohyama Y, Murase T, Fujita M. A self-assembled cage as a non-covalent protective group: regioselectivity control in the nucleophilic substitution of aryl-substituted allylic chlorides. Chem Commun 2012;48:7811–3.

[36] Ono K, et al. ON/OFF red emission from azaporphine in a coordination cage in water. J Am Chem Soc 2009;131:12526–7.

[37] Kawano M, et al. Direct crystallographic observation of a coordinatively unsaturated transition-metal complex in situ generated within a self-assembled cage. J Am Chem Soc 2006;128:6558–9.

[38] Yoshizawa M, et al. Cavity-directed synthesis of labile silanol oligomers within self-assembled coordination cages. J Am Chem Soc 2001;123:10454–9.

[39] Furusawa T, Kawano M, Fujita M. The confined cavity of a coordination cage suppresses the photocleavage of a-diketones to give cyclization products through kinetically unfavorable pathways. Agnew Chem Int Ed 2007;46:5717–9.

[40] Murase T, Nishijima Y, Fujita M. Cage-catalyzed Knoevenagel condensation under neutral conditions in water. J Am Chem Soc 2012;134:162–4.

[41] Yoshizawa M, et al. Cavity-directed synthesis within a self-assembled coordination cage: highly selective [2 + 2] cross-photodimerization of olefins. J Am Chem Soc 2003;125:3243–7.

[42] Horiuchi S, Murase T, Fujita M. Diels–Alder reactions of inert aromatic compounds within a self-assembled coordination cage. Chem Asian J 2011;6:1839–47.

[43] Yoshizawa M, et al. Alkane oxidation via photochemical excitation of a self-assembled molecular cage. J Am Chem Soc 2004;126:9172–3.

[44] Murase T, Takezawa H, Fujita M. Photo-driven anti-Markovnikov alkyne hydration in self-assembled hollow complexes. Chem Commun 2011;47:10960–2.

[45] Yamaguchi T, Fujita M. Highly selective photomediated 1,4-radical addition to *o*-quinones controlled by a self-assembled cage. Angew Chem Int Ed 2008;120:2097–9.

[46] Takezawa H, et al. Cavity-directed chromism of phthalein dyes. J Am Chem Soc 2015;137: 7043–6.

[47] Sun QF, et al. Self-assembled $M_{24}L_{48}$ polyhedra and their sharp structural switch upon subtle ligand variation. Science 2010;28:1144–7.

[48] Harris K, Fujita D, Fujita M. Giant hollow MnL2n spherical complexes: structure, functionalisation and applications. Chem Commun 2013;49:6703–12.

[49] Tominaga M, et al. Finite, spherical coordination networks that self-organize from 36 small components. Angew Chem Int Ed 2004;43:5621–5.

[50] Kamiya N, et al. Saccharide-coated $M_{12}L_{24}$ molecular spheres that form aggregates by multi-interaction with proteins. J Am Chem Soc 2007;129:3816–7.

[51] Fujita D, et al. Geometrically restricted intermediates in the self-assembly of an $M_{12}L_{24}$ cuboctahedral complex. Angew Chem Int Ed 2014;53:1–5.

[52] Sun QF, et al. Self-assembled $M_{24}L_{48}$ polyhedra and their sharp structuralsSwitch upon subtle ligand variation. Science 2010;328:1144–7.

[53] Bunzen J, et al. Self-assembly of $M_{24}L_{48}$ polyhedra based on empirical prediction. Angew Chem Int Ed 2012;51:3161–3.

[54] Suzuki K, et al. Self-assembly of an M_6L_{12} coordination cube. Chem Commun 2009;1638–40.

[55] Sun QF, et al. A sphere-in-sphere complex by orthogonal self-assembly. Angew Chem Int Ed 2011;50:10318–1032.

[56] Sun QF, Sato S, Fujita M. An $M_{18}L_{24}$ stellated cuboctahedron through post-stellation of an $M_{12}L_{24}$ core. Nat Chem 2012;4:330–3.

[57] Fujita D, et al. Protein encapsulation within synthetic molecular hosts. Nat Commun 2012;3: 1–7. 1093.

[58] Suzuki K, et al. Coronene nanophase within coordination spheres: increased solubility of C_{60}. J Am Chem Soc 2010;132:2544–5.

[59] Murase T, Sato S, Fujita M. Nanometer-sized shell molecules that confine endohedral polymerizing units. Angew Chem Int Ed 2007;46:1083–5.

[60] Kikuchi T, et al. Polymerisation of an anionic omnomer in a self-assembled $M_{12}L_{24}$ coordination sphere with cationic interior. Supramol Chem 2008;20:81–94.

[61] Suzuki K, Sato S, Fujita M. Template synthesis of precisely monodisperse silica nanoparticles within self-assembled organometallic spheres. Nature Chem 2010;2:25–9.

[62] Suzuki K, et al. The Precise synthesis and growth of core–shell nanoparticles within a self-assembled spherical template. Angew Chem Int Ed 2011;50:4858–61.

[63] Kishi N, et al. An M_2L_4 molecular capsule with an anthracene shell: encapsulation of large guests up to 1 nm. J Am Chem Soc 2011;133:11438–41.

[64] Sawada T, Matsumoto A, Fujita M. Coordination-driven folding and assembly of a short peptide into a protein-like two-nanometer-sized channel. Angew Chem Int Ed 2014;53:1–6.

Chapter 10

Metal Organic Frameworks as Nanoreactors and Host Matrices for Encapsulation

F.G. Cirujano, F.X. Llabrés i Xamena
Institute of Chemical Technology, Polytechnic University of Valencia, Superior Council for Scientific Research, Valencia, Spain

1 INTRODUCTION

Metal organic frameworks (MOFs), also known as porous coordination polymers, are a relatively new class of crystalline porous solids that has attracted much interest in recent years. These materials are constructed by inorganic nodes (also known as secondary building units, or SBUs) that are connected by polydentate organic ligands through coordination bonds, forming extended 3D crystalline networks that in many cases form an ordered pore system, as in the case of zeolites [1]. But unlike zeolites, for which only tetrahedral [TO_4] (T = Si, Al, Ge,...) building units are available for framework construction, the large variety of secondary building units and organic ligands that can be used for MOF synthesis, and the different ways to connect them (*polymorphism*), offer a virtually infinite number of possible combinations. This leads to an unprecedented control over the framework structure, making it possible to obtain MOFs with pore dimensions ranging progressively from ultramicropores (2–3 Å) to well into the mesoporous range (up to ~10 nm [2]), and with huge pore volumes and unmatched specific surface areas of thousands of square meters per gram. As an example, a MOF referred to as Nu-110 has a specific surface area (S_{BET}) of up to 7000 m^2/g and a pore volume over 4 cm^3/g [3], which gives an idea of the record-holding textural properties attainable with these compounds. Moreover, the possibility of modifying an already existing MOF by postsynthesis methods [4,5] (to introduce new functionalities or to change the chemical environment inside the pores) provides an additional dimension to the MOF synthesis and tunability.

The outstanding properties of MOFs, along with their crystallinity and porosity and the potential for tailoring both their structure and properties on demand, have attracted the interest of scientists for several applications, including

Organic Nanoreactors. http://dx.doi.org/10.1016/B978-0-12-801713-5.00010-0

gas separation and storage [6,7], sensing [8], biomedicine [9], or catalysis [10–13], among others. In particular, the excellent textural properties of certain MOFs, and the presence of a pore system with fixed and regular geometry and dimensions, make these compounds interesting candidates to act as host matrices for encapsulation, or as the confined space in which a reaction is carried out, influencing the reaction rate or the distribution of products formed (the topics of this work). However, the chapter is not presented as an exhaustive and up-to-date revision of the existing literature, since there are already excellent reviews on these topics [14–16]. Rather, we have selected some relevant examples to illustrate the possible benefits of using MOFs in these applications. As mentioned previously, we have considered two different scenarios:

1. In one case, the pore system of the MOF is the (confined) physical space in which a chemical reaction takes place, thus acting as a *nanoreactor.*
2. In another scenario, the MOF is used as the solid support in which a catalytic species is encapsulated.

2 VIRTUES AND LIMITATIONS OF MOFs AS HOST MATRICES AND NANOREACTORS

Without a doubt, the most appealing properties of MOFs as supports for encapsulation or as nanoreactors are clearly their huge specific surface areas and their strictly uniform pore systems. These properties introduce two highly desirable characteristics for the applications envisaged in this chapter. On the one hand, the huge internal surface of some MOFs, and the occurrence of specific secondary interactions with their metal ions or linkers, can effectively concentrate the reactants inside the pores of the MOFs at levels considerably higher than in the bulk solution or in the surrounding atmosphere. This can translate into a significant acceleration of the reaction rate (or, when dealing with sensing applications, for instance, into lower detection limits of a given analyte). On the other hand, the pore shape and diameter of MOFs depend on their crystalline structure, and this introduces the possibility of having size/shape-selective effects. In this way, by properly choosing the structural components and their connectivity pattern, along with the use of post-synthesis modifications, it is possible to precisely select which molecules can penetrate the pore system (and access the active sites located inside) and which molecules remain at the external surface of the MOF. In other words, MOFs can behave as *molecular sieves*. These size/shape selective properties can determine the products formed in a reaction among various alternatives. Thus, as in the case of zeolites and enzymes, shape-selective catalysis is also possible with MOFs, as has been demonstrated in several examples [17–21]. Thus, shape-selectivity imposed on the reactants, transition states, and/or products can govern the outcome of a reaction occurring inside the MOF pores. The principle of shape-selectivity, originally proposed by Csicsery [22], is shown schematically in Scheme 10.1.

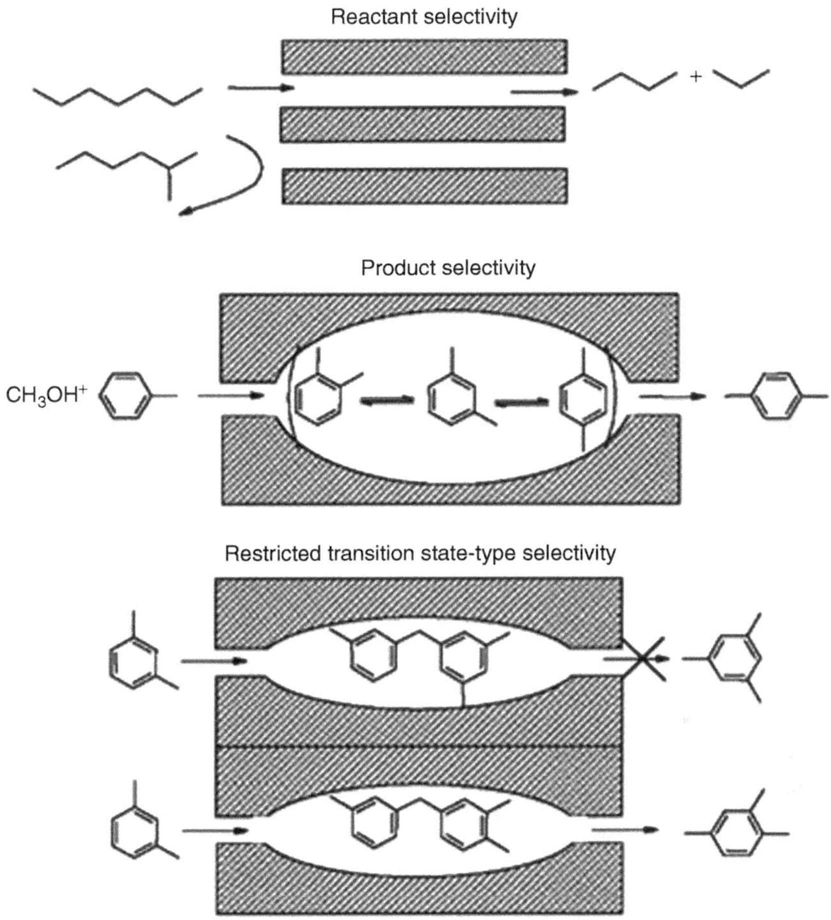

SCHEME 10.1 Shape-selective properties in porous materials according to Csicsery. *(Reproduced from Ref. [22] with permission of Elsevier. Copyright 1984.)*

Enzymes (and to some extent also zeolites) are highly shape-selective catalysts. The surroundings of their active sites usually permit a highly restricted access to specific molecules or even to precise stereoisomers. Once adsorbed, the reactants are fixed at the right place and orientation with respect to the active center of the enzyme, and they are activated at specific positions, thus allowing the strict stereo-, regio-, and chemoselectivity control necessary for the living processes. Then, according to Scheme 10.1, substrate selection in enzymes is governed by reactant selectivity, while its transformation would be controlled by transition-state selectivity. Likewise, MOFs can also exert a similar (although so far not as perfect) control of the chemical reaction through a combination of shape-selective properties and confinement effects (host–guest interactions), as we will illustrate further with some selected examples. Among the most interesting characteristics of MOFs is the possibility of introducing chiral moieties, either by using chiral linkers or through crystallization of the MOF in a chiral space group, which opens the door to asymmetric catalysis for the (shape-selective) enantioselective transformation of chiral molecules [23].

Another aspect that can represent an advantage of MOFs with respect to other potential hosts is the possible introduction of catalytic active sites either at the metallic nodes or at the organic ligands of the MOFs (and even simultaneously). This affords a means for the ready preparation of multifunctional heterogeneous catalysts when these structural functionalities of the MOFs are combined with the catalytic activity of the encapsulated species. In this way, two (or more) different active sites can be combined in the same solid in such a way that they can act in concert to exert dual activation of the reaction substrates (*synergism*) or they catalyze cascade or tandem consecutive reactions [11,24]. The high tunability of MOFs over a wide range of both, framework structures and functionalities, together with the use of post-synthesis modification approaches, allow a vast number of possibilities in engineering multifunctional MOFs with encapsulated catalytic species.

On the other hand, MOFs can also present some limitations that can delimit their application as hosts for encapsulated species or nanoreactors. Among them, the most relevant are their limited thermal and chemical stability as compared to other porous (inorganic) solids, such as zeolites. In the last years, some remarkably robust MOFs have been prepared, such as UiOs [25], MIL-101 [26], or pyrazolate-based compounds [27]. However, many MOFs are moisture sensitive and undergo amorphization or phase transformation upon standing in the ambient air even at room temperature [28,29]. Mechanical stability is also another important limitation that must be taken into account, especially when looking for applications requiring compressing or shaping the MOF, for instance, in pellets or extrudates. Indeed, some MOFs are known to collapse when submitted to mechanical compression beyond a certain pressure, or upon ball-milling [30], which produces a loss of the crystalline structure and the concomitant diminution of specific surface area and pore volume available for encapsulation or as reaction nanocavity.

3 MOFs AS NANOREACTORS

In a broad sense, a *nanoreactor* is to be understood as the nanosized container in which a chemical reaction is carried out. As discussed previously, the pores of MOFs have the right dimensions (from a few angstroms to a few nanometers) to be considered nanoreactors. However, it is evident that the confinement of the reaction in such a limited space will affect the molecular movement and the interactions between the molecules hosted inside (and between the molecules and the reactor walls), as well as the conformational flexibility of the guest. Therefore, regardless of whether or not the MOF participates actively in the reaction (by providing the active sites that catalyze the chemical transformation), it is evident that the pore system of the MOF can influence the reaction rate or the distribution of products formed, as compared with the same reaction in bulk solution and in the absence of the MOF. On one hand, secondary interactions between reactant molecules adsorbed inside the pores and the ligands or metals of the MOF walls (*host–guest interactions*) can lower the energy of the transition state. This, together with the concentration effects of the reactant molecules inside the cavities, can produce a significant acceleration of the reaction rate. But on the other hand, host–guest interactions can also determine the preferred orientation or relative conformation of the reactants, thereby facilitating a given reaction pathway among other potential alternatives and influencing the type and distribution of products formed and, thus, the selectivity (including in certain cases the regio-, chemo-, enantio- and/or diastereoselectivity). Therefore, it is not surprising that the pore structure of the MOF can largely affect the product distribution of a reaction; this can be an important benefit, as we will illustrate in the following with some selected cases.

One relevant example on the use of the MOF cavities as nanoreactor and the influence on the properties of the final products formed was reported by Uemura et al. [31]. These authors studied the polymerization of styrene in the presence of various MOFs. In particular, two different materials were considered, namely $[M_2(BDC)_2(TEDA)]$ ($M = Zn^{2+}$ or Cu^{2+}; BDC = bencene-1,4-dicarboxylate, $\mathbf{L^1}$; TEDA = triethylenediamine, $\mathbf{L^2}$), and $[Cu(PZDC)_2(4,4'\text{-}BPY)]$ (PZDC = pyrazine-2,3-dicarboxylate, $\mathbf{L^3}$; 4,4′-BPY = 4,4′-bipyridine, $\mathbf{L^4}$); both compounds having 1D channels of dimensions 7.5×7.5 Å^2 and 8.2×6.0 Å^2, respectively. Polymerization of styrene was carried out by pre-adsorbing styrene inside the MOFs, followed by radical polymerization using azobisisobutyronitrile (AIBN) as radical initiator. In spite of attaining similar loading of styrene in all cases (2.7 molecules per nm^3), polymerization was not observed for $[Cu(PZDC)_2(4,4'\text{-}BPY)]$. This was attributed to the restricted mobility of styrene inside the channels of this material (as evidenced by ^{2}H NMR spectroscopy), which was attributed in turn to the strong confinement of styrene due to the intense potential inside the channels [31]. On the contrary, polymerization of styrene inside $[M_2(BDC)_2(TEDA)]$ took place in about 70% final yield, with no significant differences observed between the Zn^{2+} and Cu^{2+} materials.

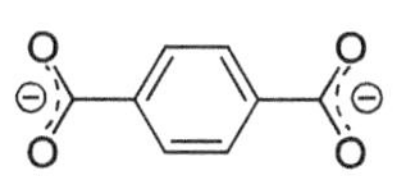

$\mathbf{L^1}$

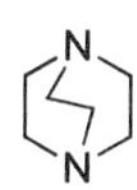

$\mathbf{L^2}$

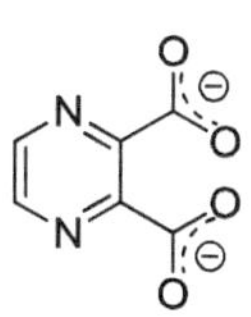

$\mathbf{L^3}$

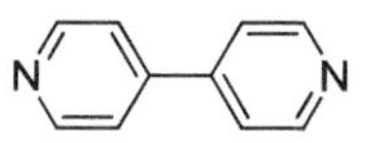

$\mathbf{L^4}$

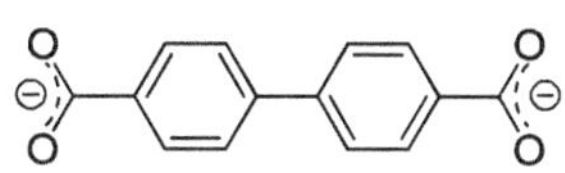

$\mathbf{L^5}$

Interestingly, after extraction of the polymer by dissolving the MOF with NaOH, the polystyrene formed was found to present a very low polydispersity (1.66) as compared to the polymer prepared in the absence of MOF (4.68). On the other hand, EPR experiments revealed the presence of a strong signal due to the propagating radical; the signal persisted even after storing the sample for one week at 343 K. The presence of this living radical species was attributed to the suppression of termination reaction and radical transfer due to the tight confinement of the growing polymer inside the MOF channels. Moreover, the reduced dimensions of the MOF channels (7.5×7.5 Å^2) did not allow the accommodation of more than one polystyrene chain (4.4×6.8 Å^2); therefore, the whole encapsulated polymer was supposed to be formed by isolated, single polystyrene chains. This particular conformation of the encapsulated polymer clearly differs from that found in the bulk polymer and is a direct consequence of the intimate contact in the host–guest composite system [32].

The same group used the [Cu(PZDC)$_2$(4,4′-BPY)] MOF as both nanoreactor and Lewis basic catalyst (arising from the carboxylate oxygen atoms of the ligands) to carry out the anionic polymerization of acidic terminal alkynes, such as methyl propiolate [33]. Interestingly, when sodium benzoate was used as a soluble model for the MOF's basic sites, no formation of poly(methyl propiolate) was observed. This suggested that the spatial confinement inside the MOF, having the precise arrangement of basic carboxylate oxygen atoms and appropriate adsorption sites for the alkyne monomers, was a crucial requirement for the success of the polymerization. This precise arrangement does not occur in solution, and this would explain the lack of any polymerization products observed when sodium benzoate was used. Meanwhile, optical and vibrational spectroscopies of the polymer recovered after extraction from the MOF micropores, showed that the conjugated polyacetylene formed had mostly a *trans* configuration of the C═C double bonds, which was obviously imposed by the reduced space inside the pores (8.2×6.0 Å^2). Other acidic terminal alkynes (cyanoacetylene and 2-ethynylpyridine) underwent polymerization like methyl propiolate, while less acidic terminal alkynes, such as phenylacetylene and 1-hexyne, or disubstituted acetylenes such as tetrolic acid methyl ester, did not polymerize inside [Cu(PZDC)$_2$(4,4′-BPY)]. Therefore, a mechanistic proposal based on the formation of acetylide species upon interaction with the basic carboxylate atoms of the MOF, followed by the subsequent stereoselective attack to form the *trans*-configured polymer chain, was proposed [33].

Another clear example of the *cage effect* introduced when MOFs are used as nanoreactors was reported by Pan et al. [34]. These authors studied the use of the MOF [Co$_3$(4,4′-BPhDC)$_3$(4,4′-BPY)] (4,4′-BPhDC = biphenyl-4,4′-dicarboxylate, **L**^{5}) as host for photochemical reactions, using *o*-methyl dibenzyl ketone (*o*-MeDBK) as a probe (see Scheme 10.2).

The photochemical decomposition of asymmetrically substituted dibenzyl ketones in confined spaces (such as zeolites), and the analysis of the product distribution can serve to determine the cage effect and the mobility of the free

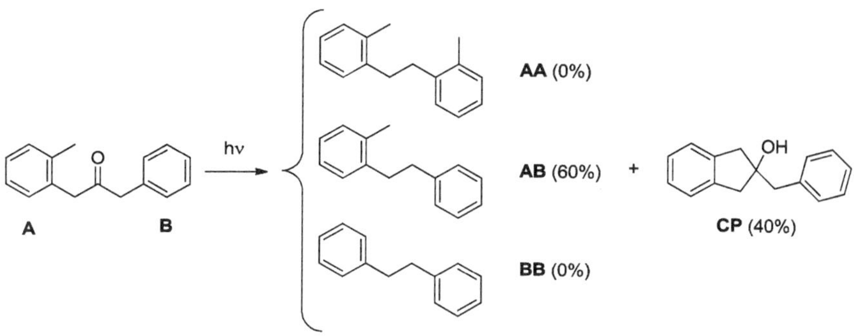

SCHEME 10.2 Product distribution in the photolysis of *o*-methyl dibenzyl ketone inside the pores of a MOF. *(Adapted from Ref. [34] with permission of Wiley-VCH. Copyright 2003.)*

radicals in these confined solids [35]. Thus, upon irradiation *o*-MeDBK undergoes decarbonylation by homolytic cleavage at the α-position, forming two different benzyl radicals (tolyl and phenyl, **A** and **B** in Scheme 10.2). When the reaction occurs in solution, these two radicals recombine randomly to give three diarylmethanes (**AA**, **AB**, and **BB**), in a statistic 1:2:1 ratio. However, when *o*-MeDBK pre-adsorbed inside the cavities of [Co_3(4,4′-BPhDC)$_3$(4,4′-BPY)] was irradiated, the limited diffusion of the radicals largely favored the formation of the diarylethane coming from the recombination of the geminate radical pair (**AB**) over the other two possible products. Thus, product **AB** was formed in 60% yield, accompanied with 40% of intramolecular hydrogen abstraction (product **CP** in Scheme 10.2). Meanwhile, the lack of formation of products **AA** and **BB** indicates that the MOF exerts 100% cage effect, allowing exclusively the recombination of the geminate benzyl radicals. For the sake of comparison, irradiation of *o*-MeDBK in the presence of the zeolite NaX gives only 78% cage effect [35].

There are several more examples reflecting how the pores of a MOF can affect the selectivity and product distribution of a reaction. For instance, *diastereoselectivity* of cycloproponation by transfer of carbene fragments from diazo compounds to olefins can be significantly enhanced if the reaction is carried out in the presence of the [$Cu_3(BTC)_2$] MOF (BTC = benzene-1,3,5-tricarboxlate), with respect to other Cu-containing catalysts [36], as shown in Table 10.1. Thus, reaction between styrene and ethyl diazoacetate (EDA) over [$Cu_3(BTC)_2$] afforded the corresponding *trans* cyclopropane in 71% diastereomeric selectivity ratio, while d.r. attained over other Cu-catalysts was less than 60% (see entries 1–4 in Table 10.1). When other combinations of (bulkier) olefins and diazo compounds were used, almost quantitative d.r. were obtained with [$Cu_3(BTC)_2$], with only the *trans* isomer being formed (entries 5–11).

Very recently, we have demonstrated [37] the ability of Cr^{3+}- and Fe^{3+}-containing terephthalate MOFs (MIL-101) to promote the benzylic oxidation of various hydrocarbons exclusively by molecular oxygen in the absence of

TABLE 10.1 Selected Catalytic Results for the Cyclopropanation Reaction Between Olefins and Diazo Compounds Catalyzed by Various Cu-Containing Catalysts

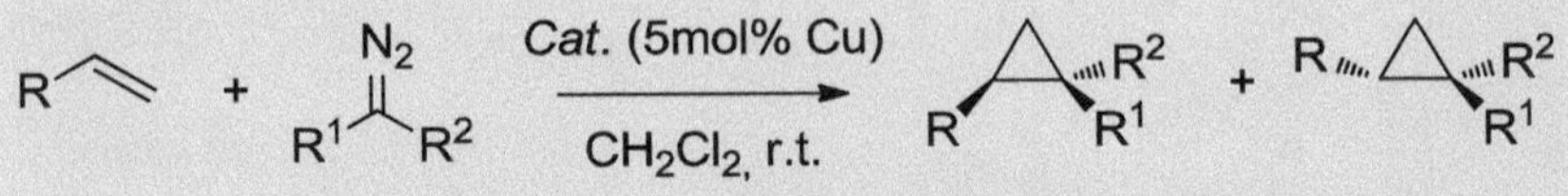

Catalyst	Olefin	Diazo compound	Yield (%)	d.r. (%)[a]
Cu-H	Styrene	EDA	36	59
Cu-USY	Styrene	EDA	32	55
Cu-MCM-41	Styrene	EDA	60	59
$[Cu_3(BTC)_2]$	Styrene	EDA	96	71
$[Cu_3(BTC)_2]$	DMHD	EDA	60	100
$[Cu_3(BTC)_2]$	Cyclohexene	EDA	99	98
$[Cu_3(BTC)_2]$	Styrene	PhEDA	85	100
$[Cu_3(BTC)_2]$	α-Methylstyrene	PhEDA	50	100
$[Cu_3(BTC)_2]$	β-Methylstyrene	PhEDA	30	100
$[Cu_3(BTC)_2]$	1-Octene	PhEDA	90	98
$[Cu_3(BTC)_2]$	Cyclooctene	PhEDA	70	98

EDA, diazoacetate; PhEDA, 2-phenyldiazoacetate; DMHD, 2,5-dimethyl-2,4-hexadiene.
[a]*Diastereomeric selectivity:* (trans-cis)/(trans+cis).
Data adapted from Ref. [36].

any other oxidizing reagent or initiator. Interestingly, the selectivity toward the desired ol/one mixture was very high, especially in the case of the Cr^{3+} compound: 87% selectivity at 30% conversion with indane as substrate. This clearly surpassed the values attained when soluble radical initiators were used: 66 and 74% selectivity at 30% indane conversion obtained with AIBN and *tert*-butyl hydroperoxide (TBHP), respectively [37]. A possible explanation for the higher selectivity [ie, (indanol + indanone)/indane ratio] was based on the assumption that the reaction takes place mainly inside the MOF (hydrophobic) cavities. In this way, the pores of the MOF should favor the adsorption of indane and desorption of the more polar indanol and indanone products, which would reduce the possibility of secondary oxidation side reactions. In this sense, isothermal adsorption measurements of individual substrates (indane, indanol, or indanone), a combination of two (indane + indanol and indane + indanone), or a combination of three substances (indane + indanol + indanone) all showed the preferential adsorption of indane in the pores of MIL-101 over the other two substrates, which provides supporting evidence of the role of the MOF cavity in the selectivity of the oxidation process.

The previous examples are just a few of many that demonstrate the high potential of MOFs as nanoreactors. To that we add the capability of some flexible MOFs (known as third-generation MOFs [38]) to reversibly change their porous structure upon application of an external stimulus (such as temperature, pressure, an electric field, or the adsorption of certain substrates), which is usually referred to as *breathing* [39] or *gate opening* effect [40]. This intriguing property of some MOFs can lead to the tuning of the adsorption properties of the solid, and thus a change in its catalytic behavior can be anticipated. Although framework flexibility of MOFs has been largely overlooked so far when studying the catalytic properties of these compounds, it can be envisaged that these properties will be considered in future developments. Then the materials should be able to adapt their pore space to the desired reaction substrate or transition state through conformational changes of their structural units, in the same way it happens in enzymes.

4 MOFs AS HOST MATRICES FOR ENCAPSULATION

Catalytic active sites of MOFs can be located at their metallic nodes [41,42] or at the organic ligands [43], or they can be incorporated a posteriori by post-synthesis modification of an existing MOF [5]. This allows tailoring the catalytic properties of the MOFs over a wide range of functionalities. But besides the "framework" active centers, certain catalytic species can also be encapsulated in the pore space of an MOF. In other words, and using the nomenclature widely applied to zeolites, we can also introduce "extra-framework" catalytic species within the MOF pores. These species are incorporated inside the structural channels and cavities of the solid in such a way that they remain entrapped and cannot leave the support. To satisfy this condition, the encapsulated species must fit within the pores of the support, and they must be larger than the windows giving access to these pores, as shown in Fig. 10.1. Therefore, not just any solid support can be used for encapsulation of a species. In general, suitable supports must necessarily possess large cavities accessible through small windows. For instance, a solid having only channels of regular and fixed dimensions will not be the most convenient choice for this purpose, unless the structure also possesses larger intersections between channels running in different directions of the crystal.

Since the target species to be encapsulated cannot physically pass through the pore apertures of the solid, specific synthesis procedures need to be used to prepare these host–guest systems, as we will illustrate in the following. Ideally, the process of encapsulation must not undermine the catalytic properties of the guest species. Rather, the presence of the support should lead to composite materials combining the properties of host and guest, ideally introducing synergic effects due to their mutual interaction. In this way, encapsulated species can benefit from, for example, increased stability, controlled size/shape, or suppression of self-deactivation pathways, as compared to the same species

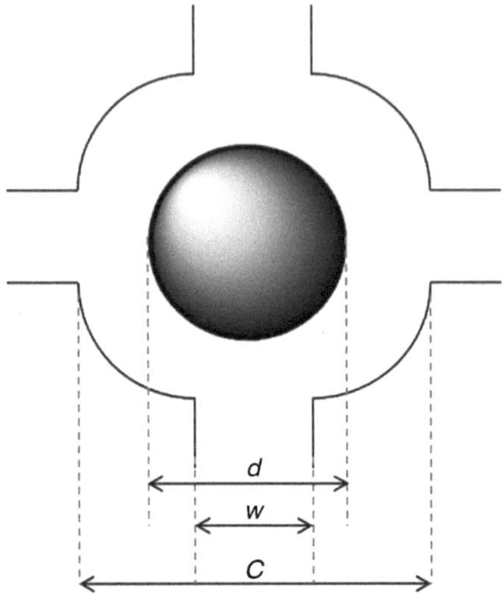

FIGURE 10.1 Schematic representation of an encapsulated species (depicted as a gray ball) inside the cavities (or channel intersections) of a porous support. In order to satisfy the condition for encapsulation, the dimensions of the imprisoned species (*d*) must be comprised between the diameter of the cavity (*C*) and the pore windows giving access to them (*w*).

in solution. Meanwhile, the solid support (host) may also introduce other benefits, such as shape-selectivity or additional catalytic centers (located at the inorganic SBUs or at the linkers), thereby leading to multifunctional materials (vide infra).

It is interesting to point out that the most popular zeolites used for encapsulation are of the FAU structure type (such as zeolites X and Y), featuring sodalite-type cages with an internal diameter of ~1.2 nm accessible through windows of 0.74 nm [1]. Therefore, species that can be encapsulated inside these materials (such that $1.2 > Ø > 0.74$ nm) will necessarily fill a large fraction of the cage volume, thus leaving little space for diffusion of incoming substrates. By contrast, the mesoporous cages present in MIL-100 type MOFs (with internal diameters of 2.5 and 2.9 nm) are accessible through microporous windows (0.55–0.86 nm) [44]. As a consequence, the size range of species that in principle can be encapsulated in these MOFs (viz., MIL-100 and other cage-like materials) is much larger than for most zeolites, while still leading to less crowded systems.

4.1 Methods of Encapsulation

The condition of encapsulation requires that, once incorporated inside the pores, the entrapped guest species cannot leave the support. This is different from simple adsorption, in which an adsorbate can freely diffuse in and out of the solid pores and is in equilibrium with the liquid or gaseous phase surrounding the solid support. This is also different from grafting, in which the adsorbed species is retained inside the pores (or at the external surface) due to the existence of covalent or electrostatic bonds between the host and the guest. Conversely, after encapsulation the species remains permanently immobilized inside the pores of the host matrix due to a physical entrapment. This type of immobilization has been coined by Stoddart as *mechanical immobilization* [45,46]. Since the size of the species that can diffuse inside the MOF is limited by the aperture of the pores, it is evident that techniques such as impregnation, ion exchange, or gas phase adsorption cannot be used for the direct encapsulation of species larger than the pore entries. Therefore, alternative and specifically designed preparation methods are required, which are essentially listed in the following sections. It is important to stress that none of the methods is intrinsically better than the others. Each has its own advantages and disadvantages, and the choice of method will depend on factors such as the stability and structural features of the MOF host, or the lability of the encapsulated guest.

4.1.1 Assembly of Preadsorbed Precursors

This preparation procedure consists of adsorbing suitable precursors inside a MOF (or another porous material) that are small enough to pass through the windows of the support. Once inside the cavities, these precursors will react or assemble in situ to form a (larger) compound that will remain imprisoned. It is therefore a bottom-up synthesis procedure carried out in the inner surface of the support. It is evident that the conditions needed for the assembly of the precursors into the target compound must be compatible with the MOF framework integrity, and this may severely delimit the type of compounds that can be prepared by using this procedure. The process may also be challenging for the assembly of molecules that require multiple steps.

This strategy has been known for a long time and was first used with other porous materials (such as zeolites) to prepare encapsulated host–guest systems [47] (see Scheme 10.3). Already in 1986, Herron coined the term "ship in a bottle" to refer to the synthesis of various metallic complexes encapsulated within zeolites [48]. The term refers to the decorative objects in which a glass bottle (or another recipient) contains a figurine inside (usually a ship) larger than the bottle neck. However, even before the coinage of this term, other groups used the same strategy to prepare metal phthalocyanines inside zeolites by treating metal-exchanged solids with 1,2-dicyanobenzene at temperatures above 200°C [49,50]. Other relevant examples include the intrazeolite encapsulation of ruthenium trisbipyridine [51], pyrylium cations [52], metal salen complexes [53], metal phthalocyanines [54], or even heteropolyacid such as $H_3PW_{12}O_{40}$ [55].

In the case of MOFs as host matrices, which may suffer from lower stability as compared with zeolites, the in situ assembly approach has been limited in general to composite systems requiring few post-synthesis operations and for which harsh reaction conditions are not necessary.

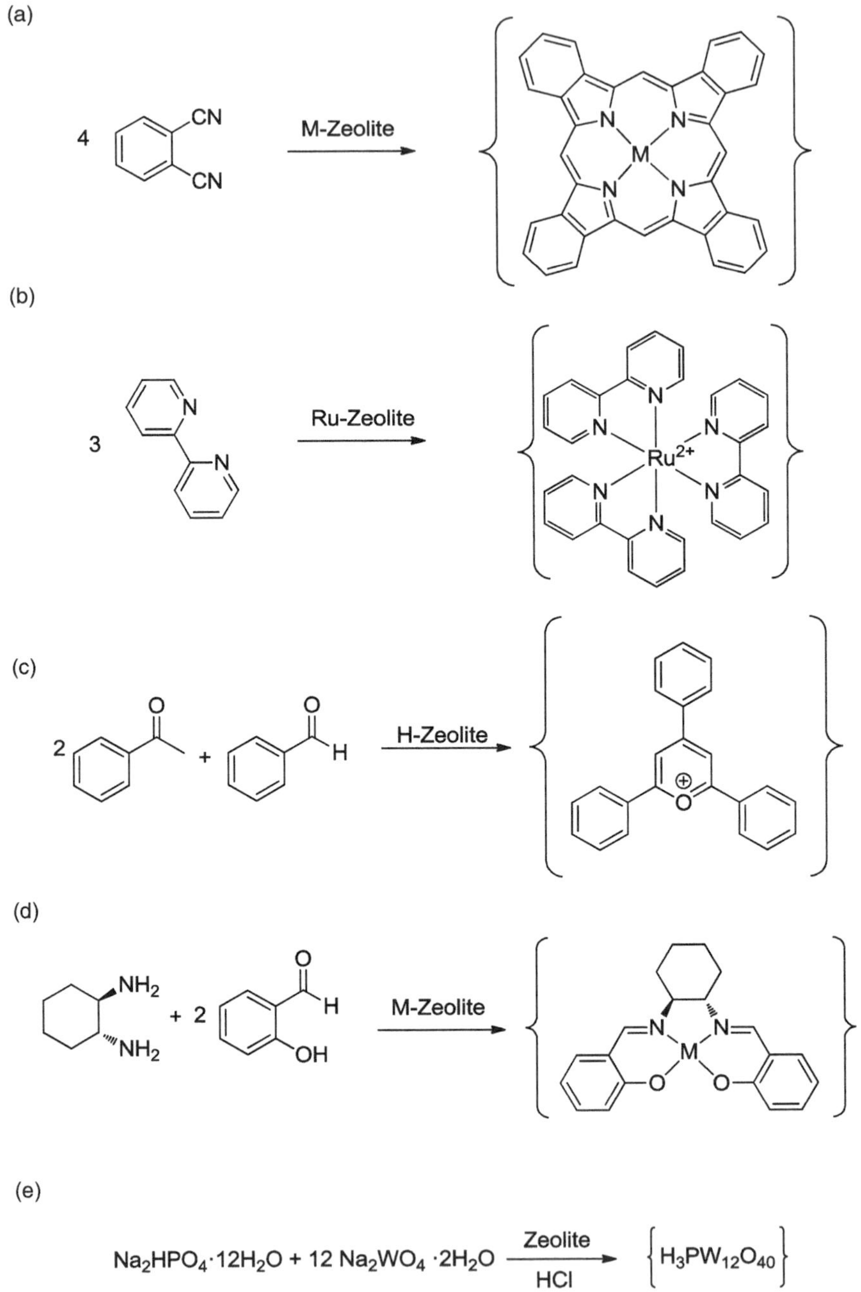

SCHEME 10.3 Some examples of encapsulated species successfully prepared by in situ assembly of precursors inside the pores of zeolites. (a) Metal phthalo-cyanines; (b) ruthenium tris-bipyridine; (c) pyrylium cations; (d) metal salen complexes; and (e) heteropoly acid. The brackets generically represent the pore system of the solid host.

Thus, in situ assembly of preadsorbed precursors is by far the approach most extensively used for preparing metal nanoparticles (MNPs) encapsulated in MOFs, or MNP@MOFs. To this end, the MOF is first loaded with a suitable metal precursor (a metal salt or an organometallic compound small enough to pass through the pore windows of the MOF) by either impregnation from a liquid solution or by adsorption from the gas phase (chemical vapor infiltration, CVI). Regardless of the method used for incorporating the precursor, this is followed by its controlled decomposition/reduction by UV irradiation, thermal treatment, or use of a suitable reducing agent (such as H_2, $NaBH_4$, hydrazine, etc.). Under appropriate conditions, MNPs are formed and remain encapsulated inside the MOF pores. In the broad sense, this method for preparing MOF-encapsulated MNPs can be considered an in situ assembly procedure, since the MNPs are formed from smaller precursor species pre-adsorbed inside the pores; thus, it is a bottom–up approach. Ideally, the whole process of MNPs encapsulation should not damage the crystalline structure of the MOF, and the maximum size of the MNPs depends on the hosting cavities. Then MOF cavities act as nanomolds for hosting and shaping the MNPs. However, in some cases, larger particles can be formed due to partial destruction of the MOF framework or to the unrestricted aggregation of the MNPs at the external surface of the solid. Indeed, Aijaz et al. developed a variant of this synthesis method specifically designed to minimize particle aggregation on the external surface by using a "double solvents" impregnation [56]. This method uses mixtures of one hydrophobic solvent (hexane) in large excess and one hydrophilic solvent (H_2O) containing the metal precursor and with a total volume close to (or slightly lower than) the pore volume of the MOF support. In this way, the metal precursor, dissolved in water, is preferentially adsorbed into the (hydrophilic) internal surface area of the MOF (as the authors showed for chromium MIL-101), thus minimizing the amount of metal precursor deposited on the outer surface. When this "directed" impregnation was followed by a mild reduction with H_2 at a relatively low temperature (200°C), the authors were able to prepare ultrafine Pt, Pd, Au, and Rh NPs inside the pores of MIL-101 without aggregation on the external surface of the solid.

In situ assembly is also the preferred method used for obtaining polymer-encapsulated MOFs (Polymer@MOF), prepared by pre-adsorbing suitable monomers inside the MOF pores followed by their in situ polymerization [31,33,57,58]. Also in this case, several precedents exist for the preparation of polymers encapsulated inside zeolites or mesoporous silica compounds [59–63]. Applications of such composites materials were long ago anticipated in the fields of electronic/ionic transport materials, optoelectronics, and sensor devices [64], and the emergence of MOFs as novel host materials will surely reactivate the interest in these composites (see for instance, Ref. [8]).

But besides MNP@MOFs and Polymer@MOFs, reports on in situ assembly of pre-adsorbed precursors leading to molecular (discrete) species inside MOFs are still very scarce. This is indeed surprising, given that several precedents exist for the preparation of molecular species encapsulated in zeolites, as commented previously [47–55]. Thus, in one of the few existing precedents to in situ assembly of precursors inside MOFs, Li et al. [65] have reported on the preparation of metal(II) phthalocyanine molecules by contacting 1,2-dicyanobenzene at 190°C with an anionic MOF previously ion-exchanged with Co(II), Ni(II), or Cu(II) cations. Co-phthalocyanine-encapsulated MOF thus prepared was tested as catalyst for the epoxidation of styrene using *tert*-butyl hydroperoxide as oxidant, attaining 72% conversion of styrene after 16 h at 60°C, with 65% selectivity to the epoxide. These results surpassed the performance of the Co-phthalocyanine in solution (38% conversion, 60% selectivity), and this was attributed to the self-deactivation of the phthalocyanine in solution due to $\pi - \pi$ stacking, which is prevented when the metal complex is encapsulated inside the MOF. The Co-pthalocyanine MOF was reused up to three cycles without a significant drop in the catalytic activity [65].

Navarro and coworkers have reported on the biophysical characterization and antitumor activity of the drug $[(\eta 6\text{-}p\text{-cymene})_2Ru_2\text{mitoxantronato}]Cl_2$ [66]. This compound was found to be too large to enter the pores of the mesoporous iron trimesate, MIL-100(Fe), featuring large cavities of about 2.9 nm internal diameter, accessible through windows of 0.86×0.86 nm (too small to allow the adsorption of the metallodrug from a solution). Nevertheless, the authors succeeded in encapsulating the drug inside the cavities of MIL-100(Fe) by following a three-step in situ assembly, as shown in Scheme 10.4. Thus, MIL-100(Fe) was first loaded with $[Ru(H_2O)_3Cl_3]$, followed by reaction with α-terpinene to form the dinuclear complex $[Ru(p\text{-cymene})Cl]_2Cl_2$, which was further reacted with mitoxantrone in EtOH to yield the target metallodrug compound.

4.1.2 Templated Growth

An alternative method for encapsulating large species inside the cavities of a porous material consists in performing the crystallization of the solid support from a solution containing the target species. Thus, the solid host is assembled around the guest species that remain imprisoned inside the pores. In this sense, the guest species acts as a "mold" or a template for the growing solid, and this is why this method is sometimes called *templated growth*. In the case of zeolite synthesis, certain "template" molecules (usually tetraalkylammonium salts or amines) can act as true structure-directing agents (SDA), in the sense that their presence during the crystallization of the zeolite determine which type of zeolite is formed among various possible structures [67]. In most cases, the SDA molecules used in the synthesis of zeolites remain trapped inside the pores and must be eliminated (usually by calcination) to make the internal pore volume available to adsorbed species. However, when the molecule that remains imprisoned inside the pores brings a new, desirable functionality to the composite host–guest system (such as a new catalytic function), the template synthesis can be a highly convenient preparative method.

However, two evident limitations of the template synthesis are that (1) the guest species has to be stable under the synthesis conditions of the host material (which, especially in the case of zeolites, can be relatively harsh in terms of pH or

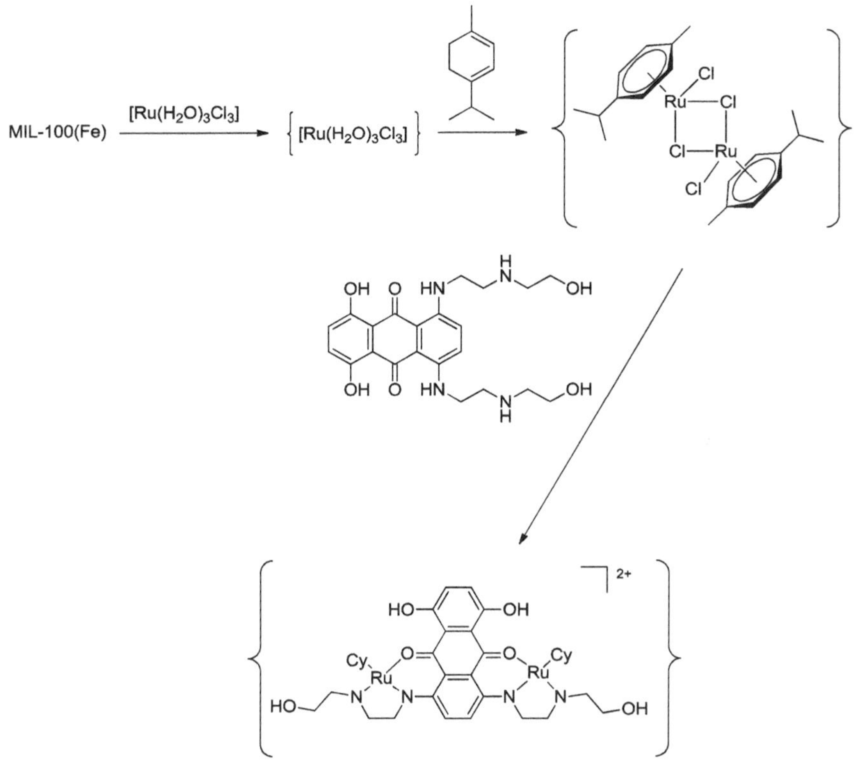

SCHEME 10.4 In situ assembly of the Ru metallodrug inside the pores of MIL-100(Fe). The brackets generically represent the pore system of the solid host. *(Adapted from Ref. [66] with permission of Royal Society of Chemistry. Copyright 2014.)*

temperature), and (2) the guest species should not disrupt the formation of the host nor alter the expected crystalline structure of the solid. This last point is critical for materials that, as in the case of zeolites and MOFs, can crystallize in various forms depending on the particular layout of the building units, TO_4 tetrahedra (T = Si, Al) in the case of zeolites, and metallic SBUs and organic linkers in the case of MOFs.

While in situ assembly strategies described in the previous section are generally preferred for preparing Metal@MOF and Polymer@MOF composites, template synthesis is the most extensively used strategy for the preparation of MOFs having discrete molecular species encapsulated. The following are two selected examples that serve to illustrate the use of these preparation procedures and their potential.

Eddaoudi and coworkers pioneered the encapsulation of metalloporphyrins inside MOFs through a template synthesis approach [68]. These authors used an indium-containing MOF with a *rho*-type topology using the 4,5-imidazoledicarboxylate ligand, denoted *rho*-ZMOF [69]. The *rho* topology is composed of truncated cuboctahedra, or α cages, connected by double eight member rings (D8Rs). In *rho*-ZMOF the α cages have internal diameter of

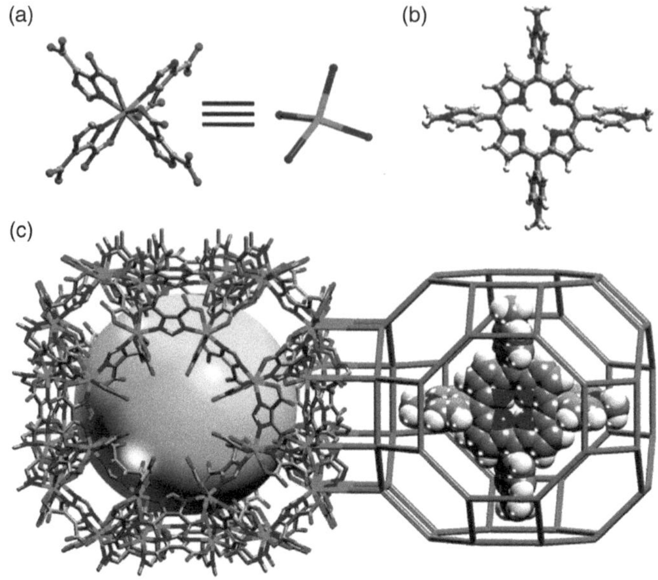

FIGURE 10.2 A *rho*-MOF containing encapsulated metal porphyrins as oxidation catalysts. *(Reproduced from Ref. [68] with permission of American Chemical Society. Copyright 2008.)*

18.2 Å, while the 8-membered ring apertures giving access to them are 9 Å in diameter [69]. When the authors performed the synthesis of *rho*-ZMOF in the presence of 5,10,15,20-tetrakis(1-methyl-4-pyridinio)porphyrin tetra(*p*-toluenesulfonate) ([H_2TMPyP][*p*-tosyl]$_4$, **L^6**), *rho*-MOF was obtained having porphyrin encapsulated inside the cavities (see Fig. 10.2). Further metallation by impreganation of the solid with methanolic solutions of various metal nitrates (Mn, Cu, Zn, or Co ions) lead to the corresponding metal porphyrins.

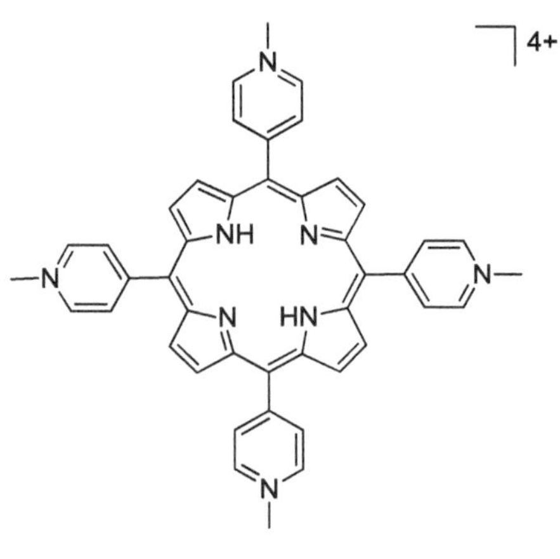

L^6

The authors demonstrated the catalytic properties of the resulting composite materials. Thus, *rho*-ZMOF with encapsulated Mn porphyrin was found to be active for cyclohexane oxidation using *tert*-butyl hydroperoxide as the oxidant, cyclohexanol and cyclohexanone being the only products detected (91.5% total yield after 24 h at 65°C and with 3.8% catalysts loading). The solid was found to retain its crystallinity, no leaching of the metalloporphyrin was detected, and it was reused for 11 catalytic cycles without significant lose of activity [68].

Several polyoxometalates (POM) have been successfully incorporated inside the pores of MOFs, including $[Cu_3(BTC)_2]$ [70,71], MIL-101(Cr) [44,72,73] and NH_2-MIL-101(Al) [74]. Thus, direct impregnation of MIL-101(Cr) with solutions containing the Keggin-type $H_3PW_{12}O_{40}$ POM readily leads to the corresponding POM/MOF material. However, dispersion of the POM inside the MOF is not optimal, since only the large cavities of MIL-101 are filled with POM, while the small cavities remain empty. This is because only the large cavities of MIL-101 have pore windows large enough to allow the diffusion of the POM. Moreover, leaching of the POM to the solution has been observed for these systems prepared by impregnation, since immobilization is only based on adsorption equilibrium. In other words, direct impregnation of MIL-101 with POM does not lead to a true encapsulated (physically imprisoned) system. Conversely, Gascon and coworkers [75] reported on the preparation of analogous POM@MIL-101 (Cr) composites in which the POM was added during the synthesis of the MOF, following a templated synthesis approach. In this way, the authors showed that, besides incorporation of the POM inside the large cavities (as in the case of direct impregnation), also the small cavities of the MIL-101 structure were occupied when the POM was added during the synthesis. As a result, much higher POM loadings (up to 50 wt%) were achieved as compared to direct impregnation (15%). Meanwhile, the 5-membered ring windows connecting both type of cavities (1 nm in diameter) were too small to allow diffusion of POM, thus avoiding agglomeration (and corresponding self-deactivation of the catalyst) and leaching to the exterior of the cavities.

The previous two examples show that it is sometimes possible to find the right conditions to allow the self-assembly of the MOF components around a large molecule that is already present in the synthesis medium, in such a way that it remains encapsulated inside the pores of the growing MOF.

As noted previously, preparation of Metal@MOFs is usually carried out by in situ assembly/reduction of suitable pre-adsorbed metal precursors. However, an alternative method for preparing Metal@MOFs involves assembling the MOF around preformed metal NPs [14], usually stabilized by suitable capping agents (such as citrate anions or polyvinylpyrrolidone, or PVP) to avoid aggregation into large particles. Thus, Tsuruoka et al. [76] added Au NPs stabilized by 11-mercaptoundecanoic acid (MUA) to a solution containing $Cu(NO_3)_2{\cdot}3H_2O$ and 1,3,5-benzenetricarboxylic acid, leading to the formation of Au NPs@[$Cu_3(BTC)_2$]. Similarly, Zhang et al. [20] succeeded in preparing the zirconium terephthalate UiO-66 MOF around preformed Pt NPs capped with PVP. Location of the Pt NPs exclusively in the internal surface of the MOF endowed the resulting material with shape-selective properties for the liquid phase hydrogenation of olefins (1-hexene, cyclooctene, *trans*-stilbene, triphenyl ethylene, and tetraphenyl ethylene), and 4-nitrophenol reduction with $NaBH_4$.

Nevertheless, a clear limitation of this method for preparing Metal@MOFs is that the substances used for stabilizing the NPs suspension (MUA or PVP in the previous examples) also remain entrapped inside the pore system of the MOF and may obstruct the interaction of the metal sites with adsorbed molecules, which can be a drawback for the potential of these materials in catalytic applications.

4.1.3 Encapsulation Through Dissociative Linker Exchange

While the previous two methods have been extensively reported for preparing encapsulated species in other porous materials (such as zeolites), there is a third approach that seems exclusive of MOFs and takes advantage of the dynamic metal–ligand association/dissociation equilibrium.

How do you put a giraffe into a refrigerator? The answer to this famous riddle is very simple: Open the refrigerator, put in the giraffe, and close the door. Similarly, the third alternative method of encapsulation in MOFs, recently proposed by Morabito et al. [77], involves adsorbing the guest under conditions that favor linker dissociation, so that short lived "open" states of the pores are formed leading to expanded apertures (*Open the refrigerator* ...). After incorporation of the guest (...*put in the giraffe*...), association of the linkers closes again the large apertures (... *and close the door*), thus trapping the guest molecules inside the MOF. To illustrate the concept, the authors reported the encapsulation of Rhodamine 6G inside a commercial zinc 2-methylimidazolate MOF (ZIF-8), as shown in Fig. 10.3.

4.2 Catalytic Species Encapsulated in MOFs

Different types of species have been successfully encapsulated inside MOFs through use of any of the three methods described above. These species include: metal nanoparticles, metal oxides, quantum dots, polyoxometalates, polymers, graphene, carbon nanotubes, biomolecules, and so on. A brief overview of the systems prepared so far is given below.

4.2.1 Metal Nanoparticles

In recent years, metal nanoparticles (MNPs) have attracted great interest in catalysis due to the unique properties associated with their small size and the high concentration of active surface sites (high surface to volume ratio). However, these MNPs are thermodynamically unstable and they tend to aggregate

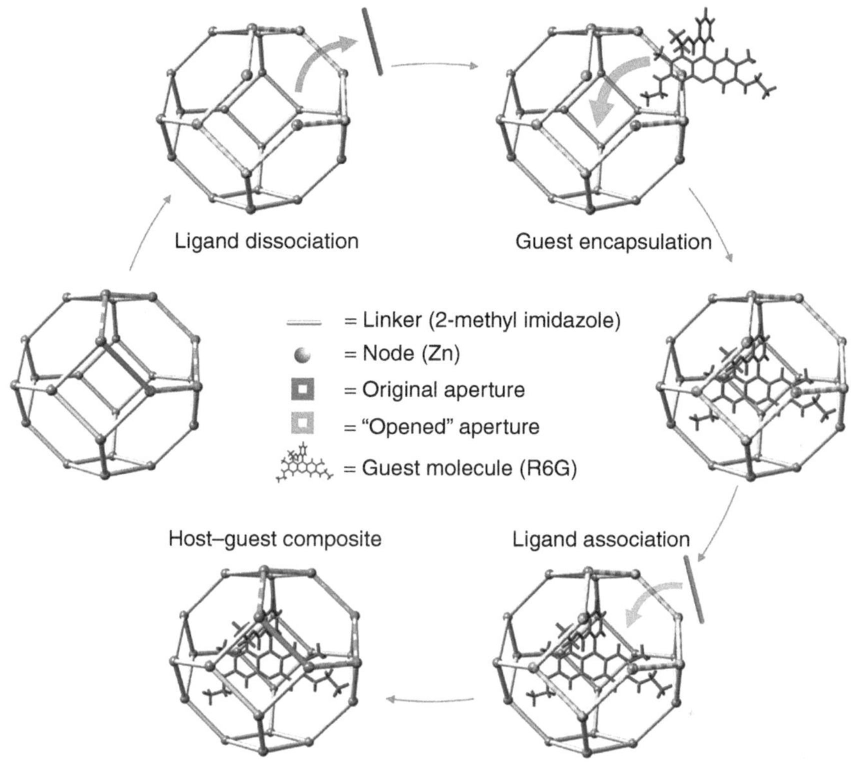

FIGURE 10.3 Encapsulation through dissociative linker exchange. *(Reproduced from Ref. [77] with permission of American Chemical Society. Copyright 2014.)*

spontaneously into larger particles. In order to prevent particle aggregation (and the concomitant decrease of surface active sites), capping agents are usually added during the liquid phase synthesis to stabilize the MNPs [78]. However, these capping agents can be detrimental to the catalytic properties of the resulting MNPs, as they represent a surface contamination which can block the catalytic active sites and hamper their interaction with the reaction substrates. Thus, an alternative way of stabilizing MNPs—without the need for any capping agent—is to employ suitable porous supports for the immobilization of these MNPs. To this end, various porous solids, including zeolites, silicas, and other organic–inorganic materials, have been successfully used [79], including also MOFs. In particular, MOFs offer a great variety of framework structures, featuring pores with tunable connectivity, size, and shape, and with various functional groups lining the pore walls, thus allowing for a rational design of the chemical environment surrounding the MNPs [14,15,80,81]. Clearly the most promising MOFs for preparing encapsulated MNPs are those having large interconnected 3D cavities accessible through small pore windows, allowing the infiltration of the metallic precursors while preventing the agglomeration of the nanoparticles due to size limitation and confinement effects. The pioneering works describing the preparation of MNPs encapsulated in MOFs from pre-adsorbed precursors were reported by the group of Fischer [82–85]. These authors used mainly CVI of organometallic precursors, as they found impregnation from solution was much less efficient. Besides the assembly of pre-adsorbed precursors, some examples exist on alternative preparation procedures, including templated synthesis using preformed capped MNPs [20] or controlled in situ reduction of metalloligands [86]. So far, several metals have been incorporated inside MOFs, including Pd, Pt, Au, Ru, and the like [14,15,80], and the application of the resulting Metal@MOF composites in catalysis is well known [15,21,87–89] for several reactions, including hydrogenation and epoxidation of olefins, alcohol and aldehyde oxidation, CO oxidation, nitroarene reduction, C–C coupling reactions (Suzuki–Miyaura, Heck, Sonogashira, Ullmann), synthesis of methyl isobutyl ketone, and so on.

When considering the potential use of MNPs@MOFs in catalysis (or the use of any other type of encapsulated species), one important issue that must be taken into account is that of MOF stability. In order to exploit the outstanding textural properties of the MOF support, it is evident that the material must withstand the synthesis conditions of metal encapsulation (infiltration/reduction/aggregation) while preserving the crystalline structure. Therefore, though much of the earlier studies on MNPs@MOFs used mainly MOF-5 and HKUST-1 as hosts, later works focused on highly thermally and chemically stable materials, such as MIL-101 [90], ZIFs [91], or UiO-type [21,92] MOFs.

4.2.2 Organic and Metal Coordination Compounds

Hereafter, some examples are presented to illustrate the large variety of molecular (discrete) organic and metallorganic compounds that have been successfully incorporated inside the pores of MOFs. Most often, these systems have been prepared by adding the species directly during the synthesis of the MOF, though some limited examples also exist on the assembly of precursors inside the pores.

Bogaerts et al. [93] have described the preparation of the chiral Jacobsen Mn(II)-salen complex ($\mathbf{L}^7$) inside the cavities of MIL-101-NH_2(Al) through a templated synthesis. This composite material was used as heterogeneous catalyst for the epoxidation of dihydronaphthalene with *meta*-chloroperoxybenzoic acid and *N*-methylmorpholine *N*-oxide. This salen complex (approximate dimensions $1.7 \times 1.2 \times 0.5$ nm) fits inside the large cages of the MOF (2.9–3.4 nm) but cannot pass through the windows. Therefore, when the complex is added during the synthesis of the MOF, it remains physically immobilized inside the cavities without the need for covalent of coordinative bonds with the support, which might result in changes of the optimal structure of the complex and decrease its high selectivity to the epoxide. Accordingly, the authors showed that the encapsulated catalysts had the same selectivity as the homogeneous complex and was completely recyclable for up to four runs, with a minimal decrease in activity and retention of the selectivity.

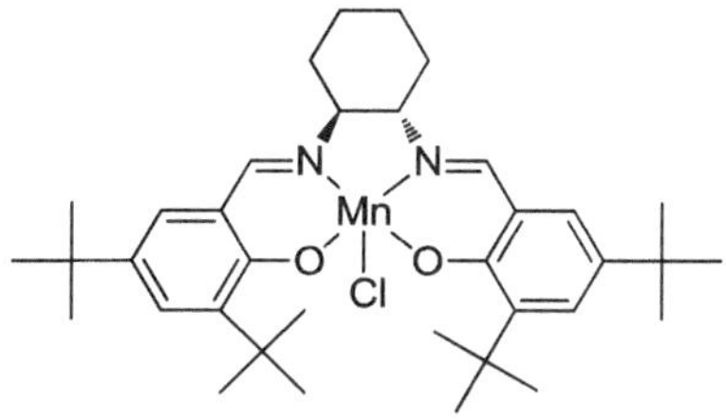

L^7

As mentioned in Section 4.1.2, polyoxomethalates (POM) can be encapsulated inside MOFs through a templated synthesis approach. Sun et al. have described the preparation inside the pores of the copper MOF $[Cu_3(BTC)_2]$ of various POM with general formula $H_nXM_{12}O_{40}]$ (X = Si, Ge, P, As; M = W, Mo) [70]. These POMs were incorporated as monomeric units inside the cuboctahedral cages of the MOF. The catalytic activity of these compounds was shown for a material containing PW_{12} species in the hydrolysis of esters in excess water, revealing a high catalytic activity and repeated reusability without activity loss. Later, Song et al. [94] prepared an analogous copper trimesate MOF having encapsulated $[CuPW_{11}O_{39}]^{5-}$ POMs. Interestingly, these authors showed that the MOF–POM interaction resulted in a synergistic stabilization of both the MOF and the POM, and this was attributed to a close matching of the POM diameter and the MOF pore size. As a result, the composite material was found to be highly active for air based oxidations, including oxidation of thiols to disulfides and H_2S to S_8.

Nevertheless, the well-known limited hydrolytic stability of the $[Cu_3(BTC)_2]$ has triggered a search for more convenient supports for encapsulated POMs, such as the metal trimesates (MIL-100) or terephthalates (MIL-101). These compounds combine larger cavities than those of $[Cu_3(BTC)_2]$ and higher thermal and chemical stability. Thus, several reports exist on the in situ encapsulation of monomeric Keggin-type POMs during the MIL-100 and MIL-101 synthesis [75,95–97], and the catalytic activity of these materials has been demonstrated for several reactions, including dehydration of carbohydrates to 5-hydroxymethylfurfural [96], or the Koevenagel condensation of benzaldehyde with ethyl cyanoacetate, esterification of *n*-butanol with acetic acid and dimethyl ether production from methanol [75].

Sometimes, the addition of a POM during the MOF synthesis can direct the formation of novel compounds that cannot otherwise be obtained without the templating effect of the POM. A clear example of this was reported by Ma et al., who described the synthesis of a sodalite-type compound named NENU-11, by reacting $CuCl_2$ and H_3BTC in the presence of the POM $Na_3PW_{12}O_{40}$. In the resulting material, the Keggin polyoxoanions $[PW_{12}O_{40}]^{3-}$ were present as noncoordinating guests inside the truncated octahedral cages [98] of NENU-11. In the absence of the POM, a MOF with another structure and compositions, $[Cu_3(BTC)_2]$ was obtained. Interestingly, this new NENU-11 material showed potential for decontamination of nerve gas, combining a good adsorption behavior of the model compound dimethyl methylphosphonate with an efficient decontamination through the hydrolysis catalyzed by the guest POM moieties.

4.2.3 Peptides and Proteins

Enzymes are the catalysts of living processes, and as such they fulfill the necessary requirements of high selectivity for reagent binding and formation of the desired products over byproducts while operating under mild conditions compatible with life. Given the high performance of enzymes, many of them have long been isolated and used as catalysts in several reactions in the biotechnology and biochemical fields (eg, in the brewery, detergency, and dairy industries). However, their low thermal stability, difficult recycling and low stability in organic solvents delimit their application in the general chemical industry. To some extent, stability and reusability of enzymes can be increased by immobilizing the enzymes onto solid supports, through covalent binding, membrane encapsulation, cross-linking, sol–gel entrapment, or encapsulation inside a porous support. Since microporous solids (such as zeolites) do contain pores bellow 1 nm (too small to accommodate enzymes inside), it is evident that enzyme encapsulation requires using mesoporous solid supports. In 1996, Balkus and Diaz reported the first immobilization of enzymes in MCM-41 mesoporous materials [99]. Since then, many other mesoporous materials based on silica (including silicas functionalized with organosilanes and PMO) have been successfully used as supports. It is evident that (mesoporous) MOFs can also play an important role as support materials for enzymes, and may also introduce interesting advantages owing to their high tunability of both pore size and functionalities.

Thus, Pisklak et al. [100] described the encapsulation of microperoxidase-11 (MP-11) inside the pores of a mesoporous MOF formed by copper paddlewheel dimeric units connected through biphenyldicarboxylate linkers forming 2D layers and pillared by 1,4-diazabicyclo[2.2.2]octane (DABCO) to form a 3D material. The mesopores of this compound (ca. 2.2 nm) are large enough to allow the adsorption of MP-11 with a specific orientation ($1.1 \times 1.7 \times 3.3$ nm). Therefore, the protein was successfully incorporated inside the MOF by adsorption from solution, up to a final concentration of 30 mmol g^{-1}. No protein leaching was observed upon prolonged stirring of the protein-encapsulated MOF into pure solvent, thus evidencing the effective encapsulation of the protein inside the pores of the MOF. The oxidative ability of the MOF-immobilized MP-11 was evaluated by monitoring the oxidation of methylene blue (MB) with H_2O_2. A conversion of 63% was attained after 1 h with the MP-11@MOF. Meanwhile, a control experiment with free MP-11 only afforded a 6% conversion, while Cu-MOF alone produced no change in the MB concentration under the same experimental conditions.

Some years later, Ma and coworkers [101] also described the immobilization of MP-11 inside another mesoporous MOF, in this case using a Tb MOF

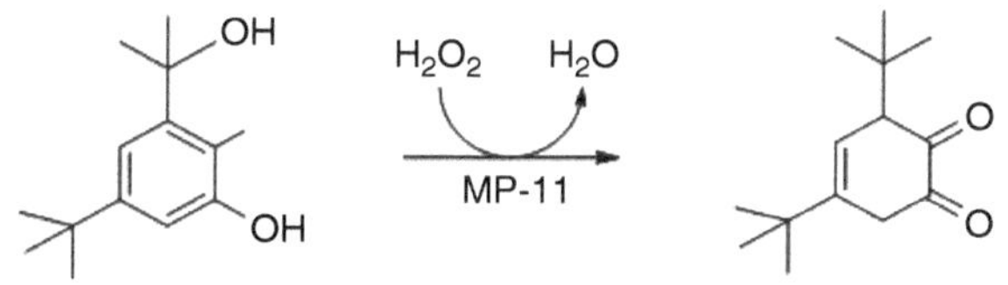

SCHEME 10.5 Oxidation of 3,5-di-*t*-butyl-catechol with H_2O_2 to the corresponding *o*-quinone product catalyzed by MP-11@Tb-mesoMOF. *(Adapted from Ref. [101] with permission of American Chemical Society. Copyright 2011.)*

with nanoscopic cages of 3.9 and 4.7 nm in diameter. A bathochromic shift of the Soret band from 400 nm in the free protein to 410 nm in the MOF-encapsulated protein was indicative of the interactions between the trapped MP-11 molecules and the hydrophobic nanoscopic cages of the MOF. The catalytic activity of MP-11@Tb-mesoMOF was assessed by monitoring the oxidation of 3,5-di-*t*-butyl-catechol with H_2O_2 to the corresponding *o*-quinone product (see Scheme 10.5). The results obtained with the MP-11@Tb-mesoMOF were compared with free MP-11 and MP-11 immobilized in a mesoporous MCM-41 silica.

Free MP-11 showed a high oxidation activity at short reaction times, but started to lose activity very quickly due to aggregation in solution. Although MP-11@Tb-mesoMOF afforded a lower oxidation rate with respect to the free protein, the activity was maintained without deactivation. Therefore, the final catechol conversion attained over MP-11@Tb-mesoMOF (48.7%) was higher than with the free protein (12.3%). For comparison, the MP-11@MCM-41 suffered from a clear protein leaching to the solution (absent in the case of the protein-MOF compound), which also resulted in a low final conversion (17%).

The aforementioned Tb-mesoMOF contains two types of mesoporous cages, arranged in an MTN topology: small cages of 3.9 nm and large cages of 4.7 nm, interconnected through 5- and 6-membered rings with free diameters of 1.3 and 1.7 nm, respectively. Therefore, cytochrome c (Cyt c), whose dimensions are 2.6 × 3.2 × 3.3 nm, should not be able in principle to pass through the pores of this MOF. Nevertheless, according to Ma et al., Cyt c was also able to enter the pores of the above, despite being larger than the access pore sizes [102]. To explain this finding, the authors suggested that the protein can undergo conformational changes to pass through the relatively narrow pores of the MOF, which would require partial unfolding of the protein's tertiary structure. Indeed, fluorescence suggested that Cyt c actually undergoes conformational changes during encapsulation, adopting a conformation that is different from both the native protein and that of the GdnHCl-denatured protein. Therefore, a protein transport mechanism reminiscent of protein translocation into cellular organelles was proposed, based on surface adsorption, partial unfolding to cross the pore, and migration into the mesoporous cage (see Fig. 10.4).

Later, Rambabu et al. [103] also encapsulated Cyt c inside the pores of a Mn-containing 2,6-naphthalene dicarboxylate MOF. The preparation procedure

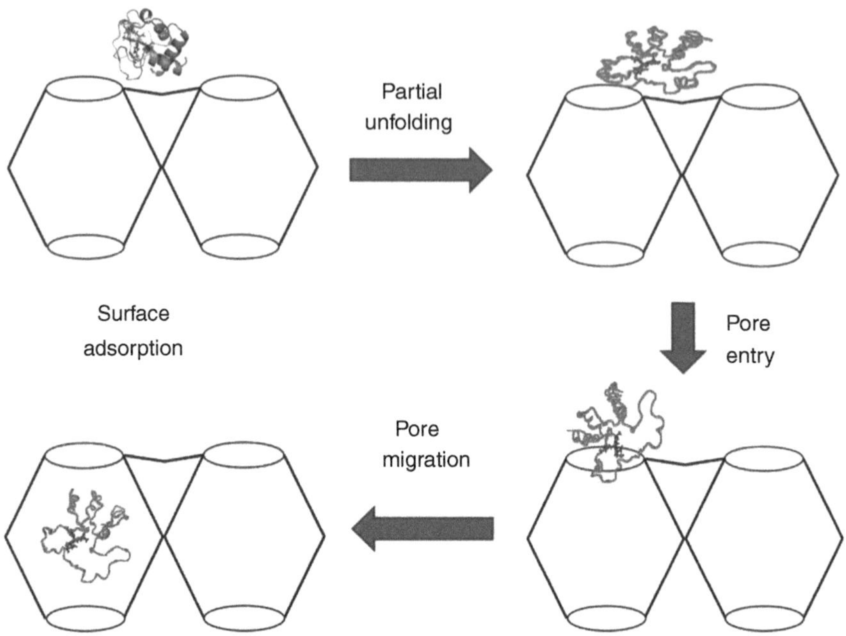

FIGURE 10.4 **Proposed mechanism for the translocation of Cyt c into the cavities of Tb-mesoMOF.** *(Reproduced from Ref. [102] with permission of American Chemical Society. Copyright 2012.)*

simply consisted in contacting desolvated MOF with a saturated solution of Cyt c in HEPES buffer at 37°C for 72 h. According to the authors, and in line with the results reported by Ma and coworkers, Cyt c gets unfolded after coming in contact with the MOF due to surface adsorption and gets entrapped after partial unfolding (as revealed by the presence of a strong absorption band at 285 nm corresponding to tryptophan of the Cyt c in partial unfolded state). Interestingly, due to the adoption of this partial unfolded state and exposure of the tryptophan residue, the material functioned as a fluorogenic probe for anion recognition, and in particular, as a sulfate selective fluorescence chemosensor.

4.2.4 Quantum Dots

Applications of quantum dots encapsulated inside MOFs (QD@MOF composites) are in the fields of photocatalysis, light harvesting, sensing, imaging, and gas storage. A very recent review of the subject can be found in Ref. [104]. An illustrative example was reported by Long et al. [105]. These authors prepared a Zn^{2+}-containing homochiral MOF with D-camphorate and bipyridine ligands. Luminescent CdTe QDs were incorporated inside the cavities of the MOF through a templated synthesis method similar to those described previously for MNP (see Section 4.1.2). Interestingly, the homochiral pore structure of the MOF endowed the resulting QD@MOF composite with enantioselective sensing properties by luminescence quenching. Thus, when this MOF, with a red fluorescence emission centered at 615 nm (λ_{exc} = 365 nm), was contacted with solutions containing six enantiomers (D- and L-tartaric acids, D- and L-dimethyl tartrates, D- and L-mandelic acids), only L-tartaric acid caused a color change of the fluorescence emission from red to blue after 4 h, while the others did not cause any obvious color change at all. Moreover, the authors also observed a linear change in the emission intensity upon contacting different solutions of tartaric acid with enantiomeric excess varying from −100% to 100%, which was attributed to the different extent of the selective QD quenching by the L enantiomer.

4.3 The Benefits of Encapsulation for Catalysis

In general, the presence of the (narrow) pore system of the solid support can introduce diffusion limitations, and therefore a decrease of the accessibility of the encapsulated species to reactants, as compared with the same species in bulk solution or simply grafted on the external surface of a solid, where they are fully accessible. This can be a serious drawback for some of the final applications envisaged for the host–guest composite system, such as sensing or catalysis, for which it is necessary to attain an intimate contact between the encapsulated species and a target molecule (viz., the analyte or the reaction substrate). Moreover, expensive and demanding preparation procedures are usually needed to prepare encapsulated systems (sometimes requiring multistep syntheses), in sharp contrast with the more simple operations commonly used for deposition or grafting. Therefore, it is evident that encapsulation will only be a practical solution when the advantages of the final host–guest composite system can compensate for the possible drawbacks and the necessary efforts taken to prepare it. Otherwise, other alternative heterogeneization procedures, such as deposition or grafting of the active species on the external surface of a solid support, can become preferred over encapsulation. Therefore, all these considerations should be properly taken into account when engineering a catalyst based on an encapsulated species inside an MOF (and in general in any other porous material).

After a thorough revision of the examples found in the literature, the benefits of encapsulation of active species in MOFs mainly consist in: (1) increased stability and prevention of self-deactivation/aggregation of the encapsulated species, (2) development of shape-selective properties upon encapsulation, and (3) possibility of designing multifunctional catalytic systems.

4.3.1 Increased Stability and Avoidance of Self-Deactivation/Aggregation

The restricted space available inside the MOF cavities can prevent the aggregation of the species hosted inside beyond a certain size limit. This is of paramount importance for species with a strong particle size–activity relationship, such as MNPs [106] for which the catalytic activity may sharply decreases when size of the particle increases. As an example, El-Shall et al. [107] reported on the preparation of Pd, Cu, and bimetallic PdCu MNPs inside the pores of MIL-101(Cr). The materials thus prepared were found to be active for CO oxidation. Among the catalysts tested, Pd@MIL-101 with 2.9 wt% showed the highest catalytic activity with a full conversion at 107°C. The authors proposed that small (2–3 nm) Pd MNPs inside the pores were responsible for most of the observed catalytic activity, while larger (4–6 nm) NPs at the external were surface of the MOF were much less active.

Aggregation can also be highly detrimental for the catalytic activity of molecular species, such as metal coordination complexes. In some cases, aggregation can lead to undesired interference between the active sites of the individual complexes, or even to mutual decomposition or self-deactivation. Once again, the size of the MOF cavities can be small enough to prevent having more than one molecule present per cavity, in such a way that aggregation of the encapsulated species is precluded by site isolation. We have already mentioned previously (Section 4.2.2) the preparation of the chiral Jacobsen Mn(II)-salen complex ($\mathbf{L}^7$) inside the cavities of MIL-101-NH_2(Al) [93], which is a clear example of the benefits of encapsulation through molecule isolation.

Nevertheless, it is relevant to stress here that the effectiveness of the MOF framework to avoid particle aggregation is limited by the resistance and integrity of the framework. In this way, depending on the particular synthesis conditions or post-synthesis treatments employed, it is possible that partial destruction of the MOF skeleton lead to pores with larger dimensions than the normal cavities, wherein aggregation of the host species can take place beyond the size expected from the ideal structure.

4.3.2 Shape-Selective Heterogeneous Catalysts

As noted previously (Section 2), by changing the structural components and their connectivity pattern, it is possible to design the pore dimensions, shape, and connectivity of MOFs. And this opens the possibility of shape-selective properties. It is evident that only the molecules that can cross the pore windows of the MOF will be able to reach and interact with the species encapsulated inside. This is a clear advantage over other composites systems, in which the catalytic species is dispersed (supported or grafted) on the surface of a support that is either nonporous or lacks a regular pore system (such as carbon or amorphous silica), precluding any possibility for control on the molecular traffic. When properly exploited, shape-selective properties of a catalyst may be crucial in determining which molecules can be transformed and which cannot, which products are produced among various possibilities and which are not, and also which specific functional groups of a given molecule can react and which cannot.

We have recently reported on the preparation of Pd NPs encapsulated in zirconium-containing UiO-66 and UiO-67 materials (with terephthalate and biphenyldicarboxylate ligands, respectively). The procedure used consisted in the gas phase loading of a Pd precursor, $[Pd(\eta^3\text{-}C_3H_5)(\eta^5\text{-}C_5H_5)]$, followed by the controlled decomposition by UV irradiation [21]. This method led to the formation of small NPs (1–2 nm) located mainly inside the pore cages of

TABLE 10.2 Turnover Frequencies (TOF) Obtained for AP and BP Hydrogenation Over the Pd-Containing Materials

Catalyst	TOF_{AP} (h^{-1})	TOF_{BP} (h^{-1})	TOF_{AP}/TOF_{BP}
Pd/UiO-66	168.6	78.9	2.12
Pd@UiO-67	55.42	8.6	6.44
Pd@UiO-66	3.2	nil	n.d.

Data adapted from Ref. [21] with permission of *Wiley-VCH Verlag GmbH & Co. KGaA, Weinheim. Copyright 2015.*

the MOF, which are accessible through triangular windows of about 6 and 7 Å in UiO-66 and UiO-67, respectively. These triangular windows effectively control which molecules can access and interact with the Pd NPs located inside the cages of the MOF. Therefore, the material showed shape-selective properties, as we demonstrated by performing the Pd-catalyzed hydrogenation of carbonyl compounds with different steric hindrance (acetophenone and benzophenone, or AP and BP, respectively). For comparison, we also evaluated a Pd/UiO-66 material containing NPs mostly located at the external surface of the MOF, and thus devoid of any shape-selectivity. The turnover frequencies (TOFs) of AP and BP hydrogenation, and the calculated TOF ratios, measured for the different catalysts are shown in Table 10.2.

As can be seen, AP is hydrogenated much faster than BP over Pd/UiO-66 ($TOF_{AP}/TOF_{BP} = 2.12$). Since this sample only contains NPs at the external surface of the MOF, the different reduction rates of AP and BP provide an indication of the relative intrinsic reactivity of the two molecules over Pd-supported UiO-type materials. However, when the reaction was carried out in the presence of Pd@UiO-67, a further decrease of the relative hydrogenation rate of BP with respect to AP was observed ($TOF_{AP}/TOF_{BP} = 6.44$), which reflects the slower diffusion of BP through the channel system of UiO-67. and thus demonstrates the shape-selective properties of this material. In the case of UiO-66, in which the pore windows are considerable smaller than in UiO-67, hydrogenation of AP is considerable slower ($TOF_{AP} = 3.2\ h^{-1}$), while BP is virtually not converted at all, as schematically represented in Fig. 10.5.

4.3.3 Multifunctional Heterogeneous Catalysts

It is well established that MOFs themselves can function as heterogeneous catalysts [10–13,108] due to the presence (either in the pristine MOF or introduced afterwards by post-synthesis modifications) of active sites located at the metallic nodes and/or functional organic linkers. Therefore, combination of two (or more) different "framework" active sites (such as acid/base), or a "framework" active site with encapsulated species located inside the pores, affords a means for preparing multifunctional catalysts. In this way, sequential catalytic transformation (usually referred to as tandem or cascade processes [24,109,110]) can

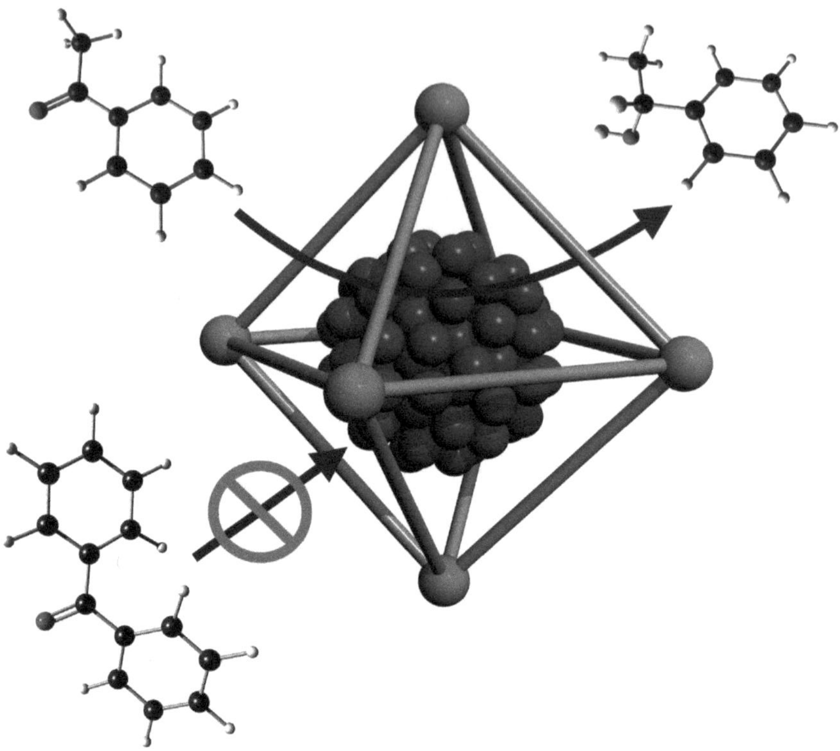

FIGURE 10.5 **Acetophenone can reach the Pd NPs located inside the cages of UiO-66 and it is converted into the corresponding alcohol.** On the contrary, benzophenone cannot enter the pores and it thus does not react. *(Reproduced from Ref. [21] with permission of Wiley-VCH Verlag GmbH & Co. KGaA, Weinheim. Copyright 2015.)*

be performed with a single material and in a single reaction vessel (one-pot), in which the product of a reaction catalyzed by one of the active sites is the substrate of a second reaction catalyzed by the other active site. This represents a remarkable process intensification, which translates into clear economic and environmental benefits; *viz.*, avoidance of the expensive, time-consuming, and energy-consuming separation and purification of intermediate products and reduction of waste products and disposal expenses. We and others have recently reviewed the preparation of multifunctional catalysts based on MOFs, and so the reader is referred to the original papers [11,87]. Among these precedents, materials combining framework active sites (either metal nodes or organic linkers) with encapsulated species represent an important class of MOF-based multifunctional catalysts. So far, in most of the existing examples of multifunctional MOF catalysts, encapsulated species consist of MNPs, whose catalytic activity is combined with coordinatively unsaturated metal sites with Lewis acid properties. In this way, several sequential transformations have been achieved, such as the one-pot synthesis of methyl isobutyl ketone (MIBK) from acetone and H_2 [111], tandem (acid-catalyzed) isomerization of citronellal to isopulegol and hydrogenation to menthol [89], or tandem nitroarene reduction followed by reductive amination and cyclization for the synthesis of secondary amines, quinolines, pyrroles, and arylpyrrolidines [88], to name only a few examples.

5 CONCLUSIONS AND PERSPECTIVES

Throughout this chapter, we have reviewed the work done on MOFs as either nanoreactors or as host matrices for encapsulating species inside their pores. The review is far from exhaustive, but we have rather preferred to illustrate the possibilities of these materials through some selected examples taken from the literature. We outlined the advantages of MOFs as nanoreactors and encapsulation systems, these advantages basically consisting in the MOFs' textural properties: regular and tunable pores and huge specific surface areas, together with easy control of the material composition and functionalization. Instead of concealing the weaknesses of MOFs to make them appear more appealing to the reader, their limitations have been instead highlighted. Thus, when assessing the potential application of these materials, special attention must be paid to the thermal, chemical, and mechanical stability of these compounds, which in certain cases can be rather limited when compared with other alternative (inorganic) counterparts.

The examples reported in the chapter have been roughly divided in two parts, one dealing with the MOF as the confined space in which a chemical process takes place (nanoreactor), the other with the MOF as the solid support used to incorporate exogenous species through encapsulation. Concerning encapsulation, we first present a short review of the existing methods for preparing encapsulated species inside the MOF pores, and the various types of encapsulated systems prepared so far. The chapter ends by illustrating the possible benefits of using MOF-based composites:(1) increased stability of the encapsulated species, (2) shape-selectivity, and (3) multifunctionality.

The reader must take into account that the research on MOFs is rather recent. While a handful of materials that nowadays would be considered as MOFs were already prepared in the 1960s, the enormous interest in MOFs started roughly at the end of the 1990s, that is, less than 20 years ago. This is a very short trajectory as compared to other porous solids, such as zeolites or clays, with more than 50 years of intensive research. At the beginning of this period, MOFs were usually regarded with caution, sometimes even with skepticism. But their potential went on to be amply demonstrated for a range of applications, including adsorption, sensing, catalysis (the main focus of this chapter), and biomedicine. Every day, new materials with improved properties are prepared, and their application to new processes is described, thus pushing back the frontiers of our knowledge of these materials and paving the way for novel applications in years to come. We can envisage a bright future for MOFs and composite materials such as those described in this chapter. We hope that the examples described here, along with the original reports listed in the reference section, will help the readers (be they experienced researchers or newcomers to this field) as a valuable guide to the hottest research and discovery in this fast-evolving field of nanoreactors and encapsulated systems.

ACKNOWLEDGMENTS

The authors thank the Spanish Ministry of Science and Innovation (projects Consolider-Ingenio 2010 (CSD-2009-0050), the SEVERO OCHOA program (project SEV-2012-0267), and project MAT2014-52085-C2-1-P) and the Generalitat Valenciana (project AICO/2015/065) for financial support. The European Union's Horizon 2020 research and innovation program under the Marie Skolodowska-Curie, grant agreement number 641887, is also acknowledged.

REFERENCES

[1] Meier WM, Olson DH, Baerlocher C. Atlas of zeolite structure types. Zeolites 1996;17: 1–229.

[2] Deng H, Grunder S, Cordova KE, Valente C, Furukawa H, Hmadeh M, et al. Large-pore apertures in a series of metal-organic frameworks. Science 2012;356:1018–23.

[3] Farha OK, Eryazici I, Jeong NC, Hauser BG, Wilmer CE, Sarjeant AA, et al. Metal-organic framework materials with ultrahigh surface areas: is the sky the limit? J Am Chem Soc 2012;134:15016–21.

[4] Burrows AD. Post-synthetic modification of MOFs. In: Llabrés i Xamena FX, Gascon J, editors. Metal organic frameworks as heterogeneous catalysts. Cambridge: The Royal Society of Chemistry; 2013. p. 31–75.

[5] Wang ZQ, Cohen SM. Postsynthetic modification of metal-organic frameworks. Chem Soc Rev 2009;38:1315–29.

[6] Suh MP, Park HJ, Prasad TK, Lim D-W. Hydrogen storage in metal-organic frameworks. Chem Rev 2012;112:782–835.

[7] Sumida K, Rogow DL, Mason JA, McDonald TM, Bloch ED, Herm ZR, et al. Carbon dioxide capture in metal-organic frameworks. Chem Rev 2012;112:724–81.

[8] Kreno LE, Leong K, Farha OK, Allendorf MD, Van Duyne RP, Hupp JT. Metal-organic framework materials as chemical sensors. Chem Rev 2012;112:1105–25.

[9] Horcajada P, Gref R, Baati T, Allan PK, Maurin G, Couvreur P, et al. Metal-organic frameworks in biomedicine. Chem Rev 2012;112:1232–68.

[10] Corma A, Garcia H, Llabrés i Xamena FX. Engineering metal organic frameworks for catalysis. Chem Rev 2010;110:4606–55.

[11] Gascon J, Corma A, Kapteijn F, Llabrés i Xamena FX. Metal organic framework catalysis: Quo vadis? ACS Catal 2014;4:361–78.

[12] Llabrés i Xamena FX, Gascon J. Metal organic frameworks as heterogeneous catalysts. In: Spivey JJ, editor. RSC catalysis series. 1st ed. Cambridge: The Royal Society of Chemistry; 2013.

[13] Lee JY, Farha OK, Roberts J, Scheidt KA, Nguyen ST, Hupp JT. Metal-organic framework materials as catalysts. Chem Soc Rev 2009;38:1450–9.

[14] Rösler C, Fischer RA. Metal-organic frameworks as hosts for nanoparticles. CrystEngComm 2015;17:199–217.

[15] Dhakshinamoorthy A, Garcia H. Catalysis by metal nanoparticles embedded on metal–organic frameworks. Chem Soc Rev 2012;41:5262–84.

[16] Juan-Alcaniz J, Ramos-Fernandez EV, Kapteijn F, Gascon J. MOFs as nano-reactors. In: Llabrés i Xamena FX, Gascon J, editors. Metal organic frameworks as heterogeneous catalysts. Cambridge: The Royal Society of Chemistry; 2013. p. 310–43.

[17] Llabrés i Xamena FX, Abad A, Corma A, Garcia H. MOFs as catalysts: activity, reusability and shape-selectivity of a Pd-containing MOF. J Catal 2007;250:294–8.

[18] Horike S, Dinca M, Tamaki K, Long JR. Size-selective lewis acid catalysis in a microporous metal-organic framework with exposed Mn^{2+} coordination sites. J Am Chem Soc 2008;130:5854–5.

[19] Roberts JM, Fini BM, Sarjeant AA, Farha OK, Hupp JT, Scheidt KA. Urea metal-organic frameworks as effective and size-selective hydrogen-bond catalysts. J Am Chem Soc 2012;134:3334–7.

[20] Zhang W, Lu G, Cui C, Liu Y, Li S, Yan W, et al. A family of metal-organic frameworks exhibiting size-selective catalysis with encapsulated noble-metal nanoparticles. Adv Mater 2014;26:4056–60.

[21] Luz I, Rösler C, Epp K, Llabrés i Xamena FX, Fischer RA. Pd@UiO-66-type MOFs prepared by chemical vapor infiltration as shape-selective hydrogenation catalysts. Eur J Inorg Chem 2015;2015:3904–12.

[22] Csicsery SM. Shape-selective catalysis in zeolites. Zeolites 1984;4:202–300.

[23] Falkowski JM, Liu S, Lin W. Asymmetric catalysis with chiral metal organic frameworks. In: Llabrés i Xamena FX, Gascon J, editors. Metal organic frameworks as heterogeneous catalysts. Cambridge: The Royal Society of Chemistry; 2013. p. 344–64.

[24] Climent MJ, Corma A, Iborra S. Heterogeneous catalysts for the one-pot synthesis of chemicals and fine chemicals. Chem Rev 2011;111:1072–133.

[25] Cavka JH, Jakobsen S, Olsbye U, Guillou N, Lamberti C, Bordiga S, et al. A new zirconium inorganic building brick forming metal organic frameworks with exceptional stability. J Am Chem Soc 2008;130:13850–1.

[26] Ferey G, Mellot-Draznieks C, Serre C, Millange F, Dutour J, Surble S, et al. A chromium terephthalate-based solid with unusually large pore volumes and surface area. Science 2005;309:2040–2.

[27] Colombo V, Galli S, Choi HJ, Han GD, Maspero A, Palmisano G, et al. High thermal and chemical stability in pyrazolate-bridged metal–organic frameworks with exposed metal sites. Chem Sci 2011;2:1311–9.

[28] Huang LM, Wang HT, Chen JX, Wang ZB, Sun JY, Zhao DY, et al. Synthesis, morphology control, and properties of porous metal-organic coordination polymers. Microporous Mesoporous Mater 2003;58:105–14.

[29] Hausdorf S, Wagler J, Mossig R, Mertens FORL. Proton and water activity-controlled structure formation in zinc carboxylate-based metal organic frameworks. J Phys Chem A 2008;112:7567–76.

[30] Van de Voorde B, Stassen I, Bueken B, Vermoortele F, de Vos DE, Ameloot R, et al. Improving the mechanical stability of zirconium based metal–organic frameworks by incorporation of acidic modulators. J Mater Chem A 2015;3:1737–42.

[31] Uemura T, Kitagawa K, Horike S, Kawamura T, Kitagawa S, Mizuno M, et al. Radical polymerisation of styrene in porous coordination polymers. Chem Commun 2005;5968–70.

[32] Uemura T, Horike S, Kitagawa K, Mizuno M, Endo K, Bracco S, et al. Conformation and molecular dynamics of single polystyrene chain confined in coordination nanospace. J Am Chem Soc 2008;130:6781–8.

[33] Uemura T, Kitaura R, Ohta Y, Nagaoka M, Kitagawa S. Nanochannel-promoted polymerization of substituted acetylenes in porous coordination polymers. Angew Chem, Int Ed 2006;45:4112–6.
[34] Pan L, Liu H, Lei X, Huang X, Olson DH, Turro NJ, et al. RPM-1: a recyclable nanoporous material suitable for ship-in-bottle synthesis and large hydrocarbon sorption. Angew Chem Int Ed 2003;42:542–6.
[35] Turro NJ. From boiling stones to smart crystals: supramolecular and magnetic isotope control of radical-radical reactions in zeolites. Acc Chem Res 2000;33:637.
[36] Corma A, Iglesias M, Llabrés i Xamena FX, Sanchez F. Cu and Au metal–organic frameworks bridge the gap between homogeneous and heterogeneous catalysts for alkene cyclopropanation reactions. Chem Eur J 2010;16:9789–95.
[37] Santiago-Portillo A, Navalón S, Cirujano FG, Llabrés i Xamena FX, Alvaro M, Garcia H. MIL-101 as reusable solid catalyst for autoxidation of benzylic 2 hydrocarbons in the absence of additional oxidizing reagent. ACS Catal 2015;5:3216–24.
[38] Kitagawa S, Kitaura R, Noro S. Functional porous coordination polymers. Angew Chem, Int Ed 2004;43:2334–75.
[39] Loiseau T, Serre C, Huguenard C, Fink G, Taulelle F, Henry M, et al. A rationale for the large breathing of the porous aluminum terephthalate (MIL-53) upon hydration. Chem Eur J 2004;10:1373–82.
[40] Nijen N, Wu H, Canepa P, Marti A, Balkus KJ, Thonhauser T, et al. Tuning the gate opening pressure of metal-organic frameworks (MOFs) for the selective separation of hydrocarbons. J Am Chem Soc 2012;134:15201–4.
[41] Valvekens P, Vermoortele F, de Vos DE. Metal–organic frameworks as catalysts: the role of metal active sites. Catal Sci Technol 2013;3:1435–45.
[42] Vermoortele F, Valvekens P, de Vos DE. Catalysis at the metallic nodes of MOFs. In: Llabrés i Xamena FX, Gascon J, editors. Metal organic frameworks as heterogeneous catalysts. The Royal Society of Chemistry; 2013. p. 268–88.
[43] Mondloch JE, Farha OK, Hupp JT. Catalysis at the organic ligands. In: Llabrés i Xamena FX, Gason J, editors. Metal organic frameworks as heterogeneous catalysts. The Royal Society of Chemistry; 2013. p. 289–309.
[44] Ferey G, Serre C, Mellot-Draznieks C, Millange F, Surble S, Dutour J, et al. A hybrid solid with giant pores prepared by a combination of targeted chemistry, simulation, and powder diffraction. Angew Chem, Int Ed 2004;43:6296–301.
[45] Raymo FM, Stoddart JF. Interlocked macromolecules. Chem Rev 1999;99:1643–63.
[46] Balzani V, Credi A, Raymo FM, Stoddart JF. Artificial molecular machines. Angew Chem, Int Ed 2000;39:3348–91.
[47] Corma A, Garcia H. Supramolecular host-guest systems in zeolites prepared by ship-in-a-bottle synthesis. Eur J Inorg Chem 2004;2004:1143–64.
[48] Herron N. A cobalt oxygen carrier in zeolite Y. A molecular "ship in a bottle". Inorg Chem 1986;25:4714–7.
[49] Zakharov VY, Zahharova OM, Romanovskii RE. Determination of dispersion of phthalocyanine in zeolite matrix. React Kinet Catal Lett 1977;6:133–7.
[50] Diggruber H, Plath PJ, Schulz-Ekloff G, Mohl M. Study on the structure, stability and propene oxidation capability of faujasite-encaged cobalt chelate complexes. J Mol Catal 1984;24:115–26.
[51] Cosa G, Chretien MN, Galletero MS, Fornes V, Garcia H, Scaiano JC. Photocatalytic activity of a multicomponent system assembled within zeolites: case of 2,4,6-triphenylpyrylium or ruthenium tris(bipyridyl) photosensitizers and titanium dioxide relays within zeolite Y. J Phys Chem B 2002;106:2460–7.

[52] Corma A, Fornes V, Garcia H, Miranda MA, Primo J, Sabater MJ. Photoinduced electron transfer within zeolite cavities: *cis*-stilbene isomerization photosensitized by 2,4,6-triphenylpyrylium cation imprisoned inside zeolite Y. J Am Chem Soc 1994;116: 2276–80.

[53] Sabater MJ, Corma A, Domenech A, Fornes V, Garcia H. Chiral salen manganese complex encapsulated within zeolite Y: a heterogeneous enantioselective catalyst for the epoxidation of alkenes. Chem Commun 1997;1285–6.

[54] Balkus KJ, Khanmamedova AK, Dixon KM, Bedioui F. Oxidations catalyzed by zeolite ship-in-a-bottle complexes. Appl Catal A 1996;143:159–73.

[55] Sulikowski B, haber J, Kubacka A, Pamin K, Olejniczak Z, Ptaszyfiski J. Novel "ship-in-the-bottle" type catalyst: evidence for encapsulation of 12-tungstophosphoric acid in the supercage of synthetic faujasite. Catal Lett 1996;39:27–31.

[56] Aijaz A, Karkamkar A, Choi YJ, Tsumori N, Ronnebro E, Autrey T, et al. Immobilizing highly catalytically active Pt nanoparticles inside the pores of metal-organic framework: a double solvents approach. J Am Chem Soc 2012;134:13926–9.

[57] Uemura T, Hiramatsu D, Kubota Y, Takata M, Kitagawa S. Topotactic linear radical polymerization of divinylbenzenes in porous coordination polymers. Angew Chem, Int Ed 2007;46:4987–90.

[58] Liu B, Shioyama H, Akita T, Xu Q. Metal-organic framework as a template for porous carbon synthesis. J Am Chem Soc 2008;130:5390–1.

[59] Zecchina A, Llabrés i Xamena FX, Paze C, Turnes Palomino G, Bordiga S, Otero Arean C. Alkyne polymerization on the titanosilicate molecular sieve ETS-10. Phys Chem Chem Phys 2001;3:1228–31.

[60] Cardin DJ, Constantine SP, Gilbert A, Lay AK, Alvaro M, Galletero MS, et al. Polymerization of alkynes in the channels of mesoporous materials containing Ni and Zn cations: almost complete filling of the voids. J Am Chem Soc 2001;123:3141–2.

[61] Llabrés i Xamena FX, Otero Arean C, Spera S, Merlo E, Zecchina A. Formaldehyde oligomerization on silicalite: an FTIR and NMR study. Catal Lett 2004;95:51–5.

[62] Alvaro M, Cardin DJ, Colquhoun HM, Garcia H, Gilbert A, Lay AK, et al. Complete filling of zeolite frameworks with polyalkynes formed in situ by transition-metal ion catalysts. Chem Mater 2005;17:2546–51.

[63] Enzel P, Bein T. Inclusion of polyanlline filaments in zeolite molecular sieves. J Phys Chem 1989;93:6270–2.

[64] Ozin GA, Kuperman A, Stein A. Advanced zeolite materials science. Angew Chem, Int Ed Engl 1989;28:359–76.

[65] Li B, Zhang Y, Ma D, Ma T, Shi Z, Ma S. Metal-cation-directed de novo assembly of a functionalized guest molecule in the nanospace of a metal-organic framework. J Am Chem Soc 2014;136:1202–5.

[66] Rojas S, Quartapelle-Procopio E, Carmona FJ, Romero MA, Navarro JAR, Barea E. Biophysical characterisation, antitumor activity and MOF encapsulation of a half-sandwich ruthenium(II) mitoxantronato system. J Mater Chem B 2014;2:2473–7.

[67] Szostak R. Molecular sieves: principles of synthesis and identification. New York: Van Nostrand Reinhold; 1989.

[68] Alkordi MH, Liu YL, Larsen RW, Eubank JF, Eddaoudi M. Zeolite-like metal-organic frameworks as platforms for applications: on metalloporphyrin-based catalysts. J Am Chem Soc 2008;130:12639–41.

[69] Liu YL, Kravtsov VC, Larsen R, Eddaoudi M. Molecular building blocks approach to the assembly of zeolite-like metal-organic frameworks (ZMOFs) with extra-large cavities. Chem Commun 2006;1488–90.

[70] Sun C-Y, Liu S-X, Liang D-D, Shao K-Z, Ren Y-H, Su Z-M. Highly stable crystalline catalysts based on a microporous metal-organic framework and polyoxometalates. J Am Chem Soc 2009;131:1883–8.

[71] Wee LH, Bajpe SR, Janssens N, Hermans I, Houthoofd K, Kirschhock CEA, et al. Convenient synthesis of Cu3(BTC)2 encapsulated Keggin heteropolyacid nanomaterial for application in catalysis. Chem Commun 2010;46:8186–8.

[72] Maksimchuk NV, Timofeeva MN, Melgunov MS, Shmakov AN, Chesalov YA, Dybtsev DN, et al. Heterogeneous selective oxidation catalysts based on coordination polymer MIL-101 and transition metal-substituted polyoxometalates. J Catal 2008;257:315–23.

[73] Bromberg L, Diao Y, Wu H, Speakman SA, Hatton TA. Chromium(III) terephthalate metal organic framework (MIL-101): HF-free synthesis, structure, polyoxometalate composites, and catalytic properties. Chem Mater 2012;24:1664–75.

[74] Ramos-Fernandez EV, Pieters C, van der Linden B, Juan-Alcaniz J, Serra-Crespo P, Verhoeven MWGM, et al. Highly dispersed platinum in metal organic framework NH2-MIL-101(Al) containing phosphotungstic acid—characterization and catalytic performance. J Catal 2012;289:42–52.

[75] Juan-Alcaniz J, Ramos-Fernandez EV, Lafont U, Gascon J, Kapteijn F. Building MOF bottles around phosphotungstic acid ships: one-pot synthesis of bi-functional polyoxometalate-MIL-101 catalysts. J Catal 2010;269:229–41.

[76] Tsuruoka T, Kawasaki H, Nawafune H, Akamatsu K. Controlled self-assembly of metal-organic frameworks on metal nanoparticles for efficient synthesis of hybrid nanostructures. ACS Appl Mater Interfaces 2011;3:3788–91.

[77] Morabito JV, Chou L-Y, Li Z, Manna CM, Petroff CA, Kyada RJ, et al. Molecular encapsulation beyond the aperture size limit through dissociative linker exchange in metal-organic framework crystals. J Am Chem Soc 2014;136:12540–3.

[78] Cushing BL, Kolesnichenko VL, O'Connor CJ. Recent advances in the liquid-phase syntheses of inorganic nanoparticles. Chem Rev 2004;104:3893–946.

[79] White RJ, Luque R, Budarin VL, Clark JH, Macquarrie DJ. Supported metal nanoparticles on porous materials. Methods and applications. Chem Soc Rev 2009;38:481–94.

[80] Meilikhov M, Yusenko K, Esken D, Turner S, Van Tendeloo G, Fischer RA. Metals@MOFs—loading MOFs with metal nanoparticles for hybrid functions. Eur J Inorg Chem 2010;2010:3701–14.

[81] Moon HR, Limb D-W, Suh MP. Fabrication of metal nanoparticles in metal-organic frameworks. Chem Soc Rev 2013;42:1807–24.

[82] Hermes S, Fischer RA, Khodeir L, Schroeder F, Schmid R, Schroeter MK, et al. Metal@MOF: perspectives of metal organic open frameworks as novel host matrices for imbedding functional nanoparticles. Abstracts of Papers of the American Chemical Society 2006; 231:83-INOR.

[83] Hermes S, Schroder F, Amirjalayer S, Schmid R, Fischer RA. Loading of porous metal-organic open frameworks with organometallic CVD precursors: inclusion compounds of the type LnM (a)@MOF-5. J Mater Chem 2006;16:2464–72.

[84] Hermes S, Schroter MK, Schmid R, Khodeir L, Muhler M, Tissler A, et al. Metal@MOF: loading of highly porous coordination polymers host lattices by metal organic chemical vapor deposition. Angew Chem, Int Ed 2005;44:6237–41.

[85] Muller M, Hermes S, Kaehler K, van den Berg MWE, Muhler M, Fischer RA. Loading of MOF-5 with Cu and ZnO nanoparticles by gas-phase infiltration with organometallic precursors: properties of Cu/ZnO@MOF-5 as catalyst for methanol synthesis. Chem Mater 2008;20:4576–87.

[86] Chen L, Chen H, Luque R, Li Y. Metalorganic framework encapsulated Pd nanoparticles: towards advanced heterogeneous catalysts. Chem Sci 2014;5:3708–14.

[87] Juan-Alcaniz J, Gascon J, Kapteijn F. Metal–organic frameworks as scaffolds for the encapsulation of active species: state of the art and future perspectives. J Mater Chem 2012;22:10102–18.

[88] Cirujano FG, Leyva-Pérez A, Corma A, Llabrés i Xamena FX. MOFs as multifunctional catalysts: synthesis of secondary arylamines, quinolines, pyrroles, and arylpyrrolidines over bifunctional MIL-101. Chem Cat Chem 2013;5:538–49.

[89] Cirujano FG, Llabres i Xamena FX, Corma A. MOFs as multifunctional catalysts: one-pot synthesis of menthol from citronellal over a bifunctional MIL-101 catalyst. Dalton Trans 2012;41:4249–54.

[90] Hermannsdorfer J, Friedrich M, Miyajima N, Alburquerque RQ, Kummel S, Kempe R. Ni/Pd@MIL-101: synergistic catalysis with cavity-conform Ni/Pd nanoparticles. Angew Chem, Int Ed 2012;51:11473–7.

[91] Esken D, Turner S, Lebedev OI, Van Tendeloo G, Fischer RA. Au@ZIFs: stabilization and encapsulation of cavity-size matching gold clusters inside functionalized aeolite imidazolate frameworks, ZIFs. Chem Mater 2010;22:6393–401.

[92] Guo Z, Xiao C, Maligal-Ganesh RV, Zhou L, Goh TW, Li X, et al. Pt nanoclusters confined within metal-organic framework cavities for chemoselective cinnamaldehyde hydrogenation. ACS Catal 2014;4:1340–8.

[93] Bogaerts T, Van Yperen-De Deyne A, Liu Y-Y, Lynen F, Van Speybroeck V, Van der Voort P. Mn-salen@MIL101(Al): a heterogeneous, enantioselective catalyst synthesized using a 'bottle around the ship' approach. Chem Commun 2013;49:8021–3.

[94] Song J, Luo Z, Britt DK, Furukawa H, Yaghi OM, Hardcastle K, et al. A multiunit catalyst with synergistic stability and reactivity: a polyoxometalate metal organic framework for aerobic decontamination. J Am Chem Soc 2011;133:16839–46.

[95] Canioni R, Roch-Marchal C, Secheresse F, Horcajada P, Serre C, Hardi-Dan M, et al. Stable polyoxometalate insertion within the mesoporous metal organic framework MIL-100(Fe). J Mater Chem 2011;21:1226–33.

[96] Zhang Y, Degirmenci V, Li C, Hensen EJM. Phosphotungstic acid encapsulated in metal-organic framework as catalysts for carbohydrate dehydration to 5-hydroxymethylfurfural. Chem Sus Chem 2011;4:59–64.

[97] Juan-Alcaniz J, Goesten MG, Ramos-Fernandez EV, Gascon J, Kapteijn F. Towards efficient polyoxometalate encapsulation in MIL-100(Cr): influence of synthesis conditions. New J Chem 2012;36:977–87.

[98] Ma F-J, Liu S-X, Sun CY, Liang DD, Ren G-J, Wei F, et al. A sodalite-type porous metal-organic framework with polyoxometalate templates: adsorption and decomposition of dimethyl methylphosphonate. J Am Chem Soc 2011;133:4178–81.

[99] Diaz JF, Balkus KJ. Enzyme immobilization in MCM-41 molecular sieve. J Mol Catal B 1996;2:115–26.

[100] Pisklak TJ, Macias M, Coutinho DH, Huang RS, Balkus KJ. Hybrid materials for immobilization of MP-11 catalyst. Top Catal 2006;38:269–78.

[101] Lykourinou V, Chen Y, Wang X-S, Meng L, Hoang T, Ming L-J, et al. Immobilization of MP-11 into a mesoporous metal-organic framework, MP-11@mesoMOF: a new platform for enzymatic catalysis. J Am Chem Soc 2011;133:10382–5.

[102] Chen Y, Lykourinou V, Vetromille C, Hoang T, Ming L-J, Larsen RW, et al. How can proteins enter the interior of a MOF? Investigation of cytochrome c translocation into a MOF consisting of mesoporous cages with microporous windows. J Am Chem Soc 2012;134:13188–91.

[103] Rambabu D, Pooja, Pradeep CP, Dhir A. A cytochrome C encapsulated metal organic framework as a bio-material for sulfate ion recognition. J Mater Chem A 2014;2:8628–31.

[104] Aguilera-Sigalat J, Bradshaw D. Synthesis and applications of metal-organic framework–quantum dot(QD@MOF) composites. Coord Chem Rev 2016;307:267–91.

[105] Long Z, Jia J, Wang S, Kou L, Hou X, Sepaniak MJ. Visual enantioselective probe based on metal organic framework incorporating quantum dots. Microchem J 2013;110:764–9.

[106] Campbell CT. The energetics of supported metal nanoparticles: relationships to sintering rates and catalytic activity. Acc Chem Res 2013;46:1712–9.

[107] El-Shall MS, Abdelsayed V, Khder AERS, Hassan HMA, El-Kaderi HM, Reich TE. Metallic and bimetallic nanocatalysts incorporated into highly porous coordination polymer MIL-101. J Mater Chem 2009;19:7625–31.

[108] Farrusseng D, Aguado S, Pinel C. Metal-organic frameworks: opportunities for catalysis. Angew Chem, Int Ed 2009;48:7502–13.

[109] Climent MJ, Corma A, Iborra S, Santos LL. Multisite solid catalyst for cascade reactions: the direct synthesis of benzodiazepines from nitro compounds. Chem Eur J 2009;15:8834–41.

[110] Climent MJ, Corma A, Iborra S, Velty A. Designing the adequate base solid catalyst with Lewis or Bronsted basic sites or with acid–base pairs. J Mol Catal A-Chem 2002;182-183:327–42.

[111] Pan Y, Yuan B, Li Y, He D. Multifunctional catalysis by Pd@MIL-101: one-step synthesis of methyl isobutyl ketone over palladium nanoparticles deposited on a metal–organic framework. Chem Commun 2010;46:2280–2.

Chapter 11

Bionanoreactors: From Confined Reaction Spaces to Artificial Organelles

Viktoriia Postupalenko, Tomaz Einfalt, Mihai Lomora, Ionel A. Dinu, Cornelia G. Palivan
Department of Chemistry, University of Basel, Basel, Switzerland

1 INTRODUCTION

The combination of biomolecules with synthetic systems — such as copolymers, inorganic particles, or solid surfaces — for the development of new complex hybrid materials/devices is a key challenge in the life and nanosciences [1]. Polymer supramolecular structures for the development of hybrid biomaterials have received a good deal of attention because their properties can be controlled to support the combination with biomolecules while preserving the stability and mechanical robustness, which characterize the original materials [2]. Thus, inspired by natural biosynthetic processes, researchers are using compartmentalization of active molecules in supramolecular assemblies to design chemical processes that can be conducted within confined reaction spaces [3,4]. In this chapter, we focus on presenting polymer supramolecular assemblies with sizes in the nanometer range and containing biomolecules that can act in the protected space provided by the assembly and without the necessity of being released, as is the case for conventional drug-delivery systems [5].

The design of bionanoreactors is based on encapsulation/insertion of active compounds (proteins, enzymes, mimics) inside polymeric supramolecular assemblies, such as dendrimers, capsules, and vesicles, in which they can act in situ while being protected from proteolytic attack [3,6]. These supramolecular assemblies should allow the passage of products or substrates to the inner spaces where the active compounds are located in order to support reactions confined within their structures. Various approaches to providing accessibility to the inner reaction space of the polymer assemblies are presented, even though not all have been used in the design of nanoreactors, in order to give an overview of the variety of solutions that can be expected to advance this field of science.

Organic Nanoreactors. http://dx.doi.org/10.1016/B978-0-12-801713-5.00011-2

Polymer assemblies have been used in the development of various reaction spaces, ranging from model nanoreactors [6,7] up to artificial organelles [4,8], either by changing the chemical nature of the polymer assembly or of the active molecules. Model bionanoreactors have provided details of the feasibility of conducting reactions in confined spaces, either as single enzyme mimics [6,9], or for the combination of multiple enzymes encapsulated/inserted in specific regions of vesicles [10,11]. More recently, bionanoreactors have been developed for translational applications, such as detoxification of free radicals associated with oxidative stress [12], production of antibiotics [13], or polymerization in a restricted space [14].

A further step has been achieved by designing bionanoreactors to function as artificial organelles, because of the retention of their activity upon cellular uptake [8]. Such systems, which have a high potential for therapeutic applications, are motivating further efforts to diversify the variety of artificial organelles to truly mimic natural ones, and to offer solutions for pathological conditions. In a biomimetic approach, the concept of multicompartmentalized polymersomes has been introduced for reactions between different inner compartments encapsulated in giant structures that can serve as models for simple cellular compartments [11,15]. We include a summary of requirements needed for successful applications and models or translational examples in the development of bionanoreactors.

2 POLYMERS AS BUILDING BLOCKS FOR NANOREACTORS

Polymers are macromolecules based on a large number of repeating units (monomers) covalently bound in a chain-like molecular architecture with a variety of compositions, structures, and properties. A large diversity of synthetic strategies and monomer combinations can be used to generate linear or nonlinear polymers, such as star-shaped, comb (brush), or branched polymers, and dendrimers (Fig. 11.1).

Depending on the chemical nature of the monomer, either *homopolymers* or *copolymers* can be formed (Fig. 11.2). *Homopolymers* are composed of one single type of repeating unit throughout the whole polymer backbone. When the monomers are charged, the resulting polymers are either anionic or cationic, and are called polyelectrolytes. Moreover, multilayer polyelectrolyte capsules can be formed by alternate adsorption of positively and negatively charged

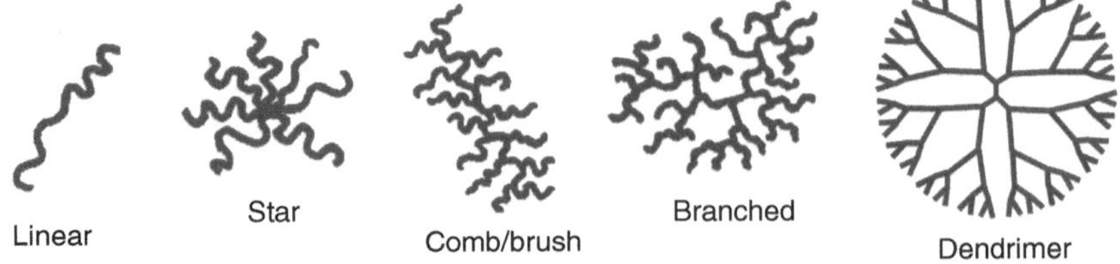

FIGURE 11.1 Various topologies for polymers.

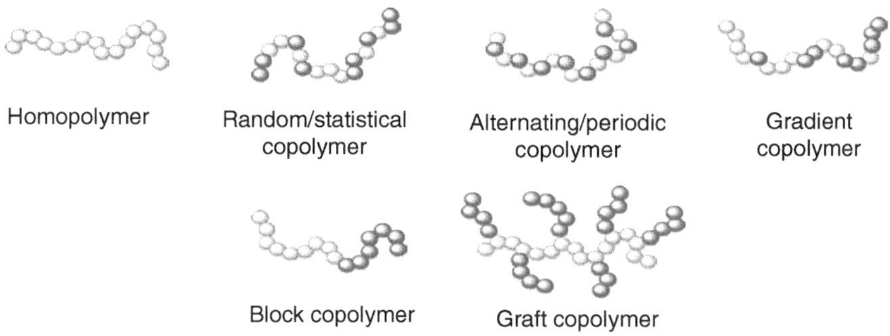

FIGURE 11.2 Schematic representation of polymers with different composition. *(Adapted from Ref. [18] with permission of Nature Publishing Group.)*

polymers; examples include poly(dimethyldiallylammonium chloride) (PDADMAC), poly(styrene sulfonate) (PSS), poly(allylamine hydrochloride) (PAH), poly(ethylenimine) (PEI), poly(L-lysine) (PLL), or dextran sulfate [16,17].

As a result of their intrinsic globular architecture, dendrimers are the only type of nonlinear polymer which can be used to engineer nanoreactors. Dendrimers are polymer systems monodispersed in size and architecture, with properties and functionality that can be adjusted by synthesis and postmodification. Previously, amphiphilic linear-dendritic AB and ABA block copolymers containing dendritic poly(benzyl ether) and linear poly(ethylene oxide) or poly(ethylene glycol) of various molecular weights were applied for the production of nanoreactors [19]. A five-generation dendrimer based on poly(amidoamine) (PAMAM) modified by capping the end groups with carboxybetaine acrylamide (CBAA) was used to generate nanoreactors [19,20].

In contrast, the presence of different chemical repeating units in a polymer chain leads to the formation of *copolymers*. By manipulating their ratio and position in the chain sequence, a large variety of structures can be produced, such as random (statistical), alternating (periodic), gradient, block and graft copolymers (Fig. 11.2) [18]. In nature, there are copolymers (e.g., peptides, proteins, DNA, and RNA) with defined roles in metabolic processes (signaling pathways, transcription of genes, etc.). They have precisely defined sequences of repeating units, specific architectures, and molecular weights, all of this optimized for their roles in biological systems [18]. The distinct sequence of a copolymer is achieved by solid-state synthesis (one monomer is added after another), or in nature by templated enzymatic reactions (e.g., translation and transcription). Of particular interest for the development of nanoreactors are *block copolymers* based on homopolymeric blocks covalently connected into a final structure. The near infinite variety of possible combinations of chemical and physical properties, functional groups, and shapes offered by block copolymers renders them attractive building blocks for nanoreactors. The different physical and chemical properties of the monomers in the individual blocks enable the creation of amphiphilic macromolecules in which hydrophobic and hydrophilic blocks are connected to each

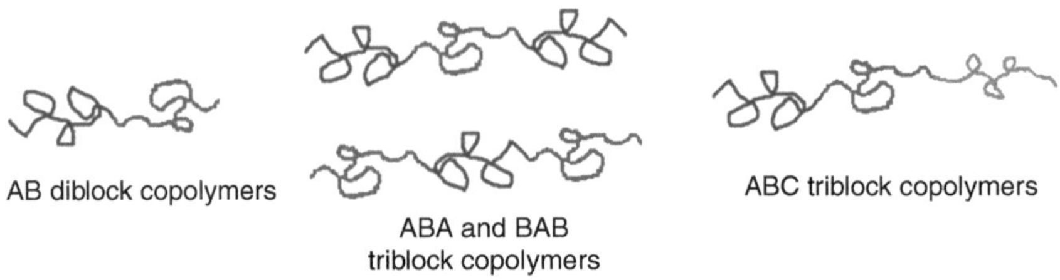

FIGURE 11.3 **Examples of linear block copolymer architectures.**

other. Diblock copolymers result from a combination of a hydrophilic (A) block and a hydrophobic (B) block, whereas triblock copolymers are formed by two hydrophilic blocks flanking one hydrophobic block (ABA or ABC) (Fig. 11.3).

Self-assembly of amphiphilic block copolymers in aqueous media generates supramolecular assemblies with different architectures (vesicles, micelles, tubes, worm-like structures); the vesicles of these architectures are some of the best candidates for the design of nanoreactors [21–23]. The self-assembly process, and therefore the supramolecular assemblies, are strongly affected by the molecular properties of each block and of the overall copolymer chain, such as molecular weight, polydispersity, and hydrophobic to hydrophilic block ratio. Another approach to generating supramolecular assemblies is to use oppositely charged pairs of polyelectrolytes — such as polyethylene poly(ethylene glycol)-*block*-poly(α,β-aspartic acid) (PEG-*b*-P(Asp)) as aniomer and poly(ethylene glycol)-*block*-poly[(5-aminopentyl)-α,β-aspartamide] (PEG-*b*-P(Asp-AP)) or poly[(8-aminooctyl)-α,β-aspartamide] (Homo-P(Asp-C8)) — as catiomers to obtain polyion complex vesicles (PICsomes) [24,25].

Various synthetic techniques have been used to obtain homopolymers and copolymers with defined chemical compositions and low polydispersity: (1) living anionic or cationic polymerization [26,27], (2) controlled ring-opening polymerization (CROP) [28], and (3) controlled radical polymerization (CRP), which includes: (a) atom transfer radical polymerization, (b) reversible addition-fragmentation chain transfer polymerization (RAFT), and (c) nitroxide mediated radical polymerization (NMP) [29–31].

Living ionic polymerization is frequently used to synthesize amphiphilic block copolymers in the absence of irreversible chain transfer and chain termination, where all polymer chains are instantaneously initiated and grow simultaneously [30]. The presence of counterions in the reaction medium associated with the propagating chains preserves the electroneutrality of the entire system. However, these living polymerization methods are affected significantly by the nature of the solvent and the presence of water, carbon dioxide, and/or impurities. Moreover, they have limited applications for the synthesis of block copolymers with functional groups as side chains.

CROP techniques are in general restricted to variations in solution conditions and less to bulk properties [28,30]. Although widely used for the synthesis of polymers with a well-defined size and composition, they require an appropriate catalyst in order to proceed, and to obtain polymers with controllable properties.

As the catalysts used in CROP are based on heavy metals, it is necessary to completely remove them when biomedical and pharmaceutical applications are intended [32].

Recent developments in CRP provide a compatible route to the synthesis of block copolymers with functional side groups [29]. CRP is less affected by the presence of impurities, and provides better conditions for chain growth based on a rapid and dynamic equilibrium between dormant chains and propagating radicals [30].

The amphiphilic block copolymers most commonly used for the development of nanoreactors consist of hydrophilic blocks, such as poly(acrylic acid) (PAA), poly(ethylene oxide) (PEO), poly(2-methyl oxazoline) (PMOXA), or poly[L-isocyanoalanine(2-thiophen-3-yl-ethyl)amide] (PIAT), and a hydrophobic block such as polystyrene (PS), polybutadiene (PB), or poly(dimethylsiloxane) (PDMS) [32–35]. Bionanoreactors have been obtained mainly by using polymer vesicles (polymersomes), but there are also a few reports based on other supramolecular assemblies (Table 11.1).

3 3D POLYMER SUPRAMOLECULAR ASSEMBLIES

3.1 Supramolecular Structures as Spaces for the Generation of Nanoreactors

Several polymer supramolecular structures such as dendrimeric structures (Fig. 11.1), polymersomes, PICsomes, and layer-by-layer (LbL) capsules (Fig. 11.4) have been used to provide defined reaction spaces, where catalytically active molecules can perform their activity [19,25,56,71]. Such polymer 3D supramolecular assemblies are generated by various methods: (1) specific synthesis (dendrimers), (2) self-assembly (polymersomes and PICsomes), and (3) adsorption of polymer layers on colloidal templates (LbL capsules) [25,37,56,71].

3.1.1 Dendrimer Structures

Dendrimers are nanosized, radially branched symmetric molecules with well-defined, homogeneous, and monodispersed morphology that have a symmetric core, an internal shell, and an external shell [72]. The increased number of terminal groups on their surfaces allows multiple ligand attachment sites, thus increasing the probability of interaction with guest molecules. However, the special conditions needed for dendrimer synthesis remain a major drawback in their development. Dendrimers are obtained either by divergent or convergent synthesis: that is, growth of the branched structure either starts from the core of the dendrimer to which the macromolecular branches are attached (divergent) or from the branches into the core (convergent). As a consequence of their highly branched structures, dendrimers possess an overall globular shape, which can reach up to 15 nm in diameter [72]. The large number of terminal groups on their surfaces provides multiple ligand attachment sites and a high probability of interaction. The globular structures of dendrimers contain inner cavities that

TABLE 11.1 Nanoreactor-Forming Polymers

Polymer	Nanostructure	Active molecule	Potential application	References
PAMAM	Dendrimers	Laccase	Model for therapeutics	[36]
PAMAM	Dendrimers	Hemin	Model for therapeutics	[20]
PBE-*b*-PEO/PBE-*b*-PEG	Dendrimers	Laccase	Model for biosensing	[37]
PEO-*b*-PEE, PEO-*b*-PB	Polymersomes	HB, Myoglobin	Therapeutics (oxygen carrier)	[38]
PEO-*b*-PCL, PEO-*b*-PLA	Polymersomes	HB	Model for therapeutics (oxygen carrier)	[39,40]
PS-*b*-PAA	Polymersomes	α-Chymotrypsin, trypsin, cytochrome c	Model for therapeutics	[41,42]
PEG-*b*-PPS-*b*-PEG	Polymersomes	GOx	Model for drug delivery, biosensing, diagnostics	[43]
PS-*b*-PIAT	Polymersomes	CalB lipase, GOx, HRP, chloroperoxidase	Therapeutics, biosensing, diagnostics, catalysis, conducting materials	[10,44–46]
PMOXA-*b*-PDMS-*b*-PMOXA	Polymersomes	SOD, HRP, trypsin, nucleoside hydrolase, thymidine phosphorylase, CAT, LPO, Rose Bengal-bovine serum albumin conjugate	Therapeutics, biosensing, diagnostics, photodynamic therapy	[8,12,13, 47–59]
PNVP-*b*-PDMS-*b*-PNVP	Polymersomes	Rose Bengal-bovine serum albumin conjugate	Photodynamic therapy	[48]
PEG-*b*-PSBA	Polymersomes	Myoglobin	Biosensing	[60]
PEG-*b*-PDEAEM-*s*-PDMIEM	Polymersomes	Myoglobin	Model for biosensing	[61]
PEO-*b*-PNBOC	Polymersomes	Alkaline phosphatase	Model for theranostics	[62]
PEO-*b*-P(DEA-*s*-CMA)	Polymersomes	GOx, Hemoglobin	Model for artificial organelles, biomedical applications	[63]

PEG-*b*-PLA	Polymersomes	CAT	Model for therapeutics	[64]
Homo-P(Asp-C8)/PEG-*b*-PAsp	PICsomes	β-Galactosidase	Model for biomedical applications	[25]
PSS/PAH	Capsules	HRP, urease, α-chymotrypsin	Model for drug delivery, diagnostics, catalysis	[65–67]
Dextran sulfate/protamine	Capsules	HRP	Model for biosensing and bioseparation	[68]
Dx/DEAE-Dx	Capsules	HRP, GOx, β-glucosidase	Model for biosensing	[69]
PSS/PDADMAC/lipids	Capsules	HRP	Model for biomedicine, sensors	[70]

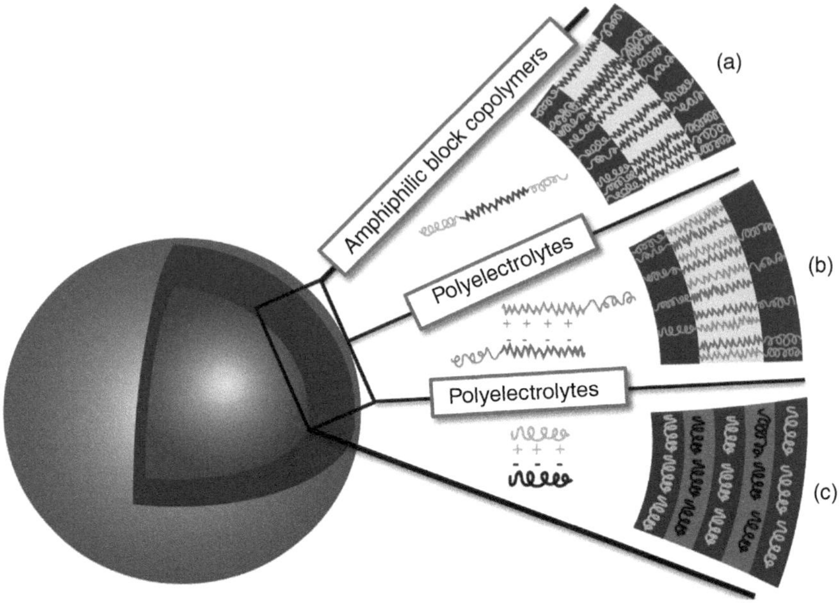

FIGURE 11.4 Schematic representation of supramolecular structures that can serve as nanoreactor compartments: (a) polymersomes, (b) PICsomes, and (c) LbL capsules.

can be exploited for the insertion of catalytic biomacromolecules, and one of their greatest advantages is the possibility of encapsulating active molecules inside their internal cavities [73]. However, up to now only a low encapsulation efficiency of one enzyme per nanostructure has been achieved [37].

3.1.2 Polymeric Vesicles (Polymersomes)

As biological systems have always been an inspiration for scientists, intracellular compartments (such as lysosomes or mitochondria) also have their artificial equivalents in polymer vesicles, called polymersomes. Polymersomes are spherical compartments with a bi- or monolayer membrane, generated by self-assembly of di- or triamphiphilic block copolymers in diluted aqueous conditions. To favor the formation of structures such as polymersomes, it is necessary to have a hydrophilic fraction of the copolymer mass of 25–40%, and polymer concentration above the critical micellar concentration. Other parameters that affect the self-assembly process, and therefore the final architecture of the polymer supramolecular assemblies, are the molecular weight of the copolymer (M_w), block lengths, solubility, and glass transition temperature (T_g) [21,22]. The relative mass or volume fraction of each block is a key parameter in the formation of a self-assembled structure with a certain membrane curvature, and ultimately, with a specific architecture. The M_w of the copolymer (and thus the block lengths) dictates the membrane thickness and polymersome properties, such as membrane fluidity, stability, and permeability [21,74].

Various methods have been used for polymersome preparation, including: (1) direct dissolution, (2) film rehydration, (3) cosolvent, and (4) electroformation. The direct dissolution method is based on slow addition of water to the solid copolymer, whereas in the film rehydration method a preformed copolymer film is hydrated using an aqueous solution under stirring. The cosolvent method makes use of emulsion templates or the dissolution of block copolymers in a suitable organic solvent, followed by the injection of this solution into an aqueous phase (cosolvent or phase-inversion technique). The electroformation method is based on the application of an alternating current of a defined frequency to the deposited polymer, which favors the formation of giant vesicles (GUVs). The direct dissolution and film rehydration techniques are preferable to the organic solvent based method, because organic solvents might be problematic in further bioapplications of the systems in combination with biomacromolecules [75].

Depending on the type of polymer chosen and the block length, polymersomes usually have diameters of 70–400 nm. The higher molecular weight of the amphiphilic block copolymers, compared to phospholipids, results in a larger membrane thickness (up to 30 nm) than that of liposome membranes (3–5 nm) [21,74]. Thicker membranes are characterized by higher robustness and lower permeability [21,76]. The size of polymersomes, as well as the thickness and permeability of their membranes are key parameters in the design of a nanoreactor, and they strongly depend on the chemical properties and the length of each polymer block [77].

One of the first and most used amphiphilic block copolymer for the development of polymer nanoreactors is poly(2-methyloxazoline)-*block*-poly(dimethylsiloxane)-*block*-poly(2-methyloxazoline) (PMOXA-PDMS-PMOXA). As the membrane is permeable only to oxygen species [6,56], it is necessary to render it permeable to other low molecular weight molecules in order to support the in situ catalytic activity of the nanoreactors [6,56]. Other amphiphilic block polymers (such as polystyrene-*block*-poly[L-isocyanoalanine (2-thiophen-3-yl-ethyl)amide] (PS-*b*-PIAT) and poly(*N*-vinyl pyrrolidone)-*block*-poly(dimethylsiloxane)-*block* -poly(N-vinyl pyrrolidone) (PNVP-*b*-PDMS-*b*-PNVP)) that are used to form nanoreactors have a porous membrane that is permeable to low molecular weight molecules [46,47,78]. Amphiphilic block copolymers, such as poly(ethylene glycol)-*block*-poly(styrene boronic acid) (PEG-*b*-PSBA), poly(ethylene glycol)-*block*-[(diethylaminoethyl methacrylate)-*stat*-(3,4-dimethylmaleimidoethyl methacrylate)] (PEG-*b*-PDEAEM-*s*-PDMIEM) or poly(ethylene oxide)-*block* -(2-nitrobenzyloxycarbonyl aminoethyl methacrylate) (PEO-*b*-PNBOC), self-assemble in polymersomes with a membrane that is rendered permeable "on demand" by changes in the environment (pH, etc.) (Fig. 11.5) [61,62]. Another example of a stimuli-responsive polymersome membrane is based on a photo-cross-linkable diblock copolymer poly(ethylene oxide)-*block*-poly[(diethylaminoethyl methacrylate)-*stat*-7-(2-methacryloyloxyethoxy)-4-methyl coumarin] (PEO-*b*-P(DEA-*s*-CMA)). This was used to develop a pH-responsive polymersome where a cross-linked coumarin-rich matrix contributed

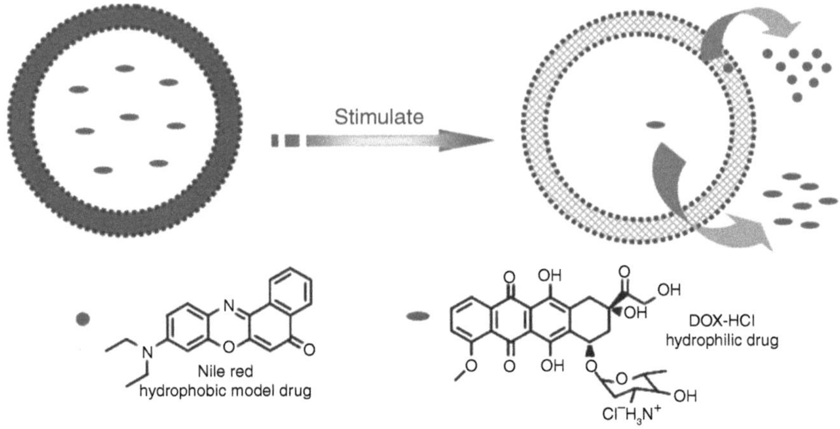

FIGURE 11.5 Polymersome nanocarriers for light regulated co-release of both hydrophobic (Nile red) and hydrophilic (Doxorubicin-HCl) drugs were created by Wang et al. [62]. Once the vesicles were irradiated with UV, the membrane were permeabilized by a change in hydrophobicity, and a co-release of the encapsulated model drugs was observed. *(Adapted from Ref. [62] with permission of Wiler-VCH.)*

to the stability of the system, while coumarin-poor patches functioned as pH-triggered tunnels for low molecular weight molecules [63].

3.1.3 PICsomes

PICsomes are hollow structures delimited by a semipermeable membrane formed by electrostatic interaction between oppositely charged block- and homoionomers [79–81]. For example, for the preparation of PICsomes, PEG block aniomers and homocatiomers were combined by simple mixing in an aqueous environment, filtering, and vortexing [25,82]. This approach avoids the use of organic solvents, which could influence the stability and activity of encapsulated biomacromolecules. The size of these nanostructures is usually in the range of 70–125 nm. An advantage of PICsomes is their stability with respect to high salt concentrations and temperatures, up to 150 mM NaCl and 70°C, respectively. In addition PICsomes based on PEG-*b*-P(Asp) can possess tunable permeability to low molecular weight molecules (Fig. 11.6).

3.1.4 LbL Capsules

LbL capsules are spherical compartments with walls formed by adsorption of alternating layers of positively and negatively charged polyelectrolytes [83]. Assembly of LbL capsules is achieved by consecutive deposition of complementary polymers onto colloidal particles (LbL deposition technique), followed by removal of the colloidal template by addition of acids [16,84–86]. $CaCO_3$ nanoparticles are commonly used as templates because they have low toxicity and are biodegradable. LbL capsules are an interesting platform for the design of nanoreactors because their permeability can be selectively modified

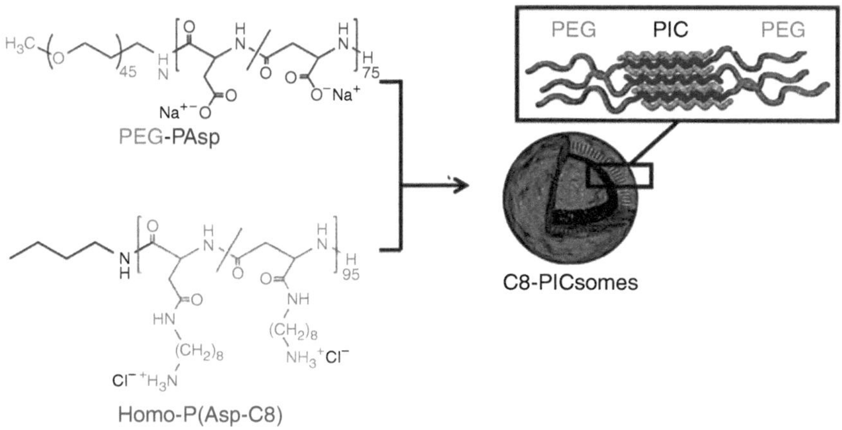

FIGURE 11.6 Schematic representation of the formation of salt-and thermal resistant PICsomes. PICsomes are created by elongation of the aliphatic spacer in the pendant group of the homocatiomer Homo-P(Asp-C8) paired with a block aniomer, PEG-*b*-P(Asp). *(Adapted from Ref. [25] with permission of American Chemical Society. Copyright 2014.)*

by different physical (light, mechanical, magnetic, ultrasound, temperature) or chemical (solvent, pH, ionic strength) conditions [85]. However, their large size (generally >1 μm) limits their development for bioapplications, and only a few LbL nanoreactors with submicron size have been reported [87].

3.2 Accessibility of 3D Polymer Supramolecular Assemblies

The accessibility of polymer supramolecular assemblies to substrates in order to allow in situ reactions is a key parameter for a nanoreactor design. In addition, such permeability also allows the release of reaction products, and therefore supports the overall function of the nanoreactor.

Dendrimers have an architecture which is intrinsically accessible, because their branched structure allows the passage of substrates and products. In the case of PICsomes, the use of polyelectrolytes, which self-assemble into intrinsically porous systems, favors accessibility to the inner cavity where the active compounds are located [25]. Their membranes are permeable to small molecules, such as Tokyo-Green-β-galactoside, while encapsulated active biomacromolecules, such as trypsin, cannot escape from the vesicular assemblies [25]. In the case of LbL capsules, permeabilization of the shell is tuned by its polymer composition, thickness, solvent, pH, or ionic strength [88,89].

Some copolymers, such as PS-*b*-PIAT, self-assemble into polymersomes with an intrinsically porous membrane [78], whereas others require membrane permeabilization in order to form nanoreactors. For example, PMOXA-*b*-PDMS-*b*-PMOXA or PMOXA-*b*-PDMS membranes are permeable to oxygen and reactive oxygen species [9], but not to small molecules [90–92]. In the case of impermeable membranes, the following procedures can be used to provide accessibility: (1) permeabilization by addition of organic solvents,

(2) incorporation of stimuli-responsive chains/chemical modifications, (3) blending lipids with copolymers followed by lipid extraction, and (4) insertion of membrane proteins or biopores. Not all these membrane permeabilization techniques have been used to form nanoreactors, but they could feature in future nanoreactor developments.

1. Permeabilization with organic solvents
   Permeabilization of polymer membranes is increased by ethanol or mixtures of specific organic solvents. For example, a small amount ($\leq$ 0.5% v/v) of ethanol [92] or mixtures of ethanol and dimethyl sulfoxide (1:1 v/v) [90] in the system has no influence on the membrane permeability of PMOXA-*b*-PDMS-*b*-PMOXA copolymers, and therefore much higher amounts of solvents are necessary to destabilize the membrane. However, this permeabilization method is not appropriate when bioapplications are intended. Likewise, a copolymer membrane based on poly(ethylene glycol)-*block*-poly(D,L-lactic-*stat*-glycolic acid) (PEG-*b*-PLGA), which exhibits 18-fold less permeability to protons than a liposome membrane (based on a mixture of soybean phospholipids and cholesterol) was rendered more permeable for proton diffusion when increasing amounts of 1,4-dioxane (5–20%) were added (Fig. 11.7) [93]. Upon removal of the 1,4-dioxane by dialysis, the proton diffusion returned to

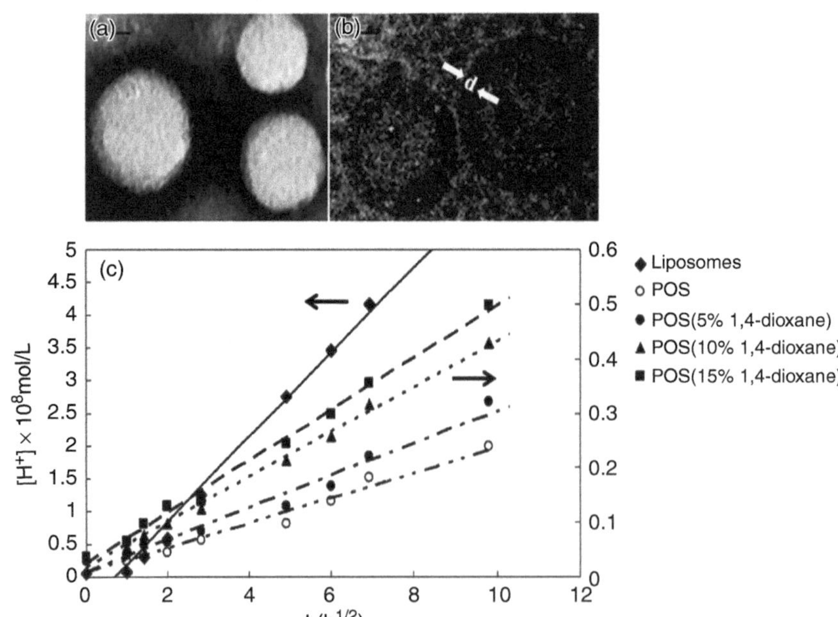

FIGURE 11.7 (a) TEM and (b) cryo-TEM micrographs of polymer vesicles based on PEG–PLGA, and (c) a comparison of proton gradient across the vesicular structures of lipids (liposomes) and polymersomes in the presence of increasing amount of 1,4-dioxane. *(Adapted from Ref. [93] with permission of Royal Society of Chemistry.)*

the initial value for nonpermeabilized polymersomes, though there are no data available on the influence of higher amounts (≥20%) of organic solvents on the morphology of polymersomes.

2. Incorporation of stimuli-responsive chains/chemical modification of polymersomes
Poration is induced in polymersome membranes by incorporating copolymer blocks, which respond to stimuli, and result in the formation of pores in the membrane when a specific stimulus is present. A typical example involves polymersomes containing poly(lactone) or poly(caprolactone) (PCL) blocks, in which gradually poration of the membrane is induced, simply by hydrolytic scission of the blocks [94]. Another strategy to render membranes permeable is chemical modification of polymersomes of PMOXA-*b*-PDMS or poly(ethylene oxide)-block -polybutadiene (PEO-*b*-PB) copolymers with a hydroxyalkylphenone: under UV irradiation, ketyl and alcohol radicals are formed, and induce a porous structure in the membrane [14,95].
In another example, a photo-cross-linkable diblock PEO-*b*-P(DEA-*s*-CMA) copolymer based vesicles, encapsulating glucose oxidase (GOx) and hemoglobin, represents a pH-responsive nanoreactor in which a cross-linked coumarin-rich matrix contributes to the stability of the system, while coumarin-poor patches function as pH-triggered tunnels [63].
3. Lipid-copolymer hybrids
In lipid-copolymer membranes, poration is induced by removing lipids with detergents. Direct removal of lipids is not possible because the entire architecture could be destabilized. Therefore, prior to phospholipid removal, the structure has to be cross-linked using a chemical initiator to increase its stability. For example, vesicular structures based on PEO-*b*-PB mixed with POPC lipids were first cross-linked and then permeabilized by adding detergents [96]. Extraction of lipids was facilitated by Triton X, in low concentrations (10 mM) that did not disturb the polymer remaining structure.
4. Insertion of membrane proteins and biopores in polymer membranes
Functional insertion of membrane proteins, polypeptides, and ion-transporters into polymer membranes presents a biomimetic strategy for membrane permeabilization, with the approach resembling cellular membranes [33]. In this respect, the research group of Wolfgang Meier performed pioneering work in the reconstitution of membrane proteins (such as OmpF) in block copolymer membranes [54], followed by successful development of polymer nanoreactors [6]. OmpF allows free diffusion of molecules less than 600 Da into the inner aqueous cavity of the polymersomes. Through the OmpF pore, substrates penetrate the membrane and reach the encapsulated active molecules and products are released after in situ reaction. It is intriguing that membrane proteins are capable of maintaining their functionality in membranes thicker (10–20 nm) than lipid membranes (3–5 nm). In recent experiments, copolymer membranes, especially those based on PMOXA-*b*-PDMS-*b*-PMOXA, were shown to be very flexible and possess high fluidity, higher than previously

obtained by simulations [74,97], and could adapt conformational changes in order to fit the proteins and provide a fluid environment (i.e., one sufficient to maintain their functionality) [74]. Insertion of membrane proteins should preserve their functionality and not perturb the membrane, with the final goal of obtaining stable and entirely functional nanoreactors [90]. Polymer membranes that have been successfully permeabilized by inserting various membrane proteins are summarized in Table 11.2, as are biopores.

3.3 Encapsulation/Insertion of Active Compounds: Design of Bionanoreactors

Encapsulation/insertion of active compounds (such as enzymes, enzyme mimics, and proteins) inside supramolecular structures is a crucial step in the generation of nanoreactors. This offers several advantages for the encapsulated enzyme, such as higher stability and catalytic rates. In addition the encapsulation of enzymes provides a simple way to engineer "active surfaces" by enzyme immobilization, because nanoreactors can be attached to surfaces by connecting functional groups [53]. Efficiency is pivotal in the design of nanoreactors with high activity, and requires optimization of the enzyme entrapment process. For example, the efficiency of polymersomes can be tuned by the block copolymer composition (hydrophilic/hydrophobic block ratio), as reported for PMOXA-*b*-PDMS-*b*-PMOXA nanoreactors [50]. A plethora of catalytically active biomolecules, such as enzymes or enzyme mimics, have been inserted/encapsulated in order to prove the concept of a nanoreactor, or to produce a functional nanoreactor (Fig. 11.8) [8,13,25,46,47,53,59,71]. Single enzymes such as horseradish peroxidase (HRP), β-lactamase, or glucose oxidase (GOx), have been used to prove that previously listed nanoarchitectures can be used as nanoreactors in combination with enzymes [6,52,109]. To maximize the encapsulation efficiency, mimics with small molecular weight were used to replace enzymes, such as Cu/Zn superoxide dismutase, inside the nanoreactor [110]. Recently, more advanced catalytic reactions were introduced by encapsulating enzymes that can act in tandem within the cavities (e.g., superoxide dismutase (SOD) + catalase (CAT) or SOD + lactoperoxidase (LPO)) [8,11,109].

3.4 Activity and Stability of 3D Nanoarchitectures

Encapsulation of enzymes in various polymer superstructures (dendrimers, polymersomes, PICsomes, or LbL capsules) provides significant advantages, such as higher stability and preserved catalytic activity over longer periods of time than bulk conditions. The process of encapsulation/entrapment of biomolecules in various 3D supramolecular assemblies should not affect the stability of the final nanoreactor structure (which is driven by the chemical nature of the copolymer used for their formation). If necessary cross-linking procedures can be used to increase the stability. However, it is also necessary to determine if

TABLE 11.2 Permeabilized Membranes Formed by Inserting Membrane Proteins and Biopores

Polymer composition used for polymersome formation	Membrane protein/biopore	Proved functionality	Nano-/microreactors	References
PMOXA-*b*-PDMS-*b*-PMOXA	Gramicidin	H^+, Na^+, K^+	No	[90]
PEO-*b*-PB	Dendritic helical pore	H^+, H_2O	No	[98]
PEG-*b*-PDMS	Calcimycin	Ca^{2+}	Yes	[99]
PMOXA-*b*-PDMS-*b*-PMOXA	Ionomycin	Ca^{2+}	No	[92]
PMOXA-*b*-PDMS-*b*-PMOXA	Asalocid A, *N*,*N*-dicyclohexyl-*N*′,*N*′-dioctadecyl-3-oxapentane-1,5-diamide, Alamethicin	Ca^{2+}	Yes	[100]
PEtOz-*b*-PDMS-*b*-PEtOz	Bacteriorhodopsin	H^+ gradient	No	[101]
PMOXA-*b*-PDMS-*b*-PMOXA	Complex I	Electron transfer	No	[102]
PMOXA-*b*-PDMS-*b*-PMOXA, PEO-*b*-PB	Aquaporin	H_2O	No	[91,103,104]
PEO-*b*-PB	α-Hemolysin	Calcein	No	[105]
PMOXA-*b*-PDMS-*b*-PMOXA, DNA-PIB	OmpF	Molecules up to 600–700 Da	Yes	[6,12,13,52, 59,106]
PMOXA-*b*-PDMS-*b*-PMOXA	LamB	DNA translocation	No	[107]
PMOXA-*b*-PDMS-*b*-PMOXA	Tsx	Inosine, adenosine, guanosine, 2-fluoroadenosine	Yes	[59]
PMOXA-*b*-PDMS-*b*-PMOXA, PIB-*b*-PEO-*b*-PIB	FhuA	Various dyes	Yes	[55,108]

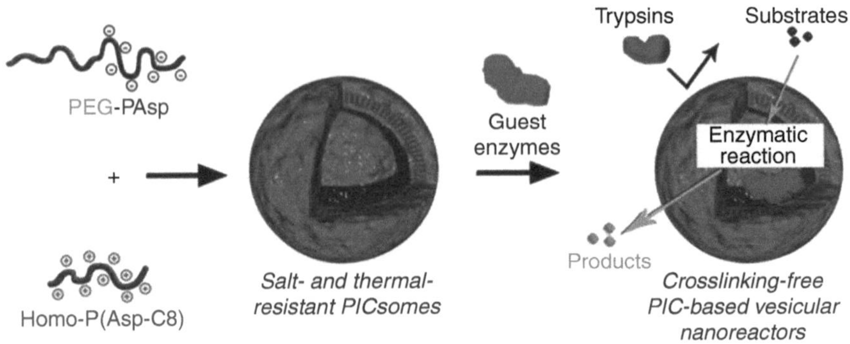

FIGURE 11.8 Schematic representation of semipermeable PICsomes encapsulating β-galactosidase served as nanoreactors. *(Reprinted from Ref. [25] with permission of American Chemical Society. Copyright 2014.)*

the procedures affect the activity of the active compound [61,96]. Therefore, polymer nanoarchitectures have the dual role of providing a stable environment for encapsulated biomacromolecules while leaving their activities unhindered. For example, nanoreactors containing encapsulated penicillin acylase have been shown to remain active and stable in buffer for more than 7 days at room temperature [13], and a thymidine phosphorylase-encapsulating nanoreactor was active in serum for 4 days at 37°C [58]. When neuraminidase, an enzyme with an essential role in combating influenza infections, was encapsulated in LbL capsules, it maintained its activity inside capsules taken up by endothelial cells without rupture for 5 days [111]. Other polymer nanoreactors with an entrapped acid phosphatase enzyme were active at pH 4–7, but inactive outside this pH range [52]. Moreover, a maximum stability of 48 h in cells, along with retained activity, has been reported for a simple artificial organelle, the artificial peroxisome, based on two entrapped enzymes (SOD coupled with either LPO or CAT) inside PMOXA-*b*-PDMS-*b*-PMOXA polymersomes [8].

There are also examples of nanoreactors which have increased enzyme activity when encapsulated compared with bulk conditions. PICsomes that are stable to salt and elevated temperatures have been successfully developed using PEG-*b*-P(Asp) and are able to encapsulate trypsin without altering the original enzymatic activity [25]. Furthermore, these specifically designed PICsomes maintained their vesicular architecture at salt concentrations of 300 mM NaCl and temperatures up to 70°C, with only a subtle increase in their size. For example, encapsulated nucleoside hydrolase has a substrate turnover number (k_{cat}) up to six times higher than the free enzyme [59]. Other enzymes, such as GOx and HRP, showed an 85% increase in activity after encapsulation in polyelectrolyte capsules [112]. In addition, laccase entrapped in linear-dendritic based nanoreactors showed superior activity compared with its activity in bulk, and also increased temperature stability (up to 70°C) [19]. However, the intrinsic branched architecture of dendrimers offers insufficient fluidity and space for entrapping larger biomolecules.

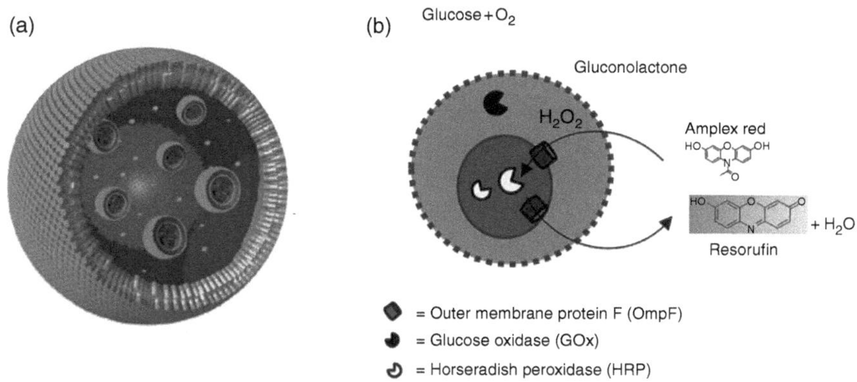

FIGURE 11.9 **Design of enzymatically active multicompartment polymersomes.** (a) Multicompartment system consists of PEO-*b*-PB vesicles loaded with cytosolic enzymes, enzymatic substrates, and semipermeable PS-*b*-PIAT nanoreactors encapsulating distinctive enzymes [3]. (b) Cascade reaction between the separately encapsulated HRP and GOx enzymes can take place, due to the structure of the system and the embedding of the porin OmpF in the inner membrane compartment [11]. *(Adapted from Refs. [3,11] with permission of Royal Society of Chemistry and Wiler-VCH.)*

3.5 Multicompartment Structures

Multicompartmentalization is a step further toward mimicking natural processes in complex spaces, which are observed in eukaryotic cells [113]. To mimic biological multicompartments micron-sized PEO-*b*-PB polymersomes containing inside smaller PS-*b*-PIAT polymersomes have been engineered (Fig. 11.9a). Such systems involve a cofactor-dependent cascade reaction between the enzyme loaded in PS-*b*-PIAT polymersomes and "cytosolic" enzymes encapsulated in the big PEO-*b*-PB compartment [109]. A similar intercompartmental enzymatic cascade reaction has been reported in polymersome-in-polymersome multicompartment architectures based on encapsulation of small PMOXA-*b*-PDMS-*b*-PMOXA polymersomes in PS-*b*-PIAT micrometer-size vesicles. GOx was encapsulated inside the inner PMOXA-*b*-PDMS-*b*-PMOXA polymersomes, while HRP was encapsulated in the larger PS-*b*-PIAT. In addition to enzymes, OmpF has been employed as a passive protein channel to allow the exchange of substrates through the membranes of inner polymersomes (Fig. 11.9b). It has to be emphasized that these systems present only basic models of eukaryotic cell mimicry, but their size has prevented further medical applications [11]. Therefore, further studies on complex systems, which combine the multicompartment concept with more interacting biomacromolecules (DNA, proteins), are required.

3.6 Requirements for Various Applications of Nanoreactors

In order to combine a polymer 3D supramolecular assembly with proteins and/or enzymes to form a confined reaction space at the nanoscale, a complex system of requirements must be fulfilled. The criteria used to design nanoreactors in various domains depend both on individual components — polymer assembly and biologically active molecule — and the intended application. The general requirements for efficient nanoreactors are: (1) stability of the polymer nanostructures and protection of the biocatalyst from degradation, (2) accessibility of substrates and products to the catalytic space, (3) efficient encapsulation/insertion of the bioactive molecule within the polymer nanostructure. In addition, the requirements for designing nanoreactors with specific properties need to be adjusted according to the application domain.

In order to be applied in medicine, there are supplementary requirements regarding the chemical nature and properties of polymers (including toxicity, biocompatibility, biodegradability, charge, flexibility, and density) [114], as well as the properties of the generated nanoreactors (size [115], shape [116], mechanical properties [114], and surface chemistry [117]).

Biocompatibility and biodegradability. Toxicity is an important criterion that must be taken into account when using polymers to build a nanoreactor for medical applications. Only polymers that are biocompatible and biodegradable and do not accumulate in the body are good candidates for use in generating nanoreactors [33,118]. Therefore, either the nanostructure-forming polymers should be biodegradable or the supramolecular assembly should break into individual polymer chains that must be small enough to be excreted from the body. However, the nanoreactors must be sufficiently robust to protect the bioactive molecule, even after cellular uptake, and thus their biodegradability needs to be delayed. Long-circulating nanoassemblies can be produced by using PMOXA-*b*-PDMS-*b*-PMOXA block copolymers that form vesicles with stability up to 48 h after uptake by various cell lines [9].

Surface chemistry. Another key point in generating nanoreactors relates to their surface chemistry. In this respect, nanoreactors with "stealth" behavior have been used in order to increase their circulation time within the body, positively influencing the efficacy, and the lifetime of entrapped active species [58,117]. Although PEG remains the standard polymer for the steric stabilization of liposomes, because of its protein repellent properties [119,120], alternative strategies have recently gained greater attention, including the use of PMOXA and poly(2-methacryloyloxyethyl phosphorylcholine) (PMPC) [121,122]. The surface charge of the supramolecular assembly influences the circulation time of the whole polymeric system inside the body, but depending on the charge sign and density it can also produce severe "toxicity". For example, positive charges are well known to be problematic for *in vivo* applications, whereas a slightly negative charged surface increases the circulation time and accumulation in tumors [98,117].

Functionalization. When a specific location for nanoreactors inside a body is required for effective therapy, or for sensitive detection, "active targeting" approaches are demanded. Targeting molecules exposed at the surface of nanoreactors can prolong the circulation time, reduce undesirable side effects, minimize nonspecific uptake, and allow for specific targeting to certain cells [123,124]. For this purpose, specific molecules have been used to functionalize the surface of polymer nanoreactors, such as growth factors and vitamins, peptides, sugar derivatives, and antibodies [125–129].

Nanoreactor size. The size of a nanoreactor plays an important role in cellular interactions, and determines biodistribution and blood circulation time [115]. Effective cell uptake and long-circulating nanoreactors are obtained if the diameter is in the range 10–100 nm [46,118], the lower limit corresponding to 10 nm dendrimeric structures [46,130]. In this respect, all supramolecular assemblies presented here can be designed to limit their size to the required domain by (1) modulation of their chemical composition, (2) selecting a template with a specific size (LbL capsules), or (3) selecting an adequate preparation method, such as extrusion (polymersomes).

Preparation conditions. In order to preserve in situ biological activity of the active species, methods are required to avoid organic solvents that lower the enzyme activity. In addition, conditions should mimic the natural environment that is specific for each biomolecule, such as: pH, ionic strength, temperature, or light [8,33,48,61,131].

4 APPLICATIONS OF NANOREACTORS

The advantageous properties of polymer nanoreactors in terms of payload protection and stability make them feasible for lowering the toxicity of contrast agents, the side effects of drugs, and their uncontrolled systemic distribution. These have led to novel developments of nanoreactor applications either in the blood stream or by mimicking cell organelles [4,8,9,46,57].

4.1 Diagnostic Applications

Different contrast agents, such as fluorescent dyes, inorganic and radiolabeled molecules, and metal complexes represent key components for increasing the sensitivity of modern medical diagnostics. However, most of these molecules are toxic, and require innovative administration strategies. Therefore, nanoreactors represent valuable candidates for protection against toxic contrast agents and offer the means of receptor-mediated and site-specific accumulation. At present though, there are only a few reports of polymer vesicles and microcapsules being used as nanoreactors for pure diagnostic purposes, described as follows.

OmpF-permeabilized PMOXA-*b*-PDMS-*b*-PMOXA nanoreactors have been proposed as promising candidates for self-regulated targeted diagnostics. For example, acid phosphatase, an enzyme with pH-dependent activity, was encapsulated and enabled the conversion of the insoluble fluorescent molecule ELF97 to a water-soluble nonfluorescent substrate at pH 4–6.5 within the nanoreactor, and thus created a controlled and triggered diagnostic system [52]. A similar concept was applied to reduce the toxicity of gadolinium, commonly used as contrast agent [51]. Acid phosphatase was encapsulated inside nanoreactors for the conversion of the glucose-1-phosphate substrate to phosphate ions, which then led to the precipitation of insoluble Gd-phosphate nanoparticles. Specific targeting of the nanoreactors by macrophage scavenger receptor A_1 and fluorescently labeled streptavidin demonstrated a novel approach for bimodal MRI and confocal imaging [51]. Despite the many advantages of nanoreactors used for diagnostic purposes, there are still several challenges to be met, such as improved detection sensitivity and specificity, higher encapsulation efficiency, and improved targeting.

4.2 Therapeutic Applications

Nanoreactors are viable candidates for a variety of therapeutic applications, such as: (1) enzyme replacement therapy [58], (2) prodrug administration [13,53,59], (3) target-specific subcellular therapeutic systems [46,57], (4) antioxidant systems to fight oxidative stress [9,12,49,50,132], (5) systems for photodynamic therapy [47,48], and (6) artificial organelles [8,12,63]. To the best of our knowledge, only polymersome-based nanoreactors have been investigated for therapeutic applications; other assemblies — such as PICsomes, LbL capsules, and dendrimers — are still at the stage of model systems and need further development to be suitable for therapeutic applications. For example, at present there is only one report indicating the potential utility of PICsomes as nanoreactors for biomedical applications [25].

1. In classical enzyme replacement therapy, the deficient or absent enzyme is administered intravenously, which results in a short circulation half-life, immunogenicity, and low enzyme stability [133]. In order to overcome these drawbacks, polymer nanoreactors have been designed as an attractive alternative for efficient enzyme replacement therapy. The first example was shown for mitochondrial neurogastrointestinal encephalomyopathy, a fatal genetic disorder caused by a deficiency of thymidine phosphorylase [58]. Thymidine phosphorylase was encapsulated in PMOXA-*b*-PDMS-*b*-PMOXA polymersomes decorated with the nucleoside-specific porin Tsx to permeabilize the membrane. These nanoreactors were able to retain their initial enzymatic activity without causing cytotoxicity in hepatocytes and stimulating macrophages.
2. Polymer nanoreactors based on PMOXA-*b*-PDMS-*b*-PMOXA have been applied in prodrug therapy by selective enzymatic activation of nontoxic prodrugs to increase local drug concentrations in a tumor while decreasing toxic side effects. Polymer vesicles were permeabilized with bacterial membrane porins (OmpF and Tsx) and loaded with the purine-specific nucleoside hydrolase of Trypanosoma vivax [59]. Three different prodrugs (2-fluoroadenosine, 2-chloroadenosine, and 6-methylpurine riboside) were efficiently hydrolyzed inside these nanoreactors, resulting in the release of the cytotoxic molecules at the site of action. Also, antibiotic-releasing nanoreactors based on PMOXA-*b*-PDMS-*b*-PMOXA were formed by encapsulating penicillin acylase and inserting OmpF to catalyze the conversion of prodrug molecules

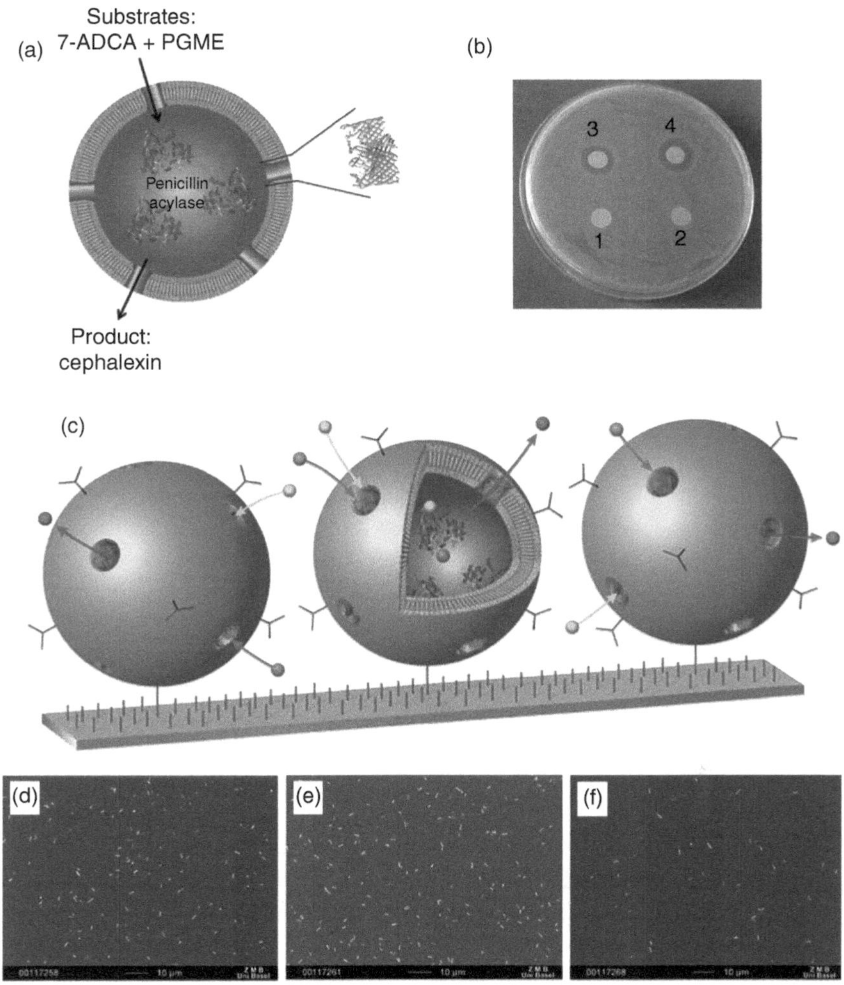

FIGURE 11.10 (a) Schematic representation of a penicillin acylase-loaded nanoreactor. (b) Picture of the MHB agar plate inoculated with bacteria. Inhibition zones: 1: control (not permeable polymersomes); 2: substrate solution; 3: nanoreactors prepared at RT; 4: nanoreactors prepared at 8°C; after incubation for 24 h at 37°C. (c) Schematic representation of enzymatically active, immobilized nanoreactors synthesizing antibiotics. SEM images of *E. coli* attached to: (d) silanized silicon surface, (e) surface with immobilized nonpermeable polymersomes encapsulating enzyme, (f) surface with immobilized nanoreactors. Scale bar: 10 mm. *(Adapted from Ref. [13,53] with permission of Royal Society of Chemistry.)*

(7-aminodesacetoxycephalosporanic acid and phenylglycine methyl ester) to cephalexin (Fig. 11.10a,b) [13]. Immobilizing such nanoreactors on surfaces of implants should increase the efficacy of antibiotics due to their localized production, and thereby reduce systemic side effects significantly (Fig. 11.10c–f) [53].

3. PMOXA-*b*-PDMS-*b*-PMOXA polymer vesicles have been functionalized with the oligonucleotide polyguanylic acid for specific recognition

by macrophage scavenger receptors and taken up by macrophages. Trypsin, an intestinal serine protease produced in the pancreas, was loaded into functionalized polymersomes and incubated with THP-1 macrophages to demonstrate intracellular trypsin activity by conversion of nonfluorescent cell-penetrating tripeptide-Rhodamine 110 conjugate (BZiPAR) into green fluorescent Rhodamine 110 [57]. Another approach was based on functionalization of PS-*b*-PIAT vesicles containing HRP with the cell penetrating peptide (Tat) in order to promote the cellular uptake into different cell lines [46]. The cellular uptake, intracellular distribution, and activity of these systems were studied inside target cells, and showed potential to serve as functional artificial organelles for different biomedical applications [46,57].

4. Disordered regulation of ROS (such as the superoxide anion radical, peroxynitrite, the hydroxyl radical and hydrogen peroxide) results in oxidative stress and plays a significant role in many diseases, including arthritis, Parkinson's disease, amyotrophic lateral sclerosis, cancer, and AIDS [134]. In normal conditions, reactive oxygen species (ROS) are regulated by cellular antioxidant mechanisms, and are rapidly scavenged by various antioxidants. In order to fight oxidative stress in pathological conditions, antioxidant nanoreactors have been designed to act in situ by detoxifying ROS species. A platform of antioxidant nanoreactors was created by self-assembly of oxygen permeable PMOXA-*b*-PDMS-*b*-PMOXA copolymers and encapsulation of SOD, which catalyzes the partitioning of the superoxide radicals [49,50], dual-enzyme mimics of superoxide dismutase, and catalase (Cu^{II}ENZm) that detoxify both superoxide radicals and H_2O_2 [9]. Nanoreactors based on the encapsulation of hemoglobin have been demonstrated to retain hemoglobin's dual functionality of peroxynitrite detoxification and oxygen transport [132]. Coencapsulation of two enzymes (SOD and LPO or CAT) that act in tandem [8,12], represented a step further in the combat against oxidative stress. In the case of polymersomes loaded with SOD and LPO enzymes, the OmpF protein was reconstituted in the vesicle membrane in order to permit the passage of substrates and products across the membrane [12]. These nanoreactors were taken up by THP-1 cells while preserving their activity inside the cells for up to 72 h. Addition of paraquat, which produces intracellular superoxide radicals, revealed the activity and high efficiency of ROS detection inside cells [12]. Therefore, such nanoreactors represent key steps in the development of artificial organelles inside cells and could be used for theranostic applications based on their production of a fluorescence product, resorufin, inside cells that enables the observation of therapeutic effects in real time [8,12].
5. Light-responsive nanoreactors, based on PMOXA-*b*-PDMS-*b*-PMOXA or PNVP-*b*-PDMS-*b*-PNVP polymer vesicles with the encapsulated photosensitizer Rose Bengal-bovine serum albumin conjugate, have been shown to serve as an efficient source of ROS "on demand" [47,48]. This system was uptaken by HeLa cells, and induced apoptosis only when irradiated with light of the appropriate wavelength (Fig. 11.11). These "Trojan horses" have been proposed as useful for photodynamic therapy.

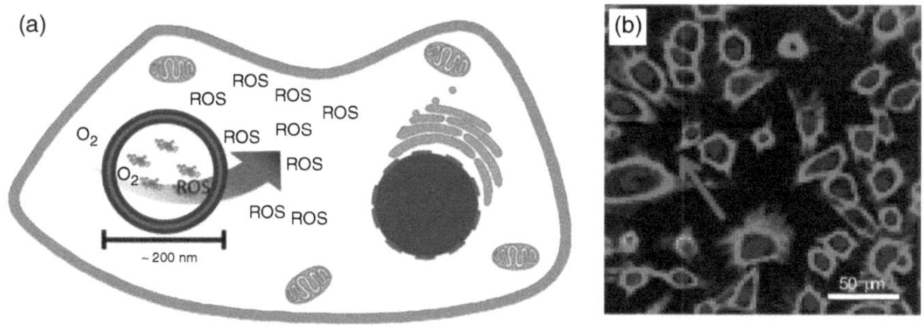

FIGURE 11.11 (a) Schematic representation of a light-responsive nanoreactor containing photosensitizers to act as cellular Trojan horses by producing ROS inside cells "on demand." (b) HeLa cells incubated with Trojan horses based on vesicles with R_h of 200 nm. Illumination time: 25 min (75 J/cm^2). The arrows indicate the presence of blebs. *(Adapted from Ref.[48] with permission of American Chemical Society. Copyright 2014.)*

6. Artificial organelles, mimicking natural peroxisomes have been obtained by coencapsulating enzymes acting in tandem with Cu/Zn-superoxide dismutase and LPO/CAT in PMOXA-*b*-PDMS-*b*-PMOXA polymersomes equipped with OmpF for permeabilization of the membrane (Fig. 11.12) [8,12]. Upon cell uptake these nontoxic, active artificial peroxisomes detoxified superoxide

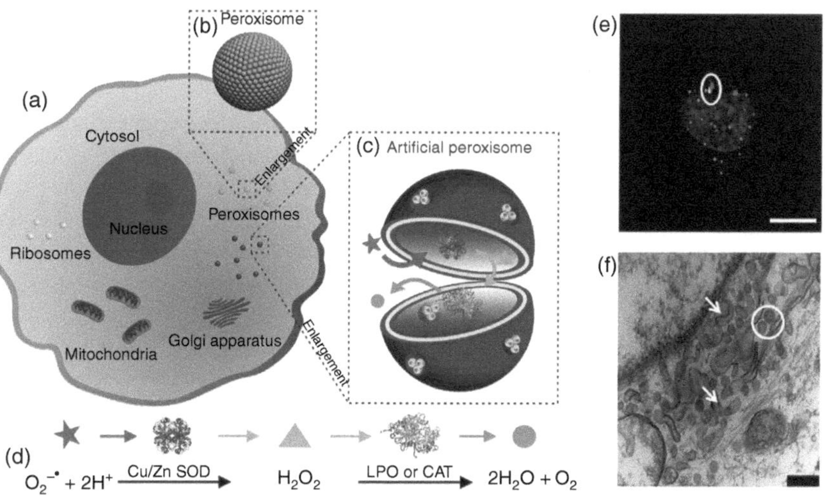

FIGURE 11.12 (a) Schematic view of a cell showing various organelles. (b) Peroxisomes are spherical, nanometer-size cell organelles that play a significant role in the regulation of ROS. (c) An artificial peroxisome based on the simultaneous encapsulation of a set of antioxidant enzymes in a polymer vesicle, with a membrane equipped with channel proteins. (d) Schematic enzymatic cascade reaction for detoxification of superoxide radicals and related H_2O_2 inside the artificial peroxisome (AP). (e) CLSM image of HeLa cells incubated for 24 h with APs [light gray (blue in the web version): nucleus stained with Hoechst, gray (green in the web version): endosome/lysosome stained with pH-rodo, dark gray (red in the web version): artificial peroxisomes] (scale bar = 5 µm). (f) TEM images of the perinuclear region of HeLa cells incubated with APs for 12 h (scale bar = 200 nm). *(Adapted from Ref. [8] with permission of American Chemical Society. Copyright 2013.)*

radicals and H_2O_2 after endosomal escape. They represent the first step in a versatile strategy to develop other "cell implants" for cell malfunction [8]. Inspired by the nuclear pore complex, polymeric nuclear envelope-like vesicles assembled from a pH-responsive and photo-cross-linkable diblock copolymer PEO-*b*-P(DEA-*s*-CMA) were loaded with glucose oxidase and hemoglobin. This system was proposed as an artificial organelle, but as it was not studied inside cells, their in vitro functionality is unclear [63]. Multicompartment structures could serve as mimics of cells but they are still at an early stage of development, and have not yet been shown to be functional in vitro [3].

5 CONCLUSIONS

To summarize, nanoreactors loaded with enzyme or enzyme mimics possess great potential for various biomedical applications by: (1) protecting enzymes from degradation and proteolysis. (2) prolonging enzymatic activity compared to the free enzyme in solution, (3) combining several enzymes with different properties (hydrophobicity, charge, etc.) that can act in tandem or synergistically, (4) enhancing reaction probabilities/efficiencies due to the confined space inside the nanoreactor, and (5) surface functionalization for specific targeting. Though increased enzyme encapsulation efficiency inside nanoreactors is desirable, it is still challenging because of intrinsic solubility of enzymes, and maintenance of their bioactivity, combined with the requirement of preserving the architecture of the supramolecular assembly upon encapsulation. The membrane of a nanoreactor enables compartmentalization in cellular environments, and by insertion of membrane proteins, selective transport of substrates/products across the nanoreactor's membrane enables the creation of artificial cellular organelles. Most of the examples presented here are basic model systems that have only been tested in vitro, but they represent key steps in the development of in vivo biosensors, new therapeutic designs, and theranostic strategies.

REFERENCES

[1] Zhang X, Tanner P, Graff A, Palivan CG, Meier W. Mimicking the cell membrane with block copolymer membranes. J Polym Sci 2012;50(12):2293–318.

[2] Baumann P, Tanner P, Onaca O, Palivan CG. Bio-decorated polymer membranes: a new approach in diagnostics and therapeutics. Polymers 2011;3(1):173–92.

[3] Peters RJRW, Marguet M, Marais S, Fraaije MW, van Hest JCM, Lecommandoux S. Cascade reactions in multicompartmentalized polymersomes. Angew Chem Int Ed 2014;53(1):146–50.

[4] Tanner P, Egli S, Balasubramanian V, Onaca O, Palivan CG, Meier W. Can polymeric vesicles that confine enzymatic reactions act as simplified organelles? FEBS Lett 2011;585(11):1699–706.

[5] Balasubramanian V, Onaca O, Enea R, Hughes DW, Palivan CG. Protein delivery: from conventional drug delivery carriers to polymeric nanoreactors. Expert Opin Drug Deliv 2010;7(1):63–78.

[6] Nardin C, Thoeni S, Widmer J, Winterhalter M, Meier W. Nanoreactors based on (polymerized) ABA-triblock copolymer vesicles. Chem Commun 2000;(15):1433–4.

[7] Vriezema DM, Comellas Aragones M, Elemans JA, Cornelissen JJ, Rowan AE, Nolte RJ. Self-assembled nanoreactors. Chem Rev 2005;105(4):1445–89.

[8] Tanner P, Balasubramanian V, Palivan CG. Aiding nature's organelles: artificial peroxisomes play their role. Nano Lett 2013;13(6):2875–83.

[9] Balasubramanian V, Onaca O, Ezhevskaya M, Van Doorslaer S, Sivasankaran B, Palivan CG. A surprising system: Polymeric nanoreactors containing a mimic with dual-enzyme activity. Soft Matter 2011;7(12):5595–603.

[10] van Dongen SF, Nallani M, Cornelissen JJ, Nolte RJ, van Hest JC. A three-enzyme cascade reaction through positional assembly of enzymes in a polymersome nanoreactor. Chemistry 2009;15(5):1107–14.

[11] Siti W, de Hoog H-PM, Fischer O, Shan WY, Tomczak N, Nallani M, Liedberg B. An intercompartmental enzymatic cascade reaction in channel-equipped polymersome-in-polymersome architectures. J Mater Chem B 2014;2(18):2733–7.

[12] Tanner P, Onaca O, Balasubramanian V, Meier W, Palivan CG. Enzymatic cascade reactions inside polymeric nanocontainers: a means to combat oxidative stress. Chem Eur J 2011;17(16):4552–60.

[13] Langowska K, Palivan CG, Meier W. Polymer nanoreactors shown to produce and release antibiotics locally. Chem Commun 2013;49(2):128–30.

[14] Spulber M, Najer A, Winkelbach K, Glaied O, Waser M, Pieles U, Meier W, Bruns N. Photoreaction of a hydroxyalkyphenone with the membrane of polymersomes: a versatile method to generate semipermeable nanoreactors. J Am Chem Soc 2013;135(24):9204–12.

[15] Marguet M, Bonduelle C, Lecommandoux S. Multicompartmentalized polymeric systems: towards biomimetic cellular structure and function. Chem Soc Rev 2013;42(2):512–29.

[16] del Mercato LL, Rivera-Gil P, Abbasi AZ, Ochs M, Ganas C, Zins I, Soennichsen C, Parak WJ. LbL multilayer capsules: recent progress and future outlook for their use in life sciences. Nanoscale 2010;2(4):458–67.

[17] del Mercato LL, Ferraro MM, Baldassarre F, Mancarella S, Greco V, Rinaldi R, Leporatti S. Biological applications of LbL multilayer capsules: from drug delivery to sensing. Adv Colloid Interface Sci 2014;207:139–54.

[18] Matyjaszewski K, Tsarevsky NV. Nanostructured functional materials prepared by atom transfer radical polymerization. Nat Chem 2009;1(4):276–88.

[19] Gitsov I. Hybrid linear dendritic macromolecules: from synthesis to applications. J Polym Sci 2008;46(16):5295–314.

[20] Wang L, Yang Q, Ma G, Lin W, Wang Z, Huang M, Chen S. Development of biocompatible PAMAM "dendrizyme" to maintain catalytic activity in biological complex medium. J Mater Chem B 2013;1(34):4259–66.

[21] Discher DE, Ahmed F. Polymersomes. Annu Rev Biomed Eng 2006;8:323–41.

[22] Lensen D, Vriezema DM, van Hest JCM. Polymeric microcapsules for synthetic applications. Macromol Biosci 2008;8(11):991–1005.

[23] Blanazs A, Armes SP, Ryan AJ. Self-assembled block copolymer aggregates: from micelles to vesicles and their biological applications. Macromol Rapid Commun 2009;30(4–5): 267–77.

[24] Kishimura A, Koide A, Osada K, Yamasaki Y, Kataoka K. Encapsulation of myoglobin in PEGylated polyion complex vesicles made from a pair of oppositely charged block ionomers: a physiologically available oxygen carrier. Angew Chem Int Ed 2007;46(32):6085–8.

[25] Chuanoi S, Anraku Y, Hori M, Kishimura A, Kataoka K. Fabrication of polyion complex vesicles with enhanced salt and temperature resistance and their potential applications as enzymatic nanoreactors. Biomacromolecules 2014;15(7):2389–97.

[26] Aoshima S, Kanaoka S. A renaissance in living cationic polymerization. Chem Rev 2009; 109(11):5245–87.
[27] Matyjaszewski K. Macromolecular engineering by controlled/living ionic and radical polymerizations. Macromol Symp. 2001;174:51–67.
[28] Dechy-Cabaret O, Martin-Vaca B, Bourissou D. Controlled ring-opening polymerization of lactide and glycolide. Chem Rev 2004;104(12):6147–76.
[29] Braunecker WA, Matyjaszewski K. Controlled/living radical polymerization: features, developments, and perspectives. Prog Polym Sci 2007;32(1):93–146.
[30] Matyjaszewski K. Atom transfer radical polymerization (ATRP): current status and future perspectives. Macromolecules 2012;45(10):4015–39.
[31] Keddie DJ. A guide to the synthesis of block copolymers using reversible-addition fragmentation chain transfer (RAFT) polymerization. Chem Soc Rev 2014;43(2):496–505.
[32] Albertsson AC, Varma IK. Recent developments in ring opening polymerization of lactones for biomedical applications. Biomacromolecules 2003;4(6):1466–86.
[33] Palivan CG, Fischer-Onaca O, Delcea M, Itel F, Meier W. Protein-polymer nanoreactors for medical applications. Chem Soc Rev 2012;41(7):2800–23.
[34] Basak S, Punetha VD, Bisht G, Bisht SS, Sahoo NG, Cho JW. Recent trends ofpolymer-protein conjugate application in biocatalysis: a review. Polym Rev 2015;55(1):163–98.
[35] Itel F, Dinu A, Tanner P, Fischer O, Palivan CG. In: Torchilin V, editor. Nanoreactors for biomedical applications, in: Torchilin V (Ed.), Handbook of nanobiomedical research. World Scientific Publishing Co., Singapore; 2014;457–508.
[36] Rahman MA, Noh H-B, Shim Y-B. Direct electrochemistry of Laccase immobilized on Au nanoparticles encapsulated-dendrimer bonded conducting polymer: application for a catechin sensor. Anal Chem 2008;80(21):8020–7.
[37] Gitsov I, Hamzik J, Ryan J, Simonyan A, Nakas JP, Omori S, Krastanov A, Cohen T, Tanenbaum SW. Enzymatic nanoreactors for environmentally benign biotransformations. 1. Formation and catalytic activity of supramolecular complexes of laccase and linear-dendritic block copolymers. Biomacromolecules 2008;9(3):804–11.
[38] Lee JCM, Bermudez H, Discher BM, Sheehan MA, Won YY, Bates FS, Discher DE. Preparation, stability, and in vitro performance of vesicles made with diblock copolymers. Biotechnol Bioeng 2001;73(2):135–45.
[39] Arifin DR, Palmer AF. Polymersome encapsulated hemoglobin: a novel type of oxygen carrier. Biomacromolecules 2005;6(4):2172–81.
[40] Rameez S, Alosta H, Palmer AF. Biocompatible and biodegradable polymersome encapsulated hemoglobin: a potential oxygen carrier. Bioconjug Chem 2008;19(5):1025–32.
[41] Chen Q, Schonherr H, Vancso GJ. Block-copolymer vesicles as nanoreactors for enzymatic reactions. Small 2009;5(12):1436–45.
[42] Hvasanov D, Wiedenmann J, Braet F, Thordarson P. Induced polymersome formation from a diblock PS-b-PAA polymer via encapsulation of positively charged proteins and peptides. Chem Commun (Camb) 2011;47(22):6314–6.
[43] Napoli A, Valentini M, Tirelli N, Muller M, Hubbell JA. Oxidation-responsive polymeric vesicles. Nat Mater 2004;3(3):183–9.
[44] Nallani M, de Hoog HP, Cornelissen JJ, Palmans AR, van Hest JC, Nolte RJ. Polymersome nanoreactors for enzymatic ring-opening polymerization. Biomacromolecules 2007;8(12): 3723–8.
[45] Vriezema DM, Hoogboom J, Velonia K, Takazawa K, Christianen PC, Maan JC, Rowan AE, Nolte RJ. Vesicles and polymerized vesicles from thiophene-containing rod-coil block copolymers. Angew Chem Int Ed Engl 2003;42(7):772–6.

[46] van Dongen SFM, Verdurmen WPR, Peters RJRW, Nolte RJM, Brock R, van Hest JCM. Cellular integration of an enzyme-loaded polymersome nanoreactor. Angew Chem Int Ed 2010;49(40):7213–6.

[47] Baumann P, Balasubramanian V, Onaca-Fischer O, Sienkiewicz A, Palivan CG. Light-responsive polymer nanoreactors: a source of reactive oxygen species on demand. Nanoscale 2013;5(1):217–24.

[48] Baumann P, Spulber M, Dinu IA, Palivan CG. Cellular trojan horse based polymer nanoreactors with light-sensitive activity. J Phys Chem B 2014;118(31):9361–70.

[49] Axthelm F, Casse O, Koppenol WH, Nauser T, Meier W, Palivan CG. Antioxidant nanoreactor based on superoxide dismutase encapsulated in superoxide-permeable vesicles. J Phys Chem B 2008;112(28):8211–7.

[50] Onaca O, Hughes DW, Balasubramanian V, Grzelakowski M, Meier W, Palivan CG. SOD antioxidant nanoreactors: influence of block copolymer composition on the nanoreactor efficiency. Macromol Biosci 2010;10(5):531–8.

[51] Broz P, Ben-Haim N, Santini F, Marsch S, Scheffler K, Meier W, Hunziker P. Polymer vesicles loaded with precipitated gadolinium nanoparticles: a novel target-specific contrast agent for magnetic resonance imaging. Eur J Nanomed 2009;2(2):43–8.

[52] Broz P, Driamov S, Ziegler J, Ben-Haim N, Marsch S, Meier W, Hunziker P. Toward intelligent nanosize bioreactors: a pH-switchable, channel-equipped, functional polymer nanocontainer. Nano Lett 2006;6(10):2349–53.

[53] Langowska K, Kowal J, Palivan CG, Meier W. A general strategy for creating self-defending surfaces for controlled drug production for long periods of time. J Mater Chem B 2014;2(29): 4684–93.

[54] Meier W, Nardin C, Winterhalter M. Reconstitution of channel proteins in (polymerized) ABA triblock copolymer membranes. Angew Chem Int Ed 2000;39(24):4599–602.

[55] Nallani M, Benito S, Onaca O, Graff A, Lindemann M, Winterhalter M, Meier W, Schwaneberg U. A nanocompartment system (synthosome) designed for biotechnological applications. J Biotechnol 2006;123(1):50–9.

[56] Nardin C, Widmer J, Winterhalter M, Meier W. Amphiphilic block copolymer nanocontainers as bioreactors. Eur Phys J E 2001;4(4):403–10.

[57] Ben-Haim N, Broz P, Marsch S, Meier W, Hunziker P. Cell-specific integration of artificial organelles based on functionalized polymer vesicles. Nano Lett 2008;8(5):1368–73.

[58] De Vocht C, Ranquin A, Willaert R, Van Ginderachter JA, Vanhaecke T, Rogiers V, Versees W, Van Gelder P, Steyaert J. Assessment of stability, toxicity and immunogenicity of new polymeric nanoreactors for use in enzyme replacement therapy of MNGIE. J Control Release 2009;137(3–4):246–54.

[59] Ranquin A, Versees W, Meier W, Steyaert J, Van Gelder P. Therapeutic nanoreactors: combining chemistry and biology in a novel triblock copolymer drug delivery system. Nano Lett 2005;5(11):2220–4.

[60] Kim KT, Cornelissen JJLM, Nolte RJM, van Hest JCM. A polymersome nanoreactor with controllable permeability induced by stimuli-responsive block copolymers. Adv Mater 2009;21(27):2787–91.

[61] Gaitzsch J, Appelhans D, Wang L, Battaglia G, Voit B. Synthetic bio-nanoreactor: mechanical and chemical control of polymersome membrane permeability. Angew Chem Int Ed 2012;51(18):4448–51.

[62] Wang X, Liu G, Hu J, Zhang G, Liu S. Concurrent block copolymer polymersome stabilization and bilayer permeabilization by stimuli-regulated "traceless" crosslinking. Angew Chem Int Ed Engl 2014;53(12):3138–42.

[63] Zhu Y, Wang F, Zhang C, Du J. Preparation and mechanism insight of nuclear envelope-like polymer vesicles for facile loading of biomacromolecules and enhanced biocatalytic activity. ACS Nano 2014;8(7):6644–54.

[64] Simone EA, Dziubla TD, Colon-Gonzalez F, Discher DE, Muzykantov VR. Effect of polymer amphiphilicity on loading of a therapeutic enzyme onto protective fulamentous and spherical polymer nanocarrlers. Biomacromolecules 2007;8(12):3914–21.

[65] Lvov Y, Antipov AA, Mamedov A, Mohwald H, Sukhorukov GB. Urease encapsulation in nanoorganized microshells. Nano Lett 2001;1(3):125–8.

[66] Petrov AI, Volodkin DV, Sukhorukov GB. Protein-calcium carbonate coprecipitation: a tool for protein encapsulation. Biotechnol Prog 2005;21(3):918–25.

[67] Shutava T, Zheng ZG, John V, Lvov Y. Microcapsule modification with peroxidase-catalyzed phenol polymerization. Biomacromolecules 2004;5(3):914–21.

[68] Balabushevich NG, Tiourina OP, Volodkin DV, Larionova NI, Sukhorukov GB. Loading the multilayer dextran sulfate/protamine microsized capsules with peroxidase. Biomacromolecules 2003;4(5):1191–7.

[69] Baumler H, Georgieva R. Coupled enzyme reactions in multicompartment microparticles. Biomacromolecules 2010;11(6):1480–7.

[70] Yashchenok AM, Delcea M, Videnova K, Jares-Erijman EA, Jovin TM, Konrad M, Mohwald H, Skirtach AG. Enzyme reaction in the pores of $CaCO_3$ particles upon ultrasound disruption of attached substrate-filled liposomes. Angew Chem Int Ed Engl 2010;49(44):8116–20.

[71] Kreft O, Javier AM, Sukhorukov GB, Parak WJ. Polymer microcapsules as mobile local pH-sensors. J Mater Chem 2007;17(42):4471–6.

[72] Gajbhiye V, Palanirajan VK, Tekade RK, Jain NK. Dendrimers as therapeutic agents: a systematic review. J Pharm Pharmacol 2009;61(8):989–1003.

[73] Hecht S, Frechet JMJ. Dendritic encapsulation of function: Applying nature's site isolation principle from biomimetics to materials science. Angew Chem Int Ed 2001;40(1):74–91.

[74] Itel F, Chami M, Najer A, Lörcher S, Wu D, Dinu IA, Meier W. Molecular organization and dynamics in polymersome membranes: a lateral diffusion study. Macromolecules 2014;47(21):7588–96.

[75] Gunkel-Grabole G, Sigg S, Lomora M, Lörcher S, Palivan CG, Meier WP. Polymeric 3D nano-architectures for transport and delivery of therapeutically relevant biomacromolecules. Biomater Sci 2015;3(1):25–40.

[76] Discher DE, Eisenberg A. Polymer vesicles. Science 2002;297(5583):967–73.

[77] Schillen K, Bryskhe K, Mel'nikova YS. Vesicles formed from a poly(ethylene oxide)-poly(propylene oxide)-poly(ethylene oxide) triblock copolymer in dilute aqueous solution. Macromolecules 1999;32(20):6885–8.

[78] Kuiper SM, Nallani M, Vriezema DM, Cornelissen JJLM, van Hest JCM, Nolte RJM, Rowan AE. Enzymes containing porous polymersomes as nano reaction vessels for cascade reactions. Org Biomol Chem 2008;6(23):4315–8.

[79] Koide A, Kishimura A, Osada K, Jang WD, Yamasaki Y, Kataoka K. Semipermeable polymer vesicle (PICsome) self-assembled in aqueous medium from a pair of oppositely charged block copolymers: physiologically stable micro-/nanocontainers of water-soluble macromolecules. J Am Chem Soc 2006;128(18):5988–9.

[80] Anraku Y, Kishimura A, Oba M, Yamasaki Y, Kataoka K. Spontaneous formation of nanosized unilamellar polyion complex vesicles with tunable size and properties. J Am Chem Soc 2010;132(5):1631–6.

[81] Kishimura A. Development of polyion complex vesicles (PICsomes) from block copolymers for biomedical applications. Polym J 2013;45(9):892–7.

[82] Anraku Y, Kishimura A, Yamasaki Y, Kataoka K. Living unimodal growth of polyion complex vesicles via two-dimensional supramolecular polymerization. J Am Chem Soc 2013;135(4):1423–9.

[83] Decher G. Fuzzy nanoassemblies: toward layered polymeric multicomposites. Science 1997;277(5330):1232–7.

[84] Chandrawati R, Odermatt PD, Chong S-F, Price AD, Staedler B, Caruso F. Triggered cargo release by encapsulated enzymatic catalysis in capsosomes. Nano Lett 2011;11(11):4958–63.

[85] Delcea M, Moehwald H, Skirtach AG. Stimuli-responsive LbL capsules and nanoshells for drug delivery. Adv Drug Del Rev 2011;63(9):730–47.

[86] Tong W, Song X, Gao C. Layer-by-layer assembly of microcapsules and their biomedical applications. Chem Soc Rev 2012;41(18):6103–24.

[87] Huang Y, Lin Y, Ran X, Ren J, Qu X. A semipermeable enzymatic nanoreactor as an efficient modulator for reversible pH regulation. Nanoscale 2014;6(19):11328–35.

[88] Antipov AA, Sukhorukov GB. Polyelectrolyte multilayer capsules as vehicles with tunable permeability. Adv Colloid Interface Sci 2004;111(1–2):49–61.

[89] She Z, Wang C, Li J, Sukhorukov GB, Antipina MN. Encapsulation of basic fibroblast growth factor by polyelectrolyte multilayer microcapsules and its controlled release for enhancing cell proliferation. Biomacromolecules 2012;13(7):2174–80.

[90] Lomora M, Garni M, Itel F, Tanner P, Spulber M, Palivan CG. Polymersomes with engineered ion selective permeability as stimuli-responsive nanocompartments with preserved architecture. Biomaterials 2015;53:406–14.

[91] Kumar M, Grzelakowski M, Zilles J, Clark M, Meier W. Highly permeable polymeric membranes based on the incorporation of the functional water channel protein Aquaporin Z. Proc Natl Acad Sci USA 2007;104(52):20719–24.

[92] Lomora M, Itel F, Dinu IA, Palivan CG. Selective ion-permeable membranes by insertion of biopores into polymersomes. Phys Chem Chem Phys 2015;17(24):15538–46.

[93] Yu Y, Jiang X, Gong S, Feng L, Zhong Y, Pang Z. The proton permeability of self-assembled polymersomes and their neuroprotection by enhancing a neuroprotective peptide across the blood–brain barrier after modification with lactoferrin. Nanoscale 2014;6(6):3250–8.

[94] Ahmed F, Discher DE. Self-porating polymersomes of PEG-PLA and PEG-PCL: hydrolysis-triggered controlled release vesicles. J Control Release 2004;96(1):37–53.

[95] Dinu MV, Spulber M, Renggli K, Wu D, Monnier CA, Petri-Fink A, Bruns N. Filling polymersomes with polymers by peroxidase-catalyzed atom transfer radical polymerization. Macromol Rapid Commun 2015;36(6):507–14.

[96] Cheng L, Cao D. Effect of tail architecture on self-assembly of amphiphiles for polymeric micelles. Langmuir 2009;25(5):2749–56.

[97] Pata V, Dan N. The effect of chain length on protein solubilization in polymer-based vesicles (polymersomes). Biophys J 2003;85(4):2111–8.

[98] Kim AJ, Kaucher MS, Davis KP, Peterca M, Imam MR, Christian NA, Levine DH, Bates FS, Percec V, Hammer DA. Proton transport from dendritic helical-pore-incorporated polymersomes. Adv Funct Mater 2009;19(18):2930–6.

[99] Picker A, Nuss H, Guenoun P, Chevallard C. Polymer vesicles as microreactors for bioinspired calcium carbonate precipitation. Langmuir 2011;27(7):3213–8.

[100] Sauer M, Haefele T, Graff A, Nardin C, Meier W. Ion-carrier controlled precipitation of calcium phosphate in giant ABA triblock copolymer vesicles. Chem Commun 2001;23: 2452–3.

[101] Choi HJ, Lee H, Montemagno CD. Toward hybrid proteo-polymeric vesicles generating a photoinduced proton gradient for biofuel cells. Nanotechnology 2005;16(9):1589–97.

[102] Graff A, Fraysse-Ailhas C, Palivan CG, Grzelakowski M, Friedrich T, Vebert C, Gescheidt G, Meier W. Amphiphilic copolymer membranes promote NADH: Ubiquinone oxidoreductase activity: towards an electron-transfer nanodevice. Macromol Chem Phys 2010;211(2): 229–38.

[103] Grzelakowski M, Cherenet MF, Shen Y-x, Kumar M. A framework for accurate evaluation of the promise of aquaporin based biomimetic membranes. J Membr Sci 2015;479:223–31.

[104] Kumar M, Habel JEO, Shen Y-x, Meier WP, Walz T. High-density reconstitution of functional water channels into vesicular and planar block copolymer membranes. J Am Chem Soc 2012;134(45):18631–7.

[105] Nallani M, Andreasson-Ochsner M, Tan C-WD, Sinner E-K, Wisantoso Y, Geifman-Shochat S, Hunziker W. Proteopolymersomes: in vitro production of a membrane protein in polymersome membranes. Biointerphases 2011;6(4):153–7.

[106] Cottenye N, Syga M-I, Nosov S, Müller AHE, Ploux L, Vebert-Nardin C. Biological-like vesicular structures self-assembled from DNA-block copolymers. Chem Commun 2012;48(20):2615–7.

[107] Graff A, Sauer M, Van Gelder P, Meier W. Virus-assisted loading of polymer nanocontainer. Proc Natl Acad Sci USA 2002;99(8):5064–8.

[108] Muhammad N, Dworeck T, Fioroni M, Schwaneberg U. Engineering of the *E. coli* outer membrane protein FhuA to overcome the hydrophobic mismatch in thick polymeric membranes. J Nanobiotechnol 2011;9.

[109] Peters RJRW, Louzao I, van Hest JCM. From polymeric nanoreactors to artificial organelles. Chem Sci 2012;3(2):335–42.

[110] Balasubramanian V, Ezhevskaya M, Moons H, Neuburger M, Cristescu C, Van Doorslaer S, Palivan C. Structural characterization of a highly active superoxide-dismutase mimic. Phys Chem Chem Phys 2009;11(31):6778–87.

[111] Liu W, Wang X, Bai K, Lin M, Sukhorukov G, Wang W. Microcapsules functionalized with neuraminidase can enter vascular endothelial cells in vitro. J R Soc Interface 2014;11(101): 20141027.

[112] Stein EW, Volodkin DV, McShane MJ, Sukhorukov GB. Real-time assessment of spatial and temporal coupled catalysis within polyelectrolyte microcapsules containing coimmobilized mucose oxiase and peroxidase. Biomacromolecules 2006;7(3):710–9.

[113] Szostak JW, Bartel DP, Luisi PL. Synthesizing life. Nature 2001;409(6818):387–90.

[114] Markovsky E, Baabur-Cohen H, Eldar-Boock A, Omer L, Tiram G, Ferber S, Ofek P, Polyak D, Scomparin A, Satchi-Fainaro R. Administration, distribution, metabolism and elimination of polymer therapeutics. J Control Release 2012;161(2):446–60.

[115] Brinkhuis RP, Stojanov K, Laverman P, Eilander J, Zuhorn IS, Rutjes FPJT, van Hest JCM. Size dependent biodistribution and SPECT imaging of In-111-labeled polymersomes. Bioconj Chem 2012;23(5):958–65.

[116] Venkataraman S, Hedrick JL, Ong ZY, Yang C, Ee PLR, Hammond PT, Yang YY. The effects of polymeric nanostructure shape on drug delivery. Adv Drug Del Rev 2011;63(14–15): 1228–46.

[117] Lee JS, Ankone M, Pieters E, Schiffelers RM, Hennink WE, Feijen J. Circulation kinetics and biodistribution of dual-labeled polymersomes with modulated surface charge in tumor-bearing mice: comparison with stealth liposomes. J Control Release 2011;155(2):282–8.

[118] Alexis F, Pridgen E, Molnar LK, Farokhzad OC. Factors affecting the clearance and biodistribution of polymeric nanoparticles. Mol Pharm 2008;5(4):505–15.

[119] Immordino ML, Dosio F, Cattel L. Stealth liposomes: review of the basic science, rationale, and clinical applications, existing and potential. Int J Nanomedicine 2006;1(3):297–315.

[120] Photos PJ, Bacakova L, Discher B, Bates FS, Discher DE. Polymer vesicles in vivo: correlations with PEG molecular weight. J Control Release 2003;90(3):323–34.
[121] Adams N, Schubert US. Poly (2-oxazolines) in biological and biomedical application contexts. Adv Drug Del Rev 2007;59(15):1504–20.
[122] LoPresti C, Massignani M, Fernyhough C, Blanazs A, Ryan AJ, Madsen J, Warren NJ, Armes SP, Lewis AL, Chirasatitsin S, Engler AJ, Battaglia G. Controlling polymersome surface topology at the nanoscale by membrane confined polymer/polymer phase separation. ACS Nano 2011;5(3):1775–84.
[123] Tanner P, Ezhevskaya M, Nehring R, Van Doorslaer S, Meier W, Palivan C. Specific His6-tag attachment to metal-functionalized polymersomes relies on molecular recognition. J Phys Chem B 2012;116(33):10113–24.
[124] Dieu LH, Wu D, Palivan CG, Balasubramanian V, Huwyler J. Polymersomes conjugated to 83-14 monoclonal antibodies: in vitro targeting of brain capillary endothelial cells. Eur J Pharm Biopharm 2014;88(2):316–24.
[125] Egli S, Nussbaumer MG, Balasubramanian V, Chami M, Bruns N, Palivan C, Meier W. Biocompatible functionalization of polymersome surfaces: a new approach to surface immobilization and cell targeting using polymersomes. J Am Chem Soc 2011;133(12):4476–83.
[126] Georgieva JV, Brinkhuis RP, Stojanov K, Weijers CA, Zuilhof H, Rutjes FP, Hoekstra D, van Hest JC, Zuhorn IS. Peptide-mediated blood-brain barrier transport of polymersomes. Angew Chem Int Ed Engl 2012;51(33):8339–42.
[127] Pourtau L, Oliveira H, Thevenot J, Wan Y, Brisson AR, Sandre O, Miraux S, Thiaudiere E, Lecommandoux S. Antibody-functionalized magnetic polymersomes: in vivo targeting and imaging of bone metastases using high resolution MRI. Adv Healthc Mater 2013;2(11):1420–4.
[128] Mu Q, Jiang G, Chen L, Zhou H, Fourches D, Tropsha A, Yan B. Chemical basis of interactions between engineered nanoparticles and biological systems. Chem Rev 2014;114(15):7740–81.
[129] Zhou H, Jiao P, Yang L, Li X, Yan B. Enhancing cell recognition by scrutinizing cell surfaces with a nanoparticle array. J Am Chem Soc 2011;133(4):680–2.
[130] Kobayashi H, Brechbiel MW. Nano-sized MRI contrast agents with dendrimer cores. Adv Drug Deliv Rev 2005;57(15):2271–86.
[131] Grafe D, Gaitzsch J, Appelhans D, Voit B. Cross-linked polymersomes as nanoreactors for controlled and stabilized single and cascade enzymatic reactions. Nanoscale 2014;6(18):10752–61.
[132] Dobrunz D, Toma AC, Tanner P, Pfohl T, Palivan CG. Polymer nanoreactors with dual functionality: simultaneous detoxification of peroxynitrite and oxygen transport. Langmuir 2012;28(45):15889–99.
[133] Jurecka A, Tylki-Szymanska A. Enzyme replacement therapy: lessons learned and emerging questions. Expert Opin Orphan Drugs 2015;3(3):293–305.
[134] Pham-Huy LA, He H, Pham-Huy C. Free radicals, antioxidants in disease and health. Int J Biomed Sci 2008;4(2):89–96.

Chapter 12

Supercritical Fluids in Nanoreactor Technology

Sodeh Sadjadi
Nuclear Science and Technology Research Institute, Tehran, Iran

1 THE CRITICAL POINT AND SUPERCRITICAL FLUIDS

The equilibrium phase boundaries between solid and liquid, solid and gas, and liquid and gas for a single component are represented by a line on a pressure–temperature diagram as shown in Fig. 12.1. The state of coexistence of all three phases in equilibrium is represented by the triple point [1]. In Fig. 12.2, one can see a surface in the three-dimension space of state variables [pressure (P), mass density (ρ) and temperature (T)] and the projections of the state surface on the planes (P, T) and (P, $1/\rho$). The three usual states of matter are separated by Coexistence Curves. If the values of (P, T) are such that the component's state is located inside a coexistence curve, then the pure component is observed under the form of two coexisting phases.

Unlike the liquid–solid coexistence curve, the liquid–vapor coexistence curve is not infinite. A maximum has been observed on this curve. Consequently, the liquid and vapor states cannot be considered as two separate states, but as the extremes of the same continuum (the dotted arrow) [2].

The discovery of what we now call the critical point came about with Cagniard de la Tour's (1822) experiments while conducting experiments involving various heated fluids and a cannon ball in a sealed cannon barrel [3]. It stimulated long discussions between Farraday and Herschel in the mid-19th century, but it is only through the works of Andrews and van der Waals that a first theoretical model explaining this transition was derived. Andrews named the top of the liquid–vapor coexistence curve the critical point, owing to its singular nature [2]. A fluid is said to be "supercritical" when its temperature and pressure exceed the critical temperature (T_c) and pressure (P_c).

Characteristic values for the gaseous, liquid, and supercritical state are listed in Table 12.1 [4]. Supercritical values for these features take place between liquids and gases. This holds true for properties of fluids, which are decisive for mass and heat transfer (diffusivity, viscosity, thermal conductivity, heat capacity) [5].

Organic Nanoreactors. http://dx.doi.org/10.1016/B978-0-12-801713-5.00012-4

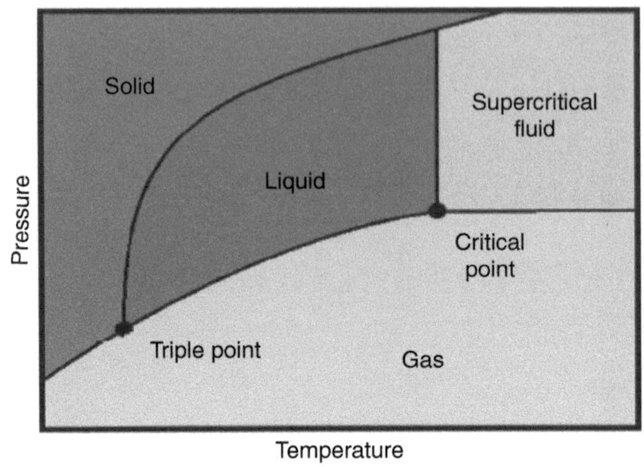

FIGURE 12.1 Pressure–temperature phase diagram. *(Reproduced from Ref. [3] with permission of Elsevier.)*

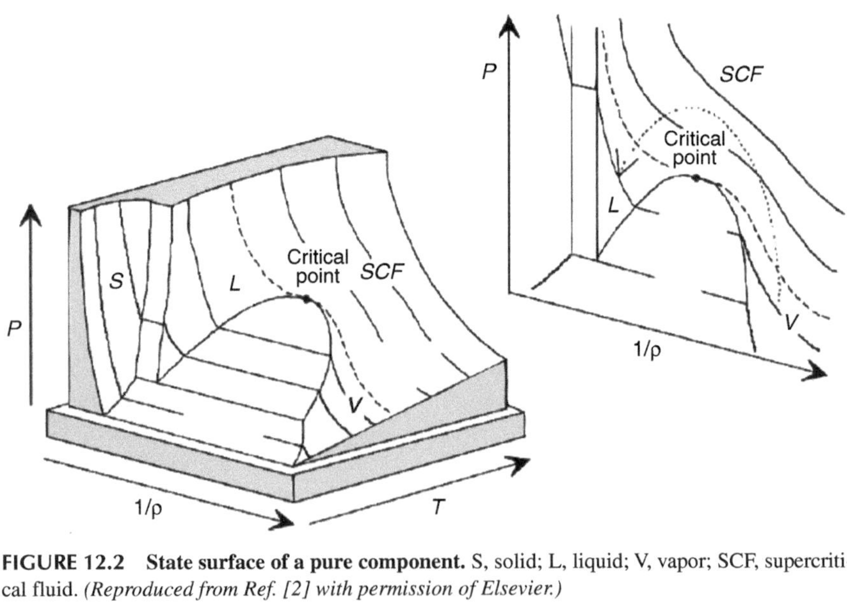

FIGURE 12.2 State surface of a pure component. S, solid; L, liquid; V, vapor; SCF, supercritical fluid. *(Reproduced from Ref. [2] with permission of Elsevier.)*

1.1 Properties of Supercritical Fluids

- Close to the critical point, the compressibility of any pure fluid goes to infinity. As a consequence of the high compressibility of fluids near the critical point, their density and density-dependent solvent properties, such as

TABLE 12.1 Characteristic Values of Gas, Liquid, and Supercritical State [4]

State of the fluid	Density (g/cm^3)	Diffusivity (cm^2/s)	Viscosity (g/cm/s)
Gas: $P = 1$ atm, $T = 15–30°C$	$(0.6–2.0) \times 10^{-3}$	0.1–0.4	$(0.6–2.0) \times 10^{-4}$
Liquid: $P = 1$ atm, $T = 15–30°C$	0.6–1.6	$(0.2–2.0) \times 10^{-5}$	$(0.2–3.0) \times 10^{-2}$
Supercritical fluid			
$P = P_C$; $T \approx T_C$	0.2–0.5	0.7×10^{-3}	$(1–3) \times 10^{-4}$
$P = 4P_C$; $T \approx T_C$	0.4–0.9	0.2×10^{-3}	$(3–9) \times 10^{-4}$

Reproduced from Ref. [4] with permission of Elsevier.

the dielectric constant and viscosity can be tuned sensitively through small changes in pressure, causing corresponding changes in dissolving power [5].

- In the supercritical state, at a constant temperature, density of a supercritical fluid (SCF) will increase with pressure. Density decreases as the temperature of the fluid rises (at a constant pressure).
- The diffusivities of SCFs are usually an order of magnitude larger than those found in the liquid phase. Diffusivity is parallel with temperature and contrary to pressure.
- Viscosity for a SCF is near that of normal gases, enhancing mass transfer. The viscosity of SCFs distinguish from liquids that temperature has a little effect on liquid viscosity while it can influence SCF viscosity in a considerable way [6].
- SCFs have also no surface tension because they are not subject to the vapor-liquid boundary [7].
- The thermophysical properties of SCFs and their consequences on SCFs dynamics and thermal transport are the subject of an excellent and very recent review by Carlès [2].

The most interesting characteristics of SCFs, on which are based all the SCFs' processes, is related to their "tunability" with pressure and temperature, especially the tunability of their solvent power [8]. The dissolving effect of a SCF is dependent on its density value. Solubility increases with increasing density (ie, with increasing pressure). The relationship with temperature is a little more complicated. Fig. 12.3 shows the solubility of a substance of low volatility in a sub- and supercritical fluid. The solubility in the SCF increases at constant pressure up to temperatures slightly below the T_c of the solvent. A further increase in temperature leads at "low" pressures to a decrease of the dissolved amount of the low-volatility substance in the subcritical liquid solvent and at "high" pressures still to an increase. "High" and "low" pressures refer to a "medium" pressure level.

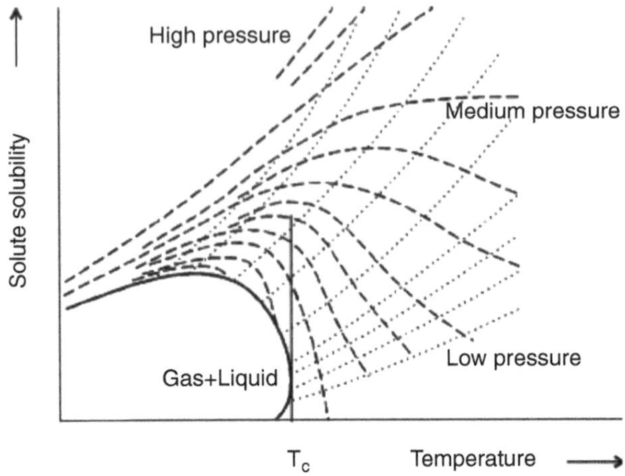

FIGURE 12.3 Variations in the solubility of a low-volatility substance (liquid or solid) in a subcritical (temperature < critical temperature, T_c) or supercritical fluid (temperature> T_c) as a function of process temperature, process pressure (segmented lines correspond to equal pressure or isobaric conditions), and solvent density (dotted lines correspond to equal density conditions). *(Reproduced from Ref. [4] with permission of Elsevier.)*

At "low" pressures, solubility of the low-volatility substance in the supercritical solvent decreases with temperature since density of the supercritical solvent decreases rapidly with increasing temperature at near critical pressures. At "high" pressures, density changes with temperature are far more moderate, so that the increase of vapor pressure is the dominating factor, while at "low" pressures loss in solvent power induced by lower density prevails. Analogous solubility behavior can be found in systems of an SCF with a solid substance or a low-volatility liquid [4].

Moreover, the fluids are commonly miscible with permanent gases (eg, N_2 or H_2) and this leads to much higher concentrations of dissolved gases than can be achieved in conventional solvents [9].

Both the capability of some SCFs to replace toxic solvents and the ability of tuning solvent characteristics for highly specific separations or reactions have attracted interest to SCFs [10]. Supercritical fluids are already applied in several processes developed to commercial scale in pharmaceutical, food, and textile industries. As research continues to investigate their capabilities, new applications of the SCF technology are developing daily. Research has shown that SCFs could be used as new reaction media for chemical and biochemical reactions, for synthesis of new materials and new catalyst supports such as aerogels, for special separation techniques such as chromatography using SCFs and extraction processes, and for particle formation and product formulation [3].

2 MICROEMULSIONS

2.1 Introduction

Microemulsions are clear, thermodynamically stable liquid mixtures of two immiscible fluids, consisting of nanometer-sized aqueous or oil droplets dispersed in the continuous phase [11]. These nanodroplets can be used as nanoreactors because of their characteristic interfacial properties allowing an intimate contact, at nanoscale level, of hydrophilic and hydrophobic domains. The relatively well understood self-assembly dynamics of these materials, and the ease of formation of enclosed structures with specified size (and, in some cases, shape) are very attractive for a comprehensive understanding of chemical reactions carried out in these media [12].

The combination of water-in-oil microemulsion and SCFs is a promising topic and may find more applications with some interesting advantages by utilizing the unique properties of SCFs. These include pressure-dependent variables such as viscosity, density, and diffusion rate, as well as the ability to readily manipulate the P–T phase behavior in the multicomponent micelle systems [13]. Much of the current research efforts in this area have been directed toward the SCF-continuous microemulsions.

2.2 SCF-Continuous Microemulsions

The ability to design water-in-SCF microemulsions offers new opportunities for protein and polymer chemistry, separation science, reaction engineering science such as providing a "nanoreactor" for the synthesis of nanoparticles and conducting chemical reactions [14].

The first observations of the micelles and microemulsions formation in SCFs (supercritical ethane and propane) were given in the pioneering work by Gale et al. [15]. As multicomponent reaction mixtures in the vicinity of a critical point can exhibit phase behavior phenomena not occurring in ordinary gas or liquid mixtures, the primary focus on these novel microemulsion systems has been on the phase behavior of water-in-SCF microemulsions. A detailed coverage of the phase behavior of this system is outside the scope of this chapter.

The phase behavior can have a significant impact on reaction outcomes (eg, reagents no longer being in contact with a catalyst due to phase separation) [16]. The phase behavior of these systems is strongly dependent on fluid pressure, in contrast to liquid systems where similar changes in pressure have little or no effect [17]. In comparison with the conventional methods to tune the properties of classical water-in-oil microemulsions (changing the composition, adding additives, changing pH, heating) [18], the properties of these systems (water-in-SCF microemulsions), can be strongly influenced by varying the pressure (changes in pressure induce large changes in fluid density).

The variations in micelle size can be easily manipulated by changing the fluid density. This property is attractive for studying nanoparticles formation or chemical reaction using water-in-SCF microemulsions as nanoreactors and provides the potential to tune the reaction environment in order to optimize reaction rate and selectivity.

The continuous phase density for reverse microemulsions has a large effect on the degree of inter-droplet attractive interactions which can directly affect the secondary structure of the microemulsion. These changes in the secondary structure can induce large changes in physical properties [19].

Ikushima et al. made a detailed study of the effects of pressure and molar ratio of water to surfactant sodium bis(2-ethylhexyl) sulfosuccinate (AOT) on the physicochemical properties such as the type of interior water molecules, micropolarity, and the structures of AOT reverse micellar aggregates in SCFs' media by using high-pressure FTIR and UV spectroscopies (see Section 2.3.2 for their results) [20].

Studies of water-in-scCO_2 reverse micelles derived from the surfactant ammonium carboxylate perfluoropolyether (PFPE) demonstrated that CO_2 can partition into the PFPE reverse micelle [21]. The important question related to the reaction rates and outcomes inside the nanoreactor is whether the pH within the reverse micelle water pool can be controlled by adjusting the CO_2 pressure. In an attempt to answer this question, Bright et al. quantitatively determined the pH within the PFPE water pools. Their results showed that the pH remains essentially constant as a function of CO_2 continuous phase pressure and micelle water loading [22].

One problem of using water-in-oil microemulsions for nanoparticle synthesis or chemical reaction is the separation and removal of solvent from products. A unique feature of the water-in-SCF microemulsions is that SCF reverse micelle phase stability is strongly dependent on the fluid pressure (density), at higher temperatures it is necessary to use higher pressures, or the density will be insufficient, the microemulsion will break, and phase separation can be accomplished [16]. So the reaction product can be obtained by a reduction in pressure without the need for extraction from reverse micelles as in the case of organic solvents [23].

The choice of surfactant depends on the composition of the continuous phase. Both surfactant mixtures and fluid mixtures can be used. For example, while neither of the two surfactant perfluoropolyether phosphate ether (PFPE-PO_4) and AOT is soluble in scCO_2, their mixtures with appropriate ratios can form stable microemulsions with water cores in scCO_2 [16].

The formation of reverse micelles has been investigated by using different SCFs such as CO_2, N_2O, CF_3Cl, SF_6, and hydrocarbon solvents (ethane, propane, *n*-butane, and *n*-pentane), scCO_2 is the preferred reaction medium among SCFs from both the economic and green chemistry perspectives [17]. However, due to the low dielectric constant ($\sim$1.33 at 200 bar, 350 K), and polarizability per volume, and not strongly engaging in van der Waals interactions, scCO_2 is generally a very poor solvent, especially for polar and high molecular weight solutes. Solubility can be increased by adding small amounts of cosolvent. The surfactants used to stabilize water-in-CO_2 reverse micelles normally consist of nonpolymeric or polymeric fluoroalkyl or fluoroether tails with nonionic or ionic head groups such as sulfonates, carboxylates, sulfates, ammonium, and others. Fluorinated tails are CO_2-philic because, like CO_2, they exhibit weak van der Waals forces [24].

The design and synthesis of low-cost, green and low toxic effective CO_2-soluble surfactants represents a major challenge and draws attention from research groups around the world [25]. So far, several fluorocarbon [26–29], silicone [30,31], phosphate fluoro surfactants [32–34], and a series of fluorinated analogs of AOT [35–37], are shown to be effective surfactants for stabilizing water droplets in $scCO_2$ [38].

For most of the CO_2-continuous microemulsions, water is used as the dispersed phase, forming the water-in-CO_2 microemulsions. Reverse micelles with ionic-liquid (ILs) cores in $scCO_2$ as the continuous phase have attracted much interest, and the systems may combine some advantages of the two fluids. ILs can dissolve large amounts of CO_2, but they are not usually measurably soluble in the $scCO_2$ phase. This system is an obvious choice for studying the effect of very small amount of polar solvent on the properties of reverse micelles (such as polarity and size) in $scCO_2$. This kind of study with the conventional polar solvent is very difficult because their solubility in $scCO_2$ continuous phase is usually not negligible [39].

Several research groups have reported ionic liquid-containing microemulsions, but literature on the ionic liquid-in-SCF microemulsion is limited. It includes ionic liquids, 1,1,3,3-tetramethylguanidinium acetate, 1,1,3,3-tetramethylguanidinium lactate, and 1,1,3,3-tetramethylguanidinium trifluoroacetate in $scCO_2$ with the aid of surfactant *N*-ethyl perfluorooctylsulfonamide [40].

Normal emulsions (ie, $scCO_2$-in-water emulsions, with water as the continuous phase) can also be obtained simply by manipulating the solubility of the surfactant (the phase in which the surfactant is most soluble will be the continuous phase), although examples for the formation of this kind of emulsion in literature are scarce and reverse micelles formed in SCFs have been used with much success as nanoreactors for a wide range of chemical syntheses and synthesizing various nanomaterials.

2.3 SCF-Continuous Microemulsions Applications as Nanoreactors

2.3.1 Formation of Nanoparticles in SCF-Continuous Microemulsions

SCF microemulsions are attracting much attention in the field of nanotechnology as a way of synthesizing inorganic and organic nanomaterials. The general method of synthesizing nanoparticles using microemulsion techniques can be divided into two cases. The first case, called the "one microemulsion method," can itself be divided into two subgroups, namely "energy triggering" and "one microemulsion plus reactant" methods. The energy triggering method generally is driven by implementing a triggering agent into the single microemulsion, which contains a reactant precursor. The "one microemulsion plus reactant" procedure is to use a system in which the micelle cores (containing a dissolved reactant) act as nanoreactors for conducting the reactions and the second reactant (precipitant) can be a gas, a solute soluble in the continuous phase. The second method, the "two microemulsion" method, is the mixing of two reverse micelles, each of them containing one reactant. The Brownian motion of the micelles leads to intermicellar collisions, and the reaction can take place by coalescence or aqueous phase exchange between the two reverse micelles [41]. The nanoparticles synthesis can be controlled by this intermicellar exchange rate, which is in turn closely related to several factors, including the size of the micelle core, the properties of the bulk solvent, and the strength of the interactions between the solvent and the surfactant tails [42]. The application of SCFs solvents may provide further control over nanoparticle characteristics through the SCF's tunable physicochemical properties by adjusting the temperature and pressure [43]. The advantages of using SCFs (especially $scCO_2$) as the continuous phase over conventional organic solvents for the synthesis of nanomaterials are that the morphologies of the products depend on the nature of the surfactant and applied solvent. Novel surfactant and CO_2 combinations can lead to novel framework structures and morphologies [23]; however, control of the morphology of synthesized particles is still a very new area. SCFs provide high diffusivity and thus accelerate reaction rates [44], and as mentioned before, nanomaterials can be simply separated from the solvent by reducing the pressure. An obvious advantage of this method could present itself if these nanomaterials were to be used as catalysts for chemical reactions; the reason is that, with all microemulsion methods, there is a possibility of leaving the surfactant residue on the catalyst, which could decrease its active surface area. Finally, as a potential substitute for traditional organic solvents, $scCO_2$ affords significant environmental benefits and can contribute to green chemistry techniques.

Many solvent properties are directly related to supercritical solvent density and will therefore depend on pressure or temperature. Though solvent density is assumed to play a leading role in the size, size distribution, and growth behavior of the particles synthesized in water-in-SCF microemulsions, it has been scarcely considered in the literature.

The effect of supercritical solvent continuous phase properties on particle growth behavior was investigated by Roberts et al. through comparison of Cu and Ag particles growth rates in supercritical alkan and in normal liquid solvents at the same conditions. Favorable properties of SCFs solvent, such as lower density and solvent power, decreased solvent interaction with the surfactant tails and led to smaller nanocrystals and faster particle growth rate, due to the increased kinetics of the intermicellar exchange mechanism [42].

Research by Wai et al. [45] on the synthesis of silver nanoparticles in $scCO_2$, showed that the pressure and the temperature dependence of particle size and size distribution in water-in-CO_2 microemulsion systems is due to variation in the density of the solvent. In this study, the density within two series of experiments were increased by increasing the pressure or decreasing the temperature. The average size of the particles decreased linearly and over a large range of values with increasing fluid density and the size distribution was larger for the particles formed at low densities. Although, at constant density (with different pressure–temperature combinations), the average size remained relatively unaffected. This linear relationship was repeatedly observed for two different water-to-surfactant molar ratio (W) values and two different concentrations of tested surfactant.

The cohesive energy density of the solvent is higher at increased solvent density, which results in better surfactant tails–solvent interactions. Therefore, smaller particle size and narrower size distributions obtained at higher densities could be attributed to weaker intermicellar interactions and dependence of the water-core size on pressure and temperature (by increasing the pressure or temperature, the size of the water core will be decreased).

In a similar fashion, this technique allowed for size control synthesis of semiconductor nanocrystals (CdS and ZnS) through manipulation of $scCO_2$ density [46].

Several research groups have shown that reverse micelle in $scCO_2$ can be employed as a "nanoreactor" for in situ synthesis of bimetallic nanoparticles [47], metal nanocrystals [42,45,48–53], metal halides [54,55], metal oxide nanoparticles [56,57], semiconductor nanomaterials [23,46,58], and nanotube-supported metallic nanoparticles [59] with good size control (Table 12.2). The formation of metallic nanometer materials have been reported by chemical reduction and photoreduction method in water-in-$scCO_2$ microemulsions [43] (Table 12.2). The two common reducing agents used in alkane-microemulsion systems are hydrazine and sodium borohydride. However, in water-in-$scCO_2$ microemulsion, both of these compounds readily react with the aqueous carbonic acid in the micelle core that is present as a result of the CO_2/H_2O equilibrium [51]. The low solubility of the conventional reducing agent in $scCO_2$ is the other important factor; it limits their use as an effective reducing agent in water-in-$scCO_2$ microemulsions systems. For example, the silver nanoparticles were synthesized by the reduction of Ag^+ ions in the water core of the AOT/PFPE-PO_4 water-in-$scCO_2$ reverse microemulsions by using $NaBH_4$ at 38°C. In this system, the reaction between $NaBH_4$ and CO_2 is unlikely, but the intensity of the characteristic Ag nanoparticle absorption band around 400 nm was not high, indicating the importance of the solubility of the reducing agent in $scCO_2$ [44]. A variety of reducing agents, such as sodium cyanoborohydride ($NaBH_3CN$), $NaBH(OAc)_3$, *N*,*N*,*N′*,*N′*-tetramethyl-*p*-phenylenediamine (TMPD), and sodium hypophosphite, have all been used to synthesize metallic nanoparticles. The reducing agent $NaBH(OAc)_3$ used in the original study by Ji et al. [51] has a limited solubility in $scCO_2$. Therefore, an addition of ethanol, for example, increases the content of the reducing agent in the $scCO_2$ system. Other

TABLE 12.2 Some Examples of Supercritical Microemulsions as Nanoreactors Which Have Been Successfully Applied to the Different Nanomaterials Synthesis

Nanomaterials	SCFs	Precursors	Reducing agent	Surfactant	Cosur-factant	*W*	Average size	References
ZnS	CO_2	$Zn(NO_3)_2$, Na_2S, F-thiol[a] stabilizer	—	F-AOT[b]	—	10 and 6	$W = 10$, 9.1–1.9 nm, $W = 6$, 7.0–1.5 nm (The average size of the nanoparticles decreases linearly with increasing fluid density from 0.86–0.99 g cm^{-3})	[46]
CdS	CO_2	$Cd(NO_3)_2$, Na_2S, F-thiol stabilizer	—	F-AOT	—	10 and 6	$W = 10$, 7.1–2.0 nm, $W = 6$, 4.0–1.3 nm (The average size of the nanoparticles decreases linearly with increasing fluid density from 0.86–0.99 g cm^{-3})	[46]

AgX (X = I, Cl, Br)	CO_2	$AgNO_3$- NaI, NaBr and NaCl	–	AOT	PFPE-PO_4	12	the calculated particle size based on the spectroscopic method: AgI:~3.4 nm AgBr: ~3.0 AgCl: particle size could not be calculated AgI particle size based on TEM: ~3–15 nm	[54]
Ag	CO_2	$AgNO_3$	Ethanolic solution of $NaBH(OAc)_3$	AOT	F-pentanol[c]	4.7	6.0 nm	[55]
AgI	CO_2	KI, $AgNO_3$	—	AOT	F-pentanol	5.6	5.7 nm	[55]
ZnS	CO_2	$Zn(NO_3)_2$, Na_2S	—	F-AOT	—	12	1.6 nm	[58]
CdS	CO_2	$Cd(NO_3)_2$, Na_2S	—	F-AOT	—	6 and 12	1.4 nm at W = 6, 1.7 nm at W = 12	[58]
Ag clusters	CO_2	$AgClO_4$, 2-propanol was added to promote the photoreaction	Photoreduction, Irradiation of water-in-$scCO_2$ microemulsions with UV light	PFPE-1, PFPE-2, PFPE-3[d]	—	10	1.7 nm Because of the aggregation during the RESS procedure, larger particles of 5–10 nm were occasionally observed in several TEM images	[43]

(Continued)

TABLE 12.2 Some Examples of Supercritical Microemulsions as Nanoreactors Which Have Been Successfully Applied to the Different Nanomaterials Synthesis (*cont.*)

Nanomaterials	SCFs	Precursors	Reducing agent	Surfactant	Cosurfactant	W	Average size	References
Au	CO_2	$HAuCl_4$	—	*N*-EtFOSA	—	For PEG-400/*N*-EtFOSA/CO_2 system, W^{corr} (the molar ratio of PEG solubilized in reverse micelles to surfactant): 0.36 For PEG-1000/*N*-EtFOSA/CO_2 system, W^{corr}:0.28	Not mentioned in the paper, from TEM image: ~ 5 nm	[50]
CuS	CO_2	$Cu(NO_3)_2$, Na_2S	—	di-HCF4[e]	—	10	4–6 nm	[23]
Cu	Ethane, isooctane cosolvent	$Cu(AOT)_2$	Hydrazine	AOT	—	7.8	9 nm	[42]
Carbon Nanotube-Supported Pt (Pt-CNT)	CO_2	CNT, Na_2PtCl_4, A small amount of hexane is necessary to increase the solubility of the AOT surfactant	H_2	AOT	—	10	8.7 nm	[59]

Ru and Pd	CO_2	$Pd(NO_3)_2$ or $RuCl_3$	H_2	Ammonium perfluorohydro-carbo-carboxyl-ate $CF_3(CF_2)_{12}COO^-NH4^+$	—	30	~3–4 nm	[48]
Bimetallic Pd–Ru nanoparticles Pd:Ru (1:1)	CO_2	$Pd(NO_3)_2$ or $RuCl_3$	H_2	Ammonium perfluorotetra-decanoate	—	30	3.95 ± 0.46 nm	[47]
Ag	CO_2	$AgNO_3$, F-thiol stabilizer	Aqueous solution of $NaCNBH_3$	F-AOT	—	6 and 10	In this work, the influence of CO_2 density, temperature, and pressure for two different *W* values and for two different concentrations of surfactant on the average size and size distribution of silver nanoparticles was tested. The average size of the metal particles varies from 1.9–13 nm	[45]
Pd	CO_2	$PdCl_2$	H_2	AOT	PFPE-PO_4	20	5–10 nm	[52]
Rh	CO_2	$RhCl_3$	H_2	AOT	PFPE-PO_4	30	3–5 nm	[53]

(Continued)

TABLE 12.2 Some Examples of Supercritical Microemulsions as Nanoreactors Which Have Been Successfully Applied to the Different Nanomaterials Synthesis (*cont.*)

Nanomaterials	SCFs	Precursors	Reducing agent	Surfactant	Cosurfactant	*W*	Average size	References
Ag and Cu	CO_2	$AgNO_3$ or $Cu(NO_3)_2$	Ethanolic solution of $NaBH_3CN$ or TMPD	AOT	PFPE-PO_4	12	Ag and Cu: 5–15 nm	[44]
Ag	CO_2	$AgNO_3$	$NaBH(OAc)_3$ (a suspension in dichloroethane)	AOT	PFPE-PO_4	12	5–15 nm	[51]
The amorphous precursor particles that was then calcined to yield TiO_2 nanoparticles	CO_2	Titanium tetraisopropoxide	—	PFPE-NH_4 and PDMAEMA-*b*-PFOMA[f]	—	5, 10,20	8–19 nm	[56]
The amorphous hydrous zirconia precursor particles that was then calcined to yield ZrO_2 nanoparticles	CO_2	Zirconyl nitrate hydrate, aqueous ammonia solution as a precipitating agent	—	PFPE	—	Not reported	Particles formed from precursors synthesized at 15.3MPa: 10–30 nm, quasispherical particles formed from precursors synthesized at	[57]

							17.3 MPa: 4–13 nm, quasi-spherical particles formed from precursors synthesized at 13.9 MPa : ellipsoidal, with lengths of 21–72 nm along the major axis and from 16– 37 nm along the minor axis	

[a] *1H,1H,2H,2H-perfluorodecanethiol.*
[b] *Sodium bis(2,2,3,3,4,4,5,5-octafluoro-1-pentyl)-2-sulfosuccinate.*
[c] *2,2,3,3,4,4,5,5-Octafluoro-1-pentanol.*
[d]

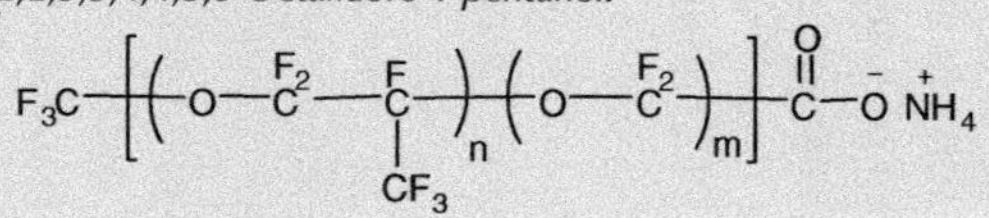

PFPE-1

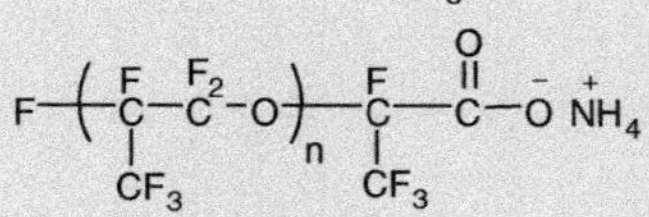

PFPE-2 (n=3) PFPE-3 (n=4)

[e] *Sodium salt of bis(2,2,3,3,4,4,5,5-octafluoro-1-pentyl)-2-sulfosuccinate.*
[f] *Poly(dimethyl amino ethyl methacrylate-block-1H,1H,2H,2H-perfluorooctyl methacrylate).*

more soluble reducing agents, including $NaBH_3CN$ and TMPD, were tested by Wai et al. [44] for synthesis of silver and copper nanoparticles in water-in-CO_2 microemulsion. TMPD is very soluble in $scCO_2$, while ethanol modifier is needed when $NaBH_3CN$ is used as a reducing agent because of its low solubility in $scCO_2$. Their results have established that the reducing agent type has a strong effect on the formation of the metal nanoparticles rate. The formation of Ag nanoparticles using $NaBH_3CN$ proceeded at a faster rate, but the size and size distribution of the synthesized particles using TMPD or $NaBH_3CN$ as a reducing agent were about the same. Diffusion and distribution of the oxidized form of the reducing agent between the water core and supercritical solvent continuous phase was suggested to be the rate-determining step for the nanoparticle synthesis. Since oxidized products of $NaBH_3CN$ are much smaller compared with $TMPD^{\cdot+}$ (the oxidized form of TMPD) it probably would diffuse faster.

Hydrogen gas exhibits high miscibility with CO_2 and is a very desirable reducing agent for producing nanometer-sized metal particles in the water-in-$scCO_2$ microemulsion [52].

Photon-based synthesis could provide an avenue to preserve the structure of microemulsions (in the chemical reduction method, the introduction of the reducing agent into the water core from outside may damage the structure of microemulsions). A photochemically induced metal nanoparticles synthesis in the water-in-$scCO_2$ microemulsion involves microemulsion formation followed by metal ion reduction through absorption of incident light by microemulsion which contains the reactants precursors [43].

Recently, poly(ethylene glycol) (PEGs) with two different molecular weights (PEG-400 and PEG-1000, which are liquid and solid at room temperature, respectively) have been reported to form nanosized domains within reverse micelles in $scCO_2$ with the aid of surfactant *N*-ethyl perfluorooctylsulfonamide (*N*-EtFOSA). These PEG domains were utilized as nanoreactors to synthesize the face-centered cubic gold nanostructures. In this reverse micelle reaction media, the role of the surfactant is twofold. The *N*-EtFOSA creates a microemulsion system consisting of water and Au^{3+} metal ions encased within an *N*-EtFOSA surfactant reverse micelle. The *N*-EtFOSA surfactant also acts as a reducing agent from it's amine head groups to form metallic Au. A continuous tuning of the size of nanosized PEG domains and nanoparticles synthesized by the CO_2 microemulsion method could be achieved by varying the density of the fluid phase, content, and the molecular weight of PEG [50].

The collection of nanoparticles entrapped within the core of the micelles from surfactants is essential to subsequently conduct catalytic reactions using these particles. Today, two of the most common methods of the collection of nanoarticles prepared in water-in-$scCO_2$ microemulsions are reducing pressure and rapid expansion method.

As mentioned before, water-in-$scCO_2$ microemulsions may be phase-separated easily for product recovery (unlike the case for conventional microemulsions) simply by depressurization. Using an in situ deposition method by reducing the pressure of the system to the cloud point of the microemulsion, nanoparticles in the microemulsion could precipitate on a solid surface (eg, on a carbon-coated copper grid for TEM). RESS technique allows collection of nanoparticles by rapidly expanding the microemulsions through a suitable nozzle into a collection solution at ambient conditions. The development of more efficient processes for collection of nanoparticles remains an important challenge since in many cases; the collected nanoparticles are tightly aggregated, resulting in a wider size distribution. Several research groups have proposed a method to prevent aggregation of nanoparticles during collection from reverse micelles using chemical stabilizers. Ligands containing thio, sulfide, and amino functional groups, with strong affinity for metal surfaces could be promising stabilizer for collecting nanoparticles with low polydispersity [11].

Supported catalysts often exhibit unique activities depending on the shape and size of nanoparticles. There are several researches, which have demonstrated the possibility to control nanoparticle size and shape during supported catalyst synthesis using water-in-scCO_2 reverse micelles as nanoreactors. For example, Wai et al. prepared three Pt onto carbon nanotubes (CNT), Pt-CNT, electrocatalysts for the reduction of oxygen using (1) water-in-scCO_2 microemulsion (Pt-CNT SCME), (2) direct scCO_2 fluid deposition, and (3) water-in-hexane microemulsion. The TEM of Pt-CNT SCME displayed a well-dispersed face center cubic platinum nanostructure on the CNT surface. Pt-CNT SCME system was found to have the highest electrocatalytic activity which is attributable to the large active surface area as measured from cyclic voltammograms (CVs) [59].

2.3.2 Chemical Reactions in Water-in-SCF Microemulsions

The use of microemulsions as media for organic reactions is a way to overcome the reagent incompatibility problems that are frequently encountered in organic synthesis. A properly formulated microemulsion can dissolve both water-soluble and water-insoluble reactants, and the large oil–water interfacial area enables contact between the two otherwise incompatible species [60]. Microemulsion droplets can catalyze or inhibit chemical reactions by compartmentalization and concentration of reactants and products [61]. The surfactant monolayer may also accelerate the reaction by attracting reagents of opposite charge situated in the water domain, thus increasing their concentration in the interfacial zone, where the reaction occurs. The presence of an oil–water interface may induce orientation of reactants in microemulsion systems, which in turn may affect the regioselectivity of organic reactions [60]. The reaction can take place inside the microemulsion droplet, at the surfactant interface, or in the continuous phase [62].

Microemulsion formation with carbon dioxide in supercritical state instead of a hydrocarbon as nonaqueous solvent is a way to take advantage of the low interfacial tension between H_2O and CO_2 that could facilitate reaction between organic substrates and hydrophilic reactants at the interface. The next point of interest is the possibility of synthesizing the nanoparticles in the fluid and subsequently conducting catalytic reactions using these nanoparticles. A remarkable advantage of water-in-scCO_2 microemulsion is the ease of separation of products and catalyst recycling by a change in phase behavior of the system at different pressures.

Several research groups have developed new procedures for tackling the problem of separating reaction products from homogeneous catalytic solutions by deposition of catalysts stabilized in water-in-oil microemulsions on solid supports. However, the deposited catalysts do not have the dynamic reaction environment inherent in microemulsion systems [63].

A wide variety of reactions has been performed in water-in-scCO_2 microemulsions. The following examples show how the properties of a fluid at conditions higher than critical (where, in particular, the specific properties are prevailing) can describe the changes observed in the catalytic reactivity or selectivity in water-in-SCF microemulsions.

Catalytic air oxidation of toluene with high conversion (> 99%) and high selectivity (> 99% to benzoic acid) was achieved based on an aqueous microemulsion containing ionic Co^{2+} and Br^- species stabilized by fluorous surfactant-like species in scCO_2–air mixture (C–H free environment). The anionic fluorinated surfactant ions chelated with Co(II), ($[CF_3(CF_2)_8COO]_2Co(II)$. nH_2O, can serve as surfactant and catalyst for the oxidation of toluene. The authors studied the effects of the different variables on the reaction conversion. They found that a critical quantity of NaBr was essential for effective catalysis. Br^- acts as ionic promoter to facilitate rapid electron transfers of the catalytic Co species. A higher conversion toward benzoic acid was observed when the amount of water added to the catalytic system was increased. The high miscibility of oxygen with scCO_2 potentially offers activity benefits, as facilitates the availability of O_2 for the catalytic sites. Preference for the oxidation process was favored at lower pressures of CO_2. Together with increased toluene diffusivity (or a greater concentration), it can help to overcome transport limitations for fast catalytic reactions. Whereas the selectivities of the reactions remained nearly constant at different scCO_2 pressures used (100, 150, and 170 bar) [64–66]. To evaluate the role of scCO_2, this reaction was also carried out in nitrogen phase using the same conditions but replacing CO_2 with N_2 gas (which cannot form a supercritical phase under the reaction conditions). The results showed that N_2 as a reaction media was far less effective than scCO_2 in terms of yields.

This new catalytic system was then tested for oxidation of various alkyl aromatic hydrocarbons (Table 12.3). The result showed that the selectivity of a conventional homogenous oxidation, which induces oxidative α–β carbon bond cleavages, could be influenced by utilizing this microemulsion method in scCO_2 (only partial oxidation took place). The different observed selectivity toward partial oxidation products can be explained in terms of solvent polarity. This approach makes the CO_2-soluble partial oxidation products less accessible to the water-soluble catalyst, preventing over-oxidations. The catalytic system was

TABLE 12.3 Partial Oxidation of Various Alkyl Aromatic Hydrocarbons[a]

Alkyl aromatic	Conversion (%)	Selectivity (%)	
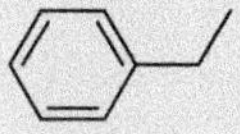	49.2	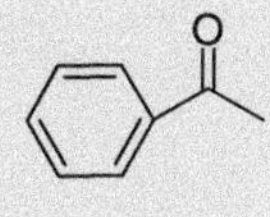 83.7	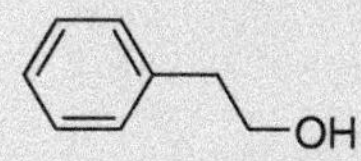 16.3
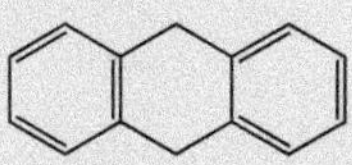	99.2	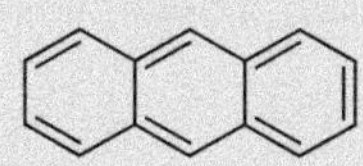 67.6	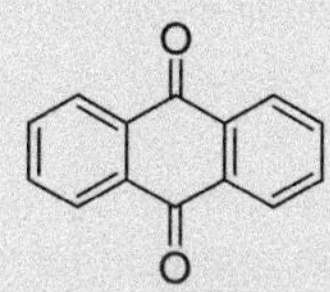 32.4
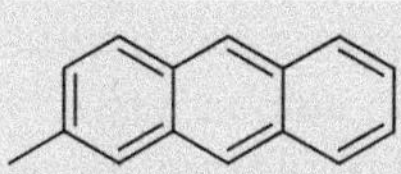	62.4	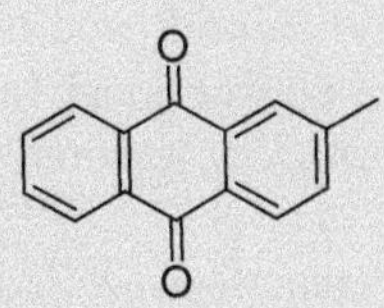 76.3	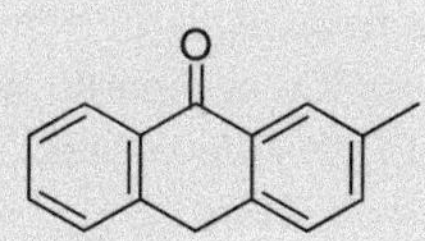 23.7

[a]*0.25 mmol Co(II) ([$CF_3(CF_2)_8COO]_2Co(II)$). nH_2O, 0.1 mL H_2O, 150 bar $scCO_2$, 10 bar O_2, 120°C, 24 h.*
Adapted from Ref. [66] with permission of Springer.

also found to be effective for the oxidation of the nonpolar bulky alkyl aromatic hydrocarbons. This is in contrast to typical homogeneous catalyst, in which low solubility of these substrates in acetic acid would significantly reduce the efficiency of the catalytic system [66]. The micellar catalytic system could be reconstructed in $scCO_2$ for the next runs of the oxidation without significant loss of catalytic activity due to solvent decarboxylation in traditional processes [65]. Nucleophilic substitution reactions (Scheme 12.1) were investigated by Johnston et al. in water-in-$scCO_2$ microemulsions [62]. The microemulsions were stabilized using an anionic perfluoropolyether ammonium carboxylate (PFPE $COO^-NH_4^+$) surfactant.

The influence of added reactants on phase behavior of the system was first studied to determine the maximum amount of the salt at which a homogeneous microemulsion will be formed. This study successively considered the effects of reaction parameters (such as temperature, and pressure) on the yield of benzyl bromide. The effect of pressure on the rate of benzyl nucleophilic substitution reaction was not significant. Upon transition from a single- into a two-phase system, the reaction yield decreased only below the cloud point

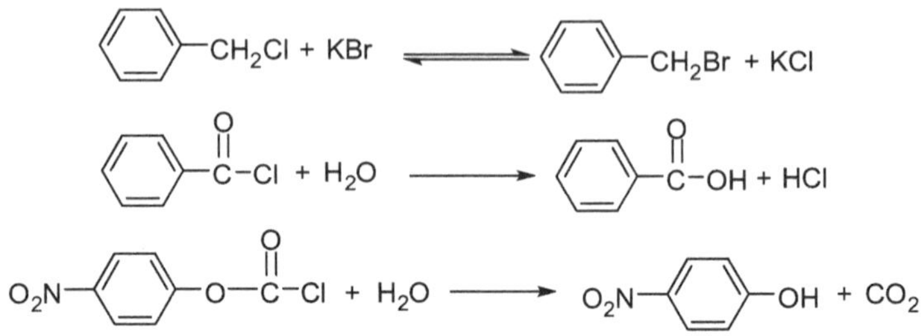

SCHEME 12.1 Nucleophilic substitution reactions. *(Adapted from Ref. [62] with permission of American Chemical Society. Copyright 1999.)*

pressure. Lowering the temperature (from 65 to 45°C) decreased the product yield. In addition to higher yields, the main advantages of performing nucleophilic substitution (the synthesis of benzyl bromide from benzyl chloride and the hydrolysis of benzoyl chloride and *p*-nitrophenyl chloroformate) reactions in $scCO_2$ compared with conventional solvent systems with anionic surfactant ($AOT/H_2O/n$-octane) were less required surfactant, omitting organic solvents and catalysts.

Using an anionic surfactant, water-in-$scCO_2$ system gave results comparable to or even greater than those obtained in a water-in-oil microemulsion stabilized by a higher concentration of cationic surfactant, and cosolvent [67], which is quite surprising since, as expected, cationic surfactants may favor enhanced product yield (a negatively charged transition state is stabilized by a cationic surfactant headgroup).

The interest in CO_2 microemulsion system for catalysis and reaction engineering is based mainly on their nature as homogeneous reaction media that provide a unique environment for homogenization of conventional heterogeneous catalysis. Various examples exist in which a colloidal rhodium particles-catalyzed hydrogenation reaction occurs slowly (during several hours or days) in organic SCF–water and water–SCF biphasic systems, in which there is an aqueous catalyst-bearing phase and a supercritical reactant phase. Wai et al. [53] have developed an effective process for rhodium-catalyzed hydrogenation of arenes in a homogeneous reaction system (ie, rhodium nanoparticles in a water-in-$scCO_2$ microemulsion). Rh nanoparticles were synthesized in the water-in-$scCO_2$ reverse microemulsions formed by the commonly used surfactant, AOT, in the presence of PFPE-PO_4 as a cosurfactant by hydrogen reduction of rhodium, and subsequently the in situ hydrogenation of arenes were catalyzed by the Rh nanoparticles in $scCO_2$. This approach gave promising results for two different CO_2-soluble (naphthalene) and water-soluble (phenol) arenes tested (Scheme 12.2). High yield (> 96%) for the desired tetralin product was achieved in short reaction time. Moreover, hydrogenation of phenol gave either cyclohexanone as a major product and cyclohexane as the minor product. Selectivity for two products cyclohexanone and cyclohexane formed in this catalytic system was 15:1, respectively. Unfortunately, Rh nanoparticles tended to lose

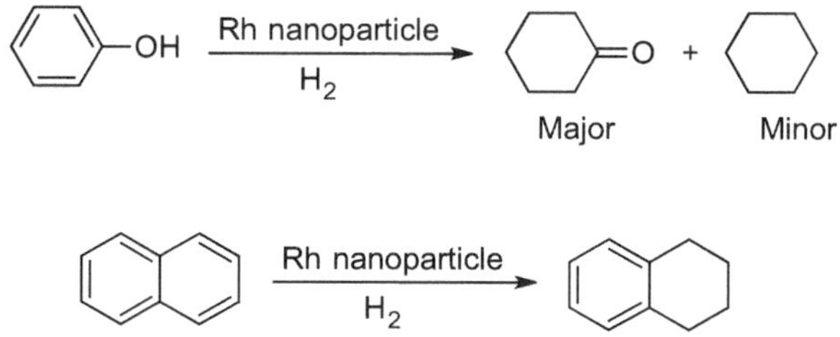

SCHEME 12.2 Catalytic hydrogenation of arenes with rhodium nanoparticles in a water-in-$scCO_2$ microemulsion. *(Adapted from Ref. [53] with permission of Royal Society of Chemistry.)*

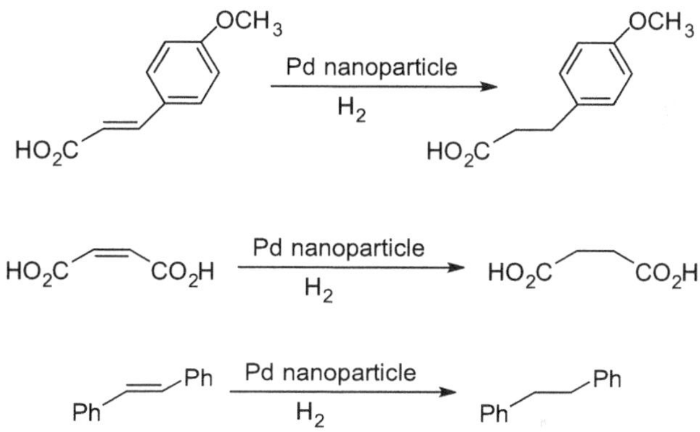

SCHEME 12.3 Hydrogenation of olefins in $scCO_2$ catalyzed by palladium nanoparticles in a water-in-CO_2 microemulsion. *(Adapted from Ref. [52] with permission of American Chemical Society.)*

their catalytic activity after a few minutes (20–30 min) of use, likely due to aggregation.

This research group prepared the stable Pd^0 nanoparticles in the same microemulsion system and demonstrated that catalytic hydrogenations of olefins (Scheme 12.3) can be achieved in $scCO_2$ by using Pd^0 nanoparticles dispersed in the fluid phase using a water-in-$scCO_2$ microemulsion. The reactions were performed by implementing liquid CO_2 (20°C) and $scCO_2$ (35 and 50°C) as the bulk solvent of the microemulsion system. The hydrogenation in $scCO_2$ was observed to be much faster than that for reaction in liquid CO_2 at similar conditions and in this range of temperature; the difference in the reaction rate between 20 and 35°C was much larger than that between 35 and 50°C. Rapid diffusion of the reactant from bulk CO_2 to the catalyst surface at the critical temperature (31.1°C) was offered as one possible explanation to this observation [52].

In a subsequent study [63], this research group took advantage of the phase behavior of $scCO_2$ and developed a simple and effective process for both separation of hydrogenation products and recycling of the catalyst. In this study, Pd^0 nanoparticles were synthesized in a water/AOT/$scCO_2$ reverse micelle system

TABLE 12.4 Catalytic Hydrogenations of Organic Compounds With Pd^0 Nanoparticles Stabilized in Water-in-CO_2 Microemulsions

Organic compounds	Time (min)	Product[a]
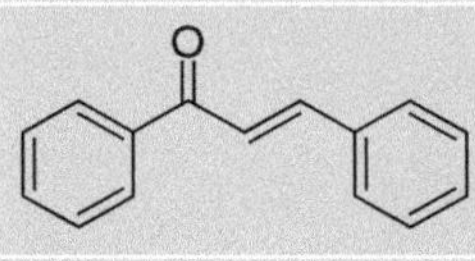	120	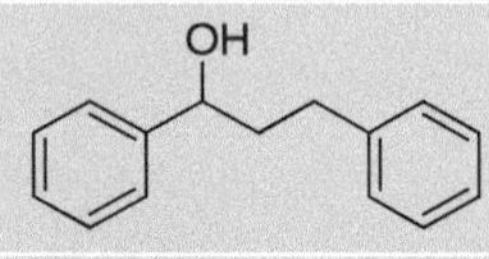
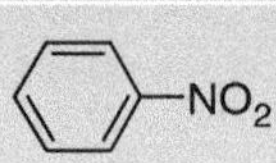	30	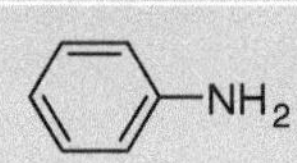
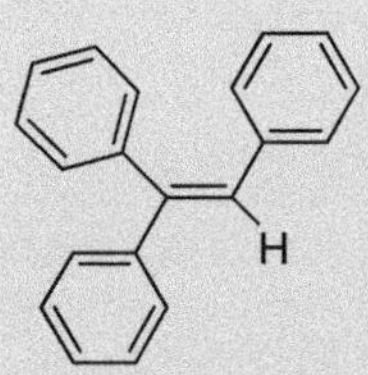	20	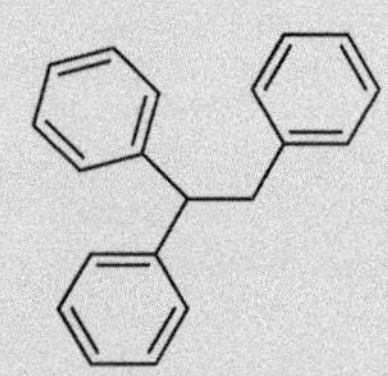

[a] *50°C and 200 atm, octanol solution containing AOT microemulsions (20 vol.%), percentage of catalyst to substrate (3–6), conversion 100%.*
Adapted from Ref. [63] with permission of American Chemical Society. Copyright 2005.

with a small amount of octanol as a cosolvent. Following depressurization, the homogeneous fluid separated into an upper CO_2 phase (containing the CO_2-soluble hydrogenated products) and a lower octanol phase (containing AOT microemulsions with Pd^0 nanoparticles). The products were then collected by flowing the upper CO_2 phase to a collection vessel. When the system was pressurized to a pressure at which a single homogeneous phase existed, it became possible for nanoparticles to redisperse into $scCO_2$ and effective catalyst recycling for the next runs of the hydrogenations. As can be seen from Table 12.4, this catalytic system showed high catalytic activities and selectivities toward hydrogenations of the double bonds of olefins, nitro groups, and keto groups (no hydrogenation of the benzene ring of these organic compounds took place).

A similar strategy has been developed for the preparation of a high-performance and reusable Pd–Ru bimetallic catalyst and its application in selective hydrogenation in $scCO_2$. Pd–Ru nanoparticles, were prepared by reduction of their salts in water-in-$scCO_2$ reverse micelles using ammonium perfluorotetradecanoate surfactant [47]. *Cis*-2-pentenenitrile molecule, which contains two reducible groups (cyanide and double bond), was selected for studies of selective hydrogenation. The bimetallic nanoparticles with different compositions were simply synthesized by using the microemulsion method in $scCO_2$ without interference from support and phase segregation to evaluate changes in catalytic activity as a function of particle composition. The optimum catalyst efficiency was achieved by the bimetallic catalysts Pd–Ru, with a Ru/Pd atomic

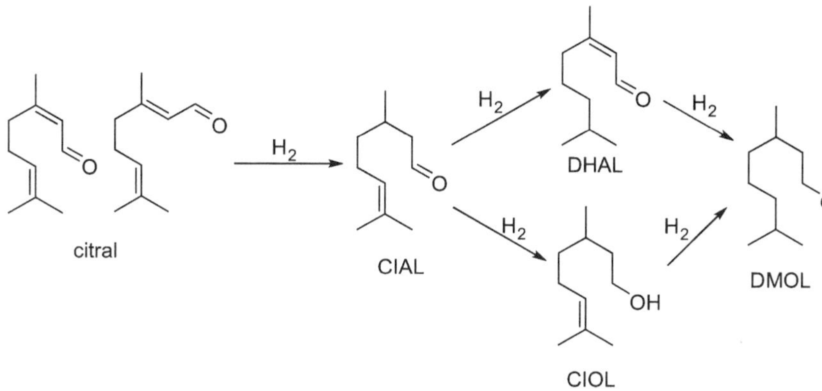

SCHEME 12.4 Reaction pathways of the citral hydrogenation. *(Adapted from Ref. [48] with permission of Elsevier.)*

ratio of 1:l. Using Pd–Ru bimetallic catalyst, high selectivity for C=C reduction was observed. The Pd–Ru catalyst had a catalytic activity much higher than either of the monometallic systems by combining the strong binding ability of Ru for nitrile with facilitated hydrogen activation and transfer ability of Pd.

Tsang and coworkers have shown in a series of papers [48,68,69] that micelle-hosted Pd nanoparticles in $scCO_2$ can be employed as a catalytic system to control the regioselectivity in hydrogenation of a citral molecule that contains three reducible groups. Exposure of citral to reducing conditions in the presence of a catalyst may result in several products, as shown in Scheme 12.4: 2,3-conjugated C=C (citronellal, CIAL), the fully saturated aldehyde (dihydrocitronellal, DHAL), the unsaturated alcohol (citronellol, CIOL), and the fully saturated alcohol (3,7-dimethyloctanol, DMOL). A micellar catalyst was obtained by self-assembly of $CF_3(CF_2)_{12}COO^-NH_4^+$ fluorosurfactant with the water-in-$scCO_2$ and reduction of metal ions dissolved in the water core of the microemulsion with a hydrogen gas. Systematic evaluation of important parameters in the hydrogenation of citral from a number of catalytic systems is presented in Table 12.5.

The results revealed that the unique micelle environment in $scCO_2$ could alter the catalytic selectivity of the hosted particles. The selectivity for hydrogenation toward the partially hydrogenated product of the 2,3-conjugated C=C (CIAL), when the reaction was performed in micellar media, can be explained by the orientation of citral at the micellar interface. This spatial orientation is mainly because of the polarity difference in citral structure, which makes the relatively hydrophilic carbonyl head group embed into the hydrophilic region of micelle, hence favoring the hydrogenation of the α,β-conjugated C=C bond that is near to the catalyst surface (selective formation of CIAL). It is believed that the hydrophobic nature of the partially hydrogenated product (more soluble in $scCO_2$ than water) would significantly reduce their contact time with the water core containing Pd nanoparticles, hence accounting for the low

TABLE 12.5 Influence of Experimental Conditions on Cital Hydrogenation[a]

Working state	W	Product distribution (%) CIAL	DHAL
Vapor phase	—	1	82
Liquid phase (cyclohexane)	—	19	66
Water-in-scCO_2 microemulsion, H_2 was added after the formation of the microemulsion	2, 5, 10	58–60	14–17
Water-in-scCO_2 microemulsion, H_2 was added after the formation of the microemulsion	30	68	11
Water-in-scCO_2 microemulsion, in buffer, H_2 was added after the formation of the microemulsion	30	67	7
No microemulsion, water remained as physical large droplets dispersing in the solvent	200	22	57
Water-in-scCO_2 microemulsion, H_2 was added prior to the formation of the microemulsion	30	30	54

[a] *500 μL Citral, 0.25 mmol Pd nitrate, 0.25 mmol F-surfactant, 140 bar CO_2 or 40 mL cyclohexane, 10 bar H_2, 40°C, 4 h. Buffered solution was prepared using a saturated aqueous solution of potassium dihydrogen phosphate.*
Adapted from Ref. [68] and [69] with permission of American Chemical Society and Springer.

selectivity to saturated aldehyde (DHAL). Interestingly, a selectivity switch appeared when H_2 was added to the mixture prior to emulsification or at high W ratios (100–200), which indeed supports the beneficial effect of the micellar environment [68,69]. The unique combination of liquid- and gas-like properties related to supercriticality is also essential for high selectivity of CITAL, since if scCO_2 was replaced by same pressure of N_2 (140 bar), low selectivity of CITAL was observed under the identical conditions over Pd and Ru catalysts (the major product was DMOL).

It is well documented that the pH inside microemulsion droplets in scCO_2 is typically 3. In order to determine whether the unusual selectivity is due to the low pH inside microemulsion droplets in scCO_2, a hydrogenation reaction was examined by addition of a phosphate buffer to raise the aqueous pH phase in waster-in-scCO_2 microemulsions to 7.5. Based on the obtained results, an increase in pH was found ineffective in changing the conversion and selectivity [68].

The influence of CO_2 pressure on catalytic activity and product distribution of the aqueous core of the microemulsion droplet with different nanoparticles guests (Pd and Ru) prepared in situ was investigated by varying the pressure of CO_2 while keeping the hydrogen pressure constant (Fig. 12.4) [48].

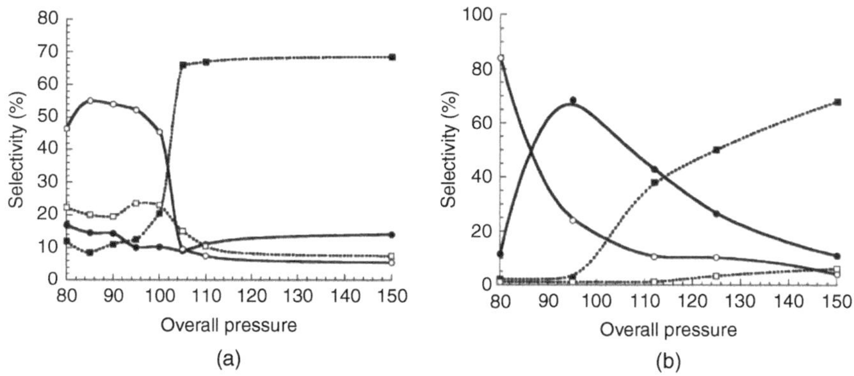

FIGURE 12.4 Product distribution curve versus total overall pressure for citral hydrogenation over (a) Ru, (b) Pd systems. [(●) DHAL, (■) CIAL, (○) DMOL, (□) CIOL]. $RuCl_2$ or $Pd(NO_3)_2$ (0.25 mmol), citral (0.5 mL), decane (0.1 mL), water (W = 30), surfactant (0.25 mmol), 40°C, 10 bar H_2 topped up with 70–140 bar CO_2 for 4 h. *(Reproduced from Ref. [48] with permission of Elsevier.)*

In the supercritical state, at the lowest pressure before the formation of microemulsion (80–85 bar), the $scCO_2$ has gas-like transport properties and citral molecules can freely access Pd or Ru nanoparticles surface, resulted in high selectivity toward DMOL (total reduction). At higher pressure (85–95 bar), a very similar product distribution was obtained over Ru. This could be attributed to the strongly bound partial hydrogenated species on the Ru surface; hence, virtually no intermediate was extracted. However, an increase in fluid pressure resulted in an increase in DHAL selectivity at the expense of DMOL selectivity over the Pd catalyst. As discussed, solvent power is directly proportional to the density of SCFs, so at higher pressure the liquid-like solubility of $scCO_2$ favors the desorption of the relatively nonpolar DHAL in citral hydrogenation over Pd nanoparticles.

At high pressure (>95 bar) corresponding to the regime of microemulsion, CIAL selectivity sharply increased and finally became the dominant product for both catalysts due to the special molecular orientation of citral guided by the micelle. Moreover, the fluid becomes more liquid-like at higher pressure, enabling solubility of chemical species from metal surface.

The alkaline fading of crystal violet in AOT micelles formed in supercritical ethane as a continuous phase has been studied by Ikushima et al. [20,70,71] (Scheme 12.5). The reaction rate of fading in this system was reported to increase compared with that in H_2O/AOT/heptane at high pressures under similar conditions.

The rate of fading was found to be accelerated by increasing either *W* or pressure. This presented a striking contrast to the fading in H_2O/AOT/heptane, in which the pseudo-first-order rate constant, *k*, decreased with pressure [72]. The authors proposed several possible factors such as the conformation

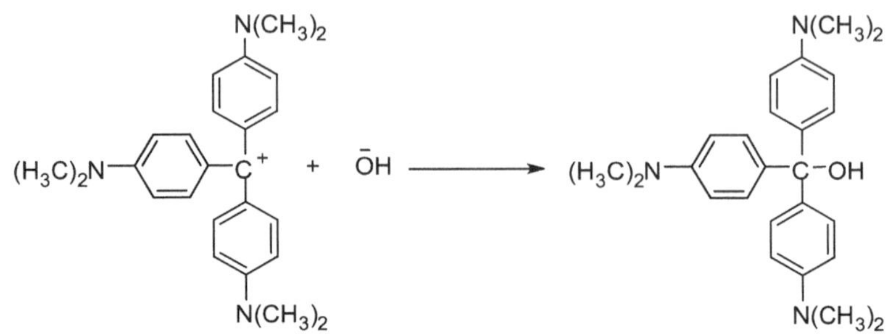

SCHEME 12.5 The alkaline fading of crystal violet (CV) by using AOT micelles formed in supercritical ethane. *(Adapted from Ref. [71] with permission of Elsevier.)*

of AOT molecule (rotational isomerism), the type of interior water molecules, and micropolarity represented by $E_T(30)$ parameter that may be responsible for the observed rate enhancement. The influences of these parameters on the rate enhancement were evaluated by dividing into the respective effects of W and pressure.

It was demonstrated that the rate enhancement with increasing W is related not only to the presence of the *trans*-like conformation of AOT molecules and of the bound-type conformation of water molecules, but to the increase in the micropolarity $E_T(30)$ value in the micelle core. The results showed that the structure of interior water and AOT molecules changed very slightly with pressure. Therefore, these two factors cannot account for the acceleration effect of pressure on the rate. It was suggested that a decrease in the size of the micelle core due to significant changes in the intermicellar and AOT-ethane interactions with increasing pressure, which causes an increase in the micropolarity inside the core, is of importance for the promotion effect by pressure [20].

2.3.3 Enzyme-Catalyzed Transformations Within Microemulsions Formed in Supercritical Fluids

The discovery that enzymes can remain active in highly apolar organic solvents, in the presence of small amounts of water, has led to the development of novel applications of such "low water" systems in chemical synthesis [73]. In comparison to an aqueous reaction medium, enzyme catalysis in low water systems has the following advantages [74]:

- Organic solvents are better solvents for many lipophilic substrates
- The reaction equilibrium shifts in desirable directions that would be rapidly hydrolyzed in aqueous solution
- The system may improve thermal stability of the enzymes
- Simpler workup, recovery and reusability of enzymes
- "Moisture-sensitive" substrates/reagents are convenient to use
- Possible control of substrate specificity, regiospecificity, and enantioselectivity.

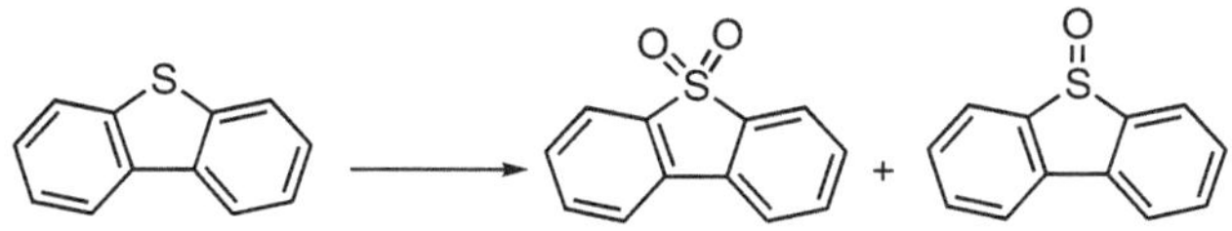

SCHEME 12.6 Biooxidation of DBT to its oxide products, DBTO, and $DBTO_2$ *(Adapted from Ref. [77].)*

On the negative side, as always with organic solvents, are the environmental problems associated with this apolar reaction medium.

SCFs are an environmentally friendly alternative to organic solvents as media for biocatalysis. A key feature of biocatalysis in SCFs is the tunability of the medium [75]. Enzymatic activity in SCFs has been proven and well documented [76]. Limiting factors, which may affect enzymatic activity in supercritical solvent systems, have been identified and are well characterized. A major limitation to the broader use of SCFs is their inability to dissolve a wide range of hydrophilic and ionic compounds, which greatly impedes their ability to carry out biotransformation with polar substrates. The interest in water-in-SCF microemulsion as reaction media stems from the fact that in such systems high concentrations of both polar and apolar molecules can be dissolved within the dispersed aqueous and continuous SCF phases, respectively.

The suitability of these "nanoreactors" for enzymatic transformations has been demonstrated recently by Ginosar and coworkers for the biooxidation of dibenzothiophenes (DBT) to produce dibenzothiophene sulfoxide (DBTO), and dibenzothiophene sulfone ($DBTO_2$) by protein catalysts in the presence of hydrogen peroxide as a cofactor (Scheme 12.6) [77,78]. This reaction was carried out in several supercritical solvent systems and in reverse emulsions/microemulsions in SCFs. In view of the fact that higher solubilities of these compounds can be achieved at supercritical conditions compared to an aqueous buffer, the potential of these reaction media is readily evident. However, nonpolar SCF reaction environments were undesirable for the protein-catalyzed biooxidation. This problem was overcome using molecular self-assembly systems of specially designed surfactants and cosolvents. The authors presented the results of variations in the bulk solvent type ($scCO_2$, scC_2H_6), surfactant [AOT, PFPE, and cetyl trimethyl ammonium bromide, (CTAB)], and protein catalysts [horseradish peroxidase (HRP), hemoglobin (HB), cytochrome c (Cyt c), and soybean peroxidase (SBP)] on the biooxidation of DBT. The acidic nature of $scCO_2$ and its negative effects on enzyme activity were underlined using a buffer in the reaction system. HB activity in PFPE reverse micelles in $scCO_2$ was very low (the product yield was less than 1%) even at the high buffer concentration. This was not the case for the scC_2H_6 (26% yield of product, even at the low buffer concentration).

It was found that the addition of cosolvent, used to improve the solubility water in $scCO_2$, suppressed the biooxidation of DBT by Cyt c. The results also

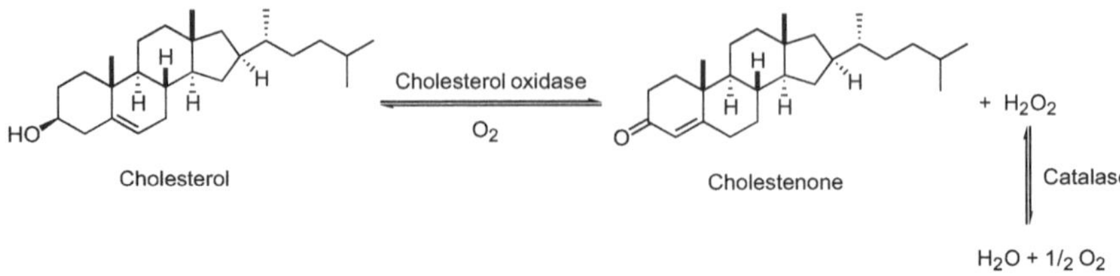

SCHEME 12.7 Enzymatic oxidation of cholesterol to cholestenone by cholesterol oxidase. *(Adapted from Ref. [80] with permission of American Chemical Society. Copyright 2000.)*

indicated that AOT or a hydrolysis product formed in the presence of AOT suppressed the biooxidation by HB in scC_2H_6. By carrying out the reaction in the presence of SBP, which is quite stable at acidic pHs [79], it was possible to form $scCO_2$/glycine buffer emulsions and microemulsions, at 2000 and 3000 psi, pH = 2.4, and at various water contents. Biooxidation of dibenzothiophenes based on emulsion catalysis in $scCO_2$ gave a product yield, which was 12 times higher than the reported yield in the aqueous media. However, biooxidation of DBT in micromulsions in SCF resulted in conversion and product yield that were comparable with those obtained in the aqueous phase reaction.

As another example, Bright and coworkers investigated the oxidation of cholesterol by cholesterol oxidase (ChOx) within the water core of reverse micelles (that had been formed in $scCO_2$ by using PFPE surfactant) in the presence of catalase, which was added in order to destroy hydrogen peroxide created within the course of the oxidation reaction (Scheme 12.7) [80]. They presented a comparative evaluation of microemulsion stabilized by AOT in isooctane [81] and microemulsion composed of PFPE in $scCO_2$ as reaction media for the oxidation reaction.

The enzymatic reaction in PFPE/$scCO_2$ reverse micelles was found to follow the classical Michaelis–Menten kinetics, whereas the activity was a strong function of the molar ratio of H_2O–PFPE. The optimum ChOx activity occurred when the molar ratio of H_2O–PFPE exceeded 12. The values of the apparent first-order rate constant ($k_{cat,app}$) were comparable with that obtained in AOT/isooctane system. In contrast, the authors observed that the best case for the apparent Michaelis constant ($K_{m,app}$) was ~twofold better than that reported for reverse micelles in liquid isooctane. There was no significant change in $k_{cat,app}$ or $K_{m,app}$ when the CO_2 pressure was changed between 100 and 260 bar.

The measured activities showed that the optimum enzyme activity occurred at near pH = 7 (W = 11), different from the conclusion reached in the earlier study by Gupte and coworkers. They had observed a pH optimum of 5 (W = 10) for AOT/isooctane reverse micelles. This observation can be rationalized on the basis that a more basic aqueous buffer was needed to offset the substantially more acidic environment encountered within the PFPE/$scCO_2$ water pool.

Hakoda et al. presented a study of lipase-catalyzed hydrolysis of triolein affected in AOT reverse micelles formed in scC_2H_6 in which the enzyme was located within the dispersed water droplets. Dynamic light-scattering experiments revealed that although the hydrodynamic diameter of reverse micelles did not change with the addition of enzyme, it increased with the increase in concentration of the triolein. They also reported on the quantitative behavior of the enzyme as a function of several factors such as temperature, ethane pressure, and reaction time. The conversion first increased to reach a maximum at 305 K (the critical temperature of ethane), and then decreased with temperature. It was found that the enzymatic reaction was very fast, reaching the maximum conversion of triolein after 1 min and then attaining the equilibrium value. At 305 K, triolein conversion was found to be independent of pressure in the range 16–37 MPa. At higher temperature (310 K), however, the conversion was small compared with that at 305 K, and it increased with increasing pressure [82].

2.3.4 Microemulsion Polymerization in Supercritical Fluids

Processes for the preparation of polymers using solvents in the supercritical state, particularly $scCO_2$, are already known in the literature [83]. Despite the potential advantages for conducting polymerization in SCFs, there are several important issues (such as solubility, and some practical aspects that are involved when they are used as polymerization solvents. For example $scCO_2$ is a poor solvent for most high molar mass polymers under mild conditions (<100°C, <350 bar). Although a change in temperature or pressure to enhance the solubility of these compounds in $scCO_2$ usually works well, it can be very expensive and it is not feasible from a practical perspective (widespread implementation in polymer manufacturing). For instance, poly(methyl acrylate) requires 2000 bar CO_2 and 100°C for a 10^5 g/mol polymer to dissolve in CO_2 [83]. The strategy of the use of CO_2-soluble surfactants to emulsify monomers into the CO_2 phase is an alternative to conduct polymerization of a hydrophilic or highly polar monomer in a supercritical solvent, while the pressure and the temperature are not impractically high.

Generally, it has been observed that a latex formed during an inverse emulsion polymerization—a reaction system comprising both micelles (10–100 nm in diameter) and large monomer droplets (partially stabilized by surfactant, of order hundreds of nanometers to microns in size) dispersed in the continuous phase—flocculates rapidly. This system is not thermodynamically stable and the monomer droplets would eventually settle from solution if stirring of the system ceased [84,85]. Microemulsion polymerization was born as an alternative process to the production of polymeric latexes with unique particle size (~10 nm in diameter), molar masses [molecular weight is very high (10^6–10^7 g/mol)] and structure [86,87]. In this case, the stability of microemulsions is usually, but not always, due to a higher concentration of surfactant than conventional emulsions.

Polymerization based on the formation of aqueous microemulsions in a SCFs (especially $scCO_2$) continuous phase could benefit from the fine control over continuous phase properties, as well as from rapid separation of solvent and from the high diffusivities inherent to SCF systems [84]. Moreover, for free radical reactions, CO_2 offers no chain transfer to solvent, thus producing higher molecular weight polymer (most solvents exhibit some degree of chain transfer, where a proton is abstracted from a solvent molecule, terminating a polymer chain) [85]. Another advantage is that to form the microemulsion, only a modest amount of surfactant may be required for CO_2 than for alkane continuous phases [62].

Despite the great potential of such methods for the preparation of polymers with specific properties, up to now very few studies have been carried out on polymerization in water-in-SCF microemulsions.

Beckman et al. presented the 2,2′-azobis-isobutyronitrile (AIBN)-initiated microemulsion polymerization of acrylamide, in microemulsions, using ethoxylated alcohols surfactant such as Brij 52 (B52) and Brij 30 (B30) in supercritical alkane (a mixture of ethane and propane), as a function of the continuous phase density and initiator concentration. It was proved that continuous phase density, and thus pressure, played a major role in regulating the reaction rate, as well as determining the product molecular weight. This appears to be due to the effect of pressure on the structure of the microemulsion. Increasing the density decreased the strength of micelle-micelle attractions, and thus the tendency for the micelles to cluster. As the extent of micelle clustering dropped, the polymer molecular weight decreased, although the rate of the polymerization increased. It was also shown that the molecular weight of the product changed but very slightly with initiator concentration. The high chain transfer reaction rates for solvent, monomer, and/or surfactant were suggested as being responsible for this behavior [84].

Regarding the potential benefits of operating a microemulsion polymerization in $scCO_2$, Ye et al. reported the AIBN-initiated microemulsion polymerization of a water-soluble vinylic monomer in $scCO_2$. The acryloxyethyltrimethyl ammonium chloride monomer and *N,N′*-methylene-bisacrylamide cross-linker were effectively emulsified in CO_2 continuous phase using an anionic phosphate fluorosurfactant (bis-[2-(F-hexyl)ethyl]phosphate sodium) [88]. Polymerization in a more confined environment, water-in-$scCO_2$ microemulsion consisting of water pools of the size of 2.47 nm, produced spherical nanosized polymeric particles (~11–15 nm) smaller than those obtainable by water-in-oil microemulsion polymerization (water pools with diameters of 5–10 nm). Particle size was found to be independent of monomer-to-cross-linker ratio and W (15, 7.3, and 9.6) at a constant temperature 65°C and CO_2 pressure of 37.2 MPa. However, in the absence of monomer, polymerization only by cross-linker resulted in smaller particle size, probably due to the high cross-linking density of the particles and their rigid structure.

3 NANOTUBES

3.1 Introduction

A promising application of hollow carbon nanostructures is their use as "nanocontainers" for various molecular species. Filling of inner cavities of the nanotubes (NTs) with different substances and carrying out a chemical reaction inside the NTs have attracted considerable attention recently. The field was opened by the encapsulation of various metals and compounds within multi-walled carbon nanotubes (MWCNTs) [89–92]. The motivation of encapsulation can be (1) the study of nanoscale chemical reactions using NTs as nanoreactors, which has attracted great interest for understanding fundamental concepts concerning the role of dimensionality and size in determining chemical properties [93], and (2) the synthesis of new materials.

CNTs and graphitized carbon nanofibers (GNFs) with their chemically inert interiors are the most common guests have been applied in the synthesis of novel nanocomposites [94]. The incorporation of guest materials into the cavity of the NTs may significantly modify the properties of the NTs, as well as alter the properties of the filling materials, implying a synergistic effect between the host and guest.

NTs can be filled with different substances including fullerenes, metals, semiconductors, organic materials, polymers, and so on, either during the synthesis process (in situ filling processes) or through subsequent opening and filling of the NTs (ex situ filling processes) [95]. Three filling methods can be distinguished: (1) wet chemistry procedures and capillarity-based physical procedures, involving filling by (2) a molten material or (3) a sublimated material [96].

Important parameters for capillary filling include the minimal inner diameter of CNT, the "wetability" of the interior surface of the NT in terms of the carbon and also in terms of the surface tension properties of the introduced liquid or solution and also the viscosity of the liquid material [97].

The capillary effect in CNTs was reported originally by Ajayan and Iijima [89], who obtained surface tension threshold values for NT wetting based on macroscopic surface tension concepts and immersion experiments of NTs in different liquids (solvents, metals, etc.) [98]. From their results, although liquids with low surface tension (<200 mN/m), including most common organic solvents would fill tubes, those with surface tension higher than 200 mN/m, such as metals, can neither wet CNTs nor initialize the capillary action [99].

SCFs, due to a variety of unique properties, have been applied as an excellent carrier fluid for the encapsulation of molecules. As mentioned before, by changing the density of the fluid through temperature and pressure variations, the solvent power of an SCF can be tuned. Another attractive feature of SCFs is that the properties lie between those of gases and liquids. They can dissolve solutes like a liquid and yet possess the low viscosity, high diffusivity, and zero surface tension characteristic of a gas; thus, an SCF can diffuse faster in a solid matrix than a liquid.

The SCF method not only provides a green route for the synthesis of NT-based composites, but also results in nanostructures that have not yet been produced by conventional methods. Furthermore, the nanocomposite properties, such as morphology and structure, can conveniently be controlled by the variation of the solvent properties [100].

Despite recent positive progress of the SCF method, more studies were dedicated to the encapsulation of different molecules within the cavity of the NTs using supercritical carrier fluids. To date, a limited number of studies exist in the literature on the effects of process parameters [such as temperature and pressure (density)] on the nanocomposite properties or chemical reactions inside the cavities of the NTs. To demonstrate the advantages of using SCFs for filling NTs with various compounds, here a short overview about encapsulation and reactions inside NTs is given with a focus on those which are expected to be interesting for nanoreactor technology. The role of SCFs also will be highlighted.

3.2 The Potential Benefits of Filling Nanotubes With Supercritical Fluids

3.2.1 Filling Nanotubes With Thermally Unstable Materials

After the encapsulation of Buckminster fullerene C_{60} into the cavity of the single-walled carbon nanotubes (SWCNTs) [101], a completely new and still increasing area of research developed. Many fullerene-related filling materials have then followed, including higher fullerenes [102], endohedral nitrogen fullerenes [103], endohedral metallofullerenes [104,105] and functionalized fullerenes [106], producing a great variety of "peapod" NTs filled with nanostructures. Filling by means of gas-phase processes is the way peapods are usually prepared. This method leads to a high-ratio filling and purity of the filled material. The principal disadvantage of this approach is the limited choice of compounds that can be encapsulated. First, the compound's vaporization or sublimation temperature should be below 1000–1200°C to ensure that it does not interact with carbon and cause closure of the NT ends. Second, many compounds such as functionalized fullerenes are not able to vaporize or sublime without decomposing, which substantially limits the number of molecules suitable for encapsulation [107].

Most SCFs are low molecular weight gases, which have relatively low critical temperatures. Operations may therefore be carried out at low or moderate temperatures [108], which is desirable in the encapsulation of molecules bearing chemical functionalities that are inclined to decompose at the temperatures generally used for gas-phase filling processes.

Recently, the Nottingham research group, led by Professor Andrei Khlobystov has been studying NTs as containers for molecules, with particular emphasis on low-temperature encapsulation of different molecules with the aid of SCFs. Taking advantage of SCFs, they have successfully encapsulated a variety of functionalized fullerenes without destroying functional groups [106]. Encapsulation at low temperatures by using SCFs offers an opportunity for studying the influence of the nature of the functional group on the NT–fullerene host–guest interactions [106], the impact of the functional group on the intermolecular spacing in NTs [109], and exploring the effects of the conditions applied on the structure of the fullerene-based peapods [110]. Some control of the assembly of molecular arrays in CNTs has also become possible by alteration of the size and geometry of the functional groups attached to the molecules inserted into the NTs [111].

It is also worth noting that there are several ways one can fill CNTs at low temperature. For example, three different methods for the filling of SWCNTs (diameter distribution of 13.6–14.9 Å) by fullerene—mechanical grinding, liquid phase filling in organic solvents, and in SCF ($scCO_2$)—have been compared [112]:

1. The grinding method by mixing SWCNTs with an equal amount of crystalline fullerene at room temperature resulted in SWCNTs partially filled with fullerene (total fullerene content in product was less than 1%) after 10–15 min. More reaction time led to irreversible changes, including the decomposition and polymerization of the fullerene and irreversible matting of SWCNTs. However, it was possible to greatly improve the filling rate (up to 20%) by storing the mixture in an inert environment (CO_2, room temperature) for several weeks.
2. The second method consisted of immersing SWCNTs in fullerene solution in toluene or CS_2 ($\sim 5 \times 10^{-3}$ M) for 7–10 days at 30–50°C. Despite the high concentration of fullerene molecules and a long processing time, a very small filling efficiency (<1%) was observed. The low filling efficiency could be attributed to the SWCNT surface and cavities' contamination with amorphous material. Reducing the concentration of the fullerene by using polar solvents (ethanol and tetrahydrofuran) did not improve the filling rate. Considering the surface tensions of all of these solvents (which are <200 mN/m) and high concentration of solvent with respect to the concentration of fullerene, it can be predicted that SWCNT cavities would be preferentially filled with solvent. The efficiency of liquid filling method depends on the CNTs inner diameters and sidewall defects. It is thermodynamically preferred for the fullerenes to displace with the trapped solvent molecules. This process will happen more efficiently in wide CNTs, where the trapped solvent molecules can diffuse between fullerene and the inner wall of NT. For small-diameter CNTs, the presence of defects on the sidewalls of CNTs may have a profound effect on diffusion of trapped solvent molecules out of NT.
3. The supercritical filling of fullerene was performed in a high-pressure reactor, at 50°C under 150 bar for 10 days. After reaction, the filled samples were sonicated in CS_2 to rinse off unfilled fullerenes. The filling rate of fullerene in SWCNTs was greatly improved by using $scCO_2$ (~30%).

In other experiments where several pressure cycles of $scCO_2$ (each cycle was performed at a pressure greater than 150 bar and subsequent depressurizing to 100 bar at 50°C) were used instead of a single cycle of a similar total length, the filling rate was dramatically improved to ~70%. The high filling efficiency is attributed to the pressure gradient between inside and outside of the SWCNTs, being assumed to drive a flow of trapped CO_2 from the inside to the outside of the tube. Supercritical CO_2 also acts as an effective cleaner, and some blockages of residual graphitic and amorphous carbon in the hollow cores of the CNTs or at the end of tubes can be blown out during the release of CO_2 and results in high filling.

Although the supercriticality has been found to be an effective factor for the high filling efficiency, the exact mechanism behind this improvement is not yet clear and some other factors could be responsible for this high filling rate too. For instance, CO_2 may prevent the formation of blocking carboxylic groups on the ends of the SWCNTs, increases the chances of C_{60} molecules that initially deposited on SWCNTs surface to be encapsulated into NTs.

3.2.2 Filling the Guest Molecules With Different Structures Into Nanotubes

For a given compound, the structure of the guest molecules may change with respect to the temperature of the encapsulation method. For example, Pagona et al. encapsulated azafullerene into SWCNTs using gas-phase high-temperature (gas-phase) and low-temperature-high-pressure (using $scCO_2$) conditions and investigated the effects of the applied temperature on the structure of azafullerene-based peapods, $C_{59}N$@SWCNTs [110]. At high temperature dissociation of azafullerene into radicals occurred and these radicals were transported through the NT. High-resolution transmission electron microscopy (HRTEM) showed several types of molecular morphology in SWCNTs, due to a variety of different stabilizing transformations forming dimers, oligomers, or monomer inside SWCNTs. However, at low temperature and applying the SCF, azafullerene could enter SWCNTs without dissociation and it existed only in dimer form.

In another example, coronene was successfully encapsulated inside SWCNTs by vapor-phase filling at high temperature (450°C) and nanoextraction from $scCO_2$ (50°C). The chemical nature of the encapsulated species could not be determined. Nevertheless, only in the material prepared at high temperature, the presence of dicoronylene, the dimerized form of coronene, on the outer walls of nanotubes was demonstrated by Raman spectroscopy [113].

3.3 Filling Nanotubes With Different Supercritical Fluids

Considerations when selecting an SCF to serve as a reaction media include critical temperature, solubilizing power, inertness, safety, environmental impact, and cost [16]. Because of the interest in developing low-temperature NT filling process, SCFs with critical temperatures around room temperature are most commonly used; among these SCFs, $scCO_2$ is the most beneficial SCF used in the filling of CNTs at low temperature. Its low critical temperature and pressure are easily accessible, and because of the small critical diameter of the CO_2 molecule, $scCO_2$ does not block NTs. Furthermore, CO_2 is chemically inert and relatively inexpensive, and the processes involving CO_2 as a solvent provide an opportunity for the recycling of CO_2.

Suitable SCFs are not limited to carbon dioxide or other SCFs with low critical temperature. The SCFs may be used individually or in combination. Some examples of different SCFs used in the filling of CNTs are mentioned here.

Supercritical ethylene (T_c = 9°C) and carbon dioxide (T_c = 31.1°C) were used as the media to take the fullerene into the SWCNTs. The filling efficiency with $scCO_2$ was much higher as compared with scC_2H_4 (~30% for $scCO_2$ and 5% for scC_2H_4). This is not very surprising, considering the small critical diameter of CO_2 in comparison with other SCFs, which facilitates its diffusion within the peapod structure. The yield of peapods could not be increased by adding organic solvents (modifiers) such as *o*-dichlorobenzene to SCFs to extend the range of their solvent power [112].

As mentioned before, CO_2 is a poor solvent for most high molecular weight under mild conditions. While the high critical temperature of many supercritical organic fluids prevents them from being used as a solvent for filling NTs with thermally unstable materials, they have great potential for the supercritical reaction to synthesize CNT-based one-dimensional new materials. Furthermore, the solvent power of high-T_c fluids is much higher than that of the low-T_c solvents and high-T_c solvents are proper solvents for high molecular weight compounds [114], so that the range of filling materials would be expanded.

Some attempts have been made recently to insert fullerenes (C_{60}, C_{70}, C_{78}, C_{84}, Gd@C_{82}, Er@C_{82}, Ho@C_{82},Y@C_{82}) into SWCNT using high-T_c supercritical fluids such as C_2H_5OH (T_c = 243°C), CH_3OH (T_c = 240°C), or $C_6H_5CH_3$ (T_c = 320°C) [115]. Nanotubes (average diameter of 14 Å) were produced by the arc discharge method. During purification, the ends were opened. Then a mixture of NTs, fullerenes, and SCFs was heated at 400°C for 10 min. In all solvents, the filling rates were proved to be more than 90%. To show the effectiveness of this supercritical filling procedure, it suffices here to note that these filling rate of fullerenes were better than those achieved by the sublimation method.

As was previously mentioned [112], the SWCNTs (diameter distribution of 13.6–14.9 Å) can be filled with fullerene in 30–70% yield using $scCO_2$ (at 50°C under 150 bar for 10 days, when the sample was exposed to several pressure cycles of $scCO_2$). Compared to $scCO_2$, the higher filling efficiency, along with shorter processing times are the major advantages of high-T_c SCFs (C_2H_5OH, CH_3OH or $C_6H_5CH_3$).

3.4 Nanotubes' Applications as Nanoreactors

3.4.1 Nanomaterials Synthesis

The nanoscale nature of the inner cavity of CNTs provides an excellent template for the precise engineering of nanomaterials synthesis, directing the growth of different nanostructures, and achieving a better size control.

Pd, Ni, and Cu nanowires and nanorods were grown through the H_2 reduction of metal precursors in $scCO_2$. Metal hexafluoroacetylacetonate, [M(hfac)$_2$·×H_2O] (M: Pd, Ni, Cu), which is very soluble in $scCO_2$, was used as precursor. TEM examinations of the products revealed that about 10% of the NTs had been filled. It was found that metal nanowires or nanorods can grow inside the interior of the tubular MWCNTs by $scCO_2$ at lower temperature and reaction time than previously employed methods [116]. Similar to many of the general approach to metal filling [chemical vapor deposition (CVD) and wet chemical processes], a side product of supercritical filling was growth of metal nanopaticles on the outer walls of NTs. Adjusting the outside or inside loading preference of metal by optimization of the reaction conditions, such as functionalization of MWCNTs, precursor concentration, CO_2 or H_2 pressure, and reaction temperature, are currently the subject of ongoing investigations.

Interestingly, Hao et al. [115], in their recent attempt to synthesize metal@SWCNTs using the supercritical method, described the first use of supercritical alcohols (scC_2H_5OH and $scCH_3OH$) for the one-step synthesis and encapsulation of pure Cu, Ni, and Ag nanowires inside the SWCNT's inner cavity without the addition of hydrogen or other reducing agents. It is believed that a low carbon number alcohol acts both as a solvent and a reducing agent at supercritical conditions [117]. Based on HRTEM study, the filling rates were proved to be more than 90%.

3.4.2 Polymerization

One of the most intriguing applications of CNT is the polymer/CNT nanocomposites. The high mechanical, electrical, and thermal properties of CNTs make them ideal candidates as fillers in light-weight polymer composites [118]. This kind of polymer/CNT composites may have some potential applications. For the last two decades, a lot of research works have been done on evaluating the potential of CNTs as filler for polymer nanocomposites. However, obtaining the uniform dispersion of CNTs in the polymer matrix is very difficult due to the insolubility of the CNTs and the inherently poor compatibility between the CNTs and the polymer, and the resulting inhomogeneous nanocomposites often have unsatisfying properties [119]. A strategy for enhancing the compatibility between NTs and polymers consists of encapsulation of polymers into the CNT's inner cavity; the polymers at the ends of the polymer@CNTs composites can contact with the bulk polymer matrix [120]. On the other hand, nanoscale polymerization reactions using the inner cavities of CNTs as a nanocontainer are of great importance because they provide an ideal geometry for studying the effects of one-dimensional confinement on polymerization.

The approach to use SCFs to transport organic molecules inside NTs seems to be very promising. There are two methods to encapsulate polymer within CNTs with the aid of SCFs. In one method, the monomer and initiator are first carried into the cavities of the CNTs with the aid of SCFs, the NTs are then taken out of the reactor, and the polymerization is carried out at a suitable temperature. In the other method, the monomer together with an initiator is filled into the NTs using SCFs, and then the temperature of the SCFs is increased and the polymerization takes place in the SCFs [121].

A few examples will be mentioned herein to illustrate the potential applications of SCFs as a processing medium enables efficient transport and polymerization of a wide range of organic molecules into NTs.

Supercritical CO_2 was used to encapsulate the polystyrene within CNTs via a two-step process. The monomer styrene together with an initiator (benzoyl peroxide) was filled into the NTs using $scCO_2$ (35°C, 12.0 MPa). After the impregnation period, the pressure was released over different times (8 and 1800 s) and the polymerization took place into an oven at 100°C for 24 h.

Perhaps the most important advantage offered by SCFs is that the filling rate can be tuned by controlling the depressurizing speed. Fast pressure release causes venting of compressed CO_2 along with some styrene monomer from the NT and reducing the filling rate [120].

Britz et al. filled the NTs and prepared the fullerene epoxide ($C_{60}O$)@SW-CNT peapods by using $scCO_2$, following their optimized previous procedure (see Ref. [112]). The filled NTs were then heated to 260°C at 10^{-6} Torr for three days to polymerize the monomer inside the CNTs. Extensive microscopic studies indicated that the topology of the polymer inside the NT was different from the bulk polymer, and that, as a consequence of confining the polymerization reaction, linear and unbranched polymers were formed [122].

Steinmetz et al. synthesized polyacetylene-filled MWCNTs by in situ polymerization. They used $scCO_2$ to fill the NTs with toluene and then added the Ziegler–Natta catalyst (toluene, facilitated infusion of the Ziegler–Natta catalyst into the CNTs). In order to assure controlled polymerization inside the NT, the catalyst outside was removed before the reaction [123]. The filled CNTs were then exposed to acetylene gas, producing polyacetylene@CNT composites.

In other work, the same group reported the encapsulation of poly(*N*-vinyl carbazole) within CNTs with the aid of $scCO_2$. The encapsulation was performed using two methods, a multistep process that consisted of filling the poly(*N*-vinyl carbazole) monomer and AIBN initiator within CNTs with the aid of $scCO_2$, and filtering and washing the filled NTs followed by polymerization in an oven at 90°C. These processes could also be made into a single step, which allowed NTs to be filled in a high-pressure reactor under $scCO_2$ at 40°C and 150 bar and then polymerized at the elevated temperature and pressure of the polymerization (70°C and 300 bar) [121].

Finally, pyrrole was impregnated into the CNTs with $scCO_2$(40°C and 120 bar), and then the polymerization was performed by oxidation in a $FeCl_3$ solution, resulting in polypyrrole@CNT composites [121].

3.4.3 Catalytic Reactions Within Nanotubes

Confined environments can be used to alter the selectivity of a reaction by influencing the orientation of the reactants, altering the mobility of trapped molecules, facilitating one reaction pathway, or selectively stabilizing the products [124].

In addition to their unique mechanical, chemical, and thermal stability, carbon nanostructures have an excellent ability to encapsulate a wide range of guest molecules that makes them particularly attractive nanocontainers and nanoreactors capable of modulating the chemical reactivity and inducing selectivity [125]. However, a major drawback of their use as nanoreactors is related to transport resistance of reactants/products within the narrow confines of a carbon nanostructure [126], which may result in significantly lowered rates of confined chemical reactions and may even preclude extraction of the products [125].

It has been demonstrated that metallic nanoparticles encapsulated within CNTs exhibit superior activity, chemoselectivity, and stereoselectivity as catalysts in chemical transformations [126,127].

Supercritical fluids play a double role in catalysis applications, acting as solvents or as reaction agents in preparation of CNT-based catalyst (Table 12.6), and as the catalytic reaction solvent. Mass transfer restrictions on the reaction rate may be alleviated in supercritical media due to their low viscosity and, thus, high diffusivity. For example, Chamberlain et al. have successfully combined the concept of CNT nanoreactors with supercritical continuous flow technology, and developed an efficient heterogeneously catalyzed hydrogenation reaction [128]. The RuNPs@SWCNT-catalyzed hydrogenation of cyclooctene was performed in gas phase (with only the reagent gases, H_2 and cyclooctene) and $scCO_2$ [128]. As might be expected, conversion was limited ($<5\%$) due to extreme confinement imposed by the narrow NTs in gas phase. Interestingly, the conversion was increased significantly when $scCO_2$ was added as a reaction solvent. This process, derived from the physical properties of $scCO_2$, allows an increase in catalyst accessibility and also the miscibility of H_2 with $scCO_2$.

4 CONCLUSIONS

As many of the examples discussed in this chapter illustrate, the unique features of SCFs can be exploited to control the behavior of many confined chemical processes in a way not possible with conventional liquid solvents. It is hoped that the results presented in this chapter will encourage and inspire researchers to use SCFs in combination with nanoreactor technology.

It is worth mentioning that SCFs cannot be considered as a simple replacement of solvents, and that the successful "transfer" of reactions to this medium depends heavily on having a specialized knowledge about the physicochemical properties of the reaction mixture and about possible chemical interactions of the reaction components with the SCFs [129]. Such knowledge is essential before we learn how to exploit these benefits in the most efficient way for new innovative processes using SCFs.

TABLE 12.6 CNT Based Catalysts Synthesized With the Aid of SCFs

Catalyst	Catalytic reaction	Important findings	References
AuNPs @GNF[a]	Oxidation of silane by hydrogen peroxide	AuNPs @GNF was identified to be the best catalysts for selective oxidation of dimethylphenylsilane to gain high conversion and considerable selectivity toward bis(dimethylphenyl) disiloxane when compared with free-standing or surface-adsorbed nanoparticles	[94]
CuNP@GNF[b]	1,3-Dipolar cycloaddition of azides to alkynes	1. The rate of formation of the 1,2,3-triazole product was higher than that obtained for CuNP/GNF 2. The regioselectivity of the reaction could not be altered by the constrained steric environment imposed by the graphitic step-edge of GNF	[127]
Rh and RhPt NP@ GNF[b]	Hydrosilylation Reaction	1. For aliphatic silane and alkyne, no confinement effect was observed. 2. For aromatic alkyne and aliphatic silan, the amount of dehydrogenative silylation products increased and the β-(Z): β-(E) isomers ratio decreased. 3. For aromatic silane and alkyne, β-(Z) regioisomer was formed.	[126]

[a]*Catalyst synthesis conditions: a mixture of thiolatestabilized AuNPs, GNF and n- pentane was prepared under an inert atmosphere of argon and then mixed with* CO_2 *(40°C and, 4000 psi), 18 h. The* CO_2 *pressure was cycled between 900 and 4000 psi three times.*
[b]*Catalyst synthesis conditions: a mixture of nanoparticles, GNF and n-hexane was impregnated into the CNTs with* CO_2 *(40°C and, 4000 psi), 18 h. The* CO_2 *pressure was cycled between 900 and 4000 psi three times.*

ACKNOWLEDGMENTS

The author is deeply grateful to coworkers at the Nuclear Science and Technology Research Institute for their helpful comments.

ABBREVIATIONS

F-thiol	1H,1H,2H,2H-Perfluorodecanethiol
F-pentanol	2,2,3,3,4,4,5,5-Octafluoro-1-pentanol
AIBN	2,2′-Azobis-isobutyronitrile
DMOL	3,7-Dimethyloctanol
$k_{cat,app}$	Apparent first-order rate constant
CNT	Carbon nanotubes
PFPE	Carboxylate perfluoropolyether
CTAB	Cetyl trimethyl ammonium bromide,
CVD	Chemical vapor deposition
ChOx	Cholesterol oxidase
CIAL	Citronellal
CIOL	Citronellol
CVs	Cyclic voltammograms
Cyt c	Cytochrome c
$DBTO_2$	Dibenzothiophene sulfone
DBTO	Dibenzothiophene sulfoxide
DBT	Dibenzothiophenes
DHAL	Dihydrocitronellal
GNFs	Graphitized carbon nanofibers
HB	Hemoglobin
$M(hfac)_2 . xH_2O$	Hexafluoroacetylacetonate
HRTEM	High-resolution transmission electron microscopy
HRP	Horseradish peroxidase
ILs	Ionic-liquid
$K_{m,app}$	Michaelis constant
$E_T(30)$	Micropolarity
MWCNTs	Multiwalled carbon nanotubes
TMPD	*N*,*N*,*N′*,*N′*-tetramethyl-*p*-phenylenediamine
***N*-EtFOSA**	*N*-ethyl perfluorooctylsulfonamide
PFPE $COO^-NH_4^+$	Perfluoropolyether ammonium carboxylate
$PFPE\text{-}PO_4$	Perfluoropolyether phosphate ether
***b*-PFOMA**	Poly(dimethyl amino ethyl methacrylate-*block*-1H,1H,2H,2H-perfluorooctyl methacrylate)
PEGs	Poly(ethylene glycol)
RESS	Rapid expansion of supercritical solution
SWCNTs	Single-walled carbon nanotubes
F-AOT	Sodium bis(2,2,3,3,4,4,5,5-octafluoro-1-pentyl)-2-sulfosuccinate
AOT	Sodium bis(2-ethylhexyl) sulfosuccinate
$NaBH_3CN$	Sodium cyanoborohydride

di-HCF4	Sodium salt of bis(2,2,3,3,4,4,5,5-octafluoro-1-pentyl)-2-sulfosuccinate
SBP	Soybean peroxidase
SCF	Supercritical fluid
SCME	Water-in-scCO_2 microemulsion
W	Water-to-surfactant molar ratio

REFERENCES

[1] York P, Kompella UB, Shekunov BY. Supercritical fluid technology for drug product development (Drugs and the Pharmaceutical Sciences), 138. Boca Raton: CRC Press; 2004. pp. 32.

[2] Carlès P. A brief review of the thermophysical properties of supercritical fluids. J Supercrit Fluids 2010;53:2–11.

[3] Knez Z, Markocic E, Leitgeb M, Primozic M, Hrncic MK, Skerget M. Industrial applications of supercritical fluids: a review. Energy 2014;77:235–43.

[4] Brunner G. Supercritical fluids: technology and application to food processing. J Food Eng 2005;67:21–33.

[5] Baiker A. Supercritical fluids in heterogeneous catalysis. Chem Rev 1999;99:453–73.

[6] Gopaliya P, Kamble PR, Kamble R, Chauhan CS. A review article on supercritical fluid chromatography. IJPRR 2014;3:59–66.

[7] http://chemwiki.ucdavis.edu/Physical_Chemistry/Physical_Properties_of_Matter/Phases_of_Matter/Supercritical_Fluids

[8] Perrut M. What is a supercritical fluid ? http://www.separex.fr/

[9] Sapkale GN, Patil SM, Surwase US, Bhatbhage PK. Supercritical fluid extraction. Int J Chem Sci 2010;8:729–43.

[10] Cansell F, Aymonier C, Loppinet-Serani A. Review on materials science and supercritical fluids. Curr Opin Solid State Mater Sci 2003;7:331–40.

[11] Liu J, Ikushima Y, Shervani Z. Environmentally benign preparation of metal nano-particles by using water-in-CO_2 microemulsions technology. Curr Opin Solid State Mater Sci 2003;7:255–61.

[12] López-Quintela MA, Tojo C, Blanco MC, Rio LG, Leis JR. Microemulsion dynamics and reactions in microemulsions. Curr Opin Colloid Interface Sci 2004;9:264–78.

[13] Matson DW, Fulton JL, Smith RD. Formation of fine particles in supercritical fluid micelle systems. Mater Lett 1987;6:31–3.

[14] Liu J, Han B, Zhang J, Li G, Zhang X, Wang J, Dong B. Formation of water-in-CO_2 microemulsions with non-fluorous surfactant Ls-54 and solubilization of biomacromolecules. Chem Eur J 2002;8:1356–60.

[15] Gale RW, Fulton JL, Smith RD. Organized molecular assemblies in the gas phase: reverse micelles and microemulsions in supercritical fluids. J. Am Chem Soc 1987;109:920–1.

[16] Leitner W, Jessop PG. Green solvents, supercritical solvents. In: Anastas PT, editor. Handbook of green chemistry, vol 4. Weinhem: Wiley-VCH; 2013.

[17] Fulton JL, Smith RD. Supercritical fluid reverse micelle systems. US5,158,704 A; 1992.

[18] Zhang J, Han B. Supercritical CO_2-continuous microemulsions and compressed CO_2-expanded reverse microemulsions. J Supercrit Fluids 2009;47:531–6.

[19] Hutchenson KW, Foster NR. Innovations in supercritical fluid science and technology. J Am Chem Soc 1995;608:1–31.

[20] Ikushima Y, Shervani Z, Saito N, Hatakeda K, Sato O, Aizawa T, Arai M. Acceleration of chemical reaction by AOT micelles under supercritical conditions. Chem Eng Sci 1999;54:2859–64.

[21] Clarke MJ, Harrison KL, Johnston KP, Howdle SM. Water in supercritical carbon dioxide microemulsions: spectroscopic investigation of a new environment for aqueous inorganic chemistry. J Am Chem Soc 1997;119:6399–406.

[22] Niemeyer ED, Bright FV. The pH within PFPE reverse micelles formed in supercritical CO_2. J Phys Chem B 1998;102:1474–8.

[23] Dong X, Potter D, Erkey C. Synthesis of CuS nanoparticles in water-in-carbon dioxide microemulsions. Ind Eng Chem Res 2002;41:4489–93.

[24] Soriano NU Jr, Venditti R, Saquing CD, Bushey D, Argyropoulos DS. Solubilizing amino acids and polypeptides in supercritical CO_2 via reverse micelle formation. Colloids Surf A 2008;315:110–6.

[25] Fei W, Lizhuang Z, Qicai W, Zhengzhong C, Yixin G, Shuquan Z. Application of surfactants in supercritical carbon dioxide. http://www.paper.edu.cn/en_releasepaper/content/5501 (2006) 1–10.

[26] Hoefling TA, Stofesky D, Reid M, Beckman EJ, Enick RM. The incorporation of a fluorinated ether functionality into a polymer of surfactant to enhance CO_2-solubility. J Supercrit Fluids 1992;5:237–41.

[27] Eastoe J, Dupont A, Steytler DC. Fluorinated surfactants in supercritical CO_2. Curr Opin Colloids Interf Sci 2003;8:267–73.

[28] Liu ZT, Erkey C. Water in carbon dioxide microemulsions with fluorinated analogues of AOT. Langmuir 2001;17:274–7.

[29] Sagisaka M, Yoda S, Takebayashi Y, Otake K, Kitiyanan B, Kondo Y, Yoshino N, Takebayashi K, Sakai H, Abe M. Preparation of a W/scCO_2 microemulsion using fluorinated surfactants. Langmuir 2003;19:220–5.

[30] Fink R, Beckman EJ. Phase behavior of siloxane-based amphiphiles in supercritical carbon dioxide. J Supercrit Fluids 2000;18:101–10.

[31] Psatthas PA, da Rocha SRP, Lee CT, Johnston KP, Lim KT, Weber S. Water-in-carbon dioxide emulsions with poly(dimethylsiloxane)-based block copolymer ionomers. Ind End Chem Res 2000;39:2655–64.

[32] Keiper JS, Simhan R, DeSimone JM, Wignal GDl, Melnichenko YB, Frielinghaus H. New phosphate fluorosurfactants for carbon dioxide. J Am Chem Soc 2002;124:1834–5.

[33] Steytler DC, Rumsey E, Thorpe M, Eastoe J, Paul A, Heenan RK. Phosphate surfactants for water-in-CO_2 microemulsions. Langmuir 2001;17:7948–50.

[34] Keiper JS, Behles JA, Bucholz TL, Simhan R, Desimone JM. Self-assembly of phosphate fluorosurfactants in carbon dioxide. Langmuir 2004;20:1065–72.

[35] Johnston KP, Harrison KL, Clarke MJ, Howdle SM, Bright FV, Carlier C, Randolph TW. Water-in-carbon dioxide microemulsions: an environment for hydrophiles including proteins. Science 1996;271:624–6.

[36] Heitz MP, Carlier C, deGrazia J, Harrison KL, Johnston KP, Randolph TW, Bright FV. Water core within perfluoropolyetherbased microemulsions formed in supercritical carbon dioxide. J Phys Chem B 1997;101:6707–14.

[37] Zielinski RG, Kline SR, Kaler EW, Rosov N. A small-angle neutron scattering study of water in carbon dioxide microemulsions. Langmuir 1997;13:3934–7.

[38] Liu J, Shervani Z, Raveendran P, Ikushima Y. Micellization of sodium bis(2-ethylhexyl) sulfosuccinate in supercritical CO_2 with fluorinated co-surfactant and its solubilization of hydrophilic species. J Supercrit Fluids 2005;33:121–30.

[39] Li J, Zhang J, Han B, Gao Y, Shen D, Wu Z. Effect of ionic liquid on the polarity and size of the reverse micelles in supercritical CO_2. Colloids Surf A 2006;279:208–12.
[40] Liu J, Cheng S, Zhang J, Feng X, Fu X, Han B. Reverse micelles in carbon dioxide with ionic-liquid domains. Angew Chem Int Ed 2007;46:3313–5.
[41] M.M.A., Wani MY, Hashim MA. Microemulsion method: a novel route to synthesize organic and inorganic nanomaterials. Arab J Chem 2012;5:397–417.
[42] Cason JP, Khambaswadkar K, Roberts CB. Supercritical fluid and compressed solvent effects on metallic nanoparticle synthesis in reverse micelles. Ind Eng Chem Res 2000;39:4749–55.
[43] Kometani N, Kaneko M, Morita T, Yonezawa Y. The formation of photolytic silver clusters in water/supercritical CO_2 microemulsions. Colloids Surf A 2008;321:301–7.
[44] Ohde H, Hunt F, Wai CM. Synthesis of silver and copper nanoparticles in a water-in-supercritical-carbon dioxide microemulsion. Chem Mater 2001;13:4130–5.
[45] Fernandez CA, Wai CM. Continuous tuning of silver nanoparticle size in a water-in-supercritical carbon dioxide microemulsion. Small 2006;2:1266–9.
[46] Fernandez CA, Wai CM. Continuous tuning of cadmium sulfide and zinc sulfide nanoparticle size in a water-in-supercritical carbon dioxide microemulsion. Chem Eur J 2007;13:5838–44.
[47] Yu KMK, Meric P, Tsang SC. Micelle-hosted bimetallic Pd–Ru nanoparticle for in situ catalytic hydrogenation in supercritical CO_2. Catal Today 2006;114:428–33.
[48] Meric P, Yu KMK, Kong ATS, Tsang SC. Pressure-dependent product distribution of citral hydrogenation over micelle-hosted Pd and Ru nanoparticles in supercritical carbon dioxide. J Catal 2006;237:330–6.
[49] Kitchens CL, Roberts CB. Copper nanoparticle synthesis in compressed liquid and supercritical fluid reverse micelle systems. Ind Eng Chem Res 2004;43:6070–81.
[50] Xue Z, Zhang J, Peng L, Li J, Mu T, Han B, Yang G. Nanosized poly(ethylene glycol) domains within reverse micelles formed in CO_2. Angew Chem Int Ed 2012;51:12325–9.
[51] Ji M, Chen X, Wai CM, Fulton JL. Synthesizing and dispersing silver nanoparticles in a water-in-supercritical carbon dioxide microemulsion. J Am Chem Soc 1999;121:2631–2.
[52] Ohde H, Wai CM, Kim H, Kim J, Ohde M. Hydrogenation of olefins in supercritical CO_2 catalyzed by palladium nanoparticles in a water-in-CO2 microemulsion. J Am Chem Soc 2002;124:4540–1.
[53] Ohde M, Ohde H, Wai CM. Catalytic hydrogenation of arenes with rhodium nanoparticles in a water-in-supercritical CO_2 microemulsion. Chem Commun 2002;2388–9.
[54] Ohde H, Rodriguez JM, Ye X-R, Wai CM. Synthesizing silver halide nanoparticles in supercritical carbon dioxide utilizing a water-in-CO_2 microemulsion. Chem Commun 2000;18:2353–4.
[55] Liu J, Raveendran P, Shervani Z, Ikushima Y, Hakuta Y. Synthesis of Ag and AgI quantum dots in AOT-stabilized water-in-CO_2 microemulsions. Chem Eur J 2005;11:1854–60.
[56] Lim KT, Hwang HS, Lee MS, Lee GD, Hongb S-S, Johnston KP. Formation of TiO_2 nanoparticles in water-in-CO_2 microemulsions. Chem Commun 2002;1528–9.
[57] Lee M-H, Lin H-Y, Thomas JL. A novel supercritical CO_2 synthesis of amorphous hydrous zirconia nanoparticles, and their calcination to zirconia. J Am Ceram Soc 2006;89:3624–30.
[58] Ohde H, Mariko Bailey F, Kim H, Wai CM. Water-in-CO_2 microemulsions as nanoreactors for synthesizing CdS and ZnS nanoparticles in supercritical CO_2. Nano Lett 2002;2:721–4.
[59] Shimizu K, Cheng IF, Wang JS, Yen CH, Yoon B, Wai CM. Water-in-supercritical CO_2 microemulsion for synthesis of carbon-nanotube-supported Pt electrocatalyst for the oxygen reduction reaction. Energy Fuels 2008;22:2543–9.
[60] Sjoblom J. Emulsions and emulsion stability: surfactant science series, 61. Boca Raton: CRC Press; 2005.

[61] Häger M, Currie F, Holmberg K. Organic reactions in microemulsions. Top Curr Chem 2003;227:53–74.

[62] Jacobson GB, Lee JCT, Johnston KP. Organic synthesis in water/carbon dioxide microemulsions. J Org Chem 1999;64:1201–6.

[63] Ohde M, Ohde H, Wai CM. Recycling nanoparticles stabilized in water-in-CO_2 microemulsions for catalytic hydrogenations. Langmuir 2005;21:1738–44.

[64] Zhu J, Robertson A, Tsang SC. Aqueous emulsion containing fluorous cobalt species in supercritical CO_2 for catalytic air oxidation of toluene. Chem Commun 2002;2044–5.

[65] Tsang SC, Zhu J, Yu KMK. Selective catalytic oxidation of alkylaromatic molecules by nanosize water droplets containing Co^{2+} species in supercritical carbon dioxide fluid. J Exp Nanosci 2006;1:435–56.

[66] Tsang SC, Zhu J, Yu KMK. A new oxidation catalyst system using fluorous cobalt(II) species in water-supercritical carbon dioxide (C–H free environment). Catal Lett 2006;106:123–6.

[67] Martin CA, McCrann PM, Angelos GH, Jaeger DA. Reactions in microemulsion media: nucleophilic displacement reaction of benzyl chloride with bromide ion. Tetrahedron Lett 1982;23:4651–4.

[68] Meric P, Yu KMK, Tsang SC. Micelle-hosted palladium nanoparticles catalyze citral molecule hydrogenation in supercritical carbon dioxide. Langmuir 2004;20:8537–45.

[69] Meric P, Yu KMK, Tsang SC. Molecular guided catalytic hydrogenation by micelle-hosted Pd nanoparticle in supercritical CO_2. Catal Lett 2004;95:39–43.

[70] Ikushima Y, Saito N, Hatakeda K, Arai M. Water/AOT/ethane microemulsion under supercritical conditions as a reaction medium. J Supercrit Fluids 1998;13:217–24.

[71] Ikushima Y, Shervani Z, Saito N, Arai M. The promotion of alkaline fading of crystal violet by AOT micelles dispersed in supercritical ethane medium. J Colloid Interf. Sci 1997;191:177–83.

[72] Tamura K, Nii N. Fluorescence polarization study of aerosol-OT reversed micelles at high pressures. J Phys Chem 1989;93:4825–9.

[73] Holmes JD, Steytler DC, Rees GD, Robinson BH. Bioconversions in a water-in-CO_2 microemulsion. Langmuir 1998;14:6371–6.

[74] Gupta MN. Enzyme function in organic solvents. Eur J Biochem 1992;203:25–32.

[75] Cantone S, Hanefeld U, Basso A. Biocatalysis in non-conventional media—ionic liquids, supercritical fluids and the gas phase. Green Chem 2007;9:954–71.

[76] Mesiano AJ, Beckman EJ, Russell A. Supercritical biocatalysis. J Chem Rev 1999;99: 623–34.

[77] Stanescu MA, Ginosar DM, Bala GA, Anderson RP. Biocatalytic treatment of organosulfur compounds in emulsions in supercritical fluids. Fuel Chem Div Preprints 2002;47:524–6.

[78] Ginosar DM, Bala GA, Anderson RP, Fox SL, Stanescu MA. Bioprocessing of high sulfur crudes via application of critical fluid biocatalysis. Bechtel Bwxt Idaho: Idaho National Engineering and Environmental Laboratory; 2002. LLC INEEL/EXT-02-00711.

[79] McEldoon JP, Dordick JS. Unusual thermal stability of soybean peroxidase. Biotechnol Prog 1996;12:555–8.

[80] Kane MA, Baker GA, Pandey S, Bright FV. Performance of cholesterol oxidase sequestered within reverse micelles formed in supercritical carbon dioxide. Langmuir 2000;16:4901–5.

[81] Gupte A, Nagarajan R, Kilara A. Enzymic oxidation of cholesterol in reverse micelles. Ind Eng Chem Res 1995;34:2910–22.

[82] Hakoda M, Shiragami N, Enomoto A, Nakamura K. Measurements of hydrodynamic diameter of AOT reverse micelles containing lipase in supercritical ethane and its enzymatic reaction. Bioprocess Biosyst Eng 2003;25:243–7.

[83] Kendall JL, Canelas DA, Young JL, DeSimone JM. Polymerizations in supercritical carbon dioxide. Chem Rev 1999;99:543–63.

[84] Beckman EJ, Smith RD. Microemulsion polymerization within near- and supercritical continuous phases: effect of microemulsion structure on reaction characteristics. J Supercrit Fluids 1990;3:205–13.

[85] Kemmere MF, Meyer T. Supercritical carbon dioxide in polymer reaction engineering. Weinheim: WILEY-VCH; 2005.

[86] Katime I, Arellano J, Mendizábal E, Flores J. Cationic microemulsion polymerization of alkyl acrylates. Mater Sci Appl 2010;1:292–300.

[87] Chern C-S. Encyclopedia of polymer science and technology: microemulsion polymerization, vol. 7. New York: John Wiley & Sons; 2002.

[88] Ye W-J, Keiper JS, DeSimone JM. Polymeric nanoparticles from supercritical CO_2 microemulsion polymerization. Chin J Polym Sci 2006;24:95–101.

[89] Ajayan PM, Iijima S. Capillarity-induced filling of carbon nanotubes. Nature 1993;361: 333–4.

[90] Seraphin S, Zhou D, Jiao J, Withers JC, Loufty R. Yttrium carbide in nanotubes. Nature 1993;362:503.

[91] Ajayan PM, Ebbesen TW, Ichihashi T, Iijima S, Tanigaki K, Hiura H. Opening carbon nanotubes with oxygen and implications for filling. Nature 1993;362:522–5.

[92] Saito Y, Yoshikawa T. Bamboo-shaped carbon tube filled partially with nickel. J Cryst Growth 1993;134:154–6.

[93] Wang XB, Liu ZM, Hu PA, Liu YQ, Han BX, Zhu DB. Nanofluids in carbon nanotubes using supercritical CO_2: a first step towards a nanochemical reaction. Appl Phys A 2005;80:637–9.

[94] Torre AL, Giménez-López MdC, Fay MW, Rance GA, Solomonsz WA, Chamberlain TW, Brown PD, Khlobystov AN. Assembly, growth, and catalytic activity of gold nanoparticles in hollow carbon nanofibers. ACS Nano 2012;6:2000–7.

[95] Weissker U, Hampel S, Leonhardt A, B¨uchner B. Carbon nanotubes filled with ferromagnetic materials. Materials 2010;3:4387–427.

[96] Bhushan B. Handbook of nanotechnology, part 3 introduction to carbon nanotubes, vol. 1. New York: Springer; 2004. p. 43–11.

[97] Monthioux M. Carbon meta-nanotubes: synthesis, properties and applications. Filled carbon nanotubes. Hoboken, NJ: John Wiley & Sons; 2012. chapter 5.

[98] Ugarte D, Stöckli T, Bonard JM, Châtelain A, de Heer WA. Filling carbon nanotubes. Appl Phys A 1998;67:101–5.

[99] Dai L. Intelligent macromolecules for smart devices: from materials synthesis to device applications (Engineering Materials and Processes). New York: Springer; 2004.

[100] Liu BZ, Han B. Synthesis of carbon-nanotube composites using supercritical fluids and their potential applications. Adv Mater 2009;21:825–9.

[101] Smith BW, Monthioux M, Luzzi DE. Encapsulated C_{60} in carbon nanotubes. Nature 1998;396:323–4.

[102] Shimada T, Ohno Y, Okazaki T, Sugai T, Suenaga K, Kishimoto S, Mizutani T, Inoue T, Taniguchi R, Fukui N, Okubo H, Shinohara H. Transport properties of C_{78}, C_{90} and Dy@C_{82} fullerenes-nanopeapods by field effect transistors. Physica E 2004;21:1089–92.

[103] Simon F, Kuzmany H, Rauf H, Pichler T, Bernardi J, Peterlik H, Korecz L, Fülöp F, Jànossy A. Low temperature fullerene encapsulation in single-wall carbon nanotubes: synthesis of N@C_{60}@SWCNT. Chem Phys Lett 2004;383:362–7.

[104] Chiu PW, Gu G, Kim GT, Philipp G, Roth S, Yang SF, Yang S. Temperature-induced change from p to n conduction in metallofullerene nanotube peapods. Appl Phys Lett 2001;79:3845–7.

[105] Okazaki T, Suenaga K, Hirahara K, Bandow S, Iijima S, Shinohara H. Electronic and geometric structures of metallofullerenes peapods. Physica B 2002;323:97–9.
[106] Britz DA, Khlobystov AN, Wang J, O'Neil AS, Poliakoff M, Ardavan A, Briggs GAD. Selective host-guest interaction of single-walled carbon nanotubes with functionalized fullerenes. Chem Commun 2004;176–7.
[107] Marulanda JM, editor. One-dimensional crystals inside single-walled carbon nanotubes: growth, structure and electronic properties. In: Electronic properties of carbon nanotubes; InTech: in Croatia (European Union); 2011 [chapter 8].
[108] Jessop PG, Walter Leitner W. Chemical synthesis using supercritical fluids. Weinhem, Chichester, New York: Wiley-VCH; 1999.
[109] Chamberlain TW, Camenisch A, Champness NR, Briggs GAD, Benjamin SC, Ardavan A, Khlobystov AN. Toward controlled spacing in one-dimensional molecular chains: alkyl-chain-functionalized fullerenes in carbon nanotubes. J Am Chem Soc 2007;129:8609–14.
[110] Pagona G, Rotas G, Khlobystov AN, Chamberlain TW, Porfyrakis K, Tagmatarchis N. Azafullerenes encapsulated within single-walled carbon nanotubes. J Am Chem Soc 2008;130:6062–3.
[111] Chamberlain TW, Pfeiffer R, Peterlik H, Kuzmany H, Zerbetto F, Melle-Franco M, Staddon L, Champness NR, Briggs GAD, Khlobystov AN. Polyarene-functionalized fullerenes in carbon nanotubes: towards controlled geometry of molecular chains. Small 2008;4:2262–70.
[112] Khlobystov AN, Britz DA, Wang J, O'Neil SA, Poliakoffb M, Briggs GAD. Low temperature assembly of fullerene arrays in single-walled carbon nanotubes using supercritical fluids. J Mater Chem 2004;14:2852–7.
[113] Botka B, Füstös V, Klupp G, Kocsis D, Székely E, Utczás M, Simándi B, Botos A, Hackl R, Kamarás K. Low-temperature encapsulation of coronene in carbon nanotubes. Phys Status Solidi B 2012;249:2432–5.
[114] Martinez JL. Supercritical fluid extraction of nutraceuticals and bioactive compounds. Boca Raton: CRC Press; 2007.
[115] Hao J, Lian Y, Guan L, Yue D, Guo X, Zhao S, Zhao Y, Ibrahim K, Wang J, Qian H, Dong J, Yuan H, Xing G, Sun B. Supercritical synthesis and characterization of SWNT-based one dimensional nanomaterials. Nanoscale 2011;3:3103–8.
[116] Ye X-R, Lin Y, Wang C, Wai CM. Supercritical fluid fabrication of metal nanowires and nanorods templated by multiwalled carbon nanotubes. Adv Mater 2003;15:316–9.
[117] Kim J, Kim D, Veriansyah B, Kang JW, Kim J-D. Metal nanoparticle synthesis using supercritical alcohol. Mater Lett 2009;63:1880–2.
[118] Bal S, Samal SS. Carbon nanotube reinforced polymer composites: a state of the art. Bull Mater Sci 2007;30:379–86.
[119] Wu C-S, Liao H-T. Study on the preparation and characterization of biodegradable polylactide/multi-walled carbon nanotubes nanocomposites. Polymer 2007;48:4449–58.
[120] Liu Z, Dai X, Xu J, Han B, Zhang J, Wang Y, Huang Y, Yang G. Encapsulation of polystyrene within carbon nanotubes with the aid of supercritical CO_2. Carbon 2004;42:458–60.
[121] Steinmetz J, Kwon S, Lee H-J, Abou-Hamad E, Almairac R, Goze-Bac C, Kim H, Park Y-W. Polymerization of conducting polymers inside carbon nanotubes. Chem Phys Lett 2006;431:139–44.
[122] Britz DA, Khlobystov AN, Porfyrakis K, Ardavanc A, Briggs GAD. Chemical reactions inside single-walled carbon nano test-tubes. Chem Commun 2005;37–9.
[123] Steinmetz J, Lee H-J, Kwon S, Lee D-S, Goze-Bac C, Abou-Hamad E, Kim H, Park Y-W. Routes to the synthesis of carbon nanotube–polyacetylene composites by Ziegler–Natta polymerization of acetylene inside carbon nanotubes. Curr Appl Phys 2007;7:39–41.

[124] Dawn S, Salpage SR, Koscher BA, Bick A, Wibowo AC, Pellechia PJ, Shimizu LS. Applications of a bis-urea phenylethynylene self-assembled nanoreactor for [2 + 2] photodimerizations. J Phys Chem A 2014;118(45):10563–74.

[125] Solomonsz WA, Rance GA, Harris BJ, Khlobystov AN. Competitive hydrosilylation in carbon nanoreactors: probing the effect of nanoscale confinement on selectivity. Nanoscale 2013;5:12200–5.

[126] Solomonsz WA, Rance GA, Suyetin M, Torre AL, Bichoutskaia E, Khlobystov AN. Controlling the regioselectivity of the hydrosilylation reaction in carbon nanoreactors. Chem Eur J 2012;18:13180–7.

[127] Rance GA, Solomonsz WA, Khlobystov AN. Click chemistry in carbon nanoreactors. Chem Commun 2013;49:1067–9.

[128] Chamberlain TW, Earley JH, Anderson DP, Khlobystov AN, Bourne RA. Catalytic nanoreactors in continuous flow: hydrogenation inside single-walled carbon nanotubes using supercritical CO_2. Chem Commun 2014;50:5200–2.

[129] Sun Y-P. Supercritical fluid technology in materials science and engineering: syntheses, properties, and applications. New York: Marcel Dekker, Inc.; 2002.

Chapter 13

Pyrene: The Guest of Honor

Nicolas P.E. Barry*, Bruno Therrien†
*Department of Chemistry, University of Warwick, Coventry, UK
†Institut de Chimie, Université de Neuchâtel, Neuchâtel, Switzerland

1 INTRODUCTION

1.1 Scope

Mimicking nature possibly remains one of the biggest challenges for chemists, and one of nature's remarkable achievements is the ability of enzymes to catalyze reactions in a confined environment. The lock-and-key model developed by Fischer [1,2] to explain the specificity between a substrate and the active site of an enzyme opened-up new avenues in supramolecular and host–guest chemistry [3,4].

Pioneered by Cram, Lehn, and Pedersen [5–7], winners of the Nobel Prize in chemistry in 1987, the preparation of molecular hosts to accommodate guest molecules is an extremely active field of research. To design new hosts, a good understanding of the interactions between the host and the guest is crucial. The utilization of a model guest molecule, such as pyrene and its derivatives, has therefore been proved to be of huge interest to the community, not only for exploring the properties of various host–guest systems, but also for comparing the hosting abilities of different molecular and supramolecular hosts. The chemical structure of pyrene and pyrene derivatives and their unique and tunable chemical and physical properties certainly make these compounds the most commonly studied family of guest molecules.

This chapter introduces some key structural features of pyrene, and the unique properties that make this polycyclic aromatic hydrocarbon molecule a guest of choice in supramolecular and host–guest chemistry. Based on these characteristics, a brief listing of some techniques that can be easily used by students and new researchers entering the field to study the dynamic process of a host–pyrene system is provided. The illustrative encapsulation of pyrene molecules in a number of organic, inorganic, and polymeric hosts is also reviewed, with an emphasis on work published from 2008 to 2014, in order to introduce some innovative and emergent directions in supramolecular chemistry, but without omitting some earlier important contributions when necessary.

Organic Nanoreactors. http://dx.doi.org/10.1016/B978-0-12-801713-5.00013-6

1.2 Pyrene: Structure and Characterization

Pyrene is the smallest peri-fused polycyclic aromatic hydrocarbon and is obtained during the combustion of organic compounds [8]. The pyrene molecule is highly symmetrical (point group D_{2h}) [9] and despite having 16 π electrons and not following the Hückel's $4n + 2$ rule, it is aromatic [10]. These particular features provide to pyrene interesting electronic properties [11], which are highly exploited in the characterization of host–guest systems. Among the numerous techniques employed to study host–guest interactions, NMR, UV-vis and fluorescence spectroscopy are the most frequently used, and an overview of the corresponding spectroscopic data of pyrene is needed before other considerations.

^1H and $^{13}C\{^1H\}$ NMR chemical shifts of pyrene in common deuterated solvents (CD_2Cl_2, $CDCl_3$, C_6D_6, $(CD_3)C_6D_5$, $(NO_2)C_6D_5$, and $(CD_3)_2SO$) are reported in Table 13.1, together with pyrene's atom-labeling scheme (according to IUPAC rules) [12]. The diffusion coefficients (D) of pyrene in these deuterated solvents are also reported in Table 13.1, owing to the importance of diffusion ordered NMR spectroscopy (DOSY) to studying host–guest systems in

TABLE 13.1 ^1H and $^{13}C\{^1H\}$ NMR Chemical Shifts (ppm) and Diffusion Coefficient D (10^{-10} m^2/s) of Pyrene (21°C) in Selected Deuterated Solvents

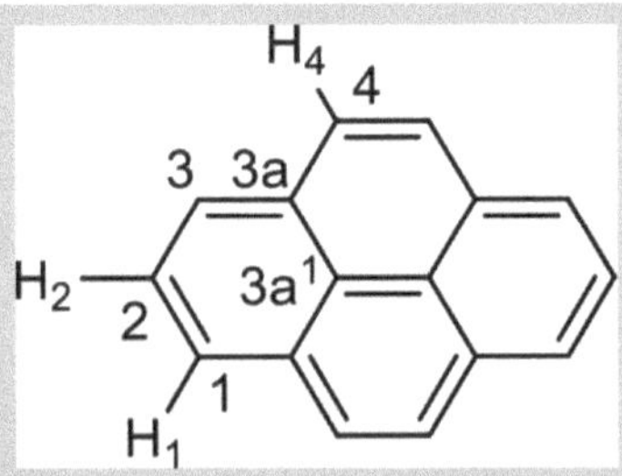

	CD_2Cl_2	$CDCl_3$	C_6D_6	$(CD_3)C_6D_5$	$(NO_2)C_6D_5$	$(CD_3)_2SO$
C_1	125.5	125.1	125.7	125.0	125.8	125.0
C_2	126.4	126.0	126.4	125.9	126.7	126.2
C_{3a}	131.7	131.3	131.9	131.6	131.9	130.6
C_{3a1}	125.1	124.8	125.6	124.8	125.2	123.8
C_4	127.9	127.6	128.0	127.6	128.1	127.3
H_1	8.22	8.22	7.90	7.87	8.18	8.27
H_2	8.02	8.04	7.73	7.72	8.03	8.05
H_4	8.11	8.11	7.78	7.76	8.07	8.16
D	19.95	15.85	12.25	14.79	8.91	8.98

solution. The resonance of the H_2 protons gives rise to an upfield triplet, while the singlet associated with the resonance of the H_4 protons is located between H_1 and H_2. The most downfield signal is the doublet assigned to the H_1 protons. This pattern is found in all solvents studied. Succession from upfield to downfield signals attributed to carbons C_{3a1}, C_1, C_2, C_4, and C_{3a} are observed in all $^{13}C\{^1H\}$ NMR spectra. More information about experimental and theoretical NMR chemical shifts can be found in the literature [13].

The absorption and fluorescence spectra of pyrene in the UV-vis region are well described in the literature [14]. Different methods for theoretical study of gas-phase or in-solvent excitation and emission properties of pyrene have been used, and these results have been compared with experimental data. In 1963, Jackson and coworkers recorded the UV-vis spectrum of pyrene in the gas-phase as well as in isooctane [15]. Tanaka did a similar investigation in *n*-heptane in 1965 [16], and Mangle and Topp completed this study in 1986 with gas-phase values for absorption and emission spectra of the two lowest singlet states (S_1 and S_2) and the corresponding spectral shifts in a series of apolar solvents [17]. More recently, ab initio and time-dependent density functional methods have been applied to discuss the nature of the different excited states and their relative positions on the energy scale [18,19]. These absorption and emission characteristics make pyrene a molecule of choice for the design of biological probes [20,21], photonic devices [22,23], or liquid crystalline materials [24–29]. Furthermore, the facile functionalization of pyrene with ethynyl groups can extend the π-conjugation system, leading to bathochromic shifts of the emission bands toward the visible region [30–33]. As an example, the UV-vis spectrum of pyrene in dichloromethane is shown in Fig. 13.1.

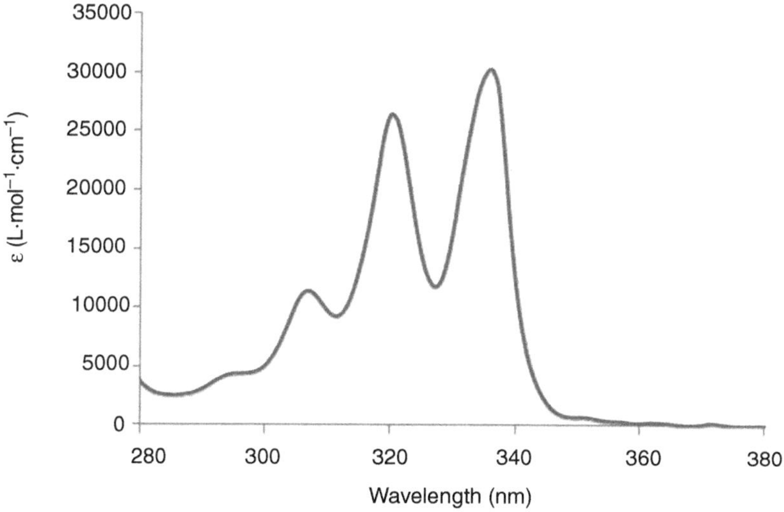

FIGURE 13.1 **UV-vis spectrum of pyrene (10^{-4} M) in dichloromethane.**

Without interactions with potential host molecules and in diluted solutions to avoid excimeric formations, pyrene presents in solution an intense and anisotropic fluorescence, as well as a high fluorescence quantum yield [34–37]. Direct evidence of ground-state interactions of pyrene with potential host molecules is provided by the emission spectra. The vibrational structure of the emission spectrum of pyrene is constituted by five fine peaks, named I_1, I_2, I_3, I_4, and I_5 (Fig. 13.2) [38]. An increase of the intensity of peak I_1 is observed in polar solvents, while I_3 is solvent insensitive. Thus, the evolution of the ratio of intensities I_1/I_3 gives information on the evolution of the polarity of the environment close to molecular pyrene, and consequently on the encapsulation of this guest in a host molecular or supramolecular object [39]. This sensitivity of pyrene, and of peri-fused polycyclic aromatic hydrocarbon molecules in general, to the polarity of the environment is a photophysic property that is extensively used to study host–guest interactions [40].

Another noticeable photochemical property of pyrene is its excimer emission in more concentrated solutions. The ground-state dimer formation of pyrene molecules leads to a large characteristic band at ca. 480 nm, which is of interest in determining a stacking of two or more pyrene molecules in solution [41]. The consequences of such excimeric emission in supramolecular chemistry will be detailed in the following sections of this chapter.

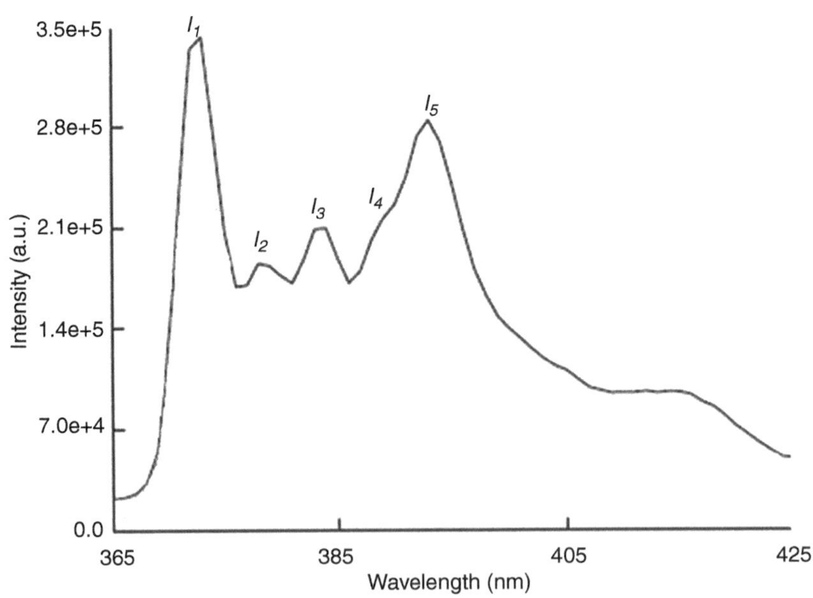

FIGURE 13.2 Fluorescence emission spectra of 0.5 μM of pyrene in a 1 M aqueous NaOH solution. Excitation wavelength: 337 nm. *(Adapted from Ref. [38] with permission of Springer.)*

1.3 Binding Constants in a Nutshell

A host–guest system can be defined as the formation of a complex obtained from the inclusion of one or more guest molecules (depending on the stoichiometry) into a host molecular or supramolecular object. A new researcher entering the field of supramolecular chemistry might be very soon confronted by the question of determining the strength of the interactions between the two entities, or in other words, to the quantification of the free energy (ΔG) associated with this particular host–guest interaction. A quantitative analysis of the host–guest complex can be estimated by the determination of the equilibrium constant "K," the most straightforward way of measuring the free energy difference:

$$\Delta G = -RT \quad \ln K$$

A binding constant is a particular case of general equilibrium constants, which measures the bonding affinity between two or more molecules at equilibrium. Its determination not only provides information on the affinity between host and guest molecules, but can also be related to other chemical parameters of either the host or the guest molecule (eg, basicity, electronegativity). Calculating the binding constants in host–guest chemistry gives a fundamental understanding of the chemical equilibrium, leading to the rationalization of the supramolecular interactions.

The difficulty in the evaluation of binding constants in supramolecular chemistry lies in the fact that it is usually not possible to determine *directly* the concentration of the complex of interest. Biochemistry techniques such as electrophoreses or filter-binding assays [42] (used for calculating the affinity of an enzyme to a substrate or the affinity of a ligand/drug to a receptor) cannot be applied to host–guest systems. Instead, spectroscopic methods (eg, NMR and UV-vis spectroscopy) are generally used to *indirectly* detect complexation. Irrespective of the method though, the determination of binding constants always requires the comparison between experimental data and theoretical models, and a number of binding models are now well described and easily available [43]:

The general $m{:}n$ host–guest (H–G) binding system is defined by the Eq. (13.1):

$$m\mathrm{H} + n\mathrm{G} \overset{\beta_{mn}}{\Leftrightarrow} \mathrm{H}_m\mathrm{G}_n \tag{13.1}$$

where β_{mn} is the overall stability constant defined according to Eq. (13.2):

$$\beta_{mn} = \frac{[\mathrm{H}]_m[\mathrm{G}]_n}{[\mathrm{H}]^m[\mathrm{G}]^n} \tag{13.2}$$

In the case of a $1{:}n$ host–guest system, the stepwise equilibrium binding constant K_n is defined by Eq. (13.3):

$$K_n = \frac{[\mathrm{HG}]_n}{[\mathrm{HG}_{n-1}][\mathrm{G}]} = \frac{k_n}{k_{-n}} \tag{13.3}$$

where k_n and k_{-n} are the forward and backward rate of kinetics of complexation, respectively.

In the case of a 1:1 equilibrium (the most observed and studied system in host–guest chemistry), the binding constant is defined according to Eq. (13.4):

$$K = K_1 = \frac{[\mathrm{HG}]}{[\mathrm{H}][\mathrm{G}]} \tag{13.4}$$

where [H], [G], and [HG] are the free concentrations of the host, guest, and host–guest species in solution. The total concentration of the host ($[\mathrm{H}]_o$) (or of the guest, $[\mathrm{G}]_o$) being the sum of the free concentration of the host (or of the guest) with the free concentration of the host–guest complex, Eq. (13.4) becomes Eq. (13.5):

$$K = \frac{[\mathrm{HG}]}{[\mathrm{H}]_o[\mathrm{HG}]_o - [\mathrm{HG}]([\mathrm{H}]_o + [\mathrm{G}]_o) + [\mathrm{HG}]^2} \tag{13.5}$$

The resolution of this equation, after rearrangement and isolation of [HG], offers the solution in Eq. (13.6):

$$[\mathrm{HG}] = \frac{1}{2}\left\{\left([\mathrm{G}]_o + [\mathrm{H}]_o + \frac{1}{K}\right) - \sqrt{\left([\mathrm{G}]_o + [\mathrm{H}]_o + \frac{1}{K}\right)^2 + 4[\mathrm{H}]_o[\mathrm{G}]_o}\right\} \tag{13.6}$$

The only unknown in Eq. (13.6) is the binding constant K, which explains the interest of this equation for the study of 1:1 host–guest equilibria. Furthermore, such a simple equation can be easily solved with a number of freely or commercially available software packages [44–46]. In Section 2 of this chapter, we will briefly introduce the UV-vis and NMR versions of this equation.

2 TECHNIQUES USED TO STUDY HOST–PYRENE INTERACTIONS IN SOLUTION

2.1 Introduction

A host–guest study can usually be addressed through a two-step process: the characterization of the interactions between a host and a guest (*How many guest molecules can interact with this host?*) and its quantification (*How strong is this interaction?*). The determination of the host–guest stoichiometry should be performed first, seconded by the calculation of the corresponding binding constants. The most popular methods used to determine binding constants include UV-vis, fluorescence, and NMR spectroscopy.

Each method has advantages and drawbacks:

- UV-vis spectroscopy is cheap, simple, and particularly valuable when chromophoric molecules, such as pyrene, are involved. The main limitation of

UV-vis spectroscopy is that the titrations must be in a concentration range where the absorbance follows the Beer–Lambert law (absorbance < 1). It is usually suitable for the determination of association constants comprised between 10^2 and 10^5 M^{-1} [47].

- The high sensitivity of fluorescence spectroscopy and the linear proportionality between fluorescence intensity and fluorophore concentration in diluted solutions is an asset, and usually makes fluorescence the technique of choice for studying host–pyrene interactions. Nonetheless, the smallest presence of fluorophore impurities may bias the emission spectra. This method is applicable for large K values (up to 10^7 M^{-1}).
- NMR spectroscopy possesses the advantage of providing unrivaled structural information on the inclusion complex in addition to offer unbiased signals from impurities data required for the calculations of binding constants, but may suffer from the broadening of the signals upon encapsulation of the pyrene molecule into the cavity of the host. This method is usually employed to determine association constants around 10^4 M^{-1}.

Knowledge of the different techniques for the study of a host–guest system in order to benefit from the advantages of one technique while avoiding the inherent drawbacks of the others is essential. These various methods are now briefly introduced with selected examples to facilitate comprehension. The discussion is limited to the simplest cases, and only a few equations related to 1:1 equilibria are presented. There are many excellent textbooks, reviews [43,48], and software packages [44–46,49] available for use in supramolecular chemistry that show how binding constants can be obtained through computation models, some of them with emphasis on the basic equations used, others describing the software protocols. This section of the chapter does not cover these computational methods, but aims at introducing the fundamental experiments that will enable students and new researchers to collect the data needed for running these calculations in a convivial way. Some graphical methods are also briefly introduced in order to help the students entering the field to easily double-check the validity of the calculations, the human eye sometimes finding it easier to spot deviation from linearity than to look directly at hyperbolic plots.

2.2 UV-vis Spectroscopy

Different methods can be used to determine the guest-to-host ratio, or stoichiometry, of a host–guest system. Among them, the continuous variations, also called Job's method [50], and the mole ratio method [51], are certainly the easiest and most common techniques.

In the method of continuous variations, several measurements are recorded in which the concentrations of the compounds are varied stepwise, to minimize the error of the estimation of the binding process. In this method, the total concentration of [host] + [pyrene] is kept constant, but the ratio [host]/[pyrene]

varies in small iterations. The plot of the absorbance variation (ΔA) versus the pyrene molar ratio ([pyrene]/([host] + [pyrene])) leads to a curve with a maximum (X_{max}) defined by Eq. (13.7):

$$X_{max} = \frac{n}{m+n} = \frac{1}{(m/n)+1} \quad (13.7)$$

where m and n correspond to a m:n stoichiometry of the host–pyrene system [52]. Thus, in the case of a 1:1 host–guest system ($m = n = 1$), the maximum curve will occur at $X_{max} = 0.5$. For a 1:2 stoichiometry ($m = 1$, $n = 2$) the maximum of the curve will be observed at $X_{max} = 2/(1 + 2) = 0.67$. This simple relation can be applied for all values of m and n. In 2010, Peinador and coworkers used this method for investigating the complexation of pyrene with a self-assembled bipyridinium palladium metalla-cycle in water (Fig. 13.3). The corresponding Job's plot (typical of a 1:1 stoichiometry) is shown in Fig. 13.3c [53].

In the mole ratio method, the concentration of the [host] is kept constant while the concentration of [pyrene] is varied, and the absorbance of the system versus the molar ratio [pyrene]/[host] is plotted. An abrupt change in the slope will correspond to the stoichiometric ratio. With this method, the absorption of the complex is proportional to its concentration (Beer–Lambert law); thus, selecting the value of absorbance at λ_{max} offers the highest precision, although this value can be measured at any wavelength if necessary. Models for 1:1 and 1:2 stoichiometries are shown in Fig. 13.4a,b, respectively (assuming K ca. 10^4 M^{-1}; concentration range: $10^{-2} - 10^{-5}$ M). This method is particularly useful

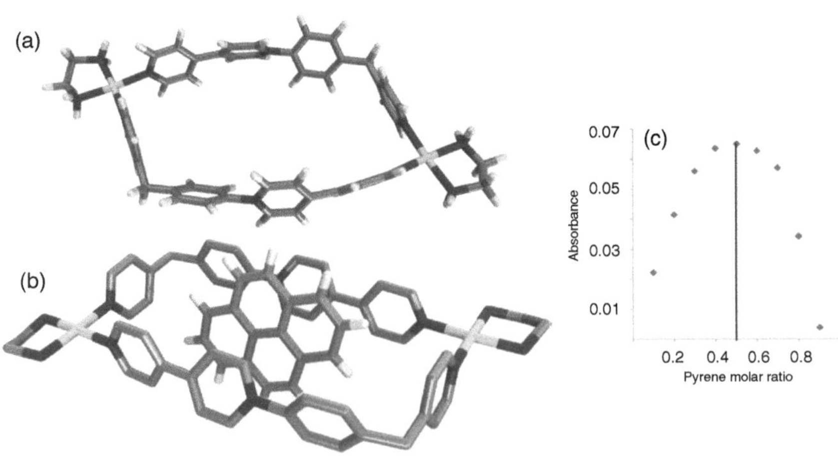

FIGURE 13.3 (a) Molecular structure of a self-assembled palladium metalla-cycle. (b) Its pyrene inclusion derivative. (c) Job plot showing the 1:1 stoichiometry. Color labeling scheme: nitrogen [dark gray (blue in the web version)], palladium [light gray (yellow in the web version)], carbon (gray), and hydrogen (white). *(Adapted from Ref. [53] with permission of American Chemical Society.)*

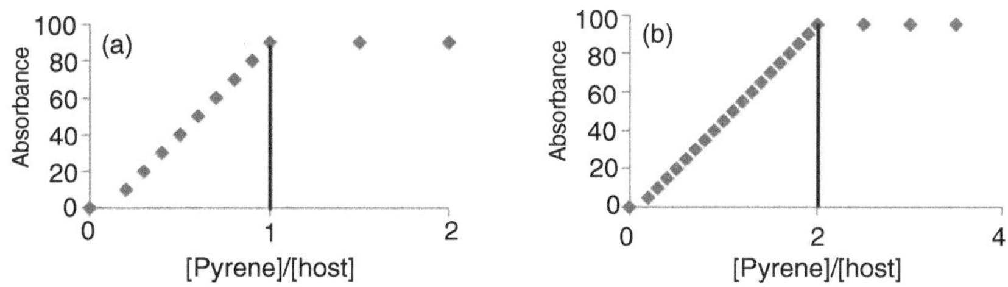

FIGURE 13.4 Plots obtained via the mole ratio method for a 1:1 (a) and a 1:2 (b) stoichiometries (assumed K ca. 10^4 M^{-1}, concentration range: $10^{-2}-10^{-5}$ M).

for determining high degrees of complexation as compared to the Job's plot method: in the method of continuous variations, the maxima are observed at $X_{max} = 0.800$, $X_{max} = 0.833$, and $X_{max} = 0.857$ for 1:4, 1:5, and 1:6 stoichiometry, which is difficult to precisely determine experimentally. In the mole ratio method, these differences are expanded to one unit [51].

Importantly, the timescale for UV-vis transitions ($<10^{-12}$ s) being smaller than the lifetimes of the guest, host, and host–guest system studied, the equilibrium monitored by UV-vis spectroscopy is always in the slow exchange region. Consequently, the observed absorbance is the sum of all the absorbances. Pyrene showing strong UV-vis bands (Fig. 13.1), the guest absorption has usually to be taken into consideration. In the case of a 1:1 equilibrium system, Eq. (13.8) must be solved [UV-vis version of Eq. (13.6)]:

$$\Delta A = \varepsilon_G[G]_0 + \varepsilon_{\Delta HG}\left(\frac{1}{2}\left\{\left([G]_0+[H]_0+\frac{1}{K}\right)-\sqrt{\left([G]_0+[H]_0+\frac{1}{K}\right)^2+4[H]_0[G]_o}\right\}\right) \tag{13.8}$$

2.3 Fluorescence Spectroscopy

Fluorescence spectroscopy is a popular and highly sensitive technique (down to the nanomolar detection) for studying host–guest systems, and very large association constants (up to 10^7 M^{-1}) can be estimated by this method. At low concentration, the fluorescence intensity is linearly proportional to the concentration of the fluorophore. The observed fluorescence can be described by Eq. (13.9):

$$F = 2.3I_0\phi\varepsilon bc \tag{13.9}$$

where F is the fluorescence intensity, I_0 is the intensity of the excitation radiation, ε is the molar absorptivity at the excitation wavelength, b is the path length (cm), c is the molar concentration and ϕ is the quantum yield [54].

Pyrene being fluorescently active, it is advised to use this technique to study host–pyrene systems in which the molecular or supramolecular host is fluorescent-inactive or silent. Upon encapsulation in the host molecule, the fluorescence of pyrene is expected to be reduced, offering a particularly powerful and rapid method of complexation detection. Fluorescence quenching may encompass a number of mechanisms of molecular interactions, including molecular rearrangements, ground-state complex formation, excited-state reactions, energy transfer and collisional (or dynamic) quenching. Excimeric formation (one of the dynamic quenching mechanisms) is of significant importance in the case of host–pyrene interactions, as will be illustrated in the following sections of this chapter. As a note of warning, it should be emphasized that the choice of solvent is also of crucial importance, owing to the sensitivity of pyrene to the polarity of the environment.

When the host is fluorescent-inactive and the gradual addition of host upon a pyrene solution results in both dynamic and static quenching, the Stern–Volmer description of the observed fluorescence intensity ratio F/F_0 can be defined by Eq. (13.10):

$$\frac{F}{F_0} = \frac{1 + \left(\phi^{\text{host–pyrene}} / \phi^{\text{pyrene}}\right) K[\text{H}]}{1 + K[\text{H}]} \tag{13.10}$$

where $\phi^{\text{host-pyrene}}/\phi^{\text{pyrene}}$ is the fluorescence quantum yield ratio between the host–pyrene system and the free analyte.

Despite its limitations [48], the Benesi–Hildebrand method can be considered for rapidly determining the stoichiometry and binding constants of a host–pyrene system by fluorescence spectroscopy; this method, still commonly employed for such determination, is based on Eq. (13.11) [55–59]:

$$\frac{1}{F - F_0} = \frac{1}{K(F_n - F_0)[\text{H}]^n} + \frac{1}{F_n - F_0} \tag{13.11}$$

The plot of $1/(F-F_0)$ versus $1/[\text{H}]^n$ will only give a straight line as well as correct standard errors and correlation coefficients for the right 1:n value (the right n stoichiometry). Fig. 13.5 shows an example of the use of this double reciprocal plot of the Benesi–Hildebrand equation for the determination of the stoichiometry (1:1 in this case) of a host–guest system constructed with star-shaped cholic acid derivative and pyrene [40]. Although very simple and useful for students not proficient with computational software packages, it must be emphasized that graphical methods such as this one should not be used instead of recent software able to deal with nonlinear problems. We believe that they are nonetheless useful methods for checking the outputs of initial calculations.

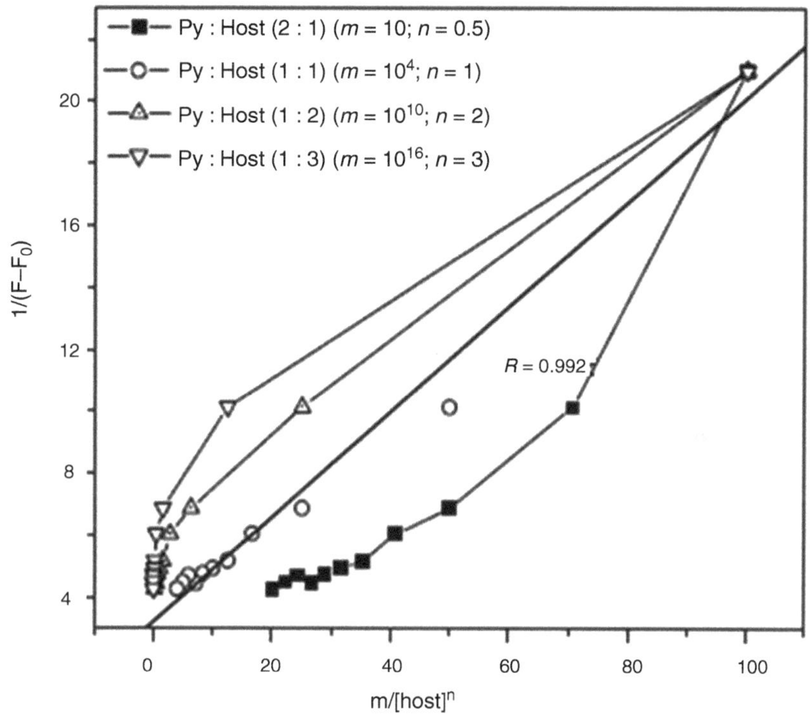

FIGURE 13.5 Double reciprocal plots of Benesi-Hildebrand equation for the determination of the stoichiometry of a host–guest system, Py = pyrene. *(Reproduced from Ref. [40] with permission of American Chemical Society. Copyright 2008.)*

2.4 NMR Spectroscopy and Diffusion Coefficients

Two cases should be envisaged when NMR spectroscopy is used to study host–guest systems: either the exchange between the signals of free and host–guest systems is fast as compared to the NMR time scale, or this exchange process is slow. Under a fast exchange regime, the chemical shifts of the chosen signals are followed upon encapsulation, and at least one proton of both the free and complex species must be shifted during the formation of the host–guest system: this phenomenon, known as complexation-induced shift (CIS) effect, is particularly useful in the case of host–pyrene systems. The encapsulation of a pyrene molecule in a host cavity generally leads to an easily observable CIS phenomenon due to a "ring current" effect via electronic shielding of pyrene from the host [60]. If the species are in fast exchange, the observed NMR resonance δ will be the weighted average of the unbound host and host–guest molecules, and for a simple 1:1 host–guest system, the equation that needs to be solved is therefore the NMR version of Eq. (13.6):

$$\Delta\delta = \frac{\delta_{\Delta HG}}{[H]_0}\left(\frac{1}{2}\left\{\left([G]_0 + [H]_0 + \frac{1}{K}\right) - \sqrt{\left([G]_0 + [H]_0 + \frac{1}{K}\right)^2 + 4[H]_0[G]_0}\right\}\right) \quad (13.12)$$

In the second configuration (slow exchange), the concentrations of host, pyrene, and host–guest complex can be directly determined by the integration ratio of the NMR signals. Host, pyrene, and the host–pyrene system should possess discrete NMR signals, and the association constant of a 1:1 host–guest system can be directly calculated from Eq. (13.4). This case of slow exchange is scarce but exists, and examples can be found in the literature [61,62].

NMR spectroscopy allows a direct determination of the diffusion coefficient (*D*) from different techniques such as DOSY (diffusion ordered spectroscopy) [63], PFGSE (pulsed field gradient spin echo experiment) [64] PFGSTE (pulsed field gradient stimulated echo experiment) [65] PFGLED (pulsed field gradient longitudinal eddy current delay sequence) [66], or BPPLED (bipolar pulse pairs longitudinal eddy current delay sequence) [67]. These NMR experiments may offer information on the constitution of the system. In a tied host–guest system, the diffusion coefficient of the guest molecule should be exactly the same as for the host molecule; this perfect encapsulation of one or several pyrene molecules within a carceplex host molecule is not common, although examples of fully encapsulated guest molecules into host objects can be found in the literature [68,69].

DOSY NMR is a powerful tool for studying host–guest systems: an example of DOSY experiments of a 1:1 host–guest system built with an arene ruthenium metalla-rectangle and pyrene as guest molecule is shown in Fig. 13.6 [70]. It is clear from the DOSY spectra that the signals of the guest and of the host species are diffusing at almost the same coefficient (about 6.3×10^{-10} m^2/s) for the 1:1 mixture of pyrene and metalla-rectangle, while the increase of the pyrene ratio leads to a differentiation of the host and guest diffusion coefficients, along with a chemical shift of both the pyrene and the metalla-rectangle signals.

Usually, the diffusion coefficient observed for the guest molecule (D_{obs}) is the average molar fraction (X) of $D_{host\text{–}guest}$ and D_{guest} [71]:

$$D_{obs} = X_{pyrene} D_{pyrene} + X_{host\text{–}pyrene} D_{host\text{–}pyrene} \tag{13.13}$$

If diffusion experiments provide an experimental proof of the encapsulation of a guest molecule (such as pyrene) into the cavity of a host molecule, they can also be directly used for the determination of association constants. Such determination of stability constant by diffusion experiments was fully reviewed by Cohen and coworkers in 2005 and we would advise the readers to read this article before using diffusion NMR spectroscopy [72]. This method is now commonly employed in the study of host–guest interactions and in estimating the corresponding binding constants [73]. However, to the best of our knowledge no example has been presented so far in the literature with pyrene as guest molecule. Fielding reported in 2001 a prediction of the diffusion coefficients behavior of two idealized host–guest systems that can be useful to compare with experimental data, see Fig. 13.7 [71].

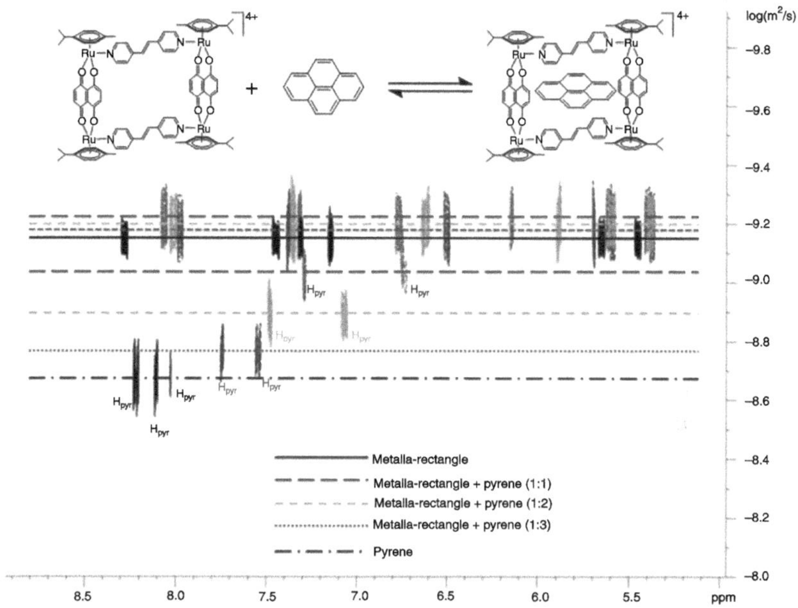

FIGURE 13.6 DOSY ^{1}H NMR spectra of pyrene, metalla-rectangle, and of additional amounts of pyrene (1 to 3 mol equiv) to a 4.0 mM solution of metalla-rectangle. *(Adapted from Ref. [70] with permission of WILEY-VCH.)*

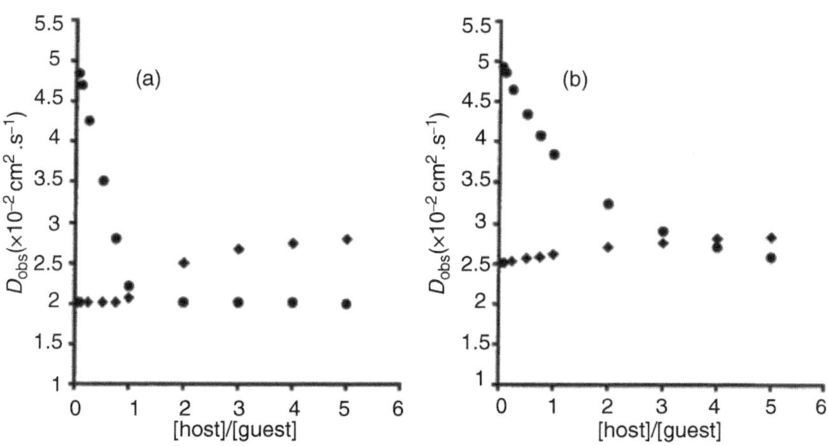

FIGURE 13.7 Predicted diffusion behavior of two idealized host–guest systems. (a) $K = 10^{5}$ M^{-1}; (b) $K = 5 \times 10^{2}$ M^{-1} (• guest or small molecule species; ♦ host or larger molecule). *(Reproduced from Ref. [71] with permission of American Chemical Society. Copyright 2001.)*

3 PYRENE AND ORGANIC SUPRAMOLECULAR HOSTS

3.1 A Brief History and Some Examples of Organic Host Species

Supramolecular chemistry is based on a "bottom-up" approach to designing molecular aggregates, or supramolecules, from organic molecules. Host–guest systems with high selectivity, efficiency, and responsiveness are key elements of such an approach and have played a major role in the development of innovative materials [74]. The landmark discovery [75] of the macrocycle polyethers and crown ethers by Pedersen in 1967 [6] offered the first generation of hosts in host–guest chemistry (Fig. 13.8) [76]. Crown ethers have found applications in artificial molecular machines [77], drug-delivery systems [78] and supramolecular polymers [79]. A remarkable property of such macrocycle polyethers is their ability to encapsulate guest molecules such as paraquat derivatives [80] and secondary ammonium salts [81], a feature that has been exploited for synthesizing the first pseudorotaxanes [82,83]. A pseudorotaxane consists of a threadlike molecule containing recognition sites and a macrocycle (eg, crown ether) that encircles the molecular thread. In order to prevent dethreading of the ring, the threadlike molecule may be terminated by bulky groups to give a rotaxane. These self-assembled rotaxane structures are themselves precursors to catenanes, mechanically interlocked molecular architectures consisting of two or more interlocked macrocycles.

The discovery of crown ethers by Pedersen was rapidly followed by the report by Lehn and coworkers of the preparation of the *N,N′*-diazamacrobicyclicpolyethers (cryptands) in 1969 (Fig. 13.8) [84]. Although the synthesis of a diazabicyclic "katapinand" by Park and Simmons in 1968 [85,86] is considered to be the first cryptand, the actual word "cryptand" was introduced by Lehn in 1969 to describe bicyclic ligands [87].

In the 1970s Cram began to develop cavitands [7]. Cavitands are structurally preorganized, which enhance the van der Waals contacts between the enforced cavities and the guest molecules. In 1983, Cram introduced the word "carceplex" to define molecular systems that can only release the entrapped guest by rupture of a covalent bond [88]. Calixarenes are macrocycles or cyclic oligomers based on the hydroxyalkylation product of a phenol and an aldehyde, and are a particular case of cavitands, which have been developed in the early 1980s (Fig. 13.8) [89].

Cyclodextrins (CD) are another widely used class of organic host molecules. They are composed of sugar molecules (ranging from 5 to more than 150 rings) bound together in a cyclic oligosaccharide fashion (Fig. 13.8b shows the molecular structure of a β-cyclodextrin). The most commonly used cyclodextrins contain 6-, 7-, and 8-membered sugar units and are named α-, β-, and γ-CD, respectively.

Comprehensively reviewing the pyrene-organic host systems that have been reported in the last decades is beyond the scope of this chapter. Furthermore, the interactions between pyrene and "classic" host organic

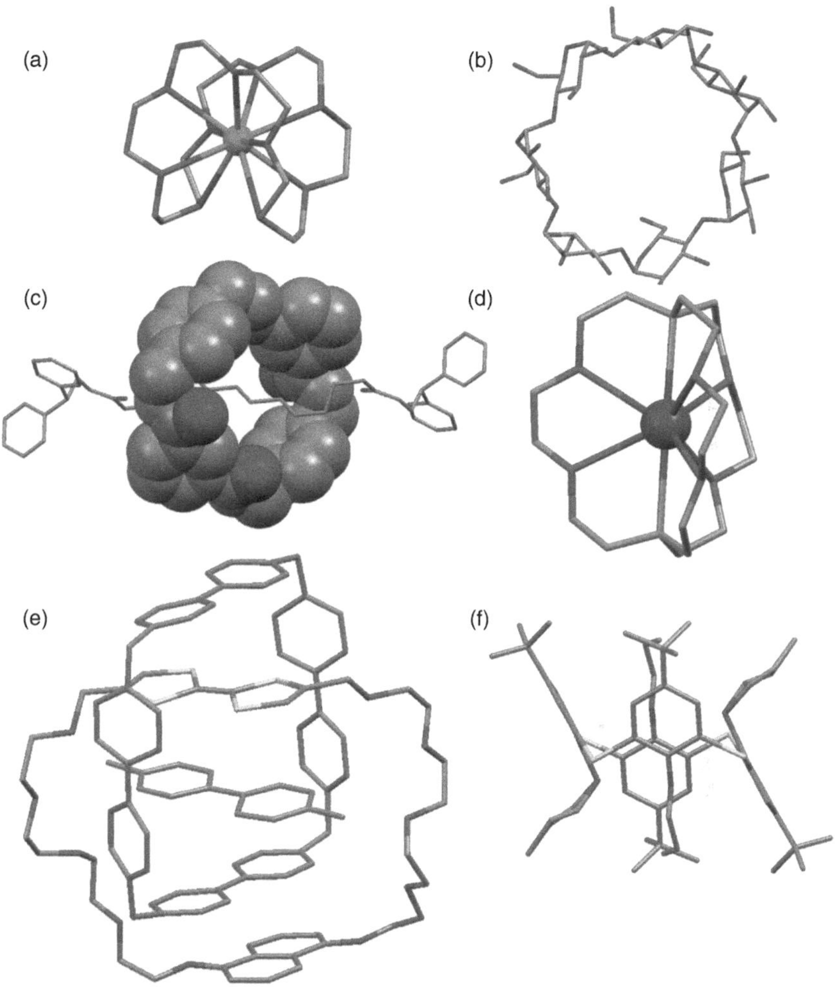

FIGURE 13.8 Molecular structures of some common organic host species. (a) Encapsulation of a Ba^{2+} ion in a 30-crown-10-ether. (b) Crystal structure of a β-cyclodextrin. (c) Crystal structure of a rotaxane. (d) Encapsulation of a K^{+} ion in a cryptand. (e) Crystal structure of a catenane. (f) Crystal structure of a thiacalix[4]arene. *(Part a: Adapted from Ref. [90], CCDC 638697. Part b: Adapted from Ref. [91], CCDC 762697. Part c: Adapted from Ref. [92], CCDC 213301. Part d: Adapted from Ref. [93], CCDC 241667. Part e: Adapted from Ref. [94], CCDC 917010. Part f: Adapted from Ref. [95], CCDC 193202)*

supramolecules, such as cyclodextrins and cucurbiturils, have been explored and rationalized as early as the 1980s [96,97] and we do not feel the need to expand on the matter (see Table 13.2 for some recent examples of host–pyrene systems). In the following sections, we will focus on selected examples in order to highlight some of the unique spectroscopic properties of pyrene,

TABLE 13.2 Some Recently Reported Pyrene-Organic Host Systems and the Techniques Used to Study the Related Host–Guest Equilibria, the Stoichiometries of the Systems, and the Binding Constants

Organic host	Spectroscopy	Stoichiometry	Binding constants	Years	References
Cucurbit[6]uril	Fluorescence	1:1 Linear double-reciprocal plot	Nonlinear fitting of the data (Eq. (13.10), Sigma Plot [44])	2011	[98]
Cucurbit[8]uril	Fluorescence	1:1 Job's plot	Isothermal titration calorimetry	2013	[99]
Heptakis-(6-amino)-(6-deoxy)- β-CD	Fluorescence	1:2 Job's plot	Nonlinearized version for Benesi–Hildebrand method	2002	[100]
γ-Cyclodextrin	Fluorescence	1:1, 1:2, 2:2 Fluorescence deconvolution	Fitting routine with the software Scientist (Micromath v.2.02) [46]	2003	[101]
β-Cyclodextrin	Fluorescence	1:2 Job plot	Benesi–Hildebrand double-reciprocal plot	2005	[102,103]
Polypeptide-grafted cyclodextrin bioconjugates	Fluorescence	1:1 and 1:2 Excimer/monomer emission bands ratio	nd[a]	2012	[104]
Cyclophane (ExBox)	^{1}H NMR	1:1 Job plot	Isothermal titration calorimetry	2014	[105]
Cyclophane (ExBox)	^{1}H NMR	1:1 (mixture)	NMR data fitting routine with the software Dynafit [106]	2012	[107]
Diphenyl disulfide cyclophane	Fluorescence	1:1 Job plot	Benesi–Hildebrand method	2013	[108]
Cryptand	Cyclic voltammetry	1:1	nd[a]	2014	[109]

[a]nd, not determined

along with host and inclusion complex structural, thermodynamic, and kinetic parameters that are easily accessible via the utilization of pyrene as a guest molecule.

3.2 Pyrene–Cavitand Interactions

Deep-cavity cavitands that dimerize into capsules via the hydrophobic effect, in the presence of a suitable guest molecule and in aqueous solution, have been developed by the Gibb group [110,111]. Such complexes possess hydrophilic outer coats, hydrophobic rims that favor self-assembly, and deep hydrophobic pockets (up to 1 nm wide to 2 nm long). They have been used to drive the formation of high-definition assemblies with a number of guest molecules, including steroid and hydrocarbon molecules [112]. Reactions within the capsule (eg, selective oxidation of substrates) and potential applications in hydrocarbon gas separation [111] have been also achieved or demonstrated [113].

Despite their elegant architectures and the promising properties of deep-cavity cavitands, the addition of cosolvents (needed for the encapsulation of guest molecules) weakens the hydrophobic effect and denatures these assemblies [114]. To provide information on the formation/dissociation dynamics of these deep-cavity cavitands host–guest complexes, Gibb, Bohne, and coworkers measured in real time the kinetics of the formation of 1:1 and 2:1 host–guest complexes formed between such a cavitand (so-called octa-acid (OA), (Fig. 13.9) and pyrene as guest molecule [115]. Upon encapsulation, the emission spectrum of pyrene is red shifted (5–6 nm) indicating either a pyrene excimeric formation or an interaction with the cavitand. However, the absence of the characteristic excimeric band at 480 nm in the emission spectra ruled out the excimer hypothesis and suggested monomeric pyrene encapsulation. To confirm this monomeric complexation, time-resolved fluorescence measurements were carried out. The fluorescence lifetime of the pyrene molecule was found to be longer in the presence of OA than in the absence (361 ± 1 ns vs. 130 ± 1 ns). Concurrently, the quenching rate constant for pyrene in the host was found to be dramatically lower than in free solution (aqueous borate buffer; $5.9 \times 10^5 M^{-1}s^{-1}$ vs. $1.0 \times 10^{5-1}s^{-1}$) [115]. This confirmed the entrapment of pyrene in the hydrophobic cavity of the cavitand. Interestingly, no other pyrene emission with intermediate lifetime was observed, indicating a transient nature of the 1:1 complex. Upon addition of a second OA cavitand, a similar iodide-quenching rate constant for singlet-excited-state pyrene was found.

The overall binding constant of the reaction between OA and pyrene was first calculated by binding isotherms obtained from steady-state fluorescence experiments. The overall reaction, shown in Fig. 13.9c, can be summarized by Eq. (13.14), a particular case of Eq. (13.1):

$$\text{Pyrene} + 2\text{OA} \overset{\beta_{21}}{\Leftrightarrow} \text{OA} - \text{Pyrene} - \text{OA} \qquad (13.14)$$

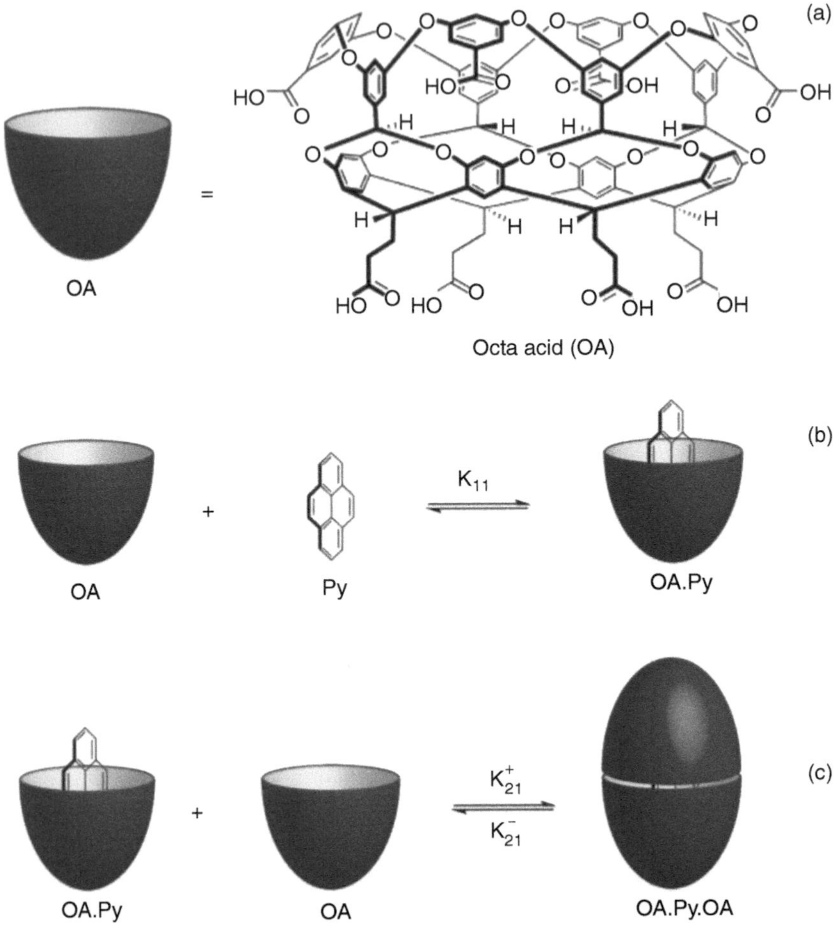

FIGURE 13.9 (a) Molecular structure of the deep-cavity cavitand OA. (b) Cartoon representations of the formation of the 1:1 OA-Py system and (c) of the 2:1 OA-Py-OA host–guest system. *(Adapted from Ref. [115] with permission of American Chemical Society. Copyright 2012.)*

To determine the overall binding constant (β_{21}), the data from the fluorescence titration of pyrene with OA were fitted numerically using the software Scientist 3 from Micromath [45], the independent variable being the total concentration of OA; the equilibrium concentrations of free OA, free pyrene, 2:1 complex, and the fluorescence emission intensity being defined as dependent variables. The resulting binding isotherm was found to fit the expected 2:1 binding model and gave an overall β_{21} value of $3.19 \pm 0.06 \times 10^{12}$ M^{-2}. Due to the transient nature of the 1:1 host–guest system, the binding constant (K_{11}) of the reaction shown in Fig. 13.9b could only be calculated from kinetic studies. Stopped-flow experiments demonstrated that the reaction between pyrene and

OA occurs within the first millisecond of the mixing time, and the dependence of the offset intensity change measured by stopped-flow with the concentration of OA revealed a 1:1 binding isotherm. The data were also fitted using the software Scientist 3, and a binding constant K_{11} of $4.5 \pm 0.6 \times 10^5\ M^{-1}$ was found.

This example of utilization of pyrene as guest molecule highlights some particularly interesting host–guest features:

- The high fluorescence emission of pyrene makes it a useful fluorescence probe to study multistepped host–guest reactions (in the previous example 1:1 and 2:1 additions), by time-resolved fluorescence, steady-state fluorescence, and fluorescence kinetic studies.
- The access of small molecules (eg, O^2 and I^-) to pyrene encapsulated in a host-cavity, without release of the entrapped pyrene, allows quenching-rate constant measurement for singlet-excited-state pyrene inside the capsule (in the previous example, such access occurred via fast opening/closing of capsular complexes).
- In host–guest chemistry, the guest usually dictates the kinetics of the association, particularly in the case of multistepped reactions.

Another remarkable property of the pyrene fluorescence is that a change of polarity in the vicinity of molecular pyrene leads to a dramatic modification of the intensity of the 0–0 vibronic band (I_1 peak), without affecting the intensity of the I_3 peak. The I_1/I_3 ratio of the pyrene emission spectrum is therefore an extensively used probe parameter to determine the micropolarity of the environment in which it resides.

This property of pyrene emission has been recently used to monitor the micropolarity of the interior of the deep-cavity cavitand OA [116]. In borate buffer, pyrene (10^{-5} M) exhibits a dual-emission resulting from monomeric emission (Fig. 13.10b; bands from 380 to 430 nm) and excimeric emission (broad, characteristic band at 480 nm). Addition of 2 mol equiv of OA to this solution leads to a dramatic displacement of the latter weak emission with an intense monomeric emission (Fig. 13.10a).

The excited-state S_1 lifetime of pyrene included in OA was found to be 350 ns, suggesting that pyrene resides as a single molecule within the OA capsule [117]. The I_1/I_3 ratio after addition of 2 mol equiv OA was found to be 1.01, similar to that of benzene, indicating a closed and compact capsular structure in which water molecules from the solvent are excluded.

3.3 Pyrene-containing Receptors: The Particular Case of Pillar[*n*]arenes

As shown in the previous sections, the unique photophysic properties of pyrene have made this molecule a guest of choice for investigating the hosting abilities of a number of molecular and supramolecular host systems. Pyrene has also

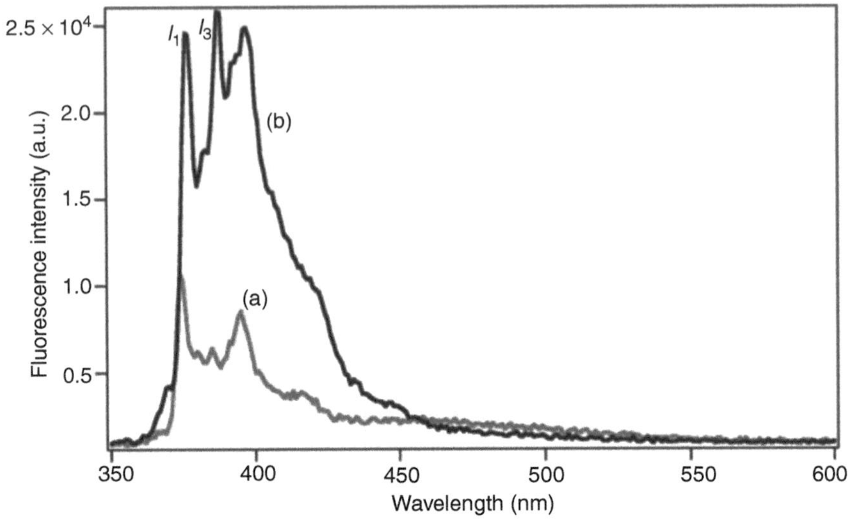

FIGURE 13.10 **Fluorescence emission spectra in borate buffer (pH ≈ 9) of (a) 1 × 10^{-5} Mpyrene and (b) pyrene + 2 mol equiv OA.** Excitation wavelength: 320 nm. *(Reproduced from Ref. [116] with permission of American Chemical Society. Copyright 2009.)*

been extensively used as a fluorescent probe for characterizing the microenvironmental conditions inside host cavities (eg, polarity, dielectric constant; see Section 5). One of the most striking facets of pyrene fluorescence is the wavelength associated with the excimer emission, generally located between 410 and 480 nm, and the intensity ratio between the excimer and monomer emission bands (I_E/I_M). These features are not only convenient for understanding conformational changes in host–pyrene recognition, but can also be of interest in the study of pyrene-containing-receptor/guest systems.

A number of examples of pyrene-containing receptors acting as molecular fluorescent sensors for the detection of cations and anions have been surveyed in two reviews by de Silva in 1997 [117] and Wallace in 2012 [118]. Although excellent, these two reviews do not cover pyrene-containing pillar[*n*]arene macrocycles. It is rare that chemists announce the arrival of a new class of macrocycles, and the report of pillar[*n*]arenes in 2008 came quite as a surprise [119]. The Lewis acid $BF_3{\cdot}OEt_2$ catalyzed condensation of 1,4-dimethoxybenzene offered the selective isolation of the symmetrical host 1,4-dimethoxypillar[5]arene, which was then deprotected with BBr_3 to yield the first pillar[5]arene host molecule (Fig. 13.11). Pillar[*n*]arenes are composed of *n* hydroquinone rings connected in their 2- and 5-positions by methylene bridges. The substitution pattern creates a cylindrical, or pillar-like structure, that has identical upper and lower rims. The facile synthesis of

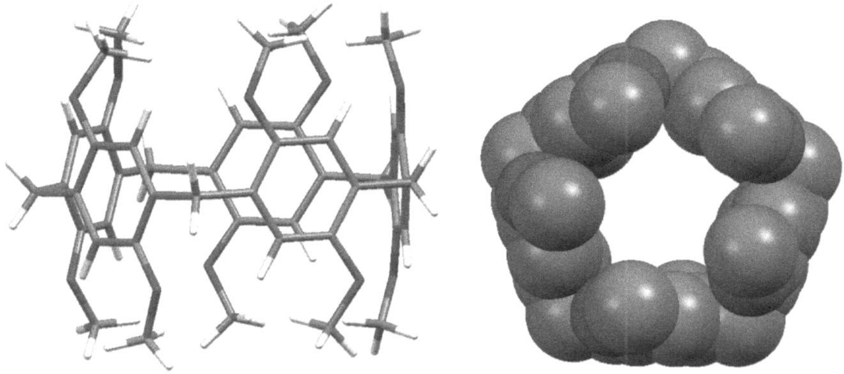

FIGURE 13.11 **Crystal structure of 1,4-dimethoxypillar[5]arene.** *(Adapted from Ref. [119], CCDC 692166.)*

pillar[*n*]arenes led to the report of a number of such molecules with *n* varying from 5 to 14 [120–123].

In the first example of a pillar[5]arene-based sensor, Stoddart reported a fluorescent copillar[5]arene that is quenched by alkanediamines binding (Fig. 13.12a) [124]. Some alkanediamines are banned substances by the World Anti-Doping Agency (eg, tuaminoheptane and methylhexanamine) [125], and their detection is of interest. Monofunctionalization of macrocycles for incorporating sensor abilities without altering the binding ability of the host is a known strategy [126]. Using a similar strategy, a pyrene-functionalized pillar-[5]arene derivative was synthesized, and its interactions with a series of alkanediamine guest molecules were investigated via fluorescence-quenching experiments in a MeCN/H_2O mixture. The 1:1 stoichiometry of the host–guest equilibria was determined by Job plot analyses. The complex between the pyrene-functionalized pillar-[5]arene (host) derivative and 1,8-diaminooctane was studied by fluorescence spectroscopy (total molar concentration = 248 μM), and the change in intensity multiplied by the mole fraction of host was plotted against the mole fraction of host (Fig. 13.12b).

Fluorescence titrations (Fig. 13.12d,e) were subsequently carried out by adding small aliquots (5 μL, 17 mM, 1:1MeCN/H_2O) of various potential guests into a solution of pyrene-functionalized pillar-[5]arene(57.2 μM, 1:1 MeCN/H_2O). The peak excitation λ (355 nm) was used, and the fluorescence intensity was monitored at the peak emission λ (395 nm). The association constant between the guests and pyrene-functionalized pillar-[5]arene was calculated using the software Dynafit [106], which is based on nonlinear least-squares regression analysis.

As shown in Fig. 13.12c, the binding constants for the C_3- to C_8-alkanediamines are all close to $10^4\ M^{-1}$. Interestingly, the smallest binding constants are

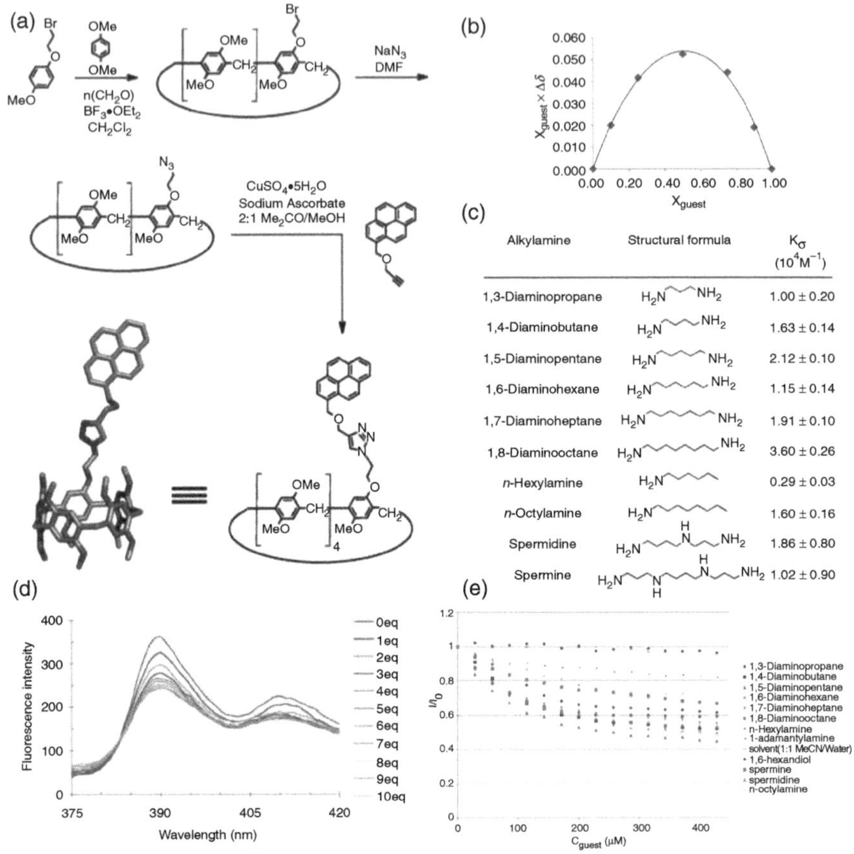

FIGURE 13.12 (a) Synthesis of pyrene-functionalized pillar-[5]arene. (b) Job plot between the host and 1,8-diaminooctane (guest) collected by measuring the fluorescence intensity of the host (1:1 MeCN/H_2O, excitation $\lambda = 355$ nm, emission $\lambda = 395$ nm) upon the change in the mole fraction of the host (X_{Host}). The plot indicates a 1:1 binding between the host and guest. (c) Association constants for pyrene-functionalized pillar-[5]arene and alkylamines complexation, calculated from fluorescence-quenching experiments. (d) Example of fluorescence-quenching experiments (57.2 μM, 1:1 MeCN/H_2O, $\lambda_{ex} = 355$ nm, $\lambda_{em} = 395$ nm). (e) Fluorescence titration of various amines and control compounds into a solution of host (57.2 μM, 1:1 MeCN/H_2O, excitation $\lambda = 355$ nm, emission $\lambda = 395$ nm). *(Adapted from Ref. [124] with permission of American Chemical Society. Copyright 2011.)*

observed for the mono terminal amine group alkylamines (*n*-hexylamine and *n*-octylamine), a fact that results from two possible orientations of the amine group of such molecules inside the cavity of the pyrene-functionalized pillar-[5]arene: either toward the pyrene molecule, or facing away from the pyrene moiety. Only the first orientation can lead to a quench of fluorescence. Furthermore, the loss of one amino group, compared to diamine derivatives, impairs the hydrogen bonding between —NH_2 groups and the oxygen atoms on the pillar rim, weakening the host–guest interactions.

4 PYRENE AND NANOSCALE HOSTS

4.1 Coordination-Driven Self-Assembly and Metalla-Assemblies

Metal complexes offer unique chemical diversities, which are quite distinct from organic compounds. This arises not only from the choice of the metal itself and its oxidation state, but also from the types and numbers of coordinated ligands and the coordination geometry of the resulting complex. Hence, coordination-driven self-assembly allows a directional-bonding approach that organizes the different building blocks following the coordination modes of the transition metal [127]. In addition to being highly directional, covalent metal–ligand bonds are also usually stronger than most of the weaker nondirectional interactions that are used in organic supramolecular chemistry (eg, hydrogen bonds, van der Waals, and hydrophobic interactions < 40 kJ mol^{-1}), which offers a unique scaffold for designing robust and rigid self-assembled metal-organic frameworks or assemblies with well-defined shapes, sizes, and geometries [128].

The structure of an inorganic metalla-assembly (discrete two-dimensional and three-dimensional polygonal and polyhedral coordination assembly) is dictated, first, by the stereochemistry of the metal (electron-poor centers), and second by the geometry of the ligands (electron-rich organic donors) connecting the metal centers. The shape of the ligands is mainly dominated by the turning angles (defined as the angles formed between the coordinating sites of a ligand, Fig. 13.13). A linear ditopic ligand, such as 4,4′-bipyridine (bpy),

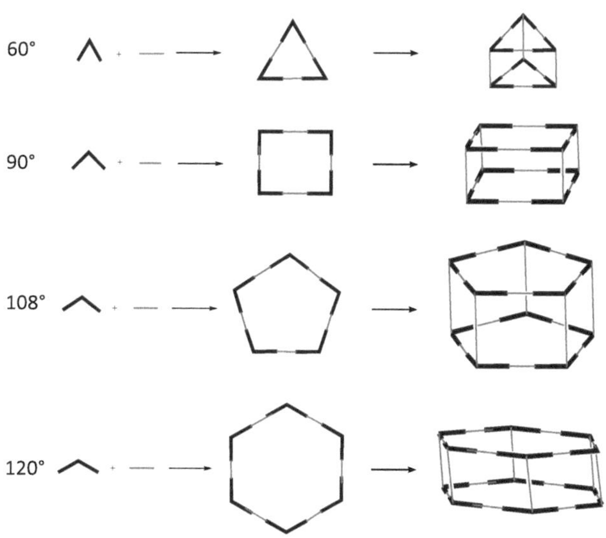

FIGURE 13.13 Some 2D and 3D structures isolable from ligands with different turning angles.

will have its two sites of coordination oriented at 180° from each other, while a ditopic building block such as 3,5-bis(pyridin-4-yl)-benzene (bpb) has a turning angle of 120° between its two pyridyl donor sites. By extension of this directional-bonding approach, building blocks with more than two coordination sites allow the design of various geometries. For example, the 2,4,6-tri(pyridin-4-yl)-1,3,5-triazine (tpt) ligand or the 5,10,15,20-tetra(pyridin-4-yl)porphyrin (tpp-2H) have an angle of 120 and 90° [115], respectively, between the adjacent pyridyl donor sites.

Over the last two decades, the groups of Fujita [129], Stang [130], Raymond [131], Lehn [132], and Mirkin [133] have pioneered the use of the directional-bonding coordination-driven approach to self-assemble a number of inorganic and organometallic architectures, and pyrene has been extensively used to probe the hosting abilities of such metalla-assemblies.

When working with discrete coordination-driven self-assemblies, standard analytical techniques are generally available, unless dealing with paramagnetic compounds (NMR unfriendly) or poorly soluble materials. Therefore, the techniques used to characterize organic host–guest systems have been also exploited to study the guest sorption of pyrene in metalla-assemblies. However, pyrene can be more than just a guest; it can serve as a template to help the formation of a specific host during the self-assembly process. This template effect has been nicely demonstrated by Fujita with a pyrazine-pillared metalla-cage [134] and more recently by Vajpayee, Stang, and Chi with a metalla-rectangle [135]. Indeed, the discrete metalla-rectangle was only obtained by addition of pyrene; otherwise an interlocked structure was isolated, with no possibility of returning back to the unlocked system. Nevertheless, pyrene was also used as standard guest molecule for several metalla-assemblies.

As shown in Section 2.4, DOSY ^{1}H NMR experiments were used to characterize a series of metalla-rectangles [70]. The same technique has been employed to confirm the encapsulation of pyrene in a series of Fe_4L_6 cages (L = diaminoterphenylene) [136], and in the metalla-prisms of the general formula $[(p\text{-cymene})_6Ru_6(3\text{-tpt})_2(OO\cap OO)_3]^{6+}$ (3-tpt = 2,4,6-tris (3-pyridyl)-1,3,5-triazine; $OO\cap OO$ = dioxido-naphthoquinonato ligands) [137]. Analogous metalla-prisms have been synthesized by Jin, $[(Cp^*)_6Ir_6(4\text{-tpt})_2(NC\cap CN)_3]^{6+}$(4-tpt = 2,4,6-tris(4-pyridyl)-1,3,5-triazine; $NC\cap CN$ = terephthal-bis-aromatic imine ligands), and the single-crystal X-ray structure analysis of a metalla-prism with pyrene sitting inside the cavity has been obtained (Fig. 13.14) [138].

Two pyrene molecules have been stacked inside the hydrophobic cavity of a palladium-based metalla-prism constructed from two 4-tpt panels, three 2,2′,6,6′-tetramethyl-4,4′-bipyridine (Me_4-bpy) pillars and six Pd(en) corners (Fig. 13.15) [139]. Inside the cavity, π–π stacking interactions between the pyrenes as well as between the pyrene and the 4-tpt panels are observed, the separation between the planar aromatic molecules being around 3.3 Å.

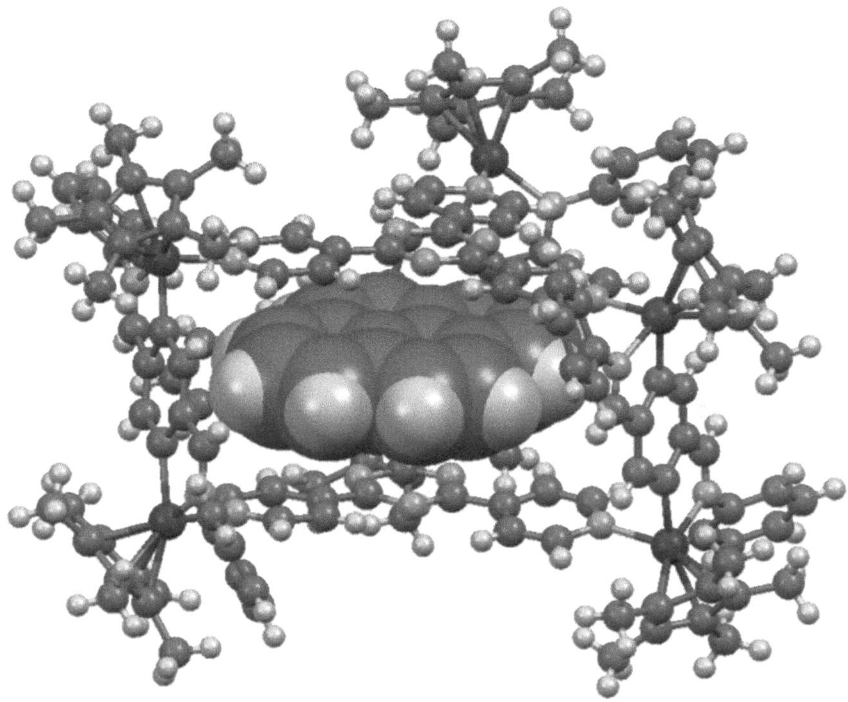

FIGURE 13.14 Crystal structure of pyrene⊂[(Cp*)$_6$Ir$_6$(4-tpt)$_2$(NC∩CN)$_3$]$^{6+}$. *(Adapted from Ref. [138], CCDC 909393)*

Interestingly, a water-soluble octahedral cage of the formula [(en)$_6$Pd$_6$(4-tpt)$_3$]$^{12+}$ was able to encapsulate pyrene together with a molecule of *N*-cyclohexylmaleimide [140]. Upon heating at 100°C for 24 h, a [2 + 2] cycloaddition reaction occurred between the two guest molecules, thus giving exclusively the endo Diels–Alder adduct. The confinement of the two guest molecules in the hydrophobic cavity controlled the stereochemistry of the product, thus to some extent mimicking enzymes' reaction pockets.

4.2 Metal-Based Nanoparticles and Metal Organic Frameworks

Metal-based nanoparticles are extremely popular because they show a wide range of applications in different fields [141]. Among the different properties of nanoparticles, their loading capacity is an important one. Indeed, knowing the amount of guest molecules trapped in a nanoparticle core and how strongly bound the guest molecules are to this nanoparticle are two essential parameters to estimate before using metal-based nanoparticles as drug-delivery vectors.

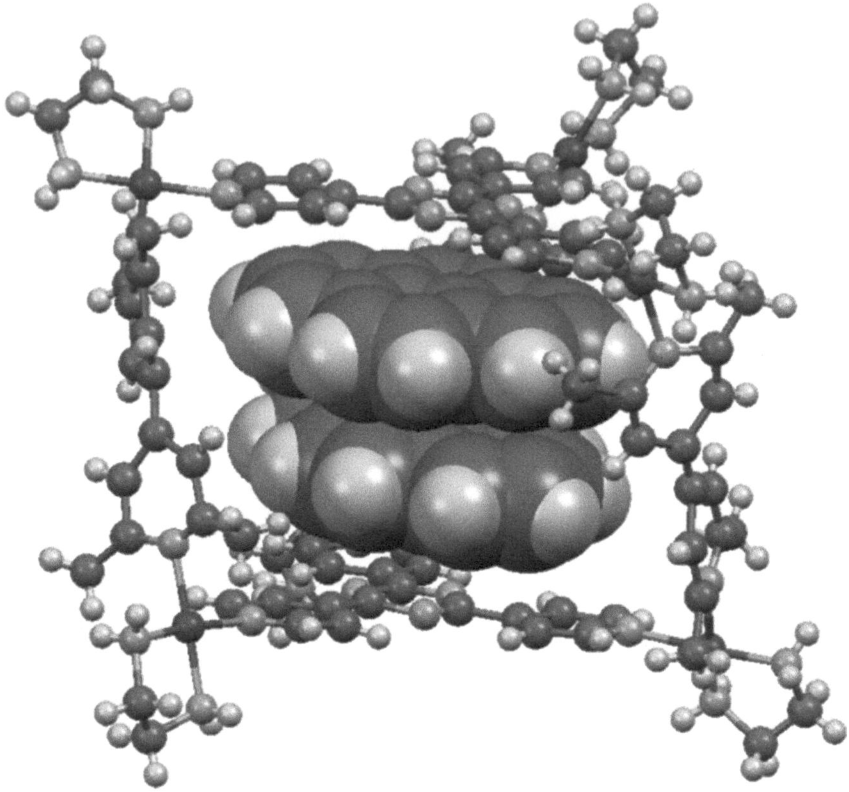

FIGURE 13.15 Crystal structure of (pyrene)$_2$⊂[(en)$_6$Pd$_6$(4-tpt)$_2$(Me$_4$-bpy)$_3$]$^{12+}$. *(Adapted from Ref. [139], CCDC 251181)*

Several techniques can be used to determine the loading capacity of nanoparticles [142] and we want here to give only a few examples in which pyrene was incorporated in the core of the metal-based nanoparticles. As opposed to discrete metalla-assemblies, nanoparticles can incorporate several guest molecules, thus giving more qualitative answers than quantitative values. For example, confocal-laser scanning microscopy has been used to follow the uptake of pyrene in nanoparticles, exploiting the fluorescence of pyrene [143]. The diameter of the fluorescence and the intensity gave an excellent approximation of the size and shape of the nanoparticles, with strong intensity being observed at the core of the nanoparticle and weak fluorescence at the outer surface. Fluorescence microscopy was also used to study zinc-nickel-ferrite nanoparticles coated with polypyrrole/polystyrene arms [144]. The size of the nanoparticles was determined, as well as the relative polarity of the inner space of the nanoparticles. The low I_1/I_3 band ratio (1.09) suggested a hydrophobic environment inside the core of the nanoparticles and the absence of water or methanol molecules within the nanoparticles.

Metal-based nanoparticles decorated with calix[4]arene receptors to sense polycyclic aromatic hydrocarbons (PAHs) have been developed by Sanchez-Cortes [145]. Each combination of host and PAH (pyrene, triphenylene, benzo[*c*]phenanthrene, coronene) gave a different signature in their surface-enhanced Raman scattering spectra, thus offering valuable information on the interaction between the calix[4]arene receptors and the PAH molecules.

Metal organic frameworks (MOFs) have been found to have industrial applications [146], among them the solid-phase extraction of polycyclic aromatic hydrocarbons (PAHs) offers an alternative to water decontamination. Pyrene is one of these PAHs that can be found as a pollutant in contaminated water. Therefore, MOFs able to detect and/or extract pyrene from water are highly desirable, and some examples of MOFs designed for the detection and extraction of pyrene can be found in the literature [147,148]. This proves one more time that pyrene is a guest of honor.

4.3 Nanoparticles Made of Polymers and Micelles

Nanoscience is a recent and exciting area of research that attracts the imagination of a large body of students, scientists, and engineers [149–151]. It also provides innovative solutions for addressing critical societal and economic issues and impact on other fields, such as medicine [152] and energy [153,154]. Nanotechnology offers unique tools to develop nanoscience and to investigate the engineering and manufacturing of materials at the atomic and molecular scales [155].

Among these nanotechnology tools, nanoparticles made of polymers (NPs) have raised considerable attention. NPs encompass a variety of nanoarchitectures, such as micelles, vesicles, multimicellar clusters, and hydrogels (Fig. 13.16) [156–162]. NPs are of particular interest as drug-delivery vectors, owing to their synthetic versatility, as well as their tunable properties (eg, thermosensitivity and pH-response). No sooner had Gref and coworkers demonstrated that nanospheres synthesized from amphiphilic copolymers composed of two biocompatible blocks (including a polyethylene glycol(PEG)-block) exhibited dramatically increased blood circulation times and low liver accumulation in mice in a seminal paper in 1994 [163], that numerous reports of PEGylated NPs with various architectures followed [164]. This remarkable research effort eventually led to PEG being listed as "generally recognized as safe" by the FDA, and to the clinical translation of a number of NPs [165]. Among them, the biodegradable poly(ethylene glycol)-poly(D,L-lactide) copolymer self-assembled into micellar NPs [Fig. 13.16 for the molecular structures

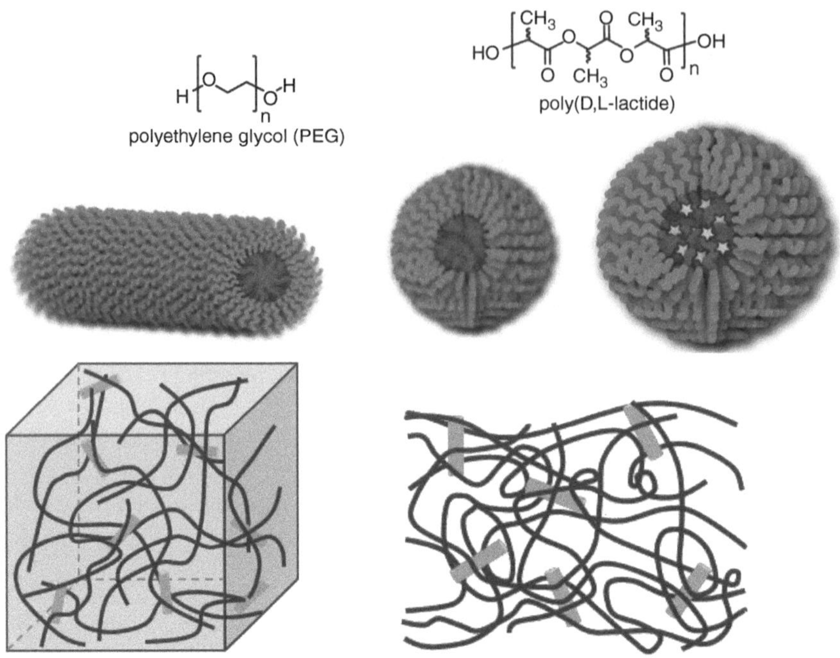

FIGURE 13.16 Molecular structures of PEG and poly(D,L-lactide), and cartoon representations of some typical self-assemblies in solution or in bulk.

of polyethylene glycol (PEG) and poly(D,L-lactide)], which can entrap paclitaxel, an organic drug used for the treatment of various cancers, including lung, ovarian, and breast. This formulation, named Genexol-PM, has been approved by the FDA for use in patients with breast cancer.

In this context, understanding the loading capabilities of newly synthesized nanoparticles made of polymers is essential, not only to control the self-assembly process, but also to understand the stimuli-responsive disassembly and the release of the cargo to a specific target. It was therefore a necessity for researchers in the nanotechnology field to look for a family of guest molecules allowing them, first, to study the physical and chemical parameters of their nanoarchitectures, second, to investigate the guest-loading behavior of these, and third, to evaluate and compare together such nanoassemblies. Pyrene and pyrene derivatives, possessing unique and well-understood chemical and physical properties, and having been used to spectroscopically probe a range of organic and inorganic molecular and supramolecular hosts, were therefore a suitable choice for such a family of guest candidates.

The encapsulation abilities of a number of micellar aggregates were tested using pyrene as a hydrophobic spectroscopic probe. A recent example is the investigation of the pH-regulated swelling and sustained release of pyrene

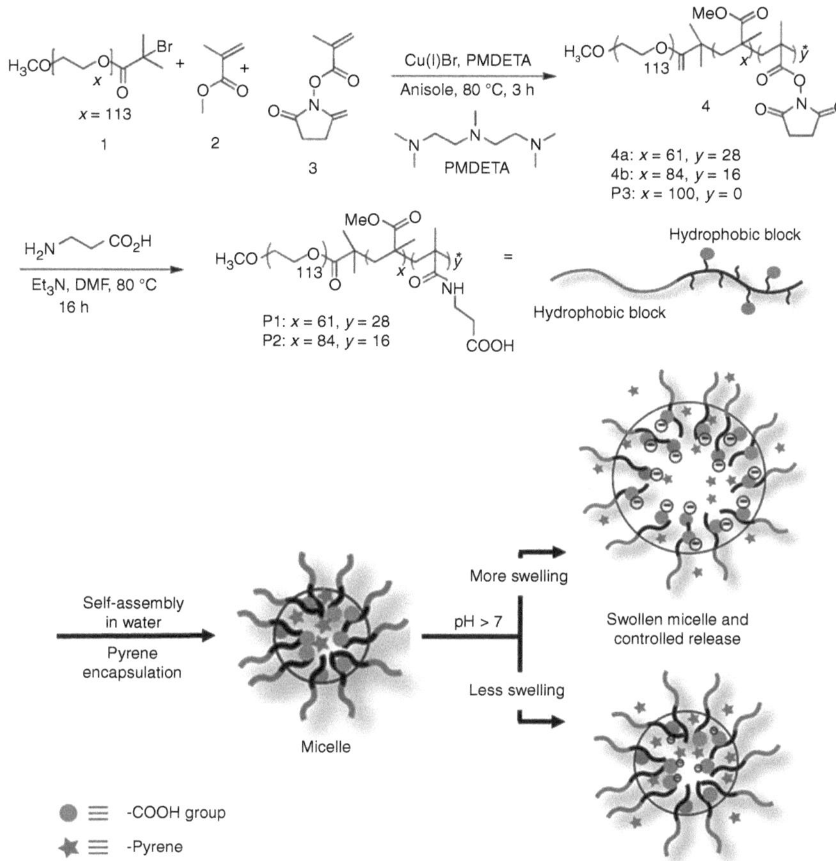

FIGURE 13.17 Schematic representation of P1 micellization and pH-dependent tunable swelling of block copolymer micelles. *(Adapted from Ref. [166] with permission of American Chemical Society. Copyright 2013.)*

from the core of functionalized poly-(ethyleneglycol)monomethylether-block-poly-(methylmethacrylate-*co*-methacryloxysuccinimide) (PEO-*b*-PMMA-*co*-PMASI) block copolymer micelles P1 (Fig. 13.17) [166].

Taking advantage of the poor solubility of pyrene in water, the authors studied the emission spectra of pyrene (10^{-6} M concentration) dissolved in aqueous solutions of block copolymer P1 (spectroscopically silent) at different concentrations. The ratio of the emission intensities of the first (I_1) and the third (I_3) pyrene vibronic bands allowed the determination of the dielectric constant of the micellar core (by comparison with the I_1/I_3 ratio in dichloromethane). Furthermore, the extinction coefficient of pyrene at 337 nm offered an estimation of the concentration of the encapsulated pyrene. The critical aggregation concentration (CAC) was also deduced from the I_1/I_3 ratio.

To investigate the pH-sensibility of P1 micellar aggregation, dynamic light scattering (DLS) experiments were performed in a pH range of 5.4–10.0. Interestingly, a reversible sevenfold increase in the hydrodynamic diameter (D_h) was observed (from ca. 20 to 142 nm) above pH = 6. To prove that this change of micellar size is due to pH-triggered micellar swelling rather than disassembly and disorganization in the micellar architecture, a naphthalene derivative donor was coencapsulated with the pyrene acceptor molecule, and Förster resonance energy transfer experiments were carried out in the same range of pH. Upon pH increase, the pyrene emission intensity gradually decreased, suggesting that the naphthalene derivative and the pyrene are allowed to distribute themselves over larger volume compared to that in the compact micelle at lower pH, and therefore demonstrating the swelling hypothesis.

This elegant study highlights three important characteristics of pyrene that can be efficiently used for studying micellar aggregation and guest-triggered pH release:

- The fine and resolved vibronic structure of the pyrene emission spectrum coupled with the low water-solubility of this aromatic molecule, make pyrene a cheap, efficient, and polymer-cost effective fluorescent probe for characterizing micellar cores (eg, dielectric constants).
- The high sensitivity of fluorescence spectroscopy and the strong emission of pyrene between 365 and 425 nm allow a direct estimation of the CAC.
- The modification of the micellar core size (eg, swelling) can be easily monitored by studying the pyrene emission.

Another remarkable feature of the pyrene fluorescence properties is that an increase in the pyrene environment polarity results in a dramatic increase in the intensity of the 0–0 vibronic band (I_1 peak), while the I_3 peak is solvent insensitive. Thus, a transition from a nonpolar to a polar environment results in a dramatic increase of the I_1/I_3 ratio in emission spectrum [167].

This property of pyrene emission has been recently used for indirectly investigating the effect of drug hydrophobicity and solution temperature on the core/shell architecture of Pluronic® F127 micelles [167]. Pluronic F127 is a ABA triblock copolymer belonging to a pool of more than 50 amphiphilic, water-soluble and polymorphic materials [(A = hydrophilic block poly(ethylene oxide), or PEO), and B = hydrophobic block poly(propylene oxide), or PPO, (Fig. 13.18)], commercially available as Pluronic (nonproprietary name "poloxamers") [168]. The physical and chemical properties of Pluronic copolymers can be finely tuned by modifying the molar mass ratio between the PEO and PPO blocks (from 1:9 to 8:2), which directly modifies the in vivo properties and interactions with cells and cell membranes, and provides high potential for the design of innovative nanomedicines and new biomaterials. The properties that make Pluronic block copolymers unique nanotechnology

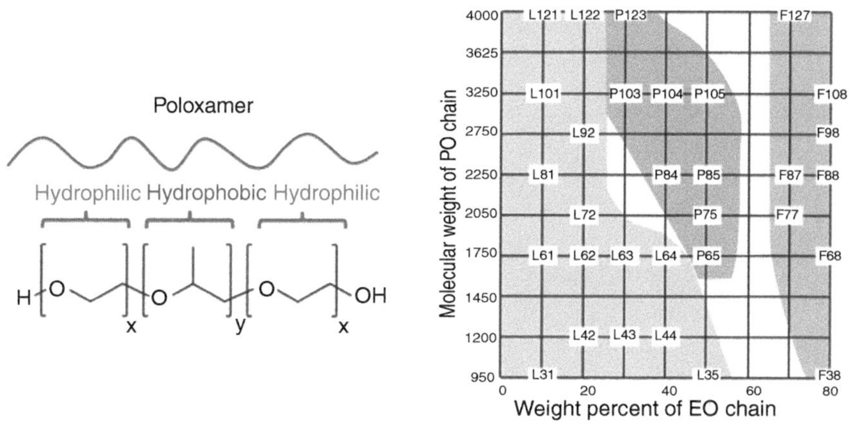

FIGURE 13.18 Molecular structure of a Pluronic triblock copolymer (left) and the Pluronic grid (right; color code: physical state of copolymers under ambient conditions: light gray (green in the web version) = liquid; dark gray (red in the web version) = paste; gray (orange in the web version) = flake). *(Adapted from Ref. [170] with permission of Royal Society of Chemistry.)*

tools and some areas of the current interest in their biochemistry have been recently reviewed [169,170].

Pyrene molecules are hydrophobic, and thus are expected to be fully encapsulated in the PPO cores of Pluronic micelles [167], which may offer valuable information on the degree of hydration of the micellar core. Using pyrene as a fluorescent probe (excitation at $\lambda = 339$ nm), the variation in the I_1/I_3 vibronic emission ratio upon gradual decrease of the solution temperature (from 60 to 10°C) in the absence or presence of coencapsulated ibuprofen (0.2 wt%), aspirin (0.5 wt%), and erythromycin (0.05 wt%) was monitored (Fig. 13.19).

In the absence of ibuprofen, aspirin, and erythromycin, the I_1/I_3 pyrene emission ratio increases with a decrease in temperature, suggesting an enhancement in the polarity (aqueous content) of the micellar core at low temperature. This change in the polarity environment of the micellar core eventually results in the release of the pyrene molecules at 8 °C via rupture of the core-shell micelle architecture, and disassembly of the structure in monomers. Interestingly, at higher drug-micelle loading, the I_1/I_3 pyrene emission ratio does not dramatically increase at low temperature, indicating a strong stabilization of the core-shell structure by the hydrophobic interactions between the ibuprofen, aspirin, or erythromycin guest molecules and the PPO micellar cores. This study demonstrates the potential of pyrene fluorescence emission for providing indirect information on the structural stability of core-shell micelles.

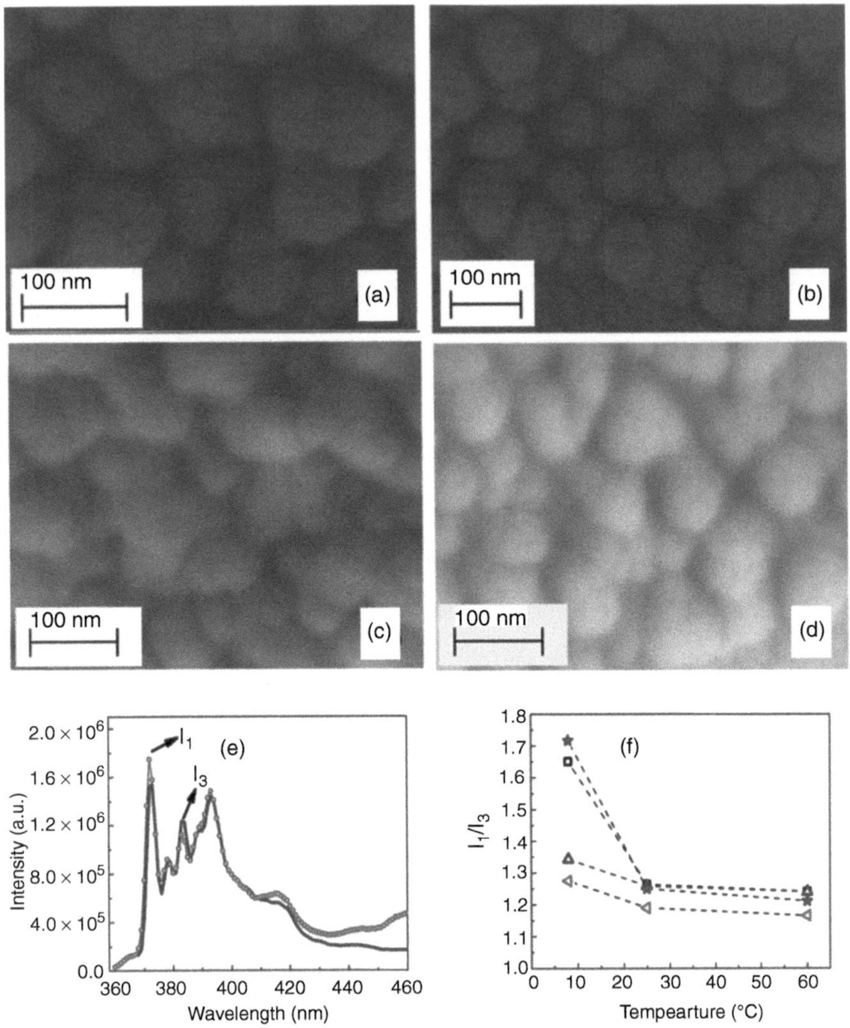

FIGURE 13.19 Cryo-SEM images of 5.26 wt% F127 micelles (a) with no encapsulated drugs and when (b) 0.1 wt% ibuprofen, (c) 0.1 wt% aspirin, and (d) 0.1 wt% erythromycin are incorporated in the micellar cores. (e) Pyrene emission spectra of 5.26 wt% F127 solutions at 40°C (solid line) and 8°C (circleline). (f) Variations of the I_1/I_3pyrene emission ratio upon temperature when no drug (squares), 0.2 wt% ibuprofen (up-triangles), 0.5 wt% aspirin (left-triangles), and 0.05 wt% erythromycin (stars) are incorporated in F127 solutions. *(Adapted from Ref. [167] with permission of American Chemical Society. Copyright 2013.)*

5 CONCLUSIONS AND CLOSING REMARKS

Nowadays, host compounds come from all different fields of chemistry, and the development of host compounds has paralleled the development of spectroscopic techniques. However, to be able to compare the hosting abilities of a new host with those of previously published compounds, it is essential to use a standard guest molecule. A literature survey clearly shows that pyrene has been systematically used to estimate the binding constants of newly designed host materials. Without a doubt, pyrene is a remarkable guest molecule. The spectroscopic features of pyrene and its availability have certainly contributed to its notoriety, and selected examples have emphasized the effectiveness of using pyrene as a primary guest molecule. In our opinion, pyrene was, and remains, the guest of honor when studying host–guest interactions.

ACKNOWLEDGMENTS

We thank the Leverhulme Trust (Early Career Fellowship ECF-2013-414 to NPEB), the Swiss National Science Foundation, and the University of Neuchatel for financial support.

REFERENCES

[1] Fischer E. Einfluss der configuration auf die wirkung der enzyme. Ber Dt Chem Ges 1894;27:2985.
[2] Kunz H. Emil Fischer—unequalled classicist, master of organic chemistry research, and inspired trailblazer of biological chemistry. Angew Chem Int Ed 2002;41:4439–51.
[3] Cram DJ, Cram JM. Design of complexes between synthetic hosts and organic guests. Acc Chem Res 1978;11:8–14.
[4] Cram DJ, Cram JM. Host-guest chemistry: complexes between organic compounds simulate the substrate selectivity of enzymes. Science 1974;183:803–9.
[5] Lehn J-M. Supramolecular chemistry—scope and perspectives molecules, supermolecules, and molecular devices (Nobel lecture). Angew Chem Int Ed 1988;27:89–112.
[6] Pedersen CJ. Cyclic polyethers and their complexes with metal salts. J Am Chem Soc 1967;89:7017–36.
[7] Moran JR, Karbach S, Cram DJ. Cavitands: synthetic molecular vessels. J Am Chem Soc 1982;104:5826–8.
[8] Wu JI, Dobrowolski MA, Cyrański MK, Merner BL, Bodwell GJ, Mo Y, Schleyer PvR. On the aromatic stabilization energy of the 4N π electron pyrene. Mol Phys 2009;107:1177–86.
[9] Hammonds M, Pathak A, Sarre PJ. TD-DFT calculations of electronic spectra of hydrogenated protonated polycyclic aromatic hydrocarbon (PAH) molecules: implications for the origin of the diffuse interstellar bands? Phys Chem Chem Phys 2009;11:4458–64.
[10] Randić M. Aromaticity of polycyclic conjugated hydrocarbons. Chem Rev 2003;103: 3449–606.
[11] Baba M, Saitoh M, Kowaka Y, Taguma K, Yoshida K, Semba Y, Kasahara S, Yamanaka T, Ohshima Y, Hsu Y-C, Lin SH. Vibrational and rotational structure and excited-state dynamics of pyrene. J Chem Phys 2009;131:224318.
[12] Moss GP. Nomenclature of fused and bridged fused ring systems. Pure & Appl Chem 1998;70:143–216.
[13] Thonhauser T, Ceresoli D, Marzari N. NMR shifts for polycyclic aromatic hydrocarbons from first-principles. Int J Quantum Chem 2009;109:3336–42.
[14] Vullev VI, Jiang H, Jones G. In: Geddes CD, Lakowicz JR, editors. Advanced concepts in fluorescence sensing: macromolecular sensing, vol. 10. New York: Springer Science + Business Media; 2005. p. 211.

[15] Becker RS, Singh IS, Jackson EA. Comprehensive spectroscopic investigation of polynuclear aromatic hydrocarbons. I. absorption spectra and state assignments for the tetracyclic hydrocarbons and their alkyl-substituted derivatives. J Chem Phys 1963;38:2144–71.

[16] Tanaka J. The electronic spectra of pyrene, chrysene, azulene, coronene and tetracene crystals. Bull Chem Soc Jpn 1965;38:86–102.

[17] Mangle EA, Topp MR. Excited-state dynamics of jet-cooled pyrene and some molecular complexes. J Phys Chem 1986;90:802–7.

[18] Kerkines ISK, Petsalakis ID, Theodorakopoulos G, Klopper W. Low-lying absorption and emission spectra of pyrene, 1,6-dithiapyrene, and tetrathiafulvalene: A comparison between ab initio and time-dependent density functional methods. J Chem Phys 2009;131:224315.

[19] Bito Y, Shida N, Toru T. Ab initio MRSD-CI calculations of the ground and the two lowest-lying excited states of pyrene. Chem Phys Lett 2000;328:310–5.

[20] Okamoto A, Kanatani K, Saito I. Pyrene-labeled base-discriminating fluorescent DNA probes for homogeneous SNP typing. J Am Chem Soc 2004;126:4820–7.

[21] Langenegger SM, Haner R. Excimer formation by interstrand stacked pyrenes. Chem Commun 2004;2792–3.

[22] Otsubo T, Aso Y, Takimiya K. Functional oligothiophenes as advanced molecular electronic materials. J Mater Chem 2002;12:2565–75.

[23] Jia W-L, Cormick TMc, Liu Q-D, Fukutani H, Motala M, Wang R-Y, Tao Y, Wang S. Diarylamino functionalized pyrene derivatives for use in blue OLEDs and complex formation. J Mater Chem 2004;14:3344–50.

[24] Sienkowska MJ, Farrar JM, Zhang F, Kusuma S, Heiney PA, Kaszynski P. Liquid crystalline behavior of tetraaryl derivatives of benzo[c]cinnoline, tetraazapyrene, phenanthrene, and pyrene: the effect of heteroatom and substitution pattern on phase stability. J Mater Chem 2007;17:1399–411.

[25] Sienkowska MJ, Monobe H, Kaszynski P, Shimizu Y. Photoconductivity of liquid crystalline derivatives of pyrene and carbazole. J Mater Chem 2007;17:1392–8.

[26] Kim YH, Yoon DK, Lee EH, Ko YK, Jung H-T. Photoluminescence properties of a perfluorinated supramolecular columnar liquid crystal with a pyrene core: effects of the ordering and orientation of the columns. J Phys Chem B 2006;110:20836–42.

[27] Kamikawa Y, Kato T. One-Dimensional chiral self-assembly of pyrene derivatives based on dendritic oligopeptides. Org Lett 2006;8:2463–6.

[28] de Halleux V, Calbert JP, Brocorens P, Cornil J, Declercq JP, Brédas JL, Geerts Y. 1,3,6,8-Tetraphenylpyrene Derivatives: towards fluorescent liquid-crystalline columns? Adv Funct Mater 2004;14:649–59.

[29] Percec V, Glodde M, Bera TK, Miura Y, Shiyanovskaya I, Singer KD, Balagurusamy VSK, Heiney PA, Schnell I, Rapp A, Spiess HW, Hudson SD, Duan H. Self-organization of supramolecular helical dendrimers into complex electronic materials. Nature 2002;417:384–7.

[30] Maeda H, Maeda T, Mizuno K, Fujimoto K, Shimizu H, Inouye M. Alkynylpyrenes as improved pyrene-based biomolecular probes with the advantages of high fluorescence quantum yields and long absorption/emission wavelengths. Chem Eur J 2006;12:824–31.

[31] Venkataramana G, Sankararaman S. Synthesis and spectroscopic investigation of aggregation through cooperative $\pi-\pi$ and C−H···O Interactions in a Novel Pyrene Octaaldehyde Derivative. Org Lett 2006;8:2739–42.

[32] Kim HM, Lee YO, Lim CS, Kim JS, Cho BR. Two-photon absorption properties of alkynyl-conjugated pyrene derivatives. J Org Chem 2008;73:5127–30.

[33] Leroy-Lhez S, Fages F. Synthesis and photophysical properties of a highly fluorescent ditopic ligand based on 1,6-bis(ethynyl)pyrene as central aromatic core. Eur J Org Chem 2005;2005:2684–8.

[34] Birks JB. Photophysics of aromatic molecules. London: Wiley-Interscience; 1970.
[35] Diring S, Camerel F, Donnio B, Dintzer T, Toffanin S, Capelli R, Muccini M, Ziessel R. Luminescent ethynyl–pyrene liquid crystals and gels for optoelectronic devices. J Am Chem Soc 2009;131:18177–85.
[36] Sagara Y, Mutai T, Yoshikawa I, Araki K. Material design for piezochromic luminescence: hydrogen-bond-directed assemblies of a pyrene derivative. J Am Chem Soc 2007;129: 1520–1.
[37] Song JH, Sailor MJ. Quenching of photoluminescence from porous silicon by aromatic molecules. J Am Chem Soc 1997;119:7381–5.
[38] Pandey S, Ali M, Bishnoi A, Azam A, Pandey S, Chawla HM. Quenching of pyrene fluorescence by calix[4]arene and calix[4]resorcinarenes. J Fluoresc 2008;18:533–9.
[39] Munoz de la Pena A, Ndou T, Zung JB, Warner IM. Stoichiometry and formation constants of pyrene inclusion complexes with *beta*- and *gamma*-cyclodextrin. J Phys Chem 1991;95:3330–4.
[40] Chen Y, Luo J, Zhu XX. Fluorescence study of inclusion complexes between star-shaped cholic acid derivatives and polycyclic aromatic fluorescent probes and the size effects of host and guest molecules. J Phys Chem B 2008;112:3402–9.
[41] Pistolis G, Malliaris A. Evidence for highly selective supramolecular formation between perylene/γ-CD and pyrene/γ-CD complexes in water. J Phys Chem B 2004;108:2846–50.
[42] Goodrich JA, Kugel JF. Binding and kinetics for molecular biologists. New York: Cold Spring Harbour Laboratory Press; 2007.
[43] Thordarson P. Binding constants and their measurement. Supramolecular chemistry: from molecules to nanomaterials. Hoboken: John Wiley & Sons; 2012.
[44] http://www.sigmaplot.co.uk/products/sigmaplot/sigmaplot-details.php.
[45] http://www.scientificsoftware-solutions.com/product.php?productid=17664.
[46] http://www.micromath.com/.
[47] Tsukube H, Furuta H, Odani A, Takeda Y, Kudo Y, Inoue Y, Liu Y, Sakamoto H, Kimura K. Comprehensive supramolecular chemistry: Introduction Atwood JL, Davies JED, MacNicol DD, Vögtle F, Lehn J.-M, Ripmeester JA, editors. New York, Pergamon; 1996, vol. 8, pp. 425.
[48] Thordarson P. Determining association constants from titration experiments in supramolecular chemistry. Chem Soc Rev 2011;40:1305–23.
[49] http://www.chemistry.unsw.edu.au/research/research-groups/thordarson-group/software
[50] Job P. Formation and stability of inorganic complexes in solution. Annal Chim Applicata 1928;9:113–203.
[51] Meyer AS, Ayres GH. The mole ratio method for spectrophotometric determination of complexes in solution. J Am Chem Soc 1957;79:49–53.
[52] Connors KA. Binding constants. Wiley: New York; 1987. pp. 24.
[53] Peinador C, Pía E, Blanco Vc, García MD, Quintela JM. Complexation of pyrene in aqueous solution with a self-assembled palladium metallocycle. Org Lett 2010;12:1380–3.
[54] Lakowicz JR. Principles of fluorescence spectroscopy. New York: Springer; 2010.
[55] Arun KT, Jayaram DT, Avirah RR, Ramaiah D. β-Cyclodextrin as a photosensitizer carrier: Effect on photophysical properties and chemical reactivity of squaraine dyes. J Phys Chem B 2011;115:7122–8.
[56] Pandey R, Kumar P, Singh AK, Shahid M, Li P-Z, Singh SK, Xu Q, Misra A, Pandey DS. Fluorescent Zinc(II) complex exhibiting "On-Off-On" switching toward Cu2+ and Ag+ Ions. Inorg Chem 2011;50:3189–97.
[57] Paul BK, Samanta A, Guchhait N. Modulation of excited-state intramolecular proton transfer reaction of 1-hydroxy-2-naphthaldehyde in different supramolecular assemblies. Langmuir 2009;26:3214–24.

[58] Shi Y, Zhang Z. Host-guest interactions in aqueous media withp-tert.-butylcalix[4]arene bearing polyoxyethylene chains. J Inclusion Phenom Mol Recognit Chem 1994;18:137–44.
[59] Neelakandan PP, Hariharan M, Ramaiah D. A supramolecular ON–OFF–ON fluorescence assay for selective recognition of GTP. J Am Chem Soc 2006;128:11334–5.
[60] Ji L, Chen W, Zheng S, Xu Z, Zhu D. Adsorption of sulfonamide antibiotics to multiwalled carbon nanotubes. Langmuir 2009;25:11608–13.
[61] Haino T, Rudkevich DM, Rebek J. Kinetically stable caviplexes in water. J Am Chem Soc 1999;121:11253–4.
[62] Sarkar S, Li S, Wayland BB. Comparison of Rh–OCH3 and Rh–CH2OH bond dissociation energetics from methanol C−H and O−H bond reactions with rhodium(II) porphyrins. J Am Chem Soc 2010;132:13569–71.
[63] Morris KF, Johnson CS. Resolution of discrete and continuous molecular size distributions by means of diffusion-ordered 2D NMR spectroscopy. J Am Chem Soc 1993;115:4291–9.
[64] Stejskal EO, Tanner JE. Spin diffusion measurements: Spin echoes in the presence of a time-dependent field gradient. J Chem Phys 1965;42:288–92.
[65] Tanner JE. Use of the stimulated echo in nmr diffusion studies. J Chem Phys 1970;52: 2523–6.
[66] Gibbs SJ, Johnson CS Jr. A PFG NMR experiment for accurate diffusion and flow studies in the presence of eddy currents. J Magn Res 1991;93:395–402.
[67] Wu DH, Chen AD, Johnson CS. An improved diffusion-ordered spectroscopy experiment incorporating bipolar-gradient pulses. J Magn Res A 1995;115:260–4.
[68] Frish L, Matthews SE, Bohmer V, Cohen Y. A pulsed gradient spin echo NMR study of guest encapsulation by hydrogen-bonded tetraurea calix[4]arene dimers. J Chem Soc Perkin Trans 1999;2:669–72.
[69] Mattsson J, Zava O, Renfrew AK, Sei Y, Yamaguchi K, Dyson PJ, Therrien B. Drug delivery of lipophilic pyrenyl derivatives by encapsulation in a water soluble metalla-cage. Dalton Trans 2010;39:8248–55.
[70] Barry NPE, Furrer J, Freudenreich J, Süss-Fink G, Therrien B. Designing the host-guest properties of tetranuclear arene ruthenium metalla-rectangles to accommodate a pyrene molecule. Eur J Inorg Chem 2010;725–8.
[71] Cameron KS, Fielding L. NMR diffusion spectroscopy as a measure of host−guest complex association constants and as a probe of complex size. J Org Chem 2001;66:6891–5.
[72] Cohen Y, Avram L, Frish L. Diffusion NMR spectroscopy in supramolecular and combinatorial chemistry: An old parameter—new insights. Angew Chem Int Ed 2005;44:520–54.
[73] Dethlefs C, Eckelmann J, Kobarg H, Weyrich T, Brammer S, Näther C, Lüning U. Determination of binding constants of hydrogen-bonded complexes by ITC, NMR CIS, and NMR diffusion experiments. Eur J Org Chem 2011;2066–74.
[74] Schneider H-J, Yatsimirsky AK. Selectivity in supramolecular host-guest complexes. Chem Soc Rev 2008;37:263–77.
[75] Pedersen CJ. The discovery of crown ethers (Noble lecture). Angew Chem Int Ed 1988;27: 1021–7.
[76] Zhang M, Yan X, Huang F, Niu Z, Gibson HW. Stimuli-responsive host–guest systems based on the recognition of cryptands by organic guests. Acc Chem Res 2014;47:1995–2005.
[77] Zhu K, Vukotic VN, Loeb SJ. Molecular shuttling of a compact and rigid h-shaped [2] rotaxane. Angew Chem Int Ed 2012;51:2168–72.
[78] Ito T, Yamaguchi T. Osmotic pressure control in response to a specific ion signal at physiological temperature using a molecular recognition ion gating membrane. J Am Chem Soc 2004;126:6202–3.

[79] Zheng B, Wang F, Dong S, Huang F. Supramolecular polymers constructed by crown ether-based molecular recognition. Chem Soc Rev 2012;41:1621–36.
[80] Huang F, Gibson HW, Bryant WS, Nagvekar DS, Fronczek FR. First pseudorotaxane-like [3]complexes based on cryptands and paraquat: self-assembly and crystal structures. J Am Chem Soc 2003;125:9367–71.
[81] Ashton PR, Chrystal EJT, Glink PT, Menzer S, Schiavo C, Spencer N, Stoddart JF, Tasker PA, White AJP, Williams DJ. Pseudorotaxanes formed between secondary dialkylammonium salts and crown ethers. Chem Eur J 1996;2:709–28.
[82] Harrison IT, Harrison S. Synthesis of a stable complex of a macrocycle and a threaded chain. J Am Chem Soc 1967;89:5723–4.
[83] Allwood BL, Shahriari-Zavareh H, Stoddart JF, Williams DJ. Complexation of paraquat and diquat by a bismetaphenylene-32-crown-10 derivative. J Chem Soc Chem Commun 1987;1058–61.
[84] Dietrich B, Lehn JM, Sauvage JP. Diaza-polyoxa-macrocycles et macrobicycles. Tetrahedron Lett 1969;10:2885–8.
[85] Simmons HE, Park CH. Macrobicyclic amines. I. Out-in isomerism of 1,(k+2)-diazabicyclo[k.l.m]alkanes. J Am Chem Soc 1968;90:2428–9.
[86] Park CH, Simmons HE. Macrobicyclic amines. II. out-out in-in Prototropy in 1, (k + 2)-diazabicyclo [k.l.m] alkaneammonium ions. J Am Chem Soc 1968;90:2429–31.
[87] Dietrich B, Lehn JM, Sauvage JP. Les cryptates. Tetrahedron Lett 1969;10:2889–92.
[88] Cram DJ. Cavitands: Organic hosts with enforced cavities. Science 1983;219:1177–83.
[89] Gutsche CD. Calixarenes: an introduction. Cambridge, UK: The Royal Society of Chemistry Publishing; 2008.
[90] Masci B, Thuery P. Versatility of {M(30-crown-10)} (M = K+, Ba2+) as a guest in UO22+ complexes of [3.1.3.1]- and [3.3.3]homooxacalixarenes. CrystEngComm 2007;9:582–90.
[91] Ramos AI, Braga TM, Silva P, Fernandes JA, Ribeiro-Claro P, Lopes MdFS, Paz FAA, Braga SS. Chloramphenicol[middle dot]cyclodextrin inclusion compounds: co-dissolution and mechanochemical preparations and antibacterial action. CrystEngComm 2013;15:2822–34.
[92] Bottari G, Dehez F, Leigh DA, Nash PJ, Pérez EM, Wong JKY, Zerbetto F. Entropy-driven translational isomerism: a tristable molecular shuttle. Angew Chem Int Ed 2003;42:5886–9.
[93] Nasri H, Ellison MK, Shang M, Schulz CE, Scheidt WR. Variable π-bonding in iron(II) porphyrinates with nitrite, CO, and tert-butyl isocyanide: Characterization of [Fe(TpivPP)(NO2)(CO)]. Inorg Chem 2004;43:2932–42.
[94] Barin G, Frasconi M, Dyar SM, Iehl J, Buyukcakir O, Sarjeant AA, Carmieli R, Coskun A, Wasielewski MR, Stoddart JF. Redox-controlled selective docking in a [2]catenane host. J Am Chem Soc 2013;135:2466–9.
[95] Lhoták P, Himl M, Stibor I, Petříčková H. Alkylation of thiacalix[4]arenes. Tetrahedron Lett 2002;43:9621–4.
[96] Rekharsky MV, Inoue Y. Complexation thermodynamics of cyclodextrins. Chem Rev 1998;98:1875–918.
[97] Lagona J, Mukhopadhyay P, Chakrabarti S, Isaacs L. The cucurbit[*n*]uril family. Angew Chem Int Ed 2005;44:4844–70.
[98] Sueldo Occello VN, Veglia AV. Cucurbit[6]uril nanocavity as an enhanced spectrofluorimetric method for the determination of pyrene. Anal Chim Acta 2011;689:97–102.
[99] Chen H, Yang H, Xu W, Tan Y. A supramolecular switch based on three binding states of a pyrene derivate: a reversible three-state switch with only two stimuli. RSC Advances 2013;3:13311–7.

[100] D'Anna F, Riela S, Lo Meo P, Gruttadauria M, Noto R. The binary pyrene/heptakis-(6-amino-6-deoxy)-β-cyclodextrin complex: a suitable chiral discriminator. Spectrofluorimetric study of the effect of some α-amino acids and esters on the stability of the binary complex. Tetrahedron Asymmetry 2002;13:1755–60.
[101] Dyck ASM, Kisiel U, Bohne C. Dynamics for the assembly of pyrene−γ-cyclodextrin host−guest complexes. J Phys Chem B 2003;107:11652–9.
[102] D'Anna F, Riela S, Gruttadauria M, Lo Meo P, Noto R. A spectrofluorimetric study of binary fluorophore–cyclodextrin complexes used as chiral selectors. Tetrahedron 2005;61: 4577–83.
[103] D'Anna F, Riela S, Meo PL, Noto R. Stability and stoichiometry of some binary fluorophore–cyclodextrin complexes. Tetrahedron 2004;60:5309–14.
[104] Lin Y-C, Wang P-I, Kuo S-W. Water-soluble, stable helical polypeptide-grafted cyclodextrin bioconjugates: synthesis, secondary and self-assembly structures, and inclusion complex with guest compounds. Soft Matter 2012;8:9676–84.
[105] Dale EJ, Vermeulen NA, Thomas AA, Barnes JC, Juríček M, Blackburn AK, Strutt NL, Sarjeant AA, Stern CL, Denmark SE, Stoddart JF. ExCage. J Am Chem Soc 2014;136: 10669–82.
[106] Kuzmič P. Program DYNAFIT for the analysis of enzyme kinetic data: application to HIV proteinase. Anal Biochem 1996;237:260–73.
[107] Barnes JC, Juríček M, Strutt NL, Frasconi M, Sampath S, Giesener MA, McGrier PL, Bruns CJ, Stern CL, Sarjeant AA, Stoddart JF. ExBox: a polycyclic aromatic hydrocarbon scavenger. J Am Chem Soc 2012;135:183–92.
[108] Hayashida O, Ichimura K, Sato D, Yasunaga T. Synthesis, guest-binding, and reduction-responsive degradation properties of water-soluble cyclophanes having disulfide moieties. J Org Chem 2013;78:5463–9.
[109] Woiczechowski-Pop A, Gligor D, Bende A, Varodi C, Bogdan E, Terec A, Grosu I. Synthesis, structure, electrochemical behaviour and electrochemical investigations on the assembling with pyrene of a novel C3 cryptand. Supramol Chem 2014;1–8.
[110] Gibb CLD, Gibb BC. Well-defined, organic nanoenvironments in water: The hydrophobic effect drives a capsular assembly. J Am Chem Soc 2004;126:11408–9.
[111] Gibb CLD, Gibb BC. Templated assembly of water-soluble nano-capsules: Inter-phase sequestration, storage, and separation of hydrocarbon gases. J Am Chem Soc 2006;128: 16498–9.
[112] Liu S, Gibb BC. High-definition self-assemblies driven by the hydrophobic effect: synthesis and properties of a supramolecular nanocapsule. Chem Commun 2008;3709–16.
[113] Gibb CLD, Sundaresan AK, Ramamurthy V, Gibb BC. Templation of the excited-state chemistry of α-(n-Alkyl) dibenzyl ketones: How guest packing within a nanoscale supramolecular capsule influences photochemistry. J Am Chem Soc 2008;130:4069–80.
[114] Liu S, Gibb BC. Solvent denaturation of supramolecular capsules assembled via the hydrophobic effect. Chem Commun 2011;47:3574–6.
[115] Tang H, de Oliveira CS, Sonntag G, Gibb CLD, Gibb BC, Bohne C. Dynamics of a supramolecular capsule assembly with pyrene. J Am Chem Soc 2012;134:5544–7.
[116] Porel M, Jayaraj N, Kaanumalle LS, Maddipatla MVSN, Parthasarathy A, Ramamurthy V. Cavitand octa acid forms a nonpolar capsuleplex dependent on the molecular size and hydrophobicity of the guest. Langmuir 2009;25:3473–81.
[117] de Silva AP, Gunaratne HQN, Gunnlaugsson T, Huxley AJM, McCoy CP, Rademacher JT, Rice TE. Signaling recognition events with fluorescent sensors and switches. Chem Rev 1997;97:1515–66.

[118] Manandhar E, Wallace KJ. Host–guest chemistry of pyrene-based molecular receptors. Inorg Chim Acta 2012;381:15–43.
[119] Ogoshi T, Kanai S, Fujinami S, Yamagishi T-a, Nakamoto Y. *para*-Bridged symmetrical pillar[5]arenes: Their lewis acid catalyzed synthesis and host–guest property. J Am Chem Soc 2008;130:5022–3.
[120] Cao D, Kou Y, Liang J, Chen Z, Wang L, Meier H. A facile and efficient preparation of pillararenes and a pillarquinone. Angew Chem Int Ed 2009;48:9721–3.
[121] Ogoshi T, Ueshima N, Akutsu T, Yamafuji D, Furuta T, Sakakibara F, Yamagishi T-a. The template effect of solvents on high yield synthesis, co-cyclization of pillar[6]arenes and interconversion between pillar[5]- and pillar[6]arenes. Chem Commun 2014;50:5774–7.
[122] Hu X-B, Chen Z, Chen L, Zhang L, Hou J-L, Li Z-T. Pillar[*n*]arenes (n = 8-10) with two cavities: synthesis, structures and complexing properties. Chem Commun 2012;48:10999–1001.
[123] Ogoshi T, Ueshima N, Sakakibara F, Yamagishi T-a, Haino T. Conversion from pillar[5]arene to pillar[6–15]arenes by ring expansion and encapsulation of C60 by pillar[*n*]arenes with nanosize cavities. Org Lett 2014;16:2896–9.
[124] Strutt NL, Forgan RS, Spruell JM, Botros YY, Stoddart JF. Monofunctionalized pillar[5] arene as a host for alkanediamines. J Am Chem Soc 2011;133:5668–71.
[125] http://list.wada-ama.org/
[126] Janshoff A, Steinem C, Michalke A, Henke C, Galla H-J. Monofunctionalized β-cyclodextrins as sensor elements for the detection of small molecules. Sensor Actuat B 2000;70:243–53.
[127] Northrop BH, Zheng Y-R, Chi K-W, Stang PJ. Self-organization in coordination-driven self-assembly. Acc Chem Res 2009;42:1554–63.
[128] Northrop BH, Yang H-B, Stang PJ. Coordination-driven self-assembly of functionalized supramolecular metallacycles. Chem Commun 2008;5896–908.
[129] Fujita M, Tominaga M, Hori A, Therrien B. Coordination assemblies from a Pd(II)-cornered square complex. Acc Chem Res 2005;38:369–78.
[130] Seidel SR, Stang PJ. High-symmetry coordination cages via self-assembly. Acc Chem Res 2002;35:972–83.
[131] Caulder DL, Raymond KN. Supermolecules by design. Acc Chem Res 1999;32:975–82.
[132] Lehn J-M. Perspectives in supramolecular chemistry—from molecular recognition towards molecular information processing and self-organization. Angew Chem Int Ed Engl 1990;29:1304–19.
[133] Holliday BJ, Mirkin CA. Strategies for the construction of supramolecular compounds through coordination chemistry. Angew Chem Int Ed 2001;40:2022–43.
[134] Kumazawa K, Biradha K, Kusukawa T, Okano T, Fujita M. Multicomponent assembly of a pyrazine-pillared coordination cage that selectively binds planar guests by intercalation. Angew Chem Int Ed 2003;42:3909–13.
[135] Vajpayee V, Song YH, Cook TR, Kim H, Lee Y, Stang PJ, Chi K-W. A unique non-catenane interlocked self-assembled supramolecular architecture and its photophysical properties. J Am Chem Soc 2011;133:19646–9.
[136] Bolliger JL, Belenguer AM, Nitschke JR. Enantiopure water-soluble [Fe4L6] cages: Host–Guest chemistry and catalytic activity. Angew Chem Int Ed 2013;52:7958–62.
[137] Freudenreich J, Furrer J, Süss-Fink G, Therrien B. Template-directed synthesis of hexanuclear arene ruthenium complexes with trigonal-prismatic architecture based on 2,4,6-tris (3-pyridyl)triazine Ligands. Organometallics 2011;30:942–51.
[138] Han Y-F, Lin Y-J, Hor TSA, Jin G-X. Efficient route to organometallic cage formation via C–H activation-directed muticomponent assembly accompanying aromatic guest encapsulation. Organometallics 2012;31:995–1000.

[139] Yoshizawa M, Nakagawa J, Kumazawa K, Nagao M, Kawano M, Ozeki T, Fujita M. Discrete Stacking of Large Aromatic Molecules within Organic-Pillared Coordination Cages. Angew Chem Int Ed 2005;44:1810–3.

[140] Nishioka Y, Yamaguchi T, Yoshizawa M, Fujita M. Unusual [2+4] and [2+2] cycloadditions of arenes in the confined cavity of self-assembled cages. J Am Chem Soc 2007;129: 7000–1.

[141] Kamat PV. Photophysical, photochemical and photocatalytic aspects of metal nanoparticles. J Phys Chem B 2002;106:7729–44.

[142] Müller RH, Mäder K, Gohla S. Solid lipid nanoparticles (SLN) for controlled drug delivery – a review of the state of the art. Eur J Pharm Biopharm 2000;50:161–77.

[143] Jang J, Lim B. Facile fabrication of inorganic-polymer core–shell nanostructures by a one-step vapor deposition polymerization. Angew Chem Int Ed 2003;42:5600–3.

[144] Głowala P, Budniak A, Krug P, Wysocka B, Berbeć S, Dec R, Dołęga I, Kacprzak K, Wojciechowski J, Kawałko J, Kípka P, Kępińska D, Kijewska K, Mazur M. Incorporation of pyrene in polypyrrole/polystyrene magnetic beads. Spect Chim Acta A 2014;131:667–73.

[145] Guerrini L, Garcia-Ramos JV, Domingo C, Sanchez-Cortes S. Sensing polycyclic aromatic hydrocarbons with dithiocarbamate-functionalized ag nanoparticles by surface-enhanced raman scattering. Anal Chem 2009;81:953–60.

[146] Mueller U, Schubert M, Teich F, Puetter H, Schierle-Arndt K, Pastre J. Metal-organic frameworks-prospective industrial applications. J Mater Chem 2006;16:626–36.

[147] Huo S-H, Yan X-P. Facile magnetization of metal-organic framework MIL-101 for magnetic solid-phase extraction of polycyclic aromatic hydrocarbons in environmental water samples. Analyst 2012;137:3445–51.

[148] Hawes CS, Nolvachai Y, Kulsing C, Knowles GP, Chaffee AL, Marriott PJ, Batten SR, Turner DR. Metal-organic frameworks as stationary phases for mixed-mode separation applications. Chem Commun 2014;50:3735–7.

[149] Müllen K. Evolution of graphene molecules: structural and functional complexity as driving forces behind nanoscience. ACS Nano 2014;8:6531–41.

[150] Zan R, Bangert U, Ramasse Q, Novoselov KS. Interaction of metals with suspended graphene observed by transmission electron microscopy. J Phys Chem 2012;3:953–8.

[151] Novoselov KS. Graphene: Materials in the flatland (Nobel lecture). Angew Chem Int Ed 2011;50:6986–7002.

[152] Barry NPE, Sadler PJ. Challenges for metals in medicine: How nanotechnology may help to shape the future. ACS Nano 2013;7:5654–9.

[153] Zhang Y, Wang Z. Research progress in nanocrystalline solar power system. J Comput Theor Nanos 2013;10:2912–5.

[154] Warren SC, Guney-Altay O, Grzybowski BA. Responsive and nonequilibrium nanomaterials. J Phys Chem Lett 2012;3:2103–11.

[155] Farokhzad OC, Langer R. Impact of nanotechnology on drug delivery. ACS Nano 2009;3: 16–20.

[156] Evans DF, Wennerstrom H. The colloidal domain. 2nd ed. New York: Wiley-VCH; 1999.

[157] Holmberg K, Jonsson BB, Kronberg B, Lindman B. Surfactants and polymers in aqueous solution. 2nd ed. New York: John Wiley & Sons; 2003.

[158] Discher DE, Eisenberg A. Polymer Vesicles. Science 2002;297:967–73.

[159] Kale TS, Klaikherd A, Popere B, Thayumanavan S. Supramolecular assemblies of amphiphilic homopolymers. Langmuir 2009;25:9660–70.

[160] Theato P, Sumerlin BS, O'Reilly RK, Epps IIITH. Stimuli responsive materials. Chem Soc Rev 2013;42:7055–6.

[161] Zhuang J, Gordon MR, Ventura J, Li L, Thayumanavan S. Multi-stimuli responsive macromolecules and their assemblies. Chem Soc Rev 2013;42:7421–35.
[162] Grubbs RB, Sun Z. Shape-changing polymer assemblies. Chem Soc Rev 2013;42:7436–45.
[163] Gref R, Minamitake Y, Peracchia MT, Trubetskoy V, Torchilin V, Langer R. Biodegradable long-circulating polymeric nanospheres. Science 1994;263:1600–3.
[164] Jokerst JV, Lobovkina T, Zare RN, Gambhir SS. Nanoparticle PEGylation for imaging and therapy. Nanomedicine 2011;6:715–28.
[165] Kamaly N, Xiao Z, Valencia PM, Radovic-Moreno AF, Farokhzad OC. Targeted polymeric therapeutic nanoparticles: design, development and clinical translation. Chem Soc Rev 2012;41:2971–3010.
[166] Basak D, Ghosh S. pH-Regulated controlled swelling and sustained release from the core functionalized amphiphilic block copolymer micelle. ACS Macro Lett 2013;2:799–804.
[167] Basak R, Bandyopadhyay R. Encapsulation of hydrophobic drugs in pluronic F127 micelles: Effects of drug hydrophobicity, solution temperature, and pH. Langmuir 2013;29:4350–6.
[168] Kabanov AV, Batrakova EV, Alakhov VY. Pluronic® block copolymers as novel polymer therapeutics for drug and gene delivery. J Control Release 2002;82:189–212.
[169] Batrakova EV, Kabanov AV. Pluronic block copolymers: Evolution of drug delivery concept from inert nanocarriers to biological response modifiers. J Control Release 2008;130:98–106.
[170] Pitto-Barry A, Barry NPE. Pluronic® block-copolymers in medicine: From chemical and biological versatility to rationalisation and clinical advances. Polym Chem 2014;5:3291–7.

Chapter 14

Nanoreactors Based on Porphyrin-Functionalized Carbon Compounds

Pietro Tagliatesta*, Barbara Floris*, Stefano Bellucci†

**Dipartimento di Scienze e Tecnologie Chimiche, Università Tor Vergata, via della ricerca scientifica, Roma, Rome, Italy*

†INFN - Istituto Nazionale di Fisica Nucleare, Laboratori Nazionali di Frascati LNF, Frascati, Latium, Rome, Italy

1 INTRODUCTION

Although the term "nanoreactor" can be dated back to the end of the past century (see, for example, [1]), since then it acquired a well-defined meaning, that of a confined environment (of any type), where chemical conversions take place, in analogy to what happens in nature [2,3]. The field grew impressively in the last years, utilizing different "cages," from micelles to vescicles, to liposomes, to the so-called polymersomes, that can even be multicompartmentalized, giving catalytic cascade reactions [4,5]. Moreover, functionalized nanoparticles (NP) can be considered nanoreactors, in that reactions occur on the nanoparticle. Thus, NP-biological composites (with NP materials including quantum dots, carbon nanotubes, viral capsides, and liposomes) were obtained [6,7].

Functionalization of nanoparticles with porphyrins and analogous macrocycles opened the way to important applications, such as, to name only a few: gelators [8], phototherapy of cancer [9], degradation of organic pollutants by singlet oxygen photosensitization [10], and photoexcited electron transfer [11]. The interest in porphyrin-based nanoreactors lies in the possibility of using a variety of metals that can enter in catalytic processes [12].

Actually, porphyrins can themselves constitute nanoreactors, from the molecular square made of four porphyrins coordinated with four rhenium centers, that encapsulate a Mn(III) porphyrins (catalyst for alkene epoxidation) [2 and references therein] to self-organized porphyrinic materials [13], to multiporphyrinic cages [14].

On the other hand, nanoreactors based on carbon occupy a special position among NPs, due to the variety of structures presented by carbon materials.

Organic Nanoreactors. http://dx.doi.org/10.1016/B978-0-12-801713-5.00014-8

Besides the well-known allotropic species, diamond, graphite, and amorphous carbon, other forms are possible and were investigated, from fullerenes (C_{60} being the most studied [15]) to single- or multiwalled carbon nanotubes (SWCNT and MWCNT) [16], to the more recent graphene [17–19] and graphene oxide (GO) [20–22]. Moreover, nowadays we can meet carbon nanodots [23,24], carbon nanoleaves [25], yolk-shell nanoparticles [26], nanorings [27], three-dimensional porous carbon materials [28], nanorods [29], or composite [30] and porous carbon [31] materials.

Possible uses of carbon NPs range from medicinal [32–34] or environmental [35–37] to analytical [38–40], technological [41–46], or catalytic [47–50] applications and were largely reviewed. The growth in utilization of carbon NPs raised some concern about possible adverse effects on health and environment [51–54].

If we consider that properties of porphyrin and analogous compounds may derive advantage from constraint in a limited environment (see, for example, [55,56]), and that carbon NPs may offer a versatile basis for nanoreactors (see, for example, [57]), the decoration of carbon nanospecies with porphyrinoid compounds should give rise to interesting nanoreactors. The aim of this chapter is to present what has been done in the field.

1.1 Graphene- and Graphene Oxide-Porphyrin Nanoreactors

Graphene is nature's ideal 2D conductor, made of a single sheet of hexagonally packed carbon atoms. Its extraordinary electronic, mechanical, thermal, optical, and optoelectronic properties [58,59] made this material the object of growing investigation of fundamental and practical aspects in optics and optoelectronics [60], as well as in photo-driven chemical conversion using solar energy [61], to cite only two fields. Different functionalizations of graphene and GO are available, with covalent and noncovalent approaches, and were reviewed few years ago [62,63].

In view of the peculiar photo- and optoelectronic properties of self-assembled architectures of porphyrins [64] and of porphyrin-based porous sheets [65], it is easy to understand why many efforts have been devoted toward functionalization of graphene and graphene-oxide with these macrocycle.

Altough graphene structure is almost planar, it should be considered as a sort of "open" nanoreactor, since in several examples it is reported that it needs the combination graphene-porphirin to be effective. Therefore, we decided to extend the review to such system, also because in several cases, without the presence of graphene (for example, with amorphous carbon nanoparticles) the reactions do not work with the same efficiency.

A 2015 study reported hybrids of cationic porphyrins with graphene, carbon nanotubes, and fullerene, discussing their properties in view of possible applications [66].

Structures of the most commonly used porphyrins in junction with graphenes are shown in Chart 14.1.

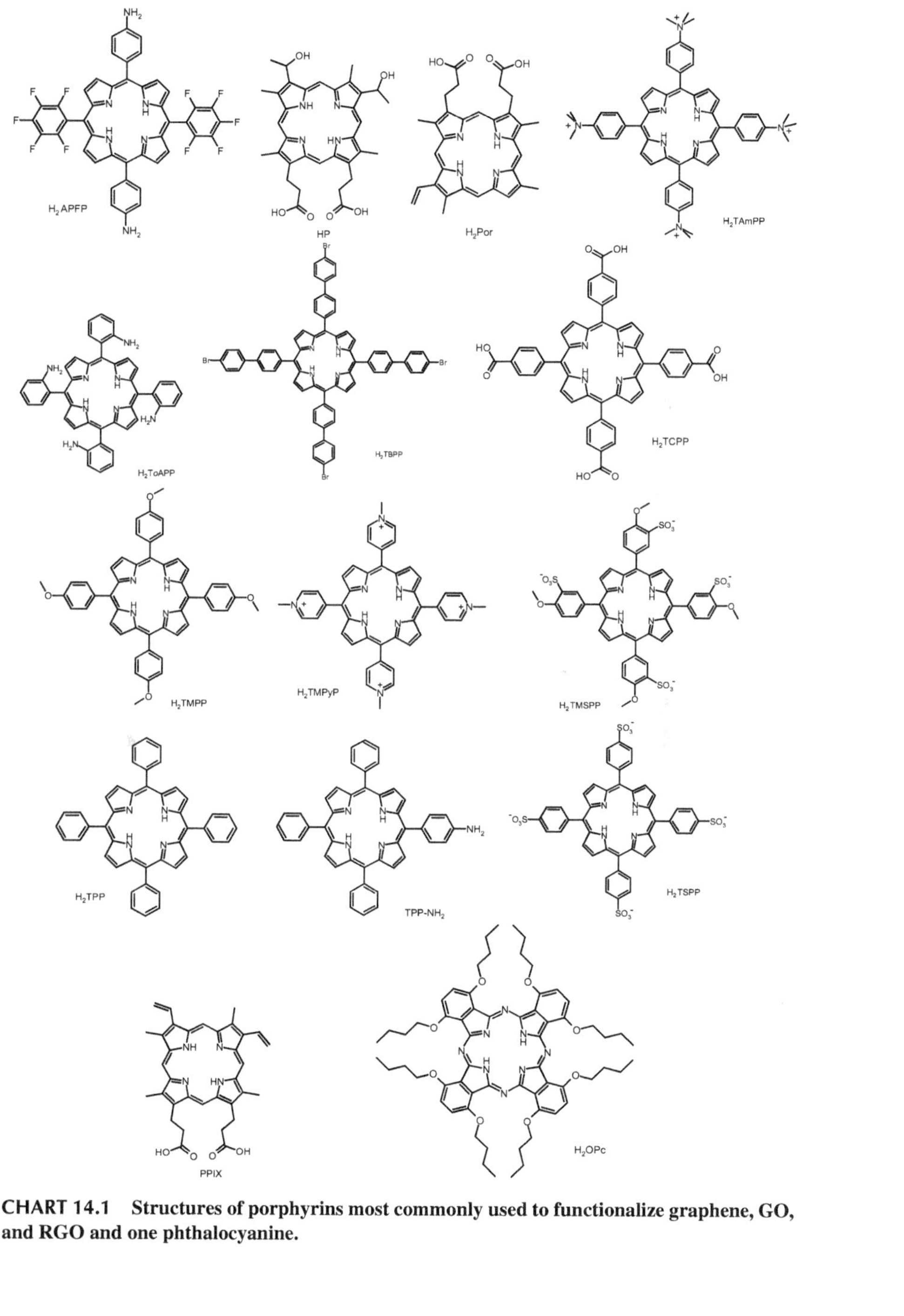

CHART 14.1 Structures of porphyrins most commonly used to functionalize graphene, GO, and RGO and one phthalocyanine.

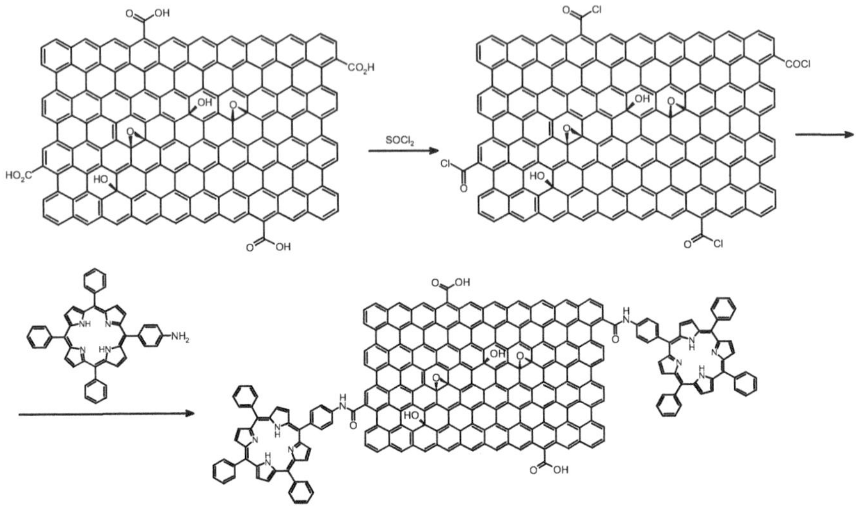

SCHEME 14.1 Synthesis of covalently linked porphyrin-graphene TPP-NH_2/GO nanohybrid. *(Adapted from Ref. [67] with permission of American Chemical Society, 2009.)*

The covalently linked porphyrin-graphene nanohybrid, Graphene-TPP, was obtained reacting the monoamino-substituted 5-(4-aminophenyl)-10,15,20-triphenylporphyrin (TPP-NH_2) with GO in DMF [67,68] (Scheme 14.1).

A different approach was the application of the Prato cycloaddition reaction [69,70] by in situ–generated azomethine ylides, thus incorporating fused pyrrolidines onto the skeleton of graphene. Thus, porphyrins and phthalocyanines were covalently linked to exfoliated graphene. The former macrocycles were attached using the formyl derivative of TPP [71], while the latter ones were introduced indirectly: exfoliated graphene was first treated with *N*-methylglycine and 4-formylbenzoic acid, and the carboxylic group of the resulting pyrrolidine ring underwent esterification with the benzyl–OH of substituted phthalocyanines [72]. More recently, this strategy was applied also to porphyrins [73]. Reduced graphene oxide (RGO) underwent 1,3-dipolar cycloaddition with *N*-methylglycine and 4-hydroxybenzaldehyde; the resulting *p*-hydroxyphenyl pyrrolidine was used in nucleophilic substitution with 5-[4-(2-bromoethoxy) phenyl]-10,15,20-triphenylporphyrin (Scheme 14.2).

Noncovalent functionalizations were obtained synthesizing porphyrinoids in the presence of sulfonated graphene, which acted as solid acid carbocatalyst [74]. The obtained porphyrinoids gave supramolecular interactions with pristine graphene. Alternatively, GO intercalated into self-assembled porphyrins J-aggregates [75].

The hybrids were investigated to understand the consequences of functionalization on properties [76], especially electrical ones [77,78]. Moreover, applications were investigated in several fields, such as optoelectronic devices [79] and electron and charge transfer processes for solar cells [80,81].

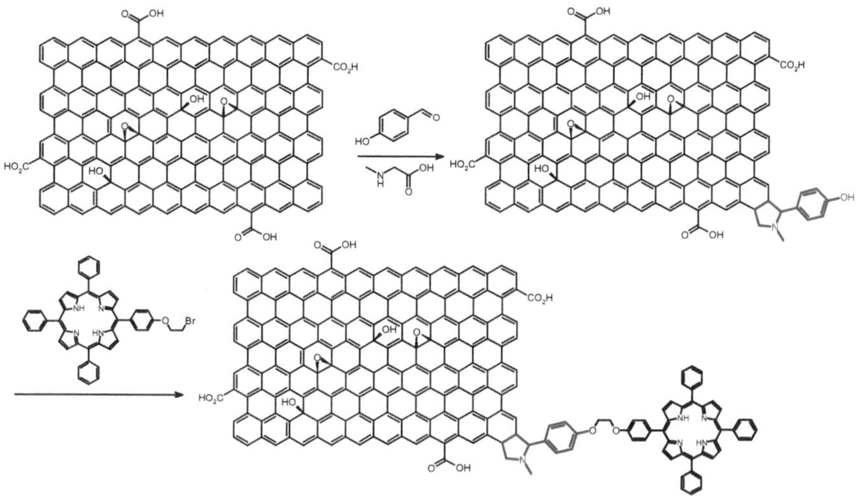

SCHEME 14.2 Two-step functionalization of RGO. *(Adapted from Ref. [73] with permission of Springer, 2014.)*

Porphyrin-graphene (GO or RGO) hybrids work as nanoreactors in catalysis (either electrochemical or photochemical catalysis), in sensing (both as chemical and biochemical sensors) or in diagnostic and therapeutic applications, as illustrated in the following section.

1.1.1 Electrochemical Nanoreactors

Electrocatalysis by porphyrins-graphene nanoreactors can occur either in oxidation or in reduction reactions. As to oxidation reaction, electropolymerization of graphene and 5-[2-(4-bromopentyloxy)phenyl]-10,15,20-triphenylporphrin composite film allowed for the electrocatalytic oxidation of indirubin [82]. More recently, electrooxidation of methanol was obtained, using as catalyst graphene noncovalently functionalized with 5,10,15,20-tetrakis(1-methylpyridinio-4-yl) porphyrin tetra(*p*-toluenesulfonate) (H_2TMPyP) [83]. Graphene and TMPyP, dispersed in water, were ultrasonicated. Subsequently, Pt nanoparticles were deposited by hydrothermal polyol treatment under ultrasonication. The electrode exhibited electrocatalytic activity for methanol oxidation and resulted more active and more stable catalyst than the Pt/graphene and commercial Pt/C catalysts. Such a nanoreactor may find important application in "direct methanol fuel cells" (DMFCs), that have particularly appealing characteristics: high power density, low operating temperature, ease of handling as a liquid fuel, low pollution emission, and potential applications in portable electronic devices and electric vehicles.

GO, both as such and as hybrid derivatives, is also emerging as a new class of carbocatalysts. The properties of these sheets have been discussed [84], with reference to the question whether GO acts as a catalyst or as a reactant, due to the large number of oxygen atoms present.

Electroreduction has been more explored. The Cobalt(II) complex of H_2T-MPyP, positively charged, could be layer-by-layer assembled with negatively charged GO and then transformed, by electrochemical reduction, into a cobalt porphyrin and electrochemically reduced graphene oxide (ERGO) functional multilayer films [85]. The resulting $[ERGO@CoTMPyP]_n$ ($n = 1, 3, 5, 7, 9, 11$) multilayer films displayed good electrocatalytic activity and stability toward hydrogen evolution reaction (HER) in alkaline media (Fig. 14.1).

Cathodic oxygen reduction reaction (ORR) is a very active area of research, since it plays a crucial role in electrochemical energy conversion in fuel cells, especially DMFCs. Typically, direct methanol fuel cells operate by oxidizing an aqueous solution of methanol to CO_2 and reducing oxygen to water. The kinetics of the ORR are very slow and require an efficient catalyst for the ORR cathode, usually Pt-based ones [86]. Replacement of the costly Pt, however, is necessary for large scale commercialization, and attention was directed toward alkaline DMFCs, that have the advantages of more facile electrode catalytic reactions, lower methanol permeability from anode to cathode, and simpler water management [87]. To this aim, pyridine-functionalized RGO was used as a building block in the assembly of metal organic framework (MOF), by reaction with iron porphyrin [86]. The graphene-iron 5,10,15,20-tetrakis(4-carboxyphenyl) porphyrin, FeTCPP, MOF showed facile 4-electron ORR (Fig. 14.2), with an enhanced catalytic activity for ORRs attributed to the combination of increased graphene porosity, a higher polarity, and the presence of the catalytic active iron-TCPP.

Several nonprecious metals porphyrins and phthalocyanines were used to prepare graphene composites in order to investigate ORR [88], with Co^{2+} being the most used metal. Cobalt tetrakis(*o*-aminophenyl)porphyrin, CoT*o*APP, was covalently anchored to graphene—besides other carbon nanomaterials—showing excellent electrocatalytic effect and efficient electrocatalytic performance for the ORR [89].

Electropolymerized films of Co(III) tetrakis(4-hydroxyphenyl)porphyrin, doped with ionic liquid (CoPor-IL/G) functionalized graphene gave O_2 electroreduction [90]. 1-Butyl-3-methylimidazolium tetrafluoroborate was used as IL. It is noteworthy that the performance of CoPor-IL/G resulted better than that of cobalt porphyrin alone, in terms of potential required and amount of undesired H_2O_2.

High activity for ORR in alkaline solution was found with glassy carbon electrodes modified with RGO-N_4 macrocycles, namely iron phthalocyanine, and cobalt 5,10,15,20-tetrakis(4-metoxyphenyl)porphyrins [91], T(OM)PP, physically adsorbed.

A cobalt 5,10,15-tetrakis(1-methylpyridinio-4-yl)porphyrin/poly(sodium *p*-styrene sulfonate)/reduced graphene oxide composite (CoTPyP/PSS-RGO), was fabricated using an in situ solvothermal synthesis method [92]. It showed excellent catalytic activity for ORR, attributed by the authors to its large electrochemically accessible surface area, to the excellent electrical conductivity of PSS-RGO, and to the synergistic effect between CoTPyP and PSS-RGO.

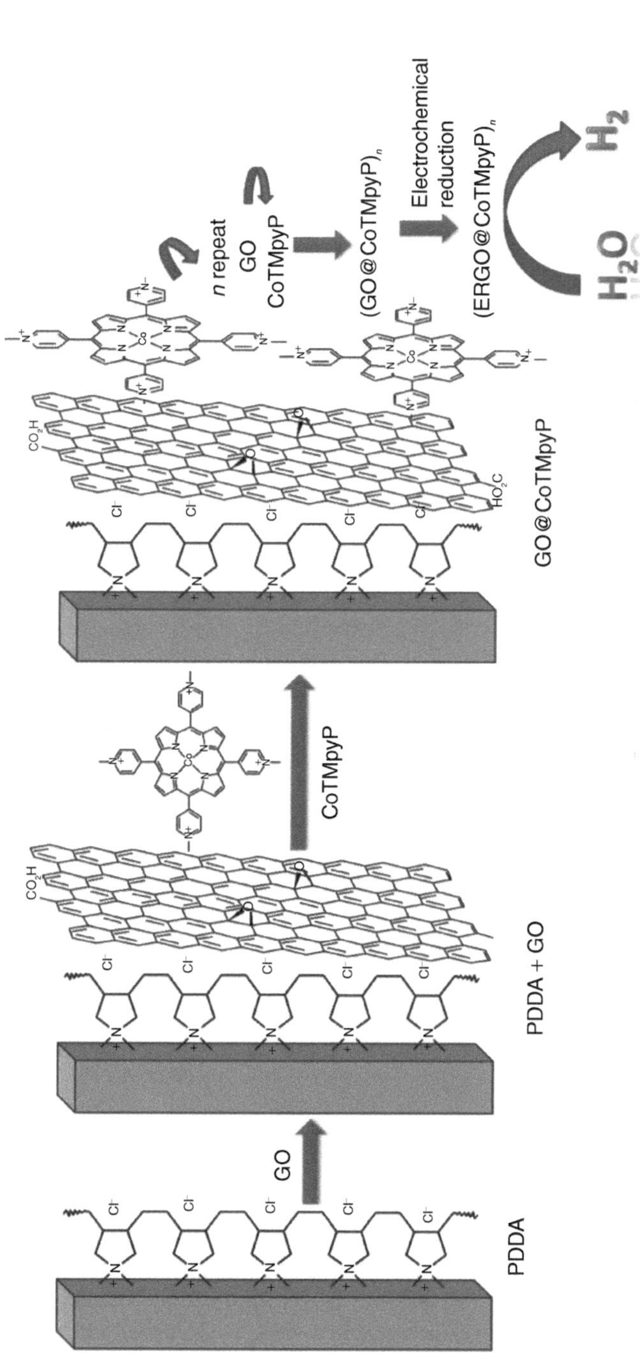

FIGURE 14.1 Construction of [ERGO@CoTMPyP]$_n$ nanoreactors and electrocatalytic HER. *(Adapted from Ref. [85] with permission of American Chemical Society, 2014.)*

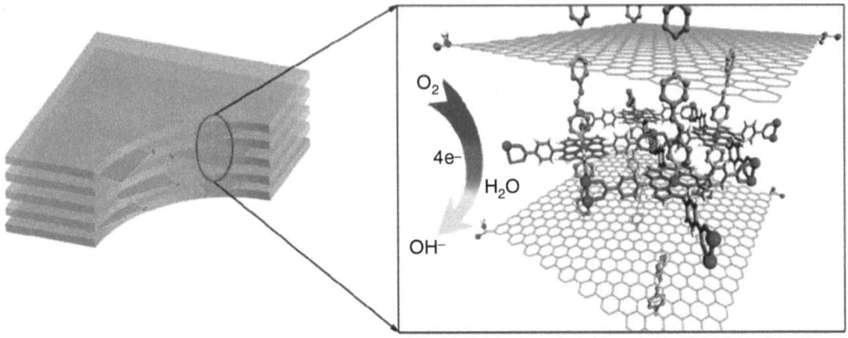

FIGURE 14.2 Graphene-Porphyrin MOF in electrocatalytic ORR. *(Reprinted from Ref. [86] with permission of American Chemical Society, Copyright 2012.)*

A different strategy to prepare the nanoreactor was to adsorb Co^{2+} ions on GO, then to mix the Co^{2+} graphene sheets with 5,10,15,20-tetrakis(4-hydroxyphenyl)porphyrin, H_2THPP. Multilayers on the GO substrate were obtained by many cycles of alternative adsorption of Co^{2+} and THPP. Finally, GO support was reduced by hydrazine in ammonia solution, yielding the RGO/(Co^{2+}–THPP)$_n$ composite, with n = number of cycles and, consequently, of Co^{2+} and porphyrin layers [93]. According to the authors, this nanoreactor shows electrocatalityc activity comparable to that of the commercial C/Pt catalysts, but better stability and increased tolerance to the crossover effect.

The same porphyrin was used with different nonnoble metal ions, cobalt(II) included, to grow M-THPP (M = Fe^{3+}, Ni^{2+}, Mn^{2+}, Co^{2+}) on poly(sodium *p*-styrene sulfonate) modified reduced graphene oxide (PSS-RGO), using an in situ solvothermal synthesis method [94] (Fig. 14.3).The authors state that, among the assembled nanoreactors, CoTHPP/PSS-RGO catalyst possesses high activity, good long-term stability, excellent tolerance for the crossover effect of methanol, and a facile 4-electron pathway for ORR, thus resulting in a promising Pt-free catalyst in an alkaline direct methanol fuel cell.

Another cobalt porphyrin, cobalt[5,15-(*p*-aminophenyl)-10,20-(pentafluorophenyl)porphyrin], CoAPFP, was covalently linked to GO and the resulting hybrid underwent electrocatalytic electrons oxygen reduction [95].

Fundamental studies are also important, helping to design more efficient catalysts. For example, the observation that axial zirconium porphyrinoids' coordination on GO led to self-organization with formation of a material with conducting properties [96]—properties that quite often are not so good when linkers are covalently bound—supported the investigation of the effects of axial coordination of the metal center on the activity of iron(III) 5,10,15,20-tetraphenylporphyrin chloride on RGOs modified with aliphatic amine groups [97]. The authors showed that axial coordination of iron in tetraphenylporphyrin environment enhanced the catalytic activity of the porphyrin in ORR and improved the selectivity of the reaction for water. On the other hand, density functional theory calculations were

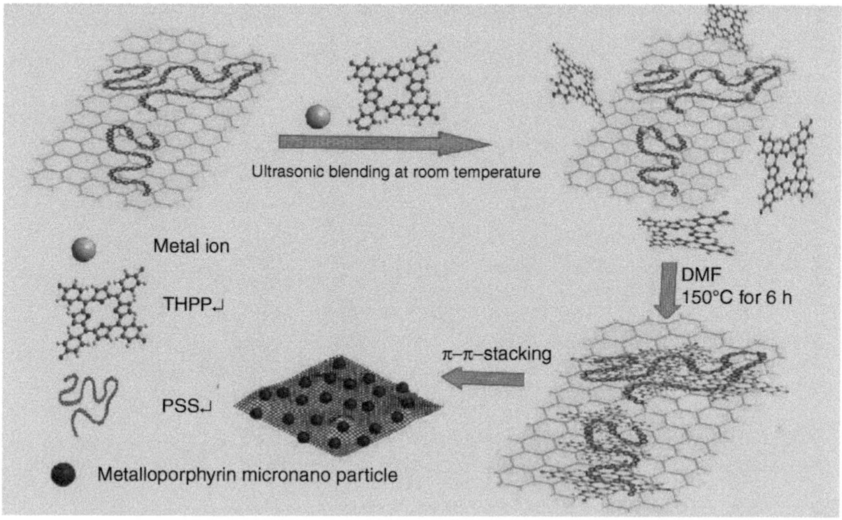

FIGURE 14.3 Preparation of CoTHPP/PSS-RGO catalyst. *(Reproduced from Ref. [94] with permission of Royal Society of Chemistry.)*

used to investigate the possibility of chemically tailoring metal-porphyrin-like active sites on carbon nanotubes and graphene for minimal overpotential in the electrochemical oxygen evolution and ORR [98].

It is interesting to note that metal porphyrinoids are so important in ORR that attempts were made to organize them in two-dimensional covalent organic polymers analogues of graphene using 5,10,15,20-tetrakis(4′-bromobiphenyl) porphyrin, which acted as efficient electrocatalysts for oxygen reduction in both alkaline and acid media [99] or in metal-functionalized porphyrin-like graphenes, that were able to perform electrochemical reduction of CO_2 and CO [100].

Using iron phthalocyanine as a model compound, different assemblies were prepared on graphene nanosheets, varying the composite architecture and investigating electrochemical oxygen reduction [101].

1.1.2 Photochemical Nanoreactors

Hydrogen production from solar water splitting has been considered as the ultimate solution to the energy and environmental problems, and graphene was found useful in improving the light-driven hydrogen generation performance [102]. Moreover, graphene's flexible structure, large surface area, high transparency, and high thermal conductivity, together with unique electrical properties, made graphene the right component of composites to be used as photocatalysts, for example in water splitting, in degradation or oxidation of organic contaminants, in photoreduction of CO_2 into renewable fuels, in elimination of toxic heavy metal ions, and in antibacterial applications [61]. Composites with porphyrinoid chromophores should improve the graphene photocatalytic ability.

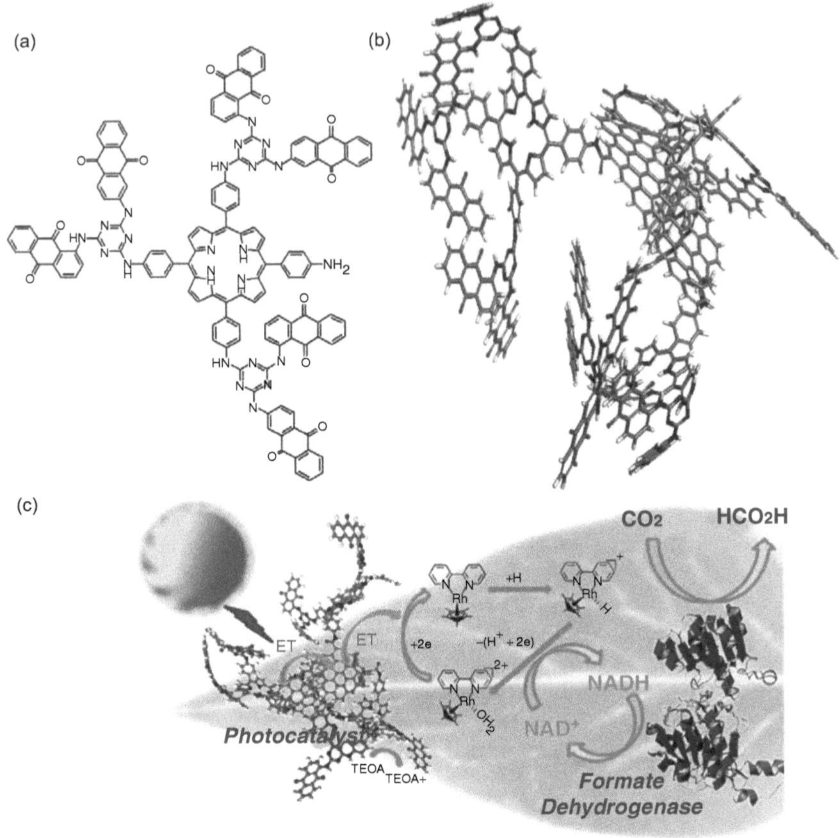

FIGURE 14.4 Multianthraquinone-substituted porphyrin graphene-based photocatalyst, formula (a), structure (b), catalyzed artificial photosynthesis of formic acid from CO_2 under visible light (c). *(Adapted from Ref. [103] with permission of American Chemical Society. Copyright 2012.)*

Thus, it was reported that methanoic acid, HCO_2H, could be obtained photocatalytically from carbon dioxide using a graphene-based photocatalyst, covalently bonded to multianthraquinone substituted porphyrin chromophore [103]. The system is quite complicated: 1-aminonaphtoquinone, 2,4,6-trichlorotriazine and 5-(4-aminophenyl)-10,15,20-triphenylporphyrin gave the multianthraquinone derivative of porphyrin, which was covalently linked to RGO, via formation of NH—CO bond (Fig. 14.4).

The photocatalyst, coupled with the enzyme formate dehydrogenase, was effective in photoregenerating NADH and in visible-light driven artificial photosynthesis of formic acid from CO_2.

Another successful development of a photocatalyst/biocatalyst is the integrated system of graphene-based visible-light active photocatalyst and sequentially coupled enzymes, reported by the same authors [104]. An isatin derivative

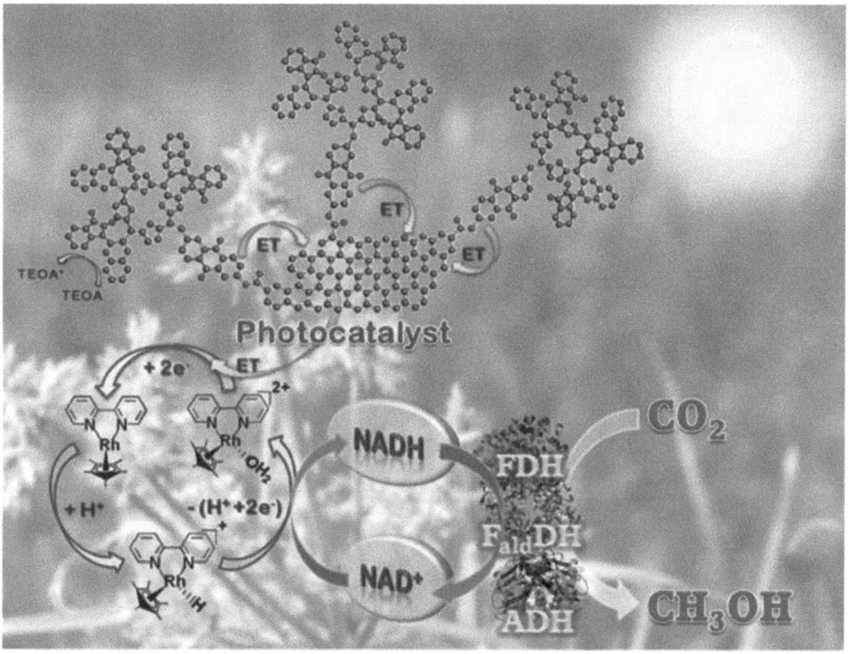

FIGURE 14.5 **Photocatalyst/biocatalyst integrated system for methanol formation from CO_2 photocatalyzed by a graphene-porphyrin nanoreactor.** *(Reprinted from Ref. [104] with permission of American Chemical Society. Copyright 2014.)*

of porphyrin, covalently attached to graphene, in conjunction with formate dehydrogenase, formaldehyde dehydrogenase, and alcohol dehydrogenase, gave a highly selective methanol formation derived from CO_2 (Fig. 14.5).

2,18-Bis(2-carboxyethyl)-8,13-divinyl-3,7,12,17-tetramethylporphyrin tin(IV) dichloride (SnPor) was used to prepare a binary nanohybrid with graphene oxide (SnPor/RGO) [105], taking advantage of the SnPor planarity that facilitated adsorption on the GO surface, as a result of π–π-stacking interactions. Meanwhile, SnPor reduced GO. The binary nanohybrid, in turn, was transformed into a ternary organic/inorganic nanohybrid with silver nanoparticles (SnPor/AgNP/RGO) under irradiation with visible light. The ternary nanohybrid showed improved photocatalytic activity in the degradation of Rhodamine B under visible light and high catalytic activity toward reduction of 4-nitrophenol (Fig. 14.6).

Zinc 5,10,15,20-tetrakis(4-pyridyl)porphyrin, ZnTPyP, was assembled to form monodimensional supramolecular nanostructures via surfactant-assisted self-assembly (SAS) method, with GO acting as a sheet-like surfactant [106]. This SAS-enhanced photocatalytic activity for the photodegradation of Rhodamine B under visible-light irradiation with respect to ZnTPyP nanoassemblies formulated via a conventional SAS, with hexadecyltrimethylammonium bromide used as surfactant.

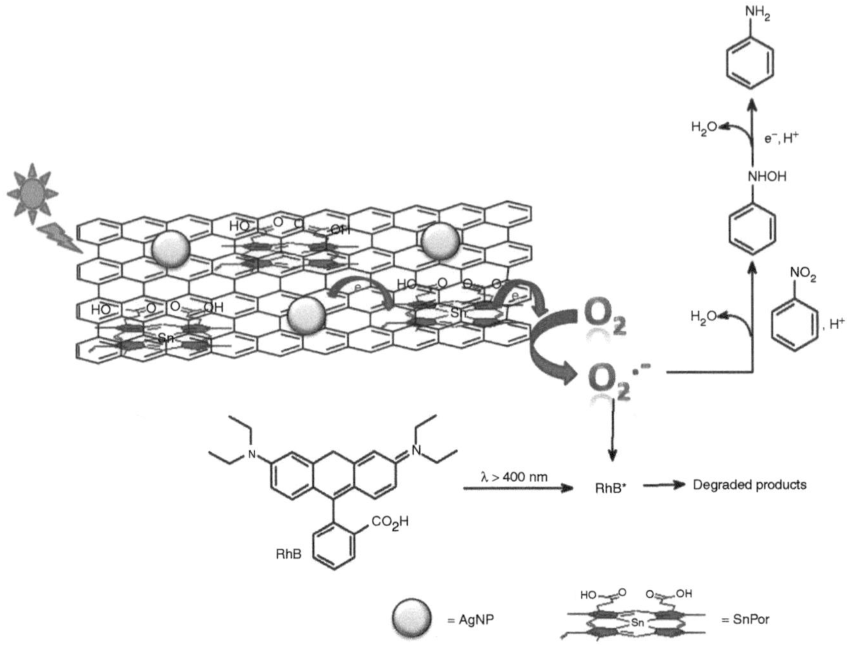

FIGURE 14.6 Photocatalytic and catalytic reduction performed by the ternary organic/inorganic nanohybrid SnPor/AgNP/RGO. *(Adapted from Ref. [105].)*

Photocatalytic degradation of Rhodamine B and methylene blue (MB) was efficiently achieved by a nanohybrid film prepared from 5,10,15,20-tetrakis(4-hydroxyphenyl)porphyrin, H_2THPP [107]. The nanoensemble exhibited enhanced visible-light photocatalytic activity compared to each moiety of the hybrid. Moreover, the photocatalyst can be easily separated and recycled for successive use with excellent stability.

Zinc 5,10,15,20-tetrakis(4-pyridyl)porphyrin was used to functionalize graphene quantum dots [108]. The nanocomposite was efficient in photocatalyzing degradation of methylene blue upon visible-light irradiation. Moreover, the nanomaterial could be recovered and reused with no variance of degradation rate. Therefore, it can be used as visible light-driven photocatalyst to decompose organic pollutants soluble in water.

A nanohybrid, synthesized by immobilizing noncovalently 5,10,15,20-tetrakis(4-hydroxyphenyl) porphyrin, H_2THPP, on RGO nanosheets, loaded with Pt nanoparticles, showed enhanced photocatalytic activity under UV-vis light irradiation to promote hydrogen evolution [109].

Although a fundamental investigation is worth mentioning that the preparation of a nanoensemble made of hematoporphyrin (as photosensitizer) and GO (as nanocarrier) is covalently linked via esterification reaction [110]. According to the authors, the hybrid structure, where GO plays the role of a nontoxic sheet-like nanocarrier, opens the way to functional nanocomposites with a high level of selective emission and efficient photosensitizing ability, utilizing unique graphene properties as a singlet oxygen generation enhancer.

Research on CO_2 photoreduction has yielded various photocatalysts that gave mostly CH_4, CO, CH_3OH, and some other single-carbon molecules as the photoreduced species. Ethyne, C_2H_2, was produced in a remarkable amount in visible light, using a catalyst made of graphene and 5,10,15,20-tetrakis(4-hydroxyphenyl) porphyrin or its Co(II) complex [111]. It was suggested that, upon illumination, the photoexcited electrons are injected into graphene from the porphyrin molecules and then transferred to the adsorbed CO_2. Due to the hydrophobicity of graphene, H_2O molecules cannot compete to obtain electrons from graphene.

Finally, it is worth mentioning the photocatalyst constructed by modifying TiO_2/RGO with 5,10,15,20-tetrakis(4-nitrophenyl)porphyrin, TNO_2PP [112]. The porphyrin sensitized TiO_2/graphene showed more than 35% higher photocatalytic efficiency than TiO_2 nanowires, under the same conditions, in decomposition of methylene blue by visible light.

1.1.3 Sensors

The sensing abilities of porphyrin-graphene nanohybrids are strictly related to electrocatalytic properties.

Excellent recent reviews on graphene-based electrochemical sensors [113] and on analytical devices in biomolecules detection and cancer diagnostics [114] appeared recently, where the superior electrochemical sensing performances of graphene-based electrodes toward the detection of various biomolecules and chemicals have been demonstrated. Also in this field, porphyrin-functionalized graphene hybrids resulted in interesting analytical applications, as will be discussed next.

Trace amounts of Cu(II) ions in water samples were easily extracted using a glassy membrane disk modified with 5-(4-aminophenyl)-10,15,20-triphenylporphyrin covalently linked to GO [115]; the same method was used for determination of Cr(III) ions [116]: complexation of metal ions by the porphyrin was the key for the success of the method. Instead, for the rapid and selective detection of Cd^{2+} ions in aqueous solution, a sensor was devised electrostatically linking a positive porphyrin (H_2TMPyP) with the negatively charged chemically converted graphene sheets [117]. Complexation of Cd(II) ions induced a large bathochromic shift of the porphyrin Soret band, thus originating an efficient optical probe.

RGO functionalized with a picket fence porphyrin, (5,10,15,20-tetrakis[2-(trimethylammoniomethylcarbonylamino)phenyl]porphyrin) iron complex, FeTMAPP, gave a nanocomposite that resulted in highly sensitive chlorite ion detection [118]. The effects of common interfering species on the sensing response were examined. Interestingly, even if present in large excess with respect to chlorite ion, neither cations, anions, sugars nor organic acids (lactic, ascorbic, uric, acetic) did interfere.

A cobalt tetraphenylporphyrin/reduced graphene oxide nanocomposite, CoTPP/RGO, exhibited high electrocatalytic activity both for oxidation and reduction of H_2O_2. As a consequence, it worked as a H_2O_2 biosensor, with good antiinterfering ability toward oxidative interferences at the oxidative potential of +0.50 V and good antiinterfering ability toward reductive interferences at the reductive potential of −0.20 V [119].

Nanoreactors prepared by π-stacking and electrostatic interactions between graphene quantum dots and iron(III) 5,10,15,20-tetrakis(1-methylpyridinio-4-yl) porphyrin acted as H_2O_2 sensor [120]. In fact, the photoluminescence of graphene quantum dots, quenched by the porphyrin was restored in presence of H_2O_2 that reacts with porphyrin, disrupting the tetrapyrrolic nucleus, with loss of iron.

5,10,15,20-Tetrakis(4-pyridyl)porphyrin was covalently linked to GO and electrochemical reduction of explosive nitroaromatic compounds was obtained by the hybrid, thus performing detection of ultra-trace explosives [121]. Comparing the sensitivity and detection limit of various electrochemical sensors toward TNT detection, the resulting porphyrin/graphene-modified GC electrode was of superior quality, allowing detection of 0.5 ppb. Detection limit for other nitroarenes were 1 (2,4-dinitrotoluene), 1 (1,3,5-trinitrobenzene) and 2 (1,3-dinitrobenzene) ppb, respectively.

A system for the detection of pesticides [122] was developed coupling an electrochemical sensor to flow injection analysis. Hemin-GO and nickel 1,4,8,11,15,18,22,25-octabutoxyphthalocyanine-GO composites showed high sensitivity and selectivity toward Carbofuran-fenol and Diuron, respectively (Scheme 14.3).

Organoposphorous insecticides play a major role in crop protection but are a threat to health. Fast and highly sensitive analytical methods are necessary for their detection. In this respect, dimethoate, $(MeO)_2P(S)SCH_2CONHMe$, was detected by manganese 5,10,15,20-tetrakis(1-methylpyridinio-4-yl)porphyrin-graphene nanohybrid on glassy carbon electrode, with linear response in the range $1 \times 10^{-6} \div 1 \times 10^{-4}$ M [123].

Hydroquinone, an important xenobiotic micropollutant, could be detected by photoelectrochemical sensing due to a porphyrin/Au nanoparticles/graphene nanocomposite: upon white-light illumination, hydroquinone was oxidized, generating photocurrent, the intensity of which increased with the increase of hydroquinone concentration [124].

Sulfonamides antibacterial drugs, largely used to cure human and animal diseases, accumulate in water. A method for their detection in tap water and in rivers has been developed using a GO/Fe_3O_4/5,10,15,20-tetrakis(4-carboxyphenyl)porphyrin nanocomposite [125]. It exhibited great magnetic solid-phase extraction capacity, adsorbing positively charged sulfonamides over its negatively charged surface, allowing preconcentration of sulfonamides and their HPLC detection in environmental water samples (Scheme 14.4).

The electrochemical oxidation of 17β-estradiol—a steroid that causes abnormal thyroid function in birds and fish—was obtained with copper *meso*-tetra(thien-2-yl)porphyrin/RGO hybrid [126].

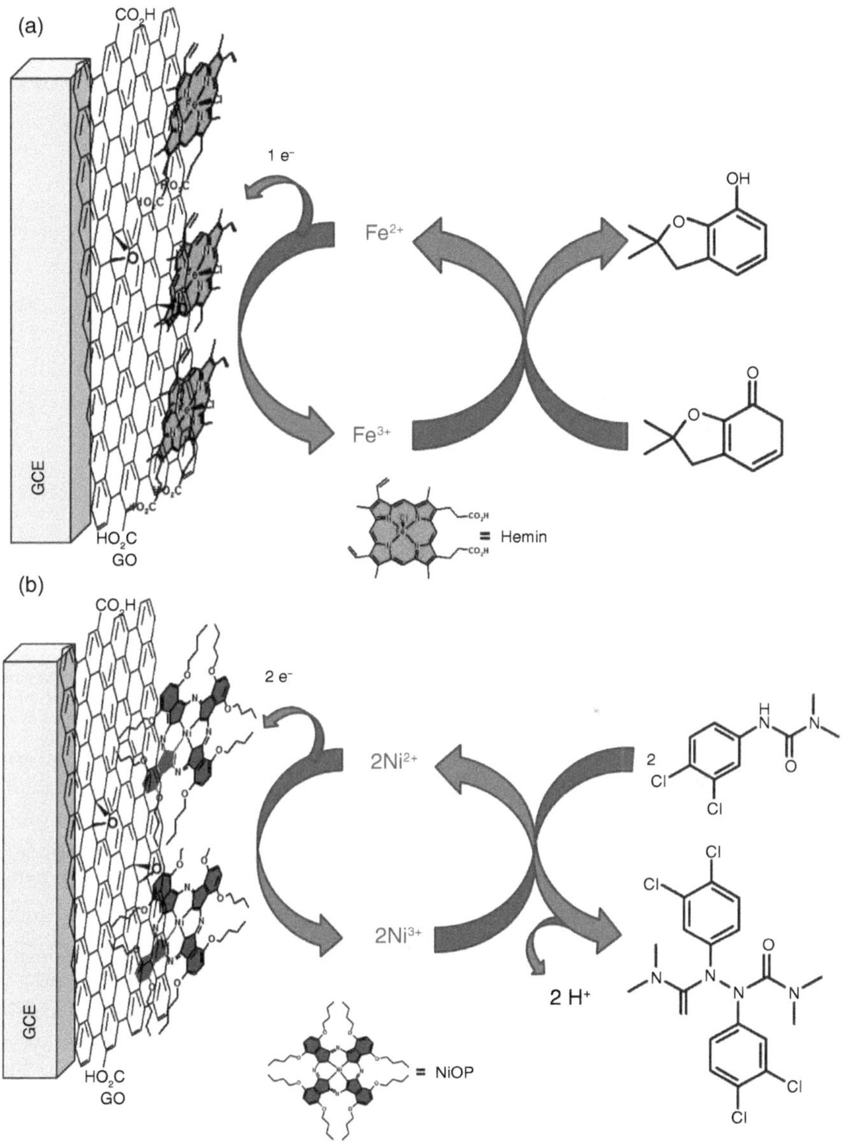

SCHEME 14.3 Hemin-GO (a) and NiOP-GO (b) sensors for electrochemical determination of carbofuran-fenol and diuron, respectively. *(Adapted from Ref. [122].)*

Several papers described the use of porphyrin-graphene hybrids as sensors for biomolecules. Thus, the electrochemical interactions between hemoglobin and H_2TPP made the hydroxypropyl-β-cyclodextrin/RGO/tetraphenylporphyrin composite a material with high biorecognition capability [127].

Electrochemical detection of cyclin A_2, which is a prognostic indicator in early-stage cancers, with ultrahigh sensitivity and selectivity was realized using

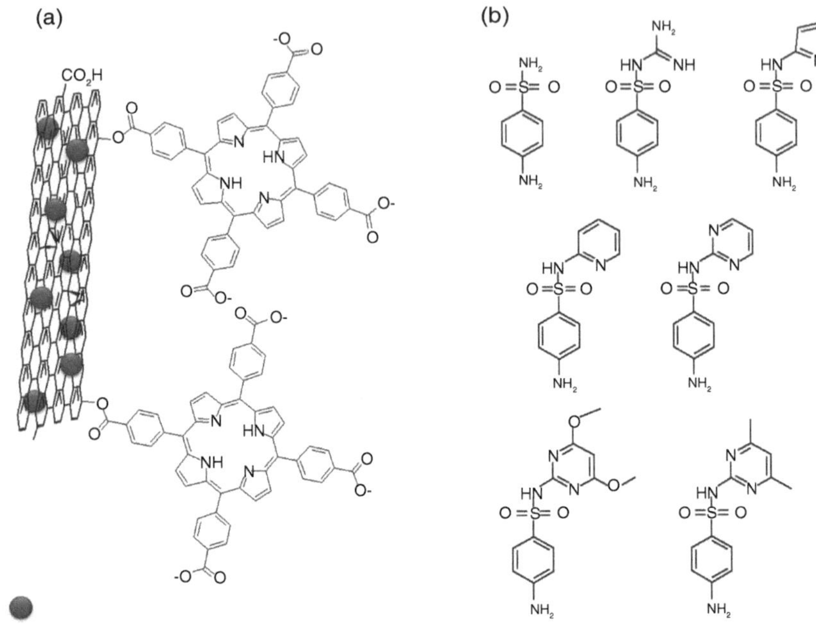

SCHEME 14.4 (a) GO/Fe_3O_4/5,10,15,20-tetrakis(4-carboxyphenyl)porphyrin nanocomposite and (b) sulfonamides detected in tap and river water samples. *(Adapted from Ref. [125] with permission of Elsevier.)*

a glassy carbon electrode modified with graphene noncovalently functionalized with 5,10,15,20-tetrakis(4-carboxyphenyl)porphyrin, one specific hexapeptide as detection probe, and a commercially available aliphatic ester to prevent non-specific binding [128]. The authors stress that this electrochemical sensor not only can detect cyclin A_2 in cancer cell extracts but can differentiate cancer cells from normal ones. The same research group, using the same strategy, obtained telomerase activity detection based on a porphyrin-graphene nanocomposite modified electrode [129]. The glassy carbon electrode modified with graphene was functionalized with a positively charged porphyrin (5,10,15,20-tetrakis[4-(*N*,*N*,*N*-trimethyl)ammoniophenyl]porphyrin, H_2TAmPP) and $Ru(bpy)_3^{2+}$ was used as signal reporter.

Highly selective and sensitive detection of dopamine was realized using porphyrin-functionalized graphene [130]. Graphene modified with the negatively charged 5,10,15,20-tetrakis(4-carboxyphenyl)porphyrin was able to interact with the positively charged dopamine via aromatic π–π-stacking and electrostatic attraction. The selective interaction allowed dopamine detection also in the presence of interferents, such as ascorbic acid and uric acid. Satisfactory results were obtained also with real human urine and serum samples.

Later, protoporphyrin IX (PPIX) was used to prepare a porphyrin-RGO nanocomposite that was used for dopamine detection, with a limit of detection of 9.45×10^{-9} M [131].

Dopamine and serotonin were simultaneously and selectively detected using a porphyrin-graphene sensor, prepared by a condensation reaction of GO and 5,15-bis(*p*-aminophenyl)-10,20-bis(pentafluorophenyl)porphyrin followed by electrochemical reduction [132].

A photoelectrochemical sensing platform was fabricated, based on the composition of protoporphyrin IX, tungsten trioxide, WO_3, and RGO on an indium tin oxide (ITO) electrode for detecting cysteine in aqueous solution, with a 25 nM detection limit [133].

Another interesting report presents the one-pot preparation of a ternary hybrid nanomaterial composed of a porphyrin-functionalized graphene, tin oxide, and gold nanoparticles [134]. This composite was applied to the simultaneous determination of epinephrine and uric acid. First, 5,10,15,20-tetrakis(4-sulfonatophenyl)porphyrin was linked to GO sheets dispersed in aqueous solution, then $HAuCl_4$ was added, followed by $SnCl_2$. The latter ensured reduction to graphene and Au nanoparticles while forming SnO_2. Reduction of GO was completed with hydrazine in alkaline medium.

Graphene modified with a selection of porphyrins and phthalocyanines was tested for the analysis of folic acid in pharmaceutical tablets [135].

1.1.4 Biosensors

Electrochemical reduction of GO nanoribbons and sequential functionalization with the water-soluble 5,10,15,20-tetrakis(1-methylpyridinio-4-yl)porphyrin gave an electrode with excellent electrocatalytic behavior toward reduction of O_2 [136]. The electrode, modified by addition of glucose oxidase (GOx), GOx/RGO nanoribbons, worked as a biosensor for electrochemical detection of glucose, based on the consumption of O_2 with enzymatic amplification.

5,10,15,20-Tetrakis(4-methoxy-3-sulfonatophenyl)porphyrin gave with graphene hybrid nanosheets that join the high conductivity and large specific surface area of graphene with photochemical electron-transfer ability and excellent electrochemical activity of porphyrins. Combined with the high affinity and specificity of an aptamer (that is, a biopolymer with a unique secondary structure that can bind specifically to its target), a simple, rapid, sensitive, and label-free electrochemical aptasensor was successfully fabricated for adenosine triphosphate (ATP) detection [137].

By the same strategy, a glassy carbon electrode modified with a graphene-porphyrin nanocomposite was prepared; with the thrombin aptamer, it produced an aptasensor for ultrasensitive determination of thrombin [138].

Loading FeTMPyP and streptavidin onto graphene, a novel horse radish peroxidase (HRP) mimic with high peroxidase activity was obtained. It was used to recognize a biotinylated molecular beacon for specific electrochemical detection of DNA, even at attomolar levels [139] (Fig. 14.7).

On the basis of the known ability of porphyrins to bind nucleic acids and of the catalytic properties of cobalt porphyrins toward oxygen reduction, a biosensor was prepared for electrocatalytic detection of short noncoding ribonucleic

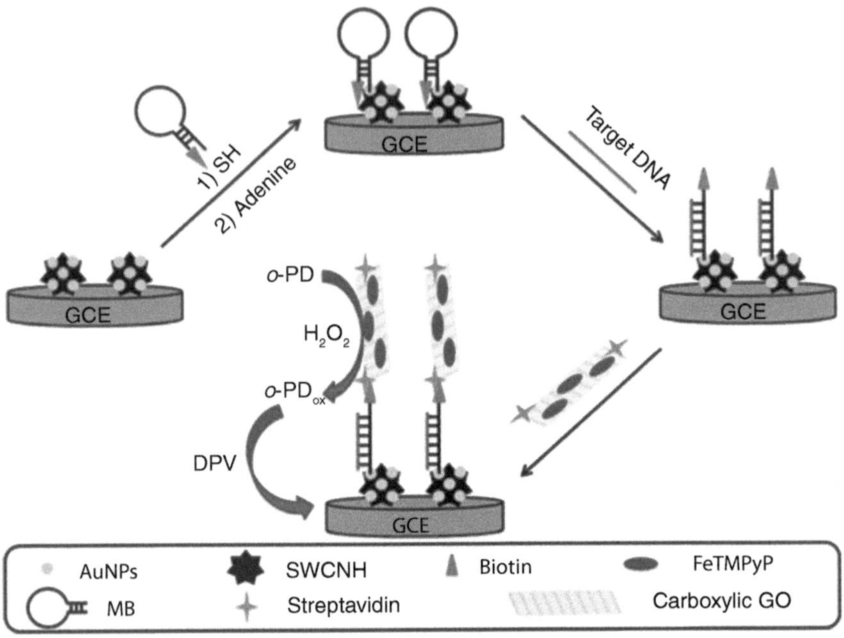

FIGURE 14.7 Graphene-supported FeTMPyP as an HRP mimicking trace label for electrochemical detection of DNA. *(Reprinted from Ref. [139] with permission of Royal Society of Chemistry.)*

acids (micro-RNA), which play an important role in translation regulation and in RNA degradation [140]. CoTCPP was physisorbed on reduced GO, and the hybrid was deposited on a glassy carbon electrode. NH_2-modified DNA probes were covalently grafted on CoTCPP/RGO/GCE. Hybridization of the probe with short micro-RNA targets led to a change in conformation of DNA strands, from a random coil structure to a well-organized double strand that influenced the electrocatalytic behavior of CoTCPP toward oxygen reduction and, consequently, allowed micro-RNA transduction.

1.1.5 Diagnostic and Therapeutic Applications

The unique two-dimensional structure of graphene and GO and their peculiar physicochemical properties have been largely explored also in the field of bioapplications, opening new horizons for diagnostic and therapeutic applications, as well illustrated in a recent review [141]. However, especially in the field of drug delivery, prudence is recommended [142,143]. The different synthetic strategies and routes available to fabricate the graphene conjugates, and the opportunities they open for the design of multifunctional tools for synthetic biology, have been discussed by Prato and coworkers [144] and can be summarized as in Fig. 14.8.

A nanosupramolecular assembly was built with a folic acid-modified β-cyclodextrin (as a target unit), an adamantanyl porphyrin (as a linker unit),

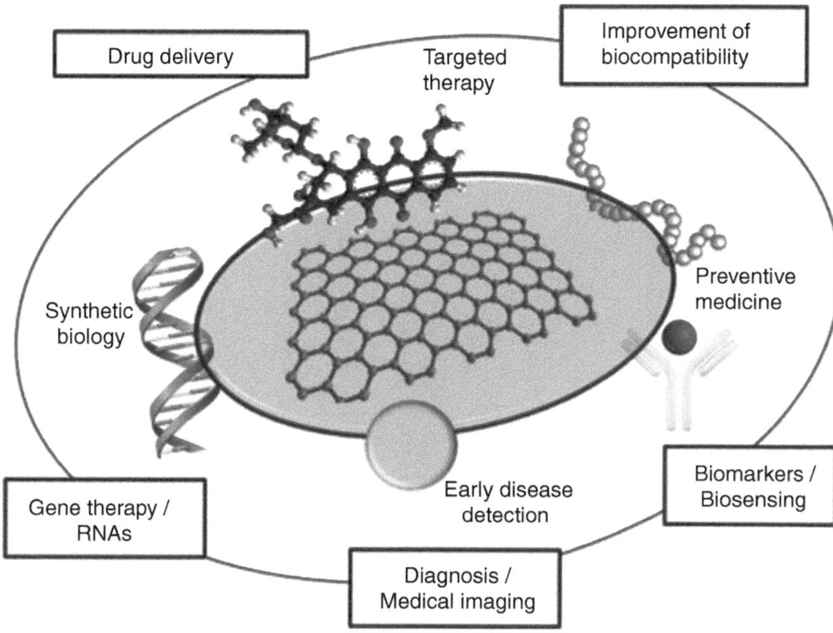

FIGURE 14.8 Graphene-conjugate possible applications for multifunctional nanomedicines aimed at personalized medicine and synthetic biology. *(Reprinted with permission from Ref. [144], from Elsevier.)*

doxorubicin (Dox), and GO (as a carrier unit) [145]. The resulting quarternary supramolecular nanoarchitecture, after association with doxorubicin, was employed as a targeted delivery system to efficiently carry the drug molecules. It exhibited, according to the authors, better drug activity and much lower toxicity than free doxorubicin in vivo.

In the attempt to develop a new enzyme-activatable GO–photosensitizer complex as a biologically tunable theranostic agent for near-infrared (NIR) fluorescence imaging and subsequent photo-induced cancer therapy, a photosensitizer (chlorin e6, Ce6) was coupled to the hyaluronic acid (HA) backbone, forming an amphiphilic HA–Ce6 conjugate [146]. Physical adsorption of HA–Ce6 on the surface of GO plates formed the noncovalent complex GO–HA–Ce6, which resulted in strong inhibition of both fluorescence and singlet oxygen generation with respect to HA–Ce6. In the presence of hyaluronidase (HAdase), HA polymer degrades, releasing fragments from GO surface, Ce6 becomes highly fluorescent and phototoxic, thus ensuring effective NIR-light-induced photothermal therapy for cancers. Thus, GO–photosensitizer complex acts as an enzyme-activatable theranostic agent.

A different approach led to the development of a theranostic platform with a single photothermal and photodynamic reagent [147]. Silicon 2,3-naphthalocyanine bis(trihexylsilyloxide) was used as a photosensitizer. The hydrophobic silicon naphthalocyanine was immobilized (via π–π-stacking) on water-dispersible graphene

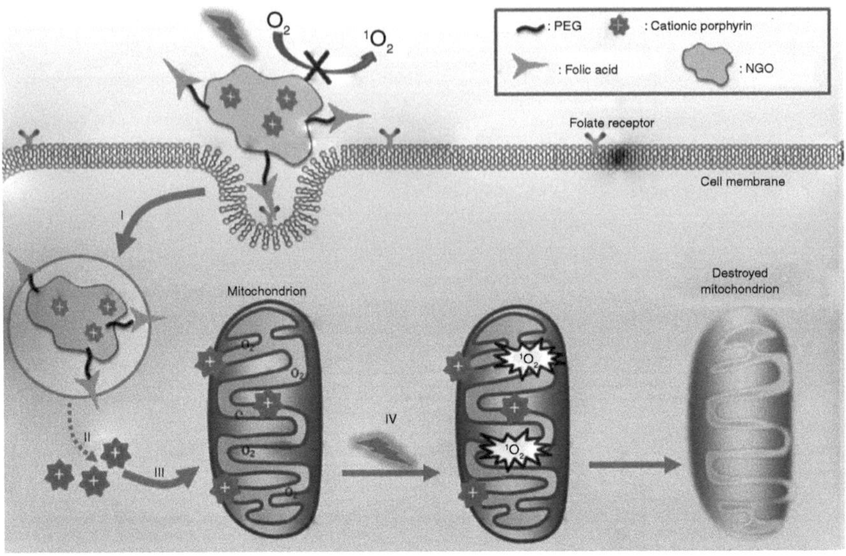

FIGURE 14.9 (a) Internalization. (b) Release of photosensitizer. (c) Localization of photosensitizers in mitochondria. (d) Oxidative damage of mitochondria. *(Reproduced from Ref. [148] with permission of American Chemical Society. Copyright 2015.)*

that was magnetic and fluorescent, contemporarily (MFG). In vitro experiments demonstrated that MFG/Si naphthalocyanine has dual modal imaging and photothermal/photodynamic capability, with killing efficacy of ca. 98% cancer cells.

A cellular and subcellular targeting photodynamic therapy nanosystem was obtained by loading a cationic porphyrin, 5-[4-(4-trimethylammoniobutoxy)phenyl]-10,15,20-triphenylporphyrin tetrabromide, on nanographene oxide (NGO) conjugated with polyethylene glycol (PEG) and functionalized with folic acid (FSA) [148]. As illustrated in Fig. 14.9, the NGO-based carrier quenches porphyrin fluorescence and singlet oxygen generation; moreover, it also internalizes the photosensitizer into mithochondria that, upon light irradiation and singlet oxygen generation, are destroyed.

Sensors for pattern recognition of neuron specific enolase (NSE) and carcinoembryonic antigen (CEA) were proposed for the screening of whole blood samples [149]. The best sensor for the assay of CEA was the one based on tetraphenylporphyrin/graphite; for the assay of NSE, the best sensor was the one based on the same porphyrin and graphene. The authors state that this is the only screening test for early detection of lung cancer, by identification of NSE and CEA in whole blood samples.

Finally, it is worth mentioning that a noninvasive photothermal therapy for brain cancer, realized with a biocompatible porphyrin-functionalized GO [150]. *Meso*-tetrakis(4-methylphenyl)porphyrin was conjugated with GO by π–π-stacking, giving a photothermal platform that induced cell death in 4 min, irradiating with a 808 nm laser (in vitro experiments).

1.2 Carbon Nanotubes-Porphyrin Nanoreactors

Within the huge amount of papers dealing with carbon nanotubes (CNTs or, simply NTs) that appeared in the literature since the seminal work by Iijima [151] and exponentially increased in the following decades [152,153], porphyrin-decorated NT represent an important part, both with the single-walled (SWCNTs) and multiwalled (MWCNTs) carbon nanotube type.

Applications of porphyrin-nanotube composites are, by and large, mainly in the photovoltaic [154–156] and optoelectronic [157,158] fields, that are out of the scope of the present chapter. Nevertheless, a number of interesting reactions have been investigated with porphyrin-CNT systems that can be regarded as nanoreactors. Moreover, being monodimensional conducting polymers, they are expected to play a role in different nanotechnology fields [159,160].

CNTs can be functionalized with porphyrins or related macrocycles either covalently or in a supramolecular manner [161–163].

As an example of covalent binding, we can mention the Suzuki coupling between an iodoaryl-SWCNT and the boronate derivative of zinc arylporphyrin shown in Scheme 14.5 [164]. The reaction, performed in *N,N*-dimethylformamide (DMF) at 100°C, with $Pd(PPh_3)_4$ as the catalyst and Cs_2CO_3 as the base, gave a cross-coupled SWCNT highly soluble in organic solvents.

A different strategy was used by Guldi and coworkers. They used click chemistry to introduce a limited amount of functionality on nanotubes (not to disrupt the conjugated π-system), but the set-up was versatile enough to

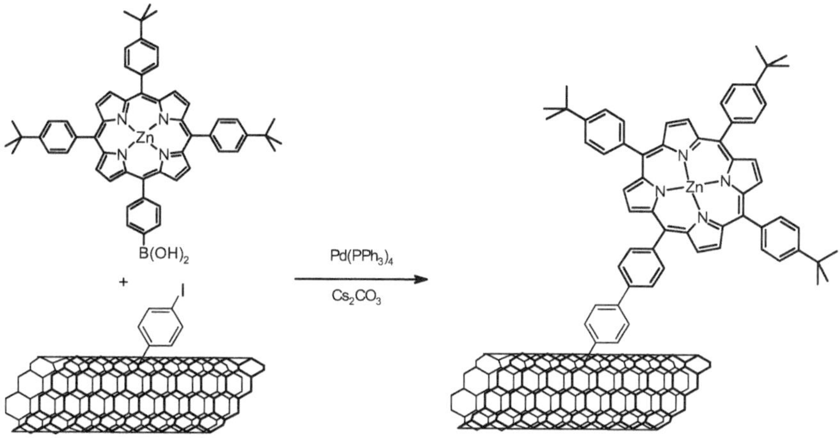

SCHEME 14.5 Example of Suzuki coupling to covalently functionalize SWCNTs. Double bonds in NT omitted for clarity. *(Adapted from Ref. [164] with permission of American Chemical Society, 2006.)*

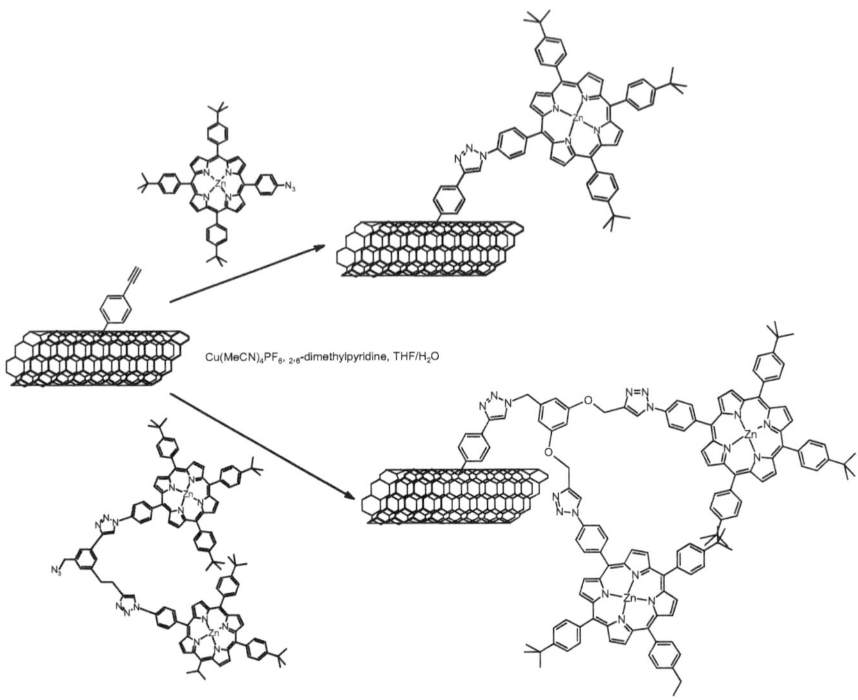

SCHEME 14.6 Covalent functionalization of SWCNTs via click chemistry. Double bonds in NT omitted for clarity. *(Adapted from Ref. [165] with permission of American Chemical Society, 2009.)*

accommodate different spacers and one or two porphyrin moiety per functional group [165] (Scheme 14.6).

On the other hand, it was reported that addition of SWCNT to a DMF solution of zinc 5,15-bispentylporphyrin, followed by sonication and filtration, resulted in a very stable complex [166].

The problem of aggregation of NT dispersions, which makes postprocessing complicated, has been addressed using porphyrins with chiral and hydrophilic substituents with consequent redissolution of aggregates in water [167].

Porphyrin-carbon nanotube (both SWCNT and MWCNT) hybrids act as nanoreactors in catalysis (chemical and electrochemical catalysis), in sensing (both as chemical and biochemical sensors) or in diagnostic and therapeutic applications, as will be discussed next.

1.2.1 Catalysis and Biocatalysis

The reaction more frequently investigated using porphyrin-CNTs nanoreactors is alkenes epoxidation.

Mn(III) 5,10,15,20-tetrakis(4-aminophenyl)porphyrin chloride, Mn(TNH$_2$PP)Cl (Scheme 14.7), covalently bound to multiwalled CNTs via amido functionality, [Mn(TNH$_2$PP)Cl-MWCNT]/NaIO$_4$, was an efficient catalyst for the

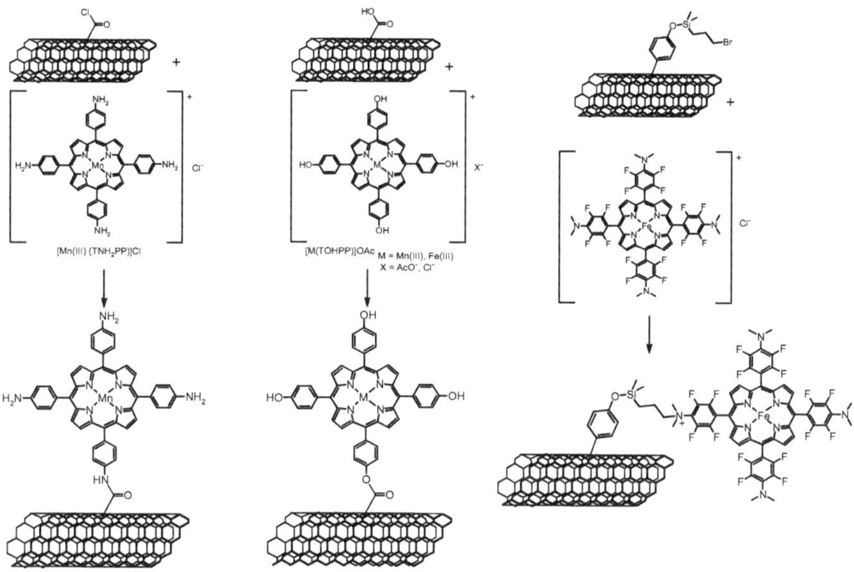

SCHEME 14.7 Porphyrins-CNTs hybrids used to catalyze alkenes epoxidation. Double bonds in NT omitted for clarity.

epoxidation of olefins with sodium periodate at room temperature, with imidazole as the axial ligand [168]. Terminal and internal alkenes—aliphatic, alicyclic, and aromatic ones—were transformed into the corresponding oxacyclopropanes with generally high yields. Recycle experiments showed that, up to seven runs, no significant yield decrease occurred. Later, Mn(III) and Fe(III) complexes of 5,10,15,20-tetrakis(4-hydroxyphenyl)porphyrin (TOHPP, Scheme 14.7) were covalently bound to CNTs by an ester functionality and were used in epoxidation of alkenes using *tert*-butyl hydroperoxide as the primary oxidant [169]. Alkene conversions were better with Fe(III) than with Mn(III) porphyrin, but selectivity was low, with high percentage formation of the corresponding aldehyde. MWCNT-[Mn(III)TOHPP]OAc was used by the same research group, with H_2O_2 as the primary oxidant and irradiation by ultrasound [170]. Results were excellent with cyclooctene (100% conversion, 100% epoxide formation), but not so good with other alkenes. Also, Fe(III) complex of TOHPP gave good results with cyclooctene, but other alkenes were not investigated [171].

Cyclooctene epoxidation was investigated with the heterogenous iron(III) complex of 5,10,15,20-tetrakis(4-dimethylaminoperfluorophenyl)porphyrin, linked to MWCNTs via an organosilyl group (Scheme 14.7), heating, or using microwaves [172]. Both systems could be recycled, although some decrease in conversion was observed in subsequent cycles. Selectivity was 100% toward oxacyclopropane in all cases.

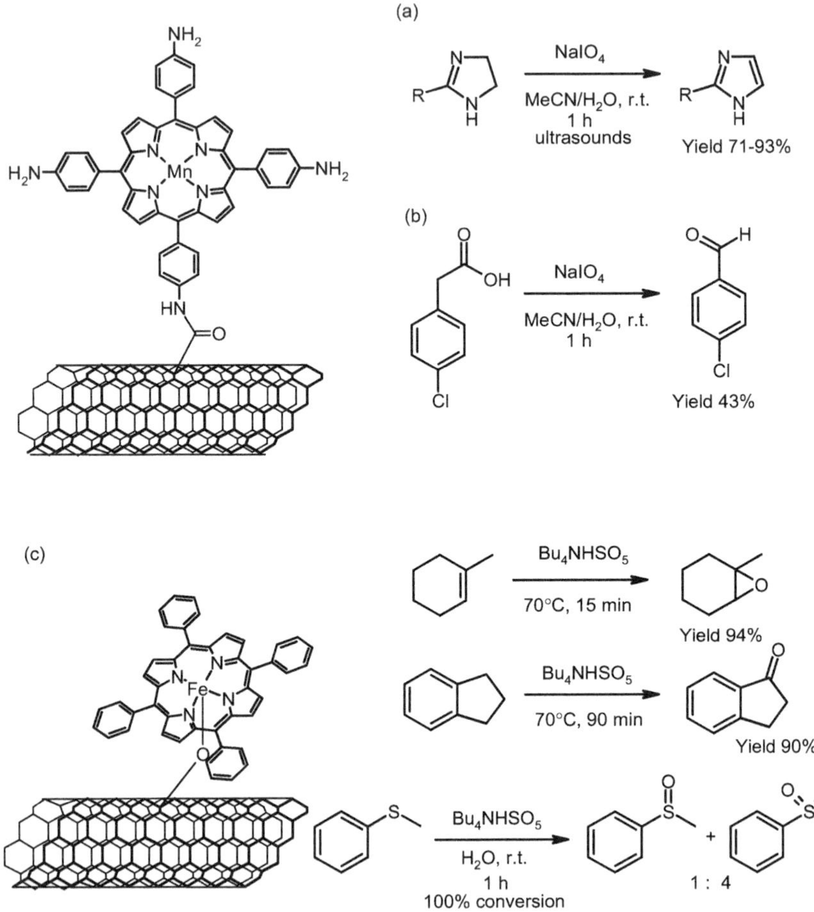

SCHEME 14.8 Examples of porphyrin-CNTs hybrids as nanoreactors in catalytic reactions. *(Adapted from Refs. [174] (a), [175] (b), [176] (c) with permission of Elsevier, Springer and Royal Society of Chemistry.)*

MWCNTs noncovalently bound to Mn(III) complexes of $TPPH_2$ and $TPyPH_2$ (Chart 14.1) were also used as catalysts in alkene epoxidation, using Oxone® ($2KHSO_5.KHSO_4.K_2SO_4$) as the primary oxidant [173]. The results were good as far as resistance to degradation of heterogeneous catalysts is considered, but disappointing in terms of alkene conversion and epoxide yield.

Other reactions that involved [Mn(TNH_2PP)Cl@MWCNT] as catalytic nanoreactor are the oxidation with sodium periodate of 2-imidazolines in MeCN [174] and the oxidative decarboxylation of α-arylcarboxylic acids with sodium periodate [175] (Scheme 14.8 a,b). On the other hand, iron(III) *meso*-tetra(2-chlorophenyl)porphyrin, coordinated to hydroxyl-functionalyzed MWCNTs, catalyzed the oxidation of alkanes, alkenes, and sulfides using tetrabutylammonium peroxomonosulfate (Bu_4NHSO_5) as oxygen source [176] (Scheme 14.8c).

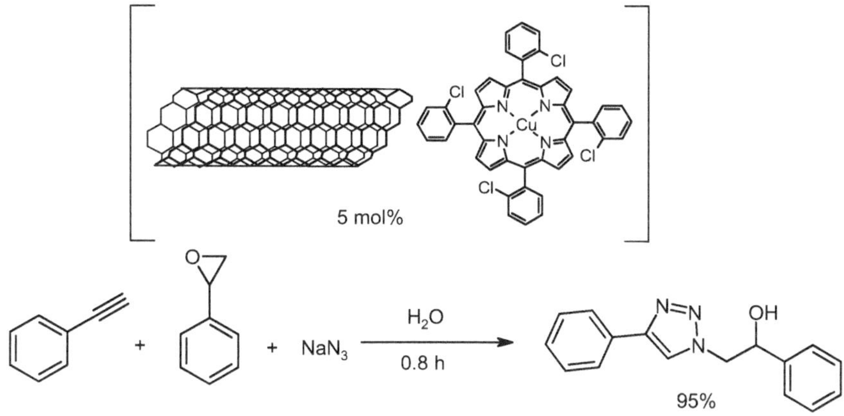

SCHEME 14.9 Example of regioelective synthesis of β-hydroxy-1,4-disubstituted 1,2,3-triazoles. *(Adapted from Ref. [177] with permission of Wiley-VCH, 2009.)*

An interesting catalytic activity was observed with copper *meso*-tetra(2-chlorophenyl)porphyrin, CuT(*o*-Cl)PP, anchored on MWCNT by microwave irradiation [177]. This nanoreactor efficiently catalyzed the click three-component synthesis of 1,2,3-triazoles in water, from various epoxides, terminal alkynes, and sodium azide (Scheme 14.9). High yields were accompanied by the easy recovery of the catalyst, which produced results robust enough to be reused ten times, with negligible copper leaching.

A biocatalytic nanoreactor was also reported, made of acrylonitrile-based copolymers bearing porphyrin pendants that, after blending with CNTs, were electrospun into nanofibers. The nanofibers, in turn, were covalently bound to catalase, maintaining it in the active conformation [178]. As a consequence, the enzyme activity and stability improved.

1.2.2 Electrochemical Nanoreactors

As mentioned for porphyrins–graphene hybrids, electrocatalysis by porphyrins-CNTs nanoreactors can be found in oxidation or in reduction reactions.

Dioxygen reduction reaction (ORR) attracted numerous investigations, with the aim of finding efficient and durable electrode catalysts to be applied in hydrogen-oxygen fuel cells. A theoretical approach was used to address the interactions of molecular dioxygen with metal porphyrins (Mn, Fe, Co, Ni) covalently linked to SWCNTs [179], with results that supported experimental evidence that ORR activity is favored by covalently bound CNT-metal porphyrin hybrids. Later, in an attempt to rationalize factors that determine the overpotential value (η^{ORR}) of electrochemical oxygen reduction on metal porphyrins-NTs hybrids, a density functional theory calculation was performed [180]. The study demonstrated that the η^{ORR} value can be lowered below that of Pt—that is the most active catalyst for ORR—coordinating an axial ligand to the Fe(III) metal ion of porphyrins and incorporating the metal macrocycle into single-walled CNTs.

Many relevant papers are too long to discuss in detail. But we can give highlights of some. A hybrid thin film on glassy carbon electrode was prepared with cobalt complex of 5,10,15,20-tetrakis(1-methylpyridinium-4-yl) porphyrin, $CoTMPyP(ClO_4)_5$, and MWCNTs [181,182]. A nanocomposite was prepared through noncovalent adsorption of iron picket-fence porphyrin (iron 5,10,15,20-tetrakis[αααα-2-trismethylammoniomethylphenyl] porphyrin) on MWCNTs [183]. A gold nanoparticles/nanocomposite self-assembled monolayer was formed on gold electrode. Besides resulting in oxygen reduction, it exhibited good response to oxygen, without interference of ascorbic and uric acids, thus showing itself useful as a sensor in the detection of dissolved oxygen. Multiwalled carbon nanotubes modified with iron tetrakis(4-methoyxphenyl) porphyrin chloride were active toward electrocatalytic oxygen reduction in acidic media [184]. Tailor-made dual electrocatalysts for the four-electrons reduction of oxygen were prepared by dissolving cobalt(II) 5,10,15,20-tetrakis(4-methoxyphenyl)porphyrin in an ionic liquid, (1-butyl-3-methylimidazolium hexafluorophosphate, $bmimPF_6$) containing Prussian Blue nanoparticles [185].

Cobalt tetrakis(2,4,6-trimethylphenyl)porphyrin-SWCNTs modified glassy carbon electrode and Nafion in aqueous solution afforded efficient electrocatalytic four-electron reduction of dioxygen to water directly [186]. Water-soluble cationic iron 5,10,15,20-tetrakis(1-methylpyridinium-4-yl)porphyrin, FeTMPyP, was assembled on SWCNTs containing 4-(pyren-1-yl)butanoic acid (PBA) via electrostatic interaction. The formed SWCNTs/PBA/FeTMPyP kept the initial structure of SWCNTs and was electrocatalytically active in reduction of O_2 and NO [187]. Metalloporphyrin films were obtained by electrochemically induced polymerization on top of a MWCNT modified with 4-(pyrrole-1-yl) benzoic acid and showed activity toward ORR better than that of without MWCNTs [188]. The combination of Co(III) protoporphyrin IX chloride (2,18-bis(2-carboxyethyl)-8,13-divinyl-3,7,12,17-tetramethylporphyrin, PPIX), MWCNTs, and tungsten oxide produced a film, deposited on glassy carbon electrode, capable of electrocatalytic oxygen reduction in acidic medium [189–191]. Cobalt 5-(4-hydroxyphenyl)-10,15,20-triphenylporphyrin was covalently grafted to functionalized MWCNTs. The nanocomposite, mixed with Nafion, was able to catalytically reduce oxygen in 0–5 pH acidic media, at room temperature [192]. Reduction of O_2 and H_2O_2 has been reported for conductive composite film containing MWCNTs coated with poly iron tetrakis(*o*-aminophenyl)porphyrin [193].

Selected compounds from a library of hanging porphyrin xanthene derivatives, containing pendant groups with various proton-donating abilities, were complexed with Co(II) and immobilized on MWCNTs, showing that catalytic reduction of oxygen to water can be achieved at the single cobalt center as far as the electron-deficient "hangman" porphyrin platforms possess a distal group capable of proton transfer [194].

Iron(II) phthalocyanine (FePc) and cobalt tetrakis(4-methoxyphenyl)porphyrin were attached to the MWCNTs' surface by π-stacking interactions producing a nanohybrid that, according to the authors, was capable of high

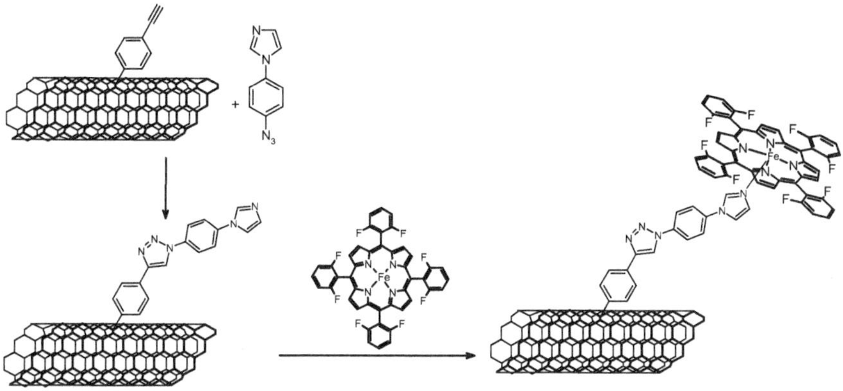

SCHEME 14.10 Bioinspired catalyst for efficient ORR. *(Adapted from Ref. [201] with permission of Wiley-VCH, 2014.)*

performance in oxygen reduction [195]. MWCNTs, drop-coated in the form of ink on a glassy-carbon electrode, were modified with a catalytically active film of iron(III) tetrakis(4-methoxyphenyl)porphyrin chloride (FeTMOPP-Cl) through a pulsed potential deposition technique [196]. ORR was investigated with different Co(II) and Fe(II) phthalocyanines supported on different types of carbon NTs (single-, double-, or multiwalled ones) [197]. Iron phthalocyanine on MWCNTs resulted in the best system in terms of catalytic activity and low overpotential (60 mV). Iron and cobalt phthalocyanine, as well as iron and cobalt tetrakis(4-methoxyphenyl)porphyrin, were adsorbed on the surface of MWCNTs, pyrolyzed (at 400 or 800°C) in flowing argon atmosphere, and tested in ORR [198]. Electrocatalytic activity is especially high with nanoreactors thermally treated at 800°C.

A single-walled carbon nanotube-Nafion® film, with cobalt (II) tetrakis(2-aminophenyl)porphyrin and palladium nanoparticles, over a glassy carbon electrode gave the four-electrons electrocatalytic reduction of O_2 to water with a reported minimal generation of H_2O_2 [199].

In a comparative investigation, an iron(III) tetraphenylporphyrin-MWCNT hybrid gave better performances than a cobalt(II) tetraphenylporphyrin-MWCNT one, not only in oxygen, but also in proton and carbon dioxide reduction reactions [200].

The conclusions from the theoretical investigations reported above [180] were experimentally confirmed when iron tetrakis(2,6-difluorophenyl)porphyrin was assembled on imidazole-functionalized carbon nanotubes (Im-CNTs) through coordination with the imidazole group (Scheme 14.10, [201]). In fact, this catalyst displayed higher ORR activity, superior stability, and excellent methanol tolerance in comparison to the commercial Pt/C catalyst.

The axial imidazole-coordinated porphyrin mimicked the active site of O_2-activating heme-containing enzymes. and the resulting nanoreactor

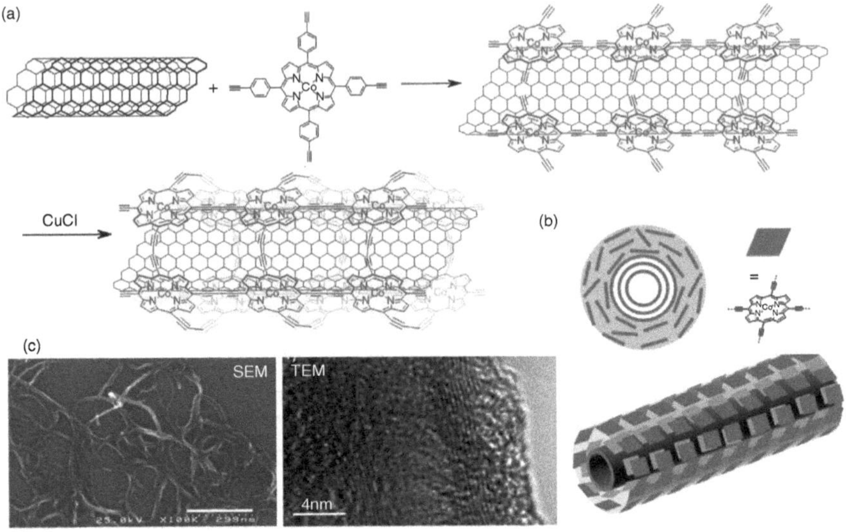

FIGURE 14.10 (a) Construction of porphyrin network on MWCNTs, (b) idealized scheme of the nanotubes coated with the porphyrin polymer, (c) SEM and TEM images of Co-porphyrin-coated MWCNTs. *(Adapted from Ref. [202] with permission of American Chemical Society. Copyright 2014.)*

demonstrated a high selectivity (99%) in the direct four-electrons reduction of dioxygen to water, while the two-electron production of H_2O_2 was minimal.

Beside development of more efficient catalysts, much research is currently devoted to finding more robust materials that maintain the peculiar properties of CNTs after functionalization. An innovative approach was that of first, absorbing cobalt(II) tetrakis(4-ethynylphenyl)porphyrin (CoPor) on the sidewalls of MWCNTs, and then dimerizing the triple bonds, with CNTs acting as a template [202] (Fig. 14.10).

The possibility of bioelectrochemical oxygen reduction was also explored. A bifunctional bioelectrocatalytic system was fabricated with the combination of Co-porphyrin supported on polyoxometalate-modified CNTs and peroxidase enzymes (horseradish and cabbage), deposited upon glassy carbon [203]. Later, the same research group used a cobalt porphyrin immobilized within the same peroxidases supported on MWCNTs [204].

Other electrocatalyzed reduction reactions were less investigated, but some of them may play an important role in the near future. For example, electrochemical reduction of carbon dioxide, such as that described with iron(III) 5,10,15,20-tetraphenylporphyrin chloride (FeTPPCl) electrostatically deposited onto carboxylated MWCNTs [205], might become a method of reducing CO_2 emission, transforming it into valuable chemicals.

As to oxidation reactions, some of them involved catalytic electro-oxidation of small inorganic molecules. For example, nitric oxide NO was electro-oxidized

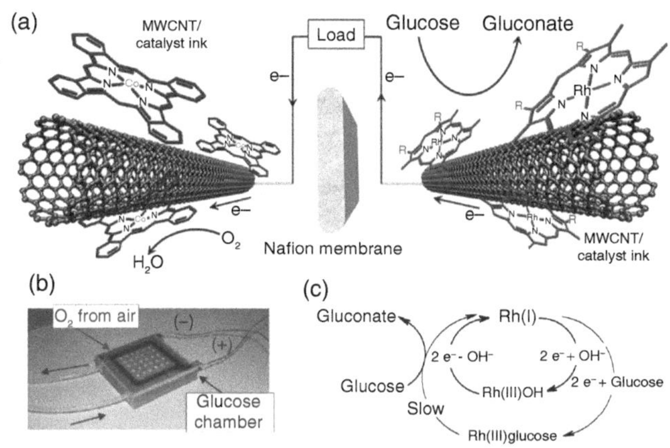

FIGURE 14.11 (a) Schematic representation of cathodic oxygen reduction and anodic glucose oxidation at MWCNT-metal phthalocyanine and porphyrin nanoreactors, respectively, (b) glucose fuel cell, (c) electrocatalytic oxidation of glucose by Rh(III). *(Adapted and reprinted from Ref. [207] with permission of American Chemical Society, Copyright 2012.)*

to nitrate ion by a nanoreactor made of cobalt (II) tetrakis(4-sulfonatophenyl) porphyrin (CoTPPS), axially coordinated to amino-modified MWCNTs [206].

Attention was devoted to electrocatalytic oxidation of glucose within the frame of constructing glucose/O_2 fuel cells. A glucose fuel cell was prepared using unique molecular catalysts at the cathode and the anode, and using a Rh(III) porphyrin on MWCNT for glucose oxidation, and cobalt(II) phthalocyanine on MWCNT for the oxygen reduction [207] (Fig. 14.11). The resulting nonenzymatic molecular catalysis-based glucose fuel cell presented a power density $P_{max} = 0.182$ mW cm^{-2} at 0.22 V and an open circuit voltage of 0.64 V.

Subsequently, Au(III) porphyrin, immobilized on MWCNT matrix, was reported by the same research group to oxidize glucose at low overpotential [208].

In the field of energy production, oxidative water splitting is an appealing process, especially if it can be achieved without noble metal catalysts. Thus, cobalt porphyrin, covalently bound to MWCNTs, gave a promising catalyst, in that it resulted in active electrochemical oxidation of water at low overpotential [209].

Finally, it is worth mentioning the oxidation of L-cysteine, obtained with iron phthalocyanines absorbed on MWCNTs [210] and the enzymatic MWCNTs electrodes linked to hemin, which showed improved electron transfer of pyrroloquinone-dependent glucose dehydrogenase [211].

1.2.3 Sensors

Carbon nanotubes modified with porphyrins constitute nanoreactors very useful for chemical sensing applications. Many results involve volatile organic compounds (VOCs) detection.

SWCNTs [212] or MWCNTs [213] functionalized with substituted free-base porphyrins allowed to measure several volatile analytes. Electropolymerization of pyrrole-substituted porphyrin monomers SWCNTs allowed to detect 1-butanol in concentrations close to the odor threshold [214].

SWCNTs-porphyrin nanosensors have been fabricated for monitoring toxic substances in the environment [215]. Free-base, Ru and Fe octaethyl- and tetraphenyl-substituted porphyrins provided good selectivity and sensitivity to various VOCs tested (acetone, butanone, methanol, ethanol). Noncovalently functionalized SWCNTs with iron tetraphenylporphyrin were used for benzene detection [216]. SWCNTs noncovalently functionalized with copper phthalocyanine and free-base porphyrins were used as sensing layers for the detection of toluene [217]. Also, MWCNTs were used as sensors for benzene, toluene, and xylene, when functionalized with metal tetraphenyl porphyrins [218]. SWCNTs-poly(tetraphenylporphyrin) hybrid was prepared and tested as a low-power chemiresistor sensor for acetone vapor [219]. A chemiresistive sensor array was fabricated from SWCNTs noncovalently functionalized with metallo *meso*-tetraphenylporphyrins (Cr(III), Mn(III), Fe(III), Co(III), Co(II), Ni(II), Cu(II), and Zn(II)) [220]. Its responses were treated by statistical analyses and allowed to classify VOCs into five classes: alkanes, aromatics, ketones, alcohols, amines. Amines' detection as an indicator of meat spoilage was achieved by the same group with the same sensor array [221].

Nonvolatile organic compounds can also be detected with CNTs-porphyrin composites. The introduction of hydroxyl groups in *meso*-tetraphenylporphyrin/CNTs systems enhanced the sensitivity of detection of explosives such as dinitrobenzenes [222].

5,10,15,20-Tetrakis(4-hydroxyphenyl)porphyrin/carbon nanotubes film was able to rapidly and simultaneously detect dihydroxybenzene isomers, taking advantage of their different oxidation potentials [223].

As to inorganic species, MWCNT-metal porphyrins hybrids were used to detect chlorate, bromate, and iodate ions (with iron(III) meso-teraphenylporphyrin [224]), sulfite ion in beverages (with picket fence [iron (III) 5,10,15,20-tetrakis(α,α,α,α-2-pivalamidopheny)porphyrin], FeTpivPP [225], Cu(II) [with 5-(4-benzylacrylamido)-10,15,20-tris(4-benzylamino) porphyrin [226]] and Hg(II) cations [with 5,10,15,20-tetrakis(3-hydroxyphenyl) porphyrin] [227]).

On a glassy carbon electrode modified with nanowires of polytetrakis (*o*-aminophenyl)porphyrin, SWCNT and Nafion as binders resulted in an efficient electrochemical detection of hydrogen peroxide [228], while detection of H_2O_2 in beverages was achieved with the picket-fence porphyrin, FeTpivPP, on MWCNTs [229].

Iron(III) meso-tetrakis (*N*-methylpyridinum-4-yl)porphyrin and gold nanoparticles were alternately assembled as cationic and anionic layers on a CNT wrapped with poly(diallyldimethylammonium chloride [230]. The

CNT-based electrode detected peroxyacetic acid with no interference by hydrogen peroxide.

Trichloroacetic acid in polluted water was successfully detected by a SWCNT-hematin composite in IL (1-butyl-3-methylimidazolium hexafluorophosphate, bmimPF_6) [231].

Analytes of biological importance can also be detected with CNT-porphyrin nanoreactors. Thus, Sudan I (1-phenylazo-2-naphthol), a food coloring agent with metabolites associated with bladder cancer, could be detected in real samples using iron(III) tetraphenylporphyrin-SWCNT system to modify a glassy carbon electrode [232]. On the other hand, the herbicide 2,4-D (2,4-dichlorophenoxyacetic acid) was selectively detected by a sensor made by modifying a carbon paste electrode with iron(III) *meso*-tetrakis(pentafluorophenyl)porphyrin/MWCNT composite [233].

Hemoglobin was determined with a sensor made of tetra-(*N*-phenylpyrazole) porphyrin/β-cyclodextrin/carbon nanotube film [234].

Dopamine, an important neurotransmitter molecule, was detected with metalloporphyrin—either Zn(II)-5,10,15,20-tetrakis(4-pyridyl)porphyrin or Co(II)-5,10,15,20-tetrakis(4-methoxyphenyl)-porphyrin—thin films on screen-printed carbon nanotube electrodes, prepared by matrix-assisted pulsed laser evaporation [235]. A few years later, an iron(III) [3,7,12,17-tetramethyl-8,13-divinylporphyrin-2,18-dipropanoato]/MWCNT composite was used to fabricate a modified glassy carbon electrode that was successfully used for the simultaneous determination of ascorbic acid, dopamine, uric acid, and nitrite ion [236].

1.2.4 Biosensors

Glucose oxidase was immobilized on a nanodendritic cobalt(II) poly[*meso*-tetrakis(2-thienyl)porphyrine]-SWCNT [237], supported on glassy carbon. The biosensor showed good glucose detection, based on oxygen reduction.

A selective sensor for reduced L-glutathione (GSH) in erythrocyte samples was obtained with iron(III) *meso*-tetrakis(1-methylpyridinio-4-yl)porphyrin absorbed on MWCNT [238]. Successively, the same research group reported the use of a supramolecular assembly of Co(II) phthalocyanine tertrasulfonate and the same cationic porphyrin on MWCNT for GSH detection [239].

Immobilization of hemin on poly(amidoamine)/MWCNT gave a composite useful as an amperometric sensor for L-tyrosine [240].

Ni(II) *meso*-tetrakis(4-sulphonatophenyl)porphyrin/carbon nanotube composite formed an electrode to detect bisphenol A, nonylphenol and ethynylestradiol, which are phenolic endocrine compounds [241].

A suspension obtained from the water-soluble anionic *meso*-tetrakis(4-carboxyphenyl)porphyrin and MWCNTs was used as an optical probe for DNA detection [242]. The same target was achieved very recently, constructing a supramolecular assembly with the water-soluble anionic 5-(4-aminophenyl)-10,15,20-tris(4-sulfonatophenyl)porphyrin, covalently linked to MWCNTs and complexed with cyclodextrins (CD, either α-CD or β-CD) [243].

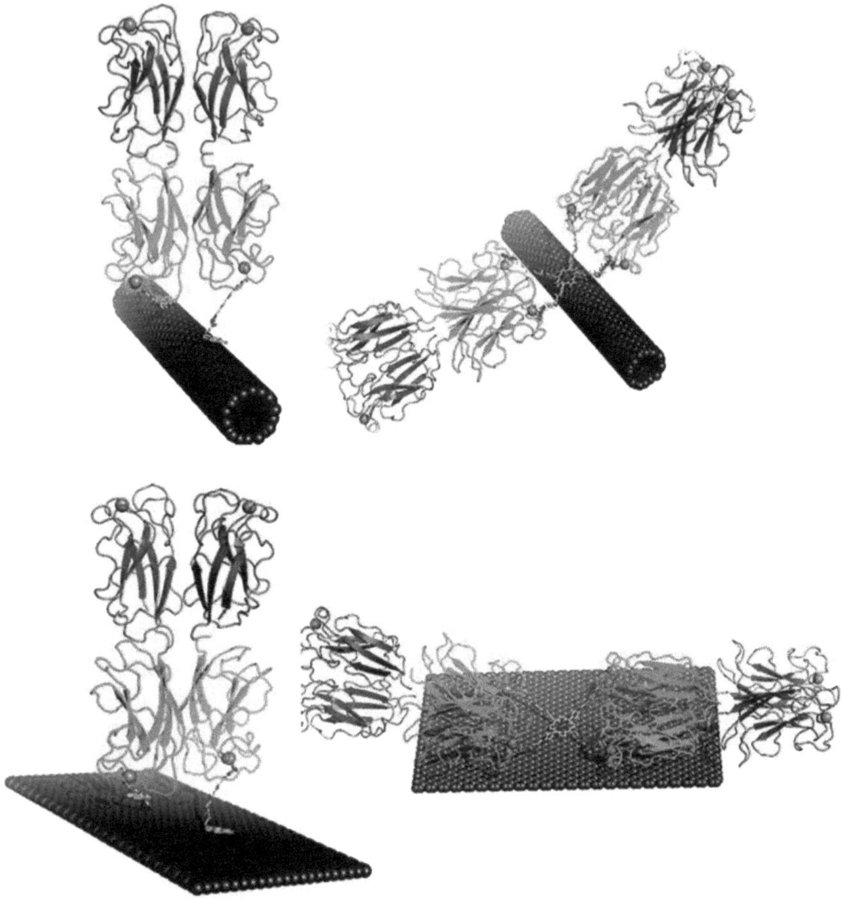

FIGURE 14.12 Computer-aided models of lectin interaction with glycoporphyrins functionalized SWCNT (above) and chemically converted graphene. *(Adapted and reprinted from Ref. [246] with permission of American Chemical Society. Copyright 2012.)*

Single-walled carbon nanotubes field-effect transistor (NTFET) devices were functionalized with two porphyrin receptor molecules, giving glycoconjugates the ability to interact with bacterial lectins [244]. The interactions between lectins and glycoconjugates were transduced as conductance changes measured by SWCNT-FET and compared to those obtained with similar chemically converted graphene (CCG-FET) devices [245,246]. SWCNT-FET showed larger response and better selectivity than CCG-FET, and the authors suggested that the difference was the result of one-dimensional and two-dimensional structural differences between the two materials, with a more favorable lectin binding to sugar moieties attached to the SWCNT surface (Fig. 14.12).

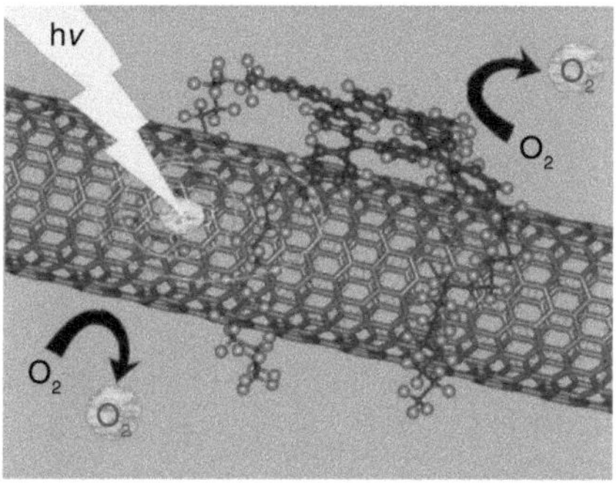

FIGURE 14.13 Oxygen singlet production by the hybrid of *meso*-tetrakis(1-methylpyridinium-4-yl)porphyrin with E-SWCNTs. *(Reprinted from Ref. [252] with permission of American Chemical Society. Copyright 2015.)*

1.2.5 Diagnostic and Therapeutic Applications

Although they cannot be strictly considered nanoreactors, CNTs became so important in therapeutic applications that we feel we can stretch the concept of nanoreactors a little. Although the are some concerns about safety, carbon nanotubes were largely used for delivery of small molecule drugs, as illustrated in a 2013 review [247], and are promising in diagnostic fields, such as photoacoustic imaging [248].

Combining CNTs with porphyrins made it possible to introduce a photosensitive moiety into the carbon nanoreactor. Thus, Protoporphyrin IX, immobilized on multiwalled carbon nanotubes, gave a conjugate that was reported as acting as a highly efficient, light-activated, antiviral agent against influenza virus [249]. The same conjugate was active against *Staphylococcus aureus* upon irradiation with visible light [250].

Hematoporphyrin monomethyl ether hyaluronic acid-derivatized carbon nanotubes were used in photodynamic therapy for both in vivo and in vitro cancer treatment [251]. According to the authors, photothermal treatment was superior to photodynamic or thermal treatment alone.

A different, very recent approach was the use of chirality enriched (6,5) single-walled carbon nanotubes (E-SWCNTs) to prepare hybrids with 5,10,15,20-tetrakis(1-methylpyridinium-4-yl)porphyrin or the corresponding 1-tetradecyl analogue. These species transfer energy directly to oxygen, promoting a triplet-singlet transition [252]. Moreover, the hybrid system of E-SWCNT with the former porphyrin synergistically generates an oxygen singlet (Fig. 14.13).

The hybrid was more efficient than porphyrin in the wavelength range used in medical therapy, thus being a very promising candidate, according to the authors, for photodynamic therapy.

1.3 Fullerene Nanoreactors

As is well known, allotropic forms of carbon fullerenes (Fig. 14.14) were detected for the first time in molecular beams and by microscopy [253] and later

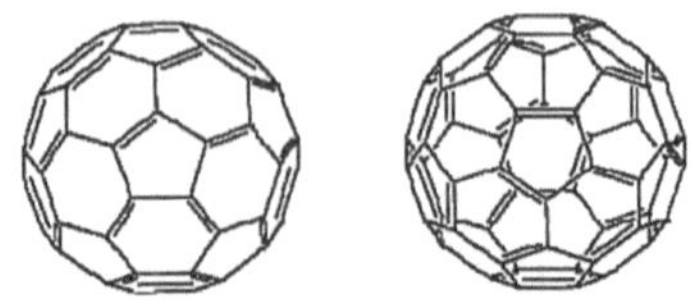

FIGURE 14.14 **The structures of C_{60} and C_{70}.**

isolated at the University of Sussex and Rice University by the Nobel laureates Kroto, Smalley, and Curl [254]. They were named in honor of Richard Buckminster "Bucky" Fuller, an architect who worked on geodesic domes whose shapes were similar to those of such materials.

The synthesis and purification of fullerenes were very difficult, and the prices of purified materials were high at the beginning. Later, a cheaper method was discovered, and grams of C_{60}, the most important fullerene, are now available at reasonable cost [255].

The production of such compounds opened large fields of research in several topics of chemistry [256], physics [257–259], and engineering [260]. The functionalization of fullerenes was the first target for the chemists working on such interesting materials. Because of the aromaticity of fullerenes, it was mandatory to accomplish the functionalization through the attack of strong electrophylic groups [256,261].

The Prato-Maggini-Scorrano reaction [262] (Scheme 14.11) is actually the most used method for the functionalization of fullerenes, but other methods (the Bingel reaction [263] and the use of diazocompounds [264]) are extensively used.

Fullerene C_{60} and its derivatives have been deeply studied, and in the last 20 years the number of publications skyrocketed [265] (see one example in Scheme 14.12). Electrochemical properties of C_{60} [266], which is able to accept up to six electrons in reduction, prompted many groups to study the possibility of using this material, coupling with suitable photons—antenna-like porphyrins and chlorins—to mimick the properties of the photosynthetic centers of bacteria

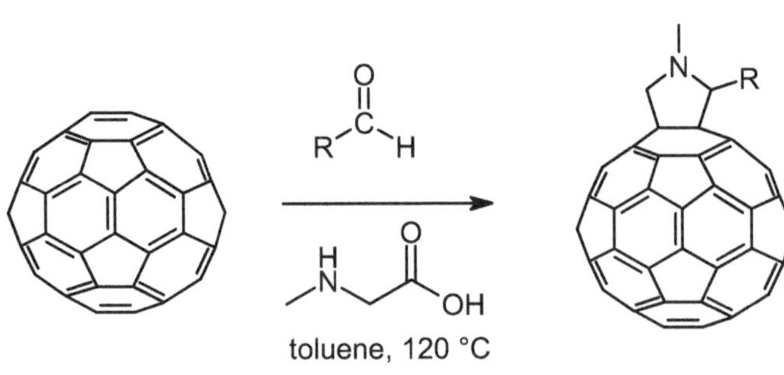

SCHEME 14.11 **Prato–Maggini–Scorrano reaction.**

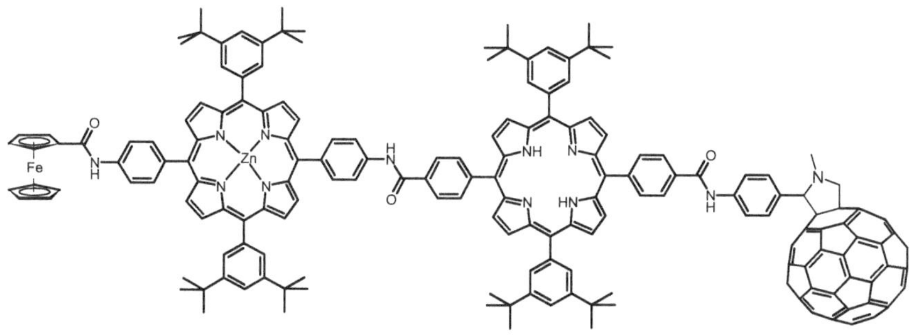

SCHEME 14.12 **Structure of the Imahori tetrad.** *(Adapted from Ref. [265].)*

and plants. Furthermore, these properties are promising for building the polymeric organic photovoltaic cells [267].

Many research groups have also been involved in the investigation of the catalytic properties of such materials and many examples of their characteristics are now present in the literature. Among all the catalytic systems, nanoreactors using fullerenes as the main components are studied and used to catalyze several reactions, even if their use is not as large as that of graphene and CNTs, probably due to the rivals' better solubility. The assembling of porphyrins and fullerenes to form nanoreactors is one of the topics that is now under investigation, and it has the possibility of reaching new and unexpected results.

Some reactions involve photochemical processes, which are catalyzed by the presence of porphyrins and enhanced by the action of $C_{60.}$

A first report was published in 2011 by Song et al. claiming the synthesis and characterization of porphyrin-fullerene dyads and the application of such compounds in the photoinduced H_2 evolution [268]. As an example, the reaction pathway to one the dyads used in the experiments is shown in Scheme 14.13.

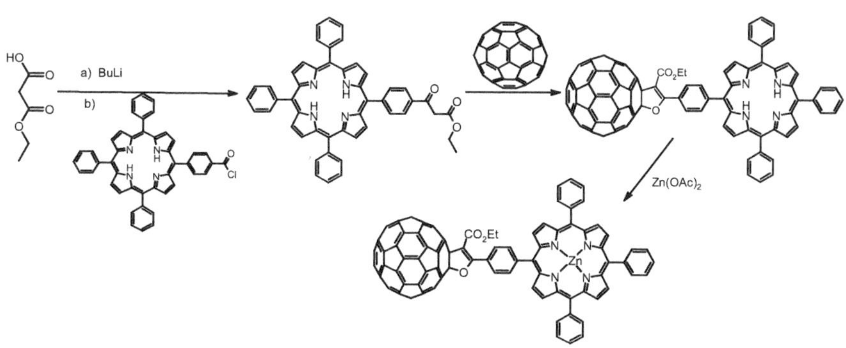

SCHEME 14.13 **Reaction pathway to obtain one of the Song dyads.** *(Adapted from Ref. [268] with permission of American Chemical Society, 2011.)*

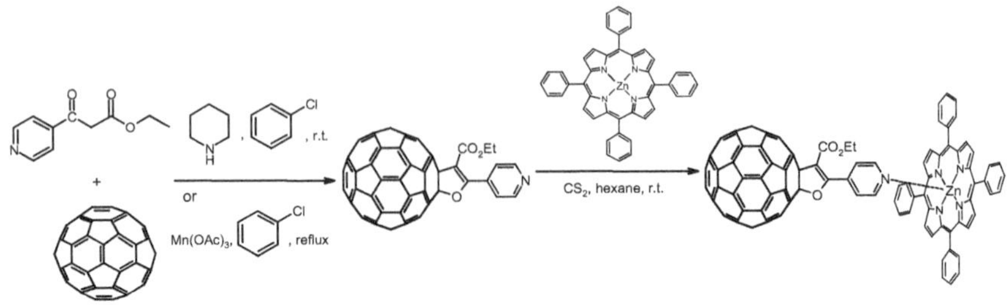

SCHEME 14.14 Reaction pathway for obtaining the C_{60} pyridine derivative and subsequent complexation of ZnTPP. *(Adapted from Ref. [268] with permission of American Chemical Society, 2011.)*

Another dyad was obtained by the axial complexation of zinc tetraphenylporphyrin by a pyridine C_{60} derivative as shown in Scheme 14.14 [268].

The fullerene-porphyrin composite has been tested as catalysts in the H_2 evolution reaction in a five-component system showing a good activity with a turnover number of 73.

The reaction is based on the electron-donating properties of EDTA, dyad of Scheme 14.14, the electron mediator MV^{+2} (methylviologen), colloidal Pt, acetic acid, and Triton X-100, irradiating with a 350 W Xe lamp equipped with a pyrex filter ($\lambda > 400$ nm).

Based on these observations, some analytical systems were built and used to detect important analytes.

Zou and coworkers assembled a porphyrin-C_{60} dyad able to work as an electrochemical sensor for sensing hydrogen peroxide and nitrite [269]. *Ortho* and *para* isomers were synthesized and tested showing good stability and reproducibility as sensors for river water and rainwater. In Scheme 14.15 the structure of one of the isomers used in this paper is reported.

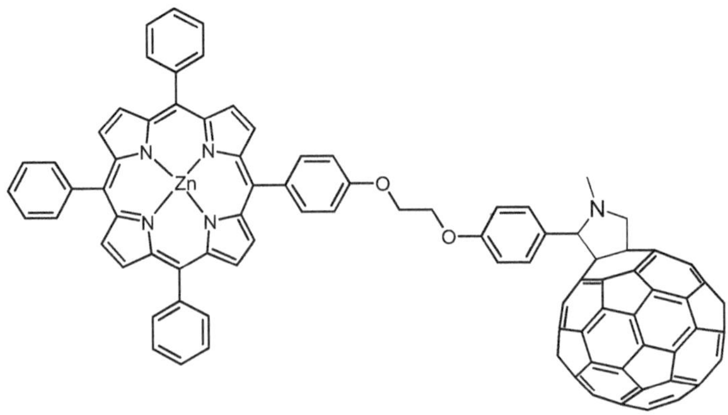

SCHEME 14.15 The structure of a dyad used to detect hydrogen peroxide and nitrite. *(Adapted from Ref. [269].)*

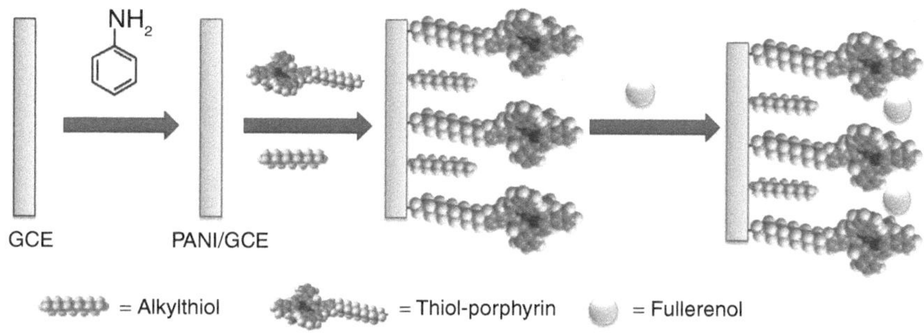

FIGURE 14.15 Deposition scheme for the preparation of the fullerene-porphyrin derivative used in DNB detection. *(Adapted with permission from Ref. [270]. Reprinted with permission from Elsevier.)*

A second report on the analytical properties of the mixture of polyaniline, fullerenols, an alkyl functionalized thiol, and a thiol-porphyrin for detecting *m*-dinitrobenzene (DNB) was published by Lu and coworkers [270]. The determination of the analyte was obtained by differential pulse voltammetry (DPV) and was directed to DNB because of its testicular toxicity, brainstem damage, and water solubility, which make such chemicals extremely dangerous.

The active electrode was prepared in three different steps: the first one was the electrodeposition of polyaniline from a solution of anyline in aqueos H_2SO_4. After that the electrode was immersed in a solution containing a thiol-porphyrin and the alkylthiol and cycled 10 times between −0.1 and 0.8 V. Finally, the electrode was stirred in fullerenols giving the active polymeric nanocomposite.

The deposition scheme of the polymeric film is shown in Fig. 14.15.

The electrochemical response to the analytes was increased by the presence of fullerenols in the active films, giving the greatest ratio between the signal and the concentration of DNB.

In reduction, three peaks appear at −0.598 V, −0714 V, and 0891 V, corresponding to the consecutive formation to nitroso, hydroxylamino, and amino-derivatives (Fig. 14.16).

A very interesting application [271] of the interaction between a manganese porphyrin and C_{60}, graphite, or diamond, is the recognition in whole-blood samples of HER-1 (epidermal growth factor receptor-1), which plays a critical role in the proliferation of cancer. HER-1 normally acts to repair tissue, in angiogenesis and fibrogenesis. The abnormal regulation of HER-1 can give the survival of cancer cells, increased transcription of genes, cell migration and metastasis [272–274].

The active sides of the sensing platform were three modified pastes formed by C_{60}, graphite, or diamond mixed in paraffin oil with the porphyrin [Mn(III)

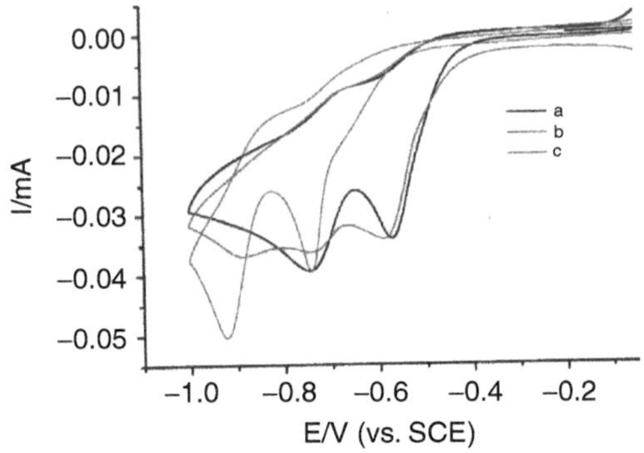

FIGURE 14.16 The electrochemical response of different modified films in PBS (pH = 7) that contained 0.5 mol L^{-1} at scan rate 50 mVs^{-1}: (a) GCE, (b) PANI-mixed-thiol/GCE, and (c) fullerenols/PANI-mixed-thiol/GCE. *(Reprinted with permission from Ref. [270], from Elsevier.)*

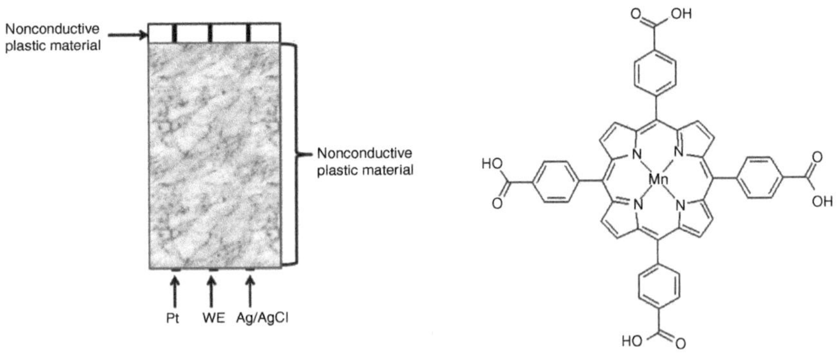

FIGURE 14.17 The electrochemical platform for HER-1 recognition in whole-blood samples. *(Reprinted from Ref. [271] with permission of Wiley.)*

tetra-4-carboxyphenylporphyrin]. Such pastes were applied to the platform reported in Fig. 14.17.

In DPV the sensitivity and linearity of the sensors are good enough and the platforms can be used continously for more than 100 tests, making them cheap and reliable for early detection of cancers.

In 2012 a Brazilian group [275] reported the use of the water-soluble silver porphyrins for detecting in water solution carbon nanoparticles with a diameter lower than 10 nm. The targets of the interaction with metalloporphyrins were the single-walled carbon nanotubes (SWCNT) and SWCNT-CO_2H or fullerenes. The photoluminescnce spectra of the porphyrin in solution change as a function of the nanoparticles that interact with the metallated macrocycle. In Fig. 14.18, the luminescence spectra of the different cosolutions of porphyrin and carbon nanoparticles are reported with the images of the cuvettes showing different colors.

The authors discussed the interaction mechanism in terms of the hydrophobic interaction.

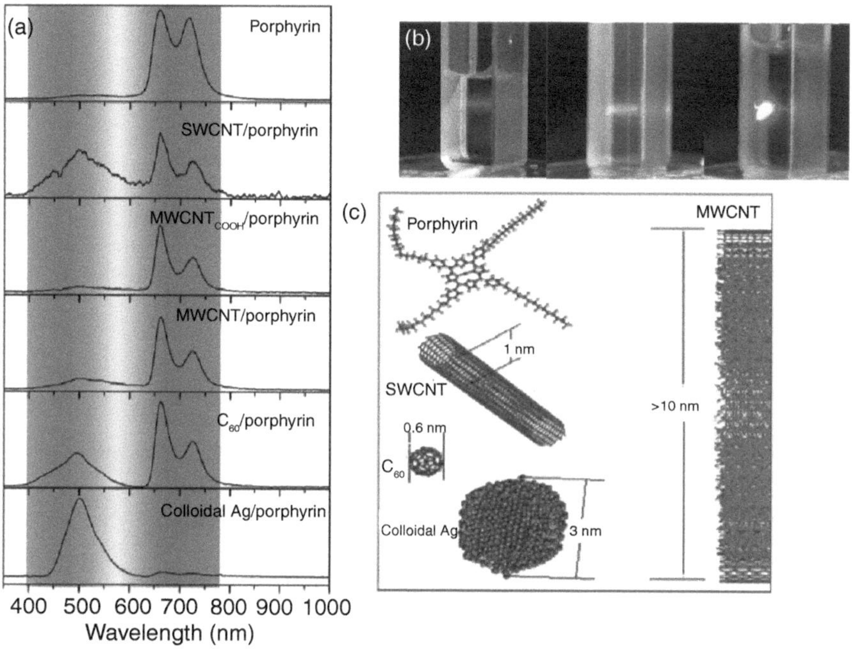

FIGURE 14.18 (a) Photoluminescence spectra of the porphyrin in several diluted nanoparticle suspensions (left). The visible range is represented to their correspondent wavelengths. (b) From left to right, a SWCNT dilute suspension, tap water, and deionized water, all containing the porphyrin, exposed to a UV laser. The light gray (yellowish in the web version) hue in the tap water comes from the excessive porphyrin that is "unreacted" to any substrate. The dark gray (red in the web version) in the solution/suspensions comes from the porphyrin luminescence. (c) Comparative size of the porphyrin to the different nanosubstrates. *(Reproduced from Ref. [275] with permission of IOP Publishing. Copyright 2012.)*

2 CONCLUSIONS

From inspection of the literature, it is evident that carbon-based nanoreactors constitute a rapid growing field, with several applications, some of them already well established, others with a promising future. Catalysis and electrocatalysis, as well as sensors and biosensors, have proved to be important applications of carbon-based nanoreactors, as have photoelectrochemical features. We feel that diagnostic and therapeutic applications will have an important future as well.

ABBREVIATIONS

CNT	Carbon nanotube
DMF	*N,N′*-Dimethylformamide
DMFC	Direct methanol fuel cell
DNB	*m*-Dinitrobenzene
DPV	Differential pulse voltammetry

EDTA	Ethylenediaminetetraacetic acid (IUPAC: *N,N,N,N*-tetra(carboxymethyl) ethane-1,2-diamine)
E-SWCNT	Chirality enriched single-walled carbon nanotubes
GCE	Glassy carbon electrode
GO	Graphene oxide
H_2APFP	5,15-(*p*-Aminophenyl)-10,20-(pentafluorophenyl)porphyrin
H_2OPc	1,4,8,11,15,18,22,25-Octabutoxyphthalocyanine
HP	Hematoporphyrin
H_2Por	2,18-Bis(2-carboxyethyl)-8,13-divinyl-3,7,12,17-tetramethylporphyrin
H_2TAmPP	5,10,15,20-Tetrakis[4-(*N,N,N*-trimethyl)ammoniophenyl]porphyrin
H_2T*o*APP	Tetrakis(*o*-aminophenyl)porphyrin
H_2TBPP	5,10,15,20-Tetrakis(4′-bromobiphenyl)porphyrin
H_2TCPP	5,10,15,20-Tetrakis(4-carboxyphenyl)porphyrin
H_2THPP	5,10,15,20-Tetrakis(4-hydroxyphenyl)porphyrin
H_2TMPP	5,10,15,20-Tetrakis(4-methoxyphenyl)porphyrin
H_2TMAPP	5,10,15,20-Tetrakis[2-(trimethylammoniomethylcarbonylamino)phenyl] porphyrin
H_2TMPyP	5,10,15,20-Tetrakis(1-methylpyridinio-4-yl)porphyrin tetra(*p*-toluenesulfonate)
H_2TMSPP	5,10,15,20-Tetrakis(4-methoxy-3-sulfonatophenyl)porphyrin
H_2TNPP	5,10,15,20-Tetrakis(4-nitrophenyl)porphyrin
H_2TOHPP	5,10,15,20-Tetrakis(4-hydroxyphenyl)porphyrin
H_2TPP	5,10,15,20-Tetraphenylporphyrin
H_2TPP-NH_2	5-(4-Aminophenyl)-10,15,20-triphenylporphyrin
H_2TSPP	5,10,15,20-Tetrakis(4-sulfonatophenyl)porphyrin
ITO	Indium tin oxide
MV^{+2}	Methylviologen (IUPAC: 1,1′-dimethyl-4,4′-bipyridinium dichloride)
NTFET	Nanotubes field-effect transistor
MWCNT	Multi-walled carbon nanotube
NP	Nanoparticle
ORR	Oxygen reduction reaction
PPIX	Protoporphyrin IX
PSS	Poly(sodium *p*-styrene sulfonate)
RGO	Reduced graphene oxide
SAS	Surfactant-assisted self-assembly
SEM	Scanning electron microscopy
SWCNT	Single-walled carbon nanotube
TEM	Transmission electron microscopy
TNT	2,4,6-Trinitrotoluene
VOC	Volatile organic compound

REFERENCES

[1] Kane RS, Cohen RE. Synthesis of doped ZnS nanoclusters within block copolymer nanoreactors. Chem Mater 1999;11:90–3.

[2] Vriezema DM, Comellas Aragonès M, Elemans JAAW, Cornelissen M, Rowan AE, Nolte RJM. Self-assembled nanoreactors. Chem Rev 2005;105:1445–89.

[3] Zarra S, Wood DM, Roberts DA, Nitschke JR. Molecular containers in complex chemical systems. Chem Soc Rev 2014;44:419–32.
[4] Cantrill S, Davey S, Johnson R, Pichon A. Catalysis in compartments. Nature Chem 2014; 6:5.
[5] Peters RJRW, Marguet M, Marais S, Fraaije MW, van Hest JCM, Lecommandoux S. Cascade Reactions in multicompartmentalized polymersomes. Angew Chem Int Ed 2014;53:146–50.
[6] Algar WR, Prasuhn DE, Stewart MH, Jennings TL, Blanco-Canosa JB, Dawson PE, Medintz IL. The controlled display of biomolecules on nanoparticles: A challenge suited to bioorthogonal chemistry. Bioconjugate Chem 2011;22:825–58.
[7] Sapsford KE, Algar WR, Berti L, Boeneman Gemmill K, Casey BJ, Oh E, Stewart MH, Medintz IL. Functionalizing nanoparticles with biological molecules: Developing chemistries that facilitate nanotechnology. Chem Rev 2013;113:1904–2074.
[8] Santhosh Babu S, Praveen VK, Ajayaghosh A. Functional π-gelators and their applications. Chem Rev 2014;114:1973–2129.
[9] Cheng L, Wang C, Feng L, Yang K, Liu Z. Functional nanomaterials for phototherapies of cancer. Chem Rev 2014;114:10869–939.
[10] Kim H, Kim W, Mackeyev Y, Lee G-S, Kim H-J, Tachikawa T, Hong S, Lee S, Kim J, Wilson LJ, Majima T, Alvarez PJJ, Choi W, Lee J. Selective oxidative degradation of organic pollutants by singlet oxygen-mediated photosensitization: tin porphyrin versus C_{60} aminofullerene systems. Environ Sci Technol 2012;46:9606–13.
[11] Gobeze HB, Das SK, D'Souza F. Femtosecond transient absorption study of supramolecularly assembled metal tetrapyrrole-TiO_2 thin films. J Phys Chem C 2014;118:16660–71.
[12] Wang X-S, Chrzanowski M, Wojtas L, Chen Y-S, Ma S. Formation of a metalloporphyrin-based nanoreactor by postsynthetic metal–ion exchange of a polyhedral-cage containing a metal–metalloporphyrin framework. Chem Eur J 2013;19:3297–301.
[13] Drain CM, Varotto A, Radivojevic I. Self-organized porphyrinic materials. Chem Rev 2009;109:1630–58.
[14] Durot S, Taesch J, Heitz V. Multiporphyrinic cages: architectures and functions. Chem Rev 2014;114:8542–78.
[15] Nierengarten JF. Chemical modification of C_{60} for materials science applications. New J Chem 2004;28:1177–1191.
[16] Guldi DM, Aminur Rahman GM, Jux N, Balbinot D, Hartnagel U, Tagmatarchis N, Prato M. Functional single-wall carbon nanotube nanohybrids-associating SWNTs with water-soluble enzyme model systems. J Am Chem Soc 2005;127:9830–9838 and references therein.
[17] Singh V, Joung D, Zhai L, Das S, Khondaker SI, Seal S. Graphene based materials: past, present and future. Progr Mater Sci 2011;56:1178–271.
[18] Wick P, Louw-Gaume AE, Kucki M, Krug HF, Kostarelos K, Fadeel B, Dawson KA, Salvati A, Vázquez E, Ballerini L, Tretiach M, Benfenati F, Flahaut E, Gauthier L, Prato M, Bianco A. Classification framework for graphene-based materials. Angew Chem Int Ed 2014;53:7714–8.
[19] Fang Q, Shen Y, Chen B. Synthesis, decoration and properties of three-dimensional graphene-based macrostructures: a review. Chem Eng J 2015;264:753–71.
[20] Perrozzi F, Prezioso S, Ottaviano L. Graphene oxide: from fundamentals to applications. J Phys Condens. Matter 2015;27:013002 (21pp).
[21] Dreyer DR, Todd AD, Bielawski CW. Harnessing the chemistry of graphene oxide. Chem Soc Rev 2014;43:5288–301.
[22] Eigler S, Hirsch A. Chemistry with graphene and graphene oxide-challenges for synthetic chemists. Angew Chem Int Ed 2014;53:7720–38.

[23] Lim SY, Shen W, Gao Z. Carbon quantum dots and their applications. Chem Soc Rev 2015;44:362–81.
[24] Amali AJ, Hoshino H, Wu C, Ando M, Xu Q. From metal–organic framework to intrinsically fluorescent carbon nanodots. Chem Eur J 2014;20:8279–82.
[25] Ye T-N, Lv L-B, Li X-H, Xu M, Chen J-S. Strongly veined carbon nanoleaves as a highly efficient metal-free electrocatalyst. Angew Chem Int Ed 2014;53:6905–9.
[26] Liu J, Qiao SZ, Hartono SB, Lu GQ (M). Monodisperse yolk–shell nanoparticles with a hierarchical porous structure for delivery vehicles and nanoreactors. Angew Chem 2010;122:5101–5.
[27] Iyer KS, Saunders M, Becker T, Evans CW, Raston CL. Nanorings of self-assembled fullerene C70 as templating nanoreactors. J Am Chem Soc 2009;131:16338–7163.
[28] Lu A-H, Hao G-P, Sun Q. Design of three-dimensional porous carbon materials: from static to dynamic skeletons. Angew Chem Int Ed 2013;52:7930–2.
[29] Gnanaprakasa TJ, Sridhar D, Beck WJ, Runge K, Potter BG Jr, Zega TJ, Deymier PA, Raghavan S, Muralidharan K. Graphene mediated self-assembly of fullerene nanorods. Chem Commun 2015;51:1858–61.
[30] Sun X, Sun H, Li H, Peng H. Developing polymer composite materials: carbon nanotubes or graphene? Adv Mater 2013;25:5153–76.
[31] Lu A-H, Hao G-P, Sun Q. Design of three-dimensional porous carbon materials: from static to dynamic skeletons. Angew Chem Int Ed 2013;52:7930–2.
[32] Modugno G, Ménard-Moyon C, Prato M, Bianco A. Carbon nanomaterials combined with metal nanoparticles for theranostic applications. British J Pharmacol 2015;172:975–91.
[33] Goenka S, Sant V, Sant S. Graphene-based nanomaterials for drug delivery and tissue engineering. J Control Release 2014;173:75–88.
[34] Liu Q, Cui Q, Li XJ, Jin L. The applications of buckminsterfullerene C_{60} and derivatives in orthopaedic research. Connect Tissue Res 2014;55:71–9.
[35] Apul OG, Karanfil T. Adsorption of synthetic organic contaminants by carbon nanotubes: a critical review. Water Res 2015;68:34–55.
[36] Yu J-G, Yu L-Y, Yang H, Liua Q, Chen X-H, Jiang X-Y, Chen X-Q, Jiao F-P. Graphene nanosheets as novel adsorbents in adsorption, preconcentration and removal of gases, organic compounds and metal ions. Sci Tot Environ 2015;502:70–9.
[37] Liu Q, Zhou Q, Jiang G. Nanomaterials for analysis and monitoring of emerging chemical pollutants. Trends Anal Chem 2014;58:10–22.
[38] Goryacheva I Yu, Speranskaya ES, Goftman VV, Tang D, De Saeger S. Synthesis and bioanalytical applications of nanostructures multiloaded with quantum dots. Trends Anal Chem 2015;66:53–62.
[39] Afreen S, Muthoosamy K, Manickam S, Hashim U. Functionalized fullerene (C_{60}) as a potential nanomediator in the fabrication of highly sensitive biosensors. Biosens Bioelectr 2015;63:354–64.
[40] Kong L, Chen W. Carbon nanotube and graphene-based bioinspired electrochemical actuators. Adv Mater 2014;26:1025–43.
[41] Lebedev N, Trammell SA, Tsoi S, Spano A, Kim JH, Xu J, Twigg ME, Schnur JM. Increasing efficiency of photoelectronic conversion by encapsulation of photosynthetic reaction center proteins in arrayed carbon nanotube electrode. Langmuir 2008;24:8871.
[42] Chen H, Guo X. Unique role of self-assembled monolayers in carbon nanomaterial-based field-effect transistors. Small 2013;9:1144–59.
[43] Algara-Siller G, Severin N, Chong SY, Björkman T, Palgrave RG, Laybourn A, Antonietti M, Khimyak YZ, Krasheninnikov AV, Rabe JP, Kaiser U, Cooper AI, Thomas A, Bojdys MJ.

Triazine-based graphitic carbon nitride: a two-dimensional semiconductor. Angew Chem Int Ed 2014;53:7450–5.

[44] Page ZA, Liu Y, Duzhko VV, Russell TP, Emrick T. Fulleropyrrolidine interlayers: tailoring electrodes to raise organic solar cell efficiency. Science 2014;346:441–4.

[45] Porro S, Accornero E, Pirri CF, Ricciardi C. Memristive devices based on graphene oxide. Carbon 2015;85:383–96.

[46] Gadipelli S, Guo ZX. Graphene-based materials: synthesis and gas sorption, storage and separation. Progr Mater Sci 2015;69:1–60.

[47] Wen Z, Ci S, Hou Y, Chen J. Facile one-pot, one-step synthesis of a carbon nanoarchitecture for an advanced multifunctonal electrocatalyst. Angew Chem Int Ed 2014;53:6496–500.

[48] Pentsak EO, Gordeev EG, Ananikov VP. Noninnocent nature of carbon support in metal/carbon catalysts: etching/pitting vs nanotube growth under microwave irradiation. ACS Catal 2014;4:3806–14.

[49] Freund H-J, Nilius N, Risse T, Schauermann S. A fresh look at an old nano-technology: catalysis. Phys Chem Chem Phys 2014;16:8148.

[50] Julkapli NM, Bagheri S. Graphene supported heterogeneous catalysts: an overview. Int J Hydr En 2015;46:948–79.

[51] Li Y, Zhang Y, Yan B. Nanotoxicity overview: nano-threat to susceptible populations. Int J Mol Sci 2014;15:3671–97.

[52] Zhang Y, Petibone D, Xu Y, Mahmood M, Karmakar A, Casciano D, Ali S, Biris AS. Toxicity and efficacy of carbon nanotubes and graphene: the utility of carbon-based nanoparticles in nanomedicine. Drug Metabol Rev 2014;46:232.

[53] Madl Amy K, Plummer Laurel E, Carosino C, Pinkerton KE. Nanoparticles, lung injury, and the role of oxidant stress. Annu Rev Physiol 2014;76:15.1–15.19.

[54] Seabra AB, Paula AJ, de Lima R, Alves OL, Durán N. Nanotoxicity of graphene and graphene oxide. Chem Res Toxicol 2014;27:159–68.

[55] Seol M-L, Choi S-J, Kim C-H, Moon D-I, Choi Y-K. Porphyrin-silicon hybrid field-effect transistor with individually addressable top-gate structure. ACS Nano 2012;6:183–9.

[56] Jiřičková M, Demel J, Kubát P, Hostomský J, Kovanda F, Lang K. Photoactive self-standing films made of layered double hydroxides with arranged porphyrin molecules. J Phys Chem C 2011;115:21700–6.

[57] Lebedeva MA, Chamberlain TW, Schröder M, Khlobystov AN. New pathway for heterogenization of molecular catalysts by noncovalent interactions with carbon nanoreactors. Chem Mater 2014;26:6461–6.

[58] Sutter P, Sutter E. Microscopy of graphene growth, processing, and properties. Adv Funct Mater 2013;23:2617–34.

[59] Iski EV, Yitamben EN, Gao L, Guisinger NP. Graphene at the atomic-scale: synthesis, characterization, and modification. Adv Funct Mater 2013;23:2554–64.

[60] Chang H, Wu H. Graphene-based nanomaterials: synthesis, properties, and optical and optoelectronic applications. Adv Funct Mater 2013;23:1984–97.

[61] Tu W, Zhou Y, Zou Z. Versatile graphene-promoting photocatalytic performance of semiconductors: basic principles, synthesis, solar energy conversion, and environmental applications. Adv Funct Mater 2013;23:4996–5008.

[62] Georgakilas V, Otyepka M, Bourlinos AB, Chandra V, Kim N, Kemp KC, Hobza P, Zboril R, Kim KS. Functionalization of graphene: covalent and non-covalent approaches, derivatives and applications. Chem Rev 2012;112:6156–214.

[63] Economopoulos SP, Tagmatarchis N. Chemical functionalization of exfoliated graphene. Chem Eur J 2013;19:12930–6.

[64] Hasobe T. Photo- and electro-functional self-assembled architectures of porphyrins. Phys Chem Chem Phys 2012;14:15975–87.
[65] Zhu G, Sun Q, Kawazoe Y, Jena P. Porphyrin-based porous sheet: optoelectronic properties and hydrogen storage. Int J Hydrogen Energy 2015;40:3689–96.
[66] Girek B, Sliwa W. Hybrids of cationic porphyrins with nanocarbons. J Incl Phenom Macrocycl Chem 2015;82:283–300.
[67] Liu ZB, Xu YF, Zhang XY, Zhang XL, Chen XL, Tian JG. Porphyrin and fullerene covalently functionalized graphene hybrid materials with large nonlinear optical properties. J Phys Chem B 2009;113:9681–6.
[68] Karousis N, Sandanayaka ASD, Hasobe T, Economopoulos SP, Sarantopoulou E, Tagmatarchis N. Graphene oxide with covalently linked porphyrin antennae: synthesis, characterization and photophysical properties. J Mater Chem 2011;21:109–17.
[69] Quintana M, Spyrou K, Grzelczak M, Browne WR, Rudolf P, Prato M. Functionalization of graphene via 1,3-dipolar cycloaddition. ACS Nano 2010;4:3527–33.
[70] Quintana M, Vazquez E, Prato M. Organic functionalization of graphene in dispersions. Acc Chem Res 2013;46:138–48.
[71] Zhang X, Hou L, Cnossen A, Coleman AC, Ivashenko O, Rudolf P, van Wees BJ, Browne WR, Feringa BL. One-pot functionalization of graphene with porphyrin through cycloaddition reactions. Chem Eur J 2011;17:8957–64.
[72] Ragoussi M-E, Malig J, Katsukis G, Butz B, Spiecker E, de La Torre G, Torres T, Guldi DM. Linking photo- and redoxactive phthalocyanines covalently to graphene. Angew Chem Int Ed 2012;51:6421–5.
[73] Wang A, Yu W, Xiao Z, Song Y, Long L, Cifuentes MP, Humphrey MG, Zhang C. A 1,3-dipolar cycloaddition protocol to porphyrin-functionalized reduced graphene oxide with a push–pull motif. Nano Res 2015;8:870–86.
[74] Mishra S, Arora S, Nagpal R, Chauhan SMS. Sulfonated graphenes catalyzed synthesis of expanded porphyrins and their supramolecular interactions with pristine graphene. J Chem Sci 2014;126:1729–36.
[75] Maraka HVR, Al-Shammari RM, Al-Attar N, Lopez SG, Keyes TE, Rice JH. Graphene oxide intercalation into self-assembled porphyrin J-aggregates. Mater Res Expr 2014;1:045038.
[76] Naumenko D, Naumenko I, Grinceviciute N, Snitka V. The interaction of iron and metal-free meso-tetra(4-sulfonatophenyl) porphines with CVD graphene: a comparative Raman spectroscopy study. J Raman Spectr 2015;46:94–99.
[77] Sreeprasad TS, Berry V. How do the electrical properties of graphene change with its functionalization? Small 2013;9:341–350.
[78] Pham VD, Lagoute J, Mouhoub O, Joucken F, Repain V, Chacon C, Bellec A, Girard Y, Rousset S. Electronic interaction between nitrogen-doped graphene and porphyrin molecules. ACS Nano 2014;8:9403–9.
[79] Yadav C, Roy S. Ultrafast all-optical universal logic gates with graphene and graphene-oxide metal porphyrin composites. J Comput Electron 2014;14:209–13.
[80] Mandal B, Sarkar S, Sarkar P. Theoretical studies on understanding the feasibility of porphyrin-sensitized graphene quantum dot solar cell. J Phys Chem C 2015;119:3400–7.
[81] Malic E, Appel H, Hofmann OT, Rubio A. Energy transfer in porphyrin-functionalized graphene. Phys Status Solidi B 2014;251:2495–8.
[82] Zhu X, Jiao Q, Zhang C, Zuo X, Xiao X, Liang Y, Nan J. Electropolymerization of graphene and 5-[*o*-(4-Bromine amyloxy) phenyl]-10,15,20-triphenylporphrin (*o*-BrPETPP) composite film electrode for the electrocatalytic oxidation of indirubin. J Electroanal Chem 2012;681:133–8.

[83] Wang R-X, Fan J-J, Fan Y-J, Zhong J-P, Wang L, Sun S-G, Shen X-C. Platinum nanoparticles on porphyrin functionalized graphene nanosheets as superior catalyst for methanol electro-oxidation. Nanoscale 2014;6:14999–5007.

[84] Su C, Loh KP. Carbocatalysts: graphene oxide and its derivatives. Acc Chem Res 2013;46:2275–85.

[85] Huang D, Lu J, Li S, Luo Y, Zhao C, Hu B, Wang M, Shen Y. Fabrication of cobalt porphyrin. Electrochemically reduced graphene oxide hybrid films for electrocatalytic hydrogen evolution in aqueous solution. Langmuir 2014;30:6990–8.

[86] Jahan M, Bao Q, Loh KP. Electrocatalytically active graphene-porphyrin MOF composite for oxygen reduction reaction. J Am Chem Soc 2012;134:6707–13 and references therein.

[87] Xiong Y, Liu QL, Zhu AM, Huang SM, Zeng QH. Performance of organic–inorganic hybrid anion-exchange membranes for alkaline direct methanol fuel cells. J Power Sources 2009;186:328–33.

[88] Choi H-J, Kumar NA, Baek J-B. Graphene supported non-precious metalmacrocycle catalysts for oxygen reduction reaction in fuel cells. Nanoscale 2015;7:6991–7000.

[89] Kim SK, Jeon S. Improved electrocatalytic effect of carbon nanomaterials by covalently anchoring with CoTAPP via diazonium salt reactions. Electrochem Commun 2012;22:141–4.

[90] Kaplan A, Soifer L, Eliyahu D, Korin E, Bettelheim A. Electrocatalytic activity towards oxygen reduction of electropolymerized cobalt porphyrin doped with ionic-liquid-functionalized graphene. J Electrochem Soc 2015;162:H481–5.

[91] Kruusenberg I, Mondal J, Matisen L, Sammelselg V, Tammeveski K. Oxygen reduction on graphene-supported MN4 macrocycles in alkaline media. Electrochem Commun 2013;33:18–22.

[92] Jiang L, Cui L, He X. Cobalt-porphyrin noncovalently functionalized grapheme as nonprecious-metal electrocatalyst for oxygen reduction reaction in an alkaline medium. J Solid State Electrochem 2015;19:497–506.

[93] Tang H, Yin H, Wang J, Yang N, Wang D, Tang Z. Molecular architecture of cobalt porphyrin multilayers on reduced graphene oxide sheets for high-performance oxygen reduction reaction. Angew Chem Int Ed 2013;52:5585–9.

[94] Jiang L, Li M, Lin L, Li Y, He X, Cui L. Electrocatalytic activity of metalloporphyrins grown *in situ* on graphene sheets toward oxygen reduction reaction in an alkaline medium. RSC Adv 2014;4:26653–61.

[95] You J-M, Han HS, Lee HK, Cho S, Jeon S. Enhanced electrocatalytic activity of oxygen reduction by cobalt-porphyrin functionalized with graphene oxide in an alkaline solution. Int J Hydrogen Ener 2014;39:4803–11.

[96] Jurow M, Manichev V, Pabon C, Hageman B, Matolina Y, Drain CM. Self-organization of Zr(IV) porphyrinoids on graphene oxide surfaces by axial metal coordination. Inorg Chem 2013;52:10576–82.

[97] Chlistunoff J, Sansiñena J-M. Effects of axial coordination of the metal center on the activity of iron tetraphenylporphyrin as a nonprecious catalyst for oxygen reduction. J Phys Chem C 2014;118:19139–49.

[98] Cheng M-J, Head-Gordon M, Bell AT. How to chemically tailor metal-porphyrin-like active sites on carbon nanotubes and graphene for minimal overpotential in the electrochemical oxygen evolution and oxygen reduction reactions. J Phys Chem C 2014;118: 29482–91.

[99] Xiang Z, Xue Y, Cao D, Huang L, Chen J-F, Dai L. Highly efficient electrocatalysts for oxygen reduction based on 2D covalent organic polymers complexed with non-precious metals. Angew Chem Int Ed 2014;53:2433–7.

[100] Tripkovic V, Vanin M, Karamad M, Björketun ME, Jacobsen KW, Thygesen KS, Rossmeisl J. Electrochemical CO_2 and CO reduction on metal-functionalized porphyrin-like graphene. J Phys Chem C 2013;117:9187–95.
[101] Zhang S, Zhang H, Hua X, Chen S. Tailoring molecular architectures of Fe phthalocyanine on nanocarbon supports for high oxygen reduction performance. J Mater Chem A 2015;3:10013–9.
[102] Xie G, Zhang K, Guo B, Liu Q, Fang L, Gong JR. Graphene-based materials for hydrogen generation from light-driven water splitting. Adv Mater 2013;25:3820–39 and references therein.
[103] Yadav RK, Baeg J-O, Oh GH, Park N-J, Kong K, Kim J, Hwang DW, Biswas SK. A photocatalyst-enzyme coupled artificial photosynthesis system for solar energy in production of formic acid from CO_2. J Am Chem Soc 2012;134:11455–61.
[104] Yadav RK, Oh GH, Park N-J, Kumar A, Kong K, Baeg J-O. Highly selective solar-driven methanol from CO_2 by a photocatalyst/biocatalyst integrated system. J Am Chem Soc 2014;136:16728–31.
[105] Li H, Zhang Y, Chang G, Liu S, Tian J, Luo Y, Asiri AM, Al-Youbi AO, Sun X. Environmentally friendly photocatalytic synthesis of porphyrin/Ag nanoparticles/reduced graphene oxide ternary nanohybrids having superior catalytic activity. ChemPlusChem 2012;77: 545–50.
[106] Guo P, Chen P, Liu M. One-dimensional porphyrin nanoassemblies assisted via graphene oxide: sheetlike functional surfactant and enhanced photocatalytic behaviors. ACS Appl Mater Interfaces 2013;5:5336–45.
[107] Chen Y, Huang Z-H, Yue M, Kanga F. Integrating porphyrin nanoparticles into a 2D graphene matrix for free-standing nanohybrid films with enhanced visible-light photocatalytic activity. Nanoscale 2014;6:978–85.
[108] Lu Q, Zhang Y, Liu S. Graphene quantum dots enhanced photocatalytic activity of zinc porphyrin toward the degradation of methylene blue under visible-light. J Mater Chem A 2015;3:8552–8.
[109] Zhu M, Li Z, Xiao B, Lu Y, Du Y, Yang P, Wang X. Surfactant assistance in improvement of photocatalytic hydrogen production with the porphyrin noncovalently functionalized graphene nanocomposite. ACS Appl Mater Interfaces 2013;5:1732–40.
[110] Grinceviciute N, Snopok B, Snitka V. Functional two-dimensional nanoarchitectures based on chemically converted graphene oxide and hematoporphyrin under the sulfuric acid treatment. Chem Eng J 2014;255:577–84.
[111] Wu T, Zou L, Han D, Li F, Zhang Q, Niu L. A carbon-based photocatalyst efficiently converting CO_2 to CH_4 and C_2H_2 in visible light. Green Chem 2014;16:2142–6.
[112] Ruan C, Zhang L, Qi Y, Xu C, Zhang X, Wan J, Peng Z, Shi J, Li X, Wang L. Synthesis of porphyrin sensitized TiO_2/graphene and its photocatalytic property under visible light. Mater Lett 2015;141:362–5.
[113] Wu S, He Q, Tan C, Wang Y, Zhang H. Graphene-based electrochemical sensors. Small 2013;9:1160–72.
[114] Zhu C, Dong S. Energetic graphene-based electrochemical analytical devices in nucleic acid, protein and cancer diagnostics and detection. Electroanalysis 2014;26:14–29.
[115] Moghimi A. Solid phase extraction of trace amount of Cu(II) using functionalized graphene. Russ J Phys Chem A 2013;87:1851–8.
[116] Karimnezhad K, Moghimi A. Separation of Cr(III) from by functionalized graphene oxide with covalently linked porphyrin (GO–H_2NP) adsorbed on surfactant coated C_{18}. Orient J Chem 2014;30:187–94.

[117] Xu Y, Zhao L, Bai H, Hong W, Li C, Shi G. Chemically converted graphene induced molecular flattening of 5,10,15,20-tetrakis(1-methyl-4-pyridinio)porphyrin and its application for optical detection of cadmium(II) ions. J Am Chem Soc 2009;131:13490–7.
[118] Tu W, Lei J, Zhang S, Ju H. Characterization, direct electrochemistry, and amperometric biosensing of graphene by noncovalent functionalization with picket-fence porphyrin. Chem Eur J 2010;16:10771–7.
[119] Zheng L, Ye D, Xiong L, Xu J, Tao K, Zou Z, Huang D, Kang X, Yang S, Xia J. Preparation of cobalt-tetraphenylporphyrin/reduced graphene oxide nanocomposite and its application on hydrogen peroxide biosensor. Anal Chim Acta 2013;768:69–75.
[120] Zhang L, Peng D, Liang R-P, Qiu J-D. Graphene quantum dots assembled with metalloporphyrins for "turn on" sensing of hydrogen peroxide and glucose. Chem Eur J 2015;21: 9343–8.
[121] Guo CX, Lei Y, Li CM. Porphyrin functionalized graphene for sensitive electrochemical detection of ultratrace explosives. Electroanalysis 2011;23:885–93.
[122] Wong A, Taboada Sotomayor MDP. Determination of carbofuran and diuron in FIA system using electrochemical sensor modified with organometallic complexes and graphene oxide. J Electroanal Chem 2014;731:163–71.
[123] Wu Y, Luo S, Chen L. Manganese porphyrin functionalized graphene and its application on dimethoate electrochemical sensor. Adv Mater Res 2014;850-851:152–5.
[124] Hu Y, Xue Z, He H, Ai R, Liu X, Lu X. Photoelectrochemical sensing for hydroquinone based on porphyrin-functionalized Au nanoparticles on graphene. Biosens Bioelectr 2013;47: 45–9.
[125] Shi P, Ye N. Investigation of the adsorption mechanism and preconcentration of sulfonamides using a porphyrin-functionalized Fe_3O_4-graphene oxide nanocomposite. Talanta 2015;143:219–25.
[126] Moraes FC, Rossi B, Donatoni MC, de Oliveira KT, Pereira EC. Sensitive determination of 17β-estradiol in river water using a graphene based electrochemical sensor. Anal Chim Acta 2015;881:37–43.
[127] Xu C, Wang X, Wang J, Hu H, Wan L. Synthesis and photoelectrical properties of β-cyclodextrin functionalized graphene materials with high bio-recognition capability. Chem Phys Lett 2010;498:162–7.
[128] Feng L, Wu L, Wang J, Ren J, Miyoshi D, Sugimoto N, Qu X. Detection of a prognostic indicator in early-stage cancer using functionalized graphene-based peptide sensors. Adv Mater 2012;24:125–31.
[129] Wu L, Wang J, Feng L, Ren J, Wei W, Qu X. Label-free ultrasensitive detection of human telomerase activity using porphyrin-functionalized graphene and electrochemiluminescence technique. Adv Mater 2012;24:2447–52.
[130] Wu L, Feng L, Ren J, Qu X. Electrochemical detection of dopamine using porphyrin-functionalized graphene. Biosens Bioelectron 2012;34:57–62.
[131] Lv M, Mei T, Zhang C, Wang X. Selective and sensitive electrochemical detection of dopamine based on water-soluble porphyrin functionalized graphene nanocomposites. RSC Adv 2014;4:9261–70.
[132] Han HS, Lee HK, You J-M, Jeong H, Jeon S. Electrochemical biosensor for simultaneous determination of dopamine and serotonin based on electrochemically reduced GO-porphyrin. Sens Actuators B 2014;190:886–95.
[133] Sun B, Zhang K, Chen L, Guo L, Ai S. A novel photoelectrochemical sensor based on PPIX-functionalized WO_3–rGO nanohybrid-decorated ITO electrode for detecting cysteine. Biosens Bioelectron 2013;44:48–51.

[134] Yang A, Xue Y, Zhao H, Li X, Yuan Z. One-pot synthesis of ternary hybrid nanomaterial composed of a porphyrin-functionalized graphene, tin oxide, and gold nanoparticles, and its application to the simultaneous determination of epinephrine and uric acid. Microchim Acta 2015;182:341–9.

[135] Georgescu R, van Staden JF, van Staden R-IS, Boscornea C. Evaluation of amperometric dot microsensors for the analysis of folic acid in pharmaceutical tablets and urine samples. J Porphyrins Phthalocyanines 2015;19:679–87.

[136] Zhang S, Tang S, Lei J, Dong H, Ju H. Functionalization of graphene nanoribbons with porphyrin for electrocatalysis and amperometric biosensing. J Electroanal Chem 2011;656: 285–8.

[137] Zhang H, Han Y, Guo Y, Dong C. Porphyrin functionalized graphene nanosheets-based electrochemical aptasensor for label-free ATP detection. J Mater Chem 2012;22:23900–5.

[138] Zhang H, Shuang S, Sun L, Chen A, Qin Y, Dong C. Label-free aptasensor for thrombin using a glassy carbon electrode modified with a graphene-porphyrin composite. Microchim Acta 2014;181:189–96.

[139] Wang Q, Lei J, Deng S, Zhang L, Ju H. Graphene-supported ferric porphyrin as a peroxidase mimic for electrochemical DNA biosensing. Chem Commun 2013;49:916–8.

[140] De Souza C, Zrig S, Wang D, Pham M-C, Piro B. Electrocatalytic miRNA detection using cobalt porphyrin-modified reduced graphene oxide. Sensors 2014;14:9984–94.

[141] Feng L, Wu L, Qu X. New horizons for diagnostics and therapeutic applications of graphene and graphene oxide. Adv Mater 2013;25:168–86.

[142] Bianco A. Graphene: safe or toxic? the two faces of the medal. Angew Chem Int Ed 2013;52:4986–97.

[143] Yang K, Li Y, Tan X, Peng R, Liu Z. Behavior and toxicity of graphene and its functionalized derivatives in biological systems. Small 2013;9:1492–503.

[144] Servant A, Bianco A, Prato M, Kostarelos K. Graphene for multi-functional synthetic biology: the last 'zeitgeist' in nanomedicine. Bioorg Med Chem Lett 2014;24:1638–49.

[145] Yang Y, Zhang Y-M, Chen Y, Zhao D, Chen J-T, Liu Y. Construction of a graphene oxide based noncovalent multiple nanosupramolecular assembly as a scaffold for drug delivery. Chem Eur J 2012;18:4208–15.

[146] Cho Y, Kim H, Choi Y. A graphene oxide–photosensitizer complex as an enzyme-activatable theranostic agent. Chem Commun 2013;49:1202–4.

[147] Gollavelli G, Ling Y-C. Magnetic and fluorescent graphene for dual modal imaging and single light induced photothermal and photodynamic therapy of cancer cells. Biomater 2014;35:4499–507.

[148] Xu J, Zeng F, Wu H, Yu C, Wu S. Dual-targeting nanosystem for enhancing photodynamic therapy efficiency. ACS Appl Mater Interfaces 2015;7:9287–96.

[149] Stefan-van Stadena R-I, Comnea-Stancua IR, Surdu-Bob CC, Stanciu-Gavan C. Pattern recognition of neuron specific enolase and carcinoembryonic antigen in whole blood samples. J Mol Recognit 2015;28:103–7.

[150] Su S, Wang J, Wei J, Martínez-Zaguilán R, Qiu J, Wang S. Efficient photothermal therapy of brain cancer through porphyrin functionalized graphene oxide. New J Chem 2015;39: 5743–5749.

[151] Iijima S. Helical microtubules of graphitic carbon. Nature 1991;354:56–8.

[152] Terrones M. Science and technology of the twenty-first century: synthesis, properties, and applications of carbon nanotubes. Annu Rev Mater Res 2003;33:419–501.

[153] De Volder MFL, Tawfick SH, Baughman RH, Hart AJ. Carbon nanotubes: present and future commercial applications. Science 2013;339:535–9.

[154] Guldi DM, Prato M. Electrostatic interactions by design. Versatile methodology towards multifunctional assemblies/nanostructured electrodes. Chem Comm 2004:2517–2525.

[155] Zhang H, Bork MA, Riedy KJ, McMillin DR, Choi JH. Understanding photophysical interactions of semiconducting carbon nanotubes with porphyrin chromophores. J Phys Chem C 2014;118:11612–9.

[156] Zhang H, Pan J, Ghimire S, Bork MA, Riccitelli MM, McMillin DR, Hyun Choi J. Regeneration of light-harvesting complexes via dynamic replacement of photodegraded chromophores. ACS Appl Mater Interf 2015;7:7833–7.

[157] Langa F, Gómez-Escalonilla MJ, de la Cruz P. Carbon nanotubes and porphyrins: an exciting combination for optoelectronic devices. J. Porphyrins Phthalocyanines 2007;11:348–58.

[158] Robertson N, McGowan CA. A comparison of potential molecular wires as components for molecular electronics. Chem Soc Rev 2003;32:96–103.

[159] Lu X, Zhang W, Wang C, Wen T-C, Wei Y. One-dimensional conducting polymer nanocomposites: Synthesis, properties and applications. Progr Polym Sci 2011;36:671–712.

[160] Egger S, Ilie A, Machida S, Nakayama T. Integration of individual nanoscale structures into devices using dynamic nanostenciling. Nano Lett 2007;7:3399–404.

[161] de Almeida Magalhães Sáfara G, Carvalho da Silva Martins D, DeFreitas-Silva G, Santos Rebouças J, Idemori YM, Righi A. Interactions of porphyrins and single walled carbon nanotubes: a fine duet. Synthetic Metals 2014;193:64–70.

[162] Kharisov BI, Kharissova OV. Coordination and organometallic compounds in the functionalization of carbon nanotubes. J Coord Chem 2014;67:3769–808.

[163] Girek B, Sliwa W. Hybrids of cationic porphyrins with nanocarbons. J Incl Phenom Macrocycl Chem 2015;82:283–300.

[164] Cheng F, Adronov A. Suzuki coupling reactions for the surface functionalization of single-walled carbon nanotubes. Chem Mater 2006;18:5389–91.

[165] Palacin T, Le Khanh H, Jousselme B, Jegou P, Filoramo A, Ehli C, Guldi DM. Efficient functionalization of carbon nanotubes with porphyrin dendrons via click chemistry. J Am Chem Soc 2009;131:15394–402.

[166] Tanaka H, Yajima T, Matsumoto T, Otsuka Y, Ogawa T. Porphyrin molecular nanodevices wired using single-walled carbon nanotubes. Adv Mater 2006;18:1411–5.

[167] Peng X, Qin H, Li L, Huang Y, Peng J, Cao Y, Komatsu N. Water redissoluble chiral porphyrin-carbon nanotube composites. J Mater Chem 2012;22:5764–9.

[168] Moghadam M, Mohammadpoor-Baltork I, Tangestaninejad S, Mirkhani V, Kargar H, Zeini-Isfahani N. Manganese(III) porphyrin supported on multi-wall carbon nanotubes: a highly efficient and reusable biomimetic catalyst for epoxidation of alkenes with sodium periodate. Polyhedron 2009;28:3816–22.

[169] Rayati S, Jafarzadeh P, Zakavi S. Catalytic activity of carbon nanotube supported iron(III) and manganese(III) porphyrins in oxidation of olefins with *tert*-butyl hydroperoxide: higher activity of the iron(III) porphyrin. Inorg Chem Commun 2013;29:40–4.

[170] Rayati S, Bohloulbandi E. Multi-wall carbon nanotube-supported manganese(III) porphyrin: an efficient and reusable catalyst for the oxidation of alkenes with H_2O_2 under ultrasonic irradiation. C R Chimie 2014;17:62–8.

[171] Rayati S. Sheybanifard Z. Ultrasound promoted green oxidation of alkenes with hydrogen peroxide in the presence of iron porphyrins supported on multi-walled carbon nanotubes. J Porphyrins Phthalocyanines 2015;19:622–30.

[172] Lipińska ME, Rebelo SLH, Freire C. Iron(III) porphyrin anchored onto organosilylated multi-walled carbon nanotubes as an active catalyst for epoxidation reactions in mild conditions. J Mater Sci 2014;49:1494–505.

[173] Karimipour G. Manganese porphyrin supported on multiwalled carbon nanotube (MWCN) as solid catalyst for alkene epoxidation. Int J Chem Eng Appl 2014;5:194–7.

[174] Kargar H, Moghadam M, Mirkhani V, Tangestaninejad S, Mohammadpoor-Baltork I, Rezaei S. Multi-wall carbon nanotube supported manganese(III) porphyrin: an efficient and reusable catalyst for oxidation of 2-imidazolines with sodium periodate. Transition Met Chem 2013;38:1–5.

[175] Araghi M, Bokaei F. Manganese(III) porphyrin supported on multi-wall carbon nanotubes: a highly efficient and reusable biomimetic catalyst for oxidative decarboxylation of α-arylcarboxylic acids and oxidation of alkanes with sodium periodate. Polyhedron 2013;53:15–9.

[176] Rezaeifard A, Jafarpour M. The catalytic efficiency of Fe-porphyrins supported on multiwalled carbon nanotubes in the heterogeneous oxidation of hydrocarbons and sulfides in water. Catal Sci Technol 2014;4:1960–9.

[177] Sharghi H, Beyzavi MH, Safavi A, Doroodmand MM, Khalifeh R. Immobilization of porphyrinatocopper nanoparticles onto activated multi-walled carbon nanotubes and a study of its catalytic activity as an efficient heterogeneous catalyst for a click approach to the three-component synthesis of 1,2,3-triazoles in water. Adv Synth Catal 2009;351:2391–410.

[178] Wan L-S, Ke B-B, Wu J, Xu Z-K. Catalase immobilization on electrospun nanofibers: effects of porphyrin pendants and carbon nanotubes. J Phys Chem C 2007;111:14091–7.

[179] Orellana W. Catalytic activity toward oxygen reduction of transition metal porphyrins covalently linked to single-walled carbon nanotubes: a density functional study. Phys Rev B 2011;84:155405/1–/1155405.

[180] Cheng M-J, Head-Gordon M, Bell AT. How to chemically tailor metal-porphyrin-like active sites on carbon nanotubes and graphene for minimal overpotential in the electrochemical oxygen evolution and oxygen reduction reactions. J Phys Chem C 2014;118:29482–91.

[181] Qu J, Shen Y, Qu X, Dong S. Preparation of hybrid thin film modified carbon nanotubes on glassy carbon electrode and its electrocatalysis for oxygen reduction. Chem Comm 2004;34–5.

[182] Qu J, Shen Y, Qu X, Dong S. Electrocatalytic reduction of oxygen at multi-walled carbon nanotubes and cobalt porphyrin modified glassy carbon electrode. Electroanal 2004;16: 1444–50.

[183] Liu Y-, Yan Y-L, Lei J, Wu F, Ju H. Functional multiwalled carbon nanotube nanocomposite with iron picket-fence porphyrin and its electrocatalytic behavior. Electrochem Comm 2007;9:2564–70.

[184] Schilling T, Bron M. Oxygen reduction at Fe–N-modified multi-walled carbon nanotubes in acidic electrolyte. Electrochim Acta 2008;53:5379–85.

[185] Yu P, Yan J, Zhao H, Su L, Zhang J, Mao L. Rational functionalization of carbon nanotube/ionic liquid bucky gel with dual tailor-made electrocatalysts for four-electron reduction of oxygen. J Phys Chem C 2008;112:2177–82.

[186] Choi A, Jeong H, Kim S, Jo S, Jeon S. Electrocatalytic reduction of dioxygen by cobalt porphyrin-modified glassy carbon electrode with single-walled carbon nanotubes and nafion in aqueous solutions. Electrochim Acta 2008;53:2579–84.

[187] Tu W, Lei J, Ju H. Noncovalent nanoassembly of porphyrin on single-walled carbon nanotubes for electrocatalytic reduction of nitric oxide and oxygen. Electrochem Comm 2008;10:766–9.

[188] Okunola A, Kowalewska B, Bron M, Kulesza PJ, Schuhmann W. Electrocatalytic reduction of oxygen at electropolymerized films of metalloporphyrins deposited onto multi-walled carbon nanotubes. Electrochim Acta 2009;54:1954–60.

[189] Dembinska B, Kulesza PJ. Multi-walled carbon nanotube-supported tungsten oxide-containing multifunctional hybrid electrocatalytic system for oxygen reduction in acid medium. Electrochim Acta 2009;54:4682–7.

[190] Kulesza PJ, Dembinska B. Development of bifunctional carbon nanotube-supported co-porphyrin electrocatalytic system for oxygen reduction. ECS Trans 2009;16:13–9.

[191] Dobrzeniecka A, Zeradjanin A, Masa J, Stroka J, Goral M, Schuhmann W, Kulesza PJ. Scanning electrochemical microscopy for investigation of multi-component bioelectrocatalytic films. ECS Trans 2011;35:33–44.

[192] Zhang W, Shaikh AU, Tsui EY, Swager TM. Cobalt Porphyrin Functionalized Carbon Nanotubes for Oxygen Reduction. Chem Mater 2009;21:3234–41.

[193] Umasankar Y, Shie J-W, Chen S-M. Electrocatalytic activity of oxygen and hydrogen peroxide reduction at Poly(iron tetra(*o*-aminophenyl)porphyrin) coated multiwalled carbon nanotube composite film. J Electrochem Soc 2009;156:K238–44.

[194] McGuire R Jr, Dogutan DK, Teets TS, Suntivich J, Shao-Horn Y, Nocera DG. Oxygen reduction reactivity of cobalt(II) hangman porphyrins. Chem Sci 2010;1:411–4.

[195] Deng L, Zhou M, Liu C, Liu L, Liu C, Dong S. Development of high performance of Co/Fe/N/CNT nanocatalyst for oxygen reduction in microbial fuel cells. Talanta 2010;81:444–8.

[196] Schilling T, Okunola A, Masa J, Schuhmann W, Bron M. Carbon nanotubes modified with electrodeposited metal porphyrins and phenanthrolines for electrocatalytic applications. Electrochim Acta 2010;55:7597–602.

[197] Morozan A, Campidelli S, Filoramo A, Jousselme B, Palacin S. Catalytic activity of cobalt and iron phthalocyanines or porphyrins supported on different carbon nanotubes towards oxygen reduction reaction. Carbon 2011;49:4839–47.

[198] Kruusenberg I, Matisen L, Tammeveski K. Oxygen electroreduction on multi-walled carbon nanotube supported metal phthalocyanines and porphyrins in acid media. Int J Electrochem Sci 2013;8:1057–66.

[199] Ahmed MS, Jeong H, You J-M, Jeon S. Electrocatalytic reduction of dioxygen at a modified glassy carbon electrode based on Nafion®-dispersed single-walled carbon nanotubes and cobalt–porphyrin with palladium nanoparticles in acidic media. Electrochim Acta 2011;56:4924–9.

[200] Zhao H, Chang Y, Liu C. Electrochemical behavior of electrodes modified with metalloporphyrin and multiwalled carbon nanotubes for the reduction of oxygen, proton and carbon dioxide. J Porphyrins Phthalocyanines 2013;17:259–63.

[201] Wei P-J, Yu G-Q, Naruta Y, Liu J-G. Covalent grafting of carbon nanotubes with a biomimetic heme model compound to enhance oxygen reduction reactions. Angew Chem Int Ed 2014;53:6659–63.

[202] Hijazi I, Bourgeteau T, Cornut R, Morozan A, Filoramo A, Leroy J, Derycke V, Jousselme B, Campidelli S. Carbon nanotube-templated synthesis of covalent porphyrin network for oxygen reduction reaction. J Am Chem Soc 2014;136:6348–54.

[203] Kulesza PJ, Kowalewska B, Karnicka K, Skunik M, Ernst A, Miecznikowski K, Ginalska G, Belcarz A. Bifunctional bioelectrocatalytic systems for oxygen reduction porphyrins and supramolecular assemblies. ECS Trans 2007;6:81–90.

[204] Kowalewska B, Skunik M, Karnicka K, Miecznikowski K, Chojak M, Ginalska G, Belcarz A, Kulesza PJ. Enhancement of bio-electrocatalytic oxygen reduction at the composite film of cobalt porphyrin immobilized within the carbon nanotube-supported peroxidase enzyme. Electrochim Acta 2008;53:2408–15.

[205] Zhao H-Z, Chang Y-Y, Liu C. Electrodes modified with iron porphyrin and carbon nanotubes: application to CO_2 reduction and mechanism of synergistic electrocatalysis. J Solid State Electrochem 2013;17:1657–64.

[206] Yan Y, Yao P, Mu Q, Wang L, Mu J, Li X, Kang S-Z. Electrochemical behavior of amino-modified multi-walled carbon nanotubes coordinated with cobalt porphyrin for the oxidation of nitric oxide. Appl Surf Sci 2011;258:58–63.

[207] Elouarzaki K, Le Goff A, Holzinger M, Thery J, Cosnier S. Electrocatalytic oxidation of glucose by rhodium porphyrin-functionalized MWCNT electrodes: application to a fully molecular catalyst-based glucose/O_2 fuel cell. J Am Chem Soc 2012;134:14078–85.

[208] Elouarzaki K, Le Goff A, Holzinger M, Agnés C, Duclairoir F, Putaux J-L, Cosnier S. From gold porphyrins to gold nanoparticles: catalytic nanomaterials for glucose oxidation. Nanoscale 2014;6:8556–60.

[209] Jia H, Sun Z, Jiang D, Du P. Covalent cobalt porphyrin framework on multiwalled carbon nanotubes for efficient water oxidation at low overpotential. Chem Mater 2015;27: 4586–4593.

[210] Gutierrez CA, Silva JF, Recio FJ, Griveau S, Bedioui F, Caro CA, Zagal JH. In search of the best iron N4-macrocyclic catalysts adsorbed on graphite electrodes and on multi-walled carbon nanotubes for the oxidation of l-cysteine by adjusting the Fe(II)/(I) formal potential of the complex. Electrocatal 2014;5:426–37.

[211] Lopez RJ, Babanova S, Artyushkova K, Atanassov P. Surface modifications for enhanced enzyme immobilization and improved electron transfer of PQQ-dependent glucose dehydrogenase anodes. Bioelectrochem 2015;105:78–87.

[212] Lvova L, Paris A, Mastroianni M, Pomarico G, Santonico M, Pennazza G, Di Natale C, Paolesse R. SWCNTs modified with porphyrin units for chemical sensing applications. A D'Amico Procedia Eng 2010;5:1043–6.

[213] Penza M, Alvisi M, Rossi R, Serra E, Paolesse R, D'Amico A, Di Natale C. Carbon nanotube films as a platform to transduce molecular recognition events in metalloporphyrins. Nanotechnol 2011;22:125502/1–/1125502.

[214] Lvova L, Mastroianni M, Pomarico G, Santonico M, Pennazza G, Di Natale C, Paolesse R. D'Amico A. Carbon nanotubes modified with porphyrin units for gaseous phase chemical sensing. Sens Actuat B 2012;170:163–71.

[215] Shirsat MD, Sarkar T, Kakoullis J Jr, Myung NV, Konnanath B, Spanias A, Mulchandani A. Porphyrin-functionalized single-walled carbon nanotube chemiresistive sensor arrays for VOCs. J Phys Chem C 2012;116:3845–50.

[216] Rushi A, Datta K, Ghosh P, Mulchandani A, Shirsat MD. Iron tetraphenyl porphyrin functionalized single wall carbon nanotubes for the detection of benzene. Mater Lett 2013;96: 38–41.

[217] Ndiaye A, Bonnet P, Pauly A, Dubois M, Brunet J, Varenne C, Guerin K, Lauron B. Noncovalent functionalization of single-wall carbon nanotubes for the elaboration of gas sensor dedicated to BTX type gases: the case of toluene. J Phys Chem C 2013;117: 20217–28.

[218] Saxena S, Verma AL. Metal-tetraphenylporphyrin functionalized carbon nanotube composites as sensor for benzene, toluene and xylene vapors. Adv Mat Lett 2014;5:472–8.

[219] Sarkar T, Srinives S, Sarkar S, Haddon RC, Mulchandani A. Single-walled carbon nanotube–poly(porphyrin) hybrid for volatile organic compounds detection. J Phys Chem C 2014;118:1602–10.

[220] Liu SF, Moh LCH, Swager TM. Single-walled carbon nanotube–metalloporphyrin chemiresistive gas sensor arrays for volatile organic compounds. Chem Mater 2015;27:3560–3.

[221] Liu SF, Petty AR, Sazama GT, Swager TM. Single-walled carbon nanotube/metalloporphyrin composites for the chemiresistive detection of amines and meat spoilage. Angew Chem Int Ed 2015;54:6554–7.

[222] Lu X, Quan Y, Xue Z, Wu B, Qi H, Liu D. Determination of explosives based on novel type of sensor using porphyrin functionalized carbon nanotubes. Colloids Surf B 2011;88: 396–401.

[223] Quan Y, Xue Z, Shi H, Zhou X, Du J, Liu X, Lu X. A high-performance and simple method for rapid and simultaneous determination of dihydroxybenzene isomers. Analyst 2012;137: 944–52.

[224] Salimi A, MamKhezri H, Hallaj R, Zandi S. Modification of glassy carbon electrode with multi-walled carbon nanotubes and iron(III)-porphyrin film: application to chlorate, bromate and iodate detection. Electrochim Acta 2007;52:6097–105.

[225] Tu W, Lei J, Jian G, Hu Z, Ju H. Noncovalent assembly of picket-fence porphyrins on nitrogen-doped carbon nanotubes for highly efficient catalysis and biosensing. Chem Eur J 2010;16:4120–6.

[226] Karimipour G, Ghaedi M, Behfar M, Andikaey Z, Kowkabi S, Orojloo AH. Synthesis and application of new porphyrin derivatives for preparation of copper selective electrodes: influence of carbon nanotube on their responses. IEEE Sensors J 2012;12:2638–47.

[227] Rajabi HR, Roushani M, Shamsipur M. Development of a highly selective voltammetric sensor for nanomolar detection of mercury ions using glassy carbon electrode modified with a novel ion imprinted polymeric nanobeads and multi-wall carbon nanotubes. J Electroanal Chem 2013;693:16–22.

[228] Jeong H, Kim S, Seol H, You J-M, Jeong ES, Kim SK, Seo K-S, Jeon S. Determination of hydrogen peroxide on modified glassy carbon electrode by polytetrakis(2-aminophenyl) porphyrin nanowire. Bull Korean Chem Soc 2009;30:2979–83.

[229] Li X-R, Wang B, Xu J-J, Chen H-Y. Noncovalent assembly of picket-fence porphyrin on carbon nanotubes as effective peroxidase-like catalysts for detection of hydrogen peroxide in beverages. Electroanalysis 2011;23:2955–63.

[230] Li J, Tu W, Lei J, Tang S, Ju H. Porphyrin-functionalized gold nanoparticles for selective electrochemical detection of peroxyacetic acid. Electrochim Acta 2011;56:3159–63.

[231] Tu W, Lei J, Ju H. Functionalization of carbon nanotubes with water-insoluble porphyrin in ionic liquid: direct electrochemistry and highly sensitive amperometric biosensing for trichloroacetic acid. Chem Eur J 2009;15:779–84.

[232] Wu Y. Electrocatalysis and sensitive determination of Sudan I at the single-walled carbon nanotubes and iron(III)-porphyrin modified glassy carbon electrodes. Food Chem 2010;121:580–4.

[233] Wong A, Del Pilar Taboada Sotomayor M. Biomimetic sensor based on 5,10,15,20-tetrakis(pentafluorophenyl)-21H,23H-porphyrin iron (III) chloride and MWCNT for selective detection of 2,4-D. Sens Actuat B 2013;181:332–9.

[234] He J-L, Zou W, Wang J-X, Zeng L-H, Cao Z, Luo D-M. Tetra-(*N*-phenylpyrazole)-porphyrin/β-cyclodextrin incorporated CNT film modified hemoglobin sensor. Adv Mater Res 2011;239-242:2831–4.

[235] Cristescu R, Popescu C, Popescu AC, Mihailescu IN, Ciucu AA, Andronie A, Iordache S, Stamatin I, Fagadar-Cosma E, Chrisey DB. Functional porphyrin thin films deposited by matrix assisted pulsed laser evaporation. Mater Sci Eng B 2010;169:106–10.

[236] Wang C, Yuan R, Chai Y, Chen S, Zhang Y, Hu F, Zhang M. Non-covalent iron(III)-porphyrin functionalized multi-walled carbon nanotubes for the simultaneous determination of ascorbic acid, dopamine, uric acid and nitrite. Electrochim Acta 2012;62:109–15.

[237] Chen W, Ding Y, Akhigbe J, Brückner C, Li CM, Lei Y. Enhanced electrochemical oxygen reduction-based glucose sensing using glucose oxidase on nanodendritic poly[meso-tetrakis(2-thienyl)porphyrinato]cobalt(II)-SWNTs composite electrodes. Biosens Bioelectron 2010;26:504–10.
[238] Luz RCS, Damos FS, Tanaka AA, Kubota LT, Gushikem Y. Electrocatalysis of reduced L-glutathione oxidation by iron(III) tetra-(*N*-methyl-4-pyridyl)-porphyrin (FeT4MPyP) adsorbed on multi-walled carbon nanotubes. Talanta 2008;76:1097–104.
[239] Luz RCS, Maroneze CM, Tanaka AA, Kubota LT, Gushikem Y, Damos FS. The electrocatalytic activity of a supramolecular assembly of CoTsPc/FeT4MPyP on multi-walled carbon nanotubes towards L-glutathione, and its determination in human erythrocytes. Microchim Acta 2010;171:169–78.
[240] Ma Q, Ai S, Yin H, Chen Q, Tang T. Towards the conception of an amperometric sensor of L-tyrosine based on Hemin/PAMAM/MWCNT modified glassy carbon electrode. Electrochim Acta 2010;55:6687–94.
[241] Liu X, Feng H, Liu X, Wong DKY. Electrocatalytic detection of phenolic estrogenic compounds at NiTPPS|carbon nanotube composite electrodes. Anal Chim Acta 2011;689:212–8.
[242] Huang CZ, Liao QG, Li YF. Non-covalent anionic porphyrin functionalized multi-walled carbon nanotubes as an optical probe for specific DNA detection. Talanta 2008;75:163–6.
[243] Ning J, Wang Y, Wu Q, Zhang X, Lin X, Zhao H. Novel supramolecular assemblies of repulsive DNA–anionic porphyrin complexes based on covalently modified multi-walled carbon nanotubes and cyclodextrins. RSC Adv 2015;5:21153–60.
[244] Vedala H, Chen Y, Cecioni S, Imberty A, Vidal S, Star A. Nanoelectronic detection of lectin-carbohydrate interactions using carbon nanotubes. Nano Lett 2011;11:170–5.
[245] Chen Y, Vedala H, Kotchey GP, Audfray A, Cecioni S, Imberty A, Vidal S, Star A. Detection of lectins using glyco-functionalized nanosensors. MRS Online Proceedings Library 2012;1451:191.
[246] Chen Y, Vedala H, Kotchey GP, Audfray A, Cecioni S, Imberty A, Vidal S, Star A. Electronic detection of lectins using carbohydrate-functionalized nanostructures: graphene versus carbon nanotubes. ACS Nano 2012;6:760–70.
[247] Wong BS, Yoong SL, Jagusiak A, Panczyk T, Ho HK, Ang WH, Pastorin G. Carbon nanotubes for delivery of small molecule drugs. Adv Drug Deliv Rev 2013;65:1964–2015.
[248] Pan D, Kim B, Wang LV, Lanza GM. A brief account of nanoparticle contrast agents for photoacoustic imaging. WIREs Nanomed Nanobiotechnol 2013;5:517–43.
[249] Banerjee I, Douaisi MP, Mondal D, Kane RS. Light-activated nanotube–porphyrin conjugates as effective antiviral agents. Nanotechnology 2012;23:340201–4018.
[250] Banerjee I, Mondal D, Martin J, Kane RS. Photoactivated antimicrobial activity of carbon nanotube-porphyrin conjugates. Langmuir 2010;26:17369–74.
[251] Shi J, Ma R, Wang L, Zhang J, Liu R, Li L, Liu Y, Hou L, Yu X, Gao J, Zhang Z. The application of hyaluronic acid-derivatized carbon nanotubes in hematoporphyrin monomethyl ether-based photodynamic therapy for in vivo and in vitro cancer treatment. Int J Nanomed 2013;8:2361–73.
[252] Sáfar GAM, Gontijo RN, Fantini C, Martins DCS, Idemori YM, Pinheiro MVB, Krambrock K. Enhanced oxygen singlet production by hybrid system of porphyrin and enriched (6,5) single-walled carbon nanotubes for photodynamic therapy. J Phys Chem C 2015;119:4344–50.
[253] Iijima S. Direct observation of the tetrahedral bonding in graphitized carbon black by high resolution electron microscopi. J Cryst Growth 1980;50:675–83.

[254] Kroto HW, Heath JR, O'Brien SC, Curl RF, Smalley RE. C60: Buckminsterfullerene. Nature 1985;318:162–3.

[255] Katz EA. Fullerene, thin films as photovoltaic material. In: Sōga T, editor. Nanostructured materials for solar energy conversion. Amsterdam: Elsevier; 2006. p. 361–443.

[256] Hirsch A, editor. Fullerenes and related structures. In: Topics in current chemistry, vol. 199. Berlin: Springer; 1998.

[257] Kim DH, Lee MY, Suh YD, Kim SH. Observation of fluorescence emission from solutions of C60 and C70 fullerenes and measurement of their excited-state lifetimes. J Am Chem Soc 1992;114:4429–30.

[258] Wasielewski MR, O'Neil MP, Lykke KR, Pellin MJ, Gruen DM. Triplet states of fullerenes C60 and C70. Electron paramagnetic resonance spectra, photophysics, and electronic structures. J Am Chem Soc 1991;113:2774–6.

[259] Arbogast JW, Foote CS. Photophysical properties of C70. J Am Chem Soc 1991;113:8886–9.

[260] Rispens MT, Hummelen JC. . In: Guldi DM, Martin N, editors. Fullerenes: from synthesis to optoelectronic properties. Dordrecht: Kluwer Academic Publisher; 2010. p. 387–435.

[261] Kadish KM, Ruoff RS. Fullerenes: chemistry, physics, and technology. New York: John Wiley & Sons; 2000.

[262] Maggini M, Scorrano G, Prato M. Addition of azomethine ylides to C60: synthesis, characterization, and functionalization of fullerene pyrrolidines. J Am Chem Soc 1993;115:9798–9.

[263] Bingel C. Cyclopropylation of Fullerenes. Ber 1993;126:1957–9.

[264] Suzuki T, Li Q, Khermani KC, Wudl F. Dihydrofulleroid H_2C_{61}: Synthesis and Properties of the Parent Fulleroid. J Am Chem Soc 1992;114:7301–2.

[265] (a) Imahori H, Sekiguchi Y, Kashiwagi Y, Sato T, Araki Y, Ito O, Yamada H, Fukuzumi S. Long-lived charge-separated state generated in a ferrocene–meso, meso-linked porphyrin trimer–fullerene pentad with a high quantum yield. Chem Eu J 2004;10:3184–3196; (b) Imahori H, Guldi DM, Tamaki K, Yoshida Y, Luo C, Sakata Y, Fukuzumi S. Charge separation in a novel artificial photosynthetic reaction center lives 380 ms. J Am Chem Soc 2001;123:6617–6628; (c) Guldi DM, Imahori H, Tamaki K, Kashiwagi Y, Yamada H, Sakata Y, Fukuzumi S. A molecular tetrad allowing efficient energy storage for 1.6 s at 163 K. J Phys Chem 2004;108:541–548.

[266] Dubois D, Jones MT, Kadish KM. Electroreduction of buckminsterfullerene, C60, in aprotic solvents: electron spin resonance characterization of singly, doubly, and triply reduced C60 in frozen solutions. J Am Chem Soc 1992;114:6446–51.

[267] Rispens MT, Hummelen JC. . In: Guldi DM, Martin N, editors. Developments in fullerene science. Dordrecht, The Netherlands: Kluwer Academic Publishers; 2010. p. 387–435.

[268] Song L-C, Liu X-F, Xie Z-J, Luo F-X, Song H-B. Synthesis and structural characterization of some new porphyrin-fullerene dyads and their application in photoinduced H_2 evolution. Inorg Chem 2011;50:11162–72.

[269] Wu H, Fan S, Jin X, Zhang H, Chen H, Dai Z, Zou X. Construction of a zinc porphyrin–fullerene-derivative based nonenzymatic electrochemical sensor for sensitive sensing of hydrogen peroxide and nitrite. Anal Chem 2014;86:6285–90.

[270] Lu X, Shan D, Yang J, Huang B, Zhou X. Determination of *m*-dinitrobenzene based on novel type of sensor using thiol-porphyrin mixed monolayer-tethered polyaniline with intercalating fullerenols. Talanta 2013;115:457–61.

[271] Moldoveanu I, Stanciu Gavan C, Stefan-van Staden R-I. Molecular recognition of HER-1 in whole-blood samples. J Mol Recognit 2014;27:653–8.

[272] Herbst RS, Bunn PA. Targeting the epidermal growth factor receptor in non-small cell lung cancer. Clin Cancer Res 2003;9:5813–24.

[273] Ettinger DS. Clinical implications of EGFR expression in the development and progression of solid tumors: focus on non-small cell lung cancer. Oncologist 2006;11:358–73.

[274] Fujino S, Enokibori T, Tezuka T, Asada Y, Inoue S, Kato H, Mori A. A comparison of epidermal growth factor receptor levels and other prognostic parameters in non-small cell lung cancer. Eur J Cancer 1996;32A:2070–4.

[275] Safar GAM, Idemori YM, Carvalho Da-Silva D, Reboucas JS, Mazzoni MSC, Righi A. Nanometrological porphyrins. Nanotechnology 2012;23:275504. 6pp.

Chapter 15

Therapeutic Nanoreactors: Toward a Better Blood Substitute

Agnes Ostafin, Kyu Bum Han
Department of Materials Science and Chemical Engineering, University of Utah, Nano Institute, Salt Lake City, UT, United States of America

1 NANOREACTORS IN BIOLOGY AND MEDICINE

Nanoscience, once a field for discovery alone, has evolved into a design engineering enterprise [1]. This is evident in the proliferation of multicomponent nanomaterials created to take advantage of unique material and interfacial features present in materials with nanoscale dimensions. Nanoreactors are a subset of such materials. These unique materials are designed as hosts for chemical and physical reactions. They can provide protection from the environment, enhance reactions, and generate physical conditions at the nanoscale that would ordinarily require high temperatures or pressures to achieve in bulk, thus allowing difficult processes to take place with minimal energy expenditure.

In biological systems, namely living cells, there are many organelles and spaces that operate thanks to a nanoreactor environment. The results are highly efficient concerted reaction sequences that cannot take place in solution. In many cases, little is understood about the architecture and physics of these spaces, and so mimicking them through nanomaterial science has been extremely challenging.

One of the simplest, and perhaps best understood, nanoreactor environments in biology is the mammalian red blood cell. The size of this donut-shaped cell is about 8 microns, and inside there are many thousands of proteins that are uniquely distributed to maximize the structural integrity and gas exchange functions. Inside there exist hemoglobin (HB) rich pockets that generate nanoenvironments to protect the molecules from undesired oxidation, modulate gas transfer, and other functions. This ability to organize molecules at the nanoscale level inside cells is vital to enabling the cell to function up to 120 days without a nucleus.

Efforts to create an artificial red blood cell have not been able to recreate the structural and functional complexity found inside red blood cells. As a result, the functional life of this material is greatly shortened in the bloodstream, and the breakdown products generate immunogenic or toxic conditions that lead

Organic Nanoreactors. http://dx.doi.org/10.1016/B978-0-12-801713-5.00015-X

to serious health problems. In this work we have taken a functional nanomaterial, a multicomponent gas-carrying nanoreactor, based on perfluorocarbon nanoemulsion, and modified it further to extend its useful duty life and reduce breakdown. As a result, we created a blood substitute nanoreactor particle that enabled the development of a new clinical methodology. This new method involves the controlled introduction and removal of the reinforced blood substitute material from circulation. This avoids unwanted metabolic processing and side effects, yet allows life-saving oxygen delivery function when it is needed. This development is an example of nanomaterial design enabling the conception of new therapeutic approaches that would not have been possible otherwise.

2 NEED FOR BLOOD SUBSTITUTES

Rising demand for blood in cases of surgery, ischemia, war, and so on, cannot be satisfied in the future given the ever-decreasing availability of donations. Fears about untested diseases, unsuitability of donated blood among some religious groups, and unavailability of needed storage and delivery facilities in third world countries amplify the urgency for developing suitable replacement products [2,3]. Recently introduced stem cell based artificial red blood cells [4], advanced perfluorocarbons (PFCs) [5], and polyhemoglobin (pHB) [6] are encouraging, but will not satisfy the needs of all types of patients. For example, those undergoing chronic transfusion would prefer iron-free products to avoid iron overload, and in war zones portability is desirable. Thus, despite past disappointments, there is continuing interest and need for developing materials that can perform this function.

3 BLOOD SUBSTITUTE MATERIALS

Viewed as a material, blood substitutes are submicron-size materials that perform a physical or chemical reaction. Physical reactions may include the solvation of gases and their evaporation. Chemical reactions may involve specific ligand binding and release regulated by local environment conditions. The key to their suitability is ensuring stable function despite varying external conditions, while minimizing adverse physiological responses. In the past, blood substitutes, generally transfused in a large quantity, have been expected to spontaneously decompose and be excreted through natural metabolism. Instead, patients suffered from acute complications and side effects [7,8] from the material toxicity, NO imbalance and accumulation of breakdown products in tissues. As a result, most clinical trials in the United States were halted [9].

4 DESIGNING A BETTER BLOOD SUBSTITUTE

Given past failures, a pragmatic approach to blood-substitute material design would to be acknowledge that the human body has not evolved to deal with large quantities of any injected foreign materials, and that a better choice may be materials that are intended for use for only short periods of time and are ultimately recoverable. We suggest retrieving an injected, breakdown-resistant blood substitute in a timely manner after its administration will reduce overall side effects yet still provide needed clinical effect. Systematic removal may also stimulate erythropoiesis for faster recovery.

While it is easy to produce a material that is short lived in the human bloodstream because it is readily excreted or processed in the liver, it is far more difficult to produce a material whose residence time in the bloodstream can be manipulated externally. Such a material must be resistant to immune recognition, enzymatic processing, cellular uptake, excretion, and other mechanisms. At the same time, it must have a physical or chemical property that allows it to be recognized and removed from circulation at will.

Nanotechnology applied to drug delivery has provided many advanced materials that are capable of being directed to a specific site in the body via chemical recognition [10]. Magnetic nanoparticles can be manipulated by external magnetic fields [11]. Techniques such as dialysis or immunofiltration have shown it is possible to selectively remove substances from circulating blood without harm to the individual. However, these capabilities have yet to be combined into a single send-and-retrieve particle system for therapeutic use. A send-and-retrieve blood-substitute particle would have a tremendous clinical impact.

5 RETRIEVABLE NANOREACTOR BLOOD SUBSTITUTES

We have synthesized several kinds of retrievable nanoreactor materials for use as potential blood substitutes. Because of the large volumes needed clinically, and the rapid circuit time of blood in the body, continuous magnetic retrieval and immunofiltration are simple not practical methods for retrieval. Therefore, we focused on high buoyant density oxygen-carrying retrievable nanoreactors. Such materials could be separated in continuous fashion using clinically accepted aphaeresis units, cell savers and similar technologies.

One of our designer materials is a $CaPO_4$-shelled retrievable nanoreactor based on a DOPA-stabilized perfluorocarbon emulsion (see Table 15.1). Others include protein, polysaccharide, and polyethylene glycol-coated retrievable nanoreactors. Each contains a perfluorooocytlbromide core surrounded by one or more lipid stabilizers and a secondary coating (Table 15.1).

Regardless of the final coating layer, the synthesis of such nanoreactor materials requires considerable knowledge about emulsion formation and stability, size clarification, precipitation seeding processes, and understanding of function. The synthesis involves three stages.

1. Formation of an emulsion containing a perfluorocarbon oil and a phosphatidic acid lipid.
2. Sizing of the emulsion through extrusion through nanoporous membranes.
3. Coating of the sized emulsion exterior in a low ionic strength environment with small amounts of coating precursors, such as calcium and phosphate ions.

TABLE 15.1 Retrievable Nanoreactors Produced

	Coating	Lipid	Core
1	$CaPO_4$[a]	DOPA,[b] lecithin[c]	PFOB[d]
2	Polylysine	DOPA, lecithin	PFOB
3	Chitosan	DOPA, lecithin	PFOB
4	PEG	DOPC,[e] PEG-PL[f]	PFOB

[a]*Calcium phosphate in form of brushite as per Schmidt et al.*
[b]*DOPA, 1,2-dioleoyl-sn-glycero-3-phosphate (sodium salt) (DOPA).*
[c]*lecithin, L-α-phosphatidylcholine from soybean.*
[d]*PFOB, perfluorooctyl bromide.*
[e]*DOPC, 1,2-dioleoyl-sn-glycero-3-phosphocholine.*
[f]*PEG-PL, 1,2-dioleoyl-sn-glycero-3-phosphoethanolamine-N-[methoxy(polyethylene glycol)-1000].*

Fig. 15.1 shows a typical transmission electron microscopy image of $CaPO_4$-coated, oxygen-carrying, retrievable nanoreactors. Several variants of the calcium phosphate coating procedure can be employed to yield relatively uniformly coated particles 120–200 nm in diameter (Fig. 15.1).

Coating of the already formed emulsion is accomplished by creating suspension conditions at the boundary of supersaturation of calcium and phosphate ions, and then controlling the rate of mineral precipitation via tiny and repeated excursions into the supersaturation range. These excursions can be made by small concentration or pH changes. A model for the seeded growth of calcium phosphate atop stable emulsion particles via ionic condensation at constant temperature was developed.

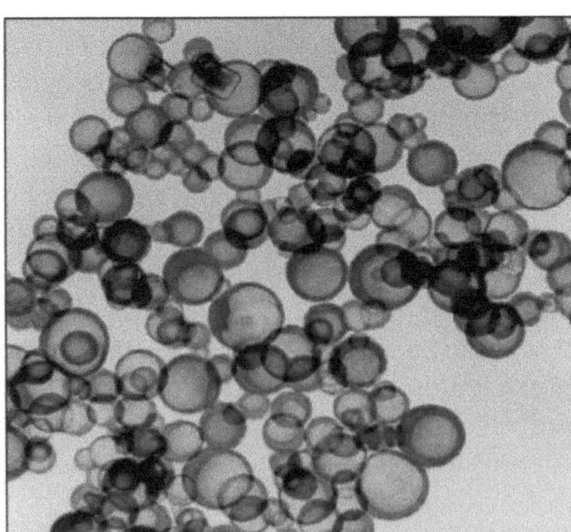

FIGURE 15.1 Transmission electron microscope image of calcium phosphate coated (dark rings) rOC. Gray interior is electron-rich perfluorocarbon liquid.

The excursions into supersaturation regime are largely localized around the periphery of the emulsion particles. The phosphatidic acid lipids used to form the emulsion are negatively charged at the pH used and create a Coulomb attraction force that attracts positively charged calcium phosphate ions. This creates a depletion zone around the particle and a region of highly concentrated calcium ions at the surface. The concentration gradient draws in more calcium from the surroundings, and reactions with phosphate ions in the highly concentrated region are enhanced. This in turn creates a gradient in phosphate ion concentration, along with a depletion zone that drives further reaction with calcium. The process continues until the reagents are depleted.

Using the aforementioned methods, retrievable nanoreactors (as shown in Fig. 15.1) can be prepared at relatively high concentrations ($> 10^9$ particles per mL), but due to the nature of ionic condensation of calcium and phosphate, the system is significantly diluted prior to coating in order to avoid end products with multiple perfluorocarbon cores fused together, or the formation of solid inorganic precipitates.

6 PERFORMANCE OF RETRIEVABLE NANOREACTOR BLOOD SUBSTITUTES

One of the first characterizations performed was measurement of the total oxygen content in mixtures of uncoated emulsion and calcium phosphate-coated retrievable nanoreactors (Fig. 15.2). The amount of dissolved oxygen in the mixtures was estimated enzymatically, an approach that allows determination

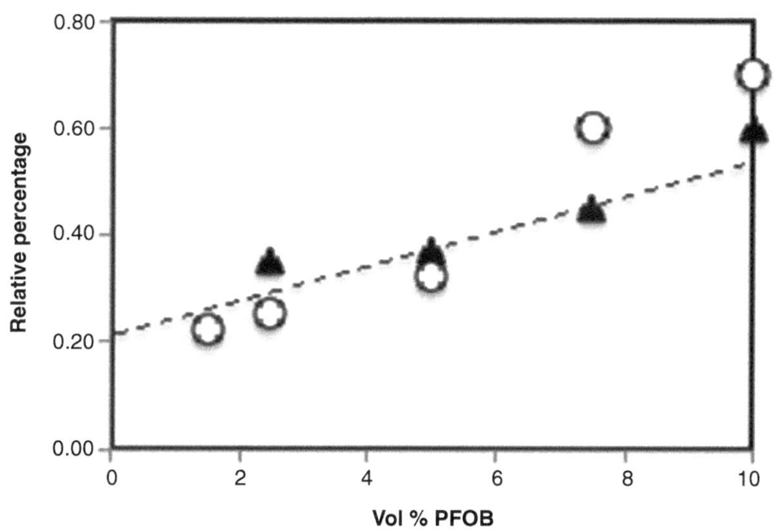

FIGURE 15.2 **Comparison of the concentration of dissolved oxygen in actual $CaPO_4$ rOC, uncoated perfluorocarbon emulsions, and calculated.**

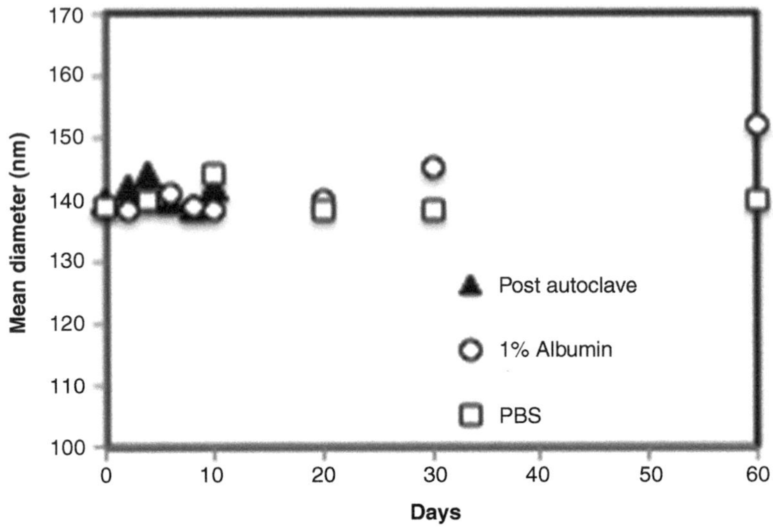

FIGURE 15.3 Mean diameter of $CaPO_4$ rOC in storage at room temperature coated ROC particles as a function of days in storage at room temperature.

of the total oxygen present both inside the particles and the surrounding solution. Dissolved oxygen electrodes in contrast measure only the dissolved oxygen in the surrounding liquid.

The suspension stability of the formulation was tested over the course of 2 months using particle size as the main indicator (Fig. 15.3). Storage in the presence of 2% albumin, or in buffer after a single autoclave cycle of 20 min, had no adverse effect on stability of the particle suspension.

The synthesis of $CaPO_4$-coated retrievable nanoreactors starts with the formation of an emulsion, and so the ternary phase diagram of the perfluoroctyl bromide, DOPA, water system was evaluated (Fig. 15.4). The initial evaluation revealed that the system exhibited three gel phases, an oil-water phase, and a water-in-oil phase. The oil-in-water phase appeared to be limited to a narrow region of the lower left corner of the phase diagram (Fig. 15.4).

To better identify the boundaries of the oil-in-water phase and how its characteristics vary with small fluctuations in composition, we extended our analyses of the ternary phase diagram to include more detailed characterization of each phase and a higher resolution map of the oil-water phase. Two lipids—DOPA, an anionic lipid, and lecithin, a zwitterionic neutral lipid—were compared to determine the effect of lipid head group. Five different characterization methods were employed: (1) visual classification of the phases, (2) buoyant density, (3) fluorescence quenching, (4) FT-IR, and (5) Cryo-TEM. The rationale for each technique is summarized in Table 15.2. A flow chart for utilizing the information obtained from these methods was developed to objectively evaluate the oil-in-water phase (Fig. 15.5).

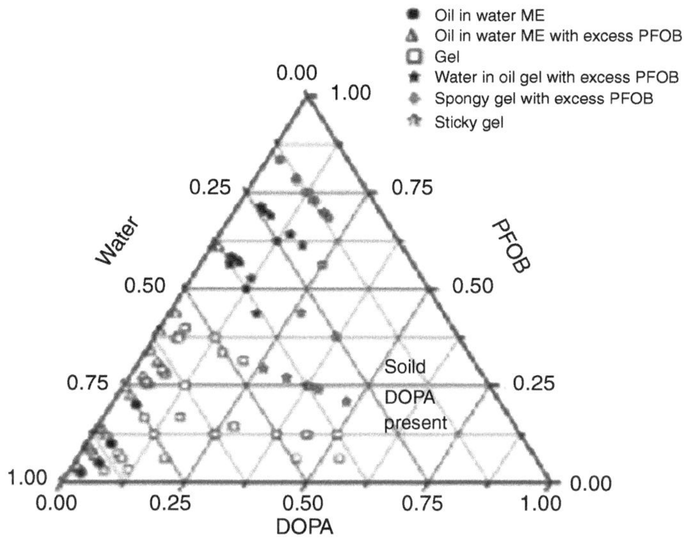

FIGURE 15.4 **Ternary phase diagram of water, PFOB, and DOPA.**

TABLE 15.2 Summary of the Techniques With Purpose and Methodology

Techniques	Purpose	Methods
Visual inspection	Coarse identification of phase boundaries	Material synthesized via titration method and the number of layers in the samples was determined visually
Buoyant density	Estimate relative oil content	The weight of the emulsion was measured in a custom pycnometer designed to handle all phases, including gels
Fluorescence quenching	Identify bicontinuous regions	Pyrene dye was dissolved in PFOB, and quencher KI in water; the amount of quenching gives insight into the phase configuration
Fourier transform infrared spectroscopy	Quantify oil content	Using ATR, Beer's law is applied to determine the concentration of oil in reference to that of pure oil
Cryogenic transmission electron microscopy	Observe morphology of each phase	The samples in each phase were frozen to form a thin ice film, which was observed by electron microscopy

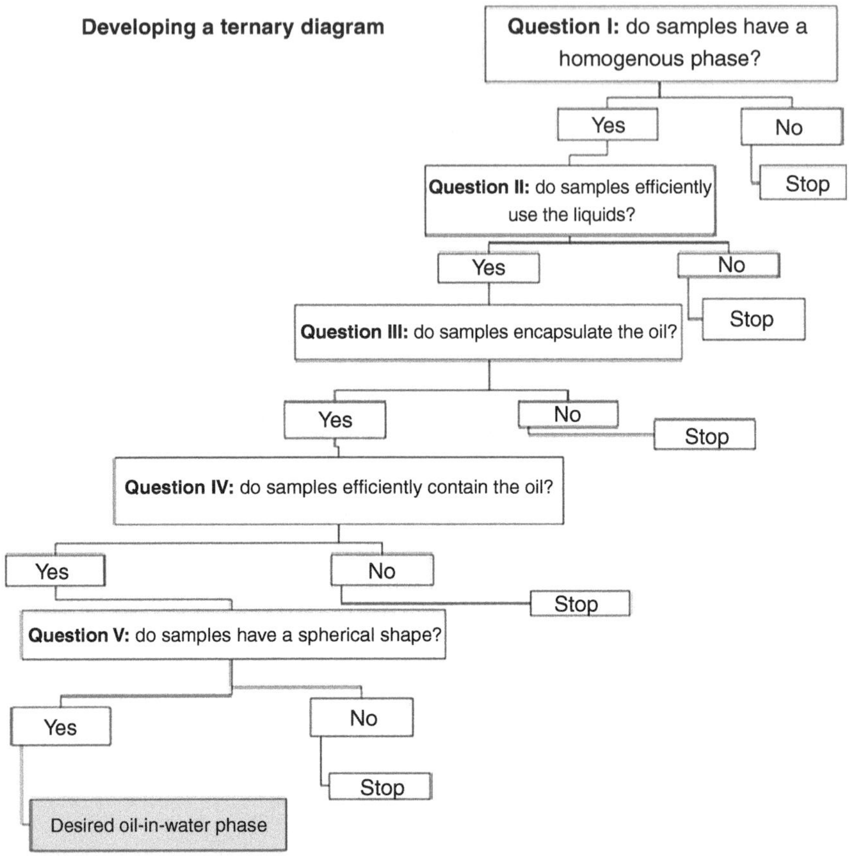

FIGURE 15.5 Flow chart to identify the desired oil-in-water phase.

A titration method was employed. The surfactant and oil were first added to a 20 mL glass vial. The mixture was stirred at 300 rpm using a magnetic stir bar (Teflon, ¾"). After 1 min, the mixture was diluted with deionized (DI) water (18.2 mΩ-cm, pH 7) dropwise within a 10 s time frame. The emulsion mixture was stirred at 1200 rpm for another 30 min, after which it was removed from the plate and allowed to relax for 6 h to ensure equilibrium. The number of visually distinct phases was counted for each composition lying along the tie lines, shown as the dashed lines in the ternary diagram of Fig. 15.6 Representative compositions for each phase were selected according to Fig. 15.7.

In general, the tie-line method provided phase diagrams (Figs. 15.9 and 15.10) similar to those typical of oil/water/surfactant systems (Fig. 15.8). Systems utilizing DOPA lipid in place of surfactant exhibited six discernible phase regions, while those with lecithin only four.

Phase A, in both Figs. 15.9 and 15.10 the desired microemulsion that yields a single homogenous oil-in-water phase. In the case of DOPA, this region is

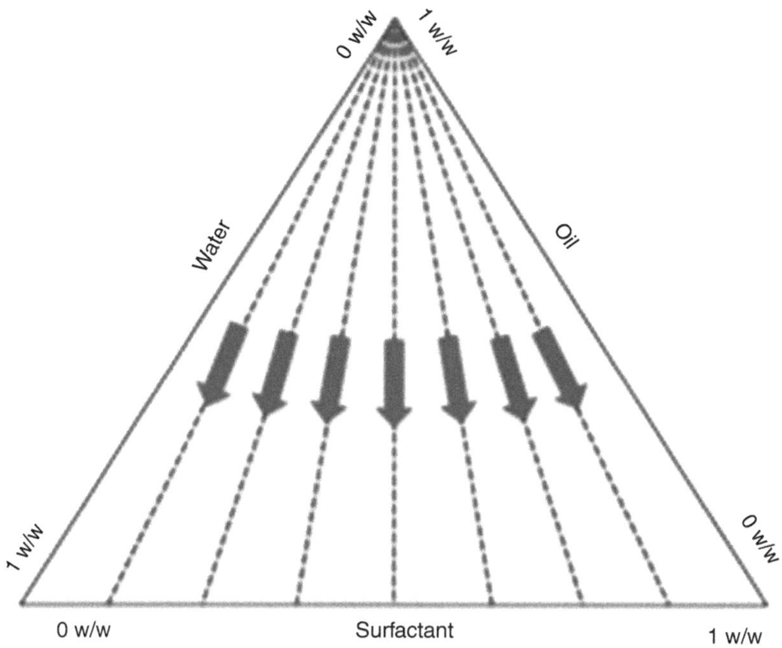

FIGURE 15.6 Schematic of the titration method used to construct the ternary diagram with the tie line [dash (blue in the web version)] . Each side represents the respective element composition in weight fractions, which was increased from 0 to 1 w/w. For example, for a water titration the samples were synthesized by following the direction [arrows (red arrows in the web version)] of low weight fraction of water to high.

very small, while lecithin occupies a much larger area of the diagram. This reflects the added stabilization offered by the larger, zwitterionic head group of lecithin.

The density of each phase was measured individually, according to Fig. 15.11, and the effective usage of perfluorooctylbromide (PFOB) and water in forming the emulsion estimated. The effective usage was defined as the amount of PFOB associated with surfactant. The selected regions ran along lines parallel to the water and surfactant axes as shown in Fig. 15.7. The detailed weight fractions for each region are described in Table 15.3. For the lecithin system, the same six weight ratio values were made allowing for a direct comparison between the two surfactants. To prepare the sample, the surfactant and oil were added to a 20 mL glass vial, while keeping the total added volume of the PFOB and the water at a constant 10 mL. The weight of the surfactants was measured with a calibrated digital scale.

A cylindrical shape vial (3.5 mL; height: 2.0 cm, diameter: 1.5 cm) with an open top was used as a pycnometer to consistently measure the weight of the various phases. The vial made working with the gels and gel-like samples

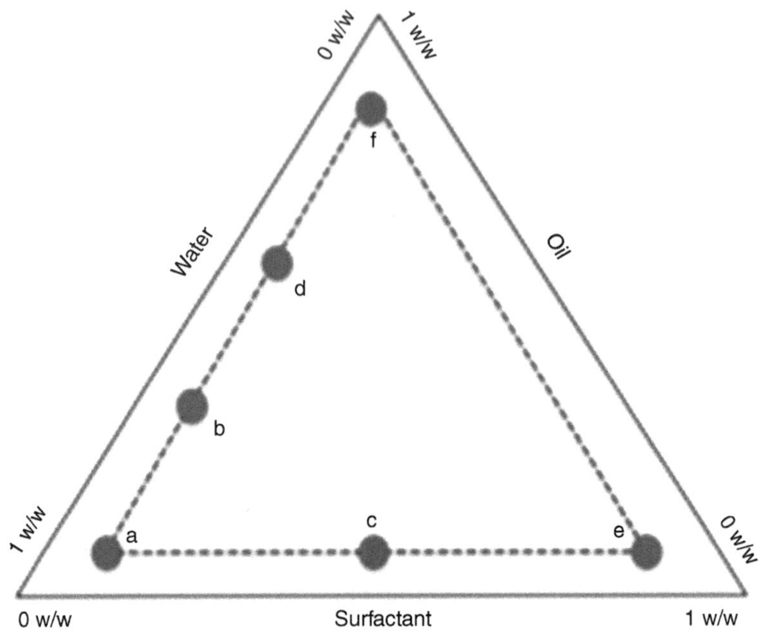

FIGURE 15.7 The ternary diagram with the selected regions [dots (red dots in the web version)] for characterization with buoyant density, fluorescence quenching, FT-IR, and cryoTEM. The line a, c, and e was parallel to the surfactant axis, and line a, b, d, and f was parallel to the water axis; the line formed by e and f was parallel to the oil.

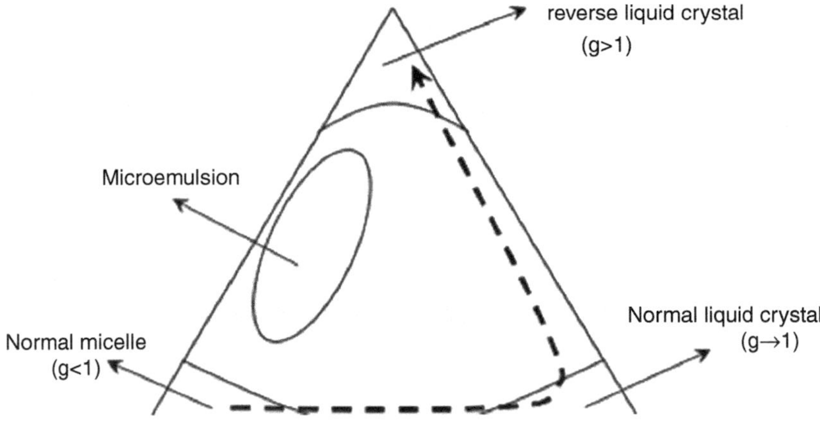

FIGURE 15.8 The typical ternary phase diagram of oil/water/surfactant. Each corner forms a type of emulsion, governed by the majority.

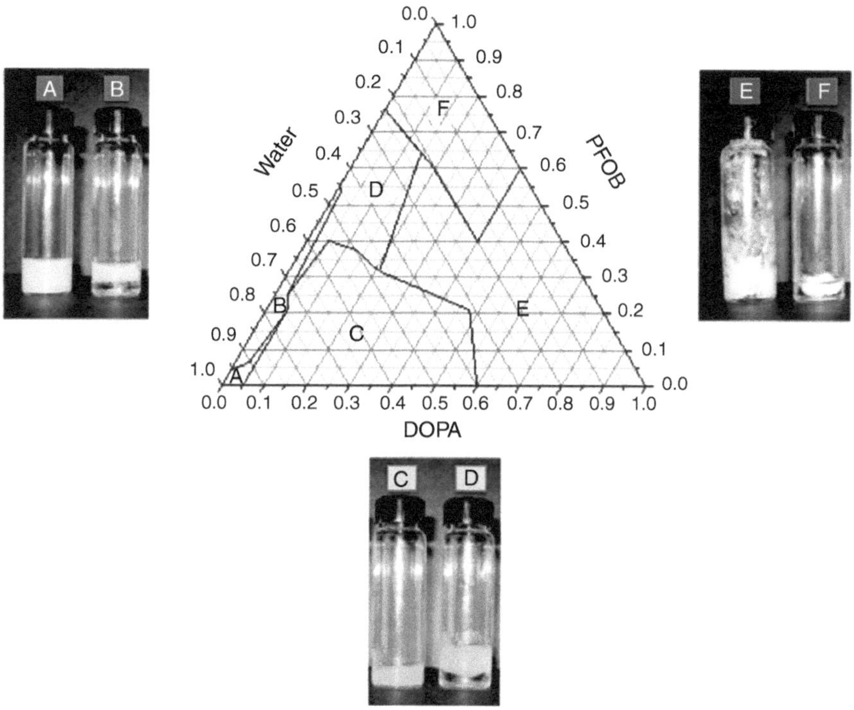

FIGURE 15.9 The ternary diagram of DOPA/PFOB/water. There were six various phases. The color of the samples was turbid or white, and some samples were viscous or had excess liquid oil or water component.

easier, and it was calibrated to obtain the buoyancy error. The correction factor is dependent on the true mass (m), and measured mass (m'), and is given by the buoyancy equation. The rule of mixtures equation was used to provide a calculated density for a given ratio of PFOB to water. This value was then compared to experimentally measured value to derive the approximate amount of PFOB that was suspended in a particular visually identifiable opaque phase. Clear phase layers were either pure PFOB (1.93 g/mL) or pure water (0.9982 g/mL). The difference between the measured and the calculated densities indicates the degree to which the oil and water are recruited in a given phase (Table 15.4). When compared to the calculated densities of the suspensions, the DOPA system showed that more than 90% of the combined liquids was used to form the emulsion in Regions a, c, and e, whereas Region f suggested poor integration. In the lecithin system, Regions a, c, and e also showed the high efficiency of material usage, while Region f was the lowest. Regions a, c, and e were all single homogenous phases.

The openness of the structures formed in each phase was evaluated with fluorescence quenching using the Cary Eclipse spectrophotometer, Varian, Palo Alto, CA and its front-face accessory (accessory solid sample holder, Varian,

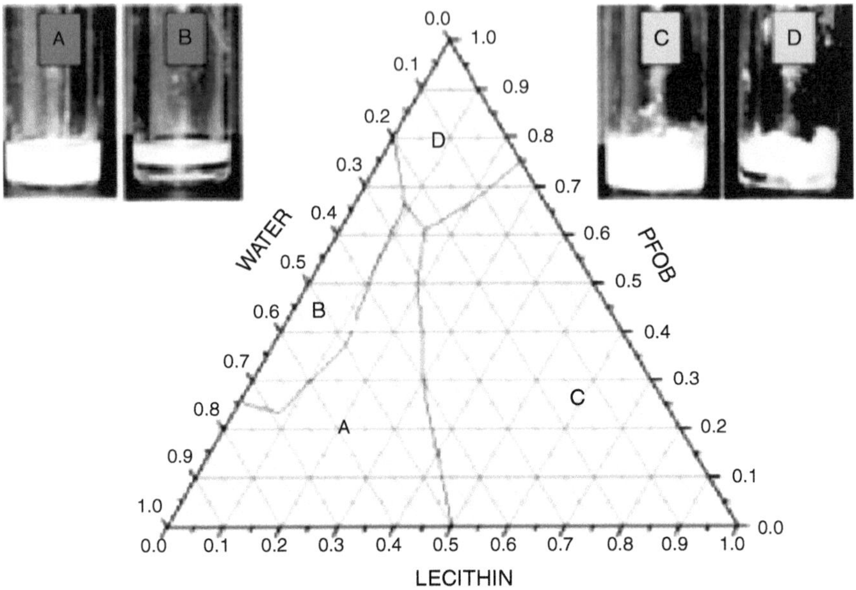

FIGURE 15.10 **The ternary diagram of soy lecithin/PFOB/water.** There were four various phases. The color of the samples was white, some samples were viscous or had excess liquid.

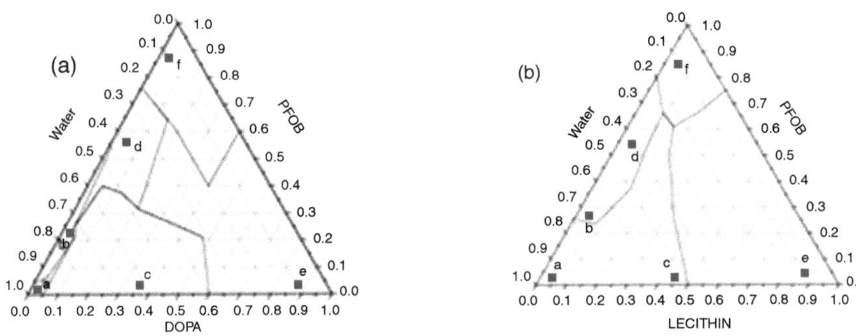

FIGURE 15.11 Points of the ternary diagrams of (a) DOPA/PFOB/water and (b) lecithin lecithin/PFOB/Water used for density, fluorescence quenching, and FTIR measurements.

Palo Alto, CA) with an angle of 62 degrees. Pyrene was selected as the fluorescent dye because it is soluble in both PFOB and water, and is easily quenched with potassium iodide (KI), which is only soluble in water.

The premise for this experiment is shown in Fig. 15.12. If the pyrene was encapsulated inside an oil/water (O/W) emulsion, it can readily diffuse across the lipid monolayer. However, KI is only soluble in aqueous environments and would remain inside of the water/oil (W/O) emulsion even though it has a high interior concentration. If the phase is an open system where molecules are free

TABLE 15.3 Compositions Corresponding to the Points in Fig 15.11

Regions	Surfactant (w/w)	PFOB (w/w)	Water (w/w)
a	0.050	0.050	0.900
b	0.025	0.255	0.750
c	0.250	0.150	0.600
d	0.100	0.500	0.400
e	0.200	0.700	0.100
f	0.600	0.200	0.200

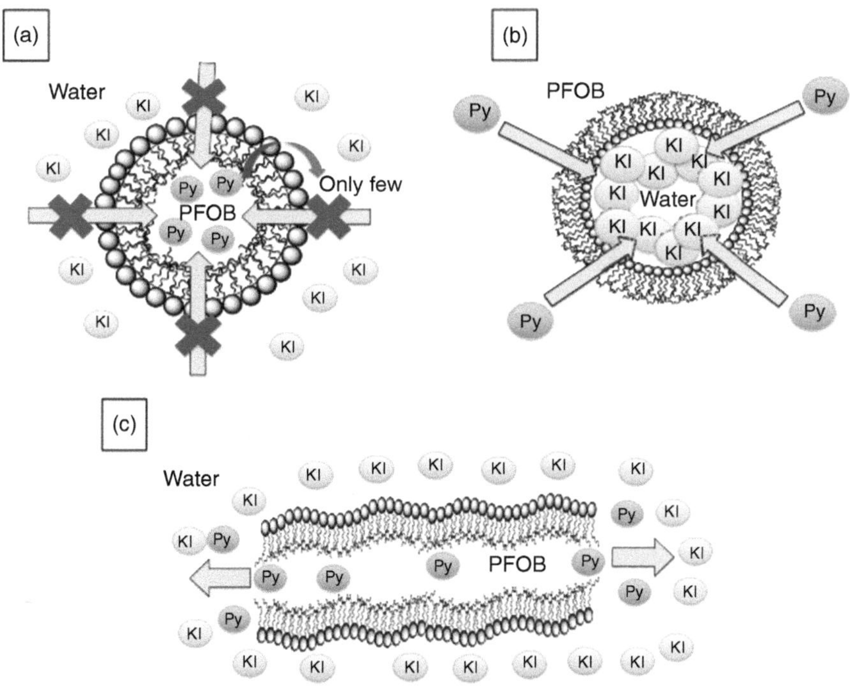

FIGURE 15.12 The schematic representing the initial idea for phases. (a) Oil-in-water (O/W), (b) water-in-oil (W/O), and (c) bicontinuous phases. The light gray and dark gray (green and blue in the web version) molecules indicate the quencher (KI) and the pyrene. (a) Oil-in-water (O/W) phase: only a few of the dye molecules transport from the oil to the water across the lipid membrane. However, KI would not be transported into the emulsion because it is not soluble in the oil. (b) Water-in-oil (W/O) phase: it would be difficult to be quenched by KI because the dye molecules would be easily trapped in the tails of the lipid molecules, and it is not soluble in the oil. (c) Bicontinuous phase: because the oil phase is not encapsulated by the surfactant membrane, the dye molecules are able to readily transport from the oil to the water, where they are quenched by KI.

TABLE 15.4 Summary of the Experimentally Measured and Calculated Densities

Regions	Composition (wt./wt.)			Measured density (g/mL)		Calculated density (g/mL)	Efficiency: formulation of emulsion (%)	
	Surfactant	PFOB	Water	DOPA	Lecithin		DOPA	Lecithin
a	0.025	0.025	0.950	0.998	0.999	1.010	98.4	98.2
b	0.025	0.225	0.750	1.003	1.007	1.123	89.3	89.7
c	0.375	0.025	0.600	1.002	1.005	1.017	98.5	98.8
d	0.025	0.575	0.400	1.125	1.180	1.395	80.6	84.6
e	0.875	0.025	0.100	1.006	1.101	1.105	91.0	99.6
f	0.025	0.875	0.100	1.156	1.004	1.761	65.7	57.0

to roam from one preferred solvent to another, then Py and KI have a greater chance of reacting. In the initial control experiments, pyrene was dissolved in water (4.51×10^{-4} mM) and titrated with KI (2.24 mM) while monitoring the fluorescence intensity. Adding KI quenched the pyrene emission linearly with its concentration. When pyrene was dissolved in PFOB, which showed that there was no significant decrease in emission intensity on addition of KI (Fig. 15.13). Fig. 15.14 shows the percentage change of fluorescence intensity.

Based on the concepts of the initial idea described previously, the regions with maximal and minimal fluorescence extinctions would be classified as the bicontinuous and W/O phases, respectively. Examination of the data suggested Regions b and d for both the DOPA and lecithin systems. For the DOPA coordination, Region a proved to be the second most resilient to quenching, a characteristic observed in Regions a and c of the lecithin experiment. This secondary region was significant because it indicated an environment that would be consistent with an oil-in-water phase.

The same samples were further analyzed with FT-IR/attenuated total reflectance (ATR) (resolution of 4 cm^{-1} and the number of scans150) [(Varian 3100 FT-IR, Varian Inc., Palo Alto, CA), and the ATR (MIRacle™ Single Reflectance ATR, Pike Technologies, Madison, WI) accessory]. Each component material—OH 3000–3600 cm^{-1} (stretching) and 1640 cm^{-1} (bending); difluoromethylene (CF_2) at 1200 cm^{-1} and 1140 cm^{-1};(P=O) at 1170 cm^{-1}; (C—N stretching) at 1375 cm^{-1}—was characterized by FT-IR to establish a reference intensity. Regions a and c of each of the ternary diagrams were the optimal regions for our desired emulsion system. Fig. 15.15 summarizes the effective recruitment of PFOB in the opaque emulsion layers of each sample a–f.

The DOPA and lecithin emulsions were viewed with coyote (Figs. 15.16 and 15.17 Fig. 15.16, 15.17). Cryo-TEM imaging allows nearest to native views of specimens. Accordingly, liquid and biological samples are preserved by being frozen into a thin film with liquid nitrogen. For low-viscosity samples, 3 μL was dropped onto a carbon grid with a pipette, but because the highly viscous samples (ie, gel-like emulsion) are not compatible with pipettes, the samples were smeared onto the transmission electron microscope (TEM) carbon grids using a pipette tip and blotted for 3 s using an IRobot vacuum-cleaning robot. After the sample was mounted, the blotted grid was submerged in liquid nitrogen and observed under a FEI Tecnai™ TEM (FEI, Hillsboro, OR) at 120 keV.

In the micrographs, a darker feature is consistent with higher density when compared to the surroundings, and because of the relatively high electron density of fluorine, an oil-in-water emulsion would appear as a black circle. If the particle were formed as a liposome, containing a water core, there would be little contrast between the inner and outer regions of the lipid layer.

If the particle were formed as a liposome, containing a water core, there would be little contrast between the inner and outer regions of the lipid layer. In both systems, Regions c to f were too viscous for a dropwise application, so they were smeared onto the grids with a pipette tip before blotting. As indicated

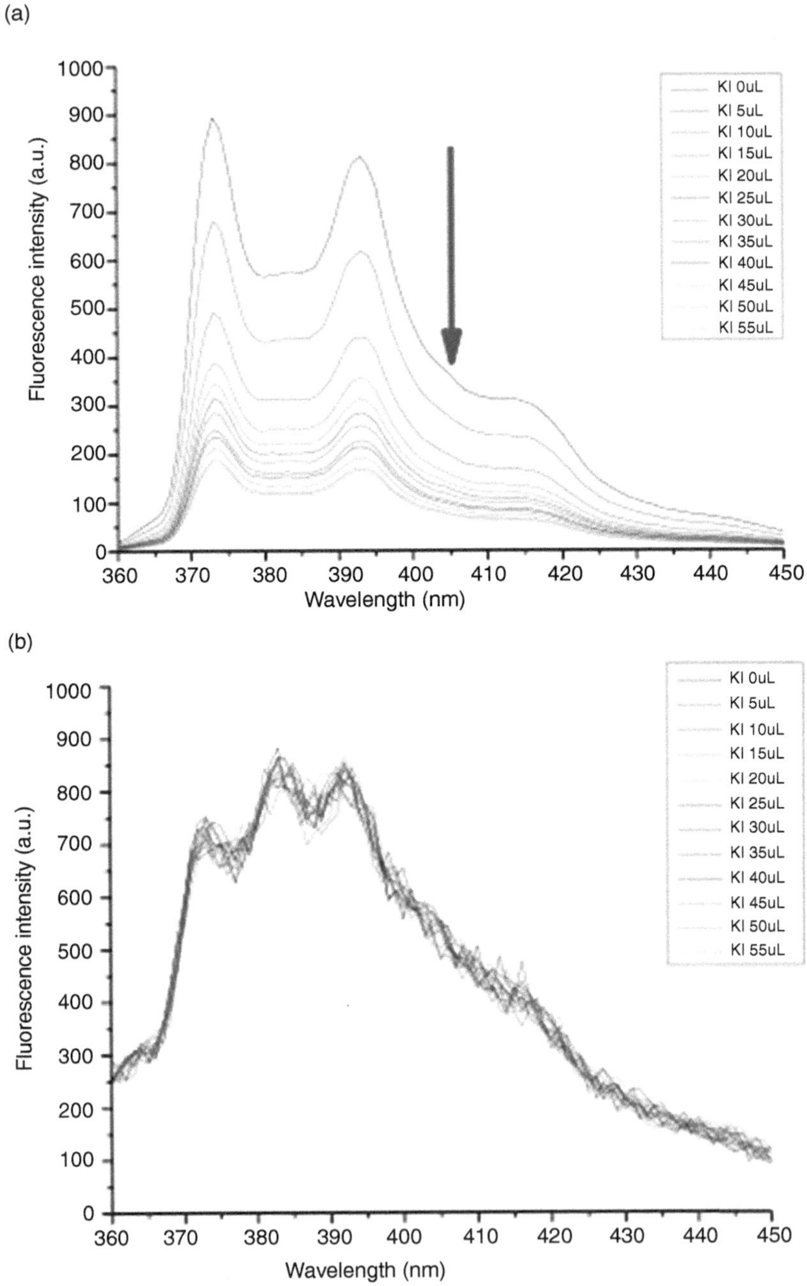

FIGURE 15.13 (a) The pyrene (4.5×10^{-4} mM in water) was quenched by the quencher, KI; (b) no quenching in PFOB.

(c)

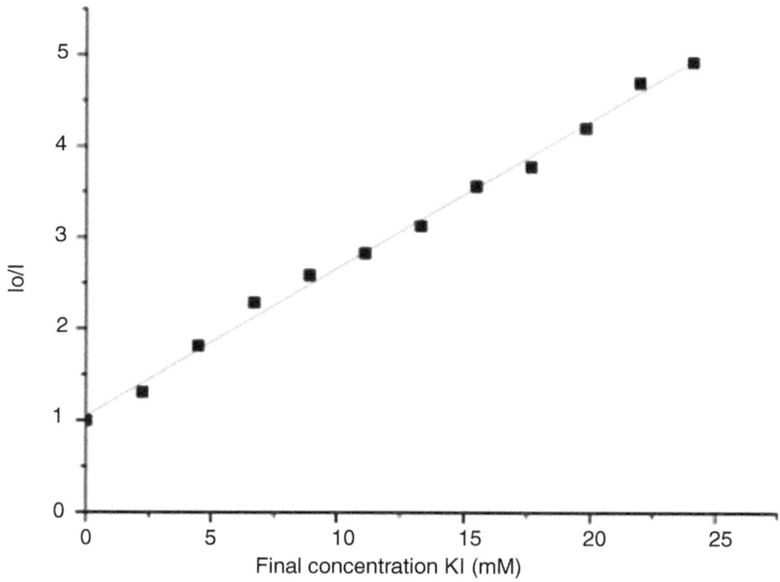

(d)

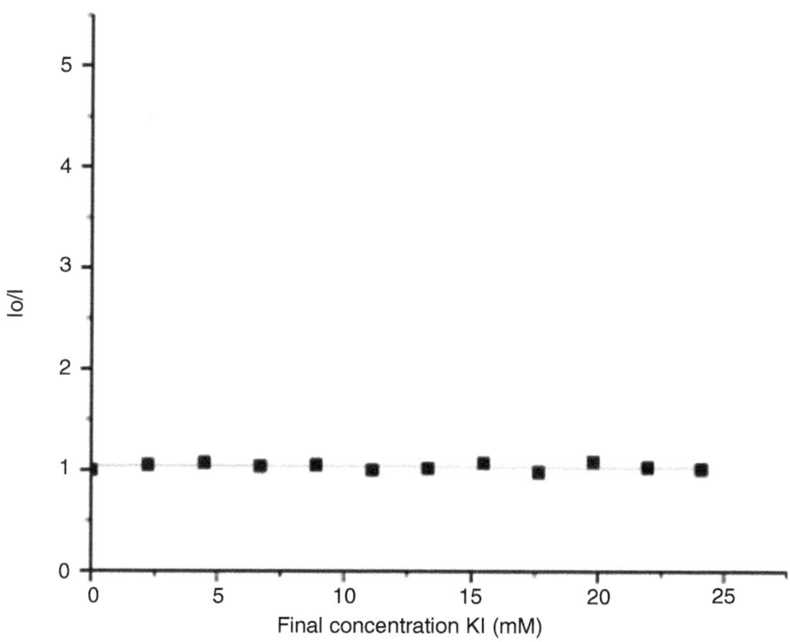

FIGURE 15.13 (*Cont.*) (c, d) Figures demonstrating that the pyrene in water were dual quenched by KI while the oil had constant fluorescence intensity. *R*-squared values for (c) 0.9949 and (d) 0.0045. Wavelengths analyzed (c) 373 nm and (d) 385 nm.

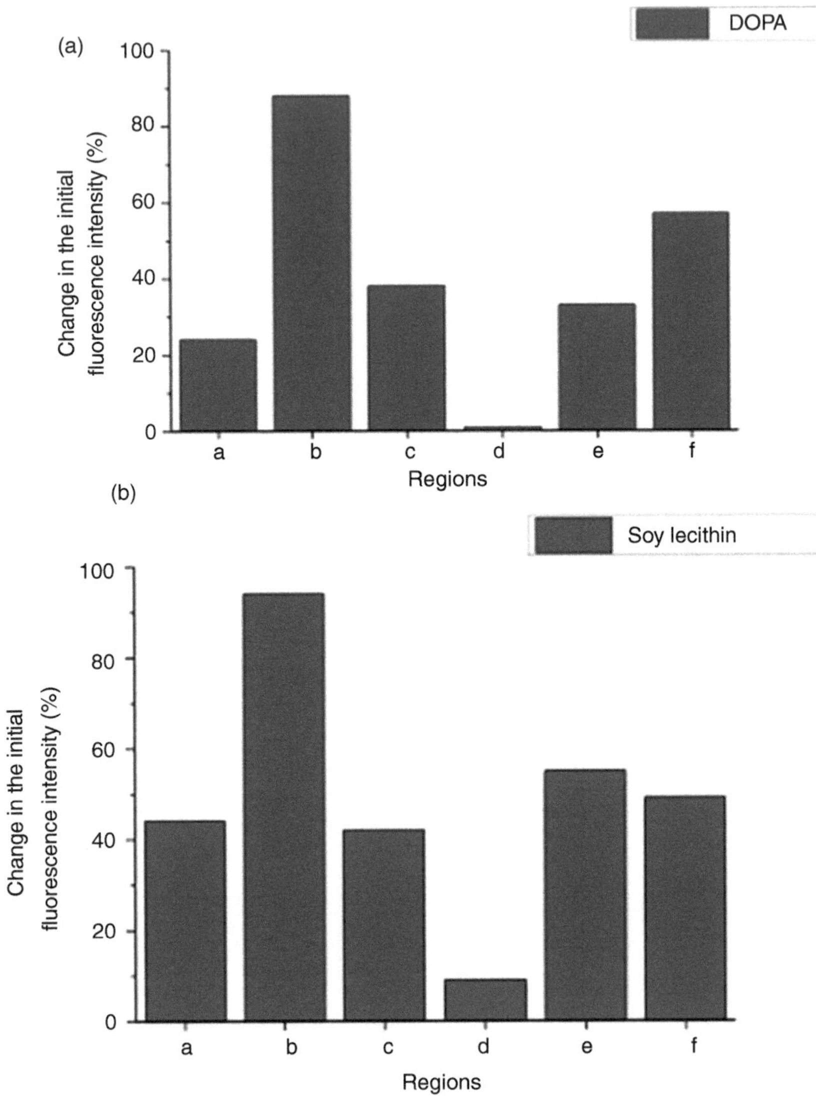

FIGURE 15.14 The change in the initial fluorescence intensity represented as a normalized percentage. Regions b and d showed the maximal and minimal fluorescent quenching in both (a) DOPA and (b) lecithin. The DOPA system demonstrated the quenching in Regions a (24%), b (88%), c (38%), d (1%), e (33%), and f (57%); the lecithin system was in Regions a (44%), b (94%), c (42%), d (9%), e (55%), and f (49%).

in Table 15.5, Region a in both the DOPA and lecithin emulsion systems exhibited a spherical morphology containing a darker core which, as explained previously, is indicative of an O/W phase. There were a wide variety of shapes in Regions b, c, and d of both systems, and the lack of contrast between multiple layers suggested the formation of a liposome and an excess of lipid. Regions

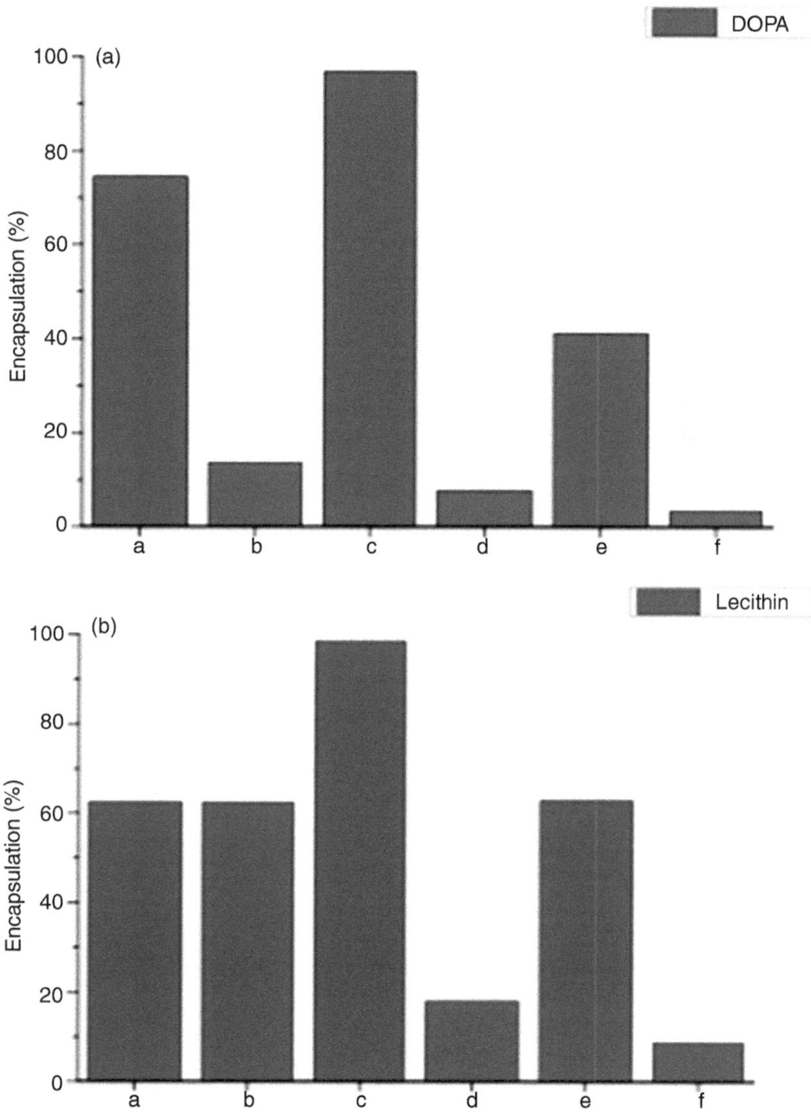

FIGURE 15.15 The calculated encapsulation of PFOB in the emulsion system by phase region. (a) DOPA emulsion system demonstrated Regions a and c encapsulate PFOB more than 50%, and (b) lecithin emulsion system mostly encapsulates PFOB except Regions d and f.

b, c, and d in the DOPA trial showed structures expected of empty liposomes. Meanwhile, in the lecithin system Regions b and c revealed few particles containing PFOB, and Region d did not have any such structure. Even though the added amount of PFOB in Region b was greater than that of a in the lecithin system, only some of the particles contained oil. In the case of lecithin, the amount of PFOB in Region c was identical with that in Region a, but Region c

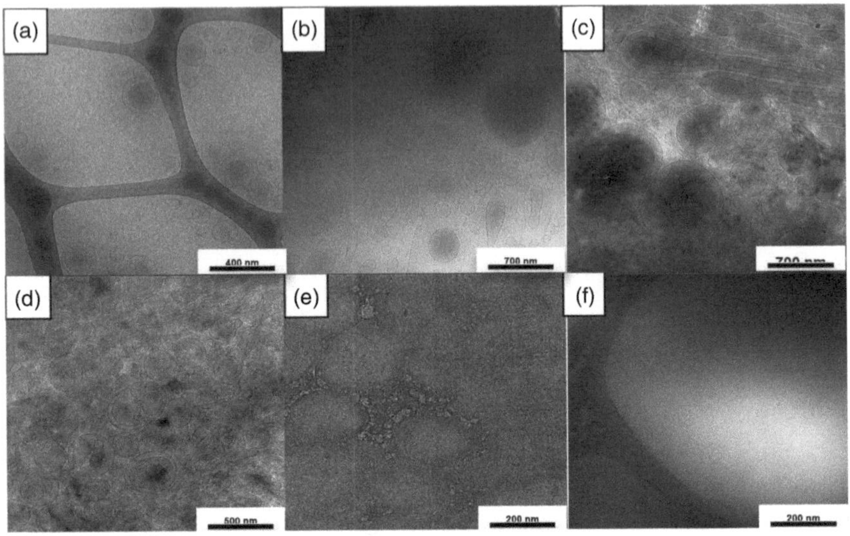

FIGURE 15.16 **The cryoTEM images of the DOPA emulsion system for each phase region from a to f.**

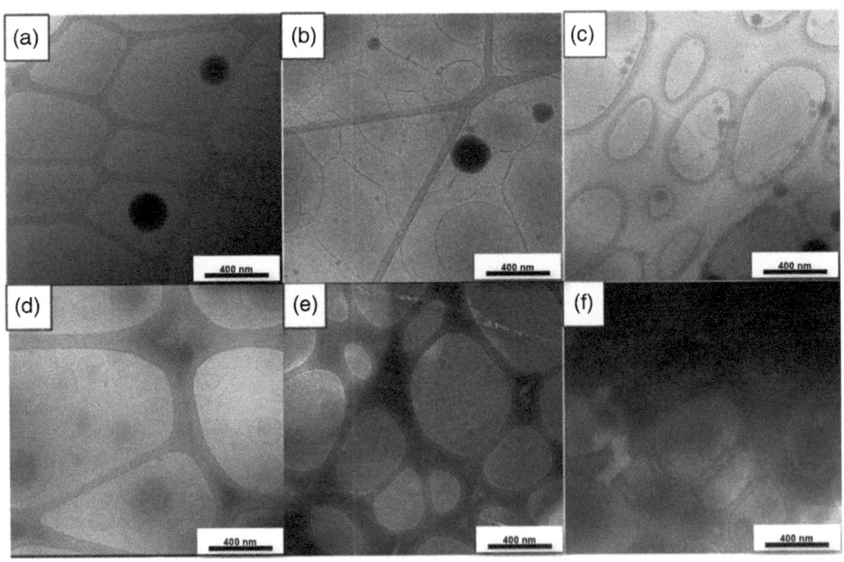

FIGURE 15.17 **The cryoTEM images of the lecithin emulsion system for each phase region from a to f.**

showed a greater number of particles with the oil core. The images of Region d show a W/O phase when taking into consideration the ternary diagram and visual observations. Regions e and f in both systems displayed a wave pattern, which implied a bicontinuous phase.

TABLE 15.5 The Summary of the Shape Characteristics Found in the Cryo-TEM Micrographs

	DOPA		Lecithin	
Regions	**Description**	**Accept?**	**Description**	**Accept?**
a	Encapsulated sphere with the oil in core	Yes	Encapsulated sphere with the oil in core	Yes
b	Encapsulated sphere; partially random shapes	No	Encapsulated sphere; partially other shapes	Yes
c	Random shapes; no oil in core	No	Random shapes; Encapsulated sphere with the oil in core	Yes
d	Random shapes; no oil in core	No	Sphere; no oil in core	No
e	Shape of a wave	No	Shape of a wave	No
f	Shape of a wave	No	Shape of a wave	No

This table describes the types of shapes and the acceptance of the morphology according to the criteria.

Considering all of the information, we conclude that phase a in both the DOPA and lecithin emulsion systems is the oil-in-water phase. There was no apparent excess liquid PFOB. The measured buoyant density was similar to the theoretical calculation. The FT-IR spectra of phase a had the highest encapsulation rate for PFOB, and performed best in preventing the quenching of pyrene by potassium iodide, indicating the pyrene was trapped in PFOB reservoirs. Finally, a spherical morphology was observed by cryoTEM, which is also consistent with the expected O/W emulsion. Phase a is 36 times larger than the same phase in the DOPA case, implying that lecithin was better at forming the oil-in-water phase compared to DOPA. This is most likely attributed to lecithin's amphiphilic head group in contrast with anionic DOPA. Because lecithin does not express a repelling force between like charges, more of the molecules are required to form the emulsion.

When looking at Region b of the lecithin system, it was discovered that it is most qualified, from the five characterization tests, to be a bicontinuous phase. This implication was confirmed by wavelike patterns observed via cryoTEM. When analyzed from a broader perspective, it was concluded that this bicontinuous phase could be the transition phase, which coexists between the oil-in-water and other phases.

In sum, even though a sample might have the visual appearance of being suitable, additional testing of samples prepared at compositions within a given

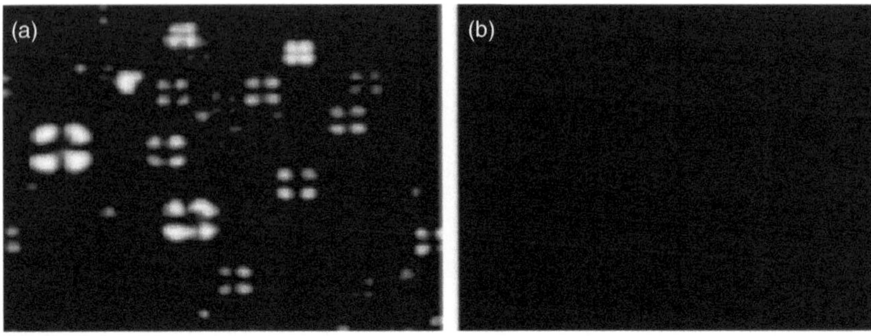

FIGURE 15.18 Maltese cross in raw emulsion (a) disappears after extrusion (b).

region of the phase diagram revealed that all did not meet the criteria—that is, (1) does not contain excess PFOB, (2) does not contain excess lipid, (3) has a uniform spherical shape. Even though the ternary diagrams demonstrated that the DOPA and lecithin emulsion systems had different numbers of phases, both DOPA and lecithin systems showed one workable phase. The zwitterionic surfactant lecithin had a greater workable region for the oil-in-water phase than DOPA, which implies that lecithin is a more reproducible system and can be used to produce a stable product.

Using a selected composition of materials, the next step is to extrude them to remove lipid multilamelarity and refine the size distribution. This can be done through high-pressure extrusion (500–600 PSI) through nanoporous membranes. Polarized microscopy was used to show that the multilamellarity disappeared after processing. Multilamellar lipid layers exhibit a characteristic Maltese cross pattern under polarized light (Fig. 15.18). In contrast, unilamellar emulsions were dark in this mode of imaging.

The final product is an opaque suspension (Fig. 15.19) with a tendency to settle. It is easy to resuspend by shaking. The particle size/size distribution is stable for as long as 6 months in room temperature storage, and stable upon autoclaving (Fig. 15.20).

The particulate suspension containing about 20% vol particulates was dialyzed against PBS to a concentration of about 10 μM free calcium, and then 1 mL of this sample was placed into 1 mL of plasma. Typically plasma contains:

- 90% water
- 8% protein
- 0.9% inorganic salts
  - Sodium 135–146 mM
  - Potassium 3.5–5.2 mM
  - Calcium 2.1–2.7 mM (1 mM protein bound, 1.2 mM ionized, and 0.2 mM complexed to anions)
  - Carbonate 23–31 mM
  - Phosphate 0.7–1.4 mM
- 1.1% organic substances

FIGURE 15.19 A photograph of a 2-L bottle filled with rOC suspension.

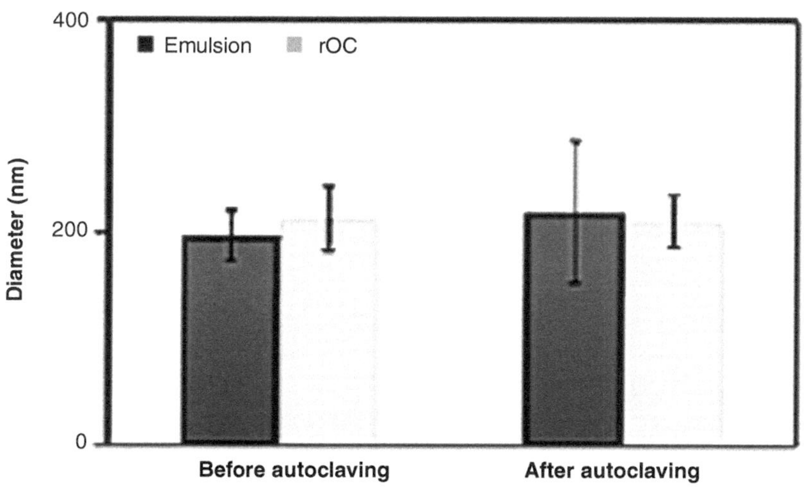

FIGURE 15.20 The rOCs are autoclave resistant.

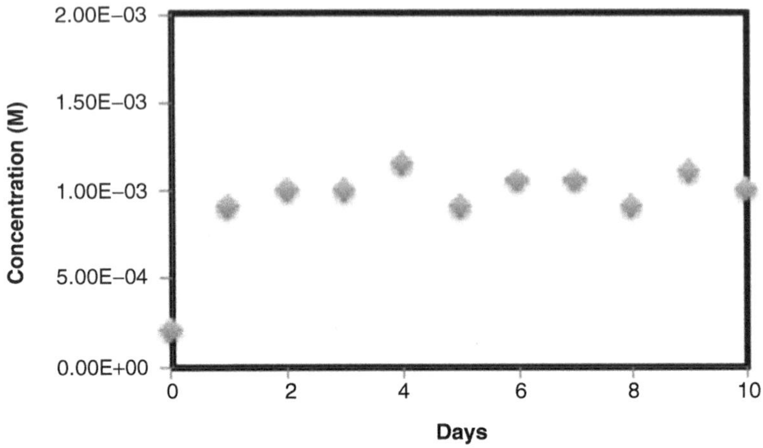

FIGURE 15.21 Free calcium levels in plasma after equivolume addition of rOC at 20% vol. First data point is rOC only, no plasma.

After addition of the retrievable nanoreactors, there was no significant change in plasma-free calcium levels (Fig. 15.21). This suggests that in plasma there is no significant leaching of calcium from the particles. Excess free calcium levels can be lethal. The particles were also tested for hemolysis of red blood cells (Fig. 15.22). While the uncoated raw emulsions did cause significant lysis (even though they were prepared in PBS at pH 7.4), the calcium phosphate-coated particles did not, showing that coating of the emulsion is an important aspect of retrievable nanoreactors' safety.

The solubility of oxygen in the calcium-coated retrievable nanoreactors was measured enzymatically [12], using a method adapted for measurements of oxygen content in oil-in-water emulsions with a precision of about 1%. The enzymatic method is based on the oxidation of glucose by molecular oxygen catalyzed by glucose oxidase. This method is commonly used for dosing glucose when oxygen in excess is present and can measure the molecular oxygen when the glucose is the excess reactant. It can be described by the following equations:

$$d\text{-glucose} + H_2O + O_2 + \text{ glucose oxidased} \rightarrow \text{gluconic acid} + H_2O_2$$

$$H_2O_2 + \text{reduced o-dianisidine(colorless)} + \text{peroxidase} \rightarrow \text{oxidized o-dianisidine (brown)}$$

$$\text{oxidized o-dianisidine(brown)} + H_2SO_4 \rightarrow \text{oxidized o-dianisidine (pink)}$$

The O_2 capacity of retrievable nanoreactors was close to that of starting emulsions, indicating no loss of PFOB during or after synthesis or blockage by

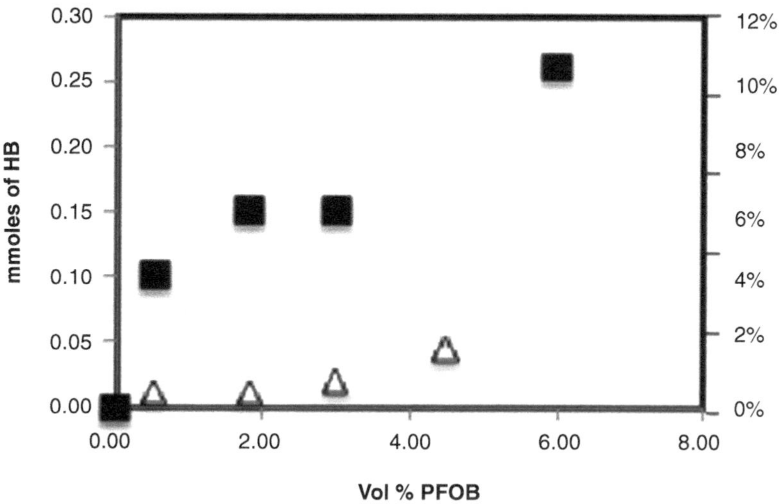

FIGURE 15.22 The rOC (open diamonds) does not hemolyze RBC, but uncoated emulsions (solid squares) do cause significant lysis.

the coating (Fig. 15.23). Due to the larger surface area/volume ratio, the rate of O_2 release from the calcium-coated retrievable nanoreactors was twice that of RBC (Fig. 15.24).

The Bohr effect of RBC was found to be unaffected by the presence of retrievable nanoreactors (Fig. 15.25). The O_2 dissociation curves of the 70% blood (30% retrievable nanoreactors) were obtained at 5 and 10% CO_2 using

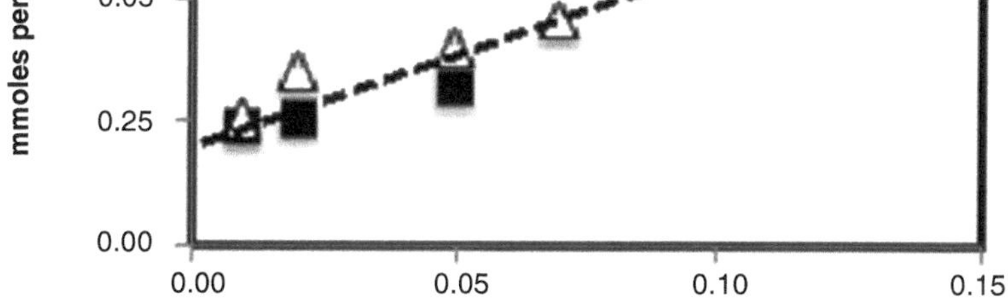

FIGURE 15.23 The O_2 capacity of rOC (solid squares) is comparable to emulsions (open diamonds), indicating there is no loss of PFOB. The dotted line is the calculated capacity, which is in good agreement with both measurements.

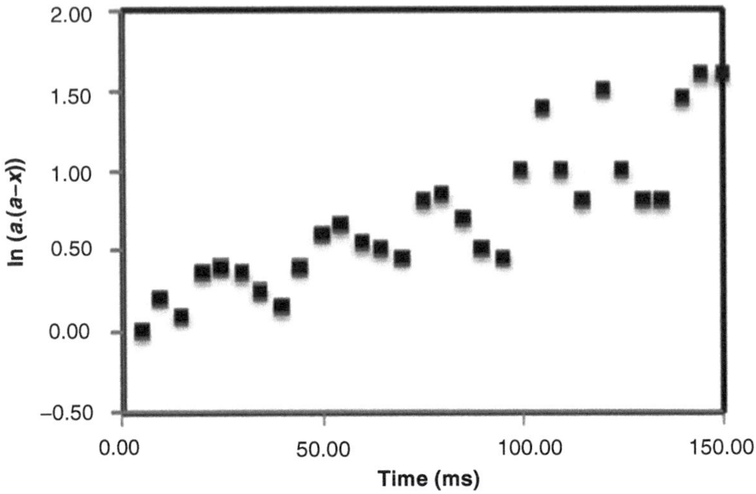

FIGURE 15.24 **The rate of O_2 offload rate from rOC is 9.8 s^{-1}; from RBC it is 4.1 s^{-1}.**

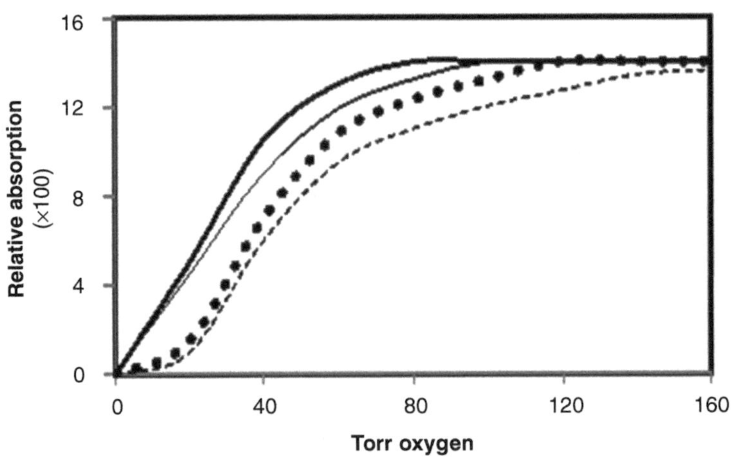

FIGURE 15.25 **Spectrophotometric O_2 dissociation diagram of 70% blood at 5 and 10% CO_2 and mixture of the same with 30% rOC.** Results show addition of rOC has little effect on O_2 affinity and Hill coefficient but the O_2 capacity has increased.

the thin layer technology developed by Mizukami and coworkers [13]. The results are shown in Fig. 15.25, which demonstrates that the addition of retrievable nanoreactors has little effect on the O_2 affinity, Hill coefficient, and Bohr effect, while the O_2 capacity is increased by addition of retrievable nanoreactors (Fig. 15.25).

Cytotoxicity in HepG2 cells (contracted with Cyprotex) yielded results suggestive of low potential for liver toxicity at working concentrations of retrievable nanoreactors compared to the positive control carbonyl cyanide 3-chlorophenylhydrazone. No significant increases in nuclear size, DNA structure, cell membrane permeability, and cytochrome C levels suggested any significant necrosis, apoptosis, or genotoxicity of retrievable nanoreactors at working concentrations. Cell proliferation declined with retrievable nanoreactors at levels 30,000 times higher than the control sample, while mitochondrial mass increased and mitochondrial membrane potential decreased, suggesting activation of metabolism at the expense of cell proliferation. Although in vitro cytotoxicity measurements are not entirely representative of physiology, these results are encouraging.

Even though HB has four positions to which oxygen can associate itself, this study was only concerned with the kinetics of the first oxygen to bind to the deoxyhemoglobin (dHB) because the other three are nonlinearly related to each other with respect to time. Therefore, a pseudo first-order reaction was assumed, and it was used to calculate the first order kinetics rate coefficient k_{a1}. By plotting the logarithm of Y, the fraction of HB absorbance at 550 nm as a function of time, k_{a1} was obtained from the slope. According to this relation, the pseudo first order rate constant showed a lower value with a thicker coating, which confirmed the idea that the coating thickness controlled the release oxygen. The amount of oxygen inside the coated particle was less than the uncoated emulsion; hence, the oxygen uptake by dHB was lower than the uncoated emulsion.

An oscilloscope trace of transmittance versus time for the HB and the calcium phosphate-coated (CaP) emulsion at pH 7.2 is shown in Fig. 15.26.With increasing coating thickness, the HB was less saturated with oxygen. From the logarithm of the absorbance versus time for the HB and CaP emulsion, the first order kinetic constant of the oxygen uptake by HB with the mean and standard deviation was estimated.

In this discussion, it was assumed that all of the oxygen inside the particle was released into the continuous phase. In other pharmaceutical studies, the Hixson–Crowell cube root law has been widely used to estimate the amount of drug released, and it assumes a spherical particle that is constantly suspended by a liquid [14]. By treating the oxygen inside the emulsion as a drug, the Hixson–Crowell cube root law was applied to the emulsion system in this study. This model estimates the amount of drug release by changes in surface area and diameter of the particle. In order for this system to be useful, the water layer surrounding the particle must be sufficiently thick when compared to the particle diameter; in other words, the particle should be smaller than 500 μm in diameter. By applying the Noyes–Whitney equation to the Hixson–Crowell law, the change of particle size in radius as a function of time can be estimated.

Even though the measured oxygen was 76 μM, excluding the dissolved oxygen in the continuous phase of the emulsion system, the measured oxygen uptake indicated that the number of particles determined the overall oxygen

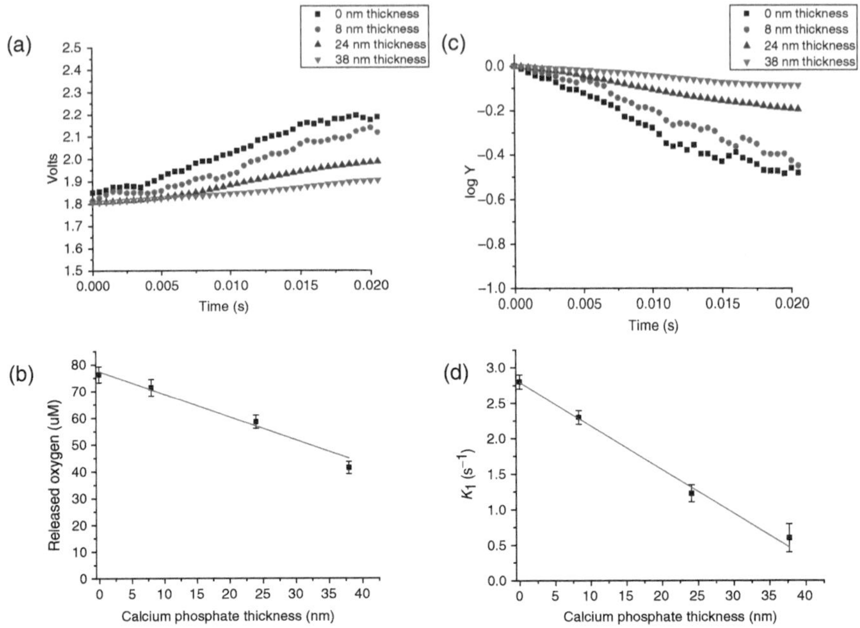

FIGURE 15.26 (a) The oscilloscope trace of transmittance versus time for the HB and the calcium phosphate-coated (CaP) emulsion at pH 7.2. With increasing coating thickness, the HB was less saturated with oxygen. (b) The oxygen uptake by Hb in molar units at various coating thickness. (c) The logarithm of the absorbance versus time. (d) The first order kinetic constant of the oxygen uptake by HB.

concentration in the system. However, the individual particle contained pure PFOB, which has an oxygen content of 4.5 mM at 4300 ft. when in equilibrium with air. Using this concept, the diffusion coefficient could be estimated by assuming all oxygen was released from the particle. The estimated diffusion coefficient was 3.28×10^{-5} cm^2/s, which was lower than in PFOB (3.5×10^{-5} cm^2/s), but greater than in water (2.5×10^{-5}cm^2/s) and plasma (2.2×10^{-5}cm^2/s) [15]. Furthermore, the estimated diffusion coefficient was greater than the diffusion coefficient of oxygen uptake by the red blood cell (2.1×10^{-7}cm^2/s) [16] (Table 15.6).

This result is due to the following: (1) the nanoemulsion has a smaller size (280 nm) than the cell (8–10 μm); the larger particle decreases the overall interfacial surface area, which provides the less open gate flow of oxygen from the particle, (2) the emulsion has a monolayer of lipid while the cell has a bilayer; thus, the emulsion has a shorter barrier for gas transfer, (3) the concentration difference of oxygen between inside and outside the emulsion is high because the fluid outside of the emulsion was deoxygenated, and (4) the oxygen concentration inside the emulsion was very high due to the solubility of oxygen in PFOB.

The thickness of the calcium phosphate coating was determined by the difference between the uncoated and the coated emulsion mean diameters. The

TABLE 15.6 The Measured Oxygen in the Uncoated and Coated Emulsion Particles, the Estimated Diffusion Coefficient, and Relative Change of Diffusion Coefficient From the Uncoated Emulsion With Various Coating Thicknesses

Calcium phosphate coating thickness (nm)	Measured oxygen in the system (μM)	Estimated diffusion coefficient (cm^2/s)	Relative change of diffusion coefficient from 0 nm coating thickness
0	76.0	3.28×10^{-5}	NA
8	68.4	3.12×10^{-5}	Decreased by 5.9%
24	60.8	3.08×10^{-5}	Decreased by 4.1%
38	45.6	3.10×10^{-5}	Decreased by 5.5%

The measured oxygen refers to the results from fast mixing.

coated emulsion increased the particle size due to its thickness, but the space for the dissolved oxygen in the liquid core was unchanged from the uncoated emulsion. Assuming that the coating layer is only around the perfluorocarbon, and that the HB is free-floating, the oxygen (ie, "drug") was released in a limited range around the particles. The estimated diffusion coefficient of the coated emulsion was 3.10×10^{-5} cm^2/s ($\pm 2.31 \times 10^{-7}$) on average. The thicker the coating, the slower the oxygen release from the particle. The greater particle size, the slower the release of oxygen to the surroundings. The oxygen released was from 76 to 45 μM. Zero-nanometer thickness is synonymous with the uncoated emulsion.

Previously, the Roughton model for oxygen diffusion and reaction with HB was used to simulate the oxygenation of red blood cells. Later, this model was improved by using the Laplace transform [17]. These theoretical approaches focused on the understanding of the diffusion coefficient via the bilayer membrane in the red blood cells. The Roughton model used oxygen outside of the red blood cell that diffused into it and interacted with HB. This model was not ideal for this study, but the fundamental concept of diffusion through the membrane was considered. The fundamental kinetics reaction was also included. In this study, the Hixson–Crowell law, which focuses on the surface area of the particle, was used to determine the amount of oxygen released. This technique was used because of the small particle size and the increase in overall surface area and concentration with a smaller individual particle. Even though the emulsion is not a solid particle, it can be treated as such because (1) the emulsion in this study is small (~280 nm) and (2) the difference in oxygen concentration between the inside and outside of the particle was high so that the monolayer boundary had little effect. Therefore, the Hixson–Crowell equation is more suitable than the Roughton model in this study because the Hixson–Crowell equation estimates drug release from the solubility of the solid.

The advantage of this method, combined with the deoxygenated condition, was less error by directly observing the reaction of the media in a shorter period of time. Because the experimental setup was in an extremely high concentration gradient, the relative change of diffusion coefficient was more significant to explain that the emulsion surface was coated by calcium phosphate and the diffusion coefficient of calcium phosphate was constant even though the thickness varied.

Other studies have found that the oxygen diffusion coefficient through an artificial lecithin bilayer membrane was 2.10×10^{-5} cm^2/s at 25°C [18,19]. The calculated value of the oxygen diffusion coefficient through this study's emulsion particle was 3.28×10^{-5} cm^2/s at 20°C, which was greater. This increase in diffusion coefficient is thought to be a result of either a lecithin monolayer (versus a bilayer), or a large concentration gradient between the particle and surrounding fluid. Nevertheless, this calculation proved that the Hickson–Crowell model would be suitable to estimate the diffusion coefficient. Furthermore, the oxygen diffusion coefficient through the calcium phosphate coating was lower, at 3.10 (+/−0.02) $\times$ 10^{-5} cm^2/s, significantly less than the uncoated emulsion. This was expected, as the coated surface would add an additional barrier for the oxygen molecules to overcome.

7 SUMMARY

This study described the encapsulation of the nanoemulsion with calcium phosphate, and its performance regarding stability and the release oxygen. The nanoemulsion was made with the desired phase, as determined from the ternary diagram results. Then the emulsion surface was coated with calcium phosphate by optimizing the reaction conditions, such as pH and ion activity. The stability of the coated emulsion was analyzed in various conditions, such as elevated temperatures, physical perturbation, and time. The coated emulsion showed improved stability across all environments and treatments when compared with the uncoated one. The uncoated and the coated emulsions exhibited release oxygen, with a thicker calcium phosphate coating slowing the rate and amount at which oxygen is released. Not discussed is the separation of these particles from blood. The high density of perfluorocarbon liquid relative to other components in blood renders density gradient separation very efficient for this purpose. Such methods are used in modern aphaeresis systems, widely available in the clinic.

ACKNOWLEDGMENTS

Research reported in this chapter was supported by the National Heart, Lung, and Blood Institute of the National Institutes of Health under award number R41HL095250. The content is solely the responsibility of the authors and does not necessarily represent the official views of the National Institutes of Health.

REFERENCES

[1] Ostafin A, Landfester K, Sridar L, Moukasian A. Nanoreactor engineering for life science and medicine. Norwood, MA: Artech House Publishers; 2009.

[2] Koch C, Li L, Sessler D, et al. Duration of red-cell storage and complications after cardiac surgery. New Eng J Med 2008;358:1229–39.

[3] Ness P, Cushing M. Oxygen therapeutics-pursuit of an alternative to the donor red blood cell. Arch Pathol Lab Med 2007;131:734–41.

[4] Lapillonne H, Kobari L, Mazurier C, et al. Red blood cells generation from human induced pluripotent stem cells: perspectives for transfusion medicine. Haematologica 2010;95:1651–9.

[5] Cohn C, Cushing M. Oxygen therapeutics: perfluorocarbons and blood substitute safety. Crit Care Clin 2009;25:399–414.

[6] Chang T. Artificial cells. Singapore: World Scientific Press; 2007.

[7] Chen J, Scerbo M, Kramer G. Review of blood substitutes: examining the history, clinical trail results and ethics of hemoglobin-based oxygen carriers. Clinics 2009;64:803–13.

[8] Silverman T, Landow K, Ko S. et al. Review of publicly available reports of adverse events associated with HBOCs, research and review, CBER, FDA, April 2008.

[9] Fergusson D, McIntyre L. The future of clinical trials evaluating blood substitutes. JAMA 2008;299:2324–6.

[10] Koo OM, Rubinstein I, Onyuksel H. Role of nanotechnology in targeted drug delivery and imaging: a concise review. Nanomed-Nanotechnol 2005;1(3):193–212.

[11] Shapiro B. Towards dynamic control of magnetic fields to focus magnetic carriers to targets deep inside the body. J Magn Mater 2009;321(10):1594.

[12] Freire M, Dias A, Coutinho A, Coelho M, Marrucho I. Enzymatic method for determining oxygen solubility in perfluorocarbon emulsions. Fluid Phase Equilibr 2005;231:109–13.

[13] Mizukami H, Beaudoin AG, Bartnicki, Adams B. Hysteresis- like behavior of oxygen association-dissociation curves of sickle cells determined by a new method. Soc Exp Biol Med 1977;154:304–9.

[14] Siepmann J, Siepmann F. Mathematical modeling of drug dissolution. Int J Pharm 2013;453(1):12–24.

[15] Moolman F. "Mass transfer considerations," in oxygen carriers for a novel bio-artificial liver support system, Moolman F, editor. Germany: Verlag Dr. Muller; 2003, p. 130.

[16] Roughton F. Diffusion and simultaneous chemical reaction velocity in hemoglobin solutions and red cell suspensions. Prog Biophys Biochem 1959;9:55.

[17] Mizukami H, Rice S. Transient effects on the initial rate of oxygenation of red blood cells. Bull Math Biol 1982;44:119–34.

[18] Fischkoff S, Vanderkooi J. Oxygen diffusion in biological and artificial membranes determined by the fluorochrome pyrene. J Gen Physiol 1975;65:663–76.

[19] Cussler E. Diffusion: mass transfer in fluid systems. Cambridge, England: Cambridge University Press; 1997.

Subject Index

D

E

N

O

P

R

S

T

U

V

W

Kolibri 360 GmbH
Ferdinand-Jühlke-Straße 7
99095 Erfurt, Deutschland
produktsicherheit@kolibri360.de